A THEORETICAL INTRODUCTION TO NUMERICAL ANALYSIS

Victor S. Ryaben'kii
Semyon V. Tsynkov

Chapman & Hall/CRC is an imprint of the
Taylor & Francis Group, an informa business

Chapman & Hall/CRC
Taylor & Francis Group
6000 Broken Sound Parkway NW, Suite 300
Boca Raton, FL 33487-2742

Printed in the United States of America on acid-free paper
10 9 8 7 6 5 4 3 2 1

International Standard Book Number-10: 1-58488-607-2 (Hardcover)
International Standard Book Number-13: 978-1-58488-607-5 (Hardcover)

Visit the Taylor & Francis Web site at
http://www.taylorandfrancis.com

and the CRC Press Web site at
http://www.crcpress.com

Contents

Preface

This book introduces the key ideas and concepts of numerical analysis. The discussion focuses on how one can represent different mathematical models in a form that enables their efficient study by means of a computer. The material learned from this book can be applied in various contexts that require the use of numerical methods. The general methodology and principles of numerical analysis are illustrated by specific examples of the methods for real analysis, linear algebra, and differential equations. The reason for this particular selection of subjects is that these methods are proven, provide a number of well-known efficient algorithms, and are used for solving different applied problems that are often quite distinct from one another.

The contemplated readership of this book consists of beginning graduate and senior undergraduate students in mathematics, science and engineering. It may also be of interest to working scientists and engineers. The book offers a first mathematical course on the subject of numerical analysis. It is carefully structured and can be read in its entirety, as well as by selected parts. The portions of the text considered more difficult are clearly identified; they can be skipped during the first reading without creating any substantial gaps in the material studied otherwise. In particular, more difficult subjects are discussed in Sections 2.3.1 and 2.3.3, Sections 3.1.3 and 3.2.7, parts of Sections 4.2 and 9.7, Section 10.5, Section 12.2, and Chapter 14.

Hereafter, numerical analysis is interpreted as a mathematical discipline. The basic concepts, such as discretization, error, efficiency, complexity, numerical stability, consistency, convergence, and others, are explained and illustrated in different parts of the book with varying levels of depth using different subject material. Moreover, some ideas and views that are addressed, or at least touched upon in the text, may also draw the attention of more advanced readers. First and foremost, this applies to the key notion of the saturation of numerical methods by smoothness. A given method of approximation is said to be saturated by smoothness if, because of its design, it may stop short of reaching the intrinsic accuracy limit (unavoidable error) determined by the smoothness of the approximated solution and by the discretization parameters. If, conversely, the accuracy of approximation self-adjusts to the smoothness, then the method does not saturate. Examples include algebraic vs. trigonometric interpolation, Newton-Cotes vs. Gaussian quadratures, finite-difference vs. spectral methods for differential equations, etc.

Another advanced subject is an introduction to the method of difference potentials in Chapter 14. This is the first account of difference potentials in the educational literature. The method employs discrete analogues of modified Calderon's potentials and boundary projection operators. It has been successfully applied to solving a variety of direct and inverse problems in fluids, acoustics, and electromagnetism.

This book covers three semesters of instruction in the framework of a commonly used curriculum with three credit hours per semester. Three semester-long courses can be designed based on Parts I, II, and III of the book, respectively. Part I includes interpolation of functions and numerical evaluation of definite integrals. Part II covers direct and iterative solution of consistent linear systems, solution of overdetermined linear systems, and solution of nonlinear equations and systems. Part III discusses finite-difference methods for differential equations. The first chapter in this part, Chapter 9, is devoted to ordinary differential equations and serves an introductory purpose. Chapters 10, 11, and 12 cover different aspects of finite-difference approximation for both steady-state and evolution partial differential equations, including rigorous analysis of stability for initial boundary value problems and approximation of the weak solutions for nonlinear conservation laws. Alternatively, for the curricula that introduce numerical differentiation right after the interpolation of functions and quadratures, the material from Chapter 9 can be added to a course based predominantly on Part I of the book.

A rigorous mathematical style is maintained throughout the book, yet very little use is made of the apparatus of functional analysis. This approach makes the book accessible to a much broader audience than only mathematicians and mathematics majors, while not compromising any fundamentals in the field. A thorough explanation of the key ideas in the simplest possible setting is always prioritized over various technicalities and generalizations. All important mathematical results are accompanied by proofs. At the same time, a large number of examples are provided that illustrate how those results apply to the analysis of individual problems.

This book has no objective whatsoever of describing as many different methods and techniques as possible. On the contrary, it treats only a limited number of well-known methodologies, and only for the purpose of exemplifying the most fundamental concepts that unite different branches of the discipline. A number of important results are given as exercises for independent study. Altogether, many exercises supplement the core material; they range from elementary to quite challenging.

Some exercises require computer implementation of the corresponding techniques. However, no substantial emphasis is put on issues related to programming. In other words, any computer implementation serves only as an illustration of the relevant mathematical concepts and does not carry an independent learning objective. For example, it may be useful to have different iteration schemes implemented for a system of linear algebraic equations. By comparing how their convergence rates depend on the condition number, one can subsequently judge the efficiency from a mathematical standpoint. However, other efficiency issues, e.g., runtime efficiency determined by the software and/or computer platform, are not addressed as there is no direct relation between them and the mathematical analysis of numerical methods.

Likewise, no substantial emphasis is put on any specific applications. Indeed, the goal is to clearly and concisely present the key mathematical concepts pertinent to the analysis of numerical methods. This provides a foundation for the subsequent specialized training. Subjects such as computational fluid dynamics, computational acoustics, computational electromagnetism, etc., are very well addressed in the literature. Most corresponding books require some numerical background from the reader, the background of precisely the kind that the current text offers.

Acknowledgments

This book has a Russian language prototype [Rya00] that withstood two editions: in 1994 and in 2000. It serves as the main numerical analysis text at Moscow Institute for Physics and Technology. The authors are most grateful to the rector of the Institute at the time, Academician O. M. Belotserkovskii, who has influenced the original concept of this textbook.

Compared to [Rya00], the current book is completely rewritten. It accommodates the differences that exist between the Russian language culture and the English language culture of mathematics education. Moreover, the current textbook includes a very considerable amount of additional material.

When writing Part III of the book, we exploited the ideas and methods previously developed in [GR64] and [GR87].

When writing Chapter 14, we used the approach of [Rya02, Introduction].

We are indebted to all our colleagues and friends with whom we discussed the subject of teaching the numerical analysis. The book has greatly benefited from all those discussions. In particular, we would like to thank S. Abarbanel, K. Brushlinskii, V. Demchenko, A. Chertock, L. Choudov, L. Demkowicz, A. Ditkowski, R. Fedorenko, G. Fibich, P. Gremaud, T. Hagstrom, V. Ivanov, C. Kelley, D. Keyes, A. Kholodov, V. Kosarev, A. Kurganov, C. Meyer, N. Onofrieva, I. Petrov, V. Pirogov, L. Strygina, E. Tadmor, E. Turkel, S. Utyuzhnikov, and A. Zabrodin. We also remember the late K. Babenko, O. Lokutsievskii, and Yu. Radvogin.

We would like to specially thank Alexandre Chorin of UC Berkeley and David Gottlieb of Brown University who read the manuscript prior to publication.

A crucial and painstaking task of proofreading the manuscript was performed by the students who took classes on the subject of this book when it was in preparation. We are most grateful to L. Bilbro, A. Constantinescu, S. Ernsberger, S. Grove, A. Peterson, H. Qasimov, A. Sampat, and W. Weiselquist. All the imperfections still remaining are a sole responsibility of the authors.

It is also a pleasure for the second author to thank Arje Nachman and Richard Albanese of the US Air Force for their consistent support of the second author's research work during and beyond the period of time when the book was written.

And last, but not least, we are very grateful to the CRC Press Editor, Sunil Nair, as well as to the company staff in London and in Florida, for their advice and assistance.

Finally, our deepest thanks go to our families for their patience and understanding without which this book project would have never been completed.

V. Ryaben'kii, Moscow, Russia
S. Tsynkov, Raleigh, USA

August 2006

Chapter 1

Introduction

Modern numerical mathematics provides a theoretical foundation behind the use of electronic computers for solving applied problems. A mathematical approach to any such problem typically begins with building a model for the phenomenon of interest (situation, process, object, device, laboratory/experimental setting, etc.). Classical examples of mathematical models include definite integrals, equation of a pendulum, the heat equation, equations of elasticity, equations of electromagnetic waves, and many other equations of mathematical physics. For comparison, we should also mention here a model used in formal logics — the Boolean algebra.

Analytical methods have always been considered a fundamental means for studying the mathematical models. In particular, these methods allow one to obtain closed form exact solutions for some special cases (for example, tabular integrals). There are also classes of problems for which one can obtain a solution in the form of a power series, Fourier series, or some other expansion. In addition, a certain role has always been played by approximate computations. For example, quadrature formulae are used for the evaluation of definite integrals.

The advent of computers in the middle of the twentieth century has drastically increased our capability of performing approximate computations. Computers have essentially transformed approximate computations into a dominant tool for the analysis of mathematical models. Analytical methods have not lost their importance, and have even gained some additional "functionality" as components of combined analytical/computational techniques and as verification tools. Yet sophisticated mathematical models are analyzed nowadays mostly with the help of computers. Computers have dramatically broadened the applicability range of mathematical methods in many traditional areas, such as mechanics, physics, and engineering. They have also facilitated a rapid expansion of the mathematical methods into various non-traditional fields, such as management, economics, finance, chemistry, biology, psychology, linguistics, ecology, and others.

Computers provide a capability of storing large (but still finite) arrays of numbers, and performing arithmetic operations with these numbers according to a given program that would run with a fast (but still finite) execution speed. Therefore, computers may only be appropriate for studying those particular models that are described by finite sets of numbers and require no more than finite sequences of arithmetic operations to be performed. Besides the arithmetic operations per se, a computer model can also contain comparisons between numbers that are typically needed for the automated control of subsequent computations.

In the traditional fields, one frequently employs such mathematical models as functions, derivatives, integrals, and differential equations. To enable the use of computers, these original models must therefore be (approximately) replaced by the new models that would only be based on finite arrays of numbers supplemented by finite sequences of arithmetic operations for their processing (i.e., finite algorithms). For example, a function can be replaced by a table of its numerical values; the derivative

$$\frac{df}{dx} = \lim_{h \to 0} \frac{f(x+h) - f(x)}{h}$$

can be replaced by an approximate formula, such as

$$f'(x) \approx \frac{f(x+h) - f(x)}{h},$$

where h is fixed (and small); a definite integral can be replaced by its integral sum; a boundary value problem for the differential equation can be replaced by the problem of finding its solution at the discrete nodes of some grid, so that by taking a suitable (i.e., sufficiently small) grid size an arbitrary desired accuracy can be achieved. In so doing, among the two methods that could seem equivalent at first glance, one may produce good results while the other may turn out completely inapplicable. The reason can be that the approximate solution it generates would not approach the exact solution as the grid size decreases, or that the approximate solution would turn out overly sensitive to the small round-off errors.

The subject of numerical analysis is precisely the theory of those models and algorithms that are applicable, i.e., that can be efficiently implemented on computers. This theory is intimately connected with many other branches of mathematics: Approximation theory and interpolation of functions, ordinary and partial differential equations, integral equations, complexity theory for functional classes and algorithms, etc., as well as with the theory and practice of programming languages. In general, both the exploratory capacity and the methodological advantages that computers deliver to numerous applied areas are truly unparalleled. Modern numerical methods allow, for example, the computation of the flow of fluid around a given aerodynamic configuration, e.g., an airplane, which in most cases would present an insurmountable task for analytical methods (like a non-tabular integral).

Moreover, the use of computers has enabled an entirely new scientific methodology known as computational experiment, i.e., computations aimed at verifying the hypotheses, as well as at monitoring the behavior of the model, when it is not known ahead of time what may interest the researcher. In fact, computational experiment may provide a sufficient level of feedback for the original formulation of the problem to be noticeably refined. In other words, numerical computations help accumulate the vital information that eventually allows one to identify the most interesting cases and results in a given area of study. Many remarkable observations, and even discoveries, have been made along this route that empowered the development of the theory and have found important practical applications as well.

Computers have also facilitated the application of mathematical methods to non-traditional areas, for which few or no "compact" mathematical models, such as differential equations, are readily available. However, other models can be built that lend themselves to the analysis by means of a computer. A model of this kind can often be interpreted as a direct numerical counterpart (such as encoding) of the object of interest and of the pertinent relations between its elements (e.g., a language or its abridged subset and the corresponding words and phrases). The very possibility of studying such models on a computer prompts their construction, which, in turn, requires that the rules and guiding principles that govern the original object be clearly and unambiguously identified. On the other hand, the results of computer simulations, e.g., a machine translation of the simplified text from one language to another, provide a practical criterion for assessing the adequacy of the theories that constitute the foundation of the corresponding mathematical model (e.g., linguistic theories).

Furthermore, computers have made it possible to analyze probabilistic models that require large amounts of test computations, as well as the so-called imitation models that describe the object or phenomenon of interest without simplifications (e.g., functional properties of a telephone network).

The variety of problems that can benefit from the use of computers is huge. For solving a given problem, one would obviously need to know enough specific detail. Clearly, this knowledge cannot be obtained ahead of time for all possible scenarios.

Therefore, the purpose of this book is rather to provide a systematic perspective on those fundamental ideas and concepts that span across different applied disciplines and can be considered established in the field of numerical analysis. Having mastered the material of this book, one should encounter little or no difficulties when receiving subsequent specialized training required for the successful work in a given research or industrial field. The general methodology and principles of numerical analysis are illustrated in the book by "sampling" the methods designed for mathematical analysis, linear algebra, and differential equations. The reason for this particular selection is that the aforementioned methods are most mature, lead to a number of well-known, efficient algorithms, and are extensively used for solving various applied problems that are often quite distant from one another.

Let us mention here some of the general ideas and concepts that require the most thorough attention in every particular setting. These general ideas acquire a concrete interpretation and meaning in the context of each specific problem that needs to be solved on a computer. They are the discretization of the problem, conditioning of the problem, numerical error, and computational stability of a given algorithm. In addition, comparison of the algorithms along different lines obviously plays a central role when selecting a specific method. The key criteria for comparison are accuracy, storage, and operation count requirements, as well as the efficiency of utilization of the input information. On top of that, different algorithms may vary in how amenable they are to parallelization — a technique that allows one to conduct computations simultaneously on multi-processor computer platforms.

In the rest of the Introduction, we provide a brief overview of the foregoing notions and concepts. It helps create a general perspective on the subject of numerical mathematics, and establishes a foundation for studying the subsequent material.

1.1 Discretization

Let $f(x)$ be a function of the continuous argument $x \in [0, 1]$. Assume that this function provides (some of) the required input data for a given problem that needs to be approximately solved on a computer. The value of the function f at every given x can be either measured or obtained numerically. Then, to store this function in the memory of a computer, one may need to approximately characterize it with a table of values at a finite set of points: $x_1, x_2, \ldots, x_n$. This is an elementary example of discretization: The problem of storing the function defined on the interval $[0, 1]$, which is a continuum of points, is replaced by the problem of storing a table of its discrete values at the subset of points $x_1, x_2, \ldots, x_n$ that all belong to this interval.

Let now $f(x)$ be sufficiently smooth, and assume that we need to calculate its derivative at a given point x. The problem of exactly evaluating the expression

$$f'(x) = \lim_{h \to 0} \frac{f(x+h) - f(x)}{h}$$

that contains a limit can be replaced by the problem of computing an approximate value of this expression using one of the following formulae:

$$f'(x) \approx \frac{f(x+h) - f(x)}{h}, \tag{1.1}$$

$$f'(x) \approx \frac{f(x) - f(x-h)}{h}, \tag{1.2}$$

$$f'(x) \approx \frac{f(x+h) - f(x-h)}{2h}. \tag{1.3}$$

Similarly, the second derivative $f''(x)$ can be replaced by the finite formula:

$$f''(x) \approx \frac{f(x+h) - 2f(x) + f(x-h)}{h^2}. \tag{1.4}$$

One can show that all these formulae become more and more accurate as h becomes smaller; this is the subject of Exercise 1, and the details of the analysis can be found in Section 9.2.1. Moreover, for every fixed h, each formula (1.1)–(1.4) will only require a finite set of values of f and a finite number of arithmetic operations. These formulae are examples of discretization for the derivatives $f'(x)$ and $f''(x)$.

Let us now consider a boundary value problem:

$$\begin{gathered} \frac{d^2y}{dx^2} - x^2 y = \cos x, \qquad 0 \le x \le 1, \\ y(0) = 2, \qquad y(1) = 3, \end{gathered} \tag{1.5}$$

where the unknown function $y = y(x)$ is defined on the interval $0 \le x \le 1$. To construct a discrete approximation of problem (1.5), let us first partition the interval

[0, 1] into N equal sub-intervals of size $h = N^{-1}$. Instead of the continuous function $y(x)$, we will be looking for a finite set of its values $y_0, y_1, \dots, y_N$ on the grid $x_k = kh$, $k = 0, 1, \dots, N$. At the interior nodes of this grid: x_k, $k = 1, 2, \dots, N-1$, we can approximately replace the second derivative $y''(x)$ by expression (1.4). After substituting into the differential equation of (1.5) this yields:

$$\frac{y_{k+1} - 2y_k + y_{k-1}}{h^2} - (kh)^2 y_k = \cos(kh), \qquad k = 1, 2, \dots, N-1. \tag{1.6}$$

Furthermore, the boundary conditions at $x = 0$ and at $x = 1$ from (1.5) translate into:

$$y_0 = 2, \qquad y_N = 3. \tag{1.7}$$

The system of $N+1$ linear algebraic equations (1.6), (1.7) contains exactly as many unknowns $y_0, y_1, \dots, y_N$, and renders a discrete counterpart of the boundary value problem (1.5). One can, in fact, show that the finer the grid, i.e., the larger the N, the more accurate will the approximation be that the discrete solution of problem (1.6), (1.7) provides for the continuous solution of problem (1.5). Later, this fact will be formulated and proven rigorously.

Let us denote the continuous boundary value problem (1.5) by M_∞, and the discrete boundary value problem (1.6), (1.7) by M_N. By taking $N = 2, 3, \dots$, we associate an infinite sequence of discrete problems $\{M_N\}$ with the continuous problem M_∞. When computing the solution to a given problem M_N for any fixed N, we only have to work with a finite array of numbers that specify the input data, and with a finite set of unknown quantities $y_0, y_1, y_2, \dots, y_N$. It is, however, the entire infinite sequence of finite discrete models $\{M_N\}$ that plays the central role from the standpoint of numerical mathematics. Indeed, as those models happen to be more and more accurate, we can always choose a sufficiently large N that would guarantee any desired accuracy of approximation.

In general, there are many different ways of transitioning from a given continuous problem M_∞ to the sequence $\{M_N\}$ of its discrete counterparts. In other words, the approximation (1.6), (1.7) of the boundary value problem (1.5) is by no means the only one possible. Let $\{M_N\}$ and $\{M'_N\}$ be two sequences of approximations, and let us also assume that the computational costs of obtaining the discrete solutions of M_N and M'_N are the same. Then, a better method of discretization would be the one that provides the same accuracy of approximation with a smaller value of N.

Let us also note that for two seemingly equivalent discretization methods M_N and M'_N, it may happen that one will approximate the continuous solution of problem M_∞ with an increasingly high accuracy as N increases, whereas the other will yield "an approximate solution" that would bear less and less resemblance to the continuous solution of M_∞. We will encounter situations like this in Part III of the book, where we also discuss how the corresponding difficulties can be partially or fully overcome.

Exercises

1. Let $f(x)$ have as many bounded derivatives as needed. Show that the approximation error of formulae (1.1), (1.2), (1.3), and (1.4), is $\mathcal{O}(h)$, $\mathcal{O}(h)$, $\mathcal{O}(h^2)$, and $\mathcal{O}(h^2)$.

1.2 Conditioning

Speaking in most general terms, for any given problem one can basically identify the input data and the output result(s), i.e., the solution, so that the former determine the latter. In this book, we will mostly analyze problems for which the solution exists and is unique. If, in addition, the solution depends continuously on the data, i.e., if for a vanishing perturbation of the data the corresponding perturbation of the solution will also be vanishing, then the problem is said to be *well-posed*.

A somewhat more subtle characterization of the problem, on top of its well-posedness, is known as the *conditioning*. It has to do with quantifying the sensitivity of the solution, or some of its key characteristics, to perturbations of the input data. This sensitivity may vary strongly for different problems that could otherwise look very similar. If it is "low" (weak), then the problem is said to be *well conditioned;* if, conversely, the sensitivity is "high" then the problem is *ill conditioned.* The notions of low and high are, of course, problem-specific. We emphasize that the concept of conditioning pertains to both continuous and discrete problems. Typically, not only do ill conditioned problems require excessively accurate definition of the input data, but also appear more difficult for computations.

Consider, for example, the quadratic equation $x^2 - 2\alpha x + 1 = 0$ for $|\alpha| > 1$. It has two real roots that can be expressed as functions of the argument α: $x_{1,2} = \alpha \pm \sqrt{\alpha^2 - 1}$. We will interpret α as the datum in the problem, and $x_1 = x_1(\alpha)$ and $x_2 = x_2(\alpha)$ as the corresponding solution. Clearly, the sensitivity of the solution to the perturbations of α can be characterized by the magnitude of the derivatives $\frac{dx_{1,2}}{d\alpha} = 1 \pm \frac{\alpha}{\sqrt{\alpha^2-1}}$. Indeed, $\Delta x_{1,2} \approx \frac{dx_{1,2}}{d\alpha}\Delta\alpha$. We can easily see that the derivatives $\frac{dx_{1,2}}{d\alpha}$ are small for large $|\alpha|$, but they become large when α approaches 1. We can therefore conclude that the problem of finding the roots of $x^2 - 2\alpha x + 1 = 0$ is well conditioned when $|\alpha| \gg 1$, and ill conditioned when $|\alpha| = \mathcal{O}(1)$. We should also note that conditioning can be improved if, instead of the original quadratic equation, we consider its equivalent $x^2 - \frac{1+\beta^2}{\beta}x + 1 = 0$, where $\beta = \alpha + \sqrt{\alpha^2 - 1}$. In this case, $x_1 = \beta$ and $x_2 = \beta^{-1}$; the two roots coincide for $|\beta| = 1$, or equivalently, $|\alpha| = 1$. However, the problem of evaluating $\beta = \beta(\alpha)$ is still ill conditioned near $|\alpha| = 1$.

Our next example involves a simple ordinary differential equation. Let $y = y(t)$ be the concentration of some substance at the time t, and assume that it satisfies:

$$\frac{dy}{dt} - 10y = 0.$$

Let us take an arbitrary t_0, $0 \le t_0 \le 1$, and perform an approximate measurement of the actual concentration $y_0 = y(t_0)$ at this moment of time, thus obtaining:

$$y\Big|_{t=t_0} = y_0^*.$$

Our overall task will be to determine the concentration $y = y(t)$ at all other moments of time t from the interval $[0, 1]$.

If we knew the quantity $y_0 = y(t_0)$ exactly, then we could have used the exact formula available for the concentration:

$$y(t) = y_0 e^{10(t-t_0)}. \tag{1.8}$$

We, however, only know the approximate value $y_0^* \approx y_0$ of the unknown quantity y_0. Therefore, instead of (1.8), the next best thing is to employ the approximate formula:

$$y^*(t) = y_0^* e^{10(t-t_0)}. \tag{1.9}$$

Clearly, the error $y^* - y$ of the approximate formula (1.9) is given by:

$$y^*(t) - y(t) = (y_0^* - y_0)e^{10(t-t_0)}, \quad 0 \le t \le 1.$$

Assume now that we need to measure y_0^* to the the accuracy δ, $|y_0^* - y_0| < \delta$, that would be sufficient to guarantee an initially prescribed tolerance ε for determining $y(t)$ everywhere on the interval $0 \le t \le 1$, i.e., would guarantee the error estimate:

$$|y^*(t) - y(t)| < \varepsilon, \quad 0 \le t \le 1.$$

It is easy to see that $\max_{0 \le t \le 1} |y^*(t) - y(t)| = |y^*(1) - y(1)| = |y_0^* - y_0| e^{10(1-t_0)}$. This yields the following constraint that the accuracy δ of measuring y_0 must satisfy:

$$\delta \le \varepsilon e^{-10(1-t_0)}. \tag{1.10}$$

Let y_0 be measured at the moment of time $t_0 = 0$. Then, inequality (1.10) would imply that this measurement has to be e^{10} times, i.e., thousands of times, more accurate than the required guaranteed accuracy of the result ε. In other words, the answer $y(t)$ appears quite sensitive to the error in specifying the input data y_0, and the problem is ill conditioned.

On the other hand, if y_0 were to be measured at $t_0 = 1$, then $\delta = \varepsilon$, and it would be sufficient to conduct the measurement with a considerably lower accuracy than the one required in the case of $t_0 = 0$. This problem is well conditioned.

Exercises

1. Consider the problem of computing $y(x) = \frac{1+x}{1-x}$ as a function of x, for $x \in (1/2, 1)$ and also for $x \in (-1, 0)$. On which of the two intervals is this problem better conditioned with respect to the perturbations of x?
2. Let $y = \sqrt{2} - 1$. Equivalently, one can write $y = (\sqrt{2} + 1)^{-1}$. Which of the two formulae is more sensitive to the error when $\sqrt{2}$ is approximated by a finite decimal fraction? **Hint.** Compare absolute values of derivatives for the functions $(x-1)$ and $(x+1)^{-1}$.

1.3 Error

In any computational problem, one needs to find the solution given some appropriate input data. If the solution can be obtained with an ideal accuracy, then there is no error. Typically, however, there is a certain error content in every feasible numerical solution. This error may be attributed to (at least) three different mechanisms.

First, the input data are often specified with some degree of uncertainty that, in turn, will generate uncertainty in the corresponding output. Then, the solution to the problem of interest may only be obtained with an error called *unavoidable error.*

Second, even if we eliminate the foregoing uncertainty by fixing the input data, and subsequently compute the solution using an approximate method, then we still won't find the solution that would exactly correspond to the specified data. There will be *error due to the choice of an approximate computational procedure*.

Third, the chosen approximate method is not implemented exactly either, because of *round-off errors* that arise when performing computations on a real machine.

Therefore, the overall error in the solution consists of unavoidable error, the error of the method, and round-off error. We will now illustrate these concepts.

1.3.1 Unavoidable Error

Assume that we need to find the value y of some function $y = f(x)$ for a given $x = x_0$. The quantity x_0, as well as the relation f itself that associates the value of the function with every given value of its argument, are considered the input data of the problem, whereas the quantity $y = y(x_0)$ will be its solution.

Now let the function $f(x)$ be known approximately rather than exactly, say, $f(x) \approx \sin x$, and suppose that $f(x)$ may differ from $\sin x$ by no more than a specified $\varepsilon > 0$:

$$\sin x - \varepsilon \leq f(x) \leq \sin x + \varepsilon. \tag{1.11}$$

Let the value of the argument $x = x_0$ be also measured approximately: $x = x_0^*$, so that regarding the actual x_0 we can only say that

$$x_0^* - \delta \leq x_0 \leq x_0^* + \delta, \tag{1.12}$$

where $\delta > 0$ characterizes the accuracy of the measurement.

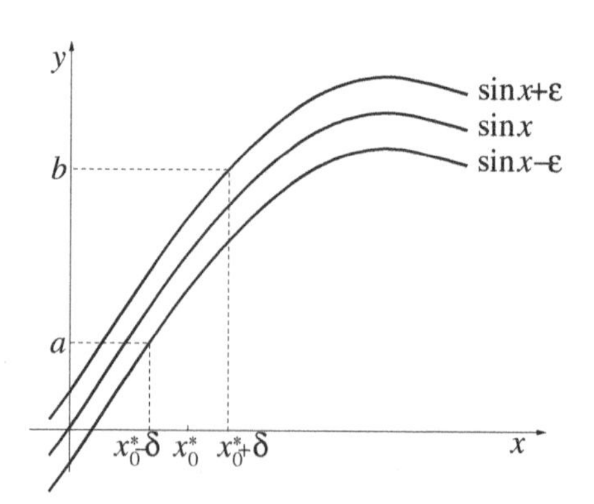

FIGURE 1.1: Unavoidable error.

One can easily see from Figure 1.1 that any point on the interval $[a, b]$ of variable y, where $a = \sin(x_0^* - \delta) - \varepsilon$ and $b = \sin(x_0^* + \delta) + \varepsilon$, can serve in the capacity of $y = f(x_0)$. Clearly, by taking an arbitrary $y^* \in [a, b]$ as the approximate value of $y = f(x_0)$, one can always guarantee the error estimate:

$$|y - y^*| \leq |b - a|. \tag{1.13}$$

For the given uncertainty in the input data, see formulae (1.11) and (1.12), this estimate cannot be considerably improved. In fact, the best error estimate that one can guarantee is obtained by choosing y^* exactly in the middle of the interval $[a, b]$:

$$y^* = y^*_{\text{opt}} = (a+b)/2.$$

From Figure 1.1 we then conclude that

$$|y - y^*| \leq |b-a|/2. \tag{1.14}$$

This inequality transforms into an exact equality when $y(x_0) = a$ or when $y(x_0) = b$.

As such, the quantity $|b-a|/2$ is precisely the unavoidable (or irreducible) error, i.e., the minimum error content that will always be present in the solution and that cannot be "dodged" no matter how the approximation y^* is actually chosen, simply because of the uncertainty that exists in the input data. For the optimal choice of the approximate solution y^*_{opt} the smallest error (1.14) can be guaranteed; otherwise, the appropriate error estimate is (1.13).

We see, however, that the optimal error estimate (1.14) is not that much better than the general estimate (1.13). We will therefore stay within reason if we interpret any arbitrary point $y^* \in [a, b]$, rather than only y^*_{opt}, as an approximate solution for $y(x_0)$ obtained within the limits of the unavoidable error. In so doing, the quantity $|b-a|$ shall replace $|b-a|/2$ of (1.14) as the estimate of the unavoidable error.

Along with the simplest illustrative example of Figure 1.1, let us consider another example that would be a little more realistic and would involve one of the most common problem formulations in numerical analysis, namely, that of reconstructing a function of continuous argument given its tabulated values at some discrete set of points. More precisely, let the values $f(x_k)$ of the function $f = f(x)$ be known at the equidistant grid nodes $x_k = kh$, $h > 0$, $k = 0, \pm 1, \pm 2, \ldots$. Let us also assume that the first derivative of $f(x)$ is bounded everywhere: $|f'(x)| \leq 1$, and that together with $f(x_k)$, this is basically all the information that we have about $f(x)$. We need to be able to obtain the (approximate) value of $f(x)$ at an arbitrary "intermediate" point x that does not necessarily coincide with any of the nodes x_k.

A large variety of methods have been developed in the literature for solving this problem. Later, we will consider interpolation by means of algebraic (Chapter 2) and trigonometric (Chapter 3) polynomials. There are other ways of building the approximating polynomials, e.g., the least squares fit, and there are other types of functions that can be used as a basis for the approximation, e.g., wavelets. Each specific method will obviously have its own accuracy. We, however, are going to show that *irrespective of any particular technique* used for reconstructing $f(x)$, there will always be error due to incomplete specification of the input data. This error merely reflects the uncertainty in the formulation; it is unavoidable and cannot be suppressed by any "smart" choice of the reconstruction procedure.

Consider the simplest case $f(x_k) = 0$ for all $k = 0, \pm 1, \pm 2, \ldots$. Clearly, the function $f_1(x) \equiv 0$ has the required trivial table of values, and also $|f_1'(x)| \leq 1$. Along with $f_1(x)$, it is easy to find another function that would satisfy the same constraints, e.g., $f_2(x) = \frac{h}{\pi}\sin\left(\frac{\pi x}{h}\right)$. Indeed, $f_2(x_k) = 0$, and $|f_2'(x)| = \left|\cos\left(\frac{\pi x}{h}\right)\right| \leq 1$. We therefore see that there are at least two different functions that cannot be told apart based on the available information. Consequently, the error $\max_x |f_1(x) - f_2(x)| = \mathcal{O}(h)$ is unavoidable when reconstructing the function $f(x)$, given its tabulated values $f(x_k)$ and the fact that its first derivative is bounded, no matter what specific reconstruction methodology may be employed.

For more on the notion of the unavoidable error in the context of reconstructing continuous functions from their discrete values see Section 2.2.4 of Chapter 2.

1.3.2 Error of the Method

Let $y^* = \sin x_0^*$. The number y^* belongs to the interval $[a, b]$; it can be considered a non-improvable approximate solution of the first problem analyzed in Section 1.3.1. For this solution, the error satisfies estimate (1.13) and is unavoidable. The point $y^* = \sin x_0^*$ has been selected among other points of the interval $[a, b]$ only because it is given by the formula convenient for subsequent analysis.

To evaluate the quantity $y^* = \sin x_0^*$ on a computer, let us use Taylor's expansion for the function $\sin x$:

$$\sin x = x - \frac{x^3}{3!} + \frac{x^5}{5!} - \dots.$$

Thus, for computing y^* one can take one of the following approximate expressions:

$$\begin{aligned} y^* &\approx y_1^* = x_0^*, \\ y^* &\approx y_2^* = x_0^* - \frac{(x_0^*)^3}{3!}, \\ \dots\dots & \quad \dots \quad \dots \quad \dots \\ y^* &\approx y_n^* = \sum_{k=1}^{n} (-1)^{k-1} \frac{(x_0^*)^{2k-1}}{(2k-1)!}. \end{aligned} \tag{1.15}$$

By choosing a specific formulae (1.15) for the approximate evaluation of y^*, we select our method of computation. The quantity $|y^* - y_n^*|$ is then known as *the error of the computational method.* In fact, we are considering a family of methods parameterized by the integer n. The larger the n the smaller the error, see (1.15); and by taking a sufficiently large n we can always make sure that the associated error will be smaller than any initially prescribed threshold.

It, however, does not make sense to drive the computational error much further down than the level of the unavoidable error. Therefore, the number n does not need to be taken excessively large. On the other hand, if n is taken too small so that the error of the method appears much larger than the unavoidable error, then one can say that the chosen method does not fully utilize the information about the solution that is contained in the input data, or equivalently, loses a part of this information.

1.3.3 Round-off Error

Assume that we have fixed the computational method by selecting a particular n in (1.15), i.e., by setting $y^* \approx y_n^*$. When calculating this y_n^* on an actual computer, we will, generally speaking, obtain a different value $\tilde{y}_n^*$ due to rounding. Rounding is an intrinsic feature of the floating-point arithmetic on computers, as they only operate with numbers that can be represented as finite binary fractions of a given fixed length. As such, all other real numbers (e.g., infinite fractions) may only be stored approximately in the computer memory, and the corresponding approximation procedure is known as rounding. The error $|y_n^* - \tilde{y}_n^*|$ is called *the round-off error.*

This error shall not noticeably exceed the error of the computational method. Otherwise, a loss of the overall accuracy will be incurred due to the round-off error.

Exercises[1]

1. Assume that we need to calculate the value $y = f(x)$ of some function $f(x)$, while there is an uncertainty in the input data x^*: $x^* - \delta \le x \le x^* + \delta$.

   How does the corresponding unavoidable error depend on x^* and on δ for the following functions:

   a) $f(x) = \sin x$;

   b) $f(x) = \ln x$, where $x > 0$?

   For what values of x^*, obtained by approximately measuring the "loose" quantity x with the accuracy δ, can one guarantee only a one-sided upper bound for $\ln x$ in problem b)? Find this upper bound.

2. Let the function $f(x)$ be defined by its values sampled on the grid $x_k = kh$, where $h = 1/N$ and $k = 0, \pm 1, \pm 2, \ldots$. In addition to these discrete values, assume that $\max_x |f''(x)| \le 1$.

   Prove that as the available input data are incomplete, they do not, generally speaking, allow one to reconstruct the function at an arbitrary given point x with accuracy better than the unavoidable error $\varepsilon(h) = h^2/\pi^2$.

   Hint. Show that along with the function $f(x) \equiv 0$, which obviously has all its grid values equal to zero, another function, $\varphi(x) = \left(h^2/\pi^2\right)\sin(N\pi x)$, also has all its grid values equal to zero, and satisfies the condition $\max_x |\varphi''(x)| \le 1$, while $\max_x |f(x) - \varphi(x)| = h^2/\pi^2$.

3. Let $f = f(x)$ be a function, such that the absolute value of its second derivative does not exceed 1. Show that the approximation error for the formula:

   $$f'(x) \approx \frac{f(x+h) - f(x)}{h}$$

   will not exceed h.

4. Let $f = f(x)$ be a function that has bounded second derivative: $\forall x: \; |f''(x)| \le 1$. For any x, the value of the function $f(x)$ is measured and comes out to be equal to some $f^*(x)$; in so doing we assume that the accuracy of the measurement guarantees the following estimate:

   $$|f(x) - f^*(x)| \le \varepsilon, \qquad \varepsilon > 0.$$

   Suppose now that we need to approximately evaluate the first derivative $f'(x)$.

[1] Hereafter, we will be using the symbol * to indicate the increased level of difficulty for a given problem.

a) How shall one choose the parameter h so that to minimize the guaranteed error estimate of the approximate formula:

$$f'(x) \approx \frac{f^*(x+h) - f^*(x)}{h}.$$

b) Show that given the existing uncertainty in the input data, the unavoidable error of evaluating $f'(x)$ is at least $\mathcal{O}(\sqrt{\varepsilon})$, no matter what specific method is used.

Hint. Consider two functions, $f(x) \equiv 0$ and $f^*(x) = \varepsilon \sin(x/\sqrt{\varepsilon})$. Clearly, the absolute value of the second derivative for either of these two functions does not exceed 1. Moreover, $\max_x |f(x) - f^*(x)| \leq \varepsilon$. At the same time,

$$\left| \frac{df^*}{dx} - \frac{df}{dx} \right| = \left| \sqrt{\varepsilon} \cos \frac{x}{\sqrt{\varepsilon}} \right| = \mathcal{O}(\sqrt{\varepsilon}).$$

By comparing the solutions of sub-problems a) and b), verify that the specific approximate formula for $f'(x)$ given in a) yields the error of the irreducible order $\mathcal{O}(\sqrt{\varepsilon})$; and also show that the unavoidable error is, in fact, exactly of order $\mathcal{O}(\sqrt{\varepsilon})$.

5. For storing the information about a linear function $f(x) = kx + b$, $\alpha \leq x \leq \beta$, that satisfies the inequalities: $0 \leq f(x) \leq 1$, we use a table with six available cells, such that one of the ten digits: $0, 1, 2, \ldots, 9$, can be written into each cell.

   What is the unavoidable error of reconstructing the function, if the foregoing six cells of the table are filled according to one of the following recipes?

   a) The first three cells contain the first three digits that appear right after the decimal point when the number $f(\alpha)$ is represented as a normalized decimal fraction; and the remaining three cells contain the first three digits after the decimal point in the normalized decimal fraction for $f(\beta)$.

   b) Let $\alpha = 0$ and $\beta = 10^{-2}$. The first three cells contain the first three digits in the normalized decimal fraction for k, the fourth cell contains either 0 or 1 depending on the sign of k, and the remaining two cells contain the first two digits after the decimal point in the normalized decimal fraction for b.

   c)* Show that irrespective of any specific strategy for filling out the aforementioned six-cell table, the unavoidable error of reconstructing the linear function $f(x) = kx + b$ is always at least 10^{-3}.

   Hint. Build 10^6 different functions from the foregoing class, such that the maximum modulus of the difference between any two of them will be at least 10^{-3}.

1.4 On Methods of Computation

Suppose that a mathematical model is constructed for studying a given object or phenomenon, and subsequently this model is analyzed using mathematical and computational means. For example, under certain assumptions the following problem:

$$\begin{aligned} &\frac{d^2y}{dt^2} + y = 0, \quad t \geq 0, \\ &y(0) = 0, \quad \left.\frac{dy}{dt}\right|_{t=0} = 1, \end{aligned} \tag{1.16}$$

can provide an adequate mathematical model for small oscillations of a pendulum, where $y(t)$ is the pendulum displacement from its equilibrium at the time t.

A study of harmonic oscillations based on this mathematical model, i.e., on the Cauchy problem (1.16), can benefit from a priori knowledge about the physical nature of the object of study. In particular, one can predict, based on physical reasoning, that the motion of the pendulum will be periodic. However, once the mathematical model (1.16) has been built, it becomes a separate and independent object that can be investigated using any available mathematical tools, including those that have little or no relation to the physical origins of the problem. For example, the numerical value of the solution $y = \sin t$ to problem (1.16) at any given moment of time $t = z$ can be obtained by expanding $\sin z$ into the Taylor series:

$$\sin z = z - \frac{z^3}{3!} + \frac{z^5}{5!} - \dots,$$

and subsequently taking its appropriate partial sum. In so doing, representation of the function $\sin t$ as a power series hardly admits any tangible physical interpretation.

In general, when solving a given problem on the computer, many different methods, or different algorithms, can be used. Some of them may prove far superior to others. In subsequent parts of the book, we are going to describe a number of established, robust and efficient, algorithms for frequently encountered classes of problems in numerical analysis. In the meantime, let us briefly explain how the algorithms may differ.

Assume that for computing the solution y to a given problem we can employ two algorithms, A_1 and A_2, that yield the approximate solutions $y_1^* = A_1(X)$ and $y_2^* = A_2(X)$, respectively, where X denotes the entire required set of the input data. In so doing, a variety of situations may occur.

1.4.1 Accuracy

The algorithm A_2 may be more accurate than the algorithm A_1, that is:

$$|y - y_1^*| \gg |y - y_2^*|.$$

For example, let us approximately evaluate $y = \sin x\big|_{x=0.1}$ using the expansion:

$$y_n^* = \sum_{k=1}^{n} (-1)^{k-1} \frac{x^{2k-1}}{(2k-1)!}. \tag{1.17}$$

The algorithm A_1 will correspond to taking $n=1$ in formula (1.17), and the algorithm A_2 will correspond to taking $n=2$ in formula (1.17). Then, obviously,

$$|\sin 0.1 - y_1^*| \gg |\sin 0.1 - y_2^*|.$$

1.4.2 Operation Count

Both algorithms may provide the same accuracy, but the computation of $y_1^* = A_1(X)$ may require many more arithmetic operations than the computation of $y_2^* = A_2(X)$. Suppose, for example, that we need to find the value of

$$y = 1 + x + x^2 + \ldots + x^{1023} \qquad \left(\text{clearly, } y = \frac{1-x^{1024}}{1-x}\right)$$

for $x = 0.99$. Let A_1 be the algorithm that would perform the computations directly using the given formula, i.e., by raising 0.99 to the powers $1, 2, \ldots, 1023$ one after another, and subsequently adding the results. Let A_2 be the algorithm that would perform the computations according to the formula:

$$y = \frac{1-0.99^{1024}}{1-0.99}.$$

The accuracy of these two algorithms is the same — both are absolutely accurate provided that there are no round-off errors. However, the first algorithm requires considerably more arithmetic operations, i.e., it is computationally more expensive. Namely, for successively computing

$$x, \quad x^2 = x \cdot x, \quad \ldots, \quad x^{1023} = x^{1022} \cdot x,$$

one will have to perform 1022 multiplications. On the other hand, to compute 0.99^{1024} one only needs 10 multiplications:

$$0.99^2 = 0.99 \cdot 0.99, \quad 0.99^4 = 0.99^2 \cdot 0.99^2, \quad \ldots, \quad 0.99^{1024} = 0.99^{512} \cdot 0.99^{512}.$$

1.4.3 Stability

The algorithms, again, may yield the same accuracy, but $A_1(X)$ may be computationally unstable, whereas $A_2(X)$ may be stable. For example, to evaluate $y = \sin x$ with the prescribed tolerance $\varepsilon = 10^{-3}$, i.e., to guarantee $|y - y^*| \le 10^{-3}$, let us employ the same finite Taylor expansion as in formula (1.17):

$$y_1^* = y_1^*(x) = \sum_{k=1}^{n} (-1)^{k-1} \frac{x^{2k-1}}{(2k-1)!}, \tag{1.18}$$

where $n = n(\varepsilon)$ is to be chosen to ensure that the inequality

$$|y - y_1^*| \le 10^{-3}$$

will hold. The first algorithm A_1 will compute the result directly according to (1.18). If $|x| \leq \pi/2$, then by noticing that the following inequality holds already for $n = 5$:

$$\frac{1}{(2n-1)!}\left(\frac{\pi}{2}\right)^{2n-1} \leq 10^{-3},$$

we can reduce the sum (1.18) to

$$y_1^* = x - \frac{x^3}{3!} + \frac{x^5}{5!} - \frac{x^7}{7!}.$$

Clearly, the computations by this formula will only be weakly sensitive to round-off errors when evaluating each term on the right-hand side. Moreover, as for $|x| \leq \pi/2$, those terms rapidly decay when the power grows, there is no room for the cancellation of significant digits, and the algorithm A_1 will be computationally stable.

Consider now $|x| \gg 1$; for example, $x = 100$. Then, for achieving the prescribed accuracy of $\varepsilon = 10^{-3}$, the number n should satisfy the inequality:

$$\frac{100^{2n-1}}{(2n-1)!} \leq 10^{-3},$$

which yields an obvious conservative lower bound for n: $n > 49$. This implies that the terms in sum (1.18) become small only for sufficiently large n. At the same time, the first few leading terms in this sum will be very large. A small relative error committed when computing those terms will result in a large absolute error; and since taking a difference of large quantities to evaluate a small quantity $\sin x$ ($|\sin x| \leq 1$) is prone to the loss of significant digits (see Section 1.4.4), the algorithm A_1 in this case will be computationally unstable.

On the other hand, in the case of large x a stable algorithm A_2 for evaluating $\sin x$ is also easy to build. Let us represent a given x in the form $x = l\pi + z$, where $|z| \leq \pi/2$ and l is integer. Then,

$$\sin x = (-1)^l \sin z,$$

$$y_2^* = A_2(x) = (-1)^l \left(z - \frac{z^3}{3!} + \frac{z^5}{5!} - \frac{z^7}{7!}\right).$$

This algorithm has the same stability properties as the algorithm A_1 for $|x| \leq \pi/2$.

1.4.4 Loss of Significant Digits

Most typically, numerical instability manifests itself through a strong amplification of the small round-off errors in the course of computations. A key mechanism for the amplification is the loss of significant digits, which is a purely computer-related phenomenon that only occurs because the numbers inside a computer are represented as finite (binary) fractions (see Section 1.3.3). If computers could operate with infinite fractions (no rounding), then this phenomenon would not take place.

Consider two real numbers a and b represented in a computer by finite fractions with m significant digits after the decimal point:

$$a = 0.a_1a_2a_3\ldots a_m,$$
$$b = 0.b_1b_2b_3\ldots b_m.$$

We are assuming that both numbers are normalized and that they have the same exponent that we are leaving out for simplicity. Suppose that these two numbers are close to one another, i.e., that the first k out of the total of m digits coincide:

$$a_1 = b_1,\ a_2 = b_2,\ \ldots,\ a_k = b_k.$$

Then the difference $a-b$ will only have $m-k<m$ significant digits (provided that $a_{k+1} > b_{k+1}$, which we, again, assume for simplicity):

$$a-b = 0.\underbrace{0\ldots0}_{k}c_{k+1}\ldots c_m.$$

The reason for this reduction from m to $m-k$, which is called *the loss of significant digits,* is obvious. Even though the actual numbers a and b may be represented by the fractions much longer than m digits, or by infinite fractions, the computer simply has no information about anything beyond digit number m. Even if the result $a-b$ is subsequently normalized:

$$a-b = 0.c_{k+1}\ldots c_m\underbrace{c_{m+1}\ldots c_{m+k}}_{\text{artifacts}}\cdot\beta^{-k},$$

where β is the radix, or base ($\beta = 2$ for all computers), then the digits from c_{m+1} through c_{m+k} will still be completely artificial and will have nothing to do with the true representation of $a-b$.

It is clear that the loss of significant digits may lead to a very considerable degradation of the overall accuracy. The error once committed at an intermediate stage of the computation will not disappear and will rather "propagate" further and contaminate the subsequent results. Therefore, when organizing the computations, it is not advisable to compute small numbers as differences of large numbers. For example, suppose that we need to evaluate the function $f(x) = 1-\cos x$ for x which is close to 1. Then $\cos x$ will also be close to 1, and significant digits could be lost when computing $A_1(x) = 1-\cos x$. Of course, there is an easy fix for this difficulty. Instead of the original formula we should use $f(x) = A_2(x) = 2\sin^2\frac{x}{2}$.

The loss of significant digits may cause an instability even if the original continuous problem is well conditioned. Indeed, assume that we need to compute the value of the function $f(x) = \sqrt{x}-\sqrt{x-1}$. Conditioning of this problem can be judged by evaluating the maximum ratio of the resulting relative error in the solution over the eliciting relative error in the input data:

$$\sup_{\Delta x}\frac{|\Delta f|/|f|}{|\Delta x|/|x|} \approx |f'(x)|\frac{|x|}{|f|} = \frac{1}{2}\left|\frac{1}{\sqrt{x}} - \frac{1}{\sqrt{x-1}}\right|\frac{|x|}{|\sqrt{x}-\sqrt{x-1}|} = \frac{|x|}{2\sqrt{x}\sqrt{x-1}}.$$

For large x the previous quantity is approximately equal to $\frac{1}{2}$, which means that the problem is perfectly well conditioned. Yet we can expect to incur a loss of significant digits when $x \gg 1$. Consider, for example, $x = 12345$, and assume that we are operating in a six-digit decimal arithmetic. Then:

$$\sqrt{x-1} = 111.10355529865\ldots \approx 111.104,$$
$$\sqrt{x} = 111.10805551354\ldots \approx 111.108,$$

and consequently, $A_1(x) = \sqrt{x} - \sqrt{x-1} \approx 111.108 - 111.104 = 0.004$. At the same time, the true answer is $f(x) = 0.004500214891\ldots$, which implies that our approximate computation carries an error of roughly 11%. To understand where this error is coming from, consider f as a function of two arguments: $f = f(t_1,t_2) = t_1 - t_2$, where $t_1 = \sqrt{x}$ and $t_2 = \sqrt{x-1}$. Conditioning with respect to the second argument t_2 can be estimated as follows:

$$\left|\frac{\partial f}{\partial t_2}\right| \cdot \frac{|t_2|}{|f|} = \frac{t_2}{|t_1 - t_2|},$$

and we conclude that this number is large when t_2 is close to t_1, which is precisely the case for large x. In other words, although the entire function is well conditioned, there is an intermediate stage that is ill conditioned, and it gives rise to large errors in the course of computation. This example illustrates why it is practically impossible to design a stable numerical procedure for an ill conditioned continuous problem.

A remedy to overcome the previous hurdle is quite easy to find:

$$f(x) = A_2(x) = \frac{1}{\sqrt{x} + \sqrt{x-1}} \approx \frac{1}{111.104 + 111.108} = 0.00450020701\ldots.$$

This is a considerably more accurate answer.

Yet another example is given by the same quadratic equation $x^2 - 2\alpha x + 1 = 0$ as we considered in Section 1.2. The roots $x_{1,2}(\alpha) = \alpha \pm \sqrt{\alpha^2 - 1}$ have been found to be ill conditioned for α close to 1. However, for $\alpha \gg 1$ both roots are clearly well conditioned. In particular, for $x_2 = \alpha - \sqrt{\alpha^2 - 1}$ we have:

$$\left|\frac{dx_2(\alpha)}{d\alpha}\right| \cdot \frac{|\alpha|}{|x_2|} = \frac{\alpha}{\sqrt{\alpha^2 - 1}} \longrightarrow 1, \quad \text{as} \quad \alpha \longrightarrow +\infty.$$

Nevertheless, the computation by the formula $x_2(\alpha) = \alpha - \sqrt{\alpha^2 - 1}$ will obviously be prone to the loss of significant digits for large α. A cure may be to compute $x_1(\alpha) = \alpha + \sqrt{\alpha^2 - 1}$ and then $x_2 = 1/x_1$. Note that even for the equation $x^2 - 2\alpha x - 1 = 0$, for which both roots $x_{1,2}(\alpha) = \alpha \pm \sqrt{\alpha^2 + 1}$ are well conditioned for all α, the computation of $x_2(\alpha) = \alpha - \sqrt{\alpha^2 + 1}$ is still prone to the loss of significant digits and as such, to instability.

1.4.5 Convergence

Finally, the algorithm may be either convergent or divergent. Suppose we need to compute the value of $y = \ln(1+x)$. Let us employ the power series:

$$y = \ln(1+x) = x - \frac{x^2}{2} + \frac{x^3}{3} - \frac{x^4}{4} + \dots \tag{1.19}$$

and set

$$y^*(x) \approx y_n^* = \sum_{k=1}^{n} (-1)^{k+1} \frac{x^k}{k}. \tag{1.20}$$

In doing so, we will obtain a method of approximately evaluating $y = \ln(1+x)$ that will depend on n as a parameter.

If $|x| = q < 1$, then $\lim_{n\to\infty} y_n^*(x) = y(x)$, i.e., the error committed when computing $y(x)$ according to formula (1.20) will be vanishing as n increases. If, however, $x > 1$, then $\lim_{n\to\infty} y_n^*(x) = \infty$, because the convergence radius for the series (1.19) is $r = 1$. In this case the algorithm based on formula (1.20) diverges, and cannot be used for computations.

1.4.6 General Comments

Basically, the properties of continuous well-posedness and numerical stability, as well as those of ill and well conditioning, are independent. There are, however, certain relations between these concepts.

- First of all, it is clear that no numerical method can ever fix a continuous ill-posedness.[2]
- For a well-posed continuous problem there may be stable and unstable discretizations.
- Even for a well conditioned continuous problem one can still obtain both stable and unstable discretizations.
- For an ill-conditioned continuous problem a discretization will typically be unstable.

Altogether, we can say that numerical methods cannot improve things in the perspective of well-posedness and conditioning.

In the book, we are going to discuss some other characteristics of numerical algorithms as well. We will see the algorithms that admit easy parallelization, and those that are limited to sequential computations; algorithms that automatically adapt to specific characteristics of the input data, such as their smoothness, and those that only partially take it into account; algorithms that have a straightforward logical structure, as well as the more elaborate ones.

[2] The opposite of well-posedness, when there is no continuous dependence of the solution on the data.

Exercises

1. Propose an algorithm for evaluating $y = \ln(1+x)$ that would also apply to $x > 1$.

2. Show that the intermediate stages of the algorithm A_2 from page 17 are well conditioned, and there is no danger of losing significant digits when computing:

$$f(x) = A_2(x) = \frac{1}{\sqrt{x} + \sqrt{x-1}}.$$

3. Consider the problem of evaluating the sequence of numbers $x_0, x_1, \ldots, x_N$ that satisfy the difference equations:

$$2x_n - x_{n+1} = 1 + n^2/N^2, \quad n = 0, 1, \ldots, N-1,$$

and the additional condition:

$$x_0 + x_N = 1. \tag{1.21}$$

We introduce two algorithms for computing x_n. First, let

$$x_n = u_n + cv_n, \quad n = 0, 1, \ldots, N. \tag{1.22}$$

Then, in the algorithm A_1 we define u_n, $n = 0, 1, \ldots, N$, as solution of the system:

$$2u_n - u_{n+1} = 1 + n^2/N^2, \quad n = 0, 1, \ldots, N-1, \tag{1.23}$$

subject to the initial condition:

$$u_0 = 0. \tag{1.24}$$

Consequently, the sequence v_n, $n = 0, 1, \ldots, N$, is defined by the equalities:

$$2v_n - v_{n+1} = 0, \quad n = 0, 1, \ldots, N-1, \tag{1.25}$$

$$v_0 = 1, \tag{1.26}$$

and the constant c of (1.22) is obtained from the condition (1.21). In so doing, the actual values of u_n and v_n are computed consecutively using the formulae:

$$\begin{aligned} u_{n+1} &= 2u_n - (1 + n^2/N^2), \quad n = 0, 1, \ldots, \\ v_{n+1} &= 2^{n+1}, \quad n = 0, 1, \ldots. \end{aligned}$$

In the algorithm A_2, u_n, $n = 0, 1, \ldots, N$, is still defined as solution to system (1.23), but instead of the condition (1.24) an alternative condition $u_N = 0$ is employed. The sequence v_n, $n = 0, 1, \ldots, N$, is again defined as a solution to system (1.25), but instead of the condition (1.26) we use $v_N = 1$.

   a) Verify that the second algorithm, A_2, is stable while the first one, A_1, is ("violently") unstable.

   b) Implement both algorithms on the computer and try to compare their performance for $N = 10$ and for $N = 100$.

Part I

Interpolation of Functions. Quadratures

One of the key concepts in mathematics is that of a function. In the simplest case, the function $y = f(x)$, $a \le x \le b$, can be specified in the closed form, i.e., defined by means of a finite formula, say, $y = x^2$. This formula can subsequently be transformed into a computer code that will calculate the value of $y = x^2$ for every given x. In real-life settings, however, the functions of interest are rarely available in the closed form. Instead, a finite array of numbers, commonly referred to as the table, would often be associated with the function $y = f(x)$. By processing the numbers from the table in a particular prescribed way, one should be able to obtain an approximate value of the function $f(x)$ at any point x. For instance, a table can contain several leading coefficients of a power series for $f(x)$. In this case, processing the table would mean calculating the corresponding partial sum of the series.

Let us, for example, take the function

$$y = e^x, \quad 0 \le x \le 1, \quad e^x = 1 + \frac{x}{1!} + \frac{x^2}{2!} + \ldots + \frac{x^n}{n!} + \ldots,$$

for which the power series converges for all x, and consider the table

$$1, \frac{1}{1!}, \frac{1}{2!}, \ldots, \frac{1}{n!}$$

of its $n+1$ leading Taylor coefficients, where $n > 0$ is given. The larger the n, the more accurately can one reconstruct the function $f(x) = e^x$ from this table. In so doing, the formula

$$e^x \approx 1 + \frac{x}{1!} + \frac{x^2}{2!} + \ldots + \frac{x^n}{n!}$$

is used for processing the table.

In most cases, however, the table that is supposed to characterize the function $y = f(x)$ would not contain its Taylor coefficients, and would rather be obtained by sampling the values of this function at some finite set of points $x_0, x_1, \ldots, x_n \in [a, b]$. In practice, sampling can be rendered by either measurements or computations. This naturally gives rise to the problem of reconstructing (e.g., interpolating) the function $f(x)$ at the "intermediate" locations x that do not necessarily coincide with any of the nodes $x_0, x_1, \ldots, x_n$.

The two most widely used and most efficient interpolation techniques are algebraic interpolation and trigonometric interpolation. We are going to analyze both of them. In addition, in the current Part I of the book we will also consider the problem of evaluating definite integrals of a given function when the latter, again, is specified by a finite table of its numerical values. The motivation behind considering this problem along with interpolation is that the main approaches to approximate evaluation of definite integrals, i.e., to obtaining the so-called quadrature formulae, are very closely related to the interpolation techniques.

Before proceeding further, let us also mention several books for additional reading on the subject: [Hen64, IK66, CdB80, Atk89, PT96, QSS00, Sch02, DB03].

Chapter 2

Algebraic Interpolation

Let $x_0, x_1, \ldots, x_n$ be a given set of points, and let $f(x_0), f(x_1), \ldots, f(x_n)$ be values of the function $f(x)$ at these points (assumed known). The one-to-one correspondence

x_0	x_1	$\ldots$	x_n
$f(x_0)$	$f(x_1)$	$\ldots$	$f(x_n)$

will be called *a table of values of the function* $f(x)$ *at the nodes* $x_0, x_1, \ldots, x_n$. We need to realize, of course, that for actual computer implementations one may only use the numbers that can be represented as finite binary fractions (Section 1.3.3 of the Introduction), whereas the values $f(x_j)$ do not necessarily have to belong to this class (e.g., $\sqrt{3}$). Therefore, the foregoing table may, in fact, contain rounded rather than true values of the function $f(x)$.

A polynomial $P_n(x) \equiv P_n(x, f, x_0, x_1, \ldots, x_n)$ of degree no greater than n that has the form

$$P_n(x) = c_0 + c_1 x + \ldots + c_n x^n$$

and coincides with $f(x_0), f(x_1), \ldots, f(x_n)$ at the nodes $x_0, x_1, \ldots, x_n$, respectively, is called *the algebraic interpolating polynomial.*

2.1 Existence and Uniqueness of Interpolating Polynomial

2.1.1 The Lagrange Form of Interpolating Polynomial

THEOREM 2.1

Let $x_0, x_1, \ldots, x_n$ *be a given set of distinct interpolation nodes, and let the values* $f(x_0), f(x_1), \ldots, f(x_n)$ *of the function* $f(x)$ *be known at these nodes. There is one and only one algebraic polynomial* $P_n(x) \equiv P_n(x, f, x_0, x_1, \ldots, x_n)$ *of degree no greater than n that would coincide with the given* $f(x_k)$ *at the nodes* x_k, $k = 0, 1, \ldots, n$.

PROOF We will first show that there may be no more than one interpolating polynomial, and will subsequently construct it explicitly.

Assume that there are two algebraic interpolating polynomials, $P_n^{(1)}(x)$ and $P_n^{(2)}(x)$. Then, the difference between these two polynomials, $R_n(x) = P_n^{(1)}(x) - P_n^{(2)}(x)$, is also a polynomial of degree no greater than n that vanishes at the $n+1$ points $x_0, x_1, \dots, x_n$. However, for any polynomial that is not identically equal to zero, the number of roots (counting their multiplicities) is equal to the degree. Therefore, $R_n(x) \equiv 0$, i.e., $P_n^{(1)}(x) \equiv P_n^{(2)}(x)$, which proves uniqueness.

Let us now introduce the auxiliary polynomials

$$l_k(x) = \frac{(x-x_0)(x-x_1)\dots(x-x_{k-1})(x-x_{k+1})\dots(x-x_n)}{(x_k-x_0)(x_k-x_1)\dots(x_k-x_{k-1})(x_k-x_{k+1})\dots(x_k-x_n)}.$$

It is clear that each $l_k(x)$ is a polynomial of degree no greater than n, and that the following equalities hold:

$$l_k(x_j) = \begin{cases} 1, & x_j = x_k, \\ 0, & x_j \neq x_k, \end{cases} \qquad j = 0, 1, \dots, n.$$

Then, the polynomial $P_n(x)$ given by the equality

$$\begin{aligned} P_n(x) &\stackrel{\text{def}}{=} P_n(x, f, x_0, x_1, \dots, x_n) \\ &= f(x_0)l_0(x) + f(x_1)l_1(x) + \dots + f(x_n)l_n(x) \end{aligned} \tag{2.1}$$

is precisely the interpolating polynomial that we are seeking. Indeed, its degree is no greater than n, because each term $f(x_j)l_j(x)$ is a polynomial of degree no greater than n. Moreover, it is clear that this polynomial satisfies the equalities $P_n(x_j) = f(x_j)$ for all $j = 0, 1, \dots, n$. $\square$

Let us emphasize that not only have we proven Theorem 2.1, but we have also written the interpolating polynomial explicitly using formula (2.1). This formula is known as *the Lagrange form of the interpolating polynomial.* There are other convenient forms of the unique interpolating polynomial $P_n(x, f, x_0, x_1, \dots, x_n)$. The Newton form is used particularly often.

2.1.2 The Newton Form of Interpolating Polynomial. Divided Differences

Let $f(x_a), f(x_b), f(x_c), f(x_d)$, etc., be values of the function $f(x)$ at the given nodes x_a, x_b, x_c, x_d, etc. A Newton's divided difference of order zero $f(x_k)$ of the function $f(x)$ at the point x_k is defined as simply the value of the function at this point:

$$f(x_k) = f(x_k), \quad k = a, b, c, d, \dots.$$

A divided difference of order one $f(x_k, x_m)$ of the function $f(x)$ is defined for an arbitrary pair of points x_k, x_m (x_k and x_m do not have to be neighbors, and we allow

$x_k \gtrless x_m$) through the previously introduced divided differences of order zero:

$$f(x_k, x_m) = \frac{f(x_m) - f(x_k)}{x_m - x_k}.$$

In general, a divided difference of order n: $f(x_0, x_1, \dots, x_n)$ for the function $f(x)$ is defined through the preceding divided differences of order $n-1$ as follows:

$$f(x_0, x_1, \dots, x_n) = \frac{f(x_1, x_2, \dots, x_n) - f(x_0, x_1, \dots, x_{n-1})}{x_n - x_0}. \tag{2.2}$$

Note that all the points $x_0, x_1, \dots, x_n$ in formula (2.2) have to be distinct, but they do not have to be arranged in any particular way, say, from the smallest to the largest value of x_j or vice versa.

Having defined the Newton divided differences[1] according to (2.2), we can now represent the interpolating polynomial $P_n(x, f, x_0, x_1, \dots, x_n)$ in the following *Newton form:*

$$\begin{aligned} P_n(x, f, x_0, x_1, \dots, x_n) = & f(x_0) + (x - x_0) f(x_0, x_1) + \dots \\ & + (x - x_0)(x - x_1) \dots (x - x_{n-1}) f(x_0, x_1, \dots, x_n). \end{aligned} \tag{2.3}$$

Formula (2.3) itself will be proven later. In the meantime, we will rather establish several useful corollaries that it implies.

COROLLARY 2.1
The following equality holds:

$$\begin{aligned} P_n(x, f, x_0, x_1, \dots, x_n) = & P_{n-1}(x, f, x_0, x_1, \dots, x_{n-1}) \\ & + (x - x_0)(x - x_1) \dots (x - x_{n-1}) f(x_0, x_1, \dots, x_n). \end{aligned} \tag{2.4}$$

PROOF Immediately follows from formula (2.3). ∎

COROLLARY 2.2
The divided difference $f(x_0, x_1, \dots, x_n)$ of order n is equal to the coefficient c_n in front of the term x^n in the interpolating polynomial

$$P_n(x, f, x_0, x_1, \dots, x_n) = c_n x^n + c_{n-1} x^{n-1} + \dots + c_0.$$

In other words, the following equality holds:

$$f(x_0, x_1, \dots, x_n) = c_n. \tag{2.5}$$

[1] A more detailed account of divided differences and their role in building the interpolating polynomials can be found, e.g., in [CdB80].

PROOF It is clear that the monomial x^n on the right-hand side of expression (2.3) is multiplied by the coefficient $f(x_0,x_1,\ldots,x_n)$. ☐

COROLLARY 2.3
The divided difference $f(x_0,x_1,\ldots,x_n)$ may be equal to zero if and only if the quantities $f(x_0)$, $f(x_1)$, $\ldots$, $f(x_n)$ are nodal values of some polynomial $Q_m(x)$ of degree m that is strictly less than n $(m<n)$.

PROOF If $f(x_0,x_1,\ldots,x_n)=0$, then formula (2.3) implies that the degree of the interpolating polynomial $P_n(x,f,x_0,x_1,\ldots,x_n)$ is less than n, because according to equality (2.5) the coefficient c_n in front of x^n is equal to zero. As the nodal values of this interpolating polynomial are equal to $f(x_j)$, $j=0,1,\ldots,n$, we can simply set $Q_m(x)=P_n(x)$. Conversely, as the interpolating polynomial of degree no greater than n is unique (Theorem 2.1), the polynomial $Q_m(x)$ with nodal values $f(x_0)$, $f(x_1)$, $\ldots$, $f(x_n)$ must coincide with the interpolating polynomial $P_n(x,f,x_0,x_1,\ldots,x_n)=c_nx^n+c_{n-1}x^{n-1}+\ldots+c_0$. As $m<n$, equality $Q_m(x)=P_n(x)$ implies that $c_n=0$. Then, according to formula (2.5), $f(x_0,x_1,\ldots,x_n)=0$. ☐

COROLLARY 2.4
The divided difference $f(x_0,x_1,\ldots,x_n)$ remains unchanged under any arbitrary permutation of its arguments $x_0,x_1,\ldots,x_n$.

PROOF Due to its uniqueness, the interpolating polynomial $P_n(x)$ will not be affected by the order of the interpolation nodes. Let $x'_0,x'_1,\ldots,x'_n$ be a permutation of $x_0,x_1,\ldots,x_n$; then, $\forall x$: $P_n(x,f,x_0,x_1,\ldots,x_n)=P_n(x,f,x'_0,x'_1,\ldots,x'_n)$. Consequently, along with formula (2.3) one can write

$$P_n(x,f,x_0,x_1,\ldots,x_n)=f(x'_0)+(x-x'_0)f(x'_0,x'_1)+\ldots\\ +(x-x'_0)(x-x'_1)\ldots(x-x'_{n-1})f(x'_0,x'_1,\ldots,x'_n).$$

According to Corollary 2.2, one can therefore conclude that

$$f(x'_0,x'_1,\ldots,x'_n)=c_n. \tag{2.6}$$

By comparing formulae (2.5) and (2.6), one can see that $f(x_0,x_1,\ldots,x_n)=f(x'_0,x'_1,\ldots,x'_n)$. ☐

COROLLARY 2.5
The following equality holds:

$$f(x_0,x_1,\ldots,x_n)=\frac{f(x_n)-P_{n-1}(x_n,f,x_0,x_1,\ldots,x_{n-1})}{(x_n-x_0)(x_n-x_1)\ldots(x_n-x_{n-1})}. \tag{2.7}$$

PROOF Let us set $x = x_n$ in equality (2.4); then its left-hand side becomes equal to $f(x_n)$, and formula (2.7) follows. ∎

THEOREM 2.2

The interpolating polynomial $P_n(x, f, x_0, x_1, \ldots, x_n)$ can be represented in the Newton form, i.e., equality (2.3) does hold.

PROOF We will use induction with respect to n. For $n = 0$ (and $n = 1$) formula (2.3) obviously holds. Assume now that it has already been justified for $n = 1, 2, \ldots, k$, and let us show that it will also hold for $n = k + 1$. In other words, let us prove the following equality:

$$\begin{aligned} P_{k+1}(x, f, x_0, x_1, \ldots, x_k, x_{k+1}) &= P_k(x, f, x_0, x_1, \ldots, x_k) \\ &\quad + f(x_0, x_1, \ldots, x_k, x_{k+1})(x - x_0)(x - x_1)\ldots(x - x_k). \end{aligned} \tag{2.8}$$

Notice that due to the assumption of the induction, formula (2.3) is valid for $n \le k$. Consequently, the proofs of Corollaries 2.1 through 2.5 that we have carried out on the basis of formula (2.3) will also remain valid for $n \le k$.

To prove equality (2.8), we will first demonstrate that the polynomial $P_{k+1}(x, f, x_0, x_1, \ldots, x_k, x_{k+1})$ can be represented in the form:

$$\begin{aligned} &P_{k+1}(x, f, x_0, x_1, \ldots, x_k, x_{k+1}) = P_k(x, f, x_0, x_1, \ldots, x_k) \\ &\quad + \frac{f(x_{k+1}) - P_k(x_{k+1}, f, x_0, x_1, \ldots, x_k)}{(x_{k+1} - x_0)(x_{k+1} - x_1)\ldots(x_{k+1} - x_k)}(x - x_0)(x - x_1)\ldots(x - x_k). \end{aligned} \tag{2.9}$$

Indeed, it is clear that on the right-hand side of formula (2.9) we have a polynomial of degree no greater than $k + 1$ that is equal to $f(x_j)$ at all nodes x_j, $j = 0, 1, \ldots, k + 1$. Therefore, the expression on the the right-hand side of (2.9) is actually the interpolating polynomial

$$P_{k+1}(x, f, x_0, x_1, \ldots, x_k, x_{k+1}),$$

which proves that (2.9) is a true equality. Next, by comparing formulae (2.8) and (2.9) we see that in order to justify (2.8) we need to establish the equality:

$$f(x_0, x_1, \ldots, x_k, x_{k+1}) = \frac{f(x_{k+1}) - P_k(x_{k+1}, f, x_0, x_1, \ldots, x_k)}{(x_{k+1} - x_0)(x_{k+1} - x_1)\ldots(x_{k+1} - x_k)}. \tag{2.10}$$

Using the same argument as in the proof of Corollary 2.4, and also employing Corollary 2.1, we can write:

$$\begin{aligned} P_k(x, f, x_0, x_1, \ldots, x_k) &= P_k(x, f, x_1, x_2, \ldots, x_k, x_0) \\ &= P_{k-1}(x, f, x_1, x_2, \ldots, x_k) \\ &\quad + f(x_1, x_2, \ldots, x_k, x_0)(x - x_1)(x - x_2)\ldots(x - x_k). \end{aligned} \tag{2.11}$$

Then, by substituting $x = x_{k+1}$ into (2.11), we can transform the right-hand side of equality (2.10) into:

$$\begin{aligned}&\frac{f(x_{k+1}) - P_k(x_{k+1}, f, x_0, x_1, \ldots, x_k)}{(x_{k+1}-x_0)(x_{k+1}-x_1)\ldots(x_{k+1}-x_k)}\\ &= \frac{1}{x_{k+1}-x_0}\,\frac{f(x_{k+1}) - P_{k-1}(x_{k+1}, f, x_1, \ldots, x_k)}{(x_{k+1}-x_1)\ldots(x_{k+1}-x_k)} - \frac{f(x_1, x_2, \ldots, x_k, x_0)}{x_{k+1}-x_0}.\end{aligned} \tag{2.12}$$

By virtue of Corollary 2.5, the minuend on the right-hand side of equality (2.12) is equal to:

$$\frac{1}{x_{k+1}-x_0} f(x_1, x_2, \ldots, x_k, x_{k+1}),$$

whereas in the subtrahend, according to Corollary 2.4, one can change the order of the arguments so that it would coincide with

$$\frac{f(x_0, x_1, \ldots, x_k)}{x_{k+1}-x_0}.$$

Consequently, the right-hand side of equality (2.12) is equal to

$$\frac{f(x_1, x_2, \ldots, x_{k+1}) - f(x_0, x_1, \ldots, x_k)}{x_{k+1}-x_0} \stackrel{\text{def}}{=} f(x_0, x_1, \ldots, x_{k+1}).$$

In other words, equality (2.12) coincides with equality (2.10) that we need to establish in order to justify formula (2.8). This completes the proof. □

THEOREM 2.3

Let $x_0 < x_1 < \ldots < x_n$; assume also that the function $f(x)$ is defined on the interval $x_0 \le x \le x_n$, and is n times differentiable on this interval. Then,

$$n!f(x_0, x_1, \ldots, x_n) = \left.\frac{d^n f}{dx^n}\right|_{x=\xi} \equiv f^{(n)}(\xi), \tag{2.13}$$

where ξ is some point from the interval $[x_0, x_n]$.

PROOF Consider an auxiliary function

$$\varphi(x) \stackrel{\text{def}}{=} f(x) - P_n(x, f, x_0, x_1, \ldots, x_n) \tag{2.14}$$

defined on $[x_0, x_n]$; it obviously has a minimum of $n+1$ zeros on this interval located at the nodes $x_0, x_1, \ldots, x_n$. Then, according to the Rolle (mean value) theorem, its first derivative vanishes at least at one point in between every two neighboring zeros of $\varphi(x)$. Therefore, the function $\varphi'(x)$ will have a minimum of n zeros on the interval $[x_0, x_n]$. Similarly, the function $\varphi''(x)$ vanishes at least at one point in between every two neighboring zeros of $\varphi'(x)$, and will therefore have a minimum of $n-1$ zeros on $[x_0, x_n]$.

By continuing this line of argument, we conclude that the n-th derivative $\varphi^{(n)}(x)$ will have at least one zero on the interval $[x_0, x_n]$. Let us denote this zero by ξ, so that $\varphi^{(n)}(\xi) = 0$. Next, we differentiate identity (2.14) exactly n times and subsequently substitute $x = \xi$, which yields:

$$0 = \varphi^{(n)}(\xi) = f^{(n)}(\xi) - \frac{d^n}{dx^n} P_n(x, f, x_0, x_1, \ldots, x_n)\bigg|_{x=\xi}. \tag{2.15}$$

On the other hand, according to Corollary 2.2, the divided difference $f(x_0, x_1, \ldots, x_n)$ is equal to the leading coefficient of the interpolating polynomial P_n, i.e., $P_n(x, f, x_0, x_1, \ldots, x_n) = f(x_0, x_1, \ldots, x_n)x^n + c_{n-1}x^{n-1} + \ldots + c_0$. Consequently, $\frac{d^n}{dx^n} P_n(x, f, x_0, x_1, \ldots, x_n) = n! f(x_0, x_1, \ldots, x_n)$, and therefore, equality (2.15) implies (2.13). □

THEOREM 2.4

The values $f(x_0), f(x_1), \ldots, f(x_n)$ of the function $f(x)$ are expressed through the divided differences $f(x_0)$, $f(x_0, x_1), \ldots$, $f(x_0, x_1, \ldots, x_n)$ by the formulae:

$$\begin{aligned} f(x_j) &= f(x_0) + (x_j - x_0) f(x_0, x_1) + (x_j - x_0)(x_j - x_1) f(x_0, x_1, x_2) \\ &\quad + (x_j - x_0)(x_j - x_1) \ldots (x_j - x_{n-1}) f(x_0, x_1, \ldots, x_n), \qquad j = 0, 1, \ldots, n, \end{aligned}$$

i.e., by linear combinations of the type:

$$f(x_j) = a_{j0} f(x_0) + a_{j1} f(x_0, x_1) + \ldots + a_{jn} f(x_0, x_1, \ldots, x_n), \qquad j = 0, 1, \ldots, n. \tag{2.16}$$

PROOF The result follows immediately from formula (2.3) and equalities $f(x_j) = P(x, f, x_0, x_1, \ldots, x_n)\big|_{x=x_j}$ for $j = 0, 1, \ldots, n$. □

2.1.3 Comparison of the Lagrange and Newton Forms

To evaluate the function $f(x)$ at a point x that is not one of the interpolation nodes, one can approximately set: $f(x) \approx P_n(x, f, x_0, x_1, \ldots, x_n)$.

Assume that the polynomial $P_n(x, f, x_0, x_1, \ldots, x_n)$ has already been built, but in order to try and improve the accuracy we incorporate an additional interpolation node x_{n+1} and the corresponding function value $f(x_{n+1})$. Then, to construct the interpolating polynomial $P_{n+1}(x, f, x_0, x_1, \ldots, x_{n+1})$ using the Lagrange formula (2.1) one basically needs to start from the scratch. At the same time, to use the Newton formula (2.3), see also Corollary 2.1:

$$\begin{aligned} P_{n+1}(x, f, x_0, x_1, \ldots, x_{n+1}) &= P_n(x, f, x_0, x_1, \ldots, x_n) \\ &\quad + (x - x_0)(x - x_1) \ldots (x - x_n) f(x_0, x_1, \ldots, x_{n+1}) \end{aligned}$$

one only needs to obtain the correction

$$(x - x_0)(x - x_1) \ldots (x - x_n) f(x_0, x_1, \ldots, x_{n+1}).$$

Moreover, one will immediately be able to see how large this correction is.

2.1.4 Conditioning of the Interpolating Polynomial

Let all the interpolation nodes $x_0, x_1, \ldots, x_n$ belong to some interval $a \le x \le b$. Let also the values $f(x_0)$, $f(x_1), \ldots$, $f(x_n)$ of the function $f(x)$ at these nodes be given. Hereafter, we will be using a shortened notation $P_n(x,f)$ for the interpolating polynomial $P_n(x) = P_n(x, f, x_0, x_1, \ldots, x_n)$.

Let us now perturb the values $f(x_j)$ by some quantities $\delta f(x_j)$, $j = 0, 1, \ldots, n$. Then, the interpolating polynomial $P_n(x,f)$ will change and become $P_n(x, f+\delta f)$. One can clearly see from the Lagrange formula (2.1) that $P_n(x, f+\delta f) = P_n(x,f) + P_n(x, \delta f)$. Therefore, the corresponding perturbation of the interpolating polynomial, i.e., its response to δf, will be $P_n(x, \delta f)$. For a given fixed set of $x_0, x_1, \ldots, x_n$, this perturbation depends only on δf and not on f itself. As such, one can introduce *the minimum number* L_n such that the following inequality would hold for any δf:

$$\max_{a \le x \le b} |P_n(x, \delta f)| \le L_n \max_j |\delta f(x_j)|. \tag{2.17}$$

The numbers $L_n = L_n(x_0, x_1, \ldots, x_n, a, b)$ are called *the Lebesgue constants.*[2] They provide a natural measure for the sensitivity of the interpolating polynomial to the perturbations $\delta f(x_j)$ of the interpolated function $f(x)$ at the nodes x_j. The Lebesgue constants are known to grow as n increases. Their specific behavior strongly depends on how the interpolation nodes x_j, $j = 0, 1, \ldots, n$, are located on the interval $[a, b]$.

If, for example, $n = 1$, $x_0 = a$, $x_1 = b$, then $L_1 = 1$. If, however, $x_0 \ne a$ and/or $x_1 \ne b$, then $L_1 \ge \frac{b-a}{2|x_1 - x_0|}$, i.e., if x_1 and x_0 are sufficiently close to one another, then the interpolation may appear arbitrarily sensitive to the perturbations of $f(x)$. The reader can easily verify the foregoing statements regarding L_1.

In the case of equally spaced interpolation nodes:

$$x_j = a + j \cdot h, \quad j = 0, 1, \ldots, n, \quad h = \frac{b-a}{n},$$

one can show that

$$2^n > L_n > 2^{n-2} \frac{1}{\sqrt{n}} \cdot \frac{1}{n - 1/2}. \tag{2.18}$$

In other words, the sensitivity of the interpolant to any errors committed when specifying the values of $f(x_j)$ will grow rapidly (exponentially) as n increases. Note that in practice it is impossible to specify the values of $f(x_j)$ without any error, no matter how these values are actually obtained, i.e., whether they are measured (with inevitable experimental inaccuracies) or computed (subject to rounding errors).

For a rigorous proof of inequalities (2.18) we refer the reader to the literature on the theory of approximation, in particular, the monographs and texts cited in Section 3.2.7 of Chapter 3. However, an elementary treatment can also be given, and one can easily provide a qualitative argument of why the Lebesgue constants for equidistant nodes grow exponentially as the grid dimension n increases. From the

[2]Note that the Lebesgue constant L_n corresponds to interpolation on $n+1$ nodes: $x_0, \ldots, x_n$.

Lagrange form of the interpolating polynomial (2.1) and definition (2.17) it is clear that:

$$L_n = \mathcal{O}\left(\max_{a\le x\le b}\sum_{k=0}^{n}|l_k(x)|\right) \tag{2.19}$$

(later, see Section 3.2.7 of Chapter 3, we will prove an even more precise statement). Take $k \approx n/2$ and x very close to one of the edges a or b, say, $x-a=\eta \ll h$. Then,

$$\begin{aligned}
|l_k(x)| &= \left|\frac{(x-x_0)(x-x_1)\dots(x-x_{k-1})(x-x_{k+1})\dots(x-x_n)}{(x_k-x_0)(x_k-x_1)\dots(x_k-x_{k-1})(x_k-x_{k+1})\dots(x_k-x_n)}\right| \\
&\approx \frac{\eta\cdot h^{2k-1}\cdot(2k)!/k}{(h^k k!)^2} = \eta\cdot h\cdot\frac{(2k)!}{k(k!)^2} \\
&= \eta\cdot h\cdot\frac{(2\cdot 4\cdot 6\cdot\ldots\cdot 2k)(1\cdot 3\cdot 5\cdot\ldots\cdot(2k-1))}{k(k!)^2} \\
&\approx \eta\cdot h\cdot\frac{(2\cdot 4\cdot 6\cdot\ldots\cdot 2k)^2}{(k!)^2} = \eta\cdot h\cdot\frac{2^{2k}(k!)^2}{(k!)^2} \approx \eta\cdot h\cdot 2^n.
\end{aligned}$$

The foregoing estimate for $|l_k(x)|$, along with the previous formula (2.19), do imply the exponential growth of the Lebesgue constants on uniform (equally spaced) interpolation grids. Let now $a=-1$, $b=1$, and let the interpolation nodes on $[a,b]$ be rather given by the formula:

$$x_j = -\cos\frac{(2j+1)\pi}{2(n+1)}, \quad j=0,1,\dots,n. \tag{2.20}$$

It is possible to show that placing the nodes according to (2.20) guarantees a much better estimate for the Lebesgue constants (again, see Section 3.2.7):

$$L_n \le \frac{2}{\pi}\ln(n+1)+1. \tag{2.21}$$

We therefore conclude that in contradistinction to the previous case (2.18), the Lebesgue constants may, in fact, grow slowly rather than rapidly, as they do on the non-equally spaced nodes (2.20). As such, even the high-degree interpolating polynomials in this case will not be overly sensitive to perturbations of the input data. Interpolation nodes (2.20) are known as the Chebyshev nodes. They will be discussed in detail in Chapter 3.

2.1.5 On Poor Convergence of Interpolation with Equidistant Nodes

One should not think that for any continuous function $f(x)$, $x\in[a,b]$, the algebraic interpolating polynomials $P_n(x,f)$ built on the equidistant nodes $x_j=a+j\cdot h$, $x_0=a$, $x_n=b$, will converge to $f(x)$ as n increases, i.e., that the deviation of $P_n(x,f)$ from $f(x)$ will decrease. For example, as has been shown by Bernstein, the sequence of interpolating polynomials obtained for the function $f(x) = |x|$ on equally spaced nodes diverges at every point of the interval $[a, b] = [-1, 1]$ except at $\{-1, 0, 1\}$.

The next example is attributed to Runge. Consider the function $f(x) = \frac{1}{x^2+1/4}$ on the same interval $[a, b] = [-1, 1]$; not only is this function continuous, but also has continuous derivatives of all orders. It is, however, possible to show that for the sequence of interpolating polynomials with equally spaced nodes the maximum difference $\max_{-1\le x\le 1} |f(x) - P_n(x, f)|$ will not approach zero as n increases.

Moreover, by working on Exercise 4 below, one will be able to see that the areas of no convergence for this function are located next to the endpoints of the interval $[-1, 1]$. For larger intervals the situation may even deteriorate and the sequence of interpolating polynomials $P_n(x, f)$ may diverge. In other words, the quantity $\max_{a\le x\le b} |f(x) - P_n(x, f)|$ may become arbitrarily large for large n's (see, e.g., [IK66]). Altogether, these convergence difficulties can be accounted for by the fact that on the complex plane the function $f(z) = \frac{1}{z^2+1/4}$ is not an entire function of its argument z, and has singularities at $z = \pm i/2$.

On the other hand, if, instead of the equidistant nodes, we use Chebyshev nodes (2.20) to interpolate either the Bernstein function $f(x) = |x|$ or the Runge function $f(x) = \frac{1}{x^2+1/4}$, then in both cases the sequence of interpolating polynomials $P_n(x, f)$ converges to $f(x)$ uniformly as n increases (see Exercise 5).

Exercises

1. Evaluate $f(1.14)$ by means of linear, quadratic, and cubic interpolation using the following table of values:

x	1.08	1.13	1.20	1.27	1.31
$f(x)$	1.302	1.386	1.509	1.217	1.284

Implement the interpolating polynomials in both the Lagrange and Newton form.

2. Let $x_j = j \cdot h$, $j = 0, \pm 1, \pm 2, \ldots$, be equidistant nodes with spacing h. Verify that the following equality holds:

$$f(x_{k-1}, x_k, x_{k+1}) = \frac{f(x_{k+1}) - 2f(x_k) + f(x_{k-1})}{2!h^2}.$$

3. Let $a = x_0$, $a < x_1 < b$, $x_2 = b$. Find the value of the Lebesgue constant L_2 when x_1 is the midpoint of $[a, b]$: $x_1 = (a+b)/2$. Show that if, conversely, $x_1 \to a$ or $x_1 \to b$, then the Lebesgue constant $L_2 = L_2(x_0, x_1, x_2, a, b)$ grows with no bound.

4. Plot the graphs of $f(x) = \frac{1}{x^2+1/4}$ and $P_n(x, f)$ from Section 2.1.5 (Runge example) on the computer and thus corroborate experimentally that there is no convergence of the interpolating polynomial on equally spaced nodes when n increases.

5. Use Chebyshev nodes (2.20) to interpolate $f(x) = |x|$ and $f(x) = \frac{1}{x^2+1/4}$ on the interval $[-1, 1]$, plot the graphs of each $f(x)$ and the corresponding $P_n(x, f)$ for $n = 10, 20, 40$, and 80, evaluate numerically the error $\max_{-1\le x\le 1} |f(x) - P_n(x, f)|$, and show that it decreases as n increases.

2.2 Classical Piecewise Polynomial Interpolation

High sensitivity of algebraic interpolating polynomials to the errors in the tabulated values of $f(x)$, as well as the "iffy" convergence of the sequence $P_n(x,f)$ on uniform grids, prompt the use of piecewise polynomial interpolation.

2.2.1 Definition of Piecewise Polynomial Interpolation

Let the function $f(x)$, $x \in [a, b]$, be defined by the table $\{f(x_0), f(x_1), \dots, f(x_n)\}$ of its numerical values at the nodes $\{a = x_0 < x_1 < x_2 < \dots < x_n = b\}$. To reconstruct this function in between the nodes $x_0, x_1, \dots, x_n$, one can use an auxiliary function that would coincide with a polynomial of a given low degree (say, the first, the second, the third, etc.) between every two neighboring nodes of the interpolation grid. This approach is known as *piecewise polynomial interpolation;* in particular, it may be *piecewise linear, piecewise quadratic, piecewise cubic, etc.*

In the case of piecewise linear interpolation on the interval $x_k \leq x \leq x_{k+1}$, one uses the linear interpolating polynomial $P_1(x, f, x_k, x_{k+1})$ to approximate the function $f(x)$. In the case of piecewise quadratic interpolation on the interval $x_k \leq x \leq x_{k+1}$, one can use either of the two polynomials: $P_2(x, f, x_k, x_{k+1}, x_{k+2})$ or $P_2(x, f, x_{k-1}, x_k, x_{k+1})$.

Piecewise polynomial interpolation of an arbitrary degree s is obtained similarly. There is always some flexibility in constructing the interpolant, and to approximate the function $f(x)$ on the interval $x_k \leq x \leq x_{k+1}$ one can basically use any of the polynomials $P_s(x, f, x_{k-j}, x_{k-j+1}, \dots, x_{k-j+s})$, where j is one of the integers $0, 1, \dots, s-1$. It is, however, desirable that the smaller interval $[x_k, x_{k+1}]$ be located maximally close to the middle of the larger interval $[x_{k-j}, x_{k-j+s}]$ (see Section 2.1.4). For equidistant nodes, the latter requirement translates into choosing j maximally close to $s/2$. In general, once the strategy for selecting j has been adopted, one can reconstruct $f(x)$ on $[a, b]$ in the form of a piecewise polynomial of degree s. It will be composed of the individual interpolating polynomials that correspond to different intervals $[x_k, x_{k+1}]$, $k = 0, 1, \dots, n-1$. For simplicity, we will hereafter denote the piecewise polynomial as follows:

$$P_s(x, f, x_{k-j}, x_{k-j+1}, \dots, x_{k-j+s}) = P_s(x, f_{kj}).$$

2.2.2 Formula for the Interpolation Error

Let us estimate the error

$$R_s(x) \stackrel{\text{def}}{=} f(x) - P_s(x, f_{kj}), \qquad x_k \leq x \leq x_{k+1}, \tag{2.22}$$

that arises when the function $f(x)$ is approximately replaced by the polynomial $P_s(x, f_{kj})$. To do so, we will need to exploit the following general theorem:

THEOREM 2.5

Let the function $f = f(t)$ be defined on $\alpha \leq t \leq \beta$, with a continuous derivative of order $s+1$ on this interval. Let $t_0, t_1, \dots, t_s$ be an arbitrary set of distinct points that all belong to $[\alpha, \beta]$, and let $f(t_0), f(t_1), \dots, f(t_s)$ be the values of the function $f(t)$ at these points. Finally, let $P_s(t) \equiv P_s(t, f, t_0, t_1, \dots, t_s)$ be the algebraic interpolating polynomial of degree no greater than s built for these given points and function values. Then, the interpolation error $R_s(t) = f(t) - P_s(t)$ can be represented on $[\alpha, \beta]$ as follows:

$$R_s(t) = \frac{f^{(s+1)}(\xi)}{(s+1)!}(t-t_0)(t-t_1)\dots(t-t_s), \tag{2.23}$$

where $\xi = \xi(t)$ is some point from the interval (α, β).

PROOF We first notice that formula (2.23) does hold for all nodes t_j, $j = 0, 1, \dots, s$, themselves, because on one hand $\forall t_j: \ f(t_j) - P_s(t_j) = 0$, and on the other hand, $R_s(t_j) = 0$, where $R_s(t)$ is defined by formula (2.23). Let us now take an arbitrary $\bar{t} \in [\alpha, \beta]$ that does not coincide with any of $t_0, t_1, \dots, t_s$. To prove formula (2.23) for $t = \bar{t}$, we introduce an auxiliary function:

$$\varphi(t) = f(t) - P_s(t) - k(t-t_0)(t-t_1)\dots(t-t_s) \tag{2.24}$$

and choose the parameter k so that $\varphi(\bar{t}) = 0$, which obviously implies

$$k = \frac{f(\bar{t}) - P_s(\bar{t})}{(\bar{t}-t_0)(\bar{t}-t_1)\dots(\bar{t}-t_s)}. \tag{2.25}$$

The numerator in formula (2.25) coincides with the value of the error $R_s(\bar{t})$, therefore, this formula yields:

$$R_s(\bar{t}) = k(\bar{t}-t_0)(\bar{t}-t_1)\dots(\bar{t}-t_s). \tag{2.26}$$

The auxiliary function φ of (2.24) clearly has a minimum of $s+2$ zeros on the interval $[\alpha, \beta]$ located at the points $\bar{t}, t_0, t_1, \dots, t_s$. Then, its first derivative $\varphi'(t)$ will have a minimum of $s+1$ zeros on the (open) interval (α, β), because according to the Rolle (mean value) theorem, the derivative $\varphi'(t)$ has to vanish at least once in between every two neighboring points where $\varphi(t)$ itself vanishes. Similarly, $\varphi''(t)$ will have at least s zeros on (α, β), $\varphi^{(3)}(t)$ will have at least $s-1$ zeros, etc., so that finally the derivative $\varphi^{(s+1)}(t)$ will have to have a minimum of one zero on the interval (α, β). Let us denote this zero by $\xi \in (\alpha, \beta)$, so that $\varphi^{(s+1)}(\xi) = 0$.

Next, we note that

$$\frac{d^{s+1}}{dt^{s+1}} t^{s+1} = (s+1)!$$

and that $(t-t_0)(t-t_1)\dots(t-t_s) = t^{s+1} + Q_s(t)$, where $Q_s(t)$ is a polynomial of degree no greater than s. We also note that

$$\frac{d^{s+1}}{dt^{s+1}} P_s(t) \equiv \frac{d^{s+1}}{dt^{s+1}} Q_s(t) \equiv 0.$$

Using the previous two expressions, we differentiate the function $\varphi(t)$ defined by formula (2.24) $s+1$ times and obtain:

$$\varphi^{(s+1)}(t) = f^{(s+1)}(t) - k(s+1)!.$$

Substituting $t = \xi$ into the last equality, and recalling that $\varphi^{(s+1)}(\xi) = 0$, we arrive at the following expression for k:

$$k = \frac{f^{(s+1)}(t)}{(s+1)!}.$$

Finally, by substituting k into equality (2.26) we obtain a formula for $R_s(\bar{t})$ that would actually coincide with formula (2.23) because $\bar{t} \in [\alpha, \beta]$ has been chosen arbitrarily. ∎

THEOREM 2.6

Under the assumptions of the previous theorem, the following estimate holds:

$$\max_{\alpha \le t \le \beta} |R_s(t)| \le \frac{1}{(s+1)!} \max_{\alpha \le t \le \beta} |f^{(s+1)}(t)|(\beta - \alpha)^{s+1}. \tag{2.27}$$

PROOF We first note that $\forall t \in [\alpha, \beta]$ the absolute value of each expression $t - t_0$, $t - t_1$, ..., $t - t_s$ will not exceed $\beta - \alpha$. Then, we use formula (2.23):

$$\begin{aligned} |R_s(t)| &= \frac{1}{(s+1)!} |f^{(s+1)}(\xi)(t - t_0)(t - t_1)\ldots(t - t_s)| \\ &\le \frac{1}{(s+1)!} \max_{\alpha \le t \le \beta} |f^{(s+1)}(t)|(\beta - \alpha)^{s+1}. \end{aligned} \tag{2.28}$$

As $t \in [\alpha, \beta]$ on the left-hand side of formula (2.28) is arbitrary, the required estimate (2.27) follows. ∎

Let us emphasize that we have proven inequality (2.27) for an arbitrary distribution of the (distinct) interpolation nodes $t_0, t_1, \ldots, t_s$ on the interval $[\alpha, \beta]$. For a given fixed distribution of nodes, estimate (2.27) can often be improved. For example, consider a piecewise linear interpolation and assume that the nodes t_0 and t_1 coincide with the endpoints α and β, respectively, of the interval $\alpha \le t \le \beta$. Then,

$$\begin{aligned} |R_1(t)| &= \left| \frac{f''(\xi)}{(s+1)!}(t - \alpha)(t - \beta) \right| \\ &\le \frac{1}{2} \max_{\alpha \le t \le \beta} |f''(t)| \max_{\alpha \le t \le \beta} |(t - \alpha)(t - \beta)| = \frac{1}{8} \max_{\alpha \le t \le \beta} |f''(t)|(\beta - \alpha)^2, \end{aligned}$$

which yields

$$\max_{\alpha \le t \le \beta} |R_1(t)| \le \frac{1}{8} \max_{\alpha \le t \le \beta} |f''(t)|(\beta - \alpha)^2, \tag{2.29}$$

whereas estimate (2.27) for $s=1$ transforms into

$$\max_{\alpha\le t\le\beta}|R_1(t)|\le\frac{1}{2}\max_{\alpha\le t\le\beta}|f''(t)|(\beta-\alpha)^2.$$

We will now use Theorems 2.5 and 2.6 to estimate the error (2.22) of piecewise polynomial interpolation of the function $f(x)$ on the interval $x_k\le x\le x_{k+1}$. First, let

$$\alpha=x_{k-j},\quad \beta=x_{k-j+s},\quad t_0=\alpha=x_{k-j},\quad t_1=x_{k-j+1},\ \dots,\ t_s=\beta=x_{k-j+s}.$$

Then, it is clear that

$$\max_{x_k\le x\le x_{k+1}}|R_s(x,f_{kj})|\le\max_{\alpha\le x\le\beta}|R_s(x,f_{kj})|,$$

and according to (2.27) we obtain

$$\max_{x_k\le x\le x_{k+1}}|R_s(x,f_{kj})|\le\frac{1}{(s+1)!}\max_{x_{k-j}\le x\le x_{k-j+s}}|f^{(s+1)}(x)|(x_{k-j+s}-x_{k-j})^{s+1}. \tag{2.30}$$

If the quantity $|f^{(s+1)}(x)|$ undergoes strong variations on the interval $[a,b]$, then, in order for the estimate (2.30) to guarantee some prescribed accuracy, it will be advantageous to have the grid size (distance between the neighboring nodes) and the value of $x_{k-j+s}-x_{k-j}$ smaller in those parts of $[a,b]$ where $|f^{(s+1)}(x)|$ is larger.

In the case of equidistant nodes $x_0,x_1,\dots,x_n$, estimate (2.30) implies

$$\max_{x_k\le x\le x_{k+1}}|R_s(x,f_{kj})|\le\frac{s^{s+1}}{(s+1)!}\max_{x_{k-j}\le x\le x_{k-j+s}}|f^{(s+1)}(x)|h^{s+1}, \tag{2.31}$$

where $h=(b-a)/n=x_{k+1}-x_k$ is the size of the interpolation grid. Inequality (2.31) can be recast as

$$\max_{x_k\le x\le x_{k+1}}|R_s(x,f_{kj})|\le\text{const}\cdot\max_{x_{k-j}\le x\le x_{k-j+s}}|f^{(s+1)}(x)|h^{s+1}, \tag{2.32}$$

where the key consideration is that the constant on the right-hand side of (2.32) *does not depend* on the grid size h.

To conclude this section, let us specifically mention the case of piecewise linear interpolation: $s=1$, $\alpha=x_k$, and $\beta=x_{k+1}$. Then, according to estimate (2.29), we have:

$$\max_{x_k\le x\le x_{k+1}}|R_1(x)|\le\frac{1}{8}\max_{x_k\le x\le x_{k+1}}|f''(x)|(x_{k+1}-x_k)^2=\frac{h^2}{8}\max_{x_k\le x\le x_{k+1}}|f''(x)|. \tag{2.33}$$

2.2.3 Approximation of Derivatives for a Grid Function

THEOREM 2.7

Let the function $f=f(x)$ be defined on the interval $[\alpha,\beta]$, and let it have a continuous derivative of order $s+1$ on this interval. Let $x_{k-j},x_{k-j+1},\dots,x_{k-j+s}$

be a set of interpolation nodes, such that $\alpha = x_{k-j} < x_{k-j+1} < \ldots < x_{k-j+s} = \beta$. *Then, to approximately evaluate the derivatives*

$$\frac{d^q f(x)}{dx^q}, \qquad q = 1, 2, \ldots, s,$$

of the function $f(x)$ *on the interval* $x_k \leq x \leq x_{k+1}$, *one can employ the interpolating polynomial* $P_s(x, f_{kj})$ *and set*

$$\frac{d^q f(x)}{dx^q} \approx \frac{d^q}{dx^q} P_s(x, f_{kj}), \qquad x_k \leq x \leq x_{k+1}. \tag{2.34}$$

In so doing, the approximation error will satisfy the estimate:

$$\begin{aligned} \max_{x_k \leq x \leq x_{k+1}} & \left| \frac{d^q f(x)}{dx^q} - \frac{d^q}{dx^q} P_s(x, f_{kj}) \right| \\ & \leq \frac{1}{(s-q+1)!} \max_{x_{k-j} \leq x \leq x_{k-j+s}} |f^{(s+1)}(x)| (x_{k-j+s} - x_{k-j})^{s-q+1}. \end{aligned} \tag{2.35}$$

PROOF Consider an auxiliary function $\varphi(x) \stackrel{\text{def}}{=} f(x) - P_s(x, f_{kj})$; it obviously vanishes at all $s+1$ interpolation nodes $x_{k-j}, x_{k-j+1}, \ldots, x_{k-j+s}$. Therefore, its first derivative $\varphi'(x)$ will have a minimum of s zeros on the interval $x_{k-j} \leq x \leq x_{k-j+s}$, because according to the Rolle (mean value) theorem, there is a zero of the function $\varphi'(x)$ in between any two neighboring zeros of $\varphi(x)$. Similarly, the function $\frac{d^q \varphi(x)}{dx^q}$ will have at least $s-q+1$ zeros on the interval $x_{k-j} \leq x \leq x_{k-j+s}$. This implies that the derivative $\frac{d^q f(x)}{dx^q}$ and the polynomial $\frac{d^q}{dx^q} P_s(x, f_{kj})$ of degree no greater than $s-q$ coincide at $s-q+1$ distinct points. In other words, the polynomial $P_s^{(q)}(x, f_{kj})$ can be interpreted as an interpolating polynomial of degree no greater than $s-q$ for the function $f^{(q)}(x)$ on the interval $x_{k-j} \leq x \leq x_{k-j+s}$, built on some set of $s-q+1$ interpolation nodes.

Moreover, the function $f^{(q)}(x)$ has a continuous derivative of order $s-q+1$ on $[\alpha, \beta]$:

$$\frac{d^{s-q+1}}{dx^{s-q+1}} f^{(q)}(x) = \frac{d^{s+1}}{dx^{s+1}} f(x).$$

Consequently, one can use Theorem 2.6 and, by setting $\alpha = x_{k-j}$, $\beta = x_{k-j+s}$, obtain the following estimate [cf. formula (2.27)]:

$$\begin{aligned} \max_{x_{k-j} \leq x \leq x_{k-j+s}} & \left| f^{(q)}(x) - P_s^{(q)}(x, f_{kj}) \right| \\ & \leq \frac{1}{(s-q+1)!} \max_{x_{k-j} \leq x \leq x_{k-j+s}} |f^{(s+1)}(x)| (x_{k-j+s} - x_{k-j})^{s-q+1}. \end{aligned}$$

As $\alpha = x_{k-j} \leq x_k < x_{k+1} \leq x_{k-j+s} = \beta$, it immediately yields (2.35). ∎

2.2.4 Estimate of the Unavoidable Error and the Choice of Degree for Piecewise Polynomial Interpolation

Let the function $f = f(x)$ be defined on the interval $[0,\pi]$, and let its values be known at the nodes of the uniform grid: $x_k = k\pi/n \equiv kh$, $k = 0,1,\dots,n$. Using only the tabulated values of the function $f(x_0), f(x_1),\dots,f(x_n)$, one cannot, even in principle, obtain an exact reconstruction of $f(x)$ in between the nodes, because different functions may have identical tables, i.e., may coincide at the nodes x_k, $k = 0,1,\dots,n$, and at the same time be different elsewhere. If, for example, in addition to the table of values nothing is known about the function $f(x)$ except that it is simply continuous, then one cannot guarantee any accuracy at all when reconstructing $f(x)$ at $x \neq x_k$, $k = 0,1,\dots,n$.

Assume now that $f(x)$ has a bounded derivative of the maximum order $s+1$:

$$\max_x |f^{(s+1)}(x)| \le M_s = \text{const.} \tag{2.36}$$

It is easy to find two different functions from the class characterized by $M_s = 1$:

$$f_1(x) = \frac{\sin nx}{n^{s+1}} \quad \text{and} \quad f_2(x) = -\frac{\sin nx}{n^{s+1}},$$

that would deviate from one another by the value of order h^{s+1}:

$$\max_{0\le x\le\pi} |f_1(x) - f_2(x)| = \max_{0\le x\le\pi} 2\left|\frac{\sin nx}{n^{s+1}}\right| = \frac{2}{\pi^{s+1}}h^{s+1}, \tag{2.37}$$

and for which the tables would nonetheless fully coincide (both will be trivial):

$$f_1(x_k) = f_2(x_k) = 0, \qquad k = 0,1,\dots,n.$$

We therefore conclude that given the tabulated values of the function $f(x)$, and only estimate (2.36) in addition to that, one cannot, even in theory, reconstruct the function $f(x)$ on the interval $0 \le x \le \pi$ with the accuracy better than $\mathcal{O}(h^{s+1})$. In other words, the error $\mathcal{O}(h^{s+1})$ is unavoidable when reconstructing the function $f(x)$, $0 \le x \le \pi$, using its table of values on a uniform grid with size h.

It is also clear that

$$\max_{0\le x\le\pi} \left|\frac{d^q f_1(x)}{dx^q} - \frac{d^q f_2(x)}{dx^q}\right| = 2\frac{1}{n^{s-q+1}} = \frac{2}{\pi^{s-q+1}}h^{s-q+1}, \tag{2.38}$$

which means that the unavoidable error when reconstructing the derivative $\frac{d^q f(x)}{dx^q}$ is at least $\mathcal{O}(h^{s-q+1})$.

By comparing equalities (2.37) and (2.38) with estimates of the error obtained in Sections 2.2.2 and 2.2.3 for the piecewise polynomial interpolation of the function $f(x)$ and its derivatives, we conclude that the interpolation error and the unavoidable error have the same asymptotic order (of smallness) with respect to the grid size h. If, under the condition (2.36), one still chooses to use interpolating polynomials of degree $r < s$, then the interpolation error (for the function itself) will be $\mathcal{O}(h^{r+1})$. In other words, there will be an additional loss of the order of accuracy, on top of the uncertainty-based unavoidable error $\mathcal{O}(h^{s+1})$ that is due to the specification of $f(x)$ through its discrete table of values.

On the other hand, the use of interpolation of a higher degree $r > s$ cannot increase the order of accuracy beyond the threshold set by the unavoidable error $\mathcal{O}(h^{s+1})$, and therefore cannot speed up the convergence as $h \longrightarrow 0$. As such, the degree s of piecewise polynomial interpolation is optimal for the functions that satisfy (2.36).

REMARK 2.1 The considerations of the current section pertain primarily to the asymptotic behavior of the error as $h \longrightarrow 0$. For a given fixed $h > 0$, interpolation of some degree $r < s$ may, in fact, appear more accurate than the interpolation of degree s. Besides, in practice the tabulated values $f(x_k)$, $k = 0, 1, \ldots, n$, may only be specified approximately, rather than exactly, with a finite fixed number of decimal (or binary) digits. In this case, the loss of interpolation accuracy due to rounding is going to increase as s increases, because of the growth of the Lebesgue constants (defined by formula (2.18) of Section 2.1.4). Therefore, the piecewise polynomial interpolation of high degree (higher than the third) is not used routinely. □

REMARK 2.2 Error estimate (2.32) does, in fact, imply *uniform convergence* of the interpolant $P_s(x, f_{kj})$ (a piecewise polynomial) to the interpolated function $f(x)$ with the rate $\mathcal{O}(h^{s+1})$ as the grid is refined, i.e., as $h \longrightarrow 0$. Estimate (2.33), in particular, indicates that piecewise linear interpolation converges uniformly with the rate $\mathcal{O}(h^2)$. Likewise, estimate (2.35) in the case of a uniform grid with size h will imply uniform convergence of the q-th derivative of the interpolant $P_s^{(q)}(x, f_{kj})$ to the q-th derivative of the interpolated function $f^{(q)}(x)$ with the rate $\mathcal{O}(h^{s-q+1})$ as $h \longrightarrow 0$. □

REMARK 2.3 The notion of unavoidable error as presented in this section (see also Section 1.3) illustrates the concept of *Kolmogorov diameters* for compact sets of functions (see Section 12.2.5 for more detail). Let W be a linear normed space, and let $U \subset W$. Introduce also an N-dimensional linear manifold $W^{(N)} \subset W$, for example, $W^{(N)} = \text{span}\{w_1^{(N)}, w_2^{(N)}, \ldots, w_N^{(N)}\}$, where the functions $w_n^{(N)} \in W$, $n = 1, 2, \ldots, N$, are given. The N-dimensional Kolmogorov diameter of the set U with respect to the space W is defined as:

$$\kappa_N(U, W) = \inf_{W^{(N)} \subset W} \sup_{u \in U} \inf_{w \in W^{(N)}} \|w - u\|_W. \tag{2.39}$$

This quantity tells us how accurately we can approximate an arbitrary u from a given set $U \subset W$ by selecting the optimal approximating subspace $W^{(N)}$ whose dimension N is fixed. The Kolmogorov diameter and related concepts play a fundamental role in the modern theory of approximation; in particular, for the analysis of the so-called best approximations, for the analysis of saturation of numerical methods by smoothness (Section 2.2.5), as well as in the theory of ε-entropy and related theory of transmission and processing of information. □

2.2.5 Saturation of Piecewise Polynomial Interpolation

As we have seen previously (Sections 1.3.1 and 2.2.4), reconstruction of continuous functions from discrete data is normally accompanied by the unavoidable error. This error is caused by the loss of information which inevitably occurs in the course of discretization. The unavoidable error is typically determined by the regularity of the continuous function and by the discretization parameters. Hence, it is not related to any particular reconstruction technique and rather presents a common intrinsic accuracy limit for all such techniques, in particular, for algebraic interpolation.

Therefore, an important question regarding every specific reconstruction method is whether or not it can reach the aforementioned intrinsic accuracy limit. If the accuracy of a given method is limited by its own design and does not, generally speaking, reach the level of the unavoidable error determined by the smoothness of the approximated function, then the method is said to be *saturated by smoothness*. Otherwise, if the accuracy of the method automatically adjusts to the smoothness of the approximated function, then the method does not get saturated.

Let $f(x)$ be defined on the interval $[a,b]$, and let its table of values $f(x_k)$ be known for the equally spaced nodes $x_k = a+kh$, $k=0,1,\dots,n$, $h=(b-a)/n$. In Section 2.2.2, we saw that the error of piecewise polynomial interpolation of degree s is of order $\mathcal{O}(h^{s+1})$, provided that the polynomial $P_s(x,f_{kj})$ is used to approximate $f(x)$ on the interval $x_k \leq x \leq x_{k+1}$, and that the derivative $f^{(s+1)}(x)$ exists and is bounded. Assume now that the only thing we know about $f(x)$ besides the table of values is that it has a bounded derivative of the maximum order $q+1$. If $q<s$, then the unavoidable error of reconstructing $f(x)$ from its tabulated values is $\mathcal{O}(h^{q+1})$. This is not as good as the $\mathcal{O}(h^{s+1})$ error that the method can potentially achieve, and the reason for deterioration is obviously the lack of smoothness. On the other hand, if $q>s$, then the accuracy of interpolation still remains of order $\mathcal{O}(h^{s+1})$ and does not reach the intrinsic limit $\mathcal{O}(h^{q+1})$. In other words, the order of interpolation error does not react in any way to the additional smoothness of the function $f(x)$, beyond the required $s+1$ derivatives. This is a manifestation of susceptibility of the algebraic piecewise polynomial interpolation to the saturation by smoothness.

In Chapter 3, we discuss an alternative interpolation strategy based on the use of trigonometric polynomials. That type of interpolation appears to be not susceptible to the saturation by smoothness.

Exercises

1. What size of the grid h guarantees that the error of piecewise linear interpolation for the function $f(x)=\sin x$ will never exceed 10^{-6}?

2. What size of the grid h guarantees that the error of piecewise quadratic interpolation for the function $f(x)=\sin x$ will never exceed 10^{-6}?

3. The values of $f(x)$ can be measured at any given point x with the accuracy $|\delta f| \leq 10^{-4}$. What is the optimal grid size for tabulating $f(x)$, if the function is to be subsequently reconstructed by means of a piecewise linear interpolation?

Hint. Choosing h excessively small can make the interpolation error smaller than the perturbation of the interpolating polynomial due to the perturbations in the data, see Section 2.1.4.

4.* The same question as in problem 3, but for piecewise quadratic interpolation.

5. Consider two approximation formulae for the first derivative $f'(x)$:

$$f'(x) \approx \frac{f(x+h)-f(x)}{h}, \tag{2.40}$$

$$f'(x) \approx \frac{f(x+h)-f(x-h)}{2h}, \tag{2.41}$$

and let $|f''(x)| \leq 1$ and $|f'''(x)| \leq 1$.

a) Find h such that the error of either formula will not exceed 10^{-3}.

b)* Assume that the function f itself is only known with the error δ. What is the best accuracy that one can achieve using formulae (2.40) and (2.41), and how one should properly choose h?

c)* Show that the asymptotic order of the error with respect to δ, obtained by formula (2.41) with the optimal h, cannot be improved.

2.3 Smooth Piecewise Polynomial Interpolation (Splines)

A classical piecewise polynomial interpolation of any degree s, e.g., piecewise linear, piecewise quadratic, etc., see Section 2.2, yields the interpolant that, generally speaking, is not differentiable even once at the interpolation nodes. There are, however, two alternative types of piecewise polynomial interpolants — local and nonlocal splines — that do have a given number of continuous derivatives everywhere, including the interpolation nodes.

2.3.1 Local Interpolation of Smoothness s and Its Properties

Assume that the interpolation nodes x_k and the function values $f(x_k)$ are given. Let us then specify a positive integer number s and also fix another positive integer j: $0 \leq j \leq s-1$. We will associate the interpolating polynomial $P_s(x, f_{kj}) \equiv P_s(x, f, x_{k-j}, x_{k-j+1}, \dots, x_{k-j+s})$ with every point x_k; this polynomial is built on the nodes $x_{k-j}, x_{k-j+1}, \dots, x_{k-j+s}$ using the function values $f(x_{k-j}), f(x_{k-j+1}), \dots, f(x_{k-j+s})$. *A piecewise polynomial local spline* $\varphi(x,s)$ that has continuous derivatives up to the order s is defined individually for each segment $[x_k, x_{k+1}]$ by means of the following equalities:

$$\varphi(x,s) = Q_{2s+1}(x,k), \quad x \in [x_k, x_{k+1}], \quad k = 0, \pm 1, \dots, \tag{2.42}$$

where each $Q_{2s+1}(x,k)$ is a polynomial of degree no greater than $2s+1$ that satisfies the relations:

$$\left.\frac{d^m Q_{2s+1}(x,k)}{dx^m}\right|_{x=x_k} = \left.\frac{d^m P_s(x,f_{kj})}{dx^m}\right|_{x=x_k}, \qquad m=0,1,2,\ldots,s, \tag{2.43}$$

$$\left.\frac{d^m Q_{2s+1}(x,k)}{dx^m}\right|_{x=x_{k+1}} = \left.\frac{d^m P_s(x,f_{k+1,j})}{dx^m}\right|_{x=x_{k+1}}, \qquad m=0,1,2,\ldots,s. \tag{2.44}$$

THEOREM 2.8

The polynomial $Q_{2s+1}(x,k)$ of degree no greater than $2s+1$ defined by means of equalities (2.43) and (2.44) exists and is unique.

PROOF Let $Q_{2s+1}(x,k) = c_{0,k} + c_{1,k}x + \ldots + c_{2s+1,k}x^{2s+1}$. Then, relations (2.43) and (2.44) together can be considered a system of $2s+2$ linear algebraic equations with respect to the $2s+2$ unknown coefficients $c_{0,k}$, $c_{1,k}$, ..., $c_{2s+1,k}$. This system becomes homogeneous if we replace all the right-hand sides of equalities (2.43) and (2.44) by zeros. Having this system as homogeneous means that the corresponding polynomial $Q_{2s+1}(x,k)$ would have a root of multiplicity $s+1$ at $x=x_k$ and another root of multiplicity $s+1$ at $x=x_{k+1}$. In other words, $Q_{2s+1}(x,k)$ would have a total of $2s+2$ roots counting their multiplicities. This is only possible if $Q_{2s+1}(x,k) \equiv 0$, because $Q_{2s+1}(x,k)$ is a polynomial of degree no greater than $2s+1$. Consequently, $c_{0,k} = c_{1,k} = \ldots = c_{2s+1,k} = 0$, and we conclude that the homogeneous counterpart of the linear algebraic system (2.43), (2.44) may only have a trivial solution. As such, the original inhomogeneous system (2.43), (2.44) itself will have a unique solution for any given choice of its right-hand sides. □

THEOREM 2.9

Let $f(x)$ be a polynomial of degree no greater than s. Then, the interpolant $\varphi(x,s)$ coincides with this polynomial.

PROOF We will prove the identity $\varphi(x,s) \equiv f(x)$ on the interval $[x_k, x_{k+1}]$ for an arbitrary k, i.e., for all x in between any two neighboring interpolation nodes. In other words, we will prove that $Q_{2s+1}(x,k) \equiv f(x)$. Due to the uniqueness of the interpolating polynomial, we have $P_s(x,f_{kj}) \equiv P_s(x,f_{k+1,j}) \equiv f(x)$. Then, clearly, the polynomial $f(x)$ solves system (2.43), (2.44). □

THEOREM 2.10

The piecewise polynomial interpolating function $\varphi(x,s)$ defined by equalities (2.42) assumes the given values $f(x_k)$ at the interpolation nodes x_k, $k=0,\pm1,\ldots$. Moreover, $\varphi(x,s)$ has a continuous derivative of order s everywhere on its domain.

PROOF According to equalities (2.43) and (2.44), at any given node x_k the two functions: $Q_{2s+1}(x,k-1)$ and $Q_{2s+1}(x,k)$, have derivatives of orders[3] $m=0,1,2,\dots,s$ that coincide with the corresponding derivatives of one and the same interpolating polynomial $P_s(x,f_{kj})$. By virtue of equalities (2.42), this proves the theorem. ∎

Let us now recast the polynomial $Q_{2s+1}(x,k)$ as

$$Q_{2s+1}(x,k) = P_s(x,f_{kj}) + R_{2s+1}(x,k), \tag{2.45}$$

where $R_{2s+1}(x,k)$ denotes a correction to the classical interpolating polynomial $P_s(x,f_{kj})$. Then, the following theorem holds.

THEOREM 2.11
The correction $R_{2s+1}(x,k)$ defined by (2.45) can be written in the form:

$$R_{2s+1}(x,k) = \\ (x_{k+1}-x_k)^{s+1} f(x_{k-j},x_{k-j+1},\dots,x_{k-j+s+1})\, q_{2s+1}\left(\frac{x-x_k}{x_{k+1}-x_k},k\right), \tag{2.46}$$

where $f(x_{k-j},x_{k-j+1},\dots,x_{k-j+s+1})$ is a divided difference of order $s+1$, and

$$q_{2s+1}(X,k) = \left(\frac{x_{k-j+s+1}-x_{k-j}}{x_{k+1}-x_k}\right) \times \\ \sum_{r=0}^{s}\left\{\left[\prod_{i=1}^{s}\left(X-\frac{x_{k-j+i}-x_k}{x_{k+1}-x_k}\right)\right]^{(r)}_{X=1}\right\} l_r(X), \qquad X=\frac{x-x_k}{x_{k+1}-x_k}, \tag{2.47}$$

$$l_r(X) = \frac{X^{s+1}(X-1)^r}{r!s!}\sum_{m=0}^{s-r}(-1)^m\frac{(s+m)!}{m!}(X-1)^m. \tag{2.48}$$

In formula (2.47), expression $[\dots]^{(r)}_{X=1}$ denotes a derivative of order r with respect to X evaluated for $X=1$.

REMARK 2.4 Representation (2.45) of the local piecewise polynomial interpolant with s continuous derivatives can be thought of as Newton's form of the interpolating polynomial $P_{s+1}(x,f_{kj})$:

$$P_{s+1}(x,f_{kj}) = P_s(x,f_{kj}) + f(x_{k-j},x_{k-j+1},\dots,x_{k-j+s+1})\phi_{s+1}(x,k), \\ \phi_{s+1}(x,k) = (x-x_{k-j})(x-x_{k-j+1})\dots(x-x_{k-j+s}),$$

[3] A derivative of order zero shall naturally be interpreted as the function itself.

in which the polynomial $\phi_{s+1}(x,k)$ has been replaced by another polynomial: $(x_{k+1}-x_k)^{s+1} q_{2s+1}\left(\frac{x-x_k}{x_{k+1}-x_k},k\right)$. ∎

To illustrate Theorem 2.11, we first note that according to formulae (2.45)–(2.48), in the case $s=0$, $j=0$, the local spline $\varphi(s,k)$ renders a piecewise linear interpolation.[4] As far as the most interesting case from the standpoint of applications, it is, perhaps, $s=2$, $j=1$, and $x_{k+1}-x_k=h=\text{const}$ for all k. Then, we have:

$$Q_5(x,k) = P_2(x,f_{k,1}) + \frac{h^3}{2!}\frac{f(x_{k+2})-3f(x_{k+1})+3f(x_k)-f(x_{k-1})}{h^3}\times \left(\frac{x-x_k}{h}\right)^3 \frac{x-x_{k+1}}{h}\left(3-\frac{2(x-x_k)}{h}\right). \tag{2.49}$$

The proof of Theorem 2.11 is given in Section 2.3.3; it concludes our discussion of smooth piecewise polynomial interpolation.

THEOREM 2.12

Let the function $f(x)$ be defined everywhere on the interval $[x_{k-j},x_{k-j+s+1}]$ and let its derivative $f^{(s+1)}(x)$ be bounded on this interval. Then, the approximate equalities

$$f^{(m)}(x) \approx \frac{d^m\varphi(x,s)}{dx^m}, \quad m=0,1,\ldots,s, \tag{2.50}$$

hold for all $x\in[x_k,x_{k+1}]$, and the following error estimates are guaranteed:

$$\max_{x_k\le x\le x_{k+1}} \left| f^{(m)}(x) - \frac{d^m\varphi(x,s)}{dx^m}\right| \le \text{const}\cdot(x_{k-j+s}-x_{k-j})^{s+1-m} \max_{x_{k-j}\le x\le x_{k-j+s+1}} |f^{(s+1)}(x)|. \tag{2.51}$$

PROOF First, using formulae (2.42) and (2.45), we can obtain:

$$\left| f^{(m)}(x) - \frac{d^m\varphi(x,s)}{dx^m}\right| \le |f^{(m)}(x)-P_s^{(m)}(x,f_{kj})| + |R_{2s+1}^{(m)}(x,k)|. \tag{2.52}$$

Next, we recall estimate (2.35) established in Theorem 2.7:

$$\max_{x_k\le x\le x_{k+1}} |f^{(m)}(x)-P_s^{(m)}(x,f_{kj})| \le \text{const}_1\cdot(x_{k-j+s}-x_{k-j})^{s+1-m} \max_{x_{k-j}\le x\le x_{k-j+s}} |f^{(s+1)}(x)|. \tag{2.53}$$

[4] Note that we have originally required that $s>0$ and $0\le j\le s-1$. However, a formal substitution of $s=0$, $j=0$ into (2.45)–(2.48) does yield a piecewise linear interpolant.

Then, taking into account that $\frac{d^m}{dx^m} = \frac{1}{(x_{k+1}-x_k)^m}\frac{d^m}{dX^m}$, where $X = \frac{x-x_k}{x_{k+1}-x_k}$, and using formulae (2.46) and (2.47) we obtain:

$$\left|\frac{d^m R_{2s+1}(x,k)}{dx^m}\right| = (x_{k+1}-x_k)^{s+1-m}|f(x_{k-j},x_{k-j+1},\dots,x_{k-j+s+1})|\times$$
$$\left|\frac{x_{k-j+s+1}-x_{k-j}}{x_{k+1}-x_k}\right|\left|\sum_{r=0}^{s}\left\{\left[\prod_{i=1}^{s}\left(X-\frac{x_{k-j+i}-x_k}{x_{k+1}-x_k}\right)\right]^{(r)}_{X=1}\right\}\frac{d^m l_r(X)}{dX^m}\right|. \quad (2.54)$$

Finally, according to Theorem 2.3 we can write:

$$(s+1)!f(x_{k-j},x_{k-j+1},\dots,x_{k-j+s+1}) = f^{(s+1)}(\xi),$$
$$\xi \in [x_{k-j},x_{k-j+s+1}]. \quad (2.55)$$

Then, substitution of (2.55) into (2.54) yields the estimate:

$$\left|\frac{d^m R_{2s+1}(x,k)}{dx^m}\right| \le \text{const}_2\cdot(x_{k+1}-x_k)^{s+1-m}\max_{x_{k-j}\le x\le x_{k-j+s+1}}\left|\frac{d^{s+1}f(x)}{dx^{s+1}}\right|, \quad (2.56)$$

where const_2 depends only on the quantities

$$\frac{x_{k-j+s+1}-x_{k-j}}{x_{k+1}-x_k} \quad \text{and} \quad \frac{x_{k-j+i}-x_k}{x_{k+1}-x_k}, \quad i=1,2,\dots,s \quad (2.57)$$

that appear in formula (2.54). Clearly, const_2 will remain bounded when the grid is refined as long as all ratios (2.57) remain bounded.

Then, the required estimate (2.51) follows from (2.52), (2.53), and (2.56), because $\max\limits_{x_{k-j}\le x\le x_{k-j+s}}|f^{(s+1)}(x)| \le \max\limits_{x_{k-j}\le x\le x_{k-j+s+1}}|f^{(s+1)}(x)|$ and also $(x_{k+1}-x_k)\le(x_{k-j+s}-x_{k-j})$. □

REMARK 2.5 On an equally spaced grid with size h, error estimate (2.51) means uniform convergence of the local spline $\varphi(x,s)$ and its derivatives of orders $m=1,2,\dots,s$, to the interpolated function $f(x)$ and its respective derivatives, with the rate $\mathcal{O}(h^{s+1-m})$, as the grid is refined, i.e., as $h\longrightarrow 0$. □

Next, we will discuss optimality and saturation of local splines (2.42).

REMARK 2.6 Assuming that $s+1$ is the maximum order of a bounded derivative of the function $f(x)$, one cannot reconstruct $f^{(m)}(x)$, $m=0,1,2,\dots,s$, from the tabulated values $f(x_k)$, $k=0,\pm1,\dots$, with an accuracy better than $\mathcal{O}(h^{s+1-m})$. This is an intrinsic constraint, no matter what specific reconstruction methodology is used. It is due to the lack of information contained in the values $f(x_k)$ (see Section 2.2.4). At the same time, the best allowable accuracy $\mathcal{O}(h^{s+1-m})$ is achieved by estimate (2.51).

However, if the function $f(x)$ has a bounded derivative of the maximum order $q+1$, where $q > s$, then the accuracy of obtaining $f^{(m)}(x)$ by means of local spline (2.42) still remains $\mathcal{O}(h^{s+1-m})$ and does not reach $\mathcal{O}(h^{q+1-m})$, which is the limit given by the unavoidable error. This is a manifestation of susceptibility of local splines to the saturation by smoothness. □

REMARK 2.7 To evaluate $\varphi(x,s)$ for every given x, one needs no more than $s+2$ interpolation nodes. This characteristic reflects on the local nature of formula (2.42), and it cannot be improved in the following sense. If we were to require that no more than $s+1$ interpolation nodes be used, then in order to achieve convergence of $\varphi^{(m)}(x)$ to $f^{(m)}(x)$ with the rate of $\mathcal{O}(h^{s+1-m})$ we would have had to relinquish the condition of continuity even for the first derivative of $\varphi(x,s)$, and rather resort back to the classical piecewise polynomial interpolation of degree no greater than s. □

We should also note that previously in our analysis we used the interpolation grid with no endpoints, i.e., the grid that formally contained infinitely many nodes x_k, $k=0,\pm 1,\dots$. If we were to use a finite interpolation grid instead, with a total of $n+1$ nodes: $x_0,x_1,\dots,x_n$, located on some interval $[a,b]$ so that $a=x_0<x_1<\dots<x_n=b$, then we would have had to modify formula (2.42) near the endpoints of the interval $[a,b]$ and define the new interpolant $\varphi(x)\equiv\varphi(x,s,a,b)$ by the following equalities:

$$\varphi(x,s,a,b)=\begin{cases} P_s(x,f_{jj}), & \text{if } x_0\le x\le x_j,\\ Q_{2s+1}(x,k), & \text{if } x\in[x_k,x_{k+1}] \text{ and } k=j,j+1,\dots,n+j-s-1,\\ P_s(x,f_{n+j-s,j}), & \text{if } x_{n+j-s}\le x\le x_n. \end{cases} \tag{2.58}$$

Recall that the notion of $P_s(x,f_{kj})$ was introduced in the end of Section 2.2.1.

The interpolant $\varphi(x,s,a,b)$ defined on the interval $[a,b]$ has continuous derivatives up to the order s everywhere on this interval. For this new interpolant, the results similar to those of Theorems 2.8–2.12 can be established.

REMARK 2.8 Local smooth piecewise polynomial interpolation was introduced by Ryaben'kii in 1952 for the functions specified on multi-dimensional rectangular grids with constant size; the formulae presented in this section were obtained in 1974. The multi-dimensional results and the bibliography can be found in [Rya02, Section 1.1] and in [Rya75]. □

2.3.2 Nonlocal Smooth Piecewise Polynomial Interpolation

Local splines with s continuous derivatives that we have constructed in Section 2.3.1 are realized using polynomials of degree $2s+1$. For example, a twice continuously differentiable local spline is built as a piecewise polynomial function of degree five. A natural question then arises of whether or not the same level of regularity (i.e., smoothness) can be achieved using polynomials of a lower degree.

The answer to this question is, generally speaking, affirmative. Splines $\psi(x,s)$ of minimal degree m that would guarantee the desired extent of smoothness s were constructed by Schoenberg for all positive integers s. For example, if $s=2$, then it turns out that the minimal necessary degree is $m=3$, i.e., the spline is realized by a piecewise cubic polynomial. In the literature, such splines are referred to as Schoenberg's cubic splines or simply cubic splines. The key distinction between Schoenberg's splines and local splines of Section 2.3.1 is that the lower degree of the constituent polynomials for Schoenberg's splines is basically obtained at the expense of losing their local nature. In other words, Schoenberg's splines become nonlocal, so that the coefficients of each of their polynomial pieces depend on the function values on the entire interpolation grid, rather than only at a few neighboring nodes.

Assume that a total of $n+1$ interpolation nodes: $x_0, x_1, \dots, x_n$, are given on the interval $[a,b]$ so that $a = x_0 < x_1 < \dots < x_n = b$, and assume also that the function values $f(x_k)$, $k=0,1,\dots,n$, at these nodes are known. The Schoenberg cubic spline $\psi(x,2) = \psi(x)$ is formally defined on $[a,b]$ as a function composed of individual third-degree polynomials $P_3(x,k)$ specified on each partition interval $x_k \le x \le x_{k+1}$, i.e., $\psi(x) \equiv P_3(x,k)$ for $x \in [x_k, x_{k+1}]$, $k=0,1,\dots,n-1$. The spline ψ must coincide with the original function f at the interpolation nodes: $\psi(x_k) = f(x_k)$ for all $k = 0,1,\dots,n$, and its first and second derivatives must be continuous everywhere. As ψ is a piecewise polynomial function, i.e., for $x \in [x_k, x_{k+1}]$ we have $\psi(x) = P_3(x,k) = \sum\limits_{l=0}^{3} a_{kl}(x-x_k)^l$, the condition of continuity for its derivatives everywhere on $[a,b]$ reduces to the condition of their continuity at the interpolation nodes: $\psi'(x_k - 0) = \psi'(x_k+0)$, $\psi''(x_k-0) = \psi''(x_k+0)$, $k=1,\dots,n-1$.

Consider the function $\psi''(x)$. It is piecewise linear and continuous on the interval $[a,b]$. Then, for $x \in [x_k, x_{k+1}]$ we can write:

$$\begin{aligned}\psi''(x) &= \psi''(x_k) + \frac{\psi''(x_{k+1}) - \psi''(x_k)}{h_k}(x-x_k) \\ &= \psi''(x_{k+1})\frac{x-x_k}{h_k} + \psi''(x_k)\frac{x_{k+1}-x}{h_k},\end{aligned} \tag{2.59}$$

where we denote $h_k \stackrel{\text{def}}{=} x_{k+1} - x_k$, $k=0,1,\dots,n-1$, hereafter. Moreover, according to the previously adopted notations for $P_3(x,k)$ we have: $a_{k,2} = \psi''(x_k)/2$. Then, since the second derivative of the spline is continuous, we obtain: $\psi''(x_{k+1}-0) = \psi''(x_{k+1}+0) = 2a_{k+1,2}$. Next, by integrating equality (2.59) twice we get:

$$\psi(x) = a_{k+1,2}\frac{(x-x_k)^3}{3h_k} + a_{k,2}\frac{(x_{k+1}-x)^3}{3h_k} + c_{k,1}\frac{x-x_k}{h_k} + c_{k,2}\frac{x_{k+1}-x}{h_k},$$

where the integration constants $c_{k,1}$ and $c_{k,2}$ are to be obtained from the conditions $\psi(x_k) = f(x_k)$ and $\psi(x_{k+1}) = f(x_{k+1})$, respectively. This yields:

$$a_{k+1,2}\frac{h_k^2}{3} + c_{k,1} = f(x_{k+1}) \quad \text{and} \quad a_{k,2}\frac{h_k^2}{3} + c_{k,2} = f(x_k).$$

Therefore, the cubic spline for $x \in [x_k, x_{k+1}]$ is given by the following expression:

$$\begin{aligned} \psi(x) = {} & a_{k+1,2}\frac{(x-x_k)^3}{3h_k} + a_{k,2}\frac{(x_{k+1}-x)^3}{3h_k} + \\ & \left(f(x_{k+1}) - a_{k+1,2}\frac{h_k^2}{3}\right)\frac{x-x_k}{h_k} + \left(f(x_k) - a_{k,2}\frac{h_k^2}{3}\right)\frac{x_{k+1}-x}{h_k}. \end{aligned} \tag{2.60}$$

When deriving formula (2.60) we have already used the condition $\psi(x_k) = f(x_k)$, as well as the condition of continuity of the second derivatives of the spline. We have not used the continuity of the first derivatives yet. To actually do so, we first differentiate formula (2.60) and substitute $x = x_k$ thus obtaining $\psi'(x_k+0)$. Then we write down a similar formula for the interval $[x_{k-1}, x_k]$, differentiate it and substitute $x = x_k$ thus obtaining $\psi'(x_k - 0)$. By requiring that $\psi'(x_k-0) = \psi'(x_k+0)$, we get:

$$\begin{aligned} & a_{k-1,2}\frac{h_{k-1}}{3} + a_{k,2}\frac{2h_{k-1}+2h_k}{3} + a_{k+1,2}\frac{h_k}{3} \\ & \qquad = \frac{f(x_{k+1}) - f(x_k)}{h_k} - \frac{f(x_k) - f(x_{k-1})}{h_{k-1}}. \end{aligned} \tag{2.61}$$

Equations (2.61) need to be considered for $k = 1, 2, \ldots, n-1$, which yields a total of $n-1$ linear algebraic equations with respect to $n+1$ unknown coefficients $a_{0,2}, a_{1,2}, \ldots a_{n,2}$. We therefore conclude that the cubic spline $\psi(x)$ is not yet defined unambiguously by the conditions specified previously, and two additional conditions must be provided. For example, we may set $a_{0,2} = a_{n,2} = 0$, which corresponds to $\psi''(x_0) \equiv \psi''(a) = 0$ and $\psi''(x_n) \equiv \psi''(b) = 0$. Then, system (2.61) can be written in the matrix form as follows:

$$\boldsymbol{A}\boldsymbol{a} = \boldsymbol{g}. \tag{2.62}$$

In system (2.62), $\boldsymbol{a} = [a_{1,2}, a_{2,2}, \ldots, a_{n-1,2}]^T$ is an $n-1$-dimensional vector of unknowns, $\boldsymbol{g} = [g_1, g_2, \ldots, g_{n-1}]^T$ is the right-hand side: $\boldsymbol{g} = \boldsymbol{B}\boldsymbol{f}$, where $\boldsymbol{f} = [f(x_0), f(x_1), \ldots, f(x_n)]^T$ is a given vector of dimension $n+1$, $\boldsymbol{A}$ is a square matrix of dimension $(n-1)\times(n-1)$:

$$\boldsymbol{A} = \begin{bmatrix} \frac{2h_0+2h_1}{3} & \frac{h_1}{3} & 0 & \cdots & 0 & 0 \\ \frac{h_1}{3} & \frac{2h_1+2h_2}{3} & \frac{h_2}{3} & \cdots & 0 & 0 \\ 0 & \frac{h_2}{3} & \frac{2h_2+2h_3}{3} & \cdots & 0 & 0 \\ \vdots & \ddots & \ddots & \cdots & \ddots & \vdots \\ 0 & 0 & 0 & \cdots & \frac{h_{n-2}}{3} & \frac{2h_{n-2}+2h_{n-1}}{3} \end{bmatrix},$$

and $\boldsymbol{B}$ is a matrix of dimension $(n-1)\times(n+1)$:

$$\boldsymbol{B} = \begin{bmatrix} \frac{1}{h_0} & \left(-\frac{1}{h_0} - \frac{1}{h_1}\right) & \frac{1}{h_1} & 0 & \cdots & 0 & 0 \\ 0 & \frac{1}{h_1} & \left(-\frac{1}{h_1} - \frac{1}{h_2}\right) & \frac{1}{h_2} & \cdots & 0 & 0 \\ 0 & 0 & \frac{1}{h_2} & \left(-\frac{1}{h_2} - \frac{1}{h_3}\right) & \cdots & 0 & 0 \\ \vdots & \vdots & \ddots & \ddots & \cdots & \ddots & \vdots \\ 0 & 0 & 0 & 0 & \cdots & \left(-\frac{1}{h_{n-2}} - \frac{1}{h_{n-1}}\right) & \frac{1}{h_{n-1}} \end{bmatrix}.$$

Both matrices $\boldsymbol{A}$ and $\boldsymbol{B}$ are tri-diagonal. The matrix $\boldsymbol{A}$ is strictly diagonally dominant by rows, as defined in Chapter 5 (this matrix is also symmetric). Therefore, $\boldsymbol{A}$ is non-singular, and as such, system (2.62) always has one and only one solution $\boldsymbol{a}$ for any right-hand side $\boldsymbol{g}$. This means that the spline $\psi(x)$ exists and is unique; it can be obtained from the tabulated function values $f(x_k)$, $k = 0, 1, \dots, n$, using formula (2.60). Numerically, the matrix $\boldsymbol{A}$ can be inverted by a very efficient algorithm known as the tri-diagonal elimination. This algorithm is stable, and has linear computational complexity with respect to the dimension of the matrix (see Section 5.4.2). Let us also note that once the solution $\boldsymbol{a} = [a_{1,2}, a_{2,2}, \dots, a_{n-1,2}]^T$ of system (2.62) has been obtained, the spline $\psi(x)$ can already be built. If necessary, however, the remaining three coefficients $a_{k,0}$, $a_{k,1}$, and $a_{k,3}$ of the polynomial $P_3(x,k) = \sum_{l=0}^{3} a_{kl}(x - x_k)^l$ for every $k = 0, 1, \dots, n-1$, can also be found explicitly, by merely regrouping the terms in formula (2.60).

As has been shown, the Schoenberg cubic spline basically requires two additional conditions for uniqueness. Of course, there are many other choices for these conditions, besides the previously used $\psi''(a) = \psi''(b) = 0$. For example, the spline may have a non-zero curvature at the endpoints a and b; this yields inhomogeneous conditions for its second derivatives instead of the homogeneous ones:

$$\psi''(a) \equiv 2a_{0,2} = \psi_0^{(2)}, \qquad \psi''(b) \equiv 2a_{n,2} = \psi_n^{(2)}, \tag{2.63}$$

where $\psi_0^{(2)}$ and $\psi_n^{(2)}$ are given constants. The new boundary conditions (2.63) prompt a minor change in system (2.62); namely, the first and last components of its right-hand side $\boldsymbol{g}$ will now be defined as follows:

$$\begin{aligned} g_1 &= \frac{f(x_2) - f(x_1)}{h_1} - \frac{f(x_1) - f(x_0)}{h_0} - \frac{\psi_0^{(2)} h_0}{6}, \\ g_{n-1} &= \frac{f(x_n) - f(x_{n-1})}{h_{n-1}} - \frac{f(x_{n-1}) - f(x_{n-2})}{h_{n-2}} - \frac{\psi_n^{(2)} h_{n-1}}{6}. \end{aligned}$$

Alternatively, one may specify the slope of the spline rather than its curvature at the endpoints a and b, which yields two additional conditions for the first derivatives:

$$\begin{aligned} \psi'(a) &\equiv \frac{f(x_1) - f(x_0)}{h_0} - a_{0,2}\frac{2h_0}{3} - a_{1,2}\frac{h_0}{3} = \psi_0^{(1)}, \\ \psi'(b) &\equiv \frac{f(x_n) - f(x_{n-1})}{h_{n-1}} + a_{n,2}\frac{2h_{n-1}}{3} + a_{n-1,2}\frac{h_{n-1}}{3} = \psi_n^{(1)}, \end{aligned} \tag{2.64}$$

where, again, $\psi_0^{(1)}$ and $\psi_n^{(1)}$ are some constants. Relations (2.64) can be added to system (2.61) as its new first and new last equation, thus making the overall dimension of the system $(n+1) \times (n+1)$. This modification will not alter the consistency of the system, and will not disrupt its solvability by the tri-diagonal elimination either. Finally, we may consider periodic boundary conditions for the spline:

$$f(x_0) = f(x_n), \qquad \psi'(a) = \psi'(b), \qquad \psi''(a) = \psi''(b). \tag{2.65}$$

Then, system (2.61) gets supplemented by one additional equation of the same structure (only one is needed because the other one is simply $a_{n,2} = a_{0,2}$):

$$a_{n-1,2}\frac{h_{n-1}}{3} + a_{0,2}\frac{2h_{n-1}+2h_0}{3} + a_{1,2}\frac{h_0}{3} = \frac{f(x_1)-f(x_0)}{h_0} - \frac{f(x_0)-f(x_{n-1})}{h_{n-1}},$$

and while its resulting matrix form can still be symbolically written as (2.62), the vector of unknowns will now be n-dimensional: $\boldsymbol{a} = [a_{0,2}, a_{1,2}, \dots, a_{n-1,2}]^T$, so will be the vector of the right-hand sides $\boldsymbol{g} = \boldsymbol{B}\boldsymbol{f}$; the matrix $\boldsymbol{A}$ will have dimension $n \times n$:

$$\boldsymbol{A} = \begin{bmatrix} \frac{2h_{n-1}+2h_0}{3} & \frac{h_0}{3} & 0 & \dots & 0 & \frac{h_{n-1}}{3} \\ \frac{h_0}{3} & \frac{2h_0+2h_1}{3} & \frac{h_1}{3} & \dots & 0 & 0 \\ 0 & \frac{h_1}{3} & \frac{2h_1+2h_2}{3} & \dots & 0 & 0 \\ \vdots & \ddots & \ddots & \dots & \ddots & \vdots \\ \frac{h_{n-1}}{3} & 0 & 0 & \dots & \frac{h_{n-2}}{3} & \frac{2h_{n-2}+2h_{n-1}}{3} \end{bmatrix}, \tag{2.66}$$

and the matrix $\boldsymbol{B}$ will be a square $n \times n$ matrix as well:

$$\boldsymbol{B} = \begin{bmatrix} \left(-\frac{1}{h_{n-1}} - \frac{1}{h_0}\right) & \frac{1}{h_0} & 0 & \dots & 0 & \frac{1}{h_{n-1}} \\ \frac{1}{h_0} & \left(-\frac{1}{h_0} - \frac{1}{h_1}\right) & \frac{1}{h_1} & \dots & 0 & 0 \\ 0 & \frac{1}{h_1} & \left(-\frac{1}{h_1} - \frac{1}{h_2}\right) & \dots & 0 & 0 \\ \vdots & \ddots & \ddots & \dots & \ddots & \vdots \\ \frac{1}{h_{n-1}} & 0 & 0 & \dots & \frac{1}{h_{n-2}} & \left(-\frac{1}{h_{n-2}} - \frac{1}{h_{n-1}}\right) \end{bmatrix}. \tag{2.67}$$

In so doing, the vector $\boldsymbol{f}$ whose components are given tabulated values of the function f will be n-dimensional: $\boldsymbol{f} = [f(x_0), f(x_1), \dots, f(x_{n-1})]^T$, because $f(x_n) = f(x_0)$. Notice that both the matrix $\boldsymbol{A}$ of (2.66) and the matrix $\boldsymbol{B}$ of (2.67) are symmetric and almost tri-diagonal, except for the non-zero entries in the upper-right and lower-left corners of each matrix. The matrix $\boldsymbol{A}$ of (2.66) can be efficiently inverted by a special cyclic version of the tri-diagonal elimination (see Section 5.4.3).

Other types of boundary conditions for the spline $\psi(x)$ at the endpoints a and b can be used as well.

Let us now assume that the function $f(x)$ has a bounded derivative of order $s+1$ on the interval $[a,b]$. Then, one can set the required additional conditions so that the nonlocal spline $\psi(x,s)$ with s continuous derivatives will keep the same optimal approximation properties that characterize both the classical piecewise polynomial interpolation of Section 2.2 and the local smooth piecewise polynomial interpolation of Section 2.3.1. Namely, estimates of type (2.51) will hold for $m = 0, 1, \dots, s$:

$$\max_{a \le x \le b} \left| f^{(m)}(x) - \frac{d^m \psi(x,s)}{dx^m} \right| \le \text{const} \cdot \left(\max_k h_k \right)^{s+1-m} \max_{a \le x \le b} |f^{(s+1)}(x)|.$$

In particular, provided that $f(x)$ is twice continuously differentiable on $[a,b]$, the cubic spline $\psi(x,2) \equiv \psi(x)$, along with its derivatives of orders $m = 1, 2$, will converge to the function $f(x)$ and its respective derivatives with the rate of $\mathscr{O}(h^{s+1-m})$.

On the other hand, the undesirable property of saturation by smoothness, which is inherent for both the classical piecewise polynomial interpolation and local splines, is shared by the Schoenberg splines as well, notwithstanding the loss of their local nature. Namely, the nonlocal splines of smoothness s on a uniform interpolation grid with size h will guarantee the order of error $\mathscr{O}(h^{s+1})$ for the functions that have a maximum of $s+1$ bounded derivatives, and they will not provide accuracy higher than $\mathscr{O}(h^{s+1})$ even for the functions $f(x)$ that have more than $s+1$ derivatives.

Even though the coefficients of a nonlocal spline on a given interval $[x_k, x_{k+1}]$ depend on the function values on the entire grid, it is known that in practice the influence of the remote nodes is rather weak. Nonetheless, to actually evaluate the coefficients, one needs to solve the full system (2.62). Therefore, a natural task of improving the accuracy of a Schoenberg's spline by adding a few interpolation nodes in a particular local region basically implies starting from the very beginning, i.e., writing down and then solving a new system of type (2.62). In contradistinction to that, the splines of Section 2.3.1 are particularly well suited for such local grid refinements. In doing so, the additional computational effort is merely proportional to the number of new nodes.

For further detail on the subject of splines we refer the reader to [dB01].

2.3.3 Proof of Theorem 2.11

This section can be skipped during the first reading.

The coefficients of the polynomial

$$Q_{2s+1}(x,k) = c_{0,k} + c_{1,k}x + \ldots + c_{2s+1,k}x^{2s+1} \tag{2.68}$$

are determined by solving the linear algebraic system (2.43), (2.44). The right-hand sides of the equations that compose sub-system (2.43) have the form

$$a_0^{(m)} f_{k-j} + a_1^{(m)} f_{k-j+1} + \ldots + a_s^{(m)} f_{k-j+s}, \qquad m = 0, 1, \ldots, s,$$

while those that pertain to sub-system (2.44) have the form

$$b_0^{(m)} f_{k-j+1} + b_1^{(m)} f_{k-j+2} + \ldots + b_s^{(m)} f_{k-j+s+1}, \qquad m = 0, 1, \ldots, s,$$

where $a_i^{(m)}$ and $b_i^{(m)}$, $i = 0, 1, \ldots, s$, $m = 0, 1, \ldots, s$, are some numbers that do not depend on[5] f_{k-j}, f_{k-j+1}, ..., $f_{k-j+s+1}$.

Consequently, one can say that the given quantities f_{k-j}, f_{k-j+1}, ..., $f_{k-j+s+1}$ determine the solution $c_{0,k}$, $c_{1,k}$, ..., $c_{2s+1,k}$ of system (2.43), (2.44) through the formulae of type

$$c_{r,k} = \alpha_0^{(r)} f_{k-j} + \alpha_1^{(r)} f_{k-j+1} + \ldots + \alpha_{s+1}^{(r)} f_{k-j+s+1}, \quad r = 0, 1, \ldots, 2s+1, \tag{2.69}$$

[5] Hereafter in this section, we will use the notation $f_l = f(x_l)$ for all nodes l.

where $\alpha_l^{(r)}$, $l = 0, 1, \ldots, s+1$, are, again, numbers that do not depend on f_{k-j}, f_{k-j+1}, ..., $f_{k-j+s+1}$. Next, by substituting expressions (2.69) into formula (2.68) and grouping together all the terms that contain each particular f_{k-j+l}, $l = 0, 1, \ldots, s+1$, we can recast the polynomial (2.68) as follows:

$$Q_{2s+1}(x,k) = \sum_{l=0}^{s+1} f_{k-j+l} p(x,l), \tag{2.70}$$

where $p(x,l)$, $l = 0, 1, \ldots, s+1$, are some polynomials of the argument x. Let us now replace the quantities f_{k-j+l} in formula (2.70) using formulae (2.16), see Theorem 2.4. Then, the polynomial $Q_{2s+1}(x,k)$ becomes:

$$\begin{aligned} Q_{2s+1}(x,k) &= \sum_{l=0}^{s} f(x_{k-j}, x_{k-j+1}, \ldots, x_{k-j+l}) q_l(x,k) + \\ &(x_{k+1} - x_k)^{(s+1)} f(x_{k-j}, x_{k-j+1}, \ldots, x_{k-j+s+1}) \tilde{q}_{2s+1}(x,k), \end{aligned} \tag{2.71}$$

where $q_l(x,k)$, $l = 0, 1, \ldots, s$, and $\tilde{q}_{2s+1}(x,k)$ are some polynomials independent of f_{k-1}, f_{k-j+1}, ..., $f_{k-j+s+1}$. We will exploit this independence for the purpose of finding the polynomials $q_l(x,k)$, $l = 0, 1, \ldots, s$. Let us specify f_{k-1}, f_{k-j+1}, ..., f_{k-j+s} arbitrarily, and let us specify $f_{k-j+s+1}$ as follows:

$$f_{k-j+s+1} = P_s(x, f_{kj})\Big|_{x = x_{k-j+s+1}}.$$

According to Theorem 2.9, we then have

$$Q_{2s+1}(x,k) = P_s(x, f_{kj}),$$

while Corollary 2.3 from the Newton's formula (see Section 2.1) implies that the divided difference $f(x_{k-j}, x_{k-j+1}, \ldots, x_{k-j+s+1})$ of order $s+1$ vanishes. Then, formula (2.71) transforms into:

$$P_s(x, f_{kj}) = \sum_{l=0}^{s} f(x_{k-j}, x_{k-j+1}, \ldots, x_{k-j+l}) q_l(x,k). \tag{2.72}$$

By comparing formula (2.72) with formula (2.3), we conclude that as the values f_{k-1}, f_{k-j+1}, ..., f_{k-j+s} have been chosen arbitrarily, then $q_l(x,k) = (x - x_{k-j})(x - x_{k-j+1}) \ldots (x - x_{k-j+l})$, $l = 0, 1, \ldots, s$. Equality (2.72) also implies that formula (2.71) can be recast as follows:

$$\begin{aligned} Q_{2s+1}(x,k) &= P_s(x, f_{kj}) + \\ &(x_{k+1} - x_k)^{(s+1)} f(x_{k-j}, x_{k-j+1}, \ldots, x_{k-j+s+1}) \tilde{q}_{2s+1}(x,k). \end{aligned} \tag{2.73}$$

To obtain the polynomial $\tilde{q}_{2s+1}(x,k)$ from (2.73), we specify

$$f_{k-j} = f_{k-j+1} = \ldots = f_{k-j+s} = 0, \text{ and } f_{k-j+s+1} = 1. \tag{2.74}$$

Then, equality (2.73) becomes:

$$Q_{2s+1}(x,k) = (x_{k+1}-x_k)^{(s+1)} f(x_{k-j}, x_{k-j+1}, \ldots, x_{k-j+s+1})\tilde{q}_{2s+1}(x,k). \quad (2.75)$$

Note that under conditions (2.74), the Lagrange form of the interpolating polynomial, see formula (2.1), yields

$$P_{s+1}(x,f,x_{k-j},x_{k-j+1},\ldots,x_{k-j+s+1}) = \prod_{l=0}^{s} \frac{x-x_{k-j+l}}{x_{k-j+s+1}-x_{k-j+l}}, \quad (2.76)$$

while according to its Newton's form (2.3), we have:

$$P_{s+1}(x,f,x_{k-j},x_{k-j+1},\ldots,x_{k-j+s+1}) = f(x_{k-j},x_{k-j+1},\ldots,x_{k-j+s+1}) \prod_{l=0}^{s}(x-x_{k-j+l}). \quad (2.77)$$

By comparing (2.76) and (2.77), we conclude that once conditions (2.74) hold:

$$f(x_{k-j},x_{k-j+1},\ldots,x_{k-j+s+1}) = \left[\prod_{l=0}^{s}(x_{k-j+s+1}-x_{k-j+l})\right]^{-1}.$$

Then, from formula (2.75) we find that

$$\tilde{q}_{2s+1}(x,k) = (x_{k+1}-x_k)^{-s-1} Q_{2s+1}(x,k) \prod_{l=0}^{s}(x_{k-j+s+1}-x_{k-j+l}), \quad (2.78)$$

where the expressions on the right-hand side of (2.78) are evaluated, again, under conditions (2.74).

Let us now transform formula (2.78) to the required form (2.47), (2.48), see Theorem 2.11. For that purpose, we will need to obtain an explicit expression for the polynomial $Q_{2s+1}(x,k)$ driven by the function values of (2.74). Note that due to condition (2.43), along with (2.74), the value $x = x_k$ appears to be a root of multiplicity $s+1$ for the polynomial $Q_{2s+1}(x,k)$, so that the function $Q(x) = Q_{2s+1}(x,k)/(x-x_k)^{s+1}$ is a polynomial of degree no greater than s. We can then expand $Q(x)$ with respect to the powers of $(x-x_{k+1})$:

$$Q_{2s+1}(x,k) = (x-x_k)^{s+1}Q(x) = (x-x_k)^{s+1}\frac{Q_{2s+1}(x,k)}{(x-x_k)^{s+1}} = (x-x_k)^{s+1}\sum_{r=0}^{s}\frac{1}{r!}\left[\frac{Q_{2s+1}(x,k)}{(x-x_k)^{s+1}}\right]^{(r)}_{x=x_{k+1}}(x-x_{k+1})^r. \quad (2.79)$$

Next, we can use equality (2.44) and write:

$$\frac{d^r}{dx^r}\left[\frac{Q_{2s+1}(x,k)}{(x-x_k)^{s+1}}\right]\bigg|_{x=x_{k+1}} = \sum_{l=0}^{r}\binom{r}{l}\left(\frac{d^{r-1}}{dx^{r-1}}P_s(x,f,x_{k-j+1},x_{k-j+2},\ldots,x_{k-j+s+1})\right)\bigg|_{x=x_{k+1}} \times \left[(x-x_k)^{-s-1}\right]^{(r)}_{x=x_{k+1}}, \quad (2.80)$$

where $\binom{r}{l} = \frac{r!}{l!(r-l)!}$ are the binomial coefficients. However, conditions (2.74) along with the Lagrange formula (2.1) yield [cf. formula (2.76)]:

$$P_s(x, f, x_{k-j+1}, x_{k-j+2}, \dots, x_{k-j+s+1}) = \prod_{i=1}^{s} \frac{x - x_{k-j+i}}{x_{k-j+s+1} - x_{k-j+i}}. \tag{2.81}$$

Substituting (2.81) into (2.80), we obtain:

$$\begin{aligned} \frac{d^r}{dx^r}\left[\frac{Q_{2s+1}(x,k)}{(x-x_k)^{s+1}}\right]\Bigg|_{x=x_{k+1}} &= \left[\prod_{i=1}^{s}(x_{k-j+s+1} - x_{k-j+i})\right]^{-1} \times \\ \sum_{l=0}^{r}\binom{r}{l}\left[(x-x_k)^{-s-1}\right]^{(r)}_{x=x_{k+1}} & \left[\prod_{i=1}^{s}(x - x_{k-j+i})\right]^{(r-l)}_{x=x_{k+1}}. \end{aligned} \tag{2.82}$$

Finally, we substitute (2.82) into (2.79), and subsequently (2.79) into (2.78), thus obtaining an explicit expression for $\tilde{q}_{2s+1}(x,k)$ as a double sum with respect to l and r. This sum would depend on x only as a composite function $q_{2s+1}(X,k)$ of the argument $X = (x - x_k)/(x_{k+1} - x_k)$:

$$\tilde{q}_{2s+1}(x,k) = q_{2s+1}(X,k).$$

Then, changing the summation order in the resulting formula for $q_{2s+1}(X,k)$ obtained after the last substitution into (2.78), we arrive at expressions (2.47), (2.48).

Exercises

1. Let $x_{k+1} - x_k = h = \text{const}$, and let $\varphi(x,s)$ be a local spline of given smoothness s defined by formula (2.45) for $x_k \le x \le x_{k+1}$. Explicitly write down the polynomials $l_k(x,s)$ that would enable representation of $\varphi(x,s)$ in the form [cf. the Lagrange formula (2.1)]:

$$\varphi(x,s) = \sum_k f(x_k) l_k(x,s).$$

   Analyze the cases: $s = 1$, $j = 0$ and $s = 2$, $j = 1$.

2. Let $\|f\| \stackrel{\text{def}}{=} \max\limits_{k-j \le l \le k-j+s+1} |f_l|$, and let the interpolation grid be uniform. Estimate the value of the Lebesgue constant

$$L_s = \max_{\|f\|=1} \max_{x_k \le x \le x_{k+1}} |Q_{2s+1}(x,k)|$$

   for the local spline with $s = 1$ and $j = 0$.

3.* Delineate the concluding part of the proof of Theorem 2.11, given in the last paragraph of Section 2.3.3.

4.* Prove that the nonlocal cubic spline $\psi(x)$ of (2.60) that satisfies the homogeneous boundary conditions (2.63): $\psi''(a) = \psi''(b) = 0$, delivers minimum to the functional

$$\Phi(u) = \int_a^b [u''(x)]^2 dx$$

   on the class of all interpolants $u = u(x)$, $a \le x \le b$, $u(x_k) = f(x_k)$, $k = 0, 1, \dots, n$, that have a square integrable second derivative: $u \in W_2^2[a,b]$.

5. Let $[a,b] = [-\pi,\pi]$, $f_1(x) = |x^3|$, $f_2(x) = \sin x$, and let the interpolation grid be uniform with size h. For both functions f_1 and f_2 implement on the computer the local spline (2.42), (2.58) with $s = 2$, $j = 1$, and the nonlocal cubic spline (2.60) with any type of boundary conditions: (2.63), (2.64), or (2.65). Demonstrate that in either case the convergence rate is $\mathcal{O}(h^2)$.

2.4 Interpolation of Functions of Two Variables

The problem of reconstructing a function of continuous argument from its discrete table of values can be formulated in the multi-dimensional case as well, for example, when $f = f(x,y)$, i.e., when there are two independent variables. The principal objective remains the same as in the case of one dimension, namely, to build a procedure for (approximately) evaluating the function in between the given interpolation nodes. However, in the case of two variables one can consider a much wider variety of interpolation grids. All these grids basically fall into one of the two categories: structured or unstructured.

2.4.1 Structured Grids

Typically, structured grids on the (x,y) plane are composed of rectangular cells. In other words, the nodes (x_k,y_l), $k = 0,\pm 1,\ldots$, $l = 0,\pm 1,\ldots$, are obtained as intersections of the two families of straight lines: the vertical lines $x = x_k$, $k = 0,\pm 1,\ldots$, and the horizontal lines $y = y_l$, $l = 0,\pm 1,\ldots$. In so doing, we always assume that $\forall k: \; x_{k+1} > x_k$, and $\forall l: \; y_{l+1} > y_l$. In the literature, such grids are also referred to as rectangular or Cartesian. The grid sizes $h_k^{(x)} = x_{k+1} - x_k$ and $h_l^{(y)} = y_{l+1} - y_l$ may but do not have to be constant. In the case of constant size(s), the grid is called uniform or equally spaced (in the corresponding direction). The simplest example of a uniform two-dimensional grid is a grid with square cells: $h_k^{(x)} = h_l^{(y)} = \text{const}$.

To approximately compute the value of the function f at the point $(\bar{x},\bar{y})$ that does not coincide with any of the nodes (x_k,y_l) of a structured rectangular grid, one can, in fact, use the apparatus of piecewise polynomial interpolation for the functions of one variable. To do so, we first select the parameters s (degree of interpolation) and j, as in Section 2.2. We also need to determine which cell of the grid contains the point of interest. Let us assume that $x_k < \bar{x} < x_{k+1}$ and $y_l < \bar{y} < y_{l+1}$ for some particular values of k and l. Then, we interpolate along the horizontal grid lines:

$$\bar{f}(\bar{x},y_{l-j+i}) = P_s(x, f(\cdot\,,y_{l-j+i}), x_{k-j}, x_{k-j+1},\ldots,x_{k-j+s}), \quad i = 0,1,\ldots,s.$$

and obtain the intermediate values $\bar{f}$. Having done that, we interpolate along the vertical grid lines and obtain the approximate value f:

$$f(\bar{x},\bar{y}) \approx P_s(y, \bar{f}(\bar{x},\cdot\,), y_{l-j}, y_{l-j+1},\ldots,y_{l-j+s}).$$

Clearly, the foregoing formulae can be used to approximate the function f at any point $(\bar{x},\bar{y})$ inside the rectangular grid cell $\{(x,y)|x_k < x < x_{k+1}, y_l < y < y_{l+1}\}$. For example, if we choose piecewise linear interpolation along x and y, i.e., $s=1$, then

$$f(\bar{x},\bar{y}) \approx f(x_k,y_l)\frac{(x-x_{k+1})(y-y_{l+1})}{(x_k-x_{k+1})(y_l-y_{l+1})} + f(x_{k+1},y_l)\frac{(x-x_k)(y-y_{l+1})}{(x_{k+1}-x_k)(y_l-y_{l+1})} + \\ f(x_k,y_{l+1})\frac{(x-x_{k+1})(y-y_l)}{(x_k-x_{k+1})(y_{l+1}-y_l)} + f(x_{k+1},y_{l+1})\frac{(x-x_k)(y-y_l)}{(x_{k+1}-x_k)(y_{l+1}-y_l)}. \quad (2.83)$$

Note, however, that in general the degree of interpolation does not necessarily have to be the same for both dimensions. Also note that the procedure is obviously symmetric. In other words, it does not matter whether we first interpolate along x and then along y, as shown above, or the other way around.

The piecewise polynomial interpolation on the plane, built dimension-by-dimension on a rectangular grid as explained above, inherits the key properties of the one-dimensional interpolation. For example, if the function $f = f(x,y)$ is twice differentiable, with bounded second partial derivatives, then the interpolation error of formula (2.83) on a square-cell grid with size h will be $\mathcal{O}(h^2)$. For piecewise polynomial interpolation of a higher degree, the rate of convergence will accordingly be faster, provided that the interpolated function is sufficiently smooth. On the other hand, similarly to the one-dimensional case, the two-dimensional piecewise polynomial interpolation is also prone to the saturation by smoothness.

Again, similarly to the one-dimensional case, one can also construct a smooth piecewise polynomial interpolation in two dimensions. As before, this interpolation may be either local or nonlocal. Local splines that extend the methodology of Section 2.3.1 can be built on the plane dimension-by-dimension, in much the same way as the conventional piecewise polynomials outlined previously. Their key properties will be preserved from the one-dimensional case, specifically, the relation between their degree and smoothness, the minimum number of grid nodes in each direction, the convergence rate, and susceptibility to saturation (see [Rya75] for detail).

The construction of nonlocal cubic splines can also be extended to two dimensions; in this case the splines are called bi-cubic. On a domain of rectangular shape, they can be obtained by solving multiple tri-diagonal linear systems of type (2.62) along the x and y coordinate lines of the Cartesian grid. The approximation properties of bi-cubic splines remain the same as those of the one-dimensional cubic splines.

Similar constructions, standard piecewise polynomials, local splines, and nonlocal splines, are also available for the interpolation of multivariable functions (more than two arguments). We should emphasize, however, that in general the size of the tables that would guarantee a given accuracy of interpolation for a function of certain smoothness rapidly grows as the number of arguments of the function increases. The corresponding interpolation algorithms also become more cumbersome.

2.4.2 Unstructured Grids

Unstructured grids on the (x, y) plane are typically composed of triangular cells. In so doing, it is required that any two triangles have either a common side or a common vertex, or alternatively, do not intersect at all. Unstructured grids offer a lot more flexibility in accommodating irregular geometric shapes, see Figure 2.1. In addition, they may also be more convenient when the interpolated function undergoes a strong variation in some local area.

Let, for example, the function $f = f(x, y)$ be defined on the domain schematically shown in Figure 2.1. Let us additionally assume that it varies rapidly inside the "bottleneck." In this case, a rectangular grid will not be particularly well suited for tabulating the function $f(x, y)$. One reason is that such a grid obviously cannot fit the curvilinear boundary of the domain. Another reason is that a rectangular grid cannot be refined locally, i.e., only in the bottleneck area, where most of the variation of the function $f(x, y)$ supposedly occurs. On the other hand, the unstructured triangular grid shown in Figure 2.1 is obviously free of these disadvantages. Therefore, the vertices of the triangles can be used in the capacity of interpolation nodes.

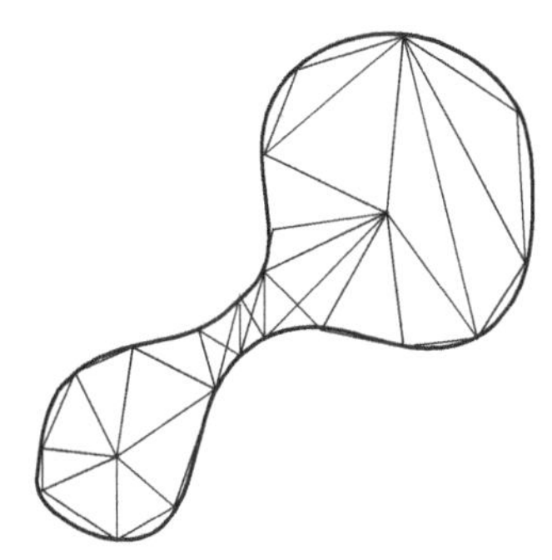

FIGURE 2.1: Unstructured grid.

However, a key difficulty in using triangular grids for interpolation is that the previously analyzed one-dimensional methods will not directly extend to this case in a dimension-by-dimension fashion. Instead, triangular grids require genuinely two-dimensional approaches to building the interpolating polynomials.

Let, for example, f_1, f_2, and f_3 be the values of the function $f(x, y)$ at the vertices (x_1, y_1), (x_2, y_2), and (x_3, y_3), respectively, of some triangle. Then, inside the triangle the unknown function $f(x, y)$ can be approximated by the linear function (a first degree polynomial of two variables):

$$f(x, y) \approx ax + by + c, \tag{2.84}$$

where the coefficients a, b, and c must satisfy the equalities:

$$\begin{aligned} ax_1 + by_1 + c &= f_1, \\ ax_2 + by_2 + c &= f_2, \\ ax_3 + by_3 + c &= f_3, \end{aligned}$$

in order for the interpolant to coincide exactly with the function values at the interpolation nodes. Geometrically, the linear function (2.84) is a plane that crosses through the three points: (x_1, y_1, f_1), (x_2, y_2, f_2), and (x_3, y_3, f_3) in the three-dimensional space. For a twice differentiable function with bounded second derivatives, the interpolation error guaranteed by this approach will be $\mathcal{O}(h^2)$, where h is the largest side of the triangle. If the linear interpolant (2.84) is built for every triangle of the grid, see Figure 2.1, then the overall interpolating function will be piecewise linear and continuous on the entire domain.

We note that linear interpolants on triangular grids are very helpful for building the finite-element approximations of partial differential equations, see Section 12.2 of Chapter 12. We also note that piecewise polynomial interpolants of higher degrees can be obtained on triangular grids as well.

Exercises

1. Prove the $\mathcal{O}\left(h^{(x)^2}+h^{(y)^2}\right)$ error estimate for the interpolation (2.83), where $h^{(x)}$ and $h^{(y)}$ are the grid sizes (constant in each direction).
2. Prove the $\mathcal{O}(h^2)$ error estimate for the interpolation (2.84), where h is the largest side of the triangle. Assume that all second order partial derivatives of $f(x,y)$ are bounded.
3. Let $f = f(x,y)$ be twice differentiable, and let all of its second partial derivatives be bounded by 1. How can one economically choose the size(s) of the rectangular interpolation grid on the plane to be able to reconstruct the function $f(x,y)$ everywhere inside the square $|x| \leq 1$, $|y| \leq 1$ with the error not exceeding 10^{-3}?
4. Let $f = f(x,y)$ be twice differentiable, let all of its second partial derivatives be bounded by 1, and let also the circles $x^2+y^2=r^2$ be level lines of the function $f = f(x,y)$ for every r. How can one economically place the interpolation nodes on the plane to achieve the same quality of the reconstructed function as in Exercise 3?
5.* Answer the question of Exercise 4, but instead of having the circles $x^2+y^2=r^2$ as level lines of f, assume that the function $f = f(x,y)$ can be represented as a product of two single-variable functions: $f = \phi(x)\psi(y)$, where $\phi(x) \neq 0$ and $\psi(y) \neq 0$.

Chapter 3

Trigonometric Interpolation

Along with the algebraic interpolation described in Chapter 2, one also uses interpolation by means of trigonometric polynomials of the type:

$$Q\left(\cos\frac{2\pi}{L}x, \sin\frac{2\pi}{L}x\right) = \sum_{k=0}^{n} a_k \cos\frac{2\pi k}{L}x + \sum_{k=1}^{n} b_k \sin\frac{2\pi k}{L}x, \tag{3.1}$$

where n is a positive integer, $L > 0$, and a_k & b_k are the coefficients. A trigonometric interpolating polynomial $Q\left(\cos\frac{2\pi}{L}x, \sin\frac{2\pi}{L}x\right)$ that would coincide with the given L-periodic function $f(x)$, $f(x+L) = f(x)$, at the equidistant interpolation nodes:

$$x_m = \frac{L}{N}m + x_0, \qquad m = 0, 1, \dots, N-1, \qquad x_0 = \text{const}, \tag{3.2}$$

can be chosen such that it will have some important advantages compared to the algebraic interpolating polynomial built on the same grid (3.2).

First, the error of the trigonometric interpolation

$$R_N(x,f) \stackrel{\text{def}}{=} f(x) - Q\left(\cos\frac{2\pi}{L}x, \sin\frac{2\pi}{L}x\right) \tag{3.3}$$

converges to zero uniformly with respect to x as $N \longrightarrow \infty$ already if the second derivative of $f(x)$ is piecewise continuous.[1] Moreover, the rate of this convergence, i.e., the rate of decay of the error (3.3) as $N \longrightarrow \infty$, automatically takes into account the smoothness of $f(x)$, i.e., increases for those functions $f(x)$ that have more derivatives. Specifically, we will prove that

$$\max_x |R_N(x,f)| = \mathcal{O}\left(\frac{M_{r+1}}{N^{r-1/2}}\right), \quad \text{where} \quad M_{r+1} = \max_x \left|\frac{d^{r+1}f(x)}{dx^{r+1}}\right|.$$

Second, it turns out that the sensitivity of the trigonometric interpolating polynomial (3.1) to the errors committed when specifying the function values $f_m = f(x_m)$ on the grid (3.2) remains "practically flat" (i.e., grows slowly) as N increases.

The foregoing two properties — automatic improvement of accuracy for smoother functions, and slow growth of the Lebesgue constants that translates into numerical stability — are distinctly different from the properties of algebraic interpolation on

[1]In fact, even less regularity may be required of $f(x)$, see Section 3.2.7.

uniform grids, see Chapter 2. It, however, turns out that algebraic interpolation of functions on an interval can also possess the same remarkable qualities. To achieve that, one shall choose the Chebyshev interpolation nodes and use Chebyshev interpolating polynomials.

When reading the book for the first time, the reader can restrict him/herself by what has been said about the contents of Chapter 3, and proceed directly to Chapter 4.

3.1 Interpolation of Periodic Functions

Let $f = f(x)$ be an L-periodic function:

$$\forall x: \; f(x+L) = f(x), \qquad L > 0, \tag{3.4}$$

defined on the grid:

$$x_m = \frac{L}{N}m + x_0, \qquad m = 0, \pm 1, \pm 2, \ldots, \qquad x_0 = \text{const}, \tag{3.5}$$

where N is a positive integer. Introducing the notation $f_m \equiv f(x_m)$, we obtain by virtue of (3.4) and (3.5):

$$\forall m: \; f_{m+N} = f_m.$$

3.1.1 An Important Particular Choice of Interpolation Nodes

THEOREM 3.1

Let $x_0 = L/(2N)$, $N = 2(n+1)$, n being a positive integer. Let f be an L-periodic function, and let f_m be its values on the grid (3.5). For a given arbitrary set of f_m, there is one and only one trigonometric interpolating polynomial:

$$Q_n\left(\cos\frac{2\pi}{L}x, \sin\frac{2\pi}{L}x, f\right) = \sum_{k=0}^{n} a_k \cos\frac{2\pi k}{L}x + \sum_{k=1}^{n+1} b_k \sin\frac{2\pi k}{L}x, \tag{3.6}$$

that satisfies the equalities:

$$Q_n\Big|_{x=x_m} = f_m, \qquad m = 0, \pm 1, \pm 2, \ldots.$$

The coefficients of this polynomial are given by the formulae:

$$a_0 = \frac{1}{N}\sum_{m=0}^{N-1} f_m, \tag{3.7}$$

$$a_k = \frac{2}{N}\sum_{m=0}^{N-1} f_m \cos k\left(\frac{2\pi}{N}m + \frac{\pi}{N}\right), \qquad k = 1,2,\dots,n, \tag{3.8}$$

$$b_k = \frac{2}{N}\sum_{m=0}^{N-1} f_m \sin k\left(\frac{2\pi}{N}m + \frac{\pi}{N}\right), \qquad k = 1,2,\dots,n, \tag{3.9}$$

$$b_{n+1} = \frac{1}{N}\sum_{m=0}^{N-1} f_m(-1)^m. \tag{3.10}$$

PROOF Let us consider a set of all real valued periodic discrete functions:

$$f_{m+N} = f_m, \qquad m = 0, \pm 1, \pm 2, \dots, \tag{3.11}$$

defined on the grid $x_m = \frac{L}{N}m + \frac{L}{2N}$. We will only be considering these functions on the grid interval $m = 0,1,\dots,N-1$, because for all other m's they can be unambiguously reconstructed by virtue of periodicity (3.11).

The entire set of these functions, supplemented by the conventional operations of addition and multiplication by real scalars, form a linear space that we will denote F_N. The dimension of this space is equal to N, because the system of N linearly independent functions (vectors) $\tilde{\psi}^{(k)} \in F_N$, $k = 1,2,\dots,N$:

$$\tilde{\psi}_m^{(k)} \stackrel{\text{def}}{=} \begin{cases} 0, & \text{if } m \neq k-1, \\ 1, & \text{if } m = k-1, \end{cases}$$

provides a basis in the space F_N. Indeed, any function $f \in F_N$, $f = \{f_m \,|\, m = 0,1,\dots,N\}$ always admits a unique representation as a linear combination of the basis functions $\tilde{\psi}^{(k)}$: $f = \sum_{k=1}^{N} f_{k-1}\tilde{\psi}^{(k)}$.

Let us now introduce a Euclidean dot (i.e., inner) product in the space F_N:

$$(f,g) = \frac{1}{N}\sum_{m=0}^{N-1} f_m g_m, \tag{3.12}$$

and show that the system of $2(n+1)$ functions: $\xi^{(k)} = \{\xi_m^{(k)}\}$, $k = 0,1,\dots,n$, and $\eta^{(k)} = \{\eta_m^{(k)}\}$, $k = 1,2,\dots,n+1$, where

$$\begin{aligned}
\xi_m^{(0)} &= \cos(0\cdot x_m) \equiv 1, \\
\xi_m^{(k)} &= \sqrt{2}\cos\left(\frac{2\pi k}{L}x_m\right), \quad k = 1,2,\dots,n, \\
\eta_m^{(k)} &= \sqrt{2}\sin\left(\frac{2\pi k}{L}x_m\right), \quad k = 1,2,\dots,n, \\
\eta_m^{(n+1)} &= \sin\left(\frac{2\pi(n+1)}{L}x_m\right) = (-1)^m,
\end{aligned} \tag{3.13}$$

forms an orthonormal basis in the space F_N (recall, $n = N/2 - 1$). The overall number of functions defined by formulae (3.13) is equal to N. Therefore, it only remains to prove the equalities:

$$(\xi^{(k)}, \xi^{(k)}) = 1, \qquad k = 0, 1, \dots, n, \tag{3.14}$$

$$(\eta^{(k)}, \eta^{(k)}) = 1, \qquad k = 1, 2, \dots, n+1, \tag{3.15}$$

$$(\xi^{(r)}, \xi^{(s)}) = 0, \qquad r \neq s, \quad r, s = 0, 1, \dots, n, \tag{3.16}$$

$$(\eta^{(r)}, \eta^{(s)}) = 0, \qquad r \neq s, \quad r, s = 1, 2, \dots, n+1, \tag{3.17}$$

$$(\xi^{(r)}, \eta^{(s)}) = 0, \qquad r = 0, 1, \dots, n, , \quad s = 1, 2, \dots, n+1. \tag{3.18}$$

To prove equalities (3.14)–(3.18), we first notice that for any N and γ the following relations hold:

$$\frac{1}{N}\sum_{m=0}^{N-1} 1 = 1, \tag{3.19}$$

$$\sum_{m=0}^{N-1} \cos l\left(\frac{2\pi}{N}m + \gamma\right) = 0, \qquad l = 1, 2, \dots, N-1, \tag{3.20}$$

$$\sum_{m=0}^{N-1} \sin l\left(\frac{2\pi}{N}m + \gamma\right) = 0, \qquad l = 1, 2, \dots, N-1. \tag{3.21}$$

Indeed, (3.19) is obviously true. To verify (3.20), we can write for any $l = 0, 1, \dots, N-1$:

$$\begin{aligned}
&\sum_{m=0}^{N-1} \cos l\left(\frac{2\pi}{N}m + \gamma\right) \\
&= \frac{1}{2}\sum_{m=0}^{N-1}\left[\exp\left\{i\left(\frac{2l\pi m}{N} + l\gamma\right)\right\} + \exp\left\{-i\left(\frac{2l\pi m}{N} + l\gamma\right)\right\}\right] \\
&= \frac{1}{2}e^{il\gamma}\sum_{m=0}^{N-1}\left(\exp\left\{i\frac{2l\pi}{N}\right\}\right)^m + \frac{1}{2}e^{-il\gamma}\sum_{m=0}^{N-1}\left(\exp\left\{-i\frac{2l\pi}{N}\right\}\right)^m \\
&= \frac{1}{2}e^{il\gamma}\frac{1 - \exp\{i2l\pi\}}{1 - \exp\{i2l\pi/N\}} + \frac{1}{2}e^{-il\gamma}\frac{1 - \exp\{-i2l\pi\}}{1 - \exp\{-i2l\pi/N\}} = 0 + 0 = 0,
\end{aligned} \tag{3.22}$$

where we have used the formula for the sum of a geometric sequence. Equality (3.21) is proven similarly.

We are now ready to show that the basis (3.13) is indeed orthonormal, i.e., that equalities (3.14)–(3.18) do hold. Equality (3.14) for $k = 0$, as well as equality (3.15) for $k = n+1$, coincide with (3.19). For $k = 1, 2, \dots, n$ both equalities (3.14) and (3.15) hold by virtue of (3.20) and (3.21):

$$(\xi^{(k)}, \xi^{(k)}) = \frac{2}{N}\sum_{m=0}^{N-1}\cos^2\left(\frac{2\pi k}{N}m + \frac{\pi k}{N}\right)$$

$$=\frac{1}{N}\sum_{m=0}^{N-1}\left[1+\cos\left(\frac{4\pi k}{N}m+\frac{2\pi k}{N}\right)\right]$$
$$=\frac{1}{N}\sum_{m=0}^{N-1}1+\sum_{m=0}^{N-1}\cos 2k\left(\frac{2\pi}{N}m+\frac{\pi}{N}\right)=1+0=1,$$

$$(\eta^{(k)},\eta^{(k)})=\frac{2}{N}\sum_{m=0}^{N-1}\sin^2\left(\frac{2\pi k}{N}m+\frac{\pi k}{N}\right)$$
$$=\frac{1}{N}\sum_{m=0}^{N-1}\left[1-\cos\left(\frac{4\pi k}{N}m+\frac{2\pi k}{N}\right)\right]=1.$$

To prove equality (3.16), we first notice that for $r,s=0,1,\dots,n$ and $r\neq s$ we always have $1\leq|r\pm s|\leq N-1$, and then use formula (3.20) to obtain:

$$(\xi^{(r)},\xi^{(s)})=\frac{2}{N}\sum_{m=0}^{N-1}\cos\left\{r\left(\frac{2\pi m}{N}+\frac{\pi}{N}\right)\right\}\cdot\cos\left\{s\left(\frac{2\pi m}{N}+\frac{\pi}{N}\right)\right\}$$
$$=\frac{1}{N}\sum_{m=0}^{N-1}\left[\cos\left\{(r+s)\left(\frac{2\pi m}{N}+\frac{\pi}{N}\right)\right\}+\cos\left\{(r-s)\left(\frac{2\pi m}{N}+\frac{\pi}{N}\right)\right\}\right]=0.$$

Equality (3.17) is proven similarly, except that instead of the trigonometric identity $2\cos\alpha\cos\beta=\cos(\alpha+\beta)+\cos(\alpha-\beta)$ that has been used when proving (3.16), one rather needs to employ another identity: $2\sin\alpha\sin\beta=\cos(\alpha+\beta)-\cos(\alpha-\beta)$. Finally, yet another trigonometric identity: $2\sin\alpha\cos\beta=\sin(\alpha+\beta)+\sin(\alpha-\beta)$ is to be used for proving formula (3.18).

Altogether, we have established that (3.13) is an orthonormal basis in the space F_N. Therefore, every function $f=\{f_m\}\in F_N$ can be represented as a linear combination of the basis functions (3.13):

$$f_m=\sum_{k=0}^{n}a_k\cos\frac{2\pi k}{L}x_m+\sum_{k=1}^{n+1}b_k\sin\frac{2\pi k}{L}x_m.$$

Calculating the dot products of both the left-hand side and the right-hand side of the previous equality with all the basis functions $\xi^{(r)}$ and $\eta^{(s)}$, $r=0,1,\dots,n$, $s=1,2,\dots,n+1$, we arrive at the equalities:

$$\begin{aligned} a_0&=(f,\xi^{(0)}),\\ a_k&=\sqrt{2}(f,\xi^{(k)}), && k=1,2,\dots,n,\\ b_k&=\sqrt{2}(f,\eta^{(k)}), && k=1,2,\dots,n,\\ b_{n+1}&=(f,\eta^{(n+1)}), \end{aligned}$$

that, according to definitions (3.13), coincide with formulae (3.7)–(3.10). □

The grid $x_m = \frac{L}{N}m + \frac{L}{2N}$, $m = 0, \pm 1, \pm 2, \ldots$, that has been used for specifying the discrete functions $f \in F_N$ in Theorem 3.1 is symmetric with respect to the origin $x = 0$,[2] so that along with the point $x = x_m$ it also contains the point $x = -x_m = x_{-(m+1)}$. Therefore, one can consider even and odd functions on this grid. The grid function $f_m \equiv f(x_m)$ is called *even* if

$$f_m = f_{-(m+1)}, \qquad m = 0, \pm 1, \pm 2, \ldots. \tag{3.23}$$

Similarly, the grid function f_m is called *odd* if

$$f_m = -f_{-(m+1)}, \qquad m = 0, \pm 1, \pm 2, \ldots. \tag{3.24}$$

THEOREM 3.2

Let f_m be an even N-periodic grid function specified at the nodes $x_m = \frac{L}{N}m + \frac{L}{2N}$, $m = 0, \pm 1, \pm 2, \ldots$, $N = 2(n+1)$. Then, the trigonometric interpolating polynomial (3.6) becomes:

$$Q_n = \sum_{k=0}^{n} a_k \cos \frac{2\pi k}{L} x, \tag{3.25}$$

where

$$a_0 = \frac{1}{n+1} \sum_{m=0}^{n} f_m, \tag{3.26}$$

$$a_k = \frac{2}{n+1} \sum_{m=0}^{n} f_m \cos k \left(\frac{\pi}{n+1} m + \frac{\pi}{2(n+1)} \right), \qquad k = 1, 2, \ldots, n. \tag{3.27}$$

PROOF For an even function f_m [see formula (3.23)] expressions (3.7) and (3.8) translate into (3.26) and (3.27), respectively, whereas expressions (3.9) and (3.10) imply that $b_k \equiv 0$. ∎

THEOREM 3.3

Let f_m be an odd N-periodic grid function specified at the nodes $x_m = \frac{L}{N}m + \frac{L}{2N}$, $m = 0, \pm 1, \pm 2, \ldots$, $N = 2(n+1)$. Then, the trigonometric interpolating polynomial (3.6) becomes:

$$Q_n = \sum_{k=1}^{n+1} b_k \sin \frac{2\pi k}{L} x, \tag{3.28}$$

[2] Note that the origin itself is not a node of this grid.

where

$$b_k = \frac{2}{n+1}\sum_{m=0}^{n} f_m \sin k\left(\frac{\pi}{n+1}m + \frac{\pi}{2(n+1)}\right), \qquad k = 1,2,\dots,n, \tag{3.29}$$

$$b_{n+1} = \frac{1}{n+1}\sum_{m=0}^{n} f_m(-1)^m. \tag{3.30}$$

PROOF For an odd function f_m [see formula (3.24)] formulae (3.7) and (3.8) imply that the coefficients a_k, $k = 0,1,\dots,n$, are equal to zero, expressions (3.9) and (3.10) then transform into (3.29) and (3.30), respectively, and the polynomial (3.6), accordingly, reduces to (3.28). □

3.1.2 Sensitivity of the Interpolating Polynomial to Perturbations of the Function Values

Let us estimate how sensitive the trigonometric interpolating polynomial (3.6) is to the errors committed when specifying the values of the grid function f_m. Assume that instead of $f = \{f_m\}$ we actually have a perturbed function $f + \delta f = \{f_m + \delta f_m\}$. Then, instead of the polynomial (3.6) we obtain a new polynomial:

$$Q_n\left(\cos\frac{2\pi}{L}x, \sin\frac{2\pi}{L}x, f + \delta f\right) = Q_n\left(\cos\frac{2\pi}{L}x, \sin\frac{2\pi}{L}x, f\right) + \delta Q_n.$$

From formulae (3.6) and (3.7)–(3.10) one can easily see that the corresponding error is given by:

$$\delta Q_n = Q_n\left(\cos\frac{2\pi}{L}x, \sin\frac{2\pi}{L}x, \delta f\right).$$

Therefore, similarly to how it has been done in Section 2.1.4 of Chapter 2, we can introduce *the Lebesgue constants:*

$$L_n \stackrel{\text{def}}{=} \sup_{\delta f \in F_N} \frac{\max_x \left|Q_n\left(\cos\frac{2\pi}{L}x, \sin\frac{2\pi}{L}x, \delta f\right)\right|}{\max_m |\delta f_m|}, \qquad n = 1,2,\dots, \tag{3.31}$$

that would naturally quantify the sensitivity of the trigonometric interpolating polynomial (3.6) to the perturbations of its input data. Obviously,

$$\max_x |\delta Q_n| \le L_n \max_m |\delta f_m|. \tag{3.32}$$

THEOREM 3.4

The Lebesgue constants of the trigonometric interpolating polynomials (3.6) satisfy the estimates

$$L_n \le 4(n+1). \tag{3.33}$$

PROOF Formulae (3.7)–(3.10) imply that

$$|a_k| \leq 2 \max_m |f_m|, \tag{3.34}$$

$$|b_k| \leq 2 \max_m |f_m|. \tag{3.35}$$

Then, from formula (3.6) and inequalities (3.34), (3.35) applied to the perturbation δf one can easily see that

$$\left| Q_n \left(\cos \frac{2\pi}{L} x, \sin \frac{2\pi}{L} x, \delta f \right) \right| \leq \sum_{k=0}^{n} |a_k| + \sum_{k=1}^{n+1} |b_k| \leq 4(n+1) \max_m |\delta f_m|. \tag{3.36}$$

Estimate (3.36) holds for any $\delta f \in F_N$ and any N. This implies (3.33). □

Let us emphasize that the growth rate of the Lebesgue constants in the case of trigonometric interpolation on equally spaced nodes, see estimate (3.33), *is considerably slower* than in the case of algebraic interpolation, see estimates (2.18) in Section 2.1.4 of Chapter 2.

3.1.3 Estimate of Interpolation Error

THEOREM 3.5
Let $f = f(x)$ be an L-periodic function that has continuous L-periodic derivatives up to the order $r > 0$ and a square integrable derivative of order $r + 1$:

$$\int_0^L \left[f^{(r+1)}(x) \right]^2 dx < \infty.$$

Let $f \in F_N$, $f = \{f_m\}$, be a table of values of this function sampled at the equally spaced grid nodes

$$x_m = \frac{L}{N} m + \frac{L}{2N}, \qquad m = 0, \pm 1, \pm 2, \ldots, \qquad N = 2(n+1),$$

and accordingly, let $Q_n \left(\cos \frac{2\pi}{L} x, \sin \frac{2\pi}{L} x, f \right)$ be the trigonometric interpolating polynomial of type (3.6). Then, for the error of the trigonometric interpolation:

$$R_n(x) \stackrel{\text{def}}{=} f(x) - Q_n \left(\cos \frac{2\pi}{L} x, \sin \frac{2\pi}{L} x, f \right),$$

the following estimate holds:

$$|R_n(x)| \leq \frac{\zeta_n}{n^{r-1/2}}, \quad \text{where} \quad \zeta_n = o(1) \quad \text{as} \quad n \longrightarrow \infty. \tag{3.37}$$

PROOF Let us represent $f(x)$ as the sum of its Fourier series:

$$f(x) = S_n(x) + \delta S_n(x),$$

where

$$S_n(x) = \frac{\alpha_0}{2} + \sum_{k=1}^{n} \alpha_k \cos\frac{2\pi kx}{L} + \sum_{k=1}^{n+1} \beta_k \sin\frac{2\pi kx}{L} \tag{3.38}$$

is a partial sum of order n and

$$\delta S_n(x) = \sum_{k=n+1}^{\infty} \alpha_k \cos\frac{2\pi kx}{L} + \sum_{k=n+2}^{\infty} \beta_k \sin\frac{2\pi kx}{L} \tag{3.39}$$

is the corresponding remainder. The coefficients α_k and β_k of the Fourier series for $f(x)$ are given by the formulae:

$$\alpha_k = \frac{2}{L}\int_0^L f(x)\cos\frac{2\pi kx}{L}dx, \qquad \beta_k = \frac{2}{L}\int_0^L f(x)\sin\frac{2\pi kx}{L}dx. \tag{3.40}$$

Let us first prove the following estimate for the remainder δS_n of (3.39):

$$|\delta S_n(x)| \leq \frac{\tilde{\zeta}_n}{n^{r+\frac{1}{2}}}, \tag{3.41}$$

where $\tilde{\zeta}_n$ is a numerical sequence such that $\tilde{\zeta}_n = o(1)$, i.e., $\lim_{n\to\infty}\tilde{\zeta}_n = 0$. Estimate (3.41) is an estimate of the convergence rate of the Fourier series for $f(x)$. Denote by A_k and B_k the Fourier coefficients of the square integrable L-periodic function $f^{(r+1)}(x)$:

$$A_k = \frac{2}{L}\int_0^L f^{(r+1)}(x)\cos\frac{2\pi kx}{L}dx, \qquad B_k = \frac{2}{L}\int_0^L f^{(r+1)}(x)\sin\frac{2\pi kx}{L}dx.$$

Then, we can integrate the previous equalities by parts $r+1$ times, employ the periodicity, and use definitions (3.40) to obtain $A_0 = 0$ and either:

$$A_k = \pm\left(\frac{2\pi k}{L}\right)^{r+1}\alpha_k, \qquad B_k = \mp\left(\frac{2\pi k}{L}\right)^{r+1}\beta_k, \tag{3.42}$$

or

$$A_k = \pm\left(\frac{2\pi k}{L}\right)^{r+1}\beta_k, \qquad B_k = \mp\left(\frac{2\pi k}{L}\right)^{r+1}\alpha_k, \tag{3.43}$$

for $k = 1, 2, \ldots$, depending on the particular value of r. Moreover, according to the Bessel inequality (see, e.g., [KF75, Section 16]) we can write:

$$\sum_{k=1}^{\infty} A_k^2 + B_k^2 \leq \frac{2}{L}\int_0^L \left[f^{(r+1)}(x)\right]^2 dx. \tag{3.44}$$

Let us then define the sequence $\mu_k \stackrel{\text{def}}{=} \left(\frac{L}{2\pi}\right)^{r+1}\sqrt{A_k^2+B_k^2}$. Inequality (3.44) implies that the series $\sum\limits_{k=1}^{\infty}\mu_k^2$ converges. Now assume for definiteness that equalities (3.42) hold; then we have:

$$|\alpha_k| = \left(\frac{L}{2\pi}\right)^{r+1}\frac{|A_k|}{k^{r+1}} \le \left(\frac{L}{2\pi}\right)^{r+1}\frac{\sqrt{A_k^2+B_k^2}}{k^{r+1}} = \frac{\mu_k}{k^{r+1}},$$

and similarly

$$|\beta_k| = \left(\frac{L}{2\pi}\right)^{r+1}\frac{|B_k|}{k^{r+1}} \le \left(\frac{L}{2\pi}\right)^{r+1}\frac{\sqrt{A_k^2+B_k^2}}{k^{r+1}} = \frac{\mu_k}{k^{r+1}}.$$

The same estimates can obviously be obtained when relations (3.43) hold instead of (3.42). We can therefore say that the Fourier coefficients (3.40) satisfy the following inequalities:

$$|\alpha_k| \le \frac{\mu_k}{k^{r+1}}, \qquad |\beta_k| \le \frac{\mu_k}{k^{r+1}}, \tag{3.45}$$

where the sequence $\mu_k = o(1)$ is such that the series $\sum\limits_{k=1}^{\infty}\mu_k^2$ converges.

Then, for the remainder (3.39) of the Fourier series we can write:

$$\begin{aligned} |\delta S_n(x)| &= \left|\sum_{k=n+1}^{\infty}\alpha_k\cos\frac{2\pi kx}{L} + \sum_{k=n+2}^{\infty}\beta_k\sin\frac{2\pi kx}{L}\right| \\ &\le \sum_{k=n+1}^{\infty}|\alpha_k| + \sum_{k=n+2}^{\infty}|\beta_k| \le \sum_{k=n+1}^{\infty}|\alpha_k|+|\beta_k| \\ &\le 2\sum_{k=n+1}^{\infty}\frac{\mu_k}{k^{r+1}} \le 2\sqrt{\sum_{k=n+1}^{\infty}\mu_k^2}\sqrt{\sum_{k=n+1}^{\infty}\frac{1}{k^{2(r+1)}}}, \end{aligned} \tag{3.46}$$

where the last estimate in (3.46) is obtained using the classical Hölder inequality (see [KF75, Section 5]):

$$\sum_{k=1}^{\infty}u_kv_k \le \sqrt{\sum_{k=1}^{\infty}u_k^2}\sqrt{\sum_{k=1}^{\infty}v_k^2},$$

which is valid if $\sum\limits_{k=1}^{\infty}u_k^2 < \infty$ and $\sum\limits_{k=1}^{\infty}v_k^2 < \infty$.

Next, we note that for $\xi \in [k-1,k]$ the following inequality always holds: $\frac{1}{k^{2(r+1)}} \le \frac{1}{\xi^{2(r+1)}}$, and consequently, $\frac{1}{k^{2(r+1)}} \le \int\limits_{k-1}^{k}\frac{d\xi}{\xi^{2(r+1)}}$. Therefore,

$$\sum_{k=n+1}^{\infty}\frac{1}{k^{2(r+1)}} \le \sum_{k=n+1}^{\infty}\int_{k-1}^{k}\frac{d\xi}{\xi^{2(r+1)}} = \int_{n}^{\infty}\frac{d\xi}{\xi^{2(r+1)}} = \frac{1}{(2r+1)n^{2r+1}}.$$

Substituting the latter estimate into inequality (3.46), we obtain:

$$|\delta S_n(x)| \le \frac{2}{\sqrt{2r+1}} \sqrt{\sum_{k=n+1}^{\infty} \mu_k^2} \frac{1}{\sqrt{n^{2r+1}}}. \quad (3.47)$$

It only remains to notice that $\tilde{\zeta}_n \stackrel{\text{def}}{=} \frac{2}{\sqrt{2r+1}}\sqrt{\sum\limits_{k=n+1}^{\infty} \mu_k^2} = o(1)$, because the series $\sum\limits_{k=1}^{\infty} \mu_k^2$ converges. Then, estimate (3.47) does imply (3.41).

Having justified estimate (3.41), we next notice that the partial sum $S_n(x)$ is, in fact, a trigonometric polynomial of type (3.6). Due to the uniqueness of the trigonometric interpolating polynomial, see Theorem 3.1, we then have

$$Q_n\left(\cos\frac{2\pi}{L}x, \sin\frac{2\pi}{L}x, S_n\right) = S_n(x). \quad (3.48)$$

Moreover, estimates (3.32), (3.33), and (3.41) together yield:

$$\left|Q_n\left(\cos\frac{2\pi}{L}x, \sin\frac{2\pi}{L}x, \delta S_n\right)\right| \le L_n|\delta S_n| = L_n\frac{\tilde{\zeta}_n}{n^{r+\frac{1}{2}}} \le 4\frac{\zeta_n}{n^{r-\frac{1}{2}}}, \quad (3.49)$$

where $\zeta_n = o(1)$ as $n \longrightarrow \infty$. Finally, by combining (3.41), (3.48), and (3.49) we obtain the following estimate for the interpolation error $R_n(x)$:

$$\begin{aligned}
|R_n(x)| &= \left|f(x) - Q_n\left(\cos\frac{2\pi}{L}x, \sin\frac{2\pi}{L}x, f\right)\right| \\
&= \left|f(x) - Q_n\left(\cos\frac{2\pi}{L}x, \sin\frac{2\pi}{L}x, S_n\right) - Q_n\left(\cos\frac{2\pi}{L}x, \sin\frac{2\pi}{L}x, \delta S_n\right)\right| \\
&= \left|(f(x) - S_n(x)) - Q_n\left(\cos\frac{2\pi}{L}x, \sin\frac{2\pi}{L}x, \delta S_n\right)\right| \\
&= \left|\delta S_n(x) - Q_n\left(\cos\frac{2\pi}{L}x, \sin\frac{2\pi}{L}x, \delta S_n\right)\right| \\
&\le |\delta S_n(x)| + \left|Q_n\left(\cos\frac{2\pi}{L}x, \sin\frac{2\pi}{L}x, \delta S_n\right)\right| \\
&\le |\delta S_n(x)| + L_n|\delta S_n(x)| \le \frac{\tilde{\zeta}_n}{n^{r+\frac{1}{2}}} + 4\frac{\zeta_n}{n^{r-\frac{1}{2}}} \le \text{const} \cdot \frac{\zeta_n}{n^{r-\frac{1}{2}}},
\end{aligned}$$

which is obviously equivalent to the required estimate (3.37). ∎

We emphasize that the rate of convergence of the trigonometric interpolating polynomials established by estimate (3.37) automatically becomes faster for smoother interpolated functions $f(x)$. In this sense, trigonometric interpolation of periodic functions appears to be *not susceptible to the saturation by smoothness*. This is a remarkable difference compared to the case of algebraic interpolation (Chapter 2), when the convergence rate is limited by the degree of the polynomial.

It shall be noted though, that there is a gap between the convergence estimate (3.37) and the estimate of the unavoidable error for reconstructing a function with $r+1$ derivatives on the grid with size $\sim 1/n$, which is $\mathcal{O}(n^{-(r+1)})$. This gap, however, is of purely technical nature. Estimate (3.33) for the Lebesgue constants that we have used when proving Theorem 3.5 can, in fact, be further improved (Section 3.2.7). Accordingly, estimate (3.37) for the interpolation error can be improved as well. An additional improvement is brought along by the Jackson inequality (Theorem 3.8). Altogether, it is shown (Section 3.2.7, page 87) that the difference between the actual interpolation error and its theoretical limit (the unavoidable error) can be driven down and made as small as a slow logarithmic factor.

3.1.4 An Alternative Choice of Interpolation Nodes

Another type of trigonometric interpolation is also important for applications.

THEOREM 3.6
Let f be an L-periodic function, and let f_m be its values on the grid:

$$x_m = \frac{L}{N}m, \qquad m = 0, \pm 1, \pm 2, \dots, \qquad N = 2n.$$

For a given arbitrary set of f_m, there is one and only one trigonometric interpolating polynomial:

$$\tilde{Q}_n\left(\cos\frac{2\pi}{L}x, \sin\frac{2\pi}{L}x, f\right) = \sum_{k=0}^{n} \tilde{a}_k \cos\frac{2\pi k}{L}x + \sum_{k=1}^{n-1} \tilde{b}_k \sin\frac{2\pi k}{L}x, \tag{3.50}$$

that satisfies the equalities:

$$\tilde{Q}_n\Big|_{x=x_m} = f_m, \qquad m = 0, \pm 1, \pm 2, \dots.$$

The coefficients of this polynomial are given by the formulae:

$$\tilde{a}_0 = \frac{1}{N}\sum_{m=0}^{N-1} f_m, \tag{3.51}$$

$$\tilde{a}_k = \frac{2}{N}\sum_{m=0}^{N-1} f_m \cos\frac{2\pi k m}{N}, \qquad k = 1, 2, \dots, n-1, \tag{3.52}$$

$$\tilde{a}_n = \frac{1}{N}\sum_{m=0}^{N-1} f_m (-1)^m, \tag{3.53}$$

$$\tilde{b}_k = \frac{2}{N}\sum_{m=0}^{N-1} f_m \sin\frac{2\pi k m}{N}, \qquad k = 1, 2, \dots, n-1. \tag{3.54}$$

PROOF It is very similar to that of Theorem 3.1, and we omit it here. ☐

For an even grid function, $f_m = f_{-m}$, formulae (3.51)–(3.54) transform into:

$$\begin{aligned}
\tilde{a}_0 &= \frac{1}{N}(f_0 + f_n) + \frac{2}{N}\sum_{m=1}^{n-1} f_m, \\
\tilde{a}_k &= \frac{2}{N}(f_0 + (-1)^k f_n) + \frac{4}{N}\sum_{m=1}^{n-1} f_m \cos\frac{2\pi km}{N}, \qquad k = 1,2,\dots,n-1, \\
\tilde{a}_n &= \frac{1}{N}(f_0 + (-1)^n f_n), \\
\tilde{b}_k &= 0, \qquad k = 1,2,\dots,n-1,
\end{aligned}$$

and the polynomial (3.50) reduces to:

$$\tilde{Q}_n\left(\cos\frac{2\pi}{L}x, \sin\frac{2\pi}{L}x, f\right) = \sum_{k=0}^{n} \tilde{a}_k \cos\frac{2\pi k}{L}x.$$

Note that the arguments which are very similar to those used when proving the key properties of the trigonometric interpolating polynomial $Q_n\left(\cos\frac{2\pi}{L}x, \sin\frac{2\pi}{L}x, f\right)$ in Theorems 3.4 and 3.5, also apply to the polynomial $\tilde{Q}_n\left(\cos\frac{2\pi}{L}x, \sin\frac{2\pi}{L}x, f\right)$ defined by formulae (3.50)–(3.54). Namely, this polynomial has slowly growing Lebesgue constants and as such, is basically stable with respect to the perturbations of the grid function f_m. Moreover, it converges to the interpolated function $f(x)$ as $n \longrightarrow \infty$ with the rate determined by the smoothness of $f(x)$, i.e., there is no saturation.

REMARK 3.1 If the interpolated function $f(x)$ has derivatives of all orders, then the rate of convergence of the trigonometric interpolating polynomials to $f(x)$ will be faster than any inverse power of n. In the literature, this type of convergence is often referred to as *spectral*. □

3.2 Interpolation of Functions on an Interval. Relation between Algebraic and Trigonometric Interpolation

Let $f = f(x)$ be defined on the interval $-1 \le x \le 1$, and let it have there a bounded derivative of order $r+1$. We have chosen this specific interval $-1 \le x \le 1$ as the domain of $f(x)$, rather than an arbitrary interval $a \le x \le b$, for the only reason of simplicity and convenience. Indeed, the transformation $x = \frac{a+b}{2} + t\frac{b-a}{2}$ renders a transition from the function $f(x)$ defined on an arbitrary interval $a \le x \le b$ to the function $F(t) \equiv f\left(\frac{a+b}{2} + t\frac{b-a}{2}\right)$ defined on the interval $-1 \le t \le 1$.

3.2.1 Periodization

According to Theorem 3.5 of Section 3.1, trigonometric interpolation is only suitable for the reconstruction of smooth periodic functions from their tables of values.

Therefore, to be able to apply it to the function $f(x)$ given on $-1 \le x \le 1$, one should first equivalently replace $f(x)$ by some smooth periodic function. However, a straightforward extension of the function $f(x)$ from its domain $-1 \le x \le 1$ to the entire real axis may, generally speaking, yield a discontinuous periodic function with the period $L = 2$, see Figure 3.1.

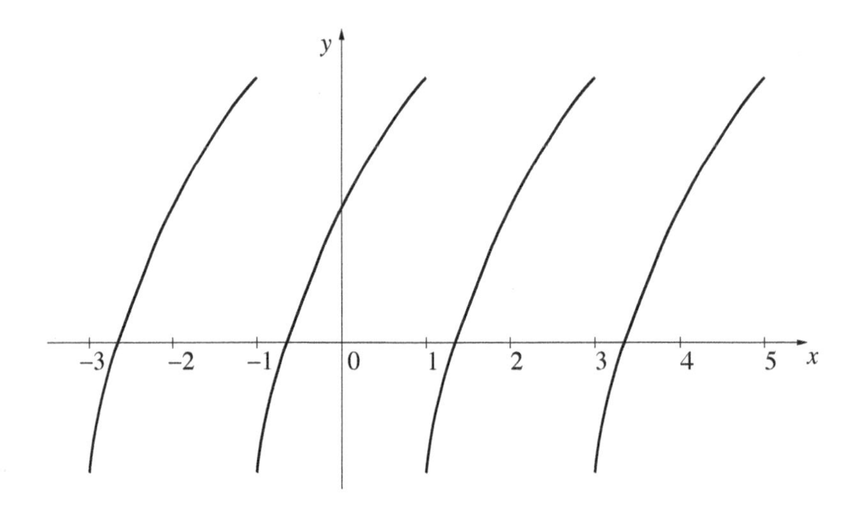

FIGURE 3.1: Straightforward periodization.

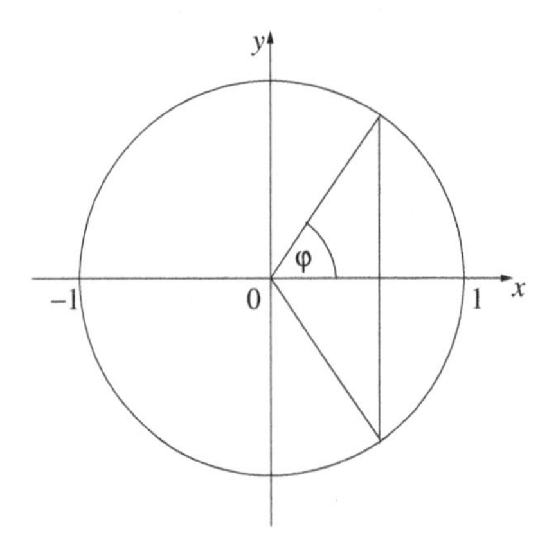

FIGURE 3.2: Periodization according to formula (3.55).

Therefore, instead of the function $f(x)$, $-1 \le x \le 1$, let us consider a new function

$$F(\varphi) = f(\cos\varphi), \quad x = \cos\varphi. \tag{3.55}$$

It will be convenient to think that the function $F(\varphi)$ of (3.55) is defined on the unit circle as a function of the polar angle φ. The value of $F(\varphi)$ is obtained by merely translating the value of $f(x)$ from the point $x \in [-1, 1]$ to the corresponding point $\varphi \in [0, \pi]$ on the unit circle, see Figure 3.2. In so doing, one can interpret the resulting function $F(\varphi)$ as even, $F(-\varphi) = F(\varphi)$, 2π-periodic function of its argument φ. Moreover, it is easy to see from definition (3.55) that the derivative $\frac{d^{r+1}F(\varphi)}{d\varphi^{r+1}}$ exists and is bounded.

3.2.2 Trigonometric Interpolation

Let us choose the following interpolation nodes:

$$\varphi_m = \frac{2\pi}{N}m + \frac{\pi}{N}, \qquad m = 0, \pm 1, \ldots, \pm n, -(n+1), \qquad N = 2(n+1). \tag{3.56}$$

According to (3.55), the values $F_m = F(\varphi_m)$ of the function $F(\varphi)$ at the nodes φ_m of (3.56) coincide with the values $f_m = f(x_m)$ of the original function $f(x)$ at the points $x_m = \cos\varphi_m$. To interpolate a 2π-periodic even function $F(\varphi)$ using its tabulated values at the nodes (3.56), one can employ formula (3.25) of Section 3.1:

$$Q_n(\cos\varphi, \sin\varphi, F) = \sum_{k=0}^{n} a_k \cos k\varphi. \tag{3.57}$$

As $F_m = f_m$ for all m, the coefficients a_k of the trigonometric interpolating polynomial (3.57) are given by formulae (3.26), (3.27) of Section 3.1:

$$\begin{aligned} a_0 &= \frac{1}{n+1}\sum_{m=0}^{n} f_m, \\ a_k &= \frac{2}{n+1}\sum_{m=0}^{n} f_m \cos k\varphi_m, \qquad k = 1, 2, \ldots, n. \end{aligned} \tag{3.58}$$

3.2.3 Chebyshev Polynomials. Relation between Algebraic and Trigonometric Interpolation

Let us use the equality $\cos\varphi = x$ and introduce the functions:

$$T_k(x) = \cos k\varphi = \cos(k \arccos x), \qquad k = 0, 1, 2, \ldots. \tag{3.59}$$

THEOREM 3.7

The functions $T_k(x)$ defined by formula (3.59) are polynomials of degree $k = 0, 1, 2, \ldots$. Specifically, $T_0(x) = 1$, $T_1(x) = x$, and all other polynomials: $T_2(x)$, $T_3(x)$, etc., can be obtained consecutively using the recursion formula

$$T_{k+1}(x) = 2xT_k(x) - T_{k-1}(x). \tag{3.60}$$

PROOF It is clear that $T_0(x) = \cos 0 = 1$ and $T_1(x) = \cos\arccos x = x$. Then, we employ a well-known trigonometric identity

$$\cos(k+1)\varphi = 2\cos\varphi\cos k\varphi - \cos(k-1)\varphi, \qquad k = 1, 2, \ldots,$$

which immediately yields formula (3.60) when $\varphi = \arccos x$. It only remains to prove that $T_k(x)$ is a polynomial of degree k; we will use induction with respect to k to do that. For $k = 0$ and $k = 1$ it has been proven directly. Let us fix some $k > 1$ and assume that for all $j = 0, 1, \ldots, k$ we have already shown that $T_j(x)$ are polynomials of degree j. Then, the expression on the right-hand side of (3.60), and as such, $T_{k+1}(x)$, is a polynomial of degree $k+1$. □

The polynomials $T_k(x)$ were first introduced and studied by Chebyshev. We provide here the formulae for a first few Chebyshev polynomials, along with their graphs, see Figure 3.3:

$$T_0(x) = 1, \qquad T_1(x) = x, \qquad T_2(x) = 2x^2 - 1, \qquad T_3(x) = 4x^3 - 3x.$$

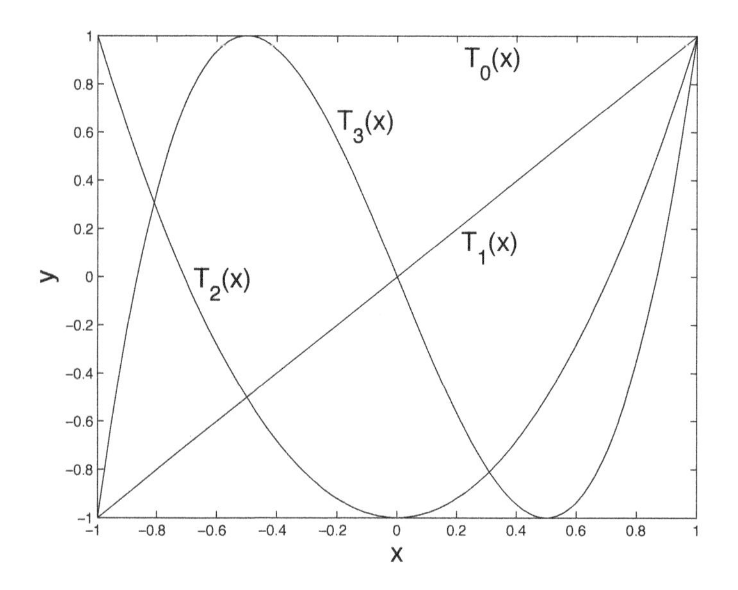

FIGURE 3.3: Chebyshev polynomials.

Next, by substituting $\varphi = \arccos x$ into the right-hand side of formula (3.57), we can recast it as a function of x, thus obtaining:

$$Q_n(\cos\varphi, \sin\varphi, F) \equiv P_n(x, f), \tag{3.61}$$

where

$$P_n(x, f) = \sum_{k=0}^{n} a_k T_k(x), \tag{3.62}$$

and

$$\begin{aligned} a_0 &= \frac{1}{n+1}\sum_{m=0}^{n} f_m = \frac{1}{n+1}\sum_{m=0}^{n} f_m T_0(x_m), \\ a_k &= \frac{2}{n+1}\sum_{m=0}^{n} f_m \cos k\varphi_m = \frac{2}{n+1}\sum_{m=0}^{n} f_m T_k(x_m). \end{aligned} \tag{3.63}$$

Therefore, we can conclude using formulae (3.61), (3.62) and Theorem 3.7, that $P_n(x,f)$ is an algebraic polynomial of degree no greater than n that coincides with the given function values $f(x_m) = f_m$ at the interpolation nodes $x_m = \cos\varphi_m$. In accordance with (3.56), these interpolation nodes can be defined by the formula:

$$x_m = \cos\varphi_m = \cos\frac{\pi(2m+1)}{2(n+1)}, \quad (3.64)$$
$$m = 0, 1, \ldots, n,$$

that basically coincides with formula (2.20) of Chapter 2. We are schematically showing the nodes (3.64) in Figure 3.4.

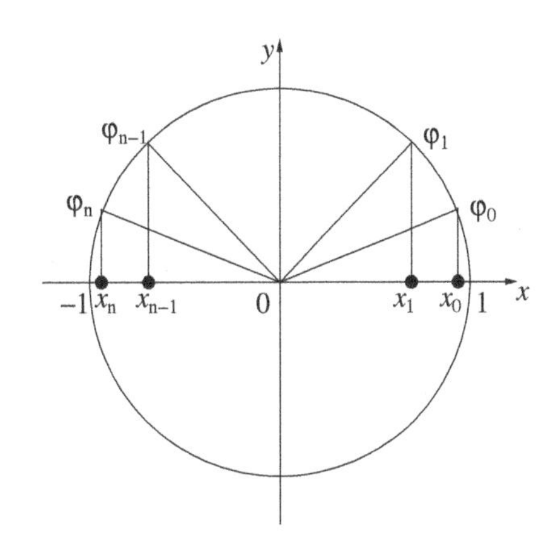

FIGURE 3.4: Chebyshev interpolation nodes.

Note that the points $\varphi_m = \frac{\pi}{n+1}m + \frac{\pi}{2(n+1)}$, $m = 0, 1, \ldots, n$, defined by formula (3.56) are actually zeros of the function $\cos(n+1)\varphi$. Accordingly, the points $x_m = \cos\varphi_m$ defined by formula (3.64) are roots of the Chebyshev polynomial $T_{n+1}(x) = \cos(n+1)\varphi$. In the literature, this particular choice of nodes for the Chebyshev grid, see Figure 3.4, is often referred to as the Chebyshev-Gauss or simply Gauss nodes (an alternative choice of nodes is discussed in Section 3.2.6).

In other words, the polynomial $P_n(x,f)$ specified by formulae (3.62), (3.63) renders algebraic interpolation of the function $f(x)$ based on its values f_m sampled at the roots x_m, see (3.64), of the Chebyshev polynomial $T_{n+1}(x)$.

3.2.4 Properties of Algebraic Interpolation with Roots of the Chebyshev Polynomial $T_{n+1}(x)$ as Nodes

Equality $Q_n(\cos\varphi, \sin\varphi, F) = P_n(x,f)$ implies that the properties of the trigonometric interpolating polynomial $Q_n(\cos\varphi, \sin\varphi, F)$ established by Theorems 3.4 and 3.5 of Section 3.1 do carry over to the algebraic interpolating polynomial $P_n(x,f)$ defined by formula (3.62). In particular, the Lebesgue constants L_n that characterize the sensitivity of the polynomial $P_n(x,f)$ to the perturbations of f_m, satisfy estimate (3.33) from Section 3.1:

$$L_n \leq 4(n+1),$$

and the interpolation error

$$R_n(x) = f(x) - P_n(x,f)$$

uniformly converges to zero as $n \longrightarrow \infty$ with the rate automatically determined by the number of derivatives $r+1$ that the function $f(x)$ has:

$$\max_{-1\leq x\leq 1} |R_n(x)| \leq \frac{\zeta_n}{n^{r-1/2}}, \quad \text{where} \quad \zeta_n = o(1), \ n \longrightarrow \infty. \quad (3.65)$$

In other words, similarly to the trigonometric interpolation (see Section 3.1.3), algebraic interpolation on the Chebyshev nodes *does not get saturated by smoothness.* In other words, the interpolation error self-adjusts to the regularity of the interpolated function without having to change anything in the construction of the method.

REMARK 3.2 Estimate (3.33) for the Lebesgue constants that we proved in Theorem 3.4 of Section 3.1 can be substantially improved. In fact, the following equality holds [see the bibliography quoted in Section 3.2.7, and cf. formula (2.21) of Section 2.1, Chapter 2]:

$$L_n = \frac{2}{\pi}\ln(n+1) + 1 - \theta_n, \qquad 0 \le \theta_n \le \frac{1}{4}. \tag{3.66}$$

Accordingly, the estimate for the interpolation error can also be improved, and instead of (3.65) we will obtain:

$$\max_{-1\le x\le 1} |R_n(x)| = o\left(\frac{\ln n}{n^{r+1/2}}\right) \quad \text{as} \quad n \longrightarrow \infty, \tag{3.67}$$

where $r+1$ is the maximum number of derivatives that the function $f(x)$ has. Further improvements of estimate (3.67) can be obtained with the help of the Jackson inequality, see Section 3.2.7. □

In contradistinction to the Chebyshev nodes (3.64), when a uniform grid is used for interpolation, the Lebesgue constants rapidly grow as n increases, see inequalities (2.18) of Chapter 2, and convergence of the interpolating polynomial to the function $f(x)$ may break down even for infinitely smooth functions, see Section 2.1.5. These are precisely the considerations that make the algebraic interpolation of high degree inappropriate, and prompt the use of piecewise polynomial or spline interpolation on uniform or arbitrary non-uniform grids (see Chapter 2).

3.2.5 An Algorithm for Evaluating the Interpolating Polynomial

To obtain the coefficients a_k of the polynomial $P_n(x, f)$ of (3.62) with the help of formulae (3.63), as well as to actually evaluate this polynomial itself at a given $x \in [-1, 1]$, one needs to be able to compute the values of the polynomials $T_k(x)$, $k = 0, 1, 2, \ldots$, for $-1 \le x \le 1$. We will show that it is appropriate to use formula (3.60) for this purpose. This formula is obviously easy to use and its computational efficacy is also apparent. We only need to demonstrate that the computations according to this formula are stable with respect to the round-off errors.

Consider a difference equation of the type:

$$y_{k+1} = 2xy_k - y_{k-1}, \tag{3.68}$$

where y_k is the unknown sequence parameterized by the integer quantity k. We will be looking for a solution of equation (3.68) in the form $y_k = q^k$, where q is a fixed number. Substituting the latter expression into the difference equation (3.68), we obtain the following algebraic equation for q:

$$q^2 - 2xq + 1 = 0.$$

It is often called the characteristic equation, and it has two roots:

$$q_{1,2} = x \pm \sqrt{x^2 - 1}.$$

Due to the linearity of equation (3.68), its general solution can be written in the form

$$y_k = c_1 q_1^k + c_2 q_2^k, \tag{3.69}$$

where c_1 and c_2 are arbitrary constants. Let us choose these constants c_1 and c_2 so that to satisfy the following conditions:

$$y_0 = T_0(x) = 1, \qquad y_1 = T_1(x) = x,$$

or equivalently,

$$c_1 + c_2 = 1, \qquad c_1 q_1 + c_2 q_2 = x.$$

This implies $c_1 = c_2 = 1/2$, and then formula (3.69) yields the following solution of equation (3.68):

$$T_k(x) = \frac{1}{2}\left(x + \sqrt{x^2 - 1}\right)^k + \frac{1}{2}\left(x - \sqrt{x^2 - 1}\right)^k. \tag{3.70}$$

According to formula (3.60), $T_k(x)$ defined by (3.70) shall be interpreted for a given fixed x as a (discrete) function of k that solves equation (3.68).

Next, note that when $|x| < 1$ the roots $q_{1,2} = x \pm \sqrt{x^2 - 1}$ of the characteristic equation are complex conjugate and have unit moduli. Consequently, the quantities q_1^k and q_2^k will remain equal to one by their absolute value as k increases. An error committed for some $k = k_0$ would cause a perturbation in the values of c_1 and c_2 that enter into formula (3.69) for $k > k_0$. However, due to the equalities $|q_1^k| = |q_2^k| = 1$, $k = 1, 2, \ldots$, this error will not get amplified as k increases. This implies numerical stability of the computations according to formula (3.60) for $|x| < 1$.

3.2.6 Algebraic Interpolation with Extrema of the Chebyshev Polynomial $T_n(x)$ as Nodes

To interpolate the function $F(\varphi) = f(\cos\varphi)$, let us now use the nodes:

$$\tilde{\varphi}_m = \frac{\pi}{n} m, \qquad m = 0, 1, \ldots, n.$$

In accordance with Theorem 3.6 of Section 3.1, and the discussion on page 72 that follows this theorem, we obtain the trigonometric interpolating polynomial:

$$\tilde{Q}_n(\cos\varphi, \sin\varphi, F) = \sum_{k=0}^{n} \tilde{a}_k \cos k\varphi,$$

$$\tilde{a}_0 = \frac{1}{2n}(f_0 + f_n) + \frac{1}{n}\sum_{m=1}^{n-1} f_m, \qquad \tilde{a}_n = \frac{1}{2n}(f_0 + (-1)^n f_n),$$

$$\tilde{a}_k = \frac{1}{n}(f_0 + (-1)^k f_n) + \frac{2}{n}\sum_{m=1}^{n-1} f_m \cos k\varphi_m, \qquad k = 1, 2, \dots, n-1.$$

Changing the variable to $x = \cos\varphi$ and denoting $\tilde{Q}_n(\cos\varphi, \sin\varphi, F) = \tilde{P}_n(x, f)$, we have:

$$\tilde{P}_n(x, f) = \sum_{k=0}^{n} \tilde{a}_k T_k(x),$$

$$\tilde{a}_0 = \frac{1}{2n}(f_0 + f_n) + \frac{1}{n}\sum_{m=1}^{n-1} f_m, \qquad \tilde{a}_n = \frac{1}{2n}(f_0 + (-1)^n f_n),$$

$$\tilde{a}_k = \frac{1}{n}(f_0 + (-1)^k f_n) + \frac{2}{n}\sum_{m=1}^{n-1} f_m T_k(\tilde{x}_m), \qquad k = 1, 2, \dots, n-1.$$

Similarly to the polynomial $P_n(x, f)$ of (3.62), the algebraic interpolating polynomial $\tilde{P}_n(x, f)$ built on the grid:

$$\tilde{x}_m = \cos\tilde{\varphi}_m = \cos\frac{\pi}{n}m, \qquad m = 0, 1, \dots, n, \tag{3.71}$$

also inherits the two foremost advantageous properties from the trigonometric interpolating polynomial $\tilde{Q}_n(\cos\varphi, \sin\varphi, F)$. They are the slow growth of the Lebesgue constants as n increases (that translates into the numerical stability with respect to the perturbations of f_m), as well convergence with the rate that automatically takes into account the smoothness of $f(x)$, i.e., no susceptibility to saturation.

Finally, we notice that the Chebyshev polynomial $T_n(x)$ reaches its extreme values on the interval $-1 \le x \le 1$ precisely at the interpolation nodes $\tilde{x}_m$ of (3.71): $T_n(\tilde{x}_m) = \cos\pi m = (-1)^m$, $m = 0, 1, \dots, n$. In the literature, the grid nodes $\tilde{x}_m$ of (3.71) are known as the Chebyshev-Gauss-Lobatto nodes or simply the Gauss-Lobatto nodes.

3.2.7 More on the Lebesgue Constants and Convergence of Interpolants

In this section, we discuss the problem of interpolation from the general perspective of approximation of functions by polynomials. Our considerations, in a substantially abridged form, follow those of [LG95], see also [Bab86]. We quote many of the fundamental results without a proof (the theorems of Jackson, Weierstrass, Faber-Bernstein, and Bernstein). The justification of these results, along with a broader and more comprehensive account of the subject, can be found in the literature on the classical theory of approximation, see, e.g., [Jac94, Ber52, Ber54, Ach92, Nat64, Nat65a, Nat65b, Lor86, Che66, Riv74]. In the numerical analysis literature, some of these issues are addressed in [Wen66]. In these books, the reader will also find references to research articles. The material of this section is more advanced, and can be skipped during the first reading.

The meaning of Lebesgue's constants L_n introduced in Chapter 2 as minimum numbers that for each n guarantee the estimate [see formula (2.17)]:

$$\max_{a\leq x\leq b} |P_n(x,\delta f)| \leq L_n \max_j |\delta f(x_j)|$$

is basically that of an operator norm. Indeed, interpolation by means of the polynomial $P_n(x,f)$ can be interpreted as a linear operator that maps the finite-dimensional space of vectors $[f_0,f_1,\dots,f_n]$ into the space $C[a,b]$ of all continuous functions $f(x)$ defined on $[a,b]$. The space $C[a,b]$ is equipped with the maximum norm $\|f\| = \max_{a\leq x\leq b} |f(x)|$. Likewise, the space of vectors $\vec{f} = [f_0,f_1,\dots,f_n]$ can also be equipped with a maximum norm, but discrete rather than continuous: $\|\vec{f}\| = \max_{0\leq j\leq n} |f_j|$. Then, L_n of (2.17) appears to be the induced norm of the foregoing linear operator:

$$L_n = \sup_{\|\vec{f}\|=1} \|P_n(x,f)\|. \tag{3.72}$$

However, for subsequent analysis it will be more convenient to use a slightly different definition of L_n — as norm of an operator that would rather map $C[a,b]$ onto itself.

DEFINITION 3.1 *The operator* $\mathscr{P}_n = \mathscr{P}_n(x_0,x_1,\dots,x_n): \ C[a,b] \longmapsto \{P_n(x)\} \subset C[a,b]$ *takes a function* $f \in C[a,b]$*, samples its values at a given set of nodes* $\{x_0,x_1,\dots,x_n\} \in [a,b]$ *thus creating the table* $\{f_0,f_1,\dots,f_n\}$*, and subsequently builds the polynomial* $P_n(x,f,x_0,x_1,\dots,x_n) \in C[a,b]$.

LEMMA 3.1
The operator $\mathscr{P}_n$ *introduced by Definition 3.1 is linear and continuous.*

PROOF The linearity of $\mathscr{P}_n$ is obvious. To show the continuity, we use the Lagrange formula (2.1) of Section 2.1, Chapter 2, and obtain:

$$|\mathscr{P}_n[f](x)| = |P_n(x,f,x_0,x_1,\dots,x_n)| \leq \sum_{k=0}^{n} |f_k||l_k(x)| \leq \|f\| \sum_{k=0}^{n} |l_k(x)|.$$

Next, we introduce a new quantity

$$\lambda_n \stackrel{\text{def}}{=} \sup_{[a,b]} \sum_{k=0}^{n} |l_k(x)| = \max_{[a,b]} \sum_{k=0}^{n} |l_k(x)|, \tag{3.73}$$

where the second equality in (3.73) holds because $[a, b]$ is a compact set, and $\sum_{k=0}^{n} |l_k(x)|$ is a continuous function. Then clearly,

$$|\mathscr{P}_n[f](x)| \leq \lambda_n \|f\|.$$

Consequently, $\mathscr{P}_n$ is a bounded operator, $\mathscr{P}_n : C[a, b] \mapsto C[a, b]$, and therefore, it is continuous. Moreover, $\|\mathscr{P}_n\| \leq \lambda_n$. ☐

DEFINITION 3.2 *The norm of the operator $\mathscr{P}_n$ introduced by Definition 3.1 is called the Lebesgue constant of the polynomial interpolation based on the nodes $x_0, x_1, \dots, x_n$:*

$$L_n = \|\mathscr{P}_n\|. \tag{3.74}$$

Recall that the operator norm on the right-hand side of formula (3.74) is given by:

$$\|\mathscr{P}_n\| = \sup_{\|f\|=1} \|\mathscr{P}_n[f](x)\| = \sup_{\|f\|=1} \|P_n(x, f)\|. \tag{3.75}$$

We have therefore formulated two alternative definitions of the Lebesgue constants — by means of formula (3.72) and by means of formulae (3.74), (3.75). We will now prove that these definitions are, in fact, equivalent. In other words, we will show that the right-hand side of formula (3.75) coincides with the right-hand side of formula (3.72). The difference between these right-hand sides is that in (3.75) the smallest upper bound is taken across the unit sphere in the space $C[a, b]$ that has infinite dimension, whereas in (3.72) it is taken across the unit sphere in the $n+1$-dimensional space of vectors $\vec{f} = [f_0, f_1, \dots, f_n]$.

LEMMA 3.2
The Lebesgue constant defined by formulae (3.74), (3.75) is the same as the Lebesgue constant defined by formula (3.72).

PROOF For every vector $\vec{f} = [f_0, f_1, \dots, f_n]$, consider a piecewise linear function defined as:

$$f(x) = f_{j+1} \frac{x - x_j}{x_{j+1} - x_j} + f_j \frac{x_{j+1} - x}{x_{j+1} - x_j} \quad \text{for} \quad x \in [x_j, x_{j+1}], \quad j = 0, 1, \dots, n-1.$$

Clearly, $f(x) \in C[a, b]$, and also if $\|\vec{f}\| = 1$ then $\|f\| = 1$. In other words, every unit vector $\vec{f} = [f_0, f_1, \dots, f_n]$ gives rise to a continuous (piecewise linear) function that belongs to the unit sphere in $C[a, b]$. Therefore, one can say that the smallest upper bound on the right-hand side of (3.75) is taken across a wider set than that on the right-hand side of (3.72). Consequently,

$$\sup_{\|f\|=1} \|P_n(x, f)\| \geq \sup_{\|\vec{f}\|=1} \|P_n(x, f)\|. \tag{3.76}$$

On the other hand, let $f(x) \in C[a, b]$ be a particular function that realizes the smallest upper bound on the right-hand side of (3.75). By construction, $\|f\| = 1$. Let us sample the values of $f(x)$ at the nodes $x_0, x_1, \dots, x_n$. This yields the table $\{f_0, f_1, \dots, f_n\}$, or equivalently, the vector $\vec{f} = [f_0, f_1, \dots, f_n]$. Assume that $\|\vec{f}\| < 1$. Then, denote $\alpha = \|\vec{f}\|^{-1} > 1$ and stretch the vector $\vec{f}$: $\vec{f} \longmapsto \alpha\vec{f} = [\alpha f_0, \alpha f_1, \dots, \alpha f_n]$ so that $\|\alpha\vec{f}\| = 1$. As the interpolation by means of the polynomials P_n is a linear operator, we obviously have $P_n(x, \alpha f) = \alpha P_n(x, f)$, and consequently, $\|P_n(x, \alpha f)\| > \|P_n(x, f)\|$. We have therefore found a unit vector $\alpha\vec{f}$, for which the norm of the corresponding interpolating polynomial will be greater than the left-hand side of (3.76). The contradiction proves that the two definitions of the Lebesgue constants are indeed equivalent. □

LEMMA 3.3

The Lebesgue constant of (3.74), (3.75) is equal to

$$L_n = \lambda_n, \tag{3.77}$$

where the quantity λ_n is defined by formula (3.73).

PROOF When proving Lemma 3.1, we have seen that $\|\mathscr{P}_n\| \le \lambda_n$. We therefore need to show that $\lambda_n \le \|\mathscr{P}_n\|$.

As has been mentioned, the function $\psi(x) \stackrel{\text{def}}{=} \sum_{k=0}^{n} |l_k(x)|$ is continuous on $[a, b]$. Consequently, $\exists x^* \in [a, b] : \psi(x^*) = \lambda_n$. Let us now consider a function $f_0 \in C[a, b]$ such that $f_0(x_k) = \operatorname{sign} l_k(x^*)$, $k = 0, 1, 2, \dots, n$, and also $\|f_0\| = 1$. For this function we have:

$$|\mathscr{P}_n[f_0](x^*)| = \left|\sum_{k=0}^{n} f_0(x_k) l_k(x^*)\right| = \left|\sum_{k=0}^{n} (\operatorname{sign} l_k(x^*))\, l_k(x^*)\right| = \sum_{k=0}^{n} |l_k(x^*)| = \psi(x^*) = \lambda_n.$$

On the other hand,

$$|\mathscr{P}_n[f_0](x^*)| \le \|\mathscr{P}_n[f_0]\| \le \|\mathscr{P}_n\| \cdot \|f_0\| = \|\mathscr{P}_n\|,$$

which implies $\lambda_n \le \|\mathscr{P}_n\|$. It only remains to construct a specific example of $f_0 \in C[a, b]$. This can be done easily by taking $f_0(x)$ as a piecewise linear function with the values $\operatorname{sign} l_k(x^*)$ at the points x_k, $k = 0, 1, 2, \dots, n$. □

We can therefore conclude that

$$L_n = \max_{a \le x \le b} \sum_{k=0}^{n} |l_k(x)|.$$

We have used a somewhat weaker form of this result in Section 2.1.4 of Chapter 2.

The Lebesgue constants of Definition 3.2 play a fundamental role when studying the convergence of interpolating polynomials. To actually see that, we will first need to introduce another key new concept and formulate some important results.

DEFINITION 3.3 *The quantity*

$$\varepsilon(f, P_n) = \min_{P_n(x)} \max_{a \le x \le b} |P_n(x) - f(x)| \tag{3.78}$$

is called the best approximation of a given function $f(x)$ by polynomials of degree no greater than n on the interval $[a, b]$.

Note that the minimum in formula (3.78) is taken with respect to all algebraic polynomials of degree no greater than n on the interval $[a, b]$, not only the interpolating polynomials. In other words, the polynomials in (3.78) do not, generally speaking, have to coincide with $f(x)$ at any given point of $[a, b]$. It is possible to show existence of a particular polynomial that realizes the best approximation (3.78). In most cases, however, this polynomial is difficult to obtain constructively. In general, polynomials of the best approximation can only be built using sophisticated iterative algorithms of non-smooth optimization. On the other hand, their theoretical properties are well studied. Perhaps the most fundamental property is given by the Jackson inequality.

THEOREM 3.8 (Jackson inequality)
Let $f = f(x)$ be defined on the interval $[a, b]$, let it be $r-1$ times continuously differentiable, and let the derivative $f^{(r-1)}(x)$ be Lipshitz-continuous:

$$\forall x_1, x_2 \in [a, b]: \quad |f^{(r-1)}(x_1) - f^{(r-1)}(x_2)| \le M|x_1 - x_2|, \qquad M > 0.$$

Then, for any $n \ge r$ the following inequality holds:

$$\varepsilon(f, P_n) < C_r \left(\frac{b-a}{2}\right)^r \frac{M}{n^r}, \tag{3.79}$$

where $C_r = \left(\frac{\pi r}{2}\right)^r \frac{1}{r!}$ are universal constants that depend neither on f, nor on n, nor on M.

The Jackson inequality [Jac94] reinforces, for sufficiently smooth functions, the result of the following classical theorem established in real analysis (see, e.g., [Rud87]):

THEOREM 3.9 (Weierstrass)
Let $f \in C[a, b]$. Then, for any $\varepsilon > 0$ there is an algebraic polynomial $P_\varepsilon(x)$ such that $\forall x \in [a, b]: \quad |f(x) - P_\varepsilon(x)| \le \varepsilon$.

A classical proof of Theorem 3.9 is based on periodization of f that preserves its continuity (the period should obviously be larger than $[a, b]$) and then on the approximation by partial sums of the Taylor series that converges uniformly. The Weierstrass theorem implies that for $f \in C[a, b]$ the best approximation defined by (3.78) converges to zero: $\varepsilon(f, P_n) \longrightarrow 0$ when $n \longrightarrow \infty$. This is basically as much as one can tell regarding the behavior of $\varepsilon(f,P_n)$ if nothing else is known about $f(x)$ except that it is continuous. On the other hand, the Jackson inequality specifies the rate of decay for the best approximation as a particular inverse power of n, see formula (3.79), provided that $f(x)$ is smooth.

Let us also note that the value of $\frac{\pi}{2}$ that enters the expression for $C_r = \left(\frac{\pi r}{2}\right)^r \frac{1}{r!}$ in the Jackson inequality (3.79) may, in fact, be replaced by smaller values:

$$K_0 = 1, \quad K_1 = \frac{\pi}{2}, \quad K_2 = \frac{\pi^2}{8}, \quad K_3 = \frac{\pi^3}{24}, \quad K_4 = \frac{5\pi^4}{384}, \ldots$$

known as the Favard constants. The Favard constants can be obtained explicitly for all $r = 0, 1, 2, \ldots$, and it is possible to show that all $K_r < \frac{\pi}{2}$. The key consideration regarding the Favard constants is that substituting them into (3.79) makes this inequality sharp.

The main result that connects the properties of the best approximation (Definition 3.3) and the quality of interpolation by means of algebraic polynomials is given by the following

THEOREM 3.10 (Lebesgue inequality)
Let $f \in C[a,b]$ and let $\{x_0, x_1, \ldots, x_n\}$ be an arbitrary set of distinct interpolation nodes on $[a,b]$. Then,

$$\varepsilon(f,P_n) \leq \|f - \mathscr{P}_n[f]\| \leq (L_n + 1)\varepsilon(f,P_n). \tag{3.80}$$

Note that according to Definition 3.1, the operator $\mathscr{P}_n$ in formula (3.80) generally speaking depends on the choice of the interpolation nodes.

PROOF It is obvious that we only need to prove the second inequality in (3.80), i.e., the upper bound. Consider an arbitrary polynomial $Q(x) \in \{P_n(x)\}$ of degree no greater than n. As the algebraic interpolating polynomial is unique (Theorem 2.1 of Chapter 2), we obviously have $\mathscr{P}_n[Q] = Q$. Next,

$$\begin{aligned}
\|f - \mathscr{P}_n[f]\| &= \|f - Q + \mathscr{P}_n[Q] - \mathscr{P}_n[f]\| \\
&\leq \|f - Q\| + \|\mathscr{P}_n[Q] - \mathscr{P}_n[f]\| = \|f - Q\| + \|\mathscr{P}_n[Q - f]\| \\
&\leq \|f - Q\| + \|\mathscr{P}_n\|\|f - Q\| = (1 + L_n)\|f - Q\|.
\end{aligned}$$

Let us now introduce $\delta > 0$ and denote by $Q_\delta(x) \in \{P_n(x)\}$ a polynomial for which $\|f - Q_\delta\| < \varepsilon(f,P_n) + \delta$. Then,

$$\|f - \mathscr{P}_n[f]\| \leq (1 + L_n)\|f - Q_\delta\| < (1 + L_n)(\varepsilon(f,P_n) + \delta).$$

Finally, by taking the limit $\delta \longrightarrow 0$, we obtain the desired inequality (3.80). □

The Lebesgue inequality (3.80) essentially provides an upper bound for the interpolation error $\|f - \mathscr{P}_n[f]\|$ in terms of a product of the best approximation (3.78) times the Lebesgue constant (3.74). Often, this estimate allows one to judge the convergence of algebraic interpolating polynomials as n increases. It is therefore clear that the behavior of the Lebesgue constants is of central importance for the convergence study.

THEOREM 3.11 (Faber-Bernstein)
For any choice of interpolation nodes $x_0, x_1, \dots, x_n$ on the interval $[a,b]$, the following inequality holds:

$$L_n > \frac{1}{8\sqrt{\pi}} \ln(n+1). \tag{3.81}$$

Theorem 3.11 shows that the Lebesgue constants always grow as the grid dimension n increases. As such, the best one can generally hope for is to be able to place the interpolation nodes in such a way that this growth will be optimal, i.e., logarithmic.

As far as the problem of interpolation is concerned, if, for example, nothing is known about the function $f \in C[a,b]$ except that it is continuous, then nothing can be said about the behavior of the error beyond the estimate given by the Lebesgue inequality (3.80). The Weierstrass theorem (Theorem 3.9) indicates that $\varepsilon(f,P_n) \longrightarrow 0$ as $n \longrightarrow \infty$, and the Faber-Bernstein theorem (Theorem 3.11) says that $L_n \longrightarrow \infty$ as $n \longrightarrow \infty$. We therefore have the uncertainty $0 \cdot \infty$ on the right-hand side of the Lebesgue inequality; and the behavior of this right-hand side is determined by which of the two processes dominates — the decay of the best approximations or the growth of the Lebesgue constants. In particular, if $\lim_{n\to\infty} L_n \varepsilon(f,P_n) = 0$, then the interpolating polynomials uniformly converge to $f(x)$.

If the function $f(x)$ is sufficiently smooth (as formulated in Theorem 3.8), then combining the Lebesgue inequality (3.80) and the Jackson inequality (3.79) we obtain the following error estimate:[3]

$$\|f - \mathscr{P}_n[f]\| < (L_n+1) C_r \left(\frac{b-a}{2}\right)^r \frac{M}{n^r}, \tag{3.82}$$

which implies that the convergence rate (if there is convergence) will depend on the behavior of L_n when n increases. If the interpolation grid is uniform (equidistant nodes), then the Lebesgue constants grow exponentially as n increases, see inequalities (2.18) of Section 2.1, Chapter 2. In this case, the limit (as $n \longrightarrow \infty$) on the right-hand side of (3.82) is infinite for any finite value of r. This does not necessarily mean that the sequence of interpolating polynomials $\mathscr{P}_n[f](x)$ diverges, because inequality (3.82) only provides an upper bound for the error. It does mean though that in this case convergence of the interpolating polynomials simply cannot be judged using the arguments based on the inequalities of Lebesgue and Jackson.

[3]Note that in estimate (3.82) the function $f(x)$ is assumed to have a maximum of $r-1$ derivatives, and the derivative $f^{(r-1)}(x)$ is required to be Lipshitz-continuous (Theorem 3.8), which basically makes $f(x)$ "almost" r times differentiable. Previously, we have used a slightly different notation and in the error estimate (3.67) the function was assumed to have a maximum of $r+1$ derivatives.

On the other hand, for the Chebyshev interpolation grid (3.64) the following theorem asserts that the asymptotic behavior of the Lebesgue constants is optimal:

THEOREM 3.12 (Bernstein)
Let the interpolation nodes $x_0, x_1, \ldots, x_n$ on the interval $[-1, 1]$ be given by roots of the Chebyshev polynomial $T_{n+1}(x)$. Then,

$$L_n < 8 + \frac{4}{\pi} \ln(n+1). \tag{3.83}$$

Therefore, according to (3.82) and (3.83), if the derivative $f^{(r-1)}(x)$ of the function $f(x)$ is Lipshitz-continuous, then the sequence of algebraic interpolating polynomials built on the Chebyshev nodes converges uniformly to $f(x)$ with the rate

$$\mathcal{O}(n^{-r} \ln(n+1)) \quad \text{as} \quad n \longrightarrow \infty.$$

Let us now recall that a Lipshitz-continuous function $f(x)$ on the interval $[-1, 1]$:

$$|f(x') - f(x'')| \leq \text{const}|x' - x''|, \quad x', x'' \in [-1, 1],$$

is absolutely continuous on $[-1, 1]$, and as such, according to the Lebesgue theorem [KF75], its derivative is integrable, i.e., exists in $L_1[-1, 1]$. In this sense, we can say that Lipshitz-continuity is "not very far" from differentiability, although this is, of course, not sufficient to claim that the derivative $f^{(r)}(x)$ is bounded. On the other hand, if $f(x)$ is r times differentiable on $[-1, 1]$ and $f^{(r)}(x)$ is bounded, then $f^{(r-1)}(x)$ is Lipshitz-continuous. Therefore, for a function with its r-th derivative bounded, the rate of convergence of Chebyshev interpolating polynomials is at least as fast as the inverse of the grid dimension raised to the power r (smoothness of the interpolated function), times an additional slowly increasing factor $\sim \ln n$. At the same time, recall that the unavoidable error of reconstructing a function with r derivatives on a uniform grid with n nodes is $\mathcal{O}(n^{-r})$. This is also true for the Chebyshev grid, because Chebyshev grid on the diameter is equivalent to a uniform grid on the circle, see Figure 3.4. Consequently, accuracy of the Chebyshev interpolating polynomials appears to be only a logarithmic factor away from the level of the unavoidable error. As such, we have shown that Chebyshev interpolation is not saturated by smoothness and practically reaches the intrinsic accuracy limit.

Altogether, we see that the type of interpolation grid may indeed have a drastic effect on convergence, which corroborates our previous observations. For the Bernstein example $f(x) = |x|$, $-1 \leq x \leq 1$ (Section 2.1.5 of Chapter 2), the sequence of interpolating polynomials constructed on a uniform grid diverges. On the Chebyshev grid we have seen experimentally that it converges. Now, using estimates (3.82) and (3.83) we can say that the rate of this convergence is at least $\mathcal{O}(n^{-1} \ln(n+1))$.

To conclude, let us also note that strictly speaking the behavior of L_n on the Chebyshev grid is only asymptotically optimal rather than optimal, because the constants in the lower bound (3.81) and in the upper bound (3.83) are different. Better values of these constants than those guaranteed by the Bernstein theorem (Theorem 3.12) have been obtained more recently, see inequality (3.66). However, there is still a gap between (3.81) and (3.66).

REMARK 3.3 Formula (3.78) that introduces the best approximation according to Definition 3.3 can obviously be recast as

$$\varepsilon(f,P_n) = \min_{P_n(x)} \|P_n(x) - f(x)\|_C,$$

where the norm on the right-hand side is taken in the sense of the space $C[a,b]$. In general, the notion of the best approximation admits a much broader interpretation, when both the norm (currently, $\|\cdot\|_C$) and the class of approximating functions (currently, polynomials $P_n(x)$) may be different. In fact, one can consider the problem of approximating a given element of the linear space by linear combinations of a pre-defined set of elements from the same space in the sense of a selected norm. This is the same concept as exploited when introducing the Kolmogorov diameters, see Remark 2.3 on page 41.

For example, consider the space $L_2[a,b]$ of all square integrable functions $f(x)$, $x \in [a,b]$, equipped with the norm:

$$\|f\|_2 \stackrel{\text{def}}{=} \left[\int_a^b f^2(x)dx\right]^{\frac{1}{2}}.$$

This space is known to be a Hilbert space. Let us take an arbitrary $f \in L_2[a,b]$ and consider a set of all trigonometric polynomials $Q_n(x)$ of type (3.6), where $L = b - a$ is the length of the interval. Similarly to the algebraic polynomials $P_n(x)$ employed in Definition 3.3, the trigonometric polynomials $Q_n(x)$ do not have to be interpolating polynomials. Then, it is known that the best approximation in the sense of L_2:

$$\varepsilon_2(f,Q_n) = \min_{Q_n(x)} \|f(x) - Q_n(x)\|_2$$

is, in fact, realized by the partial sum $S_n(x)$, see formula (3.38), of the Fourier series for $f(x)$ with the coefficients defined by (3.40). An upper bound for the actual magnitude of the L_2 best approximation is then given by estimate (3.41) for the remainder $\delta S_n(x)$ of the series, see formula (3.39):

$$\varepsilon_2(f,Q_n) \leq \frac{\zeta_n}{n^{r+\frac{1}{2}}}, \quad \text{where} \quad \zeta_n = o(1), \quad n \longrightarrow \infty,$$

where $r+1$ is the maximum smoothness of $f(x)$. Having identified what the best approximation in the sense of L_2 is, we can easily see now that both the Lebesgue inequality (Theorem 3.10) and the error estimate for trigonometric interpolation (Theorem 3.5) are, in fact, justified using the same argument. It employs uniqueness of the corresponding interpolating polynomial, the estimate for the best approximation, and the estimate of sensitivity to perturbations given by the Lebesgue constants. □

Exercises

1. Let the function $f = f(x)$ be defined on an arbitrary interval $[a, b]$, rather than on $[-1, 1]$. Construct the Chebyshev interpolation nodes for $[a, b]$, and write down the interpolating polynomials $P_n(x, f)$ and $\tilde{P}_n(x, f)$ similar to those obtained in Sections 3.2.3 and 3.2.6, respectively.

2. For the function $f(x) = \frac{1}{x^2+1/4}$, $-1 \leq x \leq 1$, construct the algebraic interpolating polynomial $P_n(x, f)$ using roots of the Chebyshev polynomial $T_{n+1}(x)$, see (3.64), as interpolation nodes. Plot the graphs of $f(x)$ and $P_n(x, f)$ for $n = 5, 10, 20, 30$, and 40. Do the same for the interpolating polynomial $P_n(x, f)$ built on the equally spaced nodes $x_k = -1 + 2k/n$, $k = 0, 1, 2, \ldots, n$ (Runge example of Section 2.1.5, Chapter 2). Explain the observable qualitative difference between the two interpolation techniques.

3. Introduce the normalized Chebyshev polynomial $\hat{T}_n(x)$ of degree n by setting $\hat{T}_n(x) = 2^{1-n}T_n(x)$.

   a) Show that the coefficient in front of x^n in the polynomial $\hat{T}_n(x)$ is equal to one.

   b) Show that the deviation $\max\limits_{-1 \leq x \leq 1} |\hat{T}_n(x)|$ of the polynomial $\hat{T}_n(x)$ from zero on the interval $-1 \leq x \leq 1$ is equal to 2^{1-n}.

   c)* Show that among all the polynomials of degree n with the leading coefficient (i.e., the coefficient in front of x^n) equal to one, the normalized Chebyshev polynomial $\hat{T}_n(x)$ has the smallest deviation from zero on the interval $-1 \leq x \leq 1$.

   d) How can one choose the interpolation nodes $t_0, t_1, \ldots, t_n$ on the interval $[-1, 1]$, so that the polynomial $(t - t_0)(t - t_1) \ldots (t - t_n)$, which is a part of the formula for the interpolation error (2.23), Chapter 2, would have the smallest possible deviation from zero on the interval $[-1, 1]$?

4. Find a set of interpolation nodes for an even 2π-periodic function $F(\varphi)$, see formula (3.55), for which the Lebesgue constants would coincide with the Lebesgue constants of algebraic interpolation on equidistant nodes.

Chapter 4

Computation of Definite Integrals. Quadratures

A definite integral $\int_a^b f(x)dx$ can only be evaluated exactly if the primitive (antiderivative) of the function $f(x)$ is available in the closed form: $F(x) = \int f(x)dx$, e.g., as a tabular integral. Then we can use the fundamental theorem of calculus (the Newton-Leibniz formula) and obtain:

$$\int_a^b f(x)dx = F(b) - F(a).$$

It is an even more rare occasion that a multiple integral, say, $\iint_\Omega f(x,y)dxdy$, where Ω is a given domain, can be evaluated exactly. Therefore, numerical methods play an important role for the approximate computation of definite integrals.

In this chapter, we will introduce several numerical methods for computing definite integrals. We will identify the methods that automatically take into account the regularity of the integrand, i.e., do not get saturated by smoothness (Section 4.2), as opposed to those methods that are prone to saturation (Section 4.1). We will also discuss the difficulties associated with the increase of dimension for multiple integrals (Section 4.4) and outline some combined analytical/numerical approaches that can be used for computing improper integrals (Section 4.3).

Formulae that are used for the approximate evaluation of definite integrals given a table of values of the integrand are called quadrature formulae.

4.1 Trapezoidal Rule, Simpson's Formula, and the Like

Recall that the value of the definite integral $\int_a^b f(x)dx$ can be interpreted geometrically as the area under the graph of $f = f(x)$.

4.1.1 General Construction of Quadrature Formulae

To compute an approximate value of the foregoing area under the graph, we first introduce a grid of distinct nodes and thus partition the interval $[a,b]$ into n subintervals:

$$a = x_0 < x_1 < \ldots < x_n = b.$$

Then, we employ a piecewise polynomial interpolation of some degree s, and replace the integrand $f = f(x)$ by the piecewise polynomial function:

$$P_s(x) \stackrel{\text{def}}{=} P_s(x, f, x_{k-j}, x_{k-j+1}, \ldots, x_{k-j+s}) \equiv P_s(x, f_{kj}), \quad \text{if} \quad x \in [x_k, x_{k+1}].$$

In other words, on every subinterval $[x_k, x_{k+1}]$, the function $P_s(x)$ is defined as a polynomial of degree no greater than s built on the $s+1$ consecutive grid nodes: $x_{k-j}, x_{k-j+1}, \ldots, x_{k-j+s}$, where j may also depend on k, see Section 2.2. Finally, we use the following approximation of the integral:

$$\begin{aligned}\int_a^b f(x)dx &\approx \int_a^b P_s(x)dx \\ &= \int_a^{x_1} P_s(x, f_{0j})dx + \int_{x_1}^{x_2} P_s(x, f_{1j})dx + \ldots + \int_{x_{n-1}}^{b} P_s(x, f_{n-1,j})dx.\end{aligned} \tag{4.1}$$

Each term on the right-hand side of this formula is the integral of a polynomial of degree no greater than s. For every $k = 0, 1, \ldots, n-1$, the polynomial is given by $P_s(x, f_{kj})$, and the corresponding integral over $[x_k, x_{k+1}]$ can be evaluated explicitly using the fundamental theorem of calculus. The resulting expression will only contain the locations of the grid nodes $x_{k-j}, x_{k-j+1}, \ldots, x_{k-j+s}$ and the function values $f(x_{k-j}), f(x_{k-j+1}), \ldots, f(x_{k-j+s})$. As such, formula (4.1) is a quadrature formula.

It is clear that the following error estimate holds for the quadrature formula (4.1):

$$\begin{aligned}\left|\int_a^b f(x)dx - \int_a^b P_s(x)dx\right| &\le \int_a^b |f(x) - P_s(x)|dx \\ &\le (b-a) \max_{a \le x \le b} |f(x) - P_s(x)|.\end{aligned} \tag{4.2}$$

THEOREM 4.1

Let the function $f(x)$ have a continuous derivative of order $s+1$ on $[a,b]$, and denote $\max_{a \le x \le b} |f^{(s+1)}(x)| = M_{s+1}$. Let also $\max_{k=0,1,\ldots,n-1} |x_{k+1} - x_k| = h$. Then,

$$\left|\int_a^b f(x)dx - \int_a^b P_s(x)dx\right| \le \text{const} \cdot |b-a| M_{s+1} h^{s+1}. \tag{4.3}$$

According to Theorem 4.1, if the integrand $f(x)$ is $s+1$ times continuously differentiable, then the quadrature formula (4.1) has order of accuracy $s+1$ with respect to the grid size h.

PROOF In Section 2.2.2, we obtained the following error estimate for the piecewise polynomial algebraic interpolation, see formula (2.32):

$$\max_{x_k \le x \le x_{k+1}} |f(x) - P_s(x, f_{kj})| \le \text{const} \cdot \max_{x_{k-j} \le x \le x_{k-j+s}} |f^{(s+1)}(x)| h^{s+1}.$$

Consequently, we can write:

$$\max_{a \le x \le b} |f(x) - P_s(x)| \le \text{const} \cdot M_{s+1} h^{s+1}.$$

Combining this estimate with inequality (4.2), we immediately arrive at the desired result (4.3). $\square$

Quadrature formulae of type (4.1) are most commonly used when the underlying interpolation is piecewise linear (the trapezoidal rule, see Section 4.1.2) or piecewise quadratic (Simpson's formula, see Section 4.1.3).

4.1.2 Trapezoidal Rule

Let $s = 1$, and let the grid be uniform:

$$x_k = a + k \cdot h, \quad k = 0, 1, \ldots, n, \quad h = \frac{b-a}{n}.$$

Then the integral $\int_{x_k}^{x_{k+1}} P_1(x)dx \equiv \int_{x_k}^{x_{k+1}} P_1(x, f_{k0})dx$ is equal to the area of the trapezoid under the graph of the linear function $P_1(x, f_{k0}) = f(x_k)(x_{k+1} - x)/h + f(x_{k+1})(x - x_k)/h$ on the interval $[x_k, x_{k+1}]$:

$$\int_{x_k}^{x_{k+1}} P_1(x)dx = \frac{b-a}{n} \frac{f_k + f_{k+1}}{2}.$$

Adding these equalities together for all $k = 0, 1, \ldots, n-1$, we transform the general formula (4.1) to:

$$\int_a^b f(x)dx \approx \int_a^b P_1(x)dx = \frac{b-a}{n} \left(\frac{f_0}{2} + f_1 + \ldots + f_{n-1} + \frac{f_n}{2} \right). \tag{4.4}$$

Quadrature formula (4.4) is called the trapezoidal rule.

Error estimate (4.3) for the general quadrature formula (4.1) can be refined in the case of the trapezoidal rule (4.4). Recall that in Section 2.2.2 we obtained the following error estimate for the piecewise linear interpolation itself, see formula (2.33):

$$\max_{x_k \le x \le x_{k+1}} |f(x) - P_1(x)| \le \frac{h^2}{8} \max_{x_k \le x \le x_{k+1}} |f''(x)|.$$

Consequently,

$$\max_{x_k \le x \le x_{k+1}} |f(x) - P_1(x)| \le \frac{h^2}{8} \max_{a \le x \le b} |f''(x)|,$$

and as the right-hand side of this inequality does not depend on what particular subinterval $[x_k, x_{k+1}]$ the point x belongs to, the same estimate holds for all $x \in [a,b]$:

$$\max_{a \le x \le b} |f(x) - P_1(x)| \le \frac{h^2}{8} \max_{a \le x \le b} |f''(x)|.$$

According to formula (4.2), the error of the trapezoidal rule built on equidistant nodes is then estimated as:

$$\left| \int_a^b f(x)dx - \int_a^b P_1(x)dx \right| \le \frac{b-a}{8} \max_{a \le x \le b} |f''(x)| h^2. \tag{4.5}$$

This estimate guarantees that the integration error will decay no slower than $\mathcal{O}(h^2)$ as $h \longrightarrow 0$. Alternatively, we say that the order of accuracy of the trapezoidal rule is second with respect to the grid size h. Furthermore, estimate (4.5) is in fact sharp in the sense that even for infinitely differentiable functions [as opposed to only twice differentiable functions that enable estimate (4.5)] the error will only decrease as $\mathcal{O}(h^2)$ and not faster. This is demonstrated by the following example.

Example

Let $f(x) = x^2$ and let $[a,b] = [0,1]$. Then, $\int_0^1 x^2 dx = \frac{1}{3}$. On the other hand, using the trapezoidal rule (4.4) we can write:

$$\begin{aligned}
\int_0^1 x^2 dx &\approx \frac{1}{n} \left(\frac{x_0^2}{2} + x_1^2 + \ldots + x_{n-1}^2 + \frac{x_n^2}{2} \right) \\
&= \frac{h^2}{n} \left[1^2 + 2^2 + \ldots + (n-1)^2 + \frac{n^2}{2} \right] \\
&= \frac{1}{n^3} \left[\frac{(n-1)n(2n-1)}{6} + \frac{n^2}{2} \right] = \frac{1}{3} + \frac{h^2}{6}.
\end{aligned}$$

Therefore, the integration error always remains equal to $h^2/6$ even though the integrand $f(x) = x^2$ has continuous derivatives of all orders rather than only up to the second order. In other words, the error is insensitive to the smoothness of $f(x)$ [beyond $f''(x)$], which means that generally speaking the trapezoidal quadrature formula is prone to the saturation by smoothness.

There is, however, one special case when the trapezoidal rule does not get saturated and rather adjusts its accuracy automatically to the smoothness of the integrand. This is the case of a smooth and periodic function $f = f(x)$ that has period $L = b - a$. For such a function, the order accuracy with respect to the grid size h will no longer be limited by $s+1=2$ and will rather increase as the regularity of $f(x)$ increases. More precisely, the following theorem holds.

THEOREM 4.2

Let $f=f(x)$ be an L-periodic function that has continuous L-periodic derivatives up to the order $r>0$ and a square integrable derivative of order $r+1$:

$$\int_a^{a+L} \left[f^{(r+1)}(x)\right]^2 dx < \infty.$$

Then the error of the trapezoidal quadrature rule (4.4) can be estimated as:

$$\left|\int_a^{a+L} f(x)dx - \int_a^{a+L} P_1(x)dx\right| \equiv \left|\int_a^{a+L} f(x)dx - h\sum_{l=0}^{n-1}\left(\frac{f_l}{2}+\frac{f_{l+1}}{2}\right)\right| \leq L\cdot\frac{\zeta_n}{n^{r+1/2}}, \quad (4.6)$$

where $\zeta_n \longrightarrow 0$ as $n \longrightarrow \infty$.

The result of Theorem 4.2 is considered well-known; a proof, for instance can be found in [DR84, Section 2.9]. Our proof below is based on a simpler argument than the one exploited in [DR84].

PROOF We first note that since the function f is integrated over a full period, we may assume without loss of generality that $a=0$. Let us represent $f(x)$ as the sum of its Fourier series (cf. Section 3.1.3):

$$f(x) = S_{n-1}(x) + \delta S_{n-1}(x), \quad (4.7)$$

where

$$S_{n-1}(x) = \frac{\alpha_0}{2} + \sum_{k=1}^{n-1} \alpha_k \cos\frac{2\pi kx}{L} + \sum_{k=1}^{n} \beta_k \sin\frac{2\pi kx}{L} \quad (4.8)$$

is a partial sum of order $n-1$ and

$$\delta S_{n-1}(x) = \sum_{k=n}^{\infty} \alpha_k \cos\frac{2\pi kx}{L} + \sum_{k=n+1}^{\infty} \beta_k \sin\frac{2\pi kx}{L} \quad (4.9)$$

is the corresponding remainder. The coefficients α_k and β_k of the Fourier series for $f(x)$ are given by the formulae:

$$\alpha_k = \frac{2}{L}\int_0^L f(x)\cos\frac{2\pi kx}{L}dx, \qquad \beta_k = \frac{2}{L}\int_0^L f(x)\sin\frac{2\pi kx}{L}dx.$$

According to (4.7), we can write:

$$\int_0^L f(x)dx = \int_0^L S_{n-1}(x)dx + \int_0^L \delta S_{n-1}(x)dx. \quad (4.10)$$

Let us first apply the trapezoidal quadrature formula to the first integral on the right-hand side of (4.10). This can be done individually for each term in the sum (4.8). First of all, for the constant component $k=0$ we immediately derive:

$$h\sum_{l=0}^{n-1}\left(\frac{\alpha_0/2}{2}+\frac{\alpha_0/2}{2}\right)=\frac{L}{2}\alpha_0=\int_0^L S_{n-1}(x)dx=\int_0^L f(x)dx.$$

For all other terms $k=1,2,\dots,n-1$, we exploit periodicity with the period L, use the definition of the grid $h=L/n$, and obtain:

$$h\sum_{l=0}^{n-1}\left(\frac{1}{2}\cos\frac{2\pi klh}{L}+\frac{1}{2}\cos\frac{2\pi k(l+1)h}{L}\right)=h\sum_{l=0}^{n-1}\cos\frac{2\pi klh}{L}$$

$$=\frac{h}{2}\sum_{l=0}^{n-1}\left(e^{i\frac{2\pi klh}{L}}+e^{-i\frac{2\pi klh}{L}}\right)=\frac{h}{2}\left(\frac{1-e^{i\frac{2\pi knh}{L}}}{1-e^{i\frac{2\pi kh}{L}}}+\frac{1-e^{-i\frac{2\pi knh}{L}}}{1-e^{-i\frac{2\pi kh}{L}}}\right)=0.$$

Analogously,

$$h\sum_{l=0}^{n-1}\left(\frac{1}{2}\sin\frac{2\pi klh}{L}+\frac{1}{2}\sin\frac{2\pi k(l+1)h}{L}\right)=0.$$

Altogether we conclude that the trapezoidal rule integrates the partial sum $S_{n-1}(x)$ given by formula (4.8) exactly:

$$h\sum_{l=0}^{n-1}\left(\frac{S_{n-1}(x_l)}{2}+\frac{S_{n-1}(x_{l+1})}{2}\right)=\int_0^L S_{n-1}(x)dx=\frac{L}{2}\alpha_0. \tag{4.11}$$

We also note that choosing the partial sum of order $n-1$, where the number of grid cells is n, is not accidental. From the previous derivation it is easy to see that equality (4.11) would no longer hold if we were to take $S_n(x)$ instead of $S_{n-1}(x)$.

Next, we need to apply the trapezoidal rule to the remainder of the series $\delta S_{n-1}(x)$ given by formula (4.9). Recall that the magnitude of this remainder, or equivalently, the rate of convergence of the Fourier series, is determined by the smoothness of the function $f(x)$. More precisely, as a part of the proof of Theorem 3.5 (page 68), we have shown that for the function $f(x)$ that has a square integrable derivative of order $r+1$, the following estimate holds, see formula (3.41):

$$\sup_{0\le x\le L}|\delta S_{n-1}(x)|\le\frac{\zeta_n}{n^{r+1/2}}, \tag{4.12}$$

where $\zeta_n=o(1)$ when $n\longrightarrow\infty$. Therefore,

$$\left|h\sum_{l=0}^{n-1}\left(\frac{\delta S_{n-1}(x_l)}{2}+\frac{\delta S_{n-1}(x_{l+1})}{2}\right)\right|\le\frac{h}{2}\sum_{l=0}^{n-1}\left(|\delta S_{n-1}(x_l)|+|\delta S_{n-1}(x_{l+1})|\right)$$

$$\le\frac{h}{2}2n\sup_{0\le x\le L}|\delta S_{n-1}(x)|=L\cdot\frac{\zeta_n}{n^{r+1/2}}.$$

It only remains now to combine the previous formula with (4.11):

$$\begin{aligned}
&\left|\int_0^L f(x)dx - h\sum_{l=0}^{n-1}\left(\frac{f_l}{2}+\frac{f_{l+1}}{2}\right)\right| \\
&= \left|\frac{L}{2}\alpha_0 - h\sum_{l=0}^{n-1}\left(\frac{S_{n-1}(x_l)}{2}+\frac{S_{n-1}(x_{l+1})}{2}\right)\right. \\
&\qquad \left. - h\sum_{l=0}^{n-1}\left(\frac{\delta S_{n-1}(x_l)}{2}+\frac{\delta S_{n-1}(x_{l+1})}{2}\right)\right| \\
&= \left|h\sum_{l=0}^{n-1}\left(\frac{\delta S_{n-1}(x_l)}{2}+\frac{\delta S_{n-1}(x_{l+1})}{2}\right)\right| \le L\cdot\frac{\zeta_n}{n^{r+1/2}},
\end{aligned}$$

which completes the proof of the theorem. □

Thus, if the integrand is periodic, then the trapezoidal quadrature rule does not get saturated by smoothness, because according to (4.6) the integration error will self-adjust to the regularity of the integrand without us having to change anything in the quadrature formula. Moreover, if the function $f(x)$ is infinitely differentiable, then Theorem 4.2 implies spectral convergence. In other words, the rate of decay of the error will be faster than any polynomial, i.e., faster than $\mathcal{O}(n^{-r})$ for any $r > 0$.

REMARK 4.1 The result of Theorem 4.2 can, in fact, be extended to a large family of quadrature formulae (4.1) called the Newton-Cotes formulae. Namely, let $n = M\cdot s$, where M and s are positive integers, and partition the original grid of n cells into M clusters of s h-size cells:

$$\tilde{x}_m = m\cdot sh, \quad m = 0,1,\ldots,M.$$

Clearly, the relation between the original grid nodes x_k, $k = 0,1,\ldots,n$, and the coarsened grid nodes $\tilde{x}_m$, $m = 0,1,\ldots,M$, is given by: $\tilde{x}_m = x_{m\cdot s}$. On each interval $[\tilde{x}_m, \tilde{x}_{m+1}]$, $m = 0,1,\ldots,M-1$, we interpolate the integrand by an algebraic polynomial of degree no greater than s:

$$f(x) \approx P_s(x, f, x_{m\cdot s}, x_{m\cdot s+1},\ldots,x_{(m+1)\cdot s}), \quad \tilde{x}_m \equiv x_{m\cdot s} \le x \le x_{(m+1)\cdot s} \equiv \tilde{x}_{m+1}.$$

This polynomial is obviously unique. In other words, we can say that if $x \in [x_k, x_{k+1}]$, $k = 0,1,\ldots,n-1$, then

$$f(x) \approx P_s(x, f, x_{k-j}, x_{k-j+1},\ldots,x_{k-j+s}), \quad \text{where} \quad j = k - [k/s]\cdot s,$$

and $[\cdot]$ denotes the integer part. The quadrature formula is subsequently built by integrating the resulting piecewise polynomial function $P_s(x)$; it clearly belongs to the class (4.1). In particular, the Simpson formula of Section 4.1.3 is obtained by taking n even, setting $s = 2$ and $M = n/2$. For periodic integrands, quadrature formulae of this type do not get saturated by smoothness, see Exercise 2. Quadrature formulae of a completely different type that do not saturate for non-periodic integrands either are discussed in Section 4.2. □

4.1.3 Simpson's Formula

This quadrature formula is obtained by replacing the integrand $f(x)$ with a piecewise quadratic interpolant. For sufficiently smooth functions $f(x)$, the error of the Simpson formula decays with the rate $\mathcal{O}(h^4)$ when $h \longrightarrow 0$, i.e., faster than the error of the trapezoidal rule.

Assume that n is even, $n = 2M$, and that the grid nodes are equidistant: $x_{k+1} - x_k = h = (b-a)/n$, $k = 0, 1, \ldots, n-1$. On every interval $[x_{2m}, x_{2m+2}]$, $m = 0, 1, \ldots, M-1$, we replace the integrand $f(x)$ by a quadratic interpolating polynomial $P_2(x, f, x_{2m}, x_{2m+1}, x_{2m+2}) \equiv P_2(x, f_{2m}, f_{2m+1}, f_{2m+2})$, which is defined uniquely and can be written, e.g., using the Lagrange formula (see Section 2.1.1):

$$f(x) \approx P_2(x, f_{2m}, f_{2m+1}, f_{2m+2}) = f_{2m}\frac{(x-x_{2m+1})(x-x_{2m+2})}{(x_{2m}-x_{2m+1})(x_{2m}-x_{2m+2})}$$
$$+ f_{2m+1}\frac{(x-x_{2m})(x-x_{2m+2})}{(x_{2m+1}-x_{2m})(x_{2m+1}-x_{2m+2})} + f_{2m+2}\frac{(x-x_{2m})(x-x_{2m+1})}{(x_{2m+2}-x_{2m})(x_{2m+2}-x_{2m+1})}.$$

Equivalently, we can say that for $x \in [x_k, x_{k+1}]$, $k = 0, 1, \ldots, n-1$, we approximate $f(x)$ by means of the quadratic polynomial $P_2(x, f_{kj}) \equiv P_2(x, f, x_{k-j}, x_{k-j+1}, x_{k-j+2})$, where $j = 0$ if k is even, and $j = 1$ if k is odd.

The polynomial $P_2(x, f_{2m}, f_{2m+1}, f_{2m+2})$ can be easily integrated over the interval $[x_{2m}, x_{2m+2}]$ with the help of the fundamental theorem of calculus, which yields:

$$\int_{x_{2m}}^{x_{2m+2}} f(x)dx \approx \int_{x_{2m}}^{x_{2m+2}} P_2(x)dx = \frac{b-a}{3n}(f_{2m} + 4f_{2m+1} + f_{2m+2}).$$

By adding these equalities together for all $m = 0, 1, \ldots, M-1$, we obtain:

$$\int_a^b f(x)dx \approx \sigma_n(f) \stackrel{\text{def}}{=} \frac{b-a}{3n}(f_0 + 4f_1 + 2f_2 + 4f_3 + \ldots + 4f_{n-1} + f_n). \tag{4.13}$$

Formula (4.13) for the approximate computation of the definite integral $\int_a^b f(x)dx$ is known as the Simpson quadrature formula.

THEOREM 4.3

Let the third derivative $f^{(3)}(x)$ of the integrand $f(x)$ be continuous on the interval $[a,b]$, and let $\max_{a \le x \le b} |f^{(3)}(x)| = M_3$. Then the error of the Simpson formula satisfies the following estimate:

$$\left| \int_a^b f(x)dx - \sigma_n(f) \right| \le \frac{(b-a)M_3}{9\sqrt{3}} h^3. \tag{4.14}$$

PROOF To estimate the integration error for the Simpson formula, we first need to recall the error estimate for the interpolation of the integrand $f(x)$ by piecewise quadratic polynomials. Let us denote the interpolation error by $R_2(x) \stackrel{\text{def}}{=} f(x) - P_2(x)$, where the function $P_2(x)$ coincides with the quadratic polynomial $P_2(x, f_{2m}, f_{2m+1}, f_{2m+2})$ if $x \in [x_{2m}, x_{2m+2}]$. Then we can use Theorem 2.5 [formula (2.23) on page 36] and write for $x \in [x_{2m}, x_{2m+2}]$:

$$R_2(x) = \frac{f^{(3)}(\xi)}{3!}(x - x_{2m})(x - x_{2m+1})(x - x_{2m+2}), \quad \xi \in (x_{2m}, x_{2m+2}).$$

Consequently,

$$|R_2(x)| \leq \frac{M_3}{6} \max_{x_{2m} \leq x \leq x_{2m+2}} |(x - x_{2m})(x - x_{2m+1})(x - x_{2m+2})| \leq \frac{M_3}{9\sqrt{3}} h^3.$$

The previous inequality holds for $x \in [x_{2m}, x_{2m+2}]$, where m can be any number: $m = 0, 1, \dots, M-1$. Hence it holds for any $x \in [a, b]$ and we therefore obtain:

$$\max_{a \leq x \leq b} |R_2(x)| \leq \frac{M_3}{9\sqrt{3}} h^3.$$

For the integration error, we thus have:

$$\begin{aligned}\left|\int_a^b f(x)dx - \sigma_n(f)\right| &= \left|\int_a^b f(x)dx - \int_a^b P_2(x)dx\right| \\ &= \left|\int_a^b (f(x) - P_2(x))dx\right| = \left|\int_a^b R_2(x)dx\right| \leq \int_a^b |R_2(x)|dx \\ &\leq \int_a^b \max_x |R_2(x)|dx \leq \frac{M_3}{9\sqrt{3}} h^3 \int_a^b dx = (b-a)\frac{M_3}{9\sqrt{3}} h^3,\end{aligned}$$

which completes the proof of the theorem. □

Theorem 4.3 implies that for the function $f(x)$ that is three times continuously differentiable on $[a, b]$ the error of the Simpson formula will decrease cubically with respect to the grid size h, i.e., with the rate $\mathcal{O}(h^3)$ as $h \longrightarrow 0$, see estimate (4.14). For a smoother integrand $f(x)$, one can, in fact, obtain an even faster convergence.

THEOREM 4.4

Let the fourth derivative $f^{(4)}(x)$ of the integrand $f(x)$ be continuous on the interval $[a, b]$, and let $\max_{a \leq x \leq b} |f^{(4)}(x)| = M_4$. Then the error of the Simpson formula satisfies the following estimate:

$$\left|\int_a^b f(x)dx - \sigma_n(f)\right| \leq \frac{(b-a)M_4}{18} h^4. \tag{4.15}$$

PROOF We will show that for each $m = 0, 1, \dots, M-1$ the following inequality holds:

$$\left| \int_{x_{2m}}^{x_{2m+2}} f(x)dx - \frac{b-a}{3n}(f_{2m} + 4f_{2m+1} + f_{2m+2}) \right| \leq \frac{M_4}{9} h^5. \tag{4.16}$$

By adding these inequalities for all $m = 0, 1, \dots, M-1$, we arrive at the desired estimate (4.15):

$$\left| \int_a^b f(x)dx - \sigma_n(f) \right| = \left| \sum_{m=0}^{M-1} \left[\int_{x_{2m}}^{x_{2m+2}} f(x)dx - \frac{b-a}{3n}(f_{2m} + 4f_{2m+1} + f_{2m+2}) \right] \right|$$

$$\leq \sum_{m=0}^{M-1} \left| \int_{x_{2m}}^{x_{2m+2}} f(x)dx - \frac{b-a}{3n}(f_{2m} + 4f_{2m+1} + f_{2m+2}) \right| \leq \frac{n}{2} \frac{M_4}{9} h^5 = \frac{(b-a)M_4}{18} h^4.$$

To prove inequality (4.16), we represent the function $f(x)$ on a neighborhood of the node x_{2m+1}: $x_{2m} \leq x \leq x_{2m+2}$, with the help of the Taylor formula:

$$f(x) = Q(x) + R(x),$$

where

$$\begin{aligned} Q(x) = & f(x_{2m+1}) + f'(x_{2m+1})(x - x_{2m+1}) \\ & + \frac{f''(x_{2m+1})}{2!}(x - x_{2m+1})^2 + \frac{f'''(x_{2m+1})}{3!}(x - x_{2m+1})^3, \end{aligned}$$

and

$$R(x) = \frac{f^{(4)}(\xi)}{4!}(x - x_{2m+1})^4, \quad x_{2m} \leq \xi \leq x_{2m+2}.$$

Next,

$$\begin{aligned} \int_{x_{2m}}^{x_{2m+2}} f(x)dx - \frac{b-a}{3n}(f_{2m} + 4f_{2m+1} + f_{2m+2}) = & \int_{x_{2m}}^{x_{2m+2}} Q(x)dx + \int_{x_{2m}}^{x_{2m+2}} R(x)dx \\ & - \frac{b-a}{3n}[(Q_{2m} + R_{2m}) + 4(Q_{2m+1} + R_{2m+1}) + (Q_{2m+2} + R_{2m+2})] \\ = & \left\{ \int_{x_{2m}}^{x_{2m+2}} Q(x)dx - \frac{b-a}{3n}(Q_{2m} + 4Q_{2m+1} + Q_{2m+2}) \right\} \\ & + \left[\int_{x_{2m}}^{x_{2m+2}} R(x)dx - \frac{b-a}{3n}(R_{2m} + 4R_{2m+1} + R_{2m+2}) \right]. \end{aligned}$$

One can verify directly that for $Q(x) \equiv 1$, $Q(x) \equiv x$, $Q(x) \equiv x^2$, and $Q(x) \equiv x^3$ the quantity in the curly brackets on the right-hand side of the previous equality is equal to zero. Consequently, this quantity is equal to zero for an arbitrary polynomial $Q(x)$ as long as its degree does not exceed 3. Therefore,

$$\left|\int_{x_{2m}}^{x_{2m+2}} f(x)dx - \frac{b-a}{3n}(f_{2m}+4f_{2m+1}+f_{2m+2})\right| \\ = \left|\int_{x_{2m}}^{x_{2m+2}} R(x)dx - \frac{b-a}{3n}(R_{2m}+4R_{2m+1}+R_{2m+2})\right|.$$

Let us now take into account that $R_{2m+1}=0$ and $|R(x)| \le \frac{M_4}{4!}h^4$. Then the right-hand side of the previous equality can be estimated as follows:

$$\int_{x_{2m}}^{x_{2m+2}} |R(x)|dx + \frac{b-a}{3n}\cdot 2\max_x |R(x)| \le \max_x |R(x)| \left[\int_{x_{2m}}^{x_{2m+2}} dx + \frac{2}{3}\frac{b-a}{n}\right] \\ \le \max_x |R(x)| \left[2\frac{b-a}{n} + \frac{2}{3}\frac{b-a}{n}\right] \le \frac{M_4}{4!}\frac{8}{3}\frac{b-a}{n}h^4 = \frac{M_4}{9}h^5.$$

Thus, we have proven inequality (4.16), and as such, estimate (4.15). □

COROLLARY 4.1

The Simpson quadrature formula on a uniform grid is exact for the integrands $f(x)$ that are algebraic polynomials of degree no higher than third.

PROOF For the cubic and lower degree polynomials, we obviously have $M_4 = \max|f^{(4)}(x)| = 0$. Consequently, the right-hand side of estimate (4.15) is zero. □

We would like to emphasize, however, that if the regularity of the integrand $f(x)$ increases further, and even if $f(x)$ becomes an infinitely differentiable function, the error of the Simpson formula will not, generally speaking, decay faster than $\mathcal{O}(h^4)$ (unless the integrand is periodic, see Remark 4.1). In other words, similarly to the trapezoidal rule, the Simpson quadrature formula is prone to the saturation by smoothness (Exercise 4). In Section 4.2, we describe alternative strategies for building numerical quadratures that will not suffer from saturation.

Note also that in practice the dimension of the grid n is typically chosen automatically in the course of computation. Namely, the computation is started with some particular n_0, after which the grid is refined (say, by a factor of two), then maybe refined again, etc., until the answer no longer changes within a prescribed tolerance.

Exercises

1. Show that the trapezoidal rule (4.4) is exact if the integrand $f(x)$ is a polynomial of degree no higher than first.

2.* Prove the result formulated in Remark 4.1.

   Hint. Show that any quadrature formula from the corresponding family can be represented as a linear combination of a number of trapezoidal formulae.

3. Show that for a nonuniform grid that satisfies $x_{k+1} - x_k \leq h = \text{const}$, the use of piecewise quadratic interpolation leads to a generalization of the Simpson formula that will only be third order accurate, i.e., will have $\mathcal{O}(M_3 h^3)$ error as $h \longrightarrow 0$, where $M_3 = \max |f'''(x)|$.

4. Construct an example of an infinitely smooth function $f(x)$, for which the error of the Simpson quadrature formula is still $\mathcal{O}(h^4)$.

4.2 Quadrature Formulae with No Saturation. Gaussian Quadratures

When we studied the interpolation of functions in Chapter 2, we first discovered that the conventional piecewise polynomial algebraic interpolation normally gets saturated by smoothness, i.e., exhibits only a fixed rate of convergence as the grid is refined irrespective of how smooth the interpolated function is. In its own turn, this property of interpolation gives rise to the saturation of the quadrature formulae by smoothness[1] that we described in Section 4.1. Later in Chapter 3, we introduced the trigonometric interpolation and showed that it does not get saturated by smoothness. In other words, it self-adjusts the accuracy of approximation to the regularity of the interpolated function, provided that the latter is periodic. In the non-periodic case, a direct analogue of the trigonometric interpolation is algebraic interpolation on Chebyshev nodes, e.g., on the Chebyshev-Gauss grid or on the Chebyshev-Gauss-Lobatto grid, see Section 3.2. Algebraic interpolation of this type does not get saturated by smoothness, and it is therefore natural to expect that if a quadrature formula is built based on the Chebyshev interpolation, then it will not be prone to saturation either. This expectation does, in fact, prove correct, and the corresponding family of quadrature formulae is commonly referred to as Gaussian quadratures. In this section, we introduce and study some examples of Gaussian quadratures. The material presented hereafter is based on the more advanced considerations of Chapter 3 and can be skipped during the first reading.

Chebyshev polynomials $T_k(x) \stackrel{\text{def}}{=} \cos(k \arccos x)$, $k = 0, 1, 2, \ldots$, were discussed in Section 3.2 in the context of interpolation. They are algebraic polynomials of degree k defined on the interval $-1 \leq x \leq 1$. An arbitrary interval $a \leq x \leq b$ can be mapped

[1] With the exception of the Newton-Cotes formulae applied to periodic functions.

onto $[-1,1]$ by a simple linear transformation: $x = \frac{a+2}{2} + t\frac{b-a}{2}$, where $-1 \leq t \leq 1$. Therefore, analyzing both interpolation and quadratures only on the interval $[-1,1]$ (as opposed to $[a,b]$) does not imply any loss of generality.

The $n+1$ roots of the Chebyshev polynomial $T_{n+1}(x)$ on the interval $[-1,1]$ are given by the formula:

$$x_m = \cos\frac{\pi(2m+1)}{2(n+1)}, \quad m = 0,1,2,\ldots,n, \tag{4.17}$$

and are referred to as the Chebyshev-Gauss nodes. The $n+1$ extrema of the Chebyshev polynomial $T_n(x)$ are given by the formula:

$$x_m = \cos\frac{\pi}{n}m, \quad m = 0,1,2,\ldots,n, \tag{4.18}$$

and are referred to as the Chebyshev-Gauss-Lobatto nodes. To employ Chebyshev polynomials in the context of numerical quadratures, we will additionally need their orthogonality property given by Lemma 4.1.

LEMMA 4.1

Chebyshev polynomials are orthogonal on the interval $-1 \leq x \leq 1$ with the weight $w(x) = 1/\sqrt{1-x^2}$, i.e., the following equalities hold for $k, l = 0,1,2,\ldots$:

$$\int_{-1}^{1} \frac{T_k(x)T_l(x)}{\sqrt{1-x^2}}dx = \begin{cases} \pi, & k = l = 0, \\ \pi/2, & k = l \neq 0, \\ 0, & k \neq l. \end{cases} \tag{4.19}$$

PROOF Let $x = \cos\varphi \Leftrightarrow \varphi = \arccos x$. Changing the variable in the integral, we obtain:

$$\int_{-1}^{1} \frac{T_k(x)T_l(x)}{\sqrt{1-x^2}}dx = \int_{\pi}^{0} \frac{\cos(k\varphi)\cos(l\varphi)}{|\sin\varphi|}(-\sin\varphi)d\varphi = \int_{0}^{\pi} \cos(k\varphi)\cos(l\varphi)d\varphi$$

$$= \frac{1}{2}\int_{0}^{\pi} [\cos(k+l)\varphi + \cos(k-l)\varphi]d\varphi = \begin{cases} \pi, & k = l = 0, \\ \pi/2, & k = l \neq 0, \\ 0, & k \neq l, \end{cases}$$

because Chebyshev polynomials are only considered for $k \geq 0$ and $l \geq 0$. ∎

Let the function $f = f(x)$ be defined for $-1 \leq x \leq 1$. We will approximate its definite integral over $[-1,1]$ taken with the weight $w(x) = 1/\sqrt{1-x^2}$ by integrating the Chebyshev interpolating polynomial $P_n(x,f)$ built on the Gauss nodes (4.17):

$$\int_{-1}^{1} \frac{f(x)}{\sqrt{1-x^2}}dx \approx \int_{-1}^{1} \frac{P_n(x,f)}{\sqrt{1-x^2}}dx. \tag{4.20}$$

The key point, of course, is to obtain a convenient expression for the integral on the right-hand side of formula (4.20) via the function values $f(x_m)$, $m = 0, 1, 2, \ldots, m$, sampled on the Gauss grid (4.17). It is precisely the introduction of the weight $w(x) = 1/\sqrt{1-x^2}$ into the integral (4.20) that enables a particularly straightforward integration of the interpolating polynomial $P_n(x, f)$ over $[-1, 1]$.

LEMMA 4.2
Let the function $f = f(x)$ be defined on $[-1,1]$, and let $P_n(x,f)$ be its algebraic interpolating polynomial on the Chebyshev-Gauss grid (4.17). Then,

$$\int_{-1}^{1} \frac{P_n(x,f)}{\sqrt{1-x^2}} dx = \frac{\pi}{n+1} \sum_{m=0}^{n} f(x_m). \tag{4.21}$$

PROOF Recall that $P_n(x,f)$ is an interpolating polynomial of degree no higher than n built for the function $f(x)$ on the Gauss grid (4.17). The grid has a total of $n+1$ nodes, and the polynomial is unique. As shown in Section 3.2.3, see formulae (3.62) and (3.63) on page 76, $P_n(x,f)$ can be represented as:

$$P_n(x,f) = \sum_{k=0}^{n} a_k T_k(x),$$

where $T_k(x)$ are Chebyshev polynomials of degree k, and the coefficients a_k are given by:

$$a_0 = \frac{1}{n+1} \sum_{m=0}^{n} f_m T_0(x_m) \quad \text{and} \quad a_k = \frac{2}{n+1} \sum_{m=0}^{n} f_m T_k(x_m), \; k = 1, \ldots, n.$$

Accordingly, it will be sufficient to show that equality (4.21) holds for all individual $T_k(x)$, $k = 0, 1, 2, \ldots, n$:

$$\int_{-1}^{1} \frac{T_k(x)}{\sqrt{1-x^2}} dx = \frac{\pi}{n+1} \sum_{m=0}^{n} T_k(x_m). \tag{4.22}$$

Let $k = 0$, then $T_0(x) \equiv 1$. In this case Lemma 4.1 implies that the left-hand side of (4.22) is equal to π, and the right-hand side also appears equal to π by direct substitution. For $k > 0$, orthogonality of the Chebyshev polynomials in the sense of formula (4.19) means that the left-hand side of (4.22) is equal to zero. Consequently, we need to show that the right-hand side is zero as well. Using trigonometric interpretation, we obtain for all $k = 1, 2, \ldots, 2n+1$:

$$\begin{aligned}
\sum_{m=0}^{n} T_k(x_m) &= \sum_{m=0}^{n} \cos\left(k \arccos\left[\cos\frac{\pi(2m+1)}{2(n+1)}\right]\right) \\
&= \sum_{m=0}^{n} \cos\left(k\frac{\pi(2m+1)}{2(n+1)}\right) = \sum_{m=0}^{n} \cos\left(k\frac{2\pi m}{2(n+1)} + k\frac{\pi}{2(n+1)}\right) = 0.
\end{aligned}$$

The last equality holds by virtue of formula (3.22), see the proof of Theorem 3.1, page 64. Thus, we have established equality (4.21). □

Lemma 4.21 allows us to recast the approximate expression (4.20) for the integral $\int_{-1}^{1} f(x)w(x)dx$ as follows:

$$\int_{-1}^{1} \frac{f(x)}{\sqrt{1-x^2}}dx \approx \frac{\pi}{n+1}\sum_{m=0}^{n} f(x_m). \tag{4.23}$$

Formula (4.23) is known as the Gaussian quadrature formula with the weight $w(x) = 1/\sqrt{1-x^2}$ on the Chebyshev-Gauss grid (4.17). It has a particularly simple structure, which is very convenient for implementation. In practice, if we need to evaluate the integral $\int_{-1}^{1} f(x)dx$ for a given $f(x)$ with no weight, we introduce a new function $g(x) = f(x)\sqrt{1-x^2}$ and then rewrite formula (4.23) as:

$$\int_{-1}^{1} f(x)dx = \int_{-1}^{1} \frac{g(x)}{\sqrt{1-x^2}}dx \approx \frac{\pi}{n+1}\sum_{m=0}^{n} g(x_m).$$

A key advantage of the Gaussian quadrature (4.23) compared to the quadrature formulae studied previously in Section 4.1 is that the Gaussian quadrature does not get saturated by smoothness. Indeed, according to the following theorem (see also Remark 4.2 right after the proof of Theorem 4.5), the integration error automatically adjusts to the regularity of the integrand.

THEOREM 4.5
Let the function $f = f(x)$ be defined for $-1 \le x \le 1$; let it have continuous derivatives up to the order $r > 0$, and a square integrable derivative of order $r+1$:

$$\int_{-1}^{1} \left[f^{(r+1)}(x)\right]^2 dx < \infty.$$

Then, the error of the Gaussian quadrature (4.23) can be estimated as:

$$\left|\int_{-1}^{1} \frac{f(x)}{\sqrt{1-x^2}}dx - \frac{\pi}{n+1}\sum_{m=0}^{n} f(x_m)\right| \le \pi\frac{\zeta_n}{n^{r-1/2}}, \tag{4.24}$$

where $\zeta_n = o(1)$ as $n \longrightarrow \infty$.

PROOF The proof of inequality (4.24) is based on the error estimate (3.65) obtained in Section 3.2.4 (see page 77) for the Chebyshev algebraic interpolation. Namely, let $R_n(x) = f(x) - P_n(x,f)$. Then, under the assumptions of the current theorem we have:

$$\max_{-1\le x\le 1}|R_n(x)| \le \frac{\zeta_n}{n^{r-1/2}}, \quad \text{where} \quad \zeta_n = o(1), \quad n \longrightarrow \infty. \tag{4.25}$$

Consequently,

$$\begin{aligned}\left|\int_{-1}^{1}\frac{f(x)}{\sqrt{1-x^2}}dx - \frac{\pi}{n+1}\sum_{m=0}^{n} f(x_m)\right| &= \left|\int_{-1}^{1}\frac{f(x)}{\sqrt{1-x^2}}dx - \int_{-1}^{1}\frac{P_n(x,f)(x)}{\sqrt{1-x^2}}dx\right| \\ &\le \int_{-1}^{1}\frac{|f(x)-P_n(x,f)|}{\sqrt{1-x^2}}dx \le \max_{-1\le x\le 1}|R_n(x)|\cdot\int_{-1}^{1}\frac{dx}{\sqrt{1-x^2}} = \pi\frac{\zeta_n}{n^{r-1/2}},\end{aligned}$$

which yields the desired estimate (4.24). □

Theorem 4.5 implies, in particular, that if the function $f = f(x)$ is infinitely differentiable on $[-1,1]$, then the Gaussian quadrature (4.23) exhibits a spectral rate of convergence as the dimension of the grid n increases. In other words, the integration error in (4.24) will decay faster than $\mathcal{O}(n^{-r})$ for any $r > 0$ as $n \longrightarrow \infty$.

REMARK 4.2 Error estimate (4.25) for the algebraic interpolation on Chebyshev grids can, in fact, be improved. According to formula (3.67), see Remark 3.2 on page 78, instead of inequality (4.25) we can write:

$$\max_{-1\le x\le 1}|R_n(x)| = o\left(\frac{\ln n}{n^{r+1/2}}\right) \quad \text{as} \quad n \longrightarrow \infty.$$

Then, the same argument as employed in the proof of Theorem 4.5 yields an improved convergence result for Gaussian quadratures:

$$\left|\int_{-1}^{1}\frac{f(x)}{\sqrt{1-x^2}}dx - \frac{\pi}{n+1}\sum_{m=0}^{n} f(x_m)\right| = o\left(\frac{\pi\ln n}{n^{r+1/2}}\right) \quad \text{as} \quad n \longrightarrow \infty,$$

where $r+1$ is the maximum number of derivatives that the integrand $f(x)$ has on the interval $-1 \le x \le 1$. However, even this is not the best estimate yet.

As shown in Section 3.2.7, by combining the Jackson inequality [Theorem 3.8, formula (3.79)], the Lebesgue inequality [Theorem 3.10, formula (3.80)], and estimate (3.83) for the Lebesgue constants given by the Bernstein theorem [Theorem 3.12], one can obtain the following error estimate for the Chebyshev algebraic interpolation:

$$\max_{-1\le x\le 1}|R_n(x)| = \mathcal{O}\left(\frac{\ln(n+1)}{n^{r+1}}\right), \quad \text{as} \quad n \longrightarrow \infty,$$

provided only that the derivative $f^{(r)}(x)$ of the function $f(x)$ is Lipshitz-continuous. Consequently, for the error of the Gaussian quadrature we have:

$$\left| \int_{-1}^{1} \frac{f(x)}{\sqrt{1-x^2}} dx - \frac{\pi}{n+1} \sum_{m=0}^{n} f(x_m) \right| = \mathcal{O}\left(\pi \frac{\ln(n+1)}{n^{r+1}} \right), \quad \text{as} \quad n \longrightarrow \infty. \tag{4.26}$$

Estimate (4.26) implies, in particular, that the Gaussian quadrature (4.23) converges at least for any Lipshitz-continuous function (page 87), and that the rate of decay of the error is not slower than $\mathcal{O}(n^{-1}\ln(n+1))$ as $n \longrightarrow \infty$. Otherwise, for smoother functions the convergence speeds up as the regularity increases, and becomes spectral for infinitely differentiable functions. □

REMARK 4.3 When proving Lemma 4.2, we have, in fact, shown that equality (4.22) holds for all Chebyshev polynomials $T_k(x)$ up to the degree $k = 2n+1$ and not only up to the degree $k = n$. This circumstance reflects on another important property of the Gaussian quadrature (4.23). It happens to be exact, i.e., it generates no error, for any algebraic polynomial of degree no higher than $2n+1$ on the interval $-1 \leq x \leq 1$. In other words, if $Q_{2n+1}(x)$ is a polynomial of degree $\leq 2n+1$, then

$$\int_{-1}^{1} \frac{Q_{2n+1}(x)}{\sqrt{1-x^2}} dx = \frac{\pi}{n+1} \sum_{m=0}^{n} Q_{2n+1}(x_m).$$

Moreover, it is possible to show that the Gaussian quadrature (4.23) is optimal in the following sense. Consider a family of the quadrature formulae on $[-1,1]$:

$$\int_{-1}^{1} f(x)w(x)dx \approx \sum_{j=0}^{n} \alpha_j f(x_j), \tag{4.27}$$

where the dimension n is fixed, the weight $w(x)$ is defined as before, $w(x) = 1/\sqrt{1-x^2}$, and both the nodes x_j and the coefficients α_j, $j = 0, 1, \ldots, n$ are to be determined. Require that this quadrature be exact for the polynomials of the highest degree possible. Then, it turns out that the corresponding degree will be equal to $2n+1$, while the nodes x_j will coincide with those of the Chebyshev-Gauss grid (4.17), and the coefficients α_j will all be equal to $\pi/(n+1)$. Altogether, quadrature (4.27) will coincide with (4.23). □

Exercises

1. Prove an analogue of Lemma 4.2 for the Gaussian quadrature on the Chebyshev-Gauss-Lobatto grid (4.18). Namely, let $\tilde{P}_n(x,f) = \sum_{k=0}^{n} \tilde{a}_k T_k(x)$ be the corresponding interpolating polynomial (constructed in Section 3.2.6). Show that

$$\int_{-1}^{1} \frac{\tilde{P}_n(x,f)}{\sqrt{1-x^2}} dx = \frac{\pi}{n} \sum_{m=0}^{n} \beta_m f(x_m),$$

where $\beta_0 = \beta_n = 1/2$ and $\beta_m = 1$ for $m = 1, 2, \ldots, m-1$.

4.3 Improper Integrals. Combination of Numerical and Analytical Methods

Even for the simplest improper integrals, a direct application of the quadrature formulae may encounter serious difficulties. For example, the trapezoidal rule (4.4):

$$\int_a^b f(x)dx \approx \frac{b-a}{n}\left(\frac{f_0}{2} + f_1 + \ldots + f_{n-1} + \frac{f_n}{2}\right)$$

will fail for the case $a = 0$, $b = 1$, and $f(x) = \cos x/\sqrt{x}$, because $f(x)$ has a singularity at $x = 0$ and consequently, $f_0 = f(0)$ is not defined. Likewise, the Simpson formula (4.13) will fail. For the Gaussian quadrature (4.23), the situation may seem a little better at a first glance, because the Chebyshev-Gauss nodes (4.17) do not include the endpoints of the interval. However, the unboundedness of the function and its derivatives will still prevent one from obtaining any reasonable error estimates. At the same time, the integral itself: $\int_0^1 \frac{\cos x}{\sqrt{x}}dx$, obviously exists, and procedures for its efficient numerical approximation need to be developed.

To address the difficulties that arise when computing the values of improper integrals, it is natural to try and employ a combination of analytical and numerical techniques. The role of the analytical part is to reduce the original problem to a new problem that would only require one to evaluate the integral of a smooth and bounded function. The latter can then be done on a computer with the help of a quadrature formula. In the previous example, we can first use the integration by parts:

$$\int_0^1 \frac{\cos x}{\sqrt{x}}dx = \cos x(2\sqrt{x})\Big|_0^1 + \int_0^1 s\sqrt{x}\sin x dx,$$

and subsequently approximate the integral on the right-hand side to a desired accuracy using any of the quadrature formulae introduced in Section 4.1 or 4.2.

An alternative approach is based on splitting the original integral into two:

$$\int_0^1 \frac{\cos x}{\sqrt{x}}dx = \int_0^c \frac{\cos x}{\sqrt{x}}dx + \int_c^1 \frac{\cos x}{\sqrt{x}}dx, \tag{4.28}$$

where $c > 0$ can be considered arbitrary as of yet. To evaluate the first integral on the right-hand side of equality (4.28), we can use the Taylor expansion:

$$\cos x = 1 - \frac{x^2}{2!} + \frac{x^4}{4!} - \frac{x^6}{6!} + \ldots.$$

Then we have:

$$\int_0^c \frac{\cos x}{\sqrt{x}}dx = \int_0^c \frac{dx}{\sqrt{x}} - \int_0^c \frac{x^{3/2}}{2!}dx + \int_0^c \frac{x^{7/2}}{4!}dx - \int_0^c \frac{x^{11/2}}{6!}dx + \dots, \tag{4.29}$$

and every integral on the right-hand side of equality (4.29) can be evaluated explicitly using the fundamental theorem of calculus. Let the tolerance $\varepsilon > 0$ be given, and assume that the overall integral needs to be approximated numerically with the error that may not exceed ε. It is then sufficient to approximate each of the two integrals on the right hand side of (4.28) with the accuracy $\varepsilon/2$. As far as the first integral, this can be achieved by taking only a few leading terms in the sum on the right-hand side of formula (4.29). The precise number of terms to be taken will, of course, depend on ε and on c. In doing so, the value of c should not be taken too large, say, $c = 1$, because in this case even if ε is not very small, the number of terms to be evaluated on the right-hand side of formula (4.29) will still be substantial. On the other hand, the value of c should not be taken too small either, say, $c \ll 1$, because in this case the first integral on the right-hand side of (4.28): $\int_0^c \frac{\cos x}{\sqrt{x}}dx$, can be evaluated very efficiently, but the integrand in the second integral $\int_c^1 \frac{\cos x}{\sqrt{x}}dx$ will have large derivatives. This will necessitate taking large values of n (grid dimension) for evaluating this integral with accuracy $\varepsilon/2$, say, using the trapezoidal rule or the Simpson formula.

A more universal and generally more efficient technique for the numerical computation of improper integrals is based on regularization, i.e., on isolation of the singularity. Unlike in the previous approach, it does not require finding a fine balance between the too small and too large values of c. Assume that we need to compute the integral

$$\int_0^1 \frac{f(x)}{\sqrt{x}}dx,$$

where the function $f(x)$ is smooth and bounded. Regularization consists of recasting this integral in an equivalent form:

$$\int_0^1 \frac{f(x)}{\sqrt{x}}dx = \int_0^1 \frac{f(x) - \phi(x)}{\sqrt{x}}dx + \int_0^1 \frac{\phi(x)}{\sqrt{x}}dx. \tag{4.30}$$

In doing so, the auxiliary function $\phi(x)$ is to be chosen so that to remove the singularity from the first integral on the right-hand side of (4.30) and enable its efficient and accurate approximation using, e.g., the Simpson formula. The second integral on the right-hand side of (4.30) will still contain a singularity, but we select $\phi(x)$ so that to be able to compute this integral analytically.

In our particular case, this goal will be achieved if $\phi(x)$ is taken in the form of just two leading terms of the Taylor expansion of $f(x)$ around $x = 0$: $\phi(x) = f(0) +$

$f'(0)x$. Substituting this expression into formula (4.30) we obtain:

$$\int_0^1 \frac{f(x)}{\sqrt{x}}dx = \int_0^1 \frac{f(x)-f(0)-f'(0)x}{\sqrt{x}}dx + f(0)\int_0^1 \frac{dx}{\sqrt{x}} + f'(0)\int_0^1 \sqrt{x}dx.$$

Returning to our original example with $f(x) = \cos x$, we arrive at the following expression for the integral:

$$\int_0^1 \frac{\cos x}{\sqrt{x}}dx = \int_0^1 \frac{\cos x - 1}{\sqrt{x}}dx + \int_0^1 \frac{dx}{\sqrt{x}}.$$

The second integral on the right-hand side is equal to 2, while for the evaluation of the first integral by the Simpson formula, say, with the accuracy $\varepsilon = 10^{-3}$, taking a fairly coarse grid $n = 4$, i.e., $h = 1/4$, already appears sufficient.

Sometimes we can also use a change of variable for removing the singularity. For instance, setting $x = t^2$ in the previous example yields:

$$\int_0^1 \frac{\cos x}{\sqrt{x}}dx = \int_0^1 \frac{\cos t^2}{t}2tdt = 2\int_0^1 \cos t^2 dt,$$

and the resulting integral no longer contains a singularity.

Altogether, a key to success is a sensible and adequate partition of the overall task into the analytical and numerical parts. The analytic derivations and transformations are to be performed by a human researcher, whereas routine number crunching is to be performed by a computer. This statement basically applies to solving any problem with the help of a computer.

Exercises

1. Propose algorithms for the approximate computation of the following improper integrals:

$$\int_0^1 \frac{\sin x}{x^{3/2}}dx, \quad \int_0^{10} \frac{\ln(1+x)}{x^{3/2}}dx, \quad \int_0^\infty e^{-x^2}dx, \quad \int_0^{\pi/2} \frac{\cos x}{\sqrt{\pi/2 - x}}dx.$$

4.4 Multiple Integrals

Computation of multiple integrals:

$$I^{(m)} = I^{(m)}(f,\Omega) = \int \cdots \int_\Omega f(x_1,\ldots,x_m)dx_1\ldots dx_m$$

over a given domain Ω of the space $\mathbb{R}^m$ of independent variables $x_1,\dots,x_m$ can be reduced to computing integrals of the type $\int_a^b g(x)dx$ using quadrature formulae that we have studied previously. However, the complexity of the corresponding algorithms and the number of arithmetic operations required for obtaining the answer with the prescribed tolerance $\varepsilon > 0$ will rapidly grow as the dimension m of the space increases. This is true even in the simplest case when the domain of integration is a cube: $\Omega = \{\boldsymbol{x} = (x_1,\dots,x_m) \,\big|\, |x_j| \le 1,\ j = 1,2,\dots,m\}$, and if Ω has a more general shape then the problem gets considerably exacerbated.

In the current section, we only provide several simple examples of how the problem of numerical computation of multiple integrals can be approached. For more detail, we refer the reader to [DR84, Chapter 5].

4.4.1 Repeated Integrals and Quadrature Formulae

We will illustrate the situation using a two-dimensional setup, $m = 2$. Consider the integral:

$$I^{(2)} = \iint\limits_{\Omega} f(x,y)dxdy$$

over the domain Ω of the Cartesian plane, which is schematically shown in Figure 4.1. Assume that this domain lies in between the graphs of two functions:

$$y = \varphi_1(x) \ \text{ and } \ y = \varphi_2(x).$$

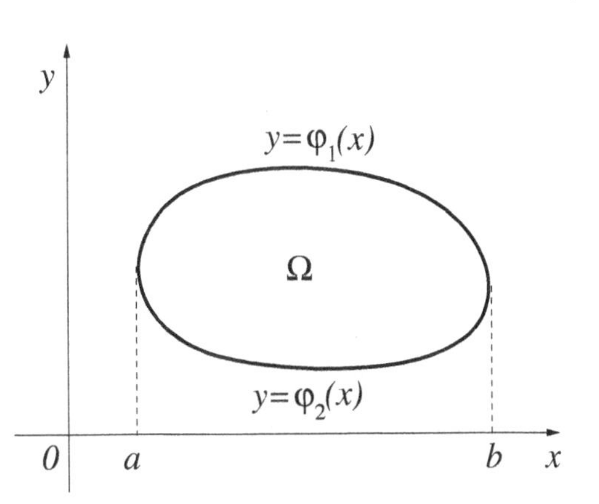

FIGURE 4.1: Domain of integration.

Then, clearly,

$$I^{(2)} = \iint\limits_{\Omega} f(x,y)dxdy = \int\limits_a^b F(x)dx,$$

where

$$F(x) = \int\limits_{\varphi_1(x)}^{\varphi_2(x)} f(x,y)dy, \quad a \le x \le b.$$

The integral $I^{(2)} = \int_a^b F(x)dx$ can be approximated with the help of a quadrature formula. For simplicity, let us use the trapezoidal rule:

$$I^{(2)} = \int\limits_a^b F(x)dx \approx \frac{b-a}{n}\left(\frac{F(x_0)}{2} + F(x_1) + \ldots + F(x_{n-1}) + \frac{F(x_n)}{2}\right), \tag{4.31}$$

$$x_j = a + j\cdot\frac{b-a}{n}, \quad j = 0,1,\dots,n.$$

In their own turn, expressions

$$F(x_j) = \int\limits_{\varphi_1(x_j)}^{\varphi_2(x_j)} f(x_j, y)dy, \quad j = 0, 1, \dots, n$$

that appear in formula (4.31) can also be computed using the trapezoidal rule:

$$\begin{aligned} F(x_j) &\approx \frac{\varphi_2(x_j) - \varphi_1(x_j)}{n} \\ &\times \left(\frac{f(x_j, y_0)}{2} + f(x_j, y_1) + \dots + f(x_j, y_{n-1}) + \frac{f(x_j, y_n)}{2} \right), \\ y_k &= \varphi_1(x_j) + k \cdot \frac{\varphi_2(x_j) - \varphi_1(x_j)}{n}, \quad k = 0, 1, \dots, n. \end{aligned} \tag{4.32}$$

In doing so, the approximation error $\varepsilon = \varepsilon(n)$ that arises when computing the integral $I^{(2)}$ by means of formulae (4.31) and (4.32) decreases as the number n increases. At the same time, the number of arithmetic operations increases as $\mathcal{O}(n^2)$. If the integrand $f = f(x, y)$ has continuous partial derivatives up to the second order on Ω, and if, in addition, the functions $\varphi_1(x)$ and $\varphi_2(x)$ are sufficiently smooth, then the properties of the one-dimensional trapezoidal rule established in Section 4.1.2 guarantee convergence of the two-dimensional quadrature (4.31), (4.32) with second order with respect to n^{-1}: $\varepsilon(n) = \mathcal{O}(n^{-2})$.

To summarize, the foregoing approach is based on the transition from a multiple integral to the repeated integral. The latter is subsequently computed by the repeated application of a one-dimensional quadrature formula. This technique can be extended to the general case of the integrals $I^{(m)}$, $m > 2$, provided that the domain Ω is specified in a sufficiently convenient way. In doing so, the number of arithmetic operations needed for the implementation of the quadrature formulae will be $\mathcal{O}(n^m)$, where n is the number of nodes in one coordinate direction. If the function $f(x_1, \dots, x_m)$ has continuous partial derivatives of order two, the boundary of the domain Ω is sufficiently smooth, and the trapezoidal rule (Section 4.1.2) is used in the capacity of the one-dimensional quadrature formula, then the rate of convergence will be quadratic with respect to n^{-1}: $\varepsilon(n) = \mathcal{O}(n^{-2})$.

In practice, if the tolerance $\varepsilon > 0$ is specified ahead of time, then the dimension n is chosen experimentally, by starting the computation with some n_0 and then refining the grid, say, by doubling the number of nodes (in each direction) several times until the answer no longer changes within the prescribed tolerance.

4.4.2 The Use of Coordinate Transformations

Let us analyze an elementary example that shows how a considerable simplification of the original formulation can be achieved by carefully taking into account the specific structure of the problem. Suppose that we need to compute the double integral $\iint_\Omega f(x, y)dxdy$, where the domain Ω is a disk: $\Omega = \{(x, y) \mid x^2 + y^2 \le R^2\}$, and the function $f(x,y)$ does not, in fact, depend on the two individual variables x and y, but rather depends only on $x^2+y^2=r^2$, so that $f(x,y) \equiv f(r)$. Then we can write:

$$I^{(2)} = \iint_{\Omega} f(x,y)dxdy = 2\pi \int_0^R f(r)rdr,$$

and the problem of approximately evaluating a double integral is reduced to the problem of evaluating a conventional definite integral.

4.4.3 The Notion of Monte Carlo Methods

The foregoing reduction of multiple integrals to repeated integrals may become problematic already for $m = 2$ if the domain Ω has a complicated shape. We therefore introduce a completely different approach to the approximate computation of multiple integrals.

Assume for simplicity that the domain Ω lies inside the square $D = \{(x,y) \mid 0 \le x \le 1, 0 \le y \le 1\}$. The function $f = f(x,y)$ is originally defined on Ω, and we extend it into $D \backslash \Omega$ and let $f(x,y) = 0$ for $(x,y) \in D \backslash \Omega$. Suppose that there is a random number generator that produces the numbers uniformly distributed across the interval $[0,1]$. This means that the probability for a given number to fall into the interval $[\alpha_1, \alpha_2]$, where $0 \le \alpha_1 < \alpha_2 \le 1$, is equal to $\alpha_2 - \alpha_1$. Using this generator, let us obtain a sequence of random pairs $P_k = (x_k, y_k)$, $k = 1, 2, \ldots$, and then construct a numerical sequence:

$$S_k = \frac{1}{k} \sum_{j=1}^{k} f(x_j, y_j).$$

It seems natural to expect that for a given $\varepsilon > 0$, the probability that the error $|I^{(2)} - S_k|$ is less than ε:

$$|I^{(2)} - S_k| < \varepsilon,$$

will approach one as the number of tests k increases. This statement can, in fact, be put into the framework of rigorous theorems.

The meaning of the Monte Carlo method becomes particularly apparent if $f(x,y) = 1$ on Ω. Then the integral $I^{(2)}$ is the area of Ω, and S_k is the ratio of the number of those points P_j, $1 \le j \le k$, that fall into Ω to the overall number k of these points. It is clear that this ratio is approximately equal to the area of Ω, because the overall area of the square D is equal to one.

The Monte Carlo method is attractive because the corresponding numerical algorithms have a very simple logical structure. It does not become more complicated when the dimension m of the space increases, or when the shape of the domain Ω becomes more elaborate. However, in the case of simple domains and special narrow classes of functions (smooth) the Monte Carlo method typically appears less efficient than the methods that take explicit advantage of the specifics of a given problem.

The Monte Carlo method has many other applications, see, e.g., [Liu01, BH02, CH06]; an interesting recent application area is in security pricing [Gla04].

Part II

Systems of Scalar Equations

In mathematics, plain numbers are often referred to as scalars; they may be real or complex. Equations or systems of equations with respect to one or several scalar unknowns appear in many branches of mathematics, as well as in applications. Accordingly, computational methods for solving these equations and systems are in the core of numerical analysis. Along with the scalar equations, one frequently encounters functional equations and systems, in which the unknown quantities are functions rather than plain numbers. A particular class of functional equations are differential equations that contain derivatives of the unknown functions. Numerical methods for solving differential equations are discussed in Part III of the book.

In the general category of scalar equations and systems one can identify a very important sub-class of *linear algebraic equations and systems*. Numerical methods for solving various systems of this type are discussed in Chapters 5, 6, and 7. In Chapter 8, we address numerical solution of the nonlinear equations and systems.

Before we actually discuss the solution of linear and nonlinear equations, let us mention here some other books that address similar subjects: [Hen64, IK66, CdB80, Atk89, PT96, QSS00, Sch02, DB03]. More specific references will be given later.

Chapter 5

Systems of Linear Algebraic Equations: Direct Methods

Systems of linear algebraic equations (or simply linear systems) arise in numerous applications; often, these systems appear to have high dimension. A linear system of order (dimension) n can be written as follows:

$$a_{11}x_1 + a_{12}x_2 + \ldots + a_{1n}x_n = f_1,$$

$$\ldots\ldots\ldots\ldots\ldots\ldots\ldots\ldots\ldots\ldots\ldots\ldots\ldots\ldots$$

$$a_{n1}x_1 + a_{n2}x_2 + \ldots + a_{nn}x_n = f_n,$$

where $a_{ij},\ i, j = 1, 2, \ldots, n$, are the coefficients, $f_i,\ i = 1, 2, \ldots, n$, are the right-hand sides (both assumed given), and $x_j,\ j = 1, 2, \ldots, n$, are the unknown quantities. A more general and often more convenient form of representing a linear system is its operator form, see Section 5.1.2.

In this chapter, we first outline different forms of consistent linear systems and illustrate those with examples (Section 5.1), and then provide a concise yet rigorous review of the material from linear algebra (Section 5.2). Next, we introduce the notion of conditioning of a linear operator (matrix) and that of the corresponding system of linear algebraic equations (Section 5.3), and subsequently describe several direct methods for solving linear systems (Sections 5.4, 5.6, and 5.7). We recall that the method is referred to as direct if it produces the exact solution of the system after a finite number of arithmetic operations.[1] (Exactness is interpreted within the machine precision if the method is implemented on a computer.) Also in this chapter we discuss what types of numerical algorithms (and why) are better suited for solving particular classes of linear systems. Additionally in Section 5.5, we establish a relation between the systems of linear algebraic equations and the problem of minimization of multi-variable quadratic functions.

Before proceeding any further, let us note that the well-known Cramer's rule is never used for computing solutions of linear systems, even for moderate dimensions $n > 5$. The reason is that it requires a large amount of arithmetic operations and may also develop computational instabilities.

[1] An alternative to direct methods are iterative methods analyzed in Chapter 6.

5.1 Different Forms of Consistent Linear Systems

A linear equation with respect to the unknowns $z_1, z_2, \ldots, z_n$ is an equation of the form

$$\alpha_1 z_1 + \alpha_2 z_2 + \ldots + \alpha_n z_n = t,$$

where $\alpha_1, \alpha_2, \ldots, \alpha_n$, and t are some given numbers (constants).

5.1.1 Canonical Form of a Linear System

Let a system of n linear algebraic equations be specified with respect to as many unknowns. This system can obviously be written in the following *canonical* form:

$$\begin{aligned} a_{11}x_1 + a_{12}x_2 + \ldots + a_{1n}x_n &= f_1, \\ \ldots\ldots\ldots\ldots\ldots\ldots\ldots\ldots \\ a_{n1}x_1 + a_{n2}x_2 + \ldots + a_{nn}x_n &= f_n. \end{aligned} \tag{5.1}$$

Both the unknowns x_j and the equations in system (5.1) are numbered by the the consecutive integers: $1, 2, \ldots, n$. Accordingly, the coefficients a_{ij} in front of the unknowns have double subscripts: $i, j = 1, 2, \ldots, n$, where i reflects on the number of the equation and j reflects on the number of the unknown. The right-hand side of equation number i is denoted by f_i for all $i = 1, 2, \ldots, n$.

From linear algebra we know that system (5.1) is *consistent,* i.e., has a solution, for any choice of the right-hand sides f_i if and only if the corresponding homogeneous system[2] only has a trivial solution: $x_1 = x_2 = \ldots = x_n = 0$. For the homogeneous system to have only the trivial solution, it is necessary and sufficient that the determinant of the system matrix $\boldsymbol{A}$:

$$\boldsymbol{A} = \begin{bmatrix} a_{11} & \ldots & a_{1n} \\ \ldots & \ldots & \ldots \\ a_{n1} & \ldots & a_{nn} \end{bmatrix} \tag{5.2}$$

differ from zero: $\det \boldsymbol{A} \neq 0$. The condition $\det \boldsymbol{A} \neq 0$ that guarantees consistency of system (5.1) for any right-hand side also implies uniqueness of the solution.

Having introduced the matrix $\boldsymbol{A}$ of (5.2), we can rewrite system (5.1) as follows:

$$\begin{bmatrix} a_{11} & \ldots & a_{1n} \\ \ldots & \ldots & \ldots \\ a_{n1} & \ldots & a_{nn} \end{bmatrix} \begin{bmatrix} x_1 \\ \vdots \\ x_n \end{bmatrix} = \begin{bmatrix} f_1 \\ \vdots \\ f_n \end{bmatrix}, \tag{5.3}$$

or, using an alternative shorter matrix notation:

$$\boldsymbol{A}\boldsymbol{x} = \boldsymbol{f}.$$

[2]Obtained by setting $f_i = 0$ for all $i = 1, 2, \ldots, n$.

5.1.2 Operator Form

Let $\mathbb{R}^n$ be a linear space of dimension n (see Section 5.2), $\boldsymbol{A}$ be a linear operator, $\boldsymbol{A}: \mathbb{R}^n \longmapsto \mathbb{R}^n$, and $\boldsymbol{f} \in \mathbb{R}^n$ be a given element of the space $\mathbb{R}^n$. Consider the equation:

$$\boldsymbol{A}\boldsymbol{x} = \boldsymbol{f} \tag{5.4}$$

with respect to the unknown $\boldsymbol{x} \in \mathbb{R}^n$. It is known that this equation is solvable for an arbitrary $\boldsymbol{f} \in \mathbb{R}^n$ if and only if the only solution of the corresponding homogeneous equation $\boldsymbol{A}\boldsymbol{x} = \boldsymbol{0}$ is trivial: $\boldsymbol{x} = \boldsymbol{0} \in \mathbb{R}^n$.

Assume that the space $\mathbb{R}^n$ consists of the elements:

$$\boldsymbol{z} = \begin{bmatrix} z_1 \\ \vdots \\ z_n \end{bmatrix}, \tag{5.5}$$

where $\{z_1, \dots, z_n\}$ are ordered sets of n numbers. When the elements of the space $\mathbb{R}^n$ are given in the form (5.5), we say that they are specified by means of *their components*. From linear algebra we know that a unique matrix of type (5.2) can be associated with any linear operator $\boldsymbol{A}$ in this case, so that for a given $\boldsymbol{x} \in \mathbb{R}^n$ the mapping $\boldsymbol{A}: \mathbb{R}^n \longmapsto \mathbb{R}^n$ will be rendered by the formula $\boldsymbol{y} = \boldsymbol{A}\boldsymbol{x}$, i.e.,

$$\begin{bmatrix} y_1 \\ \vdots \\ y_n \end{bmatrix} = \begin{bmatrix} a_{11} & \dots & a_{1n} \\ \dots & \dots & \dots \\ a_{n1} & \dots & a_{nn} \end{bmatrix} \begin{bmatrix} x_1 \\ \vdots \\ x_n \end{bmatrix}. \tag{5.6}$$

Then equation (5.4) coincides with the matrix equation (5.3) or, equivalently, with the canonical form (5.1). In this perspective, equation (5.4) appears to be no more than a mere operator interpretation of system (5.1).

However, the elements of the space $\mathbb{R}^n$ do not necessarily have to be specified as ordered sets of numbers (5.5). Likewise, the operator $\boldsymbol{A}$ does not necessarily have to be originally defined in the form (5.6) either. Thus, the operator form (5.4) of a linear system may, generally speaking, differ from the canonical form (5.1). In the next section, we provide an example of an applied problem that naturally leads to a high-order system of linear algebraic equations in the operator form. This example will be used throughout the book on multiple occasions.

5.1.3 Finite-Difference Dirichlet Problem for the Poisson Equation

Let D be a domain of square shape on the Cartesian plane: $D = \{(x,y) | 0 < x < 1,\ 0 < y < 1\}$. Assume that a Dirichlet problem for the Poisson equation:

$$\begin{aligned} -\Delta u \equiv -\left[\frac{\partial^2 u}{\partial x^2} + \frac{\partial^2 u}{\partial y^2}\right] &= f(x,y), \qquad (x,y) \in D, \\ u\big|_{\partial D} &= 0, \qquad (x,y) \in \partial D, \end{aligned} \tag{5.7}$$

needs to be solved on the domain D. In formula (5.7), ∂D denotes the boundary of D, and $f = f(x,y)$ is a given right-hand side.

To solve problem (5.7) numerically, we introduce a positive integer M, specify $h = M^{-1}$, and construct a uniform Cartesian grid on the square D:

$$(x_{m_1}, y_{m_2}) = (m_1 h, m_2 h), \quad m_1, m_2 = 0, 1, \dots, M. \tag{5.8}$$

Instead of the continuous solution $u = u(x,y)$ to the original problem (5.7), we will rather be looking for its trace, or projection, $[u]_h$ onto the grid (5.8). To compute the values u_{m_1,m_2} of this projection approximately, we replace the derivatives in the Poisson equation by the second order difference quotients at every interior node of the grid (see Section 9.2.1 for more detail), and thus obtain the following system of difference equations:

$$\begin{aligned} -\Delta^{(h)} u^{(h)}\Big|_{m_1,m_2} = &-\left(\frac{u_{m_1+1,m_2} - 2u_{m_1,m_2} + u_{m_1-1,m_2}}{h^2}\right. \\ &+ \left.\frac{u_{m_1,m_2+1} - 2u_{m_1,m_2} + u_{m_1,m_2-1}}{h^2}\right) = f_{m_1,m_2}, \\ &m_1, m_2 = 1, 2, \dots, M-1. \end{aligned} \tag{5.9}$$

The right-hand side of each equation (5.9) is defined as $f_{m_1,m_2} \stackrel{\text{def}}{=} f(m_1 h, m_2 h)$.

Clearly, the difference equation (5.9) is only valid at the interior nodes of the grid (5.8), see Figure 5.1(a). For every fixed pair of m_1 and m_2 inside the square D, it connects the values of the solution at the five points that are shown in Figure 5.1(b). These points are said to form the stencil of the finite-difference scheme (5.9).

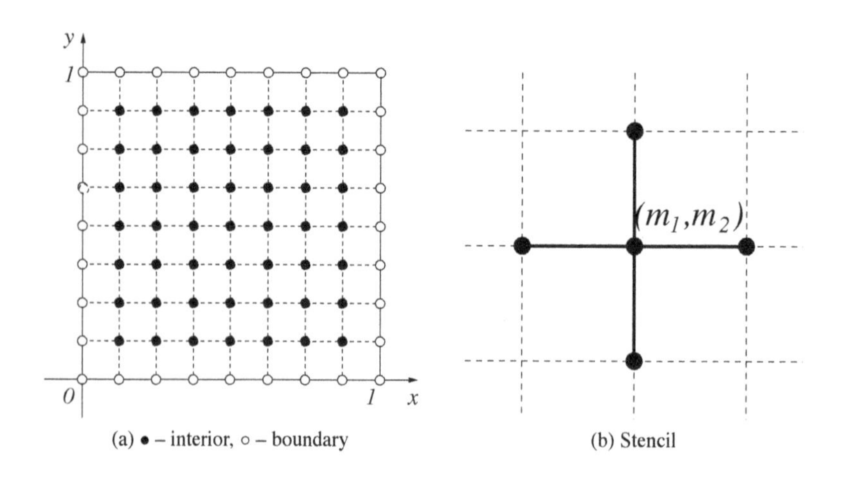

(a) • – interior, ○ – boundary

(b) Stencil

FIGURE 5.1: Finite-difference scheme for the Poisson equation.

At the boundary points of the grid (5.8), see Figure 5.1(a), we obviously cannot write the finite-difference equation (5.9). Instead, according to (5.7), we specify the homogeneous Dirichlet boundary conditions:

$$u_{m_1,m_2} = 0, \quad \text{for} \quad m_1 = 0, M \;\&\; m_2 = 0, M. \tag{5.10}$$

Substituting expressions (5.10) into (5.9), we obtain a system of $(M-1)^2$ linear algebraic equations with respect to as many unknowns: u_{m_1,m_2}, $m_1, m_2 = 1, 2, \ldots, M-1$. This system is a discretization of problem (5.7). Let us note that in general there are many different ways of constructing a discrete counterpart to problem (5.7). Its particular form (5.9), (5.10) introduced in this section has certain merits that we discuss later in Chapters 10 and 12. Here we only mention that in order to solve the Dirichlet problem (5.7) with high accuracy, i.e., in order to make the error between the exact solution $[u]_h$ and the approximate solution $u^{(h)}$ small, we should employ a sufficiently fine grid, or in other words, take a sufficiently large M. In this case, the linear system (5.9), (5.10) will be a system of high dimension.

We will now demonstrate how to convert system (5.9), (5.10) to the operator form (5.4). For that purpose, we introduce a linear space $U^{(h)} = \mathbb{R}^n$, $n \stackrel{\text{def}}{=} (M-1)^2$, that will contain all real-valued functions z_{m_1,m_2} defined on the grid $(m_1 h, m_2 h)$, $m_1, m_2 = 1, 2, \ldots, M-1$. This grid is the interior subset of the grid (5.8), see Figure 5.1(a). Let us also introduce the operator $\boldsymbol{A} \equiv -\Delta^{(h)} : U^{(h)} \longmapsto U^{(h)}$ that will map a given grid function $u^{(h)} \in U^{(h)}$ onto some $v^{(h)} \in U^{(h)}$ according to the following formula:

$$\begin{aligned} v^{(h)}\Big|_{m_1,m_2} \equiv v_{m_1,m_2} = -\Delta^{(h)} u^{(h)}\Big|_{m_1,m_2} = &-\left(\frac{u_{m_1+1,m_2} - 2u_{m_1,m_2} + u_{m_1-1,m_2}}{h^2}\right. \\ &\left. + \frac{u_{m_1,m_2+1} - 2u_{m_1,m_2} + u_{m_1,m_2-1}}{h^2}\right). \end{aligned} \tag{5.11}$$

In doing so, boundary conditions (5.10) are to be employed when using formula (5.11) for $m_1 = 1, M-1$ and $m_2 = 1, M-1$. The system of linear algebraic equations (5.9), (5.10) will then reduce to the operator form (5.4):

$$-\Delta^{(h)} u^{(h)} = f^{(h)}, \tag{5.12}$$

where $f^{(h)} = f_{m_1,m_2} \in U^{(h)} = F^{(h)}$ is a given grid function, and $u^{(h)} \in U^{(h)}$ is the unknown grid function.

REMARK 5.1 System (5.9), (5.10), as well as any other system of n linear algebraic equations with n unknowns, could be written in the canonical form (5.1). However, regardless of how it is done, i.e., how the equations and unknowns are numbered, the resulting system matrix (5.2) would still be inferior in its simplicity and readability to the specification of the system in its original form (5.9), (5.10) or in the equivalent operator form (5.12). □

REMARK 5.2 An abstract operator equation (5.4) can always be reduced to the canonical form (5.1). To do so, one needs to choose a basis $\{\boldsymbol{e}_1, \boldsymbol{e}_2, \ldots, \boldsymbol{e}_n\}$ in the space $\mathbb{R}^n$, expand the elements $\boldsymbol{x}$ and $\boldsymbol{f}$ of $\mathbb{R}^n$ with respect to this basis:

$$\boldsymbol{x} = x_1\boldsymbol{e}_1 + x_2\boldsymbol{e}_2 + \ldots + x_n\boldsymbol{e}_n, \qquad \boldsymbol{f} = f_1\boldsymbol{e}_1 + f_2\boldsymbol{e}_2 + \ldots + f_n\boldsymbol{e}_n,$$

and identify these elements with the sets of their components so that:

$$\boldsymbol{x} = \begin{bmatrix} x_1 \\ \vdots \\ x_n \end{bmatrix}, \qquad \boldsymbol{f} = \begin{bmatrix} f_1 \\ \vdots \\ f_n \end{bmatrix}.$$

Then the operator $\boldsymbol{A}$ of (5.4) will be represented by the matrix (5.2) determined by the operator itself and by the choice of the basis; this operator will act according to formula (5.6). Hence the operator equation (5.4) will reduce to the canonical form (5.3) or (5.1). We should emphasize, however, that if our primary goal is to compute the solution of equation (5.4), then replacing it with the canonical form (5.1), i.e., introducing the system matrix (5.2) explicitly, may, in fact, be an overly cumbersome and as such, ill-advised approach. □

Exercises

1. The mapping $\boldsymbol{A} : \mathbb{R}^2 \longmapsto \mathbb{R}^2$ of the space $\mathbb{R}^2 = \{\boldsymbol{x} \,|\, \boldsymbol{x} = (x_1, x_2)\}$ into itself is defined as follows. Consider a Cauchy problem for the system of two ordinary differential equations:

$$\frac{dz_1}{dt} = z_1, \quad \frac{dz_2}{dt} = z_2, \quad 0 \le t \le 1,$$
$$z_1\Big|_{t=0} = x_1, \quad z_2\Big|_{t=0} = x_2.$$

The element $\boldsymbol{y} = \boldsymbol{A}\boldsymbol{x} \in \mathbb{R}^2$ is defined as $\boldsymbol{y} = (y_1, y_2)$, $y_1 = z_1(t)\Big|_{t=1}$, $y_2 = z_2(t)\Big|_{t=1}$.

   a) Prove that the operator $\boldsymbol{A}$ is linear, i.e., $\boldsymbol{A}(\alpha\boldsymbol{u} + \beta\boldsymbol{w}) = \alpha\boldsymbol{A}\boldsymbol{u} + \beta\boldsymbol{A}\boldsymbol{w}$.

   b) Write down the matrix of the operator $\boldsymbol{A}$ if the basis in the space $\mathbb{R}^2$ is chosen as follows:

$$\boldsymbol{e}_1 = \begin{bmatrix} 1 \\ 0 \end{bmatrix}, \qquad \boldsymbol{e}_2 = \begin{bmatrix} 0 \\ 1 \end{bmatrix}.$$

   c) The same as in Part b) of this exercise, but for the basis:

$$\boldsymbol{e}_1 = \begin{bmatrix} 1 \\ 1 \end{bmatrix}, \qquad \boldsymbol{e}_2 = \begin{bmatrix} 1 \\ -1 \end{bmatrix}.$$

   d) Evaluate $\boldsymbol{x} = \boldsymbol{A}^{-1}\boldsymbol{y}$, where $\boldsymbol{y} = \begin{bmatrix} 2 \\ -1 \end{bmatrix}$.

5.2 Linear Spaces, Norms, and Operators

As we have seen, a solution of a linear system of order n can be assumed to belong to an n-dimensional linear space. The same is true regarding the error between any approximate solution of such a system and its exact solution, which is another key object of our study. In this section, we will therefore provide some background about linear spaces and related concepts. The account of the material below is necessarily brief and sketchy, and we refer the reader to the fundamental treatises on linear algebra, such as [Gan59] or [HJ85], for a more comprehensive treatment.

We recall that the space $\mathbb{L}$ of the elements $\boldsymbol{x},\boldsymbol{y},\boldsymbol{z},\ldots$ is called linear if for any two elements $\boldsymbol{x},\boldsymbol{y}\in\mathbb{L}$ their *sum* $\boldsymbol{x}+\boldsymbol{y}$ is uniquely defined in $\mathbb{L}$, so that the operation of *addition* satisfies the following properties:

1. $\boldsymbol{x}+\boldsymbol{y}=\boldsymbol{y}+\boldsymbol{x}$ (commutativity);
2. $\boldsymbol{x}+(\boldsymbol{y}+\boldsymbol{z})=(\boldsymbol{x}+\boldsymbol{y})+\boldsymbol{z}$ (associativity);
3. There is a special element $\boldsymbol{0}\in\mathbb{L}$, such that $\forall\boldsymbol{x}\in\mathbb{L}:\ \boldsymbol{x}+\boldsymbol{0}=\boldsymbol{x}$ (existence of zero);
4. For any $\boldsymbol{x}\in\mathbb{L}$ there is another element denoted $-\boldsymbol{x}$, such that $\boldsymbol{x}+(-\boldsymbol{x})=\boldsymbol{0}$ (existence of an opposite element, or inverse).

Besides, for any $\boldsymbol{x}\in\mathbb{L}$ and any scalar number α the *product* $\alpha\boldsymbol{x}$ is uniquely defined in $\mathbb{L}$ so that the operation of *multiplication by a scalar* satisfies the following properties:

1. $\alpha(\beta\boldsymbol{x})=(\alpha\beta)\boldsymbol{x}$;
2. $1\cdot\boldsymbol{x}=\boldsymbol{x}$;
3. $(\alpha+\beta)\boldsymbol{x}=\alpha\boldsymbol{x}+\beta\boldsymbol{x}$;
4. $\alpha(\boldsymbol{x}+\boldsymbol{y})=\alpha\boldsymbol{x}+\alpha\boldsymbol{y}$.

Normally, the field of scalars associated with a given linear space $\mathbb{L}$ can be either the field of all real numbers: $\alpha,\beta,\ldots\in\mathbb{R}$, or the field of all complex numbers: $\alpha,\beta,\ldots\in\mathbb{C}$. Accordingly we refer to $\mathbb{L}$ as to a real linear space or a complex linear space, respectively.

A set of elements $z_1,z_2,\ldots,z_n\in\mathbb{L}$ is called *linearly independent* if the equality $\alpha_1z_1+\alpha_2z_2+\ldots+\alpha_nz_n=\boldsymbol{0}$ necessarily implies that $\alpha_1=\alpha_2=\ldots=\alpha_n=0$. Otherwise, the set is called *linearly dependent*. The maximum number of elements in a linearly independent set in $\mathbb{L}$ is called the *dimension* of the space. If the dimension is finite and equal to a positive integer n, then the linear space $\mathbb{L}$ is also referred to as an n-dimensional *vector* space. An example is the space of all finite sequences (vectors) composed of n real numbers called components; this space is denoted by $\mathbb{R}^n$. Similarly, the space of all complex vectors with n components is denoted by $\mathbb{C}^n$. Any linearly independent set of n vectors in an n-dimensional space forms a *basis*.

To quantify the notion of the error for an approximate solution of a linear system, say in the space $\mathbb{R}^n$ or in the space $\mathbb{C}^n$, we need to be able to measure the length of a vector. In linear algebra, one normally introduces the *norm* of a vector for that purpose. Hence let us recall the definition of a normed linear space, as well as that of the norm of a linear operator acting in this space.

5.2.1 Normed Spaces

A linear space $\mathbb{L}$ is called normed if a non-negative number $\|\boldsymbol{x}\|$ is associated with every element $\boldsymbol{x} \in \mathbb{L}$, so that the following conditions (axioms) hold:

1. $\|\boldsymbol{x}\| > 0$ if $\boldsymbol{x} \neq \boldsymbol{0}$ and $\|\boldsymbol{x}\| = 0$ if $\boldsymbol{x} = \boldsymbol{0}$;
2. For any $\boldsymbol{x} \in \mathbb{L}$ and for any scalar λ (real or complex):
$$\|\lambda \boldsymbol{x}\| = |\lambda| \|\boldsymbol{x}\|;$$
3. For any $\boldsymbol{x} \in \mathbb{L}$ and $\boldsymbol{y} \in \mathbb{L}$ the triangle inequality holds:
$$\|\boldsymbol{x} + \boldsymbol{y}\| \leq \|\boldsymbol{x}\| + \|\boldsymbol{y}\|.$$

Let us consider several examples. The spaces $\mathbb{R}^n$ and $\mathbb{C}^n$ are composed of the elements $\boldsymbol{x} = \begin{bmatrix} x_1 \\ \cdots \\ x_n \end{bmatrix}$, where x_j, $j = 1, 2, \ldots, n$, are real or complex numbers, respectively. It is possible to show that the functions $\|\boldsymbol{x}\|_\infty$ and $\|\boldsymbol{x}\|_1$ defined by the equalities:

$$\|\boldsymbol{x}\|_\infty = \max_j |x_j| \tag{5.13}$$

and

$$\|\boldsymbol{x}\|_1 = \sum_{j=1}^{n} |x_j| \tag{5.14}$$

satisfy the foregoing three axioms of the norm. They are called the maximum norm and the l_1 norm, respectively. Alternatively, the maximum norm (5.13) is also referred to as the l_∞ norm or the C norm, and the l_1 norm (5.14) is sometimes called the first norm or the Manhattan norm.

Another commonly used norm is based on the notion of a scalar product or, equivalently, inner product. A scalar product $(\boldsymbol{x}, \boldsymbol{y})$ on the linear space $\mathbb{L}$ is a scalar function of a pair of vector arguments $\boldsymbol{x} \in \mathbb{L}$ and $\boldsymbol{y} \in \mathbb{L}$. This function is supposed to satisfy the following four axioms:[3]

1. $(\boldsymbol{x}, \boldsymbol{y}) = \overline{(\boldsymbol{y}, \boldsymbol{x})}$;
2. For any scalar λ: $(\lambda \boldsymbol{x}, \boldsymbol{y}) = \lambda (\boldsymbol{x}, \boldsymbol{y})$;
3. $(\boldsymbol{x} + \boldsymbol{y}, \boldsymbol{z}) = (\boldsymbol{x}, \boldsymbol{z}) + (\boldsymbol{y}, \boldsymbol{z})$;
4. $(\boldsymbol{x}, \boldsymbol{x}) \geq 0$, and $(\boldsymbol{x}, \boldsymbol{x}) > 0$ if $\boldsymbol{x} \neq 0$.

[3] The overbar in axiom 1 and later in the text denotes complex conjugation.

Note that in the foregoing definition of the scalar product the linear space $\mathbb{L}$ is, generally speaking, assumed complex, i.e., the scalars λ may be complex, and the inner product $(\boldsymbol{x},\boldsymbol{y})$ itself may be complex as well. The only exception is the inner product of a vector with itself, which is always real (axiom 4). If, however, we were to assume ahead of time that the space $\mathbb{L}$ was real, then axioms 2, 3, and 4 would remain unchanged, while axiom 1 will get simplified: $(\boldsymbol{x},\boldsymbol{y}) = (\boldsymbol{y},\boldsymbol{x})$. In other words, the scalar product on a real linear space is commutative.

The simplest example of a scalar product on the space $\mathbb{R}^n$ is given by

$$(\boldsymbol{x},\boldsymbol{y}) = x_1y_1 + x_2y_2 + \ldots + x_ny_n, \tag{5.15}$$

whereas for the complex space $\mathbb{C}^n$ we can choose:

$$(\boldsymbol{x},\boldsymbol{y}) = x_1\bar{y}_1 + x_2\bar{y}_2 + \ldots + x_n\bar{y}_n. \tag{5.16}$$

Recall that a real linear space equipped with an inner product is called *Euclidean,* whereas a complex space with an inner product is called *unitary.*

One can show that the function:

$$\|\boldsymbol{x}\|_2 = (\boldsymbol{x},\boldsymbol{x})^{\frac{1}{2}} \tag{5.17}$$

provides a norm on both $\mathbb{R}^n$ and $\mathbb{C}^n$. In the case of a real space this norm is called *Euclidean*, and in the case of a complex space it is called *Hermitian.* In the literature, the norm defined by formula (5.17) is also referred to as the l_2 norm.

In the previous examples, we have employed a standard vector form $\boldsymbol{x} = [x_1,x_2,\ldots,x_n]^T$ for the elements of the space $\mathbb{L}$ ($\mathbb{L} = \mathbb{R}^n$ or $\mathbb{L} = \mathbb{C}^n$). In much the same way, norms can be introduced on linear spaces without having to enumerate consecutively all the components of every vector. Consider, for example, the space $U^{(h)} = \mathbb{R}^n$, $n = (M-1)^2$, of the grid functions $u^{(h)} = \{u_{m_1,m_2}\}$, $m_1,m_2 = 1,2,\ldots,M-1$, that we have introduced and exploited in Section 5.1.3 for the construction of a finite-difference scheme for the Poisson equation. The maximum norm and the l_1 norm can be introduced on this space as follows:

$$\|u^{(h)}\|_\infty = \max_{m_1,m_2} |u_{m_1,m_2}|, \tag{5.18}$$

$$\|u^{(h)}\|_1 = \sum_{m_1,m_2=1}^{M-1} |u_{m_1,m_2}|. \tag{5.19}$$

Moreover, a scalar product $(u^{(h)},v^{(h)})$ can be introduced in the real linear space $U^{(h)}$ according to the formula [cf. formula (5.15)]:

$$(u^{(h)},v^{(h)}) = h^2 \sum_{m_1,m_2=1}^{M-1} u_{m_1,m_2}v_{m_1,m_2}. \tag{5.20}$$

Then the corresponding Euclidean norm is defined as follows:

$$\|u^{(h)}\|_2 = (u^{(h)},u^{(h)})^{\frac{1}{2}} = \left[h^2 \sum_{m_1,m_2=1}^{M-1} |u_{m_1,m_2}|^2\right]^{\frac{1}{2}}. \tag{5.21}$$

In general, there are many different ways of defining a scalar product on a real or complex vector space $\mathbb{L}$, besides using the simplest expressions (5.15) and (5.16). Any appropriate scalar product $(\boldsymbol{x},\boldsymbol{y})$ on $\mathbb{L}$ generates the corresponding Euclidean or Hermitian norm according to formula (5.17). It turns out that one can, in fact, describe all scalar products that satisfy axioms 1–4 on a given space and consequently, all Euclidean (Hermitian) norms. For that purpose, one first needs to fix a particular scalar product $(\boldsymbol{x},\boldsymbol{y})$ on $\mathbb{L}$, and then consider the self-adjoint linear operators with respect to this product.

Recall that the operator $\boldsymbol{B}^*: \mathbb{L} \longmapsto \mathbb{L}$ is called *adjoint* to a given linear operator $\boldsymbol{B}: \mathbb{L} \longmapsto \mathbb{L}$ if the following identity holds for any $\boldsymbol{x} \in \mathbb{L}$ and $\boldsymbol{y} \in \mathbb{L}$:

$$(\boldsymbol{Bx},\boldsymbol{y}) = (\boldsymbol{x},\boldsymbol{B}^*\boldsymbol{y}).$$

It is known that given a particular inner product on $\mathbb{L}$, one and only one adjoint operator $\boldsymbol{B}^*$ exists for every $\boldsymbol{B}: \mathbb{L} \longmapsto \mathbb{L}$. If a particular basis is selected in $\mathbb{L}$ that enables the representation of the operators by matrices (see Remark 5.2), then for a real space and the inner product (5.15), the adjoint operator is given by the transposed matrix $\boldsymbol{B}^T$ that has the entries $\mathfrak{b}_{ij} = b_{ji}$, $i,j = 1,2,\dots,n$, where b_{ij} are the entries of $\boldsymbol{B}$. For a complex space $\mathbb{L}$ and the inner product (5.16), the adjoint operator is represented by the conjugate matrix $\boldsymbol{B}^*$ with the entries $\mathfrak{b}_{ij} = \bar{b}_{ji}$, $i,j = 1,2,\dots,n$.

The operator $\boldsymbol{B}$ is called *self-adjoint* if $\boldsymbol{B} = \boldsymbol{B}^*$. Again, in the case of a real space with the scalar product (5.15) the matrix of a self-adjoint operator is symmetric: $b_{ij} = b_{ji}$, $i,j = 1,2,\dots,n$; whereas in the case of a complex space with the scalar product (5.16) the matrix of a self-adjoint operator is Hermitian: $b_{ij} = \bar{b}_{ji}$, $i,j = 1,2,\dots,n$.

The operator $\boldsymbol{B}: \mathbb{L} \longmapsto \mathbb{L}$ is called *positive definite* and denoted $\boldsymbol{B} > 0$ if $(\boldsymbol{Bx},\boldsymbol{x}) > 0$ for every $\boldsymbol{x} \in \mathbb{L}$, $\boldsymbol{x} \neq \boldsymbol{0}$. It is known that if $\boldsymbol{B} = \boldsymbol{B}^* > 0$, then the expression

$$[\boldsymbol{x},\boldsymbol{y}]_{\boldsymbol{B}} \equiv (\boldsymbol{Bx},\boldsymbol{y}) \tag{5.22}$$

satisfies all four axioms of the scalar product. Moreover, it turns out that any legitimate scalar product on $\mathbb{L}$ can be obtained using formula (5.22) with an appropriate choice of a self-adjoint positive definite operator $\boldsymbol{B} = \boldsymbol{B}^* > 0$. Consequently, any operator of this type ($\boldsymbol{B} = \boldsymbol{B}^* > 0$) generates an Euclidean (Hermitian) norm by the formula:

$$\|\boldsymbol{x}\|_{\boldsymbol{B}} = ([\boldsymbol{x},\boldsymbol{x}]_{\boldsymbol{B}})^{\frac{1}{2}}. \tag{5.23}$$

In other words, there is a one-to-one correspondence between Euclidean norms and symmetric positive definite matrices (SPD) in the real case, and Hermitian norms and Hermitian positive definite matrices in the complex case. In particular, the Euclidean norm (5.17) generated by the inner product (5.15) can be represented in the form (5.23) if we choose the identity operator for $\boldsymbol{B}$: $\boldsymbol{B} = \boldsymbol{I}$:

$$\|\boldsymbol{x}\|_2 = \|\boldsymbol{x}\|_{\boldsymbol{I}} = ([\boldsymbol{x},\boldsymbol{x}]_{\boldsymbol{I}})^{\frac{1}{2}} = (\boldsymbol{x},\boldsymbol{x})^{\frac{1}{2}}.$$

5.2.2 Norm of a Linear Operator

The operator $\boldsymbol{A} : \mathbb{L} \longmapsto \mathbb{L}$ is called *linear* if for any pair $\boldsymbol{x}, \boldsymbol{y} \in \mathbb{L}$ and any two scalars α and β it satisfies: $\boldsymbol{A}(\alpha \boldsymbol{x} + \beta \boldsymbol{y}) = \alpha \boldsymbol{A}\boldsymbol{x} + \beta \boldsymbol{A}\boldsymbol{y}$.

We introduce the norm $\|\boldsymbol{A}\|$ of a linear operator $\boldsymbol{A} : \mathbb{L} \longmapsto \mathbb{L}$ as follows:

$$\|\boldsymbol{A}\| \stackrel{\text{def}}{=} \max_{\boldsymbol{x} \in \mathbb{L},\, \boldsymbol{x} \neq \boldsymbol{0}} \frac{\|\boldsymbol{A}\boldsymbol{x}\|}{\|\boldsymbol{x}\|}, \tag{5.24}$$

where $\|\boldsymbol{A}\boldsymbol{x}\|$ and $\|\boldsymbol{x}\|$ are norms of the elements $\boldsymbol{A}\boldsymbol{x}$ and $\boldsymbol{x}$, respectively, from the linear normed space $\mathbb{L}$. The norm $\|\boldsymbol{A}\|$ given by (5.24) is said to be *induced* by the vector norm chosen in the space $\mathbb{L}$. According to formula (5.24), the norm $\|\boldsymbol{A}\|$ is the coefficient of stretching of that particular vector $\boldsymbol{x}' \in \mathbb{L}$ which is stretched at least as much as any other vector $\boldsymbol{x} \in \mathbb{L}$ under the transformation $\boldsymbol{A}$.

As before, a square matrix

$$\boldsymbol{A} = \begin{bmatrix} a_{11} & \dots & a_{1n} \\ \dots & \dots & \dots \\ a_{n1} & \dots & a_{nn} \end{bmatrix}$$

will be identified with an operator $\boldsymbol{A} : \mathbb{L} \longmapsto \mathbb{L}$ that acts in the space $\mathbb{L} = \mathbb{R}^n$ or $\mathbb{L} = \mathbb{C}^n$. Recall, either space consists of the elements (vectors) $\boldsymbol{x} = \begin{bmatrix} x_1 \\ \dots \\ x_n \end{bmatrix}$, where x_j, $j = 1, 2, \dots, n$, are real or complex numbers, respectively. The operator $\boldsymbol{A}$ maps a given $\boldsymbol{x} \in \mathbb{L}$ onto $\boldsymbol{y} = \begin{bmatrix} y_1 \\ \dots \\ y_n \end{bmatrix}$ according to the formula

$$y_i = \sum_{j=1}^{n} a_{ij} x_j, \qquad i = 1, 2, \dots, n, \tag{5.25}$$

where a_{ij} are entries of the matrix $\boldsymbol{A}$. Therefore, the norm of the matrix $\boldsymbol{A}$ can naturally be introduced according to (5.24) as the norm of the linear operator defined by formula (5.25).

THEOREM 5.1

The norms of the matrix $\boldsymbol{A} = \{a_{ij}\}$ induced by the vector norms $\|\boldsymbol{x}\|_\infty = \max_j |x_j|$ of (5.13) and $\|\boldsymbol{x}\|_1 = \sum_j |x_j|$ of (5.14), respectively, in the space $\mathbb{L} = \mathbb{R}^n$ or $\mathbb{L} = \mathbb{C}^n$, are given by:

$$\|\boldsymbol{A}\|_\infty = \max_i \sum_j |a_{ij}| \tag{5.26}$$

and

$$\|\boldsymbol{A}\|_1 = \max_j \sum_i |a_{ij}|. \tag{5.27}$$

Accordingly, the norm $\|A\|_\infty$ of (5.26) is known as the maximum row norm, and the norm $\|A\|_1$ of (5.27) is known as the maximum column norm of the matrix A. The proof of Theorem 5.1 is fairly easy, and we leave it to the reader as an exercise.

THEOREM 5.2

Let (x,y) denote a scalar product in an n-dimensional Euclidean space $\mathbb{L}$, and let $\|x\| = (x,x)^{1/2}$ be the corresponding Euclidean norm. Let A be a self-adjoint operator on $\mathbb{L}$ with respect to the chosen inner product: $\forall x,y \in \mathbb{L}:\ (Ax,y) = (x,Ay)$. Then for the induced operator norm [formula (5.24)] we have:

$$\|A\| = \max_{x\in\mathbb{R}^n,\ x\neq 0} \frac{\|Ax\|}{\|x\|} = \max_j |\lambda_j|, \tag{5.28}$$

where λ_j, $j = 1,2,\ldots,n$ are eigenvalues of the operator (matrix) A.

The set of all eigenvalues $\{\lambda_j\}$ of the matrix (operator) A is known as its spectrum. The quantity $\max_j |\lambda_j|$ on the right-hand side of formula (5.28) is called the spectral radius of the matrix (operator) A and is denoted $\rho(A)$.

PROOF From linear algebra it is known that when $A = A^*$, then an orthonormal basis exists in the space $\mathbb{L}$:

$$e_1,\ e_2,\ \ldots,\ e_n \tag{5.29}$$

that consists of the eigenvectors of the operator A:

$$Ae_j = \lambda_j e_j, \quad j = 1,2,\ldots,n.$$

Take an arbitrary $x \in \mathbb{L}$ and expand it with respect to the basis (5.29):

$$x = x_1 e_1 + x_2 e_2 + \ldots + x_n e_n.$$

Then, for the vector $Ax \in \mathbb{L}$ we can write:

$$Ax = \lambda_1 x_1 e_1 + \lambda_2 x_2 e_2 + \ldots + \lambda_n x_n e_n.$$

As the basis (5.29) is orthonormal: $(e_i,e_j) = \begin{cases} 1, & i = j, \\ 0, & i \neq j, \end{cases}$ we use the linearity of the scalar product (axioms 2 and 3) and obtain:

$$\|x\| = [x_1^2 + x_2^2 + \ldots + x_n^2]^{\frac{1}{2}},$$
$$\|Ax\| = [(\lambda_1 x_1)^2 + (\lambda_2 x_2)^2 + \ldots + (\lambda_n x_n)^2]^{\frac{1}{2}}.$$

Consequently,

$$\|Ax\| \leq \max_j |\lambda_j| [x_1^2 + x_2^2 + \ldots + x_n^2]^{\frac{1}{2}} = \max_j |\lambda_j| \|x\|.$$

Therefore, for any $\boldsymbol{x} \in \mathbb{L}$, $\boldsymbol{x} \neq \boldsymbol{0}$, the following inequality holds:

$$\frac{\|\boldsymbol{A}\boldsymbol{x}\|}{\|\boldsymbol{x}\|} \leq \max_j |\lambda_j|,$$

while for $\boldsymbol{x} = \boldsymbol{e}_k$, where $\boldsymbol{e}_k$ is the eigenvector of $\boldsymbol{A}$ that corresponds to the maximum eigenvalue: $\max_j |\lambda_j| = |\lambda_k|$, the previous inequality transforms into a precise equality:

$$\frac{\|\boldsymbol{A}\boldsymbol{e}_k\|}{\|\boldsymbol{e}_k\|} = |\lambda_k| = \max_j |\lambda_j|.$$

As such, formula (5.28) is indeed true. ∎

Note that if, for example, we were to take $\mathbb{L} = \mathbb{R}^n$ in Theorem 5.2, then the scalar product $(\boldsymbol{x}, \boldsymbol{y})$ does not necessarily have to be the simplest product (5.15). It can, in fact, be any product of type (5.22) on $\mathbb{L}$, with the corresponding Euclidean norm given by formula (5.23). Also note that the argument given when proving Theorem 5.2 easily extends to a unitary space $\mathbb{L}$, for example, $\mathbb{L} = \mathbb{C}^n$. Finally, the result of Theorem 5.2 can be generalized to the case of the so-called normal matrices $\boldsymbol{A}$ that are defined as $\boldsymbol{A}\boldsymbol{A}^* = \boldsymbol{A}^*\boldsymbol{A}$. The matrix is normal if and only if there is an orthonormal basis composed of its eigenvectors, see [HJ85, Chapter 2].

Exercises

1. Show that $\|\boldsymbol{x}\|_\infty$ of (5.13) and $\|\boldsymbol{x}\|_1$ of (5.14) satisfy all axioms of the norm.

2. Consider the space $\mathbb{R}^2$ that consists of all vectors $\boldsymbol{x} = \begin{bmatrix} x_1 \\ x_2 \end{bmatrix}$ with two real components. Represent the elements of this space by the points (x_1, x_2) on the two dimensional Cartesian plane. On this plane, draw the curve specified by the equation $\|\boldsymbol{x}\| = 1$, i.e., the unit circle, if the norm is defined as:

$$\|\boldsymbol{x}\| = \|\boldsymbol{x}\|_\infty = \max\{|x_1|, |x_2|\},$$
$$\|\boldsymbol{x}\| = \|\boldsymbol{x}\|_1 = |x_1| + |x_2|,$$
$$\|\boldsymbol{x}\| = \|\boldsymbol{x}\|_2 = (x_1^2 + x_2^2)^{1/2}.$$

3. Let $\boldsymbol{A} : \mathbb{L} \longmapsto \mathbb{L}$ be a linear operator. Prove that for any choice of the vector norm on $\mathbb{L}$, the corresponding induced operator norm $\|\boldsymbol{A}\|$ satisfies the inequality:

$$\|\boldsymbol{A}\| \geq \rho(\boldsymbol{A}),$$

where $\rho(\boldsymbol{A}) = \max_j |\lambda_j|$ is the spectral radius of the matrix $\boldsymbol{A}$.

4. Construct an example of a linear operator $\boldsymbol{A} : \mathbb{R}^2 \longmapsto \mathbb{R}^2$ that is represented by a 2×2 matrix with the eigenvalues $\lambda_1 = \lambda_2 = 1$ and such that $\|\boldsymbol{A}\|_\infty > 1000$, $\|\boldsymbol{A}\|_1 > 1000$, and $\|\boldsymbol{A}\|_2 > 1000$.

5. Let $\boldsymbol{L}$ be a vector space, and $\boldsymbol{A}$ and $\boldsymbol{B}$ be two linear operators acting on this space. Prove that $\|\boldsymbol{A}\boldsymbol{B}\| \leq \|\boldsymbol{A}\|\|\boldsymbol{B}\|$.

6. Prove Theorem 5.1.

7. For the scalar product $(\boldsymbol{x},\boldsymbol{y})$ introduced on the linear space $\mathbb{L}$, prove the Cauchy-Schwarz inequality:
$$\forall \boldsymbol{x},\boldsymbol{y} \in \mathbb{L}: \ |(\boldsymbol{x},\boldsymbol{y})|^2 \leq (\boldsymbol{x},\boldsymbol{x})(\boldsymbol{y},\boldsymbol{y}).$$
Consider two cases:
   a) The field of scalars for $\mathbb{L}$ is $\mathbb{R}$;
   b)* The field of scalars for $\mathbb{L}$ is $\mathbb{C}$.
8. Prove that the Euclidean (Hermitian) norm $\|\boldsymbol{x}\| = (\boldsymbol{x},\boldsymbol{x})^{1/2}$, where $(\boldsymbol{x},\boldsymbol{y})$ is an appropriate scalar product on the given linear space, is indeed a norm, i.e., that it satisfies all three axioms of the norm from page 126.
9. Prove Theorem 5.2 for the case of a unitary space $\mathbb{L}$.
10. Let $\mathbb{L}$ be a Euclidean space, $\boldsymbol{A}: \mathbb{L} \longmapsto \mathbb{L}$ be an arbitrary linear operator, and $\boldsymbol{B}: \mathbb{L} \longmapsto \mathbb{L}$ be an orthogonal linear operator, i.e., such that $\forall \boldsymbol{x} \in \mathbb{L}: \ (\boldsymbol{B}\boldsymbol{x},\boldsymbol{B}\boldsymbol{x}) = (\boldsymbol{x},\boldsymbol{x})$ or equivalently, $\boldsymbol{B}^T\boldsymbol{B} = \boldsymbol{I}$. Prove that $\|\boldsymbol{A}\boldsymbol{B}\|_2 = \|\boldsymbol{B}\boldsymbol{A}\|_2 = \|\boldsymbol{A}\|_2$.
11. Let $\boldsymbol{A}$ be a linear operator on the Euclidean (unitary) space $\mathbb{L}$, $\boldsymbol{A}: \mathbb{L} \longmapsto \mathbb{L}$, and $\boldsymbol{U}$ and $\boldsymbol{W}$ be orthogonal (unitary) operators on this space, i.e., $\boldsymbol{U}\boldsymbol{U}^* = \boldsymbol{I}$, $\boldsymbol{W}\boldsymbol{W}^* = \boldsymbol{I}$. Show that $\|\boldsymbol{U}\boldsymbol{A}\boldsymbol{W}\|_2 = \|\boldsymbol{A}\|_2$.
12. Let $\mathbb{L}$ be an Euclidean space and $\boldsymbol{A} = \boldsymbol{A}^*$ be a self-adjoint operator: $\boldsymbol{A}: \mathbb{L} \longmapsto \mathbb{L}$. Prove that $\|\boldsymbol{A}^2\|_2 = \|\boldsymbol{A}\|_2^2$. Construct an example showing that if $\boldsymbol{A} \neq \boldsymbol{A}^*$, then it can happen that $\|\boldsymbol{A}^2\|_2 < \|\boldsymbol{A}\|_2^2$.
13.* Let $\mathbb{L}$ be a Euclidean (unitary) space, and $\boldsymbol{A}: \mathbb{L} \longmapsto \mathbb{L}$ be an arbitrary linear operator.
   a) Prove that $\boldsymbol{A}^*\boldsymbol{A}$ is a self-adjoint non-negative definite operator on $\mathbb{L}$.
   b) Prove that
$$\|\boldsymbol{A}\|^2 = \|\boldsymbol{A}^*\boldsymbol{A}\| = \lambda_{\max}(\boldsymbol{A}^*\boldsymbol{A}),$$
   where $\lambda_{\max}(\boldsymbol{A}^*\boldsymbol{A})$ is the maximum eigenvalue of $\boldsymbol{A}^*\boldsymbol{A}$.
14. Let $\mathbb{R}^2$ be a two-dimensional space of real vectors $\boldsymbol{x} = \begin{bmatrix} x_1 \\ x_2 \end{bmatrix}$ with the inner product (5.15): $(\boldsymbol{x},\boldsymbol{y}) = x_1y_1 + x_2y_2$, and the corresponding l_2 norm (5.17): $\|\boldsymbol{x}\|_2 = (\boldsymbol{x},\boldsymbol{x})^{1/2}$. Evaluate the induced norm $\|\boldsymbol{A}\|_2$ for the following two matrices:
$$\boldsymbol{A} = \begin{bmatrix} 1 & \sqrt{6} \\ \sqrt{6} & 1 \end{bmatrix} \quad \text{and} \quad \boldsymbol{A} = \begin{bmatrix} 1 & 1 \\ 5 & 2 \end{bmatrix}.$$
Hint. Use the result of the previous Exercise.
15. Let $\mathbb{R}^2$ be a two-dimensional space of real vectors $\boldsymbol{x} = \begin{bmatrix} x_1 \\ x_2 \end{bmatrix}$ with the inner product (5.15): $(\boldsymbol{x},\boldsymbol{y}) = x_1y_1 + x_2y_2$, and let $\boldsymbol{B} = \begin{bmatrix} 1 & 1 \\ 1 & 2 \end{bmatrix}$. Verify that $\boldsymbol{B} = \boldsymbol{B}^* > 0$. Evaluate the norms $\|\boldsymbol{A}\|_{\boldsymbol{B}}$ of the two matrices from Exercise 14 induced by the vector norm (5.23): $\|\boldsymbol{x}\|_{\boldsymbol{B}} = ([\boldsymbol{x},\boldsymbol{x}]_{\boldsymbol{B}})^{1/2} = (\boldsymbol{B}\boldsymbol{x},\boldsymbol{x})^{1/2}$.
16. Let $\mathbb{L}$ be a Euclidean (unitary) space, and $\|\boldsymbol{x}\| = (\boldsymbol{x},\boldsymbol{x})^{1/2}$ be the corresponding vector norm. Let $\boldsymbol{A}: \mathbb{L} \longmapsto \mathbb{L}$ be an arbitrary linear operator on $\mathbb{L}$. Prove that $\|\boldsymbol{A}\| = \|\boldsymbol{A}^*\|$. If, additionally, $\boldsymbol{A}$ is a non-singular operator, i.e., if the inverse operator (matrix) $\boldsymbol{A}^{-1}$ exists, then prove $(\boldsymbol{A}^{-1})^* = (\boldsymbol{A}^*)^{-1}$.

5.3 Conditioning of Linear Systems

Two linear systems that look quite similar at the first glance, may, in fact, have a very different degree of sensitivity of their solutions to the errors committed when specifying the input data. This phenomenon can be observed already for the systems $\boldsymbol{Ax} = \boldsymbol{f}$ of order two:

$$\begin{aligned} a_{11}x_1 + a_{12}x_2 &= f_1, \\ a_{21}x_1 + a_{22}x_2 &= f_2. \end{aligned} \tag{5.30}$$

With no loss of generality we will assume that the coefficients of system (5.30) are normalized: $a_{i1}^2 + a_{i2}^2 = 1, i = 1, 2$. Geometrically, each individual equation of system (5.30) defines a straight line on the Cartesian plane (x_1, x_2). Accordingly, the solution of system (5.30) can be interpreted as the intersection point of these two lines as shown in Figure 5.2.

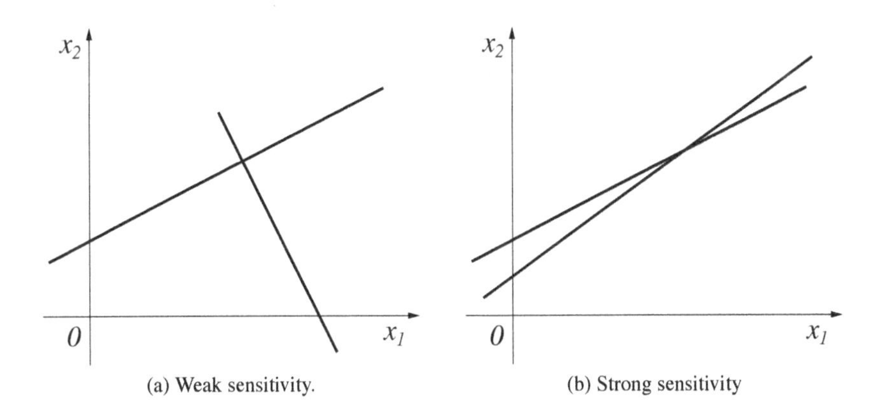

(a) Weak sensitivity.　　(b) Strong sensitivity

FIGURE 5.2: Sensitivity of the solution of system (5.30) to perturbations of the data.

Let us qualitatively analyze two antipodal situations. The straight lines that correspond to linear equations (5.30) can intersect "almost normally," as shown in Figure 5.2(a), or they can intersect "almost tangentially," as shown in Figure 5.2(b). If we slightly perturb the input data, i.e., the right-hand sides f_i, $i = 1, 2$, and/or the coefficients a_{ij}, $i, j = 1, 2$, then each line may move parallel to itself and/or tilt. In doing so, it is clear that the intersection point (i.e., the solution) on Figure 5.2(a) will only move slightly, whereas the intersection point on Figure 5.2(b) will move much more visibly. Accordingly, one can say that in the case of Figure 5.2(a) the sensitivity of the solution to perturbations of the input data is weak, whereas in the case of Figure 5.2(b) it is strong.

Quantitatively, the sensitivity of the solution to perturbations of the input data can be characterized with the help of the so-called *condition number* $\mu(\boldsymbol{A})$. We will later see that not only does the condition number determine the aforementioned sensitivity, but also that it directly influences the performance of iterative methods for solving $\boldsymbol{A}\boldsymbol{x} = \boldsymbol{f}$. Namely, it affects the number of iterations and as such, the number of arithmetic operations, required for finding an approximate solution of $\boldsymbol{A}\boldsymbol{x} = \boldsymbol{f}$ within a prescribed tolerance, see Chapter 6.

5.3.1 Condition Number

The condition number of a linear operator $\boldsymbol{A}$ acting on a normed vector space $\mathbb{L}$ is defined as follows:

$$\mu(\boldsymbol{A}) = \|\boldsymbol{A}\| \cdot \|\boldsymbol{A}^{-1}\|. \tag{5.31}$$

The same quantity $\mu(\boldsymbol{A})$ given by formula (5.31) is also referred to as the condition number of a linear system $\boldsymbol{A}\boldsymbol{x} = \boldsymbol{f}$.

Recall that we have previously identified every matrix:

$$\boldsymbol{A} = \begin{bmatrix} a_{11} & \dots & a_{1n} \\ \dots & \dots & \dots \\ a_{n1} & \dots & a_{nn} \end{bmatrix}$$

with a linear operator acting on an n-dimensional vector space $\mathbb{L}$ of the elements $\boldsymbol{x} = \begin{bmatrix} x_1 \\ \dots \\ x_n \end{bmatrix}$ (for example, $\mathbb{L} = \mathbb{R}^n$ or $\mathbb{L} = \mathbb{C}^n$). For a given $\boldsymbol{x} \in \mathbb{L}$, the operator yields $\boldsymbol{y} = \boldsymbol{A}\boldsymbol{x}$, where $\boldsymbol{y} = \begin{bmatrix} y_1 \\ \dots \\ y_n \end{bmatrix}$ is computed as follows:

$$y_i = \sum_{j=1}^{n} a_{ij} x_j, \quad i = 1, 2, \dots, n.$$

Accordingly, the definition of the condition number $\mu(\boldsymbol{A})$ by formula (5.31) also makes sense for matrices, so that one can refer to the condition number of a matrix $\boldsymbol{A}$, as well as to the condition number of a system of linear algebraic equations specified in its canonical form (5.1) rather than only in the operator form.

The norms $\|\boldsymbol{A}\|$ and $\|\boldsymbol{A}^{-1}\|$ of the direct and inverse operators, respectively, in formula (5.31) are assumed induced by the vector norm chosen in $\mathbb{L}$. Consequently, the condition number $\mu(\boldsymbol{A})$ also depends on on the choice of the norm in $\mathbb{L}$. If $\boldsymbol{A}$ is a matrix, and we use the maximum norm for the vectors from $\mathbb{L}$, $\|\boldsymbol{x}\|_\infty = \max_j |x_j|$, then we write $\mu = \mu_\infty(\boldsymbol{A})$; if we employ the first norm $\|\boldsymbol{x}\|_1 = \sum_j |x_j|$, then $\mu = \mu_1(\boldsymbol{A})$.

If $\mathbb{L}$ is a Euclidean (unitary) space with the scalar product $(\boldsymbol{x}, \boldsymbol{y})$ given, for example, by formula (5.15) or (5.16), and the corresponding Euclidean (Hermitian) norm is defined by formula (5.17): $\|\boldsymbol{x}\|_2 = (\boldsymbol{x}, \boldsymbol{x})^{1/2}$, then $\mu = \mu_2(\boldsymbol{A})$. If an alternative scalar product $[\boldsymbol{x}, \boldsymbol{y}]_{\boldsymbol{B}} = (\boldsymbol{B}\boldsymbol{x}, \boldsymbol{y})$ is introduced in $\mathbb{L}$ based on the original product $(\boldsymbol{x}, \boldsymbol{y})$ and on the operator $\boldsymbol{B} = \boldsymbol{B}^* > 0$, see formula (5.22), and if a new norm is set up accordingly by formula (5.23): $\|\boldsymbol{x}\|_{\boldsymbol{B}} = ([\boldsymbol{x}, \boldsymbol{y}]_{\boldsymbol{B}})^{1/2}$, then the corresponding condition number is denoted by $\mu = \mu_{\boldsymbol{B}}(\boldsymbol{A})$.

Let us now explain the geometric meaning of the condition number $\mu(\boldsymbol{A})$. To do so, consider the set $S \subset \mathbb{L}$ of all vectors with the norm equal to one, i.e., a unit sphere in the space $\mathbb{L}$. Among these vectors choose a particular two, $\boldsymbol{x}_{\max}$ and $\boldsymbol{x}_{\min}$, that satisfy the following equalities:

$$\|\boldsymbol{A}\boldsymbol{x}_{\max}\| = \max_{\boldsymbol{x}\in S} \|\boldsymbol{A}\boldsymbol{x}\| \quad \text{and} \quad \|\boldsymbol{A}\boldsymbol{x}_{\min}\| = \min_{\boldsymbol{x}\in S} \|\boldsymbol{A}\boldsymbol{x}\|.$$

It is easy to see that

$$\|\boldsymbol{A}\| = \|\boldsymbol{A}\boldsymbol{x}_{\max}\| \quad \text{and} \quad \|\boldsymbol{A}^{-1}\| = \frac{1}{\|\boldsymbol{A}\boldsymbol{x}_{\min}\|}.$$

Indeed, according to formula (5.24) we can write:

$$\|\boldsymbol{A}\| = \max_{\boldsymbol{x}\in\mathbb{L},\, \boldsymbol{x}\neq\boldsymbol{0}} \frac{\|\boldsymbol{A}\boldsymbol{x}\|}{\|\boldsymbol{x}\|} = \max_{\boldsymbol{x}\in\mathbb{L},\, \boldsymbol{x}\neq\boldsymbol{0}} \boldsymbol{A}\left(\frac{\boldsymbol{x}}{\|\boldsymbol{x}\|}\right) = \max_{\tilde{\boldsymbol{x}}\in S} \boldsymbol{A}\tilde{\boldsymbol{x}} = \|\boldsymbol{A}\boldsymbol{x}_{\max}\|.$$

Likewise:

$$\|\boldsymbol{A}^{-1}\| = \max_{\boldsymbol{x}\in\mathbb{L},\, \boldsymbol{x}\neq\boldsymbol{0}} \frac{\|\boldsymbol{A}^{-1}\boldsymbol{x}\|}{\|\boldsymbol{x}\|} = \max_{\tilde{\boldsymbol{x}}\in\mathbb{L},\, \tilde{\boldsymbol{x}}\neq\boldsymbol{0}} \frac{\|\tilde{\boldsymbol{x}}\|}{\|\boldsymbol{A}\tilde{\boldsymbol{x}}\|} = \left[\min_{\tilde{\boldsymbol{x}}\in\mathbb{L},\, \tilde{\boldsymbol{x}}\neq\boldsymbol{0}} \frac{\|\boldsymbol{A}\tilde{\boldsymbol{x}}\|}{\|\tilde{\boldsymbol{x}}\|}\right]^{-1} = \frac{1}{\|\boldsymbol{A}\boldsymbol{x}_{\min}\|}.$$

Substituting these expressions into the definition of $\mu(\boldsymbol{A})$ by means of formula (5.31), we obtain:

$$\mu(\boldsymbol{A}) = \frac{\max_{\boldsymbol{x}\in S} \|\boldsymbol{A}\boldsymbol{x}\|}{\min_{\boldsymbol{x}\in S} \|\boldsymbol{A}\boldsymbol{x}\|}. \tag{5.32}$$

This, in particular, implies that we always have:

$$\mu(\boldsymbol{A}) \geq 1.$$

According to formula (5.32), the condition number $\mu(\boldsymbol{A})$ is the ratio of magnitudes of the maximally stretched unit vector to the minimally stretched (i.e., maximally shrunk) unit vector. If the operator $\boldsymbol{A}$ is singular, i.e., if the inverse operator $\boldsymbol{A}^{-1}$ does not exist, then we formally set $\mu(\boldsymbol{A}) = \infty$.

The geometric meaning of the quantity $\mu(\boldsymbol{A})$ becomes particularly apparent in the case of an Euclidean space $\mathbb{L} = \mathbb{R}^2$ equipped with the norm $\|\boldsymbol{x}\|_2 = (\boldsymbol{x},\boldsymbol{x})^{1/2} = \sqrt{x_1^2 + x_2^2}$. In other words, $\mathbb{L}$ is the Euclidean plane of the variables x_1 and x_2. In this case, S is a conventional unit circle: $x_1^2 + x_2^2 = 1$. A linear mapping by means of the operator $\boldsymbol{A}$ obviously transforms this circle into an ellipse. Formula (5.32) then implies that the condition number $\mu(\boldsymbol{A})$ is the ratio of the large semiaxis of this ellipse to its small semiaxis.

THEOREM 5.3

Let $\boldsymbol{A} = \boldsymbol{A}^*$ be an operator on the vector space $\mathbb{L}$ self-adjoint in the sense of a given scalar product $[\boldsymbol{x},\boldsymbol{y}]_B$. Then,

$$\mu_B(\boldsymbol{A}) = \frac{|\lambda_{\max}|}{|\lambda_{\min}|}, \tag{5.33}$$

where $\lambda_{\max}$ and $\lambda_{\min}$ are the eigenvalues of $\boldsymbol{A}$ with the largest and smallest moduli, respectively.

PROOF Let $\{\boldsymbol{e}_1, \boldsymbol{e}_2, \dots, \boldsymbol{e}_n\}$ be an orthonormal basis in the n-dimensional space $\mathbb{L}$ composed of the eigenvectors of $\boldsymbol{A}$. Orthonormality of the basis is understood in the sense of the scalar product $[\boldsymbol{x}, \boldsymbol{y}]_B$. Let λ_j be the corresponding eigenvalues: $\boldsymbol{A}\boldsymbol{e}_j = \lambda_j \boldsymbol{e}_j$, $j = 1, 2, \dots, n$, which are all real. Also assume with no loss of generality that the eigenvalues are arranged in the descending order: $|\lambda_1| \geq |\lambda_2| \geq \dots \geq |\lambda_n|$.

An arbitrary vector $\boldsymbol{x} \in \mathbb{L}$ can be expanded with respect to this basis:

$$\boldsymbol{x} = x_1\boldsymbol{e}_1 + x_2\boldsymbol{e}_2 + \dots + x_n\boldsymbol{e}_n.$$

In doing so,

$$\boldsymbol{A}\boldsymbol{x} = \lambda_1 x_1 \boldsymbol{e}_1 + \lambda_2 x_2 \boldsymbol{e}_2 + \dots + \lambda_n x_n \boldsymbol{e}_n,$$

and because of the orthonormality of the basis:

$$\|\boldsymbol{A}\boldsymbol{x}\|_B = (|\lambda_1 x_1|^2 + |\lambda_2 x_2|^2 + \dots + |\lambda_n x_n|^2)^{1/2}.$$

Consequently, $\|\boldsymbol{A}\boldsymbol{x}\|_B \leq |\lambda_1| \|\boldsymbol{x}\|_B$, and if $\|\boldsymbol{x}\|_B = 1$, i.e., if $\boldsymbol{x} \in S$, then $\|\boldsymbol{A}\boldsymbol{x}\|_B \leq |\lambda_1|$. At the same time, for $\boldsymbol{x} = \boldsymbol{e}_1 \in S$ we have $\|\boldsymbol{A}\boldsymbol{e}_1\|_B = |\lambda_1|$. Therefore,

$$\max_{\boldsymbol{x} \in S} \|\boldsymbol{A}\boldsymbol{x}\|_B = |\lambda_1| = |\lambda_{\max}|.$$

A similar argument immediately yields:

$$\min_{\boldsymbol{x} \in S} \|\boldsymbol{A}\boldsymbol{x}\|_B = |\lambda_n| = |\lambda_{\min}|.$$

Then, formula (5.31) implies (5.33). ∎

Note that similarly to Theorem 5.2, the result of Theorem 5.3 can also be generalized to the case of normal matrices $\boldsymbol{A}\boldsymbol{A}^* = \boldsymbol{A}^*\boldsymbol{A}$.

5.3.2 Characterization of a Linear System by Means of Its Condition Number

THEOREM 5.4

Let $\mathbb{L}$ be a normed vector space (e.g., $\mathbb{L} = \mathbb{R}^n$ or $\mathbb{L} = \mathbb{C}^n$), and let $\boldsymbol{A} : \mathbb{L} \longmapsto \mathbb{L}$ be a non-singular linear operator. Consider a system of linear algebraic equations:

$$\boldsymbol{A}\boldsymbol{x} = \boldsymbol{f}, \tag{5.34}$$

where $\boldsymbol{x} \in \mathbb{L}$ *is the vector of unknowns and* $\boldsymbol{f} \in \mathbb{L}$ *is a given right-hand side. Let* $\Delta \boldsymbol{f} \in \mathbb{L}$ *be a perturbation of the right-hand side that leads to the perturbation* $\Delta \boldsymbol{x} \in \mathbb{L}$ *of the solution so that:*

$$\boldsymbol{A}(\boldsymbol{x}+\Delta\boldsymbol{x}) = \boldsymbol{f}+\Delta\boldsymbol{f}. \tag{5.35}$$

Then the relative error of the solution $\|\Delta\boldsymbol{x}\|/\|\boldsymbol{x}\|$ *satisfies the inequality:*

$$\frac{\|\Delta\boldsymbol{x}\|}{\|\boldsymbol{x}\|} \le \mu(\boldsymbol{A})\frac{\|\Delta\boldsymbol{f}\|}{\|\boldsymbol{f}\|}, \tag{5.36}$$

where $\mu(\boldsymbol{A})$ *is the condition number of the operator* $\boldsymbol{A}$*, see formula (5.31). Moreover, there are particular* $\boldsymbol{f} \in \mathbb{L}$ *and* $\Delta\boldsymbol{f} \in \mathbb{L}$ *for which (5.36) transforms into a precise equality.*

PROOF Formulae (5.34) and (5.35) imply that $\boldsymbol{A}\Delta\boldsymbol{x} = \Delta\boldsymbol{f}$, and consequently, $\Delta\boldsymbol{x} = \boldsymbol{A}^{-1}\Delta\boldsymbol{f}$. Let us also employ the expression $\boldsymbol{A}\boldsymbol{x} = \boldsymbol{f}$ that defines the original system itself. Then,

$$\frac{\|\Delta\boldsymbol{x}\|}{\|\boldsymbol{x}\|} = \frac{\|\boldsymbol{A}^{-1}\Delta\boldsymbol{f}\|}{\|\boldsymbol{x}\|} = \frac{\|\boldsymbol{A}\boldsymbol{x}\|}{\|\boldsymbol{x}\|}\frac{\|\boldsymbol{A}^{-1}\Delta\boldsymbol{f}\|}{\|\Delta\boldsymbol{f}\|}\frac{\|\Delta\boldsymbol{f}\|}{\|\boldsymbol{A}\boldsymbol{x}\|} = \frac{\|\boldsymbol{A}\boldsymbol{x}\|}{\|\boldsymbol{x}\|}\frac{\|\boldsymbol{A}^{-1}\Delta\boldsymbol{f}\|}{\|\Delta\boldsymbol{f}\|}\frac{\|\Delta\boldsymbol{f}\|}{\|\boldsymbol{f}\|}. \tag{5.37}$$

According to the definition of the operator norm, see formula (5.24), we have:

$$\frac{\|\boldsymbol{A}\boldsymbol{x}\|}{\|\boldsymbol{x}\|} \le \|\boldsymbol{A}\| \quad \text{and} \quad \frac{\|\boldsymbol{A}^{-1}\Delta\boldsymbol{f}\|}{\|\Delta\boldsymbol{f}\|} \le \|\boldsymbol{A}^{-1}\|. \tag{5.38}$$

Combining formulae (5.37) and (5.38), we obtain for any $\boldsymbol{f} \in \mathbb{L}$ and $\Delta\boldsymbol{f} \in \mathbb{L}$:

$$\frac{\|\Delta\boldsymbol{x}\|}{\|\boldsymbol{x}\|} \le \|\boldsymbol{A}\|\|\boldsymbol{A}^{-1}\|\frac{\|\Delta\boldsymbol{f}\|}{\|\boldsymbol{f}\|} = \mu(\boldsymbol{A})\frac{\|\Delta\boldsymbol{f}\|}{\|\boldsymbol{f}\|}, \tag{5.39}$$

which means that inequality (5.36) holds.

Furthermore, if $\Delta\boldsymbol{f}$ is the specific vector from the space $\mathbb{L}$ for which

$$\frac{\|\boldsymbol{A}^{-1}\Delta\boldsymbol{f}\|}{\|\Delta\boldsymbol{f}\|} = \|\boldsymbol{A}^{-1}\|,$$

and $\boldsymbol{f} = \boldsymbol{A}\boldsymbol{x}$ is the element of $\mathbb{L}$ for which

$$\frac{\|\boldsymbol{A}\boldsymbol{x}\|}{\|\boldsymbol{x}\|} \le \|\boldsymbol{A}\|,$$

then expression (5.37) coincides with inequalities (5.39) and (5.36) that transform into precise equalities for these particular $\boldsymbol{f}$ and $\Delta\boldsymbol{f}$. $\square$

Let us emphasize that the sensitivity of $\boldsymbol{x}$ to the perturbations $\Delta \boldsymbol{f}$ is, generally speaking, not the same for all vectors $\boldsymbol{f}$. It may vary quite dramatically for different right-hand sides of the system $\boldsymbol{A}\boldsymbol{x} = \boldsymbol{f}$. In fact, Theorem 5.4 describes the worst case scenario. Otherwise, for a given fixed $\boldsymbol{f}$ and $\boldsymbol{x} = \boldsymbol{A}^{-1}\boldsymbol{f}$ it may happen that $\|\boldsymbol{A}\boldsymbol{x}\|/\|\boldsymbol{x}\| \ll \|\boldsymbol{A}\|$, so that expression (5.37) will yield a much weaker sensitivity of the relative error $\|\Delta \boldsymbol{x}\|/\|\boldsymbol{x}\|$ to the relative error $\|\Delta \boldsymbol{f}\|/\|\boldsymbol{f}\|$ than provided by estimate (5.36).

To accurately evaluate the condition number $\mu(\boldsymbol{A})$, one needs to be able to compute the norms of the operators $\boldsymbol{A}$ and $\boldsymbol{A}^{-1}$. Typically, this is a cumbersome task. If, for example, the operator $\boldsymbol{A}$ is specified by means of its matrix, and we are interested in finding $\mu_\infty(\boldsymbol{A})$ or $\mu_1(\boldsymbol{A})$, then we first need to obtain the inverse matrix $\boldsymbol{A}^{-1}$ and subsequently compute $\|\boldsymbol{A}\|_\infty$, $\|\boldsymbol{A}^{-1}\|_\infty$ or $\|\boldsymbol{A}\|_1$, $\|\boldsymbol{A}^{-1}\|_1$ using formulae (5.26) or (5.27), respectively, see Theorem 5.1. Evaluating the condition number $\mu_{\boldsymbol{B}}(\boldsymbol{A})$ with respect to the Euclidean (Hermitian) norm given by an operator $\boldsymbol{B} = \boldsymbol{B}^* > 0$ could be even more difficult. Therefore, one often restricts the consideration to obtaining a satisfactory upper bound for the condition number $\mu(\boldsymbol{A})$ using the specifics of the operator $\boldsymbol{A}$ available. We will later see examples of such estimates, e.g., in Section 5.7.2.

In the meantime, we will identify a special class of matrices called matrices with diagonal dominance, for which an estimate of $\mu_\infty(\boldsymbol{A})$ can be obtained without evaluating the inverse matrix $\boldsymbol{A}^{-1}$. A matrix

$$\boldsymbol{A} = \begin{bmatrix} a_{11} & \dots & a_{1n} \\ \dots & \dots & \dots \\ a_{n1} & \dots & a_{nn} \end{bmatrix}$$

is said to have a (strict) diagonal dominance (by rows) of magnitude δ if

$$|a_{ii}| \geq \sum_{j \neq i} |a_{ij}| + \delta, \quad i = 1, 2, \dots, n, \tag{5.40}$$

where $\delta > 0$ is one and the same constant for all rows of $\boldsymbol{A}$.

THEOREM 5.5

Let $\boldsymbol{A}$ be a matrix with diagonal dominance of magnitude $\delta > 0$. Then the inverse matrix $\boldsymbol{A}^{-1}$ exists, and its maximum row norm (5.26), i.e., its norm induced by the l_∞ vector norm $\|\boldsymbol{x}\|_\infty = \max_j |x_j|$, satisfies the estimate:

$$\|\boldsymbol{A}^{-1}\|_\infty \leq \frac{1}{\delta}. \tag{5.41}$$

PROOF Take an arbitrary $\boldsymbol{f} = \begin{bmatrix} f_1 \\ \dots \\ f_n \end{bmatrix}$ and assume that the system of linear algebraic equations $\boldsymbol{A}\boldsymbol{x} = \boldsymbol{f}$ has a solution $\boldsymbol{x} = \begin{bmatrix} x_1 \\ \dots \\ x_n \end{bmatrix}$, such that $\max_j |x_j| = |x_k|$.

Let us write down equation number k from the system $\boldsymbol{Ax}=\boldsymbol{f}$:

$$a_{k1}x_1 + a_{k2}x_2 + \ldots + a_{kn}x_n = f_k.$$

Taking into account that $|x_k| \geq |x_j|$ for $j=1,2,\ldots,n$, we arrive at the following estimate:

$$|f_k| = \left|\sum_j a_{kj}x_j\right| \geq |a_{kk}||x_k| - \sum_{j\neq k} |a_{kj}||x_j|$$
$$\geq |a_{kk}||x_k| - \left(\sum_{j\neq k} |a_{kj}|\right)|x_k| = \left(|a_{kk}| - \sum_{j\neq k} |a_{kj}|\right)|x_k| \geq \delta|x_k|.$$

Consequently, $|x_k| \leq |f_k|/\delta$. On the other hand, $|x_k| = \max_j |x_j| = \|\boldsymbol{x}\|_\infty$ and $|f_k| \leq \max_i |f_i| = \|\boldsymbol{f}\|_\infty$. Therefore,

$$\|\boldsymbol{x}\|_\infty \leq \frac{1}{\delta}\|\boldsymbol{f}\|_\infty. \tag{5.42}$$

In particular, estimate (5.42) means that if $\boldsymbol{f}=\boldsymbol{0}\in\mathbb{L}$ (e.g., $\mathbb{L}=\mathbb{R}^n$ of $\mathbb{L}=\mathbb{C}^n$), then $\|\boldsymbol{x}\|_\infty = 0$, and consequently, the homogeneous system $\boldsymbol{Ax}=\boldsymbol{0}$ only has a trivial solution $\boldsymbol{x}=\boldsymbol{0}$. As such, the inhomogeneous system $\boldsymbol{Ax}=\boldsymbol{f}$ has a unique solution for every $\boldsymbol{f}\in\mathbb{L}$. In other words, the inverse matrix $\boldsymbol{A}^{-1}$ exists.

Estimate (5.42) also implies that for any $\boldsymbol{f}\in\mathbb{L}, \boldsymbol{f}\neq\boldsymbol{0}$, the following estimate holds for $\boldsymbol{x}=\boldsymbol{A}^{-1}\boldsymbol{f}$:

$$\|\boldsymbol{A}^{-1}\boldsymbol{f}\|_\infty \leq \frac{1}{\delta}\|\boldsymbol{f}\|_\infty \quad\Longrightarrow\quad \frac{\|\boldsymbol{A}^{-1}\boldsymbol{f}\|_\infty}{\|\boldsymbol{f}\|_\infty} \leq \frac{1}{\delta},$$

so that

$$\|\boldsymbol{A}^{-1}\|_\infty = \max_{\boldsymbol{f}\in\mathbb{L},\boldsymbol{f}\neq\boldsymbol{0}} \frac{\|\boldsymbol{A}^{-1}\boldsymbol{f}\|_\infty}{\|\boldsymbol{f}\|_\infty} \leq \frac{1}{\delta}.$$

□

COROLLARY 5.1

Let $\boldsymbol{A}$ *be a matrix with diagonal dominance of magnitude* $\delta > 0$. *Then,*

$$\mu_\infty(\boldsymbol{A}) = \|\boldsymbol{A}\|_\infty\|\boldsymbol{A}^{-1}\|_\infty \leq \frac{1}{\delta}\|\boldsymbol{A}\|_\infty. \tag{5.43}$$

The proof is obtained as an immediate implication of the result of Theorem 5.5.

Exercises

1. Prove that the condition numbers $\mu_\infty(\boldsymbol{A})$ and $\mu_1(\boldsymbol{A})$ of the matrix $\boldsymbol{A}$ will not change after a permutation of rows or columns.

2. Prove that for a square matrix A and its transpose A^T, the following equalities hold: $\mu_\infty(A) = \mu_1(A^T)$, $\mu_1(A) = \mu_\infty(A^T)$.
3. Show that the condition number of the operator A does not change if the operator is multiplied by an arbitrary non-zero real number.
4. Let $\mathbb{L}$ be a Euclidean space, and let $A : \mathbb{L} \longmapsto \mathbb{L}$. Show that the condition number $\mu_B = 1$ if and only if at least one of the following conditions holds:

   a) $A = \alpha I$, where $\alpha \in \mathbb{R}$;

   b) A is an orthogonal operator, i.e., $\forall x \in \mathbb{L}: [Ax, Ax]_B = [x, x]_B$.

   c) A is a composition of αI and an orthogonal operator.

5.* Prove that $\mu_B(A) = \mu_B(A_B^*)$, where A_B^* is the operator adjoint to A in the sense of the scalar product $[x, y]_B$.

6. Let A be a non-singular matrix, $\det A \neq 0$. Multiply one row of the matrix A by some scalar α, and denote the new matrix by A_α. Show that $\mu(A_\alpha) \longrightarrow \infty$ as $\alpha \longrightarrow \infty$.

7.* Prove that for any linear operator $A : \mathbb{L} \longmapsto \mathbb{L}$:

$$\mu_B(A_B^* A) = (\mu_B(A))^2,$$

where A_B^* is the operator adjoint to A in the sense of the scalar product $[x, y]_B$.

8.* Let $A = A^* > 0$ and $B = B^* > 0$ in the sense of some scalar product introduced on the linear space $\mathbb{L}$. Let the following inequalities hold for every $x \in \mathbb{L}$:

$$\gamma_1(Bx, x) \leq (Ax, x) \leq \gamma_2(Bx, x),$$

where $\gamma_1 > 0$ and $\gamma_2 > 0$ are two real numbers. Consider the operator $C = B^{-1}A$ and prove that the condition number $\mu_B(C)$ satisfies the estimate:

$$\mu_B(C) \leq \frac{\gamma_2}{\gamma_1}.$$

Remark. We will solve this problem in Section 6.1.4 as it has numerous applications.

5.4 Gaussian Elimination and Its Tri-Diagonal Version

We will describe both the standard Gaussian elimination algorithm and the Gaussian elimination with pivoting, as they apply to solving an $n \times n$ system of linear algebraic equations in its canonical form:

$$\begin{aligned} a_{11}x_1 + a_{12}x_2 + \ldots + a_{1n}x_n &= f_1, \\ \ldots\ldots\ldots\ldots\ldots\ldots\ldots\ldots\ldots\ldots \\ a_{n1}x_1 + a_{n2}x_2 + \ldots + a_{nn}x_n &= f_n. \end{aligned} \tag{5.44}$$

Recall that the Gaussian elimination procedures belong to the class of direct methods, i.e., they produce the exact solution of system (5.44) after a finite number of arithmetic operations.

5.4.1 Standard Gaussian Elimination

From the first equation of system (5.44), express x_1 through the other variables:

$$x_1 = a'_{12}x_2 + \ldots + a'_{1n}x_n + f'_1. \tag{5.45}$$

Substitute expression (5.45) for x_1 into the remaining $n-1$ equations of system (5.44) and obtain a new system of $n-1$ linear algebraic equations with respect to $n-1$ unknowns: $x_2, x_3, \ldots, x_n$. From the first equation of this updated system express x_2 as a function of all other variables:

$$x_2 = a'_{23}x_3 + \ldots + a'_{2n}x_n + f'_2, \tag{5.46}$$

substitute into the remaining $n-2$ equations and obtain yet another system of further reduced dimension: $n-2$ equations with $n-2$ unknowns. Repeat this procedure, called elimination, for $k = 3, 4, \ldots, n-1$:

$$x_k = a'_{k,k+1}x_{k+1} + \ldots + a'_{kn}x_n + f'_k, \tag{5.47}$$

and every time obtain a new smaller linear system of dimension $(n-k) \times (n-k)$. At the last stage $k = n$, no substitution is needed, and we simply solve the 1×1 system, i.e., a single scalar linear equation from the previous stage $k = n-1$, which yields:

$$x_n = f'_n. \tag{5.48}$$

Having obtained the value of x_n by formula (5.48), we can find the values of the unknowns $x_{n-1}, x_{n-2}, \ldots, x_1$ one after another by employing formulae (5.47) in the ascending order: $k = n-1, n-2, \ldots, 1$. This procedure, however, is not fail-proof. It may break down because of a division by zero. Even if no division by zero occurs, the algorithm may still end up generating a large error in the solution due to an amplification of the small round-off errors. Let us explain these phenomena.

If the coefficient a_{11} in front of the unknown x_1 in the first equation of system (5.44) is equal to zero, $a_{11} = 0$, then already the first step of the elimination algorithm, i.e., expression (5.45), becomes invalid, because $a'_{12} = -a_{12}/a_{11}$. A division by zero can obviously be encountered at any stage of the algorithm. For example, having the coefficient in front of x_2 equal to zero in the first equation of the $(n-1) \times (n-1)$ system invalidates expression (5.46). But even if no division by zero is ever encountered, and formulae (5.47) are obtained for all $k = 1, 2, \ldots, n$, then the method can still develop a computational instability when solving for $x_n, x_{n-1}, \ldots, x_1$. For example, if it happens that $a'_{k,k+1} = 2$ for $k = n-1, n-2, \ldots, 1$, while all other a'_{ij} are equal to zero, then the small round-off error committed when evaluating x_n by formula (5.48) increases by a factor of two when computing x_{n-1}, then by another factor of two when computing x_{n-2}, and eventually by a factor of 2^{n-1} when computing x_n. Already for $n = 11$ the error grows more than a thousand times.

Another potential danger that one needs to keep in mind is that the quantities f'_k can rapidly increase when k increases. If this happens, then even small relative errors committed when computing f'_k in expression (5.47) can lead to a large absolute error in the value of x_k.

Let us now formulate a sufficient condition that would guarantee the computational stability of the Gaussian elimination algorithm.

THEOREM 5.6
Let the matrix $\mathbf{A}$ *of system (5.44) be a matrix with diagonal dominance of magnitude* $\delta > 0$, *see formula (5.40). Then, no division by zero will be encountered in the standard Gaussian elimination algorithm. Moreover, the following inequalities will hold:*

$$\sum_{j=1}^{n-k} |a'_{k,k+j}| < 1, \quad k = 1, 2, \dots, n-1, \tag{5.49}$$

$$|f'_k| \leq \frac{2}{\delta} \max_j |f_j|, \quad k = 1, 2, \dots, n. \tag{5.50}$$

To prove Theorem 5.6, we will first need the following Lemma.

LEMMA 5.1
Let

$$\begin{aligned} b_{11}y_1 + b_{12}y_2 + \ldots + b_{1m}y_m &= g_1, \\ \ldots\ldots\ldots\ldots\ldots\ldots\ldots\ldots\ldots\ldots \\ b_{m1}y_1 + b_{m2}y_2 + \ldots + b_{mm}y_m &= g_m, \end{aligned} \tag{5.51}$$

be a system of linear algebraic equations with diagonal dominance of magnitude $\delta > 0$:

$$|b_{ll}| \geq \sum_{j \neq l} |b_{lj}| + \delta, \quad l = 1, 2, \dots, m. \tag{5.52}$$

Then, when reducing the first equation of system (5.51) to the form

$$y_1 = b'_{12}y_2 + \ldots + b'_{1m}y_m + g'_1 \tag{5.53}$$

there will be no division by zero. Besides, the inequality

$$\sum_{j=2}^{m} |b'_{1j}| < 1 \tag{5.54}$$

will hold. Finally, if $m > 1$, *then the variable* y_1 *can be eliminated from system (5.51) with the help of expression (5.53). In doing so, the resulting* $(m-1) \times (m-1)$ *system with respect to the unknowns* $y_2, y_3, \dots, y_m$ *will also be a system with diagonal dominance of the same magnitude* δ *as in formula (5.52).*

PROOF According to formula (5.52), for $l = 1$ we have:

$$|b_{11}| \geq |b_{12}| + |b_{13}| + \ldots + |b_{1m}| + \delta.$$

Consequently, $b_{11} \neq 0$, and expression (5.53) makes sense, where

$$b'_{1j} = -\frac{b_{1j}}{b_{11}} \quad \text{for} \quad j = 2,3,\ldots,m, \quad \text{and} \quad g'_1 = \frac{g_1}{b_{11}}.$$

Moreover,

$$\sum_{j=2}^{m} |b'_{1j}| = \frac{|b_{12}| + |b_{13}| + \ldots + |b_{1m}|}{|b_{11}|},$$

hence inequality (5.54) is satisfied.

It remains to prove the last assertion of the Lemma for $m > 1$. Substituting expression (5.53) into the equation number j $(j > 1)$ of system (5.51), we obtain:

$$(b_{j2} + b_{j1}b'_{12})y_2 + (b_{j3} + b_{j1}b'_{13})y_3 + \ldots (b_{jm} + b_{j1}b'_{1m})y_m = g'_j,$$
$$j = 2,3,\ldots,m.$$

In this system of $m-1$ equations, equation number l $(l = 1,2,\ldots,m-1)$ is the equation:

$$(b_{l+1,2} + b_{l+1,1}b'_{12})y_2 + (b_{l+1,3} + b_{l+1,1}b'_{13})y_3 + \ldots (b_{l+1,m} + b_{l+1,1}b'_{1m})y_m = g'_{l+1},$$
$$l = 1,2,\ldots,m-1. \tag{5.55}$$

Consequently, the entries in row number l of the matrix of system (5.55) are:

$$(b_{l+1,2} + b_{l+1,1}b'_{12}),\ (b_{l+1,3} + b_{l+1,1}b'_{13}),\ \ldots,\ (b_{l+1,m} + b_{l+1,1}b'_{1m}),$$

and the corresponding diagonal entry is: $(b_{l+1,l+1} + b_{l+1,1}b'_{1,l+1})$.

Let us show that there is a diagonal dominance of magnitude δ, i.e., that the following estimate holds:

$$|b_{l+1,l+1} + b_{l+1,1}b'_{1,l+1}| \geq \sum_{\substack{j=2,\\ j\neq l+1}}^{m} |b_{l+1,j} + b_{l+1,1}b'_{1j}| + \delta. \tag{5.56}$$

We will prove an even stronger inequality:

$$|b_{l+1,l+1}| - |b_{l+1,1}b'_{1,l+1}| \geq \sum_{\substack{j=2,\\ j\neq l+1}}^{m} \left[|b_{l+1,j}| + |b_{l+1,1}b'_{1j}|\right] + \delta,$$

which, in turn, is equivalent to the inequality:

$$|b_{l+1,l+1}| \geq \sum_{\substack{j=2,\\ j\neq l+1}}^{m} |b_{l+1,j}| + |b_{l+1,1}| \sum_{j=2}^{m} |b'_{1j}| + \delta. \tag{5.57}$$

Let us replace the quantity $\sum_{j=2}^{m} |b'_{1j}|$ in formula (5.57) by the number 1:

$$|b_{l+1,l+1}| \geq \sum_{\substack{j=2,\\ j\neq l+1}}^{m} |b_{l+1,j}| + |b_{l+1,1}| + \delta = \sum_{\substack{j=1,\\ j\neq l+1}}^{m} |b_{l+1,j}| + \delta. \tag{5.58}$$

According to estimate (5.54), if inequality (5.58) holds, then inequality (5.57) will automatically hold. However, inequality (5.58) is true because of estimate (5.52). Thus, we have proven inequality (5.56). ☐

PROOF OF THEOREM 5.6 We will first establish the validity of formula (5.47) and prove inequality (5.49). To do so, we will use induction with respect to k. If $k=1$ and $n>1$, formula (5.47) and inequality (5.49) are equivalent to formula (5.53) and inequality (5.54), respectively, proven in Lemma 5.1. In addition, Lemma 5.1 implies that the $(n-1)\times(n-1)$ system with respect to $x_2, x_3, \ldots, x_n$ obtained from (5.44) by eliminating the variable x_1 using (5.45) will also be a system with diagonal dominance of magnitude δ.

Assume now that formula (5.47) and inequality (5.49) have already been proven for all $k=1,2,\ldots,l$, where $l<n$. Also assume that the $(n-l)\times(n-l)$ system with respect to $x_{l+1}, x_{l+2}, \ldots, x_n$ obtained from (5.44) by consecutively eliminating the variables $x_1, x_2, \ldots, x_l$ using (5.47) is a system with diagonal dominance of magnitude δ. Then this system can be considered in the capacity of system (5.51), in which case Lemma 5.1 immediately implies that the assumption of the induction is true for $k=l+1$ as well. This completes the proof by induction.

Next, let us justify inequality (5.50). According to Theorem 5.5, the following estimate holds for the solution of system (5.44):

$$\max_j |x_j| \leq \frac{1}{\delta} \max_i |f_i|.$$

Employing this inequality along with formula (5.47) and inequality (5.49) that we have just proven, we obtain for any $k=1,2,\ldots,n$:

$$\begin{aligned} |f'_k| &= \left| x_k - \sum_{j=k+1}^{n} a'_{kj} x_j \right| \leq |x_k| + \sum_{j=k+1}^{n} |a'_{kj}||x_j| \\ &\leq \max_{1\leq j\leq n} |x_j| \left(1 + \sum_{j=k+1}^{n} |a'_{kj}|\right) \leq 2 \max_{1\leq j\leq n} |x_j| \leq \frac{2}{\delta} \max_i |f_i|. \end{aligned}$$

This estimate obviously coincides with (5.50). ☐

Let us emphasize that the hypothesis of Theorem 5.6 (diagonal dominance) provides a sufficient but not a necessary condition for the applicability of the standard Gaussian elimination procedure. There are other linear systems (5.44) that lend themselves to the solution by this method. If, for a given system (5.44), the Gaussian elimination is successfully implemented on a computer (no divisions by zero and no instabilities), then the accuracy of the resulting exact solution will only be limited by the machine precision, i.e., by the round-off errors.

Having computed the approximate solution of system (5.44), one can substitute it into the left-hand side and thus obtain the residual Δf of the right-hand side. Then, estimate

$$\frac{\|\Delta x\|}{\|x\|} \leq \mu(A)\frac{\|\Delta f\|}{\|f\|},$$

can be used to judge the error of the solution, provided that the condition number $\mu(A)$ or its upper bound is known. This is one of the ideas behind the so-called a posteriori analysis.

Computational complexity of the standard Gaussian elimination algorithm is cubic with respect to the dimension n of the system. More precisely, it requires roughly $\frac{2}{3}n^3 + 2n^2 = \mathcal{O}(n^3)$ arithmetic operations to obtain the solution. In doing so, the cubic component of the cost comes from the elimination per se, whereas the quadratic component is the cost of solving back for $x_n, x_{n-1}, \dots, x_1$.

5.4.2 Tri-Diagonal Elimination

The Gaussian elimination algorithm of Section 5.4.1 is particularly efficient for a system (5.44) of the following special kind:

$$\begin{array}{lllll} b_1x_1 + c_1x_2 & & & & = f_1, \\ a_2x_1 + b_2x_2 + & c_2x_3 & & & = f_2, \\ \quad a_3x_2 + & b_3x_3 & + \quad c_3x_4 & & = f_3, \\ \dots\dots\dots & & & & \\ & & a_{n-1}x_{n-2} + b_{n-1}x_{n-1} & + c_{n-1}x_n & = f_{n-1}, \\ & & & a_nx_{n-1} + b_nx_n & = f_n. \end{array} \tag{5.59}$$

The matrix of system (5.59) is tri-diagonal, i.e., all its entries are equal to zero except those on the main diagonal, on the super-diagonal, and on the sub-diagonal. In other words, if $A = \{\mathfrak{a}_{ij}\}$, then $\mathfrak{a}_{ij} = 0$ for $j > i+1$ and $j < i-1$. Adopting the notation of formula (5.59), we say that $\mathfrak{a}_{ii} = b_i$, $i = 1, 2, \dots, n$; $\mathfrak{a}_{i,i-1} = a_i$, $i = 2, 3, \dots, n$; and $\mathfrak{a}_{i,i+1} = c_i$, $i = 1, 2, \dots, n-1$.

The conditions of diagonal dominance (5.40) for system (5.59) read:

$$\begin{aligned} &|b_1| \geq |c_1| + \delta, \\ &|b_k| \geq |a_k| + |c_k| + \delta, \quad k = 2, 3, \dots, n-1, \\ &|b_n| \geq |a_n| + \delta. \end{aligned} \tag{5.60}$$

Equalities (5.45)–(5.48) transform into:

$$\begin{aligned} x_k &= A_kx_{k+1} + F_k, \quad k = 1, 2, \dots, n-1, \\ x_n &= F_n, \end{aligned} \tag{5.61}$$

where A_k and F_k are some coefficients. Define $A_n = 0$ and rewrite formulae (5.61) as a unified expression:

$$x_k = A_kx_{k+1} + F_k, \quad k = 1, 2, \dots, n. \tag{5.62}$$

From the first equation of system (5.59) it is clear that for $k = 1$ the coefficients in formula (5.62) are:

$$A_1 = -\frac{c_1}{b_1}, \quad F_1 = \frac{f_1}{b_1}. \tag{5.63}$$

Suppose that all the coefficients A_k and F_k have already been computed up to some fixed k, $1 \leq k \leq n-1$. Substituting the expression $x_k = A_k x_{k+1} + F_k$ into the equation number $k+1$ of system (5.59) we obtain:

$$x_{k+1} = -\frac{c_{k+1}}{b_{k+1} + a_{k+1}A_k} x_{k+2} + \frac{f_{k+1} - a_{k+1}F_k}{b_{k+1} + a_{k+1}A_k}.$$

Therefore, the coefficients A_k and F_k satisfy the following recurrence relations:

$$A_{k+1} = -\frac{c_{k+1}}{b_{k+1} + a_{k+1}A_k}, \quad F_{k+1} = \frac{f_{k+1} - a_{k+1}F_k}{b_{k+1} + a_{k+1}A_k}, \qquad k = 1, 2, \dots, n-1. \tag{5.64}$$

As such, the algorithm of solving system (5.59) gets split into two stages. At the first stage, we evaluate the coefficients A_k and F_k for $k = 1, 2, \dots, n$ using formulae (5.63) and (5.64). At the second stage, we solve back for the actual unknowns $x_n, x_{n-1}, \dots, x_1$ using formulae (5.62) for $k = n, n-1, \dots, 1$.

In the literature, one can find several alternative names for the tri-diagonal Gaussian elimination procedure that we have described. Sometimes, the term *marching* is used. The first stage of the algorithm is also referred to as the forward stage or forward marching, when the marching coefficients A_k and B_k are computed. Accordingly, the second stage of the algorithm, when relations (5.62) are applied consecutively in the reverse order is called backward marching.

We will now estimate the computational complexity of the tri-diagonal elimination. At the forward stage, the elimination according to formulae (5.63) and (5.64) requires $\mathcal{O}(n)$ arithmetic operations. At the backward stage, formula (5.62) is applied n times, which also requires $\mathcal{O}(n)$ operations. Altogether, the complexity of the tri-diagonal elimination is $\mathcal{O}(n)$ arithmetic operations. It is clear that no algorithm can be built that would be asymptotically cheaper than $\mathcal{O}(n)$, because the number of unknowns in the system is also $\mathcal{O}(n)$.

Let us additionally note that the tri-diagonal elimination is apparently the only example available in the literature of a direct method with linear complexity, i.e., of a method that produces the exact solution of a linear system at a cost of $\mathcal{O}(n)$ operations. In other words, the computational cost is directly proportional to the dimension of the system. We will later see examples of direct methods that produce the exact solution at a cost of $\mathcal{O}(n \ln n)$ operations, and examples of iterative methods that cost $\mathcal{O}(n)$ operations but only produce an approximate solution. However, no other method of computing the exact solution with a genuinely linear complexity is known.

The algorithm can also be generalized to the case of the banded matrices. Matrices of this type may contain non-zero entries on several neighboring diagonals, including the main diagonal. Normally we would assume that the number m of the non-zero diagonals, i.e., the bandwidth, satisfies $3 \leq m \ll n$. The complexity of the Gaussian elimination algorithm when applied to a banded system (5.44) is $\mathcal{O}(m^2 n)$ operations. If m is fixed and n is arbitrary, then the complexity, again, scales as $\mathcal{O}(n)$.

High-order systems of type (5.59) drew the attention of researchers in the fifties. They appeared when solving the heat equation numerically with the help of the so-called implicit finite-difference schemes. These schemes, their construction and their importance, will be discussed later in Part III of the book, see, in particular, Section 10.6.

The foregoing tri-diagonal marching algorithm was apparently introduced for the first time by I. M. Gelfand and O. V. Lokutsievskii around the late forties or early fifties. They conducted a complete analysis of the algorithm, showed that it was computationally stable, and also built its continuous "closure," see Appendix II to the book [GR64] written by Gelfand and Lokutsievskii.

Alternatively, the tri-diagonal elimination algorithm is attributed to L. H. Thomas [Tho49], and is referred to as the Thomas algorithm.

The aforementioned work by Gelfand and Lokutsievskii was one of the first papers in the literature where the question of stability of a computational algorithm was accurately formulated and solved for a particular class of problems. This question has since become one of the key issues for the entire large field of knowledge called scientific computing. Having stability is crucial, as otherwise computer codes that implement the algorithms will not execute properly. In Part III of the book, we study computational stability for the finite-difference schemes.

Theorem 5.6, which provides sufficient conditions for applicability of the Gaussian elimination, is a generalization to the case of full matrices of the result by Gelfand and Lokutsievskii on stability of the tri-diagonal marching.

Note also that the conditions of strict diagonal dominance (5.60) that are sufficient for stability of the tri-diagonal elimination can actually be relaxed. In fact, one can only require that the coefficients of system (5.59) satisfy the inequalities:

$$\begin{aligned} &|b_1| \geq |c_1|, \\ &|b_k| \geq |a_k| + |c_k|, \quad k = 2,3,\ldots,n-1, \\ &|b_n| \geq |a_n|, \end{aligned} \tag{5.65}$$

so that at least one out of the total of n inequalities (5.65) actually be strict, i.e., "$>$" rather than "$\geq$." We refer the reader to [SN89a, Chapter II] for detail.

Overall, the idea of transporting, or marching, the condition $b_1 x_1 + c_1 x_2 = f_1$ specified by the first equation of the tri-diagonal system (5.59) is quite general. In the previous algorithm, this idea is put to use when obtaining the marching coefficients (5.63), (5.64) and relations (5.62). It is also exploited in many other elimination algorithms, see [SN89a, Chapter II]. We briefly describe one of those algorithms, known as the cyclic tri-diagonal elimination, in Section 5.4.3.

5.4.3 Cyclic Tri-Diagonal Elimination

In many applications, one needs to solve a system which is "almost" tri-diagonal, but is not quite equivalent to system (5.59):

$$\begin{array}{lllll} b_1x_1 + c_1x_2 & & & + \; a_1x_n & = f_1, \\ a_2x_1 + b_2x_2 + & c_2x_3 & & & = f_2, \\ \qquad a_3x_2 + & b_3x_3 & + \quad c_3x_4 & & = f_3, \\ \multicolumn{5}{c}{\dots\dots\dots\dots\dots\dots\dots\dots\dots\dots\dots\dots} \\ & & a_{n-1}x_{n-2} + b_{n-1}x_{n-1} + c_{n-1}x_n & & = f_{n-1}, \\ c_nx_1 & & \qquad + \; a_nx_{n-1} \; + \; b_nx_n & & = f_n. \end{array} \tag{5.66}$$

In Section 2.3.2, we have analyzed one particular example of this type that arises when constructing the nonlocal Schoenberg splines; see the matrix $\boldsymbol{A}$ given by formula (2.66) on page 52. Other typical examples include the so-called central difference schemes (see Section 9.2.1) for the solution of second order ordinary differential equations with periodic boundary conditions, as well as many schemes built for solving partial differential equations in the cylindrical or spherical coordinates. Periodicity of the boundary conditions gave rise to the name *cyclic* attached to the version of the tri-diagonal elimination that we are about to describe.

The coefficients a_1 and c_n in the first and last equations of system (5.66), respectively, are, generally speaking, non-zero. Their presence does not allow one to apply the tri-diagonal elimination algorithm of Section 5.4.2 to system (5.66) directly. Let us therefore consider two auxiliary linear systems of dimension $(n-1)\times(n-1)$:

$$\begin{array}{llll} b_2u_2 + & c_2u_3 & & = f_2, \\ a_3u_2 + & b_3u_3 & + \quad c_3u_4 & = f_3, \\ \multicolumn{4}{c}{\dots\dots\dots\dots\dots\dots\dots\dots\dots\dots} \\ & & a_{n-1}u_{n-2} + b_{n-1}u_{n-1} + c_{n-1}u_n & = f_{n-1}, \\ & & \qquad a_nu_{n-1} \; + \; b_nu_n & = f_n. \end{array} \tag{5.67}$$

and

$$\begin{array}{llll} b_2v_2 + & c_2v_3 & & = -a_2, \\ a_3v_2 + & b_3v_3 & + \quad c_3v_4 & = 0, \\ \multicolumn{4}{c}{\dots\dots\dots\dots\dots\dots\dots\dots\dots\dots} \\ & & a_{n-1}v_{n-2} + b_{n-1}v_{n-1} + c_{n-1}v_n & = 0, \\ & & \qquad a_nv_{n-1} \; + \; b_nv_n & = -c_1. \end{array} \tag{5.68}$$

Having obtained the solutions $\{u_2,u_3,\dots,u_n\}$ and $\{v_2,v_3,\dots,v_n\}$ to systems (5.67) and (5.68), respectively, we can represent the solution $\{x_1,x_2,\dots,x_n\}$ to system (5.66) in the form:

$$x_i = u_i + x_1v_i, \quad i = 1,2,\dots,n, \tag{5.69}$$

where for convenience we additionally define $u_1 = 0$ and $v_1 = 1$. Indeed, multiplying each equation of system (5.68) by x_1, adding with the corresponding equation of system (5.67), and using representation (5.69), we immediately see that the equations number 2 through n of system (5.66) are satisfied. It only remains to satisfy equation number 1 of system (5.66). To do so, we use formula (5.69) for $i = 2$ and $i = n$, and substitute $x_2 = u_2 + x_1 v_2$ and $x_n = u_n + x_1 v_n$ into the first equation of system (5.66), which yields one scalar equation for x_1:

$$b_1 x_1 + c_1(u_2 + x_1 v_2) + a_1(u_n + x_1 v_n) = f_1.$$

As such, we find:

$$x_1 = \frac{f_1 - a_1 u_n - c_1 u_2}{b_1 + a_1 v_n + c_1 v_2}. \tag{5.70}$$

Altogether, the solution algorithm for system (5.66) reduces to first solving the two auxiliary systems (5.67) and (5.68), then finding x_1 with the help of formula (5.70), and finally obtaining $x_2, x_3, \dots, x_n$ according to formula (5.69). As both systems (5.67) and (5.68) are genuinely tri-diagonal, they can be solved by the original tri-diagonal elimination described in Section 5.4.2. In doing so, the overall computational complexity of solving system (5.66) obviously remains linear with respect to the dimension of the system n, i.e., $\mathcal{O}(n)$ arithmetic operations.

5.4.4 Matrix Interpretation of the Gaussian Elimination. *LU* Factorization

Consider system (5.44) written as $\boldsymbol{A}\boldsymbol{x} = \boldsymbol{f}$ or alternatively, $\boldsymbol{A}^{(0)}\boldsymbol{x} = \boldsymbol{f}^{(0)}$, where

$$\boldsymbol{A} \equiv \boldsymbol{A}^{(0)} = \begin{bmatrix} a_{11}^{(0)} & \dots & a_{1n}^{(0)} \\ \vdots & \ddots & \vdots \\ a_{n1}^{(0)} & \dots & a_{nn}^{(0)} \end{bmatrix} \quad \text{and} \quad \boldsymbol{f} \equiv \boldsymbol{f}^{(0)} = \begin{bmatrix} f_1^{(0)} \\ \vdots \\ f_n^{(0)} \end{bmatrix},$$

and the superscript "(0)" emphasizes that this is the beginning, i.e., the zeroth stage of the Gaussian elimination procedure. The notations in this section are somewhat different from those of Section 5.4.1, but we will later show the correspondence.

At the first stage of the algorithm we assume that $a_{11}^{(0)} \neq 0$ and introduce the transformation matrix:

$$\boldsymbol{T}_1 = \begin{bmatrix} 1 & 0 & 0 & \dots & 0 \\ -t_{21} & 1 & 0 & \dots & 0 \\ -t_{31} & 0 & 1 & \dots & 0 \\ \vdots & \vdots & \vdots & \ddots & \vdots \\ -t_{n1} & 0 & 0 & \dots & 1 \end{bmatrix}, \quad \text{where} \quad t_{i1} = \frac{a_{i1}^{(0)}}{a_{11}^{(0)}}, \quad i = 2, 3, \dots, n.$$

Applying this matrix is equivalent to eliminating the variable x_1 from equations number 2 through n of system (5.44):

$$\boldsymbol{T}_1 \boldsymbol{A}^{(0)} \boldsymbol{x} \equiv \boldsymbol{A}^{(1)} \boldsymbol{x} = \begin{bmatrix} a_{11}^{(0)} & a_{12}^{(0)} & \dots & a_{1n}^{(0)} \\ 0 & a_{22}^{(1)} & \dots & a_{2n}^{(1)} \\ \vdots & \vdots & \ddots & \vdots \\ 0 & a_{n2}^{(1)} & \dots & a_{nn}^{(1)} \end{bmatrix} \begin{bmatrix} x_1 \\ x_2 \\ \vdots \\ x_n \end{bmatrix} = \begin{bmatrix} f_1^{(0)} \\ f_2^{(1)} \\ \vdots \\ f_n^{(1)} \end{bmatrix} = \boldsymbol{f}^{(1)} \equiv \boldsymbol{T}_1 \boldsymbol{f}^{(0)}.$$

At the second stage, we define:

$$T_2 = \begin{bmatrix} 1 & 0 & 0 & \dots & 0 \\ 0 & 1 & 0 & \dots & 0 \\ 0 & -t_{32} & 1 & \dots & 0 \\ \vdots & & & \ddots & \vdots \\ 0 & -t_{n2} & 0 & \dots & 1 \end{bmatrix}, \quad \text{where} \quad t_{i2} = \frac{a_{i2}^{(1)}}{a_{22}^{(1)}}, \quad i = 3,4,\dots,n,$$

and eliminate x_2 from equations 3 through n, thus obtaining the system $\boldsymbol{A}^{(2)}\boldsymbol{x} = \boldsymbol{f}^{(2)}$, where

$$\boldsymbol{A}^{(2)} = \boldsymbol{T}_2\boldsymbol{A}^{(1)} = \begin{bmatrix} a_{11}^{(0)} & a_{12}^{(0)} & a_{13}^{(0)} & \dots & a_{1n}^{(0)} \\ 0 & a_{22}^{(1)} & a_{23}^{(1)} & \dots & a_{2n}^{(1)} \\ 0 & 0 & a_{33}^{(2)} & \dots & a_{3n}^{(2)} \\ \vdots & \vdots & \vdots & \ddots & \vdots \\ 0 & 0 & a_{n3}^{(2)} & \dots & a_{nn}^{(2)} \end{bmatrix} \quad \text{and} \quad \boldsymbol{f}^{(2)} = \boldsymbol{T}_2\boldsymbol{f}^{(1)} = \begin{bmatrix} f_1^{(0)} \\ f_2^{(1)} \\ f_3^{(2)} \\ \vdots \\ f_n^{(2)} \end{bmatrix}.$$

In general, at stage number k we have the transformation matrix:

$$\boldsymbol{T}_k = \begin{bmatrix} 1 & \dots & 0 & 0 & \dots & 0 \\ \vdots & \ddots & \vdots & \vdots & & \vdots \\ 0 & \dots & 1 & 0 & \dots & 0 \\ 0 & \dots & -t_{k+1,k} & 1 & \dots & 0 \\ \vdots & & \vdots & \vdots & \ddots & \vdots \\ 0 & \dots & -t_{n,k} & 0 & \dots & 1 \end{bmatrix}, \quad \text{where} \quad t_{i,k} = \frac{a_{ik}^{(k-1)}}{a_{kk}^{(k-1)}}, \quad i = k+1,\dots,n, \tag{5.71}$$

and the system $\boldsymbol{A}^{(k)}\boldsymbol{x} = \boldsymbol{f}^{(k)}$, where:

$$\boldsymbol{A}^{(k)} = \boldsymbol{T}_k\boldsymbol{A}^{(k-1)} = \begin{bmatrix} a_{11}^{(0)} & \dots & a_{1,k}^{(0)} & a_{1,k+1}^{(0)} & \dots & a_{1,n}^{(0)} \\ \vdots & \ddots & \vdots & \vdots & & \vdots \\ 0 & \dots & a_{k,k}^{(k-1)} & a_{k,k+1}^{(k-1)} & \dots & a_{k,n}^{(k-1)} \\ 0 & \dots & 0 & a_{k+1,k+1}^{(k)} & \dots & a_{k+1,n}^{(k)} \\ \vdots & & \vdots & \vdots & \ddots & \vdots \\ 0 & \dots & 0 & a_{n,k+1}^{(k)} & \dots & a_{n,n}^{(k)} \end{bmatrix}, \quad \boldsymbol{f}^{(k)} = \boldsymbol{T}_k\boldsymbol{f}^{(k-1)} = \begin{bmatrix} f_1^{(0)} \\ \vdots \\ f_k^{(k-1)} \\ f_{k+1}^{(k)} \\ \vdots \\ f_n^{(k)} \end{bmatrix}.$$

Performing all $n-1$ stages of the elimination, we arrive at the system:

$$\boldsymbol{A}^{(n-1)}\boldsymbol{x} = \boldsymbol{f}^{(n-1)},$$

where

$$\boldsymbol{A}^{(n-1)} = \boldsymbol{T}_{n-1}\boldsymbol{T}_{n-2}\dots\boldsymbol{T}_1\boldsymbol{A} \quad \text{and} \quad \boldsymbol{f}^{(n-1)} = \boldsymbol{T}_{n-1}\boldsymbol{T}_{n-2}\dots\boldsymbol{T}_1\boldsymbol{f}.$$

The resulting matrix $\boldsymbol{A}^{(n-1)}$ is upper triangular and we re-denote it by $\boldsymbol{U}$. All entries of this matrix below its main diagonal are equal to zero:

$$\boldsymbol{A}^{(n-1)} \equiv \boldsymbol{U} = \begin{bmatrix} a_{11}^{(0)} & \dots & a_{1,k}^{(0)} & a_{1,k+1}^{(0)} & \dots & a_{1,n}^{(0)} \\ \vdots & \ddots & \vdots & \vdots & & \vdots \\ 0 & \dots & a_{k,k}^{(k-1)} & a_{k,k+1}^{(k-1)} & \dots & a_{k,n}^{(k-1)} \\ 0 & \dots & 0 & a_{k+1,k+1}^{(k)} & \dots & a_{k+1,n}^{(k)} \\ \vdots & & \vdots & \vdots & \ddots & \vdots \\ 0 & \dots & 0 & 0 & \dots & a_{n,n}^{(n-1)} \end{bmatrix}. \tag{5.72}$$

The entries on the main diagonal, $a_{k,k}^{(k-1)}$, $k = 1,2,\dots,n$, are called pivots, and none of them may turn into zero as otherwise the standard Gaussian elimination procedure will fail. Solving the system $\boldsymbol{Ux} = \boldsymbol{f}^{(n-1)} \equiv \boldsymbol{g}$ is a fairly straightforward task, as we have seen in Section 5.4.1. It amounts to computing the values of all the unknowns one after another in the reverse order, i.e., starting from $x_n = f_n^{(n-1)}/a_{n,n}^{(n-1)}$, then $x_{n-1} = (f_{n-1}^{(n-2)} - a_{n-1,n}^{(n-2)}x_n)/a_{n-1,n-1}^{(n-2)}$, etc., and all the way up to x_1.

We will now show that the representation $\boldsymbol{U} = \boldsymbol{T}_{n-1}\boldsymbol{T}_{n-2}\dots\boldsymbol{T}_1\boldsymbol{A}$ implies that $\boldsymbol{A} = \boldsymbol{LU}$, where $\boldsymbol{L}$ is a lower triangular matrix, i.e., a matrix that only has zero entries above its main diagonal. In other words, we will need to demonstrate that $\boldsymbol{L} \stackrel{\text{def}}{=} (\boldsymbol{T}_{n-1}\boldsymbol{T}_{n-2}\dots\boldsymbol{T}_1)^{-1} = \boldsymbol{T}_1^{-1}\boldsymbol{T}_2^{-1}\dots\boldsymbol{T}_{n-1}^{-1}$ is a lower triangular matrix. Then, the formula

$$\boldsymbol{A} = \boldsymbol{LU} \tag{5.73}$$

will be referred to as an $\boldsymbol{LU}$ factorization of the matrix $\boldsymbol{A}$.

One can easily verify by direct multiplication that the inverse matrix for a given $\boldsymbol{T}_k$, $k = 1,2,\dots,n-1$, see formula (5.71), is:

$$\boldsymbol{T}_k^{-1} = \begin{bmatrix} 1 & \dots & 0 & 0 & \dots & 0 \\ \vdots & \ddots & \vdots & \vdots & & \vdots \\ 0 & \dots & 1 & 0 & \dots & 0 \\ 0 & \dots & t_{k+1,k} & 1 & \dots & 0 \\ \vdots & & \vdots & \vdots & \ddots & \vdots \\ 0 & \dots & t_{n,k} & 0 & \dots & 1 \end{bmatrix}. \tag{5.74}$$

It is also clear that if the meaning of the operation rendered by $\boldsymbol{T}_k$ is "take equation number k, multiply it consecutively by $t_{k+1,k},\dots,t_{n,k}$, and *subtract from* equations number $k+1$ through n, respectively," then the meaning of the inverse operation has to be "take equation number k, multiply it consecutively by the factors $t_{k+1,k},\dots,t_{n,k}$, and *add to* equations number $k+1$ through n, respectively," which is precisely what the foregoing matrix $\boldsymbol{T}_k^{-1}$ does.

We see that all the matrices $\boldsymbol{T}_k^{-1}$, $k = 1,2,\dots,n-1$, given by formula (5.74) are lower triangular matrices with the diagonal entries equal to one. Their product can be calculated directly, which yields:

$$L = T_1^{-1} T_2^{-1} \dots T_{n-1}^{-1} = \begin{bmatrix} 1 & 0 & \dots & 0 & 0 \dots 0 \\ t_{2,1} & 1 & \dots & 0 & 0 \dots 0 \\ \vdots & \vdots & \ddots & \vdots & \vdots \quad \vdots \\ t_{k,1} & t_{k,2} & \dots & 1 & 0 \dots 0 \\ t_{k+1,1} & t_{k+1,2} & \dots & t_{k+1,k} & 1 \dots 0 \\ \vdots & \vdots & & \vdots & \vdots \ddots \vdots \\ t_{n,1} & t_{n,2} & \dots & t_{n,k} & 0 \dots 1 \end{bmatrix}. \tag{5.75}$$

Consequently, the matrix $\boldsymbol{L}$ is indeed a lower triangular matrix (with all its diagonal entries equal to one), and the factorization formula (5.73) holds.

The $\boldsymbol{LU}$ factorization of the matrix $\boldsymbol{A}$ allows us to analyze the computational complexity of the Gaussian elimination algorithm as it applies to solving multiple linear systems that have the same matrix $\boldsymbol{A}$ but different right-hand sides. The cost of obtaining the factorization itself, i.e., that of computing the matrix $\boldsymbol{U}$, is cubic: $\mathcal{O}(n^3)$ arithmetic operations. This factorization obviously stays the same when the right-hand side changes. For a given right-hand side $\boldsymbol{f}$, we need to solve the system $\boldsymbol{LUx} = \boldsymbol{f}$. This amounts to first solving the system $\boldsymbol{Lg} = \boldsymbol{f}$ with a lower triangular matrix $\boldsymbol{L}$ of (5.75) and obtaining $\boldsymbol{g} = \boldsymbol{L}^{-1}\boldsymbol{f} = \boldsymbol{T}_{n-1}\boldsymbol{T}_{n-2}\dots\boldsymbol{T}_1\boldsymbol{f}$, and then solving the system $\boldsymbol{Ux} = \boldsymbol{g}$ with an upper triangular matrix $\boldsymbol{U}$ of (5.72) and obtaining $\boldsymbol{x}$. The cost of either solution is $\mathcal{O}(n^2)$ operations. Consequently, once the $\boldsymbol{LU}$ factorization has been built, each additional right-hand side can be accommodated at a quadratic cost.

In particular, consider the problem of finding the inverse matrix $\boldsymbol{A}^{-1}$ using Gaussian elimination. By definition, $\boldsymbol{AA}^{-1} = \boldsymbol{I}$. In other words, each column of the matrix $\boldsymbol{A}^{-1}$ is the solution to the system $\boldsymbol{Ax} = \boldsymbol{f}$ with the right-hand side $\boldsymbol{f}$ equal to the corresponding column of the identity matrix $\boldsymbol{I}$. Altogether, there are n columns, each adding an $\mathcal{O}(n^2)$ solution cost to the $\mathcal{O}(n^3)$ initial cost of the $\boldsymbol{LU}$ factorization that is performed only once ahead of time. We therefore conclude that the overall cost of computing $\boldsymbol{A}^{-1}$ using Gaussian elimination is also cubic: $\mathcal{O}(n^3)$ operations.

Finally, let us note that for a given matrix $\boldsymbol{A}$, its $\boldsymbol{LU}$ factorization is, generally speaking, not unique. The procedure that we have described yields a particular form of the $\boldsymbol{LU}$ factorization (5.73) defined by an additional constraint that all diagonal entries of the matrix $\boldsymbol{L}$ of (5.75) be equal to one. Instead, we could have required, for example, that the diagonal entries of $\boldsymbol{U}$ be equal to one. Then, the matrices $\boldsymbol{T}_k$ of (5.71) get replaced by:

$$\tilde{\boldsymbol{T}}_k = \begin{bmatrix} 1 \dots & 0 & 0 \dots 0 \\ \vdots \ddots & \vdots & \vdots \quad \vdots \\ 0 \dots & t_{k,k} & 0 \dots 0 \\ 0 \dots & -t_{k+1,k} & 1 \dots 0 \\ \vdots & \vdots & \vdots \ddots \vdots \\ 0 \dots & -t_{n,k} & 0 \dots 1 \end{bmatrix}, \quad \text{where} \quad t_{k,k} = \frac{1}{a_{kk}^{(k-1)}}, \tag{5.76}$$

and instead of (5.72) we have:

$$\tilde{U} = \begin{bmatrix} 1 & \dots & -a'_{1,k} & -a'_{1,k+1} & \cdots & -a'_{1,n} \\ \vdots & \ddots & \vdots & \vdots & & \vdots \\ 0 & \dots & 1 & -a'_{k,k+1} & \cdots & -a'_{k,n} \\ 0 & \dots & 0 & 1 & \dots & -a'_{k+1,n} \\ \vdots & & \vdots & \vdots & \ddots & \vdots \\ 0 & \dots & 0 & 0 & \dots & 1 \end{bmatrix}, \tag{5.77}$$

where the off-diagonal entries of the matrix $\tilde{U}$ are the same as introduced in the beginning of Section 5.4.1. The matrices $\tilde{T}_k^{-1}$ and $\tilde{L}$ are obtained accordingly; this is the subject of Exercise 1 after this section.

5.4.5 Cholesky Factorization

If A is a symmetric positive definite (SPD) matrix (with real entries), $A = A^T > 0$, then it admits a special form of LU factorization, namely:

$$A = LL^T, \tag{5.78}$$

where L is a lower triangular matrix with positive entries on its main diagonal. Factorization (5.78) is known as the Cholesky factorization. The proof of formula (5.78), which uses induction with respect to the dimension of the matrix, can be found, e.g., in [GL81, Chapter 2]. Moreover, one can derive explicit formulae for the entries of the matrix L:

$$\begin{aligned} l_{jj} &= \left(a_{jj} - \sum_{k=1}^{j-1} l_{jk}^2 \right)^{\frac{1}{2}}, \quad j = 1, 2, \dots, n, \\ l_{ij} &= \left(a_{ij} - \sum_{k=1}^{j-1} l_{ik} l_{jk} \right) \cdot l_{jj}^{-1}, \quad i = j+1, j+2, \dots, n. \end{aligned}$$

Computational complexity of obtaining the Cholesky factorization (5.78) is also cubic with respect to the dimension of the matrix: $\mathcal{O}(n^3)$ arithmetic operations. However, the actual cost is roughly one half compared to that of the standard Gaussian elimination: $n^3/3$ vs. $2n^3/3$ arithmetic operations. This reduction of the cost is the key benefit of using the Cholesky factorization, as opposed to the standard Gaussian elimination, for large symmetric matrices. The reduction of the cost is only enabled by the special SPD structure of the matrix A; there is no Cholesky factorization for general matrices, and one has to resort back to the standard LU. This is one of many examples in numerical analysis when a more efficient computational procedure can be obtained for a more narrow class of objects that define the problem.

5.4.6 Gaussian Elimination with Pivoting

The standard Gaussian procedure of Section 5.4.1 fails if a pivotal entry in the matrix appears equal to zero. Consider, for example, the following linear system:

$$\begin{aligned} x_1+2x_2+3x_3&=1,\\ 2x_1+4x_2+5x_3&=2,\\ 7x_1+8x_2+9x_3&=3. \end{aligned}$$

After the first stage of elimination we obtain:

$$\begin{aligned} x_1+2x_2+3x_3&=1,\\ 0\cdot x_1+0\cdot x_2-x_3&=0,\\ 0\cdot x_1-6x_2-12x_3&=-4. \end{aligned}$$

The pivotal entry in the second equation of this system, i.e., the coefficient in front of x_2, is equal to zero. Therefore, this equation cannot be used for eliminating x_2 from the third equation. As such, the algorithm cannot proceed any further. The problem, however, can be easily fixed by changing the order of the equations:

$$\begin{aligned} x_1+2x_2+3x_3&=1,\\ -6x_2-12x_3&=-4,\\ x_3&=0, \end{aligned}$$

and we see that the system is already reduced to the upper triangular form.

Different strategies of *pivoting* are designed to help overcome or alleviate the difficulties that the standard Gaussian elimination may encounter — division by zero and the loss of accuracy due to an instability. Pivoting exploits pretty much the same idea as outlined in the previous simple example — changing the order of equations and/or the order of unknowns in the equations. In the case $\det \boldsymbol{A}\neq 0$, the elimination procedure with pivoting guarantees that there will be no division by zero, and also improves the computational stability compared to the standard Gaussian elimination. We will describe partial pivoting and complete pivoting.

When performing partial pivoting for a non-singular system, we start with finding the entry with the maximum absolute value[4] in the first column of the matrix $\boldsymbol{A}$. Assume that this is the entry a_{i1}. Clearly, $a_{i1}\neq 0$, because otherwise $\det\boldsymbol{A}=0$. Then, we change the order of the equations in the system — the first equation becomes equation number i, and equation number i becomes the first equation. The system after this operation obviously remains equivalent to the original one. Finally, one step of the standard Gaussian elimination is performed, see Section 5.4.1. In doing so, the entries $-t_{21},-t_{31},\dots,-t_{n1}$ of the matrix $\boldsymbol{T}_1$, see formula (5.71) will all have absolute values less than one. This improves stability, because multiplication of the subtrahend by a quantity smaller than one before subtraction helps reduce the

[4]If several entries have the same largest absolute value, we take one of those.

effect of the round-off errors. Having eliminated the variable x_1, we apply the same approach to the resulting smaller system of order $(n-1)\times(n-1)$. Namely, among all the equations we first find the equation that has the maximum absolute value of the coefficient in front of x_2. We then change the order of the equations so that this coefficient moves to the position of the pivot, and only after that eliminate x_2.

Complete pivoting is similar to partial pivoting, except that at every stage of elimination, the entry with the maximum absolute value is sought for not only in the first column, but across the entire current-stage matrix:

$$\begin{bmatrix} a_{kk}^{(k-1)} & \dots & a_{kn}^{(k-1)} \\ \dots & \dots & \dots \\ a_{nk}^{(k-1)} & \dots & a_{nn}^{(k-1)} \end{bmatrix}.$$

Once the maximum entry has been determined, it needs to be moved to the position of the pivot $a_{kk}^{(k-1)}$. This is achieved by the appropriate permutations of both the equations and the unknowns. It is clear that the system after the permutations remains equivalent to the original one.

Computational complexity of Gaussian elimination with pivoting remains cubic with respect to the dimension of the system: $\mathscr{O}(n^3)$ arithmetic operations. Of course, this estimate is only asymptotic, and the actual constants for the algorithm with partial pivoting are larger than those for the standard Gaussian elimination. This is the pay-off for its improved robustness. The algorithm with complete pivoting is even more expensive than the algorithm with partial pivoting; but it is typically more robust as well. Note that for the same reason of improved robustness, the use of Gaussian elimination with pivoting can be recommended in practice, even for those systems for which the standard algorithm does not fail.

Gaussian elimination with pivoting can be put into the matrix framework in much the same way as it has been done in Section 5.4.4 for standard Gaussian elimination. Moreover, a very similar result on the ***LU*** factorization can be obtained; this is the subject of Exercise 3. To do so, one needs to exploit the so-called permutation matrices, i.e., the matrices that change the order of rows or columns in a given matrix when applied as multipliers, see [HJ85, Section 0.9]. For example, to swap the second and the third equations in the example analyzed in the beginning of this section, one would need to multiply the system matrix from the left by the permutation matrix:

$$\boldsymbol{P} = \begin{bmatrix} 1 & 0 & 0 \\ 0 & 0 & 1 \\ 0 & 1 & 0 \end{bmatrix}.$$

5.4.7 An Algorithm with a Guaranteed Error Estimate

As has already been discussed, all numbers in a computer are represented as fractions with a finite fixed number of significant digits. Consequently, many actual numbers have to be truncated, or in other words, rounded, before they can be stored in the computer memory, see Section 1.3.3. As such, when performing computations on a real machine with a fixed number of significant digits, along with the effect of inaccuracies in the input data, round-off errors are introduced at every arithmetic operation. For linear systems, the impact of these round-off errors depends on the number of significant digits, on the condition number of the matrix, as well as on the particular algorithm chosen for the solution. In work [GAKK93], a family of algorithms has been proposed that directly take into account the effect of rounding on a given computer. These algorithms either produce the result with a guaranteed accuracy, or otherwise determine in the course of computations that the system is conditioned so poorly that no accuracy can be guaranteed when computing its solution on the machine with a given number of significant digits.

Exercises

1. Write explicitly the matrices $\tilde{\boldsymbol{T}}_k^{-1}$ and $\tilde{\boldsymbol{L}}$ that correspond to the $\boldsymbol{LU}$ decomposition of $\boldsymbol{A}$ with $\tilde{\boldsymbol{T}}_k$ and $\tilde{\boldsymbol{U}}$ defined by formulae (5.76) and (5.77), respectively.
2. Compute the solution of the 2×2 system:
$$10^{-3}x+y=5, \quad x-y=6,$$
using standard Gaussian elimination and Gaussian elimination with pivoting. Conduct all computations with two significant digits (decimal). Compare and explain the results.
3.* Show that when performing Gaussian elimination with partial pivoting, the $\boldsymbol{LU}$ factorization is obtained in the form:
$$\boldsymbol{PA}=\boldsymbol{LU},$$
where $\boldsymbol{P}$ is the composition (i.e., product) of all permutation matrices used at every stage of the algorithm.
4. Consider a boundary value problem for the second order ordinary differential equation:
$$\frac{d^2u}{dx^2}-u=f(x), \quad x\in[0,1],$$
$$u(0)=0, \quad u(1)=0,$$
where $u=u(x)$ is the unknown function and $f=f(x)$ is a given right-hand side. To solve this problem numerically, we first partition the interval $0\le x\le 1$ into N equal subintervals and thus build a uniform grid of $N+1$ nodes: $x_j=j\cdot h$, $h=1/N$, $j=0,1,2,\dots,N$. Then, instead of looking for the continuous function $u=u(x)$ we will be looking for its approximate table of values $\{u_0,u_1,u_2,\dots,u_N\}$ at the grid nodes $x_0,x_1,x_2,\dots,x_N$, respectively.

   At every interior node x_j, $j=1,2,\dots,N-1$, we approximately replace the second derivative by the difference quotient (for more detail, see Section 9.2):
$$\left.\frac{d^2u}{dx^2}\right|_{x=x_j}\approx\frac{u_{j+1}-2u_j+u_{j-1}}{h^2},$$
and arrive at the following finite-difference counterpart of the original problem (a central-difference scheme):
$$\frac{u_{j+1}-2u_j+u_{j-1}}{h^2}-u_j=f_j, \quad j=1,2,\dots,N-1,$$
$$u_0=0, \quad u_N=0,$$

where $f_j \equiv f(x_j)$ is the discrete right-hand side assumed given, and u_j is the unknown discrete solution.

a) Write down the previous scheme as a system of linear algebraic equations ($N-1$ equations with $N-1$ unknowns) both in the canonical form and in an operator form with the operator specified by an $(N-1)\times(N-1)$ matrix.

b) Show that the system is tri-diagonal and satisfies the sufficient conditions of Section 5.4.2, so that the tri-diagonal elimination can be applied.

c) Assume that the exact solution u is known: $u(x) = \sin(\pi x)e^x$. Then, $f(x) = (-\pi^2\sin(\pi x) + 2\pi\cos(\pi x))e^x$. Implement on the computer the tri-diagonal elimination using double precision arithmetic. Solve the system for the right-hand side $f_j = f(x_j)$ on a sequence of grids with $N = 32, 64, 128, 256$, and 512. On each grid, calculate the relative error in the maximum norm:

$$\varepsilon_\infty(N) = \frac{\max\limits_{1\le j\le N-1} |u(x_j) - u_j|}{\max\limits_{1\le j\le N-1} |u(x_j)|}.$$

By plotting $\log_2(\varepsilon_\infty(N))$ vs. $\log_2 N$, show that every time the grid is refined by a factor of 2, the error drops by approximately a factor of 4. This should indicate the second order of grid convergence of the scheme.

5.5 Minimization of Quadratic Functions and Its Relation to Linear Systems

Consider a Euclidean space $\mathbb{R}^n$ of vectors $\boldsymbol{x}$ with n real components. Let $\boldsymbol{A}$ be a linear operator, $\boldsymbol{A}: \mathbb{R}^n \longmapsto \mathbb{R}^n, \boldsymbol{f} \in \mathbb{R}^n$ be a fixed element of the vector space $\mathbb{R}^n$, and c be a real constant. The function $F = F(\boldsymbol{x})$ of the argument $\boldsymbol{x}$ defined as:

$$F(\boldsymbol{x}) = (\boldsymbol{A}\boldsymbol{x},\boldsymbol{x}) - 2(\boldsymbol{f},\boldsymbol{x}) + c \tag{5.79}$$

is called a *quadratic function*.

Note that since $(\boldsymbol{A}\boldsymbol{x},\boldsymbol{x}) = (\boldsymbol{x},\boldsymbol{A}^*\boldsymbol{x}) = (\boldsymbol{A}^*\boldsymbol{x},\boldsymbol{x})$, the quadratic function $F(\boldsymbol{x})$ coincides with another quadratic function:

$$F(\boldsymbol{x}) = (\boldsymbol{A}^*\boldsymbol{x},\boldsymbol{x}) - 2(\boldsymbol{f},\boldsymbol{x}) + c,$$

and consequently, with the function:

$$F(\boldsymbol{x}) = \left(\frac{\boldsymbol{A}+\boldsymbol{A}^*}{2}\boldsymbol{x},\boldsymbol{x}\right) - 2(\boldsymbol{f},\boldsymbol{x}) + c.$$

Therefore, with no loss of generality we can assume that the operator $\boldsymbol{A}$ in formula (5.79) is self-adjoint. Otherwise, we can simply replace a given operator $\boldsymbol{A}$ with the corresponding self-adjoint operator $(\boldsymbol{A}+\boldsymbol{A}^*)/2$.

From now on, we will suppose that $\boldsymbol{A} = \boldsymbol{A}^*$ and $\boldsymbol{A} > 0$ in formula (5.79), i.e., $(\boldsymbol{Ax},\boldsymbol{x}) > 0$ for any $\boldsymbol{x} \in \mathbb{R}^n$, $\boldsymbol{x} \neq \boldsymbol{0}$. Given the function $F(\boldsymbol{x})$ of (5.79), let us formulate the problem of its minimization, i.e., the problem of finding a particular element $\boldsymbol{z} \in \mathbb{R}^n$ that delivers the minimum value to $F(\boldsymbol{x})$:

$$F(\boldsymbol{z}) = \min_{\boldsymbol{x} \in \mathbb{R}^n} F(\boldsymbol{x}). \tag{5.80}$$

It turns out that the minimization problem (5.80) and the problem of solving the system of linear algebraic equations:

$$\boldsymbol{Ax} = \boldsymbol{f}, \quad \text{where } \boldsymbol{A} = \boldsymbol{A}^* > 0, \tag{5.81}$$

are equivalent. We formulate this result in the form of a theorem.

THEOREM 5.7
Let $\boldsymbol{A} = \boldsymbol{A}^* > 0$. There is one and only one element $\boldsymbol{z} \in \mathbb{R}^n$ that delivers a minimum to the quadratic function $F(\boldsymbol{x})$ of (5.79). This vector $\boldsymbol{z}$ is the solution to the linear system (5.81).

PROOF As the operator $\boldsymbol{A}$ is positive definite, it is non-singular. Consequently, system (5.81) has a unique solution $\boldsymbol{z} \in \mathbb{R}^n$. Let us now show that for any $\boldsymbol{\delta} \in \mathbb{R}^n$, $\boldsymbol{\delta} \neq \boldsymbol{0}$, we have $F(\boldsymbol{z}+\boldsymbol{\delta}) > F(\boldsymbol{z})$:

$$\begin{aligned}
F(\boldsymbol{z}+\boldsymbol{\delta}) &= (\boldsymbol{A}(\boldsymbol{z}+\boldsymbol{\delta}),\boldsymbol{z}+\boldsymbol{\delta}) - 2(\boldsymbol{f},\boldsymbol{z}+\boldsymbol{\delta}) + c = \\
&= [(\boldsymbol{Az},\boldsymbol{z}) - 2(\boldsymbol{f},\boldsymbol{z}) + c] + 2(\boldsymbol{Az},\boldsymbol{\delta}) - 2(\boldsymbol{f},\boldsymbol{\delta}) + (\boldsymbol{A\delta},\boldsymbol{\delta}) \\
&= F(\boldsymbol{z}) + 2(\boldsymbol{Az}-\boldsymbol{f},\boldsymbol{\delta}) + (\boldsymbol{A\delta},\boldsymbol{\delta}) \\
&= F(\boldsymbol{z}) + (\boldsymbol{A\delta},\boldsymbol{\delta}) > F(\boldsymbol{z}).
\end{aligned}$$

Consequently, the solution $\boldsymbol{z}$ of system (5.81) indeed delivers a minimum to the function $F(\boldsymbol{x})$, because any nonzero deviation $\boldsymbol{\delta}$ implies a larger value of the function. ∎

Equivalence of problems (5.80) and (5.81) established by Theorem 5.7 allows one to reduce the solution of either problem to the solution of the other problem.

Note that linear systems of type (5.81) that have a self-adjoint positive definite matrix represent an important class of systems. Indeed, a general linear system

$$\boldsymbol{Ax} = \boldsymbol{f},$$

where $\boldsymbol{A} : \mathbb{R}^n \longmapsto \mathbb{R}^n$ is an arbitrary non-singular operator, can be easily reduced to a new system of type (5.81): $\boldsymbol{Cx} = \boldsymbol{g}$, where $\boldsymbol{C} = \boldsymbol{C}^* > 0$. To do so, one merely needs to set: $\boldsymbol{C} = \boldsymbol{A}^*\boldsymbol{A}$ and $\boldsymbol{g} = \boldsymbol{A}^*\boldsymbol{f}$. This approach, however, may only have a theoretical significance because in practice an extra matrix multiplication performed on a machine with finite precision is prone to generating large additional errors.

However, linear systems with self-adjoint matrices can also arise naturally. For example, many boundary value problems for elliptic partial differential equations can be interpreted as the Lagrange-Euler problems for minimizing certain quadratic functionals. It is therefore natural to expect that a "correct," i.e., appropriate, discretization of the corresponding variational problem will lead to a problem of minimizing a quadratic function on a finite-dimensional space. The latter problem will, in turn, be equivalent to the linear system (5.81) according to Theorem 5.7.

Exercises

1. Consider a quadratic function of two scalar arguments x_1 and x_2:

$$F(x_1,x_2) = x_1^2 + 2x_1x_2 + 4x_2^2 - 2x_1 + 3x_2 + 5.$$

Recast this function in the form (5.79) as a function of the vector argument $\boldsymbol{x} = \begin{bmatrix} x_1 \\ x_2 \end{bmatrix}$ that belongs to the Euclidean space $\mathbb{R}^2$ supplied with the inner product $(\boldsymbol{x},\boldsymbol{y}) = x_1y_1 + x_2y_2$. Verify that $\boldsymbol{A} = \boldsymbol{A}^* > 0$, and solve the corresponding problem (5.81).

2.* Recast the function $F(x_1,x_2)$ from the previous exercise in the form:

$$F(\boldsymbol{x}) = [\boldsymbol{A}\boldsymbol{x},\boldsymbol{x}]_{\boldsymbol{B}} - 2[\boldsymbol{f},\boldsymbol{x}]_{\boldsymbol{B}} + c,$$

where $[\boldsymbol{x},\boldsymbol{y}]_{\boldsymbol{B}} = (\boldsymbol{B}\boldsymbol{x},\boldsymbol{y})$, $\boldsymbol{B} = \begin{bmatrix} 1 & 0 \\ 0 & 2 \end{bmatrix}$, and $\boldsymbol{A} = \boldsymbol{A}^*$ in the sense of the inner product $[\boldsymbol{x},\boldsymbol{y}]_{\boldsymbol{B}}$.

5.6 The Method of Conjugate Gradients

Consider a system of linear algebraic equations:

$$\boldsymbol{A}\boldsymbol{x} = \boldsymbol{f}, \quad \boldsymbol{A} = \boldsymbol{A}^* > 0, \quad \boldsymbol{f} \in \mathbb{R}^n, \quad \boldsymbol{x} \in \mathbb{R}^n, \tag{5.82}$$

where $\mathbb{R}^n$ is an n-dimensional real vector space with the scalar product $(\boldsymbol{x},\boldsymbol{y})$, and $\boldsymbol{A} : \mathbb{R}^n \longmapsto \mathbb{R}^n$ is a self-adjoint positive definite operator with respect to this product. To introduce the method of conjugate gradients, we will use the equivalence of system (5.82) and the problem of minimizing the quadratic function $F(\boldsymbol{x}) = (\boldsymbol{A}\boldsymbol{x},\boldsymbol{x}) - 2(\boldsymbol{f},\boldsymbol{x})$ that was established in Section 5.5, see Theorem 5.7.

5.6.1 Construction of the Method

In the core of the method is the sequence of vectors $\boldsymbol{x}^{(p)} \in \mathbb{R}^n$, $p = 0,1,2,\ldots$, which is to be built so that it would converge to the vector that minimizes the value of $F(\boldsymbol{x})$. Along with this sequence, the method requires constructing another sequence of vectors $\boldsymbol{d}^{(p)} \in \mathbb{R}^k$, $p = 0,1,2,\ldots$, that would provide the so-called descent directions. In other words, we define $\boldsymbol{x}^{(p+1)} = \boldsymbol{x}^{(p)} + \alpha_p\boldsymbol{d}^{(p)}$, and choose the scalar α_p in order to minimize the composite function $F(\boldsymbol{x}^{(p)} + \alpha\boldsymbol{d}^{(p)})$.

Note that the original function $F(\boldsymbol{x}) = (\boldsymbol{A}\boldsymbol{x},\boldsymbol{x}) - 2(\boldsymbol{f},\boldsymbol{x}) + c$ is a function of the vector argument $\boldsymbol{x} \in \mathbb{R}^n$. Given two fixed vectors $\boldsymbol{x}^{(p)} \in \mathbb{R}^n$ and $\boldsymbol{d}^{(p)} \in \mathbb{R}^n$, we can consider the function $F(\boldsymbol{x}^{(p)} + \alpha\boldsymbol{d}^{(p)})$, which is a function of the scalar argument α. Let us find the value of α that delivers a minimum to this function:

$$\begin{aligned}\frac{dF}{d\alpha} &= (\boldsymbol{A}\boldsymbol{d}^{(p)}, (\boldsymbol{x}^{(p)} + \alpha\boldsymbol{d}^{(p)})) + (\boldsymbol{A}(\boldsymbol{x}^{(p)} + \alpha\boldsymbol{d}^{(p)}), \boldsymbol{d}^{(p)}) - 2(\boldsymbol{f}, \boldsymbol{d}^{(p)}) \\ &= (\boldsymbol{A}\boldsymbol{d}^{(p)}, \boldsymbol{x}^{(p)}) + (\boldsymbol{A}\boldsymbol{x}^{(p)}, \boldsymbol{d}^{(p)}) + 2\alpha(\boldsymbol{A}\boldsymbol{d}^{(p)}, \boldsymbol{d}^{(p)}) - 2(\boldsymbol{f}, \boldsymbol{d}^{(p)}) \\ &= 2(\boldsymbol{A}\boldsymbol{x}^{(p)}, \boldsymbol{d}^{(p)}) + 2\alpha(\boldsymbol{A}\boldsymbol{d}^{(p)}, \boldsymbol{d}^{(p)}) - 2(\boldsymbol{f}, \boldsymbol{d}^{(p)}) = 0.\end{aligned}$$

Consequently,

$$\alpha_p = \arg\min_{\alpha} F(\boldsymbol{x}^{(p)} + \alpha\boldsymbol{d}^{(p)}) = -\frac{(\boldsymbol{A}\boldsymbol{x}^{(p)} - \boldsymbol{f}, \boldsymbol{d}^{(p)})}{(\boldsymbol{A}\boldsymbol{d}^{(p)}, \boldsymbol{d}^{(p)})} = -\frac{(\boldsymbol{r}^{(p)}, \boldsymbol{d}^{(p)})}{(\boldsymbol{A}\boldsymbol{d}^{(p)}, \boldsymbol{d}^{(p)})}, \qquad (5.83)$$

where the vector quantity $\boldsymbol{r}^{(p)} \stackrel{\text{def}}{=} \boldsymbol{A}\boldsymbol{x}^{(p)} - \boldsymbol{f}$ is called *the residual.*

Consider $\boldsymbol{x}^{(p+1)} = \boldsymbol{x}^{(p)} + \alpha_p\boldsymbol{d}^{(p)}$, where α_p is given by (5.83), and introduce $F(\boldsymbol{x}^{(p+1)} + \alpha\boldsymbol{d}^{(p)})$. Clearly, $\arg\min_\alpha F(\boldsymbol{x}^{(p+1)} + \alpha\boldsymbol{d}^{(p)}) = 0$, i.e., $\frac{d}{d\alpha}F(\boldsymbol{x}^{(p+1)} + \alpha\boldsymbol{d}^{(p)})\big|_{\alpha=0} = 0$. Consequently, according to formula (5.83), we will have $(\boldsymbol{r}^{(p+1)}, \boldsymbol{d}^{(p)}) = 0$, i.e., the residual $\boldsymbol{r}^{(p+1)} = \boldsymbol{A}\boldsymbol{x}^{(p+1)} - \boldsymbol{f}$ of the next vector $\boldsymbol{x}^{(p+1)}$ will be orthogonal to the previous descent direction $\boldsymbol{d}^{(p)}$.

Altogether, the sequence of vectors $\boldsymbol{x}^{(p)}$, $p = 0, 1, 2, \ldots$, will be built precisely so that to maintain orthogonality of the residuals to all the previous descent directions, rather than only to the immediately preceding direction. Let $\boldsymbol{d}$ be given and assume that $(\boldsymbol{r}^{(p)}, \boldsymbol{d}) = 0$, where $\boldsymbol{r}^{(p)} = \boldsymbol{A}\boldsymbol{x}^{(p)} - \boldsymbol{f}$. Let also $\boldsymbol{x}^{(p+1)} = \boldsymbol{x}^{(p)} + \boldsymbol{g}$, and require that $(\boldsymbol{r}^{(p+1)}, \boldsymbol{d}) = 0$. Since $\boldsymbol{r}^{(p+1)} = \boldsymbol{A}\boldsymbol{x}^{(p+1)} - \boldsymbol{f} = \boldsymbol{A}\boldsymbol{g} + \boldsymbol{r}^{(p)}$, we obtain $(\boldsymbol{A}\boldsymbol{g}, \boldsymbol{d}) = 0$. In other words, we need to require that the increment $\boldsymbol{g} = \boldsymbol{x}^{(p+1)} - \boldsymbol{x}^{(p)}$ be orthogonal to $\boldsymbol{d}$ in the sense of the inner product $[\boldsymbol{g}, \boldsymbol{d}]_{\boldsymbol{A}} = (\boldsymbol{A}\boldsymbol{g}, \boldsymbol{d})$.

The vectors that satisfy the equality $[\boldsymbol{g}, \boldsymbol{d}]_{\boldsymbol{A}} = 0$ are often called $\boldsymbol{A}$-orthogonal or $\boldsymbol{A}$-conjugate. We therefore see that to maintain the orthogonality of $\boldsymbol{r}^{(p)}$ to the previous descent directions, the latter must be mutually $\boldsymbol{A}$-conjugate. The descent directions are directions for the one-dimensional minimization of the function F, see formula (5.83). As minimization of multi-variable functions is traditionally performed along their local gradients, the descent directions that we are going to use are called *conjugate gradients*. This is where the method draws its name from, even though none of these directions except $\boldsymbol{d}^{(0)}$ will actually be a gradient of F.

Define $\boldsymbol{d}^{(0)} = \boldsymbol{r}^{(0)} = \boldsymbol{A}\boldsymbol{x}^{(0)} - \boldsymbol{f}$, where $\boldsymbol{x}^{(0)} \in \mathbb{R}^n$ is arbitrary, and set

$$\boldsymbol{d}^{(p+1)} = \boldsymbol{r}^{(p+1)} - \beta_p\boldsymbol{d}^{(p)}, \quad p = 0, 1, 2, \ldots, \qquad (5.84)$$

where the real scalars β_p are to be chosen so that to obtain the $\boldsymbol{A}$-conjugate descent directions:

$$[\boldsymbol{d}^{(j)}, \boldsymbol{d}^{(p+1)}]_{\boldsymbol{A}} = 0, \quad j = 0, 1, \ldots, p. \qquad (5.85)$$

For $j = p$, this immediately yields the following value of β_p:

$$\beta_p = \frac{(\boldsymbol{A}\boldsymbol{d}^{(p)}, \boldsymbol{r}^{(p+1)})}{(\boldsymbol{A}\boldsymbol{d}^{(p)}, \boldsymbol{d}^{(p)})}, \quad p = 0, 1, 2, \ldots. \qquad (5.86)$$

Next, we will use induction with respect to p to show that $[\boldsymbol{d}^{(j)},\boldsymbol{d}^{(p+1)}]_A = 0$ for $j = 0,1,\ldots,p-1$. Indeed, if $p = 0$, then we only need to guarantee $[\boldsymbol{d}^{(0)},\boldsymbol{d}^{(1)}]_A = 0$, and we simply choose $\beta_0 = (\boldsymbol{A}\boldsymbol{d}^{(0)},\boldsymbol{r}^{(1)})/(\boldsymbol{A}\boldsymbol{d}^{(0)},\boldsymbol{d}^{(0)})$ as prescribed by formula (5.86). Next, assume that the desired property has already been established for all integers $0,1,\ldots,p$, and we want to justify it for $p+1$. In other words, suppose that $[\boldsymbol{d}^{(j)},\boldsymbol{d}^{(p)}]_A = 0$ for $j = 0,1,\ldots,p-1$ and for all integer p between 0 and some prescribed positive value. Then we can write:

$$[\boldsymbol{d}^{(j)},\boldsymbol{d}^{(p+1)}]_A = [\boldsymbol{d}^{(j)},(\boldsymbol{r}^{(p+1)} - \beta_p\boldsymbol{d}^{(p)})]_A = [\boldsymbol{d}^{(j)},\boldsymbol{r}^{(p+1)}]_A, \tag{5.87}$$

because $[\boldsymbol{d}^{(j)},\boldsymbol{d}^{(p)}]_A = 0$ for $j < p$.

In its own turn, the sequence $\boldsymbol{x}^{(p)}$, $p = 0,1,2,\ldots$, is built so that the residuals be orthogonal to all previous descent directions, i.e., $(\boldsymbol{d}^{(j)},\boldsymbol{r}^{(p+1)}) = 0$, $j = 0,1,2,\ldots,p$. We therefore conclude that the residual $\boldsymbol{r}^{(p+1)}$ is orthogonal to the entire linear span of the $k+1$ descent vectors: $\mathrm{span}\{\boldsymbol{d}^{(0)},\boldsymbol{d}^{(1)},\ldots,\boldsymbol{d}^{(p)}\}$. According to the assumption of the induction, all these vectors are $\boldsymbol{A}$-conjugate and as such, linearly independent. Moreover, we have:

$$\begin{aligned}
\boldsymbol{r}^{(0)} &= \boldsymbol{d}^{(0)}, \\
\boldsymbol{r}^{(1)} &= \boldsymbol{d}^{(1)} + \beta_0\boldsymbol{d}^{(0)}, \\
&\cdots\cdots\cdots\cdots\cdots\cdots \\
\boldsymbol{r}^{(p)} &= \boldsymbol{d}^{(p)} + \beta_{p-1}\boldsymbol{d}^{(p-1)},
\end{aligned} \tag{5.88}$$

and consequently, the residuals $\boldsymbol{r}^{(0)},\boldsymbol{r}^{(1)},\ldots,\boldsymbol{r}^{(p)}$ are also linearly independent, and $\mathrm{span}\{\boldsymbol{d}^{(0)},\boldsymbol{d}^{(1)},\ldots,\boldsymbol{d}^{(p)}\} = \mathrm{span}\{\boldsymbol{r}^{(0)},\boldsymbol{r}^{(1)},\ldots,\boldsymbol{r}^{(p)}\}$. In other words,

$$\boldsymbol{r}^{(p+1)} \perp \mathrm{span}\{\boldsymbol{r}^{(0)},\boldsymbol{r}^{(1)},\ldots,\boldsymbol{r}^{(p)}\}. \tag{5.89}$$

Finally, for every $j = 0,1,\ldots,p-1$ we can write starting from (5.87):

$$\begin{aligned}
[\boldsymbol{d}^{(j)},\boldsymbol{r}^{(p+1)}]_A &= (\boldsymbol{A}\boldsymbol{d}^{(j)},\boldsymbol{r}^{(p+1)}) = \frac{1}{\alpha_j}(\boldsymbol{A}(\boldsymbol{x}^{(j+1)} - \boldsymbol{x}^{(j)}),\boldsymbol{r}^{(p+1)}) \\
&= \frac{1}{\alpha_j}((\boldsymbol{r}^{(j+1)} - \boldsymbol{r}^{(j)}),\boldsymbol{r}^{(p+1)}) = 0
\end{aligned} \tag{5.90}$$

because of the relation (5.89). Consequently, formulae (5.87), (5.90) imply (5.85). This completes the induction based argument and shows that the descent directions chosen according to (5.84), (5.86) are indeed mutually $\boldsymbol{A}$-orthogonal.

Altogether, the method of conjugate gradients for solving system (5.82) can now be summarized as follows. We start with an arbitrary $\boldsymbol{x}^{(0)} \in \mathbb{R}^n$ and simultaneously build two sequences of vectors in the space $\mathbb{R}^n$: $\{\boldsymbol{x}^{(p)}\}$ and $\{\boldsymbol{d}^{(p)}\}$, $p = 0,1,2,\ldots$. The first descent direction is chosen as $\boldsymbol{d}^{(0)} = \boldsymbol{r}^{(0)} = \boldsymbol{A}\boldsymbol{x}^{(0)} - \boldsymbol{f}$. Then, for every $p = 0,1,2,\ldots$ we compute:

$$\alpha_p = -\frac{(\boldsymbol{r}^{(p)},\boldsymbol{d}^{(p)})}{(\boldsymbol{A}\boldsymbol{d}^{(p)},\boldsymbol{d}^{(p)})}, \qquad \boldsymbol{x}^{(p+1)} = \boldsymbol{x}^{(p)} + \alpha_p\boldsymbol{d}^{(p)}, \tag{5.91a}$$

$$\beta_p = \frac{(\boldsymbol{A}\boldsymbol{d}^{(p)},\boldsymbol{r}^{(p+1)})}{(\boldsymbol{A}\boldsymbol{d}^{(p)},\boldsymbol{d}^{(p)})}, \qquad \boldsymbol{d}^{(p+1)} = \boldsymbol{r}^{(p+1)} - \beta_p\boldsymbol{d}^{(p)}, \tag{5.91b}$$

while in between the stages (5.91a) and (5.91b) we also evaluate the residual: $\boldsymbol{r}^{(p+1)} = \boldsymbol{A}\boldsymbol{x}^{(p+1)} - \boldsymbol{f}$.

Note that in practice the descent vectors $\boldsymbol{d}^{(p)}$ can be eliminated, and the formulae of the method can be rewritten using only $\boldsymbol{x}^{(p)}$ and the residuals $\boldsymbol{r}^{(p)}$; this is the subject of Exercise 2.

THEOREM 5.8

There is a positive integer number $p_0 \leq n$ such that the vector $\boldsymbol{x}^{(p_0)}$ obtained by formulae (5.91) of the conjugate gradients method yields the exact solution of the linear system (5.82).

PROOF By construction of the method, all the descent directions $\boldsymbol{d}^{(p)}$ are $\boldsymbol{A}$-conjugate and as such, are linearly independent. Clearly, any linearly independent system in the n-dimensional space $\mathbb{R}^n$ has no more than n vectors. Consequently, there may be at most n different descent directions: $\{\boldsymbol{d}^{(0)}, \boldsymbol{d}^{(1)}, \ldots, \boldsymbol{d}^{(n-1)}\}$. According to formulae (5.88), the residuals $\boldsymbol{r}^{(p)}$ are also linearly independent, and we have established that each subsequent residual $\boldsymbol{r}^{(p+1)}$ is orthogonal to the linear span of all the previous ones, see formula (5.89). Therefore, there may be two scenarios. Either at some intermediate step $p_0 < n$ it so happens that $\boldsymbol{r}^{(p_0)} = \boldsymbol{0}$, i.e., $\boldsymbol{A}\boldsymbol{x}^{(p_0)} = \boldsymbol{f}$, or at the step number n we have:

$$\boldsymbol{r}^{(n)} \perp \text{span}\{\boldsymbol{r}^{(0)}, \boldsymbol{r}^{(1)}, \ldots, \boldsymbol{r}^{(n-1)}\}.$$

Hence $\boldsymbol{r}^{(n)}$ is orthogonal to all available linearly independent directions in the space $\mathbb{R}^n$, and as such, $\boldsymbol{r}^{(n)} = \boldsymbol{0}$. Consequently, $k_0 = n$ and $\boldsymbol{A}\boldsymbol{x}^{(n)} = \boldsymbol{f}$. ∎

In simple words, the method of conjugate gradients minimizes the quadratic function $F(\boldsymbol{x}) = (\boldsymbol{A}\boldsymbol{x}, \boldsymbol{x}) - 2(\boldsymbol{f}, \boldsymbol{x})$ along a sequence of linearly independent ($\boldsymbol{A}$-orthogonal) directions in the space $\mathbb{R}^n$. When the pool of such directions is exhausted (altogether there may be no more than n), the method yields a global minimum, which coincides with the exact solution of system (5.82) according to Theorem 5.7.

According to Theorem 5.8, the method of conjugate gradients is a direct method. Additionally we should note that the sequence of vectors $\boldsymbol{x}^{(0)}$, $\boldsymbol{x}^{(1)}$, ..., $\boldsymbol{x}^{(p)}$, ... provides increasingly more accurate approximations to the exact solution when the number p increases. For a given small $\varepsilon > 0$, the error $\|\boldsymbol{x} - \boldsymbol{x}^{(p)}\|$ of the approximation $\boldsymbol{x}^{(p)}$ may become smaller than ε already for $p \ll n$, see Section 6.2.2. That is why the method of conjugate gradients is mostly used in the capacity of an iterative method these days,[5] especially when the condition number $\mu(\boldsymbol{A})$ is moderate and the dimension n of the system is large.

When the dimension of the system n is large, computing the exact solution of system (5.82) with the help of conjugate gradients may require a large number of steps

[5]As opposed to direct methods that yield the exact solution after a finite number of operations, iterative methods provide a sequence of increasingly accurate approximations to the exact solution, see Chapter 6.

(5.91a), (5.91b). If the situation is also "exacerbated" by a large condition number $\mu(\boldsymbol{A})$, the computation may encounter an obstacle: numerical stability can be lost when obtaining the successive approximations $\boldsymbol{x}^{(p)}$. However, when the condition number $\mu(\boldsymbol{A})$ is not excessively large, the method of conjugate gradients has an important advantage compared to the elimination methods of Section 5.4. We comment on this advantage in Section 5.6.2.

5.6.2 Flexibility in Specifying the Operator A

From formulae (5.91a), (5.91b) it is clear that to implement the method of conjugate gradients we only need to be able to perform two fundamental operations: for a given $\boldsymbol{y} \in \mathbb{R}^n$ compute $\boldsymbol{z} = \boldsymbol{A}\boldsymbol{y}$, and for a given pair of vectors $\boldsymbol{y}$, $\boldsymbol{z} \in \mathbb{R}^n$ compute their scalar product $(\boldsymbol{y}, \boldsymbol{z})$. Consequently, the system (5.82) does not necessarily have to be specified in its canonical form. It is not even necessary to store the entire matrix $\boldsymbol{A}$ that has n^2 entries, whereas the vectors $\boldsymbol{y}$ and $\boldsymbol{z} = \boldsymbol{A}\boldsymbol{y}$ in their coordinate form are represented by only n real numbers each. In fact, the method is very well suited for the case when the operator $\boldsymbol{A}$ is defined as a procedure of obtaining $\boldsymbol{z} = \boldsymbol{A}\boldsymbol{y}$ once $\boldsymbol{y}$ is given, for example, by finite-difference formulae of type (5.11).

5.6.3 Computational Complexity

The overall computational complexity of obtaining the exact solution of system (5.82) using the conjugate gradients method is $\mathcal{O}(n^3)$ arithmetic operations. Indeed, the computation of each step by formulae (5.91a), (5.91b) clearly requires $\mathcal{O}(n^2)$ operations, while altogether no more than n steps is needed according to Theorem 5.8. We see that asymptotically the computational cost of the method of conjugate gradients is the same as that of the Gaussian elimination.

Note also that the two key elements of the method identified in Section 5.6.2, i.e., the matrix-vector multiplication and the computation of a scalar product, can be efficiently implemented on modern computer platforms that involve vector and/or multi-processor architectures. This is true regardless of whether the operator $\boldsymbol{A}$ is specified explicitly by its matrix or otherwise, say, by a formula of type (5.11) that involves grid functions. Altogether, a vector or multi-processor implementation of the algorithm may considerably reduce its total execution time.

Historically, a version of what is known as the conjugate gradients method today was introduced by C. Lanczos in the early fifties. In the literature, the method of conjugate gradients is sometimes referred to as the Lanczos algorithm.

Exercises

1. Show that the residuals of the conjugate gradients method satisfy the following relation that contains three consecutive terms:

$$\boldsymbol{r}^{(p+1)} = \alpha_p \boldsymbol{A}\boldsymbol{r}^{(p)} + \left(1 - \alpha_p \frac{\beta_{p-1}}{\alpha_{p-1}}\right) \boldsymbol{r}^{(p)} + \alpha_p \frac{\beta_{p-1}}{\alpha_{p-1}} \boldsymbol{r}^{(p-1)}.$$

2.* Show that the method of conjugate gradients (5.91) can be recast as follows:

$$\begin{aligned} \boldsymbol{x}^{(1)} &= (\boldsymbol{I} - \tau_1 \boldsymbol{A})\boldsymbol{x}^{(0)} + \tau_1 \boldsymbol{f}, \\ \boldsymbol{x}^{(p+1)} &= \gamma_{p+1}(\boldsymbol{I} - \tau_{p+1}\boldsymbol{A})\boldsymbol{x}^{(p)} + (1-\gamma_{p+1})\boldsymbol{x}^{(p-1)} + \gamma_{p+1}\tau_{p+1}\boldsymbol{f}, \end{aligned}$$

where

$$\tau_{p+1} = \frac{(\boldsymbol{r}^{(p)}, \boldsymbol{r}^{(p)})}{(\boldsymbol{A}\boldsymbol{r}^{(p)}, \boldsymbol{r}^{(p)})}, \quad \boldsymbol{r}^{(p)} = \boldsymbol{A}\boldsymbol{x}^{(p)} - \boldsymbol{f}, \quad p = 0, 1, 2, \dots,$$

$$\gamma_1 = 1, \quad \gamma_{p+1} = \left(1 - \frac{\tau_{p+1}}{\tau_p} \frac{(\boldsymbol{r}^{(p)}, \boldsymbol{r}^{(p)})}{(\boldsymbol{r}^{(p-1)}, \boldsymbol{r}^{(p-1)})} \frac{1}{\gamma_p}\right)^{-1}, \quad p = 1, 2, \dots.$$

5.7 Finite Fourier Series

Along with the universal methods, special techniques have been developed in the literature that are "fine tuned" for solving only particular, narrow classes of linear systems. Trading off the universality brings along a considerably better efficiency than that of the general methods. A typical example is given by the direct methods based on representing the solution to a linear system in the form of a finite Fourier series.

Let $\mathbb{L}$ be a Euclidean or unitary vector space of dimension n, and let us consider a linear system in the operator form:

$$\boldsymbol{A}\boldsymbol{x} = \boldsymbol{f}, \quad \boldsymbol{x} \in \mathbb{L}, \quad \boldsymbol{f} \in \mathbb{L}, \quad \boldsymbol{A} : \mathbb{L} \longmapsto \mathbb{L}, \tag{5.92}$$

where the operator $\boldsymbol{A}$ is assumed self-adjoint, $\boldsymbol{A} = \boldsymbol{A}^*$. Every self-adjoint linear operator has an orthonormal basis composed of its eigenvectors. Let us denote by $\{\boldsymbol{e}_1, \boldsymbol{e}_2, \dots, \boldsymbol{e}_n\}$ the eigenvectors of the operator $\boldsymbol{A}$ that form an orthonormal basis in $\mathbb{L}$, and by $\{\lambda_1, \lambda_2, \dots, \lambda_n\}$ the corresponding eigenvalues that are all real, so that

$$\boldsymbol{A}\boldsymbol{e}_j = \lambda_j \boldsymbol{e}_j, \quad j = 1, 2, \dots, n. \tag{5.93}$$

As we are assuming that $\boldsymbol{A}$ is non-singular, we have $\lambda_j \neq 0$, $j = 1, 2, \dots, n$. For the solution methodology that we are about to build, not only do we need the existence of the basis $\{\boldsymbol{e}_1, \boldsymbol{e}_2, \dots, \boldsymbol{e}_n\}$, but we need to explicitly know these eigenvectors and the corresponding eigenvalues of (5.93). This clearly narrows the scope of admissible operators $\boldsymbol{A}$, yet it facilitates the construction of a very efficient computational algorithm.

To solve system (5.92), we expand both its unknown solution $\boldsymbol{x}$ and its given right-hand side $\boldsymbol{f}$ with respect to the basis $\{\boldsymbol{e}_1, \boldsymbol{e}_2, \dots, \boldsymbol{e}_n\}$, i.e., represent the vectors $\boldsymbol{x}$ and $\boldsymbol{f}$ in the form of their finite Fourier series:

$$\begin{aligned} \boldsymbol{f} &= f_1 \boldsymbol{e}_1 + f_2 \boldsymbol{e}_2 + \dots + f_n \boldsymbol{e}_n, \quad \text{where} \quad f_j = (\boldsymbol{f}, \boldsymbol{e}_j), \\ \boldsymbol{x} &= x_1 \boldsymbol{e}_1 + x_2 \boldsymbol{e}_2 + \dots + x_n \boldsymbol{e}_n, \quad \text{where} \quad x_j = (\boldsymbol{x}, \boldsymbol{e}_j). \end{aligned} \tag{5.94}$$

Clearly, the coefficients x_j, although formally defined as $x_j = (\boldsymbol{x}, \boldsymbol{e}_j)$, $j = 1, 2, \dots, n$, are not known yet. To determine these coefficients, we substitute expressions (5.94) into the system (5.92), and using formula (5.93) obtain:

$$\sum_{j=1}^{n} \lambda_j x_j \boldsymbol{e}_j = \sum_{j=1}^{n} f_j \boldsymbol{e}_j. \tag{5.95}$$

Next, we take the scalar product of both the left-hand side and the right-hand side of equality (5.95) with every vector $\boldsymbol{e}_j$, $j = 1, 2, \dots, n$. As the eigenvectors are orthonormal: $(\boldsymbol{e}_i, \boldsymbol{e}_j) = \begin{cases} 1, & i = j, \\ 0, & i \neq j, \end{cases}$, we can find the coefficients x_j of the Fourier series (5.94) for the solution $\boldsymbol{x}$:

$$x_j = f_j / \lambda_j, \quad j = 1, 2, \dots, n. \tag{5.96}$$

We will illustrate the foregoing abstract scheme using the specific example of a finite-difference Dirichlet problem for the Poisson equation from Section 5.1.3:

$$-\Delta^{(h)} u^{(h)} = f^{(h)}, \quad u^{(h)} \in U^{(h)}, \quad f^{(h)} \in F^{(h)}. \tag{5.97}$$

In this case, one can explicitly write down the eigenvectors and eigenvalues of the operator $-\Delta^{(h)}$.

5.7.1 Fourier Series for Grid Functions

Consider a set of all real-valued functions $v = \{v_m\}$ defined on the grid $\boldsymbol{x}_m = mh$, $m = 0, 1, \dots, M$, $Mh = 1$, and equal to zero at the endpoints: $v_0 = v_M = 0$. The set of these functions, supplemented by the conventional component-wise operations of addition and multiplication by real numbers, forms a real vector space $\mathbb{R}^{M-1}$. The dimension of this space is indeed equal to $M - 1$ because the system of functions:

$$\tilde{\psi}_m^{(k)} = \begin{cases} 0, & m \neq k, \\ 1, & m = k, \end{cases} \qquad k = 1, 2, \dots, M - 1,$$

clearly provides a basis. This is easy to see, since any function $v = (v_0, v_1, \dots, v_m)$ with $v_0 = 0$ and $v_M = 0$ can be uniquely represented as a linear combination of the functions $\tilde{\psi}^{(1)}$, $\tilde{\psi}^{(2)}$, ..., $\tilde{\psi}^{(M-1)}$:

$$v = v_1 \tilde{\psi}^{(1)} + v_2 \tilde{\psi}^{(2)} + \dots + v_{M-1} \tilde{\psi}^{(M-1)}.$$

Let us introduce a scalar product in the space we are considering:

$$(v, w) = h \sum_{m=0}^{M} v_m w_m. \tag{5.98}$$

We will show that the following system of $M - 1$ functions:

$$\psi^{(k)} = \left\{ \sqrt{2} \sin \frac{k\pi m}{M} \right\}, \quad k = 1, 2, \dots, M - 1, \tag{5.99}$$

forms an orthonormal basis in our $M-1$-dimensional space, i.e.,

$$(\psi^{(k)},\psi^{(r)}) = \begin{cases} 0, & k \neq r, \\ 1, & k = r, \end{cases} \quad k,r = 1,2,\ldots,M-1. \tag{5.100}$$

For the proof, we first notice that

$$\begin{aligned} \sum_{m=0}^{M-1} \cos\frac{l\pi m}{M} &= \frac{1}{2}\sum_{m=0}^{M-1}\left(e^{il\pi m/M} + e^{-il\pi m/M}\right) \\ &= \frac{1}{2}\frac{1-e^{il\pi}}{1-e^{il\pi/M}} + \frac{1}{2}\frac{1-e^{-il\pi}}{1-e^{-il\pi/M}} = \begin{cases} 0, & l \text{ is even}, \\ 1, & l \text{ is odd}, \end{cases} \quad 0 < l < 2M. \end{aligned}$$

Then, for $k \neq r$ we have:

$$\begin{aligned} (\psi^{(k)},\psi^{(r)}) &= 2h\sum_{m=0}^{M} \sin\frac{k\pi m}{M}\sin\frac{r\pi m}{M} = 2h\sum_{m=0}^{M-1} \sin\frac{k\pi m}{M}\sin\frac{r\pi m}{M} \\ &= h\sum_{m=0}^{M-1}\cos\frac{(k-r)\pi m}{M} - h\sum_{m=0}^{M-1}\cos\frac{(k+r)\pi m}{M} = 0, \end{aligned}$$

and for $k = r$:

$$(\psi^{(k)},\psi^{(r)}) = h\sum_{m=0}^{M-1}\cos 0 - h\sum_{m=0}^{M-1}\cos\frac{2k\pi m}{M} = hM - h\cdot 0 = 1.$$

Therefore, the system of grid functions (5.99) is indeed orthonormal and as such, linearly independent. As the number of functions $M-1$ is the same as the dimension of the space, the functions $\psi^{(k)}$, $k = 1,2,\ldots,M-1$, do provide a basis.

Any grid function $v = (v_0, v_1, \ldots, v_M)$ with $v_0 = 0$ and $v_M = 0$ can be expanded with respect to the orthonormal basis (5.99):

$$v = c_1\psi^{(1)} + c_2\psi^{(2)} + \ldots + c_{M-1}\psi^{(M-1)},$$

or alternatively,

$$v = \sqrt{2}\sum_{k=1}^{M-1} c_k \sin\frac{k\pi m}{M}, \tag{5.101}$$

where the coefficients c_k are given by:

$$c_k = (v,\psi^{(k)}) = \sqrt{2}h\sum_{m=0}^{M} v_m \sin\frac{k\pi m}{M}. \tag{5.102}$$

It is clear that as the basis (5.99) is orthonormal, we have:

$$(v,v) = c_1^2 + c_2^2 + \ldots + c_{M-1}^2. \tag{5.103}$$

The sum (5.101) is precisely what is called the expansion of the grid function $v = \{v_m\}$ into a finite (or discrete) Fourier series. Equality (5.103) is a counterpart of the Parseval equality in the conventional theory of Fourier series.

In much the same way, finite Fourier series can be introduced for the functions defined on two-dimensional grids. Consider a uniform rectangular grid on the square $\{(x,y)|0 \le x \le 1,\ 0 \le y \le 1\}$:

$$\begin{aligned} x_{m_1} &= m_1 h, \quad m_1 = 0,1,\dots,M, \\ y_{m_2} &= m_2 h, \quad m_2 = 0,1,\dots,M, \end{aligned} \tag{5.104}$$

where $h = 1/M$ and M is a positive integer. A set of all real-valued functions $v = \{v_{m_1}, v_{m_2}\}$ on this grid that turn into zero at the boundary of the square, i.e., for $m_1 = 0, M$ and $m_2 = 0, M$, form a vector space of dimension $(M-1)^2$ once supplemented by the conventional component-wise operations of addition and multiplication by (real) scalars. An inner product can be introduced in this space:

$$(v,w) = h^2 \sum_{m_1,m_2=0}^{M} v_{m_1,m_2} w_{m_1,m_2}.$$

Then, it is easy to see that the system of grid functions:

$$\psi^{(k,l)} = \left\{ 2\sin\frac{k\pi m_1}{M}\sin\frac{l\pi m_2}{M} \right\}, \quad k,l = 1,2,\dots,M-1, \tag{5.105}$$

forms an orthonormal basis in the space that we are considering:

$$(\psi^{(k,l)}, \psi^{(r,s)}) = \begin{cases} 0, & k \ne r \text{ or } l \ne s, \\ 1, & k = r \text{ and } l = s. \end{cases} \tag{5.106}$$

This follows from formula (5.100), because:

$$\begin{aligned} (\psi^{(k,l)}, \psi^{(r,s)}) &= \sum_{m_1=0}^{M}\sum_{m_2=0}^{M} \left(2\sin\frac{k\pi m_1}{M}\sin\frac{l\pi m_2}{M}\right)\left(2\sin\frac{r\pi m_1}{M}\sin\frac{s\pi m_2}{M}\right) \\ &= \left(2\sum_{m_1=0}^{M}\sin\frac{k\pi m_1}{M}\sin\frac{r\pi m_1}{M}\right)\left(2\sum_{m_2=0}^{M}\sin\frac{l\pi m_2}{M}\sin\frac{s\pi m_2}{M}\right) \\ &= (\psi^{(k)}, \psi^{(r)})(\psi^{(l)}, \psi^{(s)}). \end{aligned}$$

Any function $v = \{v_{m_1,m_2}\}$ that is equal to zero at the boundary of the grid square can be expanded into a two-dimensional finite Fourier series [cf. formula (5.101)]:

$$v_{m_1,m_2} = 2\sum_{k,l=1}^{M-1} c_{kl}\sin\frac{k\pi m_1}{M}\sin\frac{l\pi m_2}{M}, \tag{5.107}$$

where $c_{kl} = (v, \psi^{(k,l)})$. In doing so, the Parseval equality holds [cf. formula (5.103)]:

$$(v,v) = \sum_{k,l=1}^{M-1} c_{kl}^2.$$

5.7.2 Representation of Solution as a Finite Fourier Series

By direct calculation one can easily make sure[6] that the following equalities hold:

$$-\Delta^{(h)}\psi^{(k,l)} = \frac{4}{h^2}\left(\sin^2\frac{k\pi}{2M} + \sin^2\frac{l\pi}{2M}\right)\psi^{(k,l)}, \tag{5.108}$$

where the operator $-\Delta^{(h)} : U^{(h)} \longmapsto U^{(h)}$ is defined by formulae (5.11), (5.10), and the space $U^{(h)}$ contains all discrete functions specified on the interior subset of the grid (5.104), i.e., for $m_1, m_2 = 1, 2, \dots, M-1$. Note that even though the functions $\psi^{(k,l)}$ of (5.105) are formally defined on the entire grid (5.104) rather than on its interior subset only, we can still assume that $\psi^{(k,l)} \in U^{(h)}$ with no loss of generality, because the boundary values of $\psi^{(k,l)}$ are fixed: $\psi^{(k,l)}_{m_1,m_2} = 0$ for $m_1, m_2 = 0, M$.

Equalities (5.108) imply that the grid functions $\psi^{(k,l)}$ that form an orthonormal basis in the space $U^{(h)}$, see formula (5.106), are eigenfunctions of the operator $-\Delta^{(h)} : U^{(h)} \longmapsto U^{(h)}$, whereas the numbers:

$$\lambda_{kl} = \frac{4}{h^2}\left(\sin^2\frac{k\pi}{2M} + \sin^2\frac{l\pi}{2M}\right), \quad k, l = 1, 2, \dots, M-1, \tag{5.109}$$

are the corresponding eigenvalues. As all the eigenvalues (5.109) are real (and positive), the operator $-\Delta^{(h)}$ is self-adjoint, i.e., symmetric (and positive definite). Indeed, for an arbitrary pair $v, w \in U^{(h)}$ we first use the expansion (5.107): $v = \sum_{k,l=1}^{M-1} c_{kl}\psi^{(k,l)}$, $w = \sum_{r,s=1}^{M-1} d_{rs}\psi^{(k,l)}$, and then substitute these expressions into formula (5.108), which yields [with the help of formula (5.106)]:

$$\begin{aligned}
(-\Delta^{(h)}v, w) &= \left(-\Delta^{(h)}\left(\sum_{k,l=1}^{M-1} c_{kl}\psi^{(k,l)}\right), \sum_{r,s=1}^{M-1} d_{rs}\psi^{(r,s)}\right) \\
&= \left(\sum_{k,l=1}^{M-1} c_{kl}\lambda_{kl}\psi^{(k,l)}, \sum_{r,s=1}^{M-1} d_{rs}\psi^{(r,s)}\right) = \sum_{k,l=1}^{M-1} c_{kl}\lambda_{kl}d_{kl} \\
&= \left(\sum_{k,l=1}^{M-1} c_{kl}\psi^{(k,l)}, \sum_{r,s=1}^{M-1} \lambda_{rs}d_{rs}\psi^{(r,s)}\right) \\
&= \left(\sum_{k,l=1}^{M-1} c_{kl}\psi^{(k,l)}, -\Delta^{(h)}\left(\sum_{r,s=1}^{M-1} \lambda_{rs}d_{rs}\psi^{(r,s)}\right)\right) = (v, -\Delta^{(h)}w).
\end{aligned}$$

According to the general scheme of the Fourier method given by formulae (5.94), (5.95), and (5.96), the solution of problem (5.97) can be written in the form:

$$u_{m_1,m_2} = \sum_{k,l=1}^{M-1} \frac{f_{kl}}{\lambda_{kl}} 2\sin\frac{k\pi m_1}{M}\sin\frac{l\pi m_2}{M}, \tag{5.110}$$

---

[6]See also the analysis on page 269.

where

$$f_{kl} = 2h^2 \sum_{m_1,m_2=0}^{M} f_{m_1,m_2} \sin\frac{k\pi m_1}{M} \sin\frac{l\pi m_2}{M}. \tag{5.111}$$

REMARK 5.3 In the case $M = 2^p$, where p is a positive integer, there is a special algorithm that allows one to compute the Fourier coefficients f_{kl} according to formula (5.111), as well as the solution u_{m_1,m_2} according to formula (5.110), at a cost of only $\mathcal{O}(M^2 \ln M)$ arithmetic operations, as opposed to the straightforward estimate $\mathcal{O}(M^4)$. This remarkable algorithm is called the fast Fourier transform (FFT). We describe it in Section 5.7.3. □

Let us additionally note that according to formula (5.109):

$$\lambda_{11} = \lambda_{11}(h) = \frac{8}{h^2}\sin^2\frac{\pi}{2M} = \frac{8}{h^2}\sin^2\frac{\pi h}{2}. \tag{5.112}$$

Consequently,

$$\pi^2 \leq \lambda_{11} \leq 2\pi^2. \tag{5.113}$$

At the same time,

$$\lambda_{M-1,M-1} \sim \frac{8}{h^2}. \tag{5.114}$$

Therefore, for the Euclidean condition number of the operator $-\Delta^{(h)}$ we can write with the help of Theorem 5.3:

$$\mu(-\Delta^{(h)}) = \frac{\lambda_{\max}}{\lambda_{\min}} = \frac{\lambda_{M-1,M-1}}{\lambda_{11}} = \mathcal{O}(h^{-2}), \quad \text{as} \quad h \longrightarrow 0. \tag{5.115}$$

Moreover, the spectral radius of the inverse operator $\left(-\Delta^{(h)}\right)^{-1}$, which provides its Euclidean norm, and which is equal to its maximum eigenvalue λ_{11}^{-1}, does not exceed π^{-2} according to estimate (5.113). Consequently, the following inequality (a stability estimate) always holds for the solution $u^{(h)}$ of problem (5.97):

$$\|u^{(h)}\|_2 \leq \frac{1}{\pi^2}\|f^{(h)}\|_2.$$

5.7.3 Fast Fourier Transform

We will introduce the idea of a fast algorithm in the one-dimensional context, i.e., we will show how the Fourier coefficients (5.102):

$$c_k = (v, \psi^{(k)}) = \sqrt{2}h \sum_{m=0}^{M} v_m \sin\frac{k\pi m}{M}$$

can be evaluated considerably more rapidly than at a cost of $\mathcal{O}(M^2)$ arithmetic operations. Rapid evaluation of the Fourier coefficients on a two-dimensional grid, as well as rapid summation of the two-dimensional discrete Fourier series (see Section 5.7.2) can then be performed similarly.

We begin with recasting expression (5.102) in the complex form. To do so, we artificially extend the grid function v_m antisymmetrically with respect to the point $m = M$: $v_{M+j} = v_{M-j}$, $j = 1,2,\dots,M$, so that for $m = M+1, M+2,\dots,2M$ it becomes: $v_m = v_{2M-m}$. Then we can write:

$$\begin{aligned} c_k &= \frac{h}{\sqrt{2}} \sum_{m=0}^{2M} v_m e^{ik\pi m/M} \\ &= \frac{h}{\sqrt{2}} \sum_{n=0}^{N} v_n e^{i2k\pi n/N}, \quad \text{where} \quad N \stackrel{\text{def}}{=} 2M, \end{aligned} \tag{5.116}$$

because given the antisymmetric extension of v_m beyond $m = M$, we clearly have $\sum_{m=0}^{2M} v_m \cos\frac{k\pi m}{M} = 0$. Hereafter, we will analyze formula (5.116) instead of (5.102).

First of all, it is clear that a straightforward implementation of the summation formula (5.116) requires $\mathcal{O}(N)$ arithmetic operations per one coefficient c_k even if all the exponential factors $e^{i2k\pi n/N}$ are precomputed and available. Consequently, the computation of all c_k, $k = 0,1,2,\dots,N-1$, requires $\mathcal{O}(N^2)$ arithmetic operations. Let, however, $N = N_1 \cdot N_2$, where N_1 and N_2 are positive integers. Then we can represent the numbers k and n as follows:

$$\begin{aligned} k &= k_1 N_2 + k_2, \quad k_1 = 0,1,\dots,N_1-1, \quad k_2 = 0,1,\dots,N_2-1, \\ n &= n_1 + n_2 N_1, \quad n_1 = 0,1,\dots,N_1-1, \quad n_2 = 0,1,\dots,N_2-1. \end{aligned}$$

Consequently, by noticing that $e^{i2\pi(k_1 N_2 n_2 N_1)/N} = e^{i2\pi(k_1 n_2)} = 1$, we obtain from expression (5.116):

$$\begin{aligned} c_k &= \frac{h}{\sqrt{2}} \sum_{n=0}^{N} v_n e^{i2k\pi n/N} = \frac{h}{\sqrt{2}} \sum_{n=0}^{N} v_n e^{i2\pi(k_1 N_2 + k_2)(n_1 + n_2 N_1)/N} \\ &= \frac{h}{\sqrt{2}} \sum_{n=0}^{N} v_n e^{i2\pi(k_1 N_2 n_1 + k_1 N_2 n_2 N_1 + k_2 n_1 + k_2 n_2 N_1)/N} \\ &= \frac{h}{\sqrt{2}} \sum_{n=0}^{N} v_n e^{i2\pi(k_1 N_2 n_1 + k_2 n_1 + k_2 n_2 N_1)/N} = \frac{h}{\sqrt{2}} \sum_{n=0}^{N} v_n e^{i2\pi(k n_1 + k_2 n_2 N_1)/N} \\ &= \frac{h}{\sqrt{2}} \sum_{n_1=0}^{N_1-1} \left(\sum_{n_2=0}^{N_2-1} v_{n_1+n_2N_1} e^{i2\pi(k_2 n_2 N_1)/N} \right) e^{i2\pi(k n_1)/N}. \end{aligned}$$

Therefore, the sum (5.116) can be computed in two stages. First, we evaluate the intermediate quantity:

$$\tilde{v}_{n_1,k_2} = \sum_{n_2=0}^{N_2-1} v_{n_1+n_2N_1} e^{i2\pi(k_2 n_2 N_1)/N}, \tag{5.117a}$$

and then the final result:

$$c_k = \frac{h}{\sqrt{2}} \sum_{n_1=0}^{N_1-1} \tilde{v}_{n_1,k_2} e^{i2\pi(kn_1)/N}. \qquad (5.117b)$$

The key advantage of computing c_k consecutively by formulae (5.117a)–(5.117b) rather than directly by formula (5.116), is that the cost of implementing formula (5.117a) is $\mathcal{O}(N_2)$ arithmetic operations, and the cost of subsequently implementing formula (5.117b) is $\mathcal{O}(N_1)$ arithmetic operations, which altogether yields $\mathcal{O}(N_1+N_2)$ operations. At the same time, the cost of computing a given c_k by formula (5.116) is $\mathcal{O}(N) = \mathcal{O}(N_1 \cdot N_2)$ operations. For example, if N is a large even number and $N_1 = 2$, then one basically obtains a speed-up by roughly a factor of two. The cost of computing all coefficients c_k, $k = 0,1,2,\dots,N-1$, by formula (5.116) is $\mathcal{O}(N^2)$ operations, whereas with the help of formulae (5.117a)–(5.117b) it can be reduced to $\mathcal{O}(N(N_1+N_2))$ operations.

Assume now that N_1 is a prime number, whereas N_2 is a composite number. Then N_2 can be represented as a product of two factors and a similar transformation can be applied to the sum (5.117a), for which n_1 is simply a parameter. This will further reduce the computational cost. In general, if $N = N_1 \cdot N_2 \cdots N_p$, then instead of the $\mathcal{O}(N^2)$ complexity of the original summation we will obtain an algorithm for computing c_k, $k = 0,1,2,\dots,N-1$, at a cost of $\mathcal{O}(N(N_1+N_2+\ldots+N_p))$ arithmetic operations. This algorithm is known as *the fast Fourier transform (FFT)*.

Efficient versions of the FFT have been developed for $N_i = 2,3,4$, although the algorithm can also be built for other prime factors. In practice, the most commonly used and the most convenient to implement version is the one for $N = 2^p$, where p is a positive integer. The computational complexity of the FFT in this case is

$$\mathcal{O}(N(\underbrace{2+2+\ldots+2}_{p \text{ times}})) = \mathcal{O}(N \ln N)$$

arithmetic operations,

The algorithm of fast Fourier transform was introduced in a landmark 1965 paper by Cooley and Tukey [CT65], and has since become one of the most popular computational tools for solving large linear systems obtained as discretizations of differential equations on the grid. Besides that, the FFT finds many applications in signal processing and in statistics. Practically all numerical software libraries today have FFT as a part of their standard implementation.

Exercises

1. Consider an inhomogeneous Dirichlet problem for the finite-difference Poisson equation on a grid square [cf. formula (5.11)]:

$$-\left(\frac{u_{m_1+1,m_2} - 2u_{m_1,m_2} + u_{m_1-1,m_2}}{h^2} + \frac{u_{m_1,m_2+1} - 2u_{m_1,m_2} + u_{m_1,m_2-1}}{h^2} \right) = f_{m_1,m_2},$$

$$m_1 = 1,2,\dots,M-1, \quad m_2 = 1,2,\dots,M-1, \qquad (5.118a)$$

with the boundary conditions:

$$\begin{aligned} &u_{0,m_2} = \varphi_{m_2}, \quad u_{M,m_2} = \xi_{m_2}, \quad m_2 = 1,2,\dots,M-1, \\ &u_{m_1,0} = \eta_{m_1}, \quad u_{m_1,M} = \zeta_{m_1}, \quad m_1 = 1,2,\dots,M-1, \end{aligned} \tag{5.118b}$$

where f_{m_1,m_2}, φ_{m_2}, ξ_{m_2}, η_{m_1}, and ζ_{m_1} are given functions of their respective arguments. Write down the solution of this problem in the form of a finite Fourier series (5.110).

Hint. Take an arbitrary grid function $w^{(h)} = \{w_{m_1,m_2}\}$, $m_1 = 0,1,\dots,M$, $m_2 = 0,1,\dots,M$, that satisfies boundary conditions (5.118b). Apply the finite-difference Laplace operator of (5.118a) to this function and obtain: $-\Delta^{(h)}w^{(h)} \stackrel{\text{def}}{=} g^{(h)}$. Then, consider a new grid function $v^{(h)} \stackrel{\text{def}}{==} u^{(h)} - w^{(h)}$. This function satisfies the finite difference Poisson equation: $-\Delta^{(h)}v^{(h)} = f^{(h)} - g^{(h)}$, for which the right-hand side is known, and it also satisfies the homogeneous boundary conditions: $v_{m_1,m_2} = 0$ for $m_1 = 0,M$ and $m_2 = 0,M$. Consequently, the new solution $v^{(h)}$ can be obtained using the discrete Fourier series, after which the original solution is reconstructed as $u^{(h)} = v^{(h)} + w^{(h)}$, where $w^{(h)}$ is known.

Chapter 6

Iterative Methods for Solving Linear Systems

Consider a system of linear algebraic equations:

$$Ax = f, \quad f \in \mathbb{L}, \quad x \in \mathbb{L}, \tag{6.1}$$

where $\mathbb{L}$ is a vector space and $A : \mathbb{L} \longmapsto \mathbb{L}$ is a linear operator. From the standpoint of applications, having a capability to compute its *exact* solution may be "nice," but it is typically not necessary. Quite the opposite, one can normally use an approximate solution, provided that it would guarantee the accuracy sufficient for a particular application. On the other hand, one usually cannot obtain the exact solution anyway. The reason is that the input data of the problem (the right-hand side f, as well as the operator A itself) are always specified with some degree of uncertainty. This necessarily leads to a certain unavoidable error in the result. Besides, as all the numbers inside the machine are only specified with a finite precision, the round-off errors are also inevitable in the course of computations.

Therefore, instead of solving system (6.1) by a direct method, e.g., by Gaussian elimination, in many cases it may be advantageous to use an iterative method of solution. This is particularly true when the dimension n of system (6.1) is very large, and unless a special fast algorithm such as FFT (see Section 5.7.3) can be employed, the $\mathcal{O}(n^3)$ cost of a direct method (see Sections 5.4 and 5.6) would be unbearable.

A typical iterative method (or an iteration scheme) consists of building a sequence of vectors $\{x^{(p)}\} \subset \mathbb{L}$, $p = 0, 1, 2, \ldots$, that are supposed to provide successively more accurate approximations of the exact solution x. The initial guess $x^{(0)} \in \mathbb{L}$ for an iteration scheme is normally taken arbitrarily. The notion of successively more accurate approximations can, of course, be quantified. It means that the sequence $x^{(p)}$ has to converge to the exact solution x as the number p increases: $x^{(p)} \longrightarrow x$, when $p \longrightarrow \infty$. This means that for any $\varepsilon > 0$ we can always find $p = p(\varepsilon)$ such that the following inequality:

$$\|x - x^{(p)}\| \leq \varepsilon$$

will hold for all $p \geq p(\varepsilon)$. Accordingly, by specifying a sufficiently small $\varepsilon > 0$ we can terminate the iteration process after a finite number $p = p(\varepsilon)$ of steps, and subsequently use the iteration $x^{(p)}$ in the capacity of an approximate solution that would meet the accuracy requirements for a given problem.

In this chapter, we will describe some popular iterative methods, and outline the conditions under which it may be advisable to use an iterative method rather than a direct method, or to prefer one particular iterative method over another.

6.1 Richardson Iterations and the Like

We will build a family of iterative methods by first recasting system (6.1) as follows:

$$\boldsymbol{x} = (\boldsymbol{I} - \tau \boldsymbol{A})\boldsymbol{x} + \tau \boldsymbol{f}. \tag{6.2}$$

In doing so, the new system (6.2) will be equivalent to the original one for any value of the parameter τ, $\tau > 0$. In general, there are many different ways of replacing the system $\boldsymbol{Ax} = \boldsymbol{f}$ by its equivalent of the type:

$$\boldsymbol{x} = \boldsymbol{Bx} + \boldsymbol{\varphi}, \quad \boldsymbol{x} \in \mathbb{L}, \quad \boldsymbol{\varphi} \in \mathbb{L}. \tag{6.3}$$

System (6.2) is a particular case of (6.3) with $\boldsymbol{B} = (\boldsymbol{I} - \tau \boldsymbol{A})$ and $\boldsymbol{\varphi} = \tau \boldsymbol{f}$.

6.1.1 General Iteration Scheme

The general scheme of what is known as the first order linear stationary iteration process consists of successively computing the terms of the sequence:

$$\boldsymbol{x}^{(p+1)} = \boldsymbol{Bx}^{(p)} + \boldsymbol{\varphi}, \quad p = 0, 1, 2, \dots, \tag{6.4}$$

where the initial guess $\boldsymbol{x}^{(0)}$ is specified arbitrarily. The matrix $\boldsymbol{B}$ is known as the iteration matrix. Clearly, if the sequence $\boldsymbol{x}^{(p)}$ converges, i.e., if there is a limit: $\lim_{p\to\infty} \boldsymbol{x}^{(p)} = \boldsymbol{x}$, then $\boldsymbol{x}$ is the solution of system (6.1). Later we will identify the conditions that would guarantee convergence of the sequence (6.4).

Iterative method (6.4) is first order because the next iterate $\boldsymbol{x}^{(p+1)}$ depends only on one previous iterate, $\boldsymbol{x}^{(p)}$; it is linear because the latter dependence is linear; and finally it is stationary because if we formally rewrite (6.4) as $\boldsymbol{x}^{(p+1)} = F(\boldsymbol{x}^{(p)}, \boldsymbol{A}, \boldsymbol{f})$, then the function F does not depend on p.

A particular form of the iteration scheme (6.4) based on system (6.2) is known as the stationary Richardson method:

$$\boldsymbol{x}^{(p+1)} = (\boldsymbol{I} - \tau \boldsymbol{A})\boldsymbol{x}^{(p)} + \tau \boldsymbol{f}, \quad p = 0, 1, 2, \dots. \tag{6.5}$$

Formula (6.5) can obviously be rewritten as:

$$\boldsymbol{x}^{(p+1)} = \boldsymbol{x}^{(p)} - \tau \boldsymbol{r}^{(p)}, \tag{6.6}$$

where $\boldsymbol{r}^{(p)} = \boldsymbol{Ax}^{(p)} - \boldsymbol{f}$ is the residual of the iterate $\boldsymbol{x}^{(p)}$. Instead of keeping the parameter τ constant we can allow it to depend on p. Then, departing from formula (6.6), we arrive at the so-called non-stationary Richardson method:

$$\boldsymbol{x}^{(p+1)} = \boldsymbol{x}^{(p)} - \tau_p \boldsymbol{r}^{(p)}. \tag{6.7}$$

Note that in order to actually compute the iterations according to formula (6.5) we only need to be able to obtain the vector $\boldsymbol{Ax}^{(p)}$ once $\boldsymbol{x}^{(p)} \in \mathbb{L}$ is given. This does not necessarily entail the explicit knowledge of the matrix $\boldsymbol{A}$. In other words, an iterative method of solving $\boldsymbol{Ax}=\boldsymbol{f}$ can also be realized when the system is specified in an operator form. Building the iteration sequence does not require choosing a particular basis in $\mathbb{L}$ and reducing the system to its canonical form:

$$\sum_{j=1}^{n} a_{ij}x_j = f_i, \quad i=1,2,\dots,n. \tag{6.8}$$

Moreover, when computing the terms of the sequence $\boldsymbol{x}^{(p)}$ with the help of formula (6.5), we do not necessarily need to store all n^2 entries of the matrix $\boldsymbol{A}$ in the computer memory. Instead, to implement the iteration scheme (6.5) we may only store the current vector $\boldsymbol{x}^{(p)} \in \mathbb{L}$ that has n components. In addition to the memory savings and flexibility in specifying $\boldsymbol{A}$, we will see that for certain classes of linear systems the computational cost of obtaining a sufficiently accurate solution of $\boldsymbol{Ax}=\boldsymbol{f}$ with the help of an iterative method may be considerably lower than $\mathscr{O}(n^3)$ operations, which is characteristic of direct methods.

THEOREM 6.1
Let $\mathbb{L}$ be an n-dimensional normed vector space (say, $\mathbb{R}^n$ or $\mathbb{C}^n$), and assume that the induced operator norm of the iteration matrix $\boldsymbol{B}$ of (6.4) satisfies:

$$\|\boldsymbol{B}\| = q < 1. \tag{6.9}$$

Then, system (6.3) or equivalently, system (6.1), has a unique solution $\boldsymbol{x} \in \mathbb{L}$. Moreover, the iteration sequence (6.4) converges to this solution $\boldsymbol{x}$ for an arbitrary initial guess $\boldsymbol{x}^{(0)}$, and the error of the iterate number p:

$$\boldsymbol{\varepsilon}^{(p)} \stackrel{\text{def}}{=} \boldsymbol{x} - \boldsymbol{x}^{(p)}$$

satisfies the estimate:

$$\|\boldsymbol{\varepsilon}^{(p)}\| = \|\boldsymbol{x} - \boldsymbol{x}^{(p)}\| \le q^p \|\boldsymbol{x} - \boldsymbol{x}^{(0)}\| = q^p \|\boldsymbol{\varepsilon}^{(0)}\|. \tag{6.10}$$

In other words, the norm of the error $\|\boldsymbol{\varepsilon}^{(p)}\|$ vanishes when $p \longrightarrow \infty$ at least as fast as the geometric sequence q^p.

PROOF If $\boldsymbol{\varphi} = \boldsymbol{0}$, then system (6.3) only has a trivial solution $\boldsymbol{x} = \boldsymbol{0}$. Indeed, otherwise for a solution $\boldsymbol{x} \neq \boldsymbol{0}$, $\boldsymbol{\varphi} = \boldsymbol{0}$, we could write:

$$\|\boldsymbol{x}\| = \|\boldsymbol{Bx}\| \le \|\boldsymbol{B}\|\|\boldsymbol{x}\| = q\|\boldsymbol{x}\| < \|\boldsymbol{x}\|,$$

i.e., $\|\boldsymbol{x}\| < \|\boldsymbol{x}\|$, which may not hold. The contradiction proves that system (6.3) is uniquely solvable for any $\boldsymbol{\varphi}$, and as such, so is system (6.1) for any $\boldsymbol{f}$.

Next, let $\boldsymbol{x}$ be the solution of system (6.3). Take an arbitrary $\boldsymbol{x}^{(0)} \in \mathbb{L}$ and subtract equality (6.3) from equality (6.4), which yields:

$$\boldsymbol{\varepsilon}^{(p+1)} = \boldsymbol{B}\boldsymbol{\varepsilon}^{(p)}, \quad p = 0,1,2,\dots.$$

Consequently, the following error estimate holds:

$$\|x - x^{(p)}\| = \|\varepsilon^{(p)}\| = \|B\varepsilon^{(p-1)}\| \leq q\|\varepsilon^{(p-1)}\|$$
$$\leq q^2\|\varepsilon^{(p-2)}\| \leq \ldots \leq q^p\|\varepsilon^{(0)}\| = q^p\|x - x^{(0)}\|,$$

which is equivalent to (6.10). □

REMARK 6.1 Condition (6.9) can be violated for a different choice of norm on the space $\mathbb{L}$: $\|x\|'$ instead of $\|x\|$. This, however, will not disrupt the convergence: $x^{(p)} \longrightarrow x$ as $p \longrightarrow \infty$. Moreover, the error estimate (6.10) will be replaced by

$$\|\varepsilon^{(p)}\|' \leq cq^p\|\varepsilon^{(0)}\|', \tag{6.11}$$

where c is a constant that will, generally speaking, depend on the new norm $\|\cdot\|'$, whereas the value of q will remain the same.

To justify this remark one needs to employ the equivalence of any two norms on a vector (i.e., finite-dimensional) space, see [HJ85, Section 5.4]. This important result says that if $\|\cdot\|$ and $\|\cdot\|'$ are two norms on $\mathbb{L}$, then we can always find two constants $c_1 > 0$ and $c_2 > 0$, such that $\forall x \in \mathbb{L}: \ c_1\|x\|' \leq \|x\| \leq c_2\|x\|'$, where c_1 and c_2 do not depend on x. Therefore, inequality (6.10) implies (6.11) with $c = c_2/c_1$. □

Example 1: The Jacobi Method

Let the matrix $A = \{a_{ij}\}$ of system (6.1) be diagonally dominant:

$$|a_{ii}| > \sum_{j \neq i} |a_{ij}| + \delta, \quad i = 1, 2, \ldots, n, \quad \delta > 0. \tag{6.12}$$

In the equation number i of system (6.1), we move all terms $a_{ij}x_j$, $j \neq i$, to the right-hand side and then divide this equation by a_{ii}. In doing so, we obtain a system of type (6.3) with the matrix:

$$B = \begin{bmatrix} 0 & b_{12} & b_{13} & \ldots & b_{1,n-1} & b_{1n} \\ b_{21} & 0 & b_{23} & \ldots & b_{2,n-1} & b_{2n} \\ \ldots & \ldots & \ldots & \ldots & \ldots & \ldots \\ b_{n1} & b_{n2} & b_{n3} & \ldots & b_{n,n-1} & 0 \end{bmatrix}.$$

Alternatively, if we define the diagonal $n \times n$ matrix $D = \operatorname{diag}\{a_{ii}\}$, then in the resulting system (6.3) we have $B = -D^{-1}(A - D)$ and $\varphi = D^{-1}f$.

Due to the diagonal dominance of A, one can find such a number $0 < q < 1$ that

$$\sum_{j=1}^{n} |b_{ij}| = \sum_{\substack{j=1, \\ j \neq i}}^{n} \frac{|a_{ij}|}{|a_{ii}|} \leq q < 1.$$

Consequently, the maximum norm of the iteration matrix B: $\|B\|_\infty = \max_i \sum_j |b_{ij}|$, satisfies estimate (6.9). Then, according to Theorem 6.1, the Jacobi iterations:

$$x^{(p+1)} = -D^{-1}(A - D)x^{(p)} + D^{-1}f \tag{6.13}$$

converge to the solution $\boldsymbol{x}$ of system (6.1). In the component form, the Jacobi iterative method (6.13) is written as:

$$x_i^{(p+1)} = -\sum_{\substack{j=1,\\ j\neq i}}^{n} \frac{a_{ij}}{a_{ii}} x_j^{(p)} + \frac{f_i}{a_{ii}}, \quad i = 1,2,\dots,n. \tag{6.14}$$

Let us specify $\boldsymbol{x}^{(0)}$ arbitrarily, but so that the initial error $\|\boldsymbol{\varepsilon}^{(0)}\|_\infty = \|\boldsymbol{x}-\boldsymbol{x}^{(0)}\|_\infty$ will not exceed some prescribed quantity, e.g., will not exceed one. Let us also assume that the accuracy of the approximate solution will be satisfactory if in the course of the iteration the initial error drops by three orders of magnitude, i.e., if the error becomes no greater than 10^{-3}. Then it is sufficient to choose p so that to guarantee the inequality $q^p \leq 10^{-3}$. If, for example, $q = 1/2$, then one can take $p = 10$ regardless of the value of n. The overall computational cost will then be $\mathcal{O}(10n^2) = \mathcal{O}(n^2)$ arithmetic operations, as opposed to the cubic cost $\mathcal{O}(n^3)$ of Gaussian elimination. Indeed, every matrix-vector multiplication $\boldsymbol{B}\boldsymbol{x}^{(p)}$ requires $\mathcal{O}(n^2)$ operations. Of course, the key consideration here is the actual rate of convergence determined by the value of q. We will later provide accurate estimates for the convergence rates of several iteration schemes: the Richardson method (Section 6.1.3), the Chebyshev method (Section 6.2.1), and the method of conjugate gradients (Section 6.2.2).

Example 2: The Gauss-Seidel Method

The Gauss-Seidel method is similar to the Jacobi method, except that when computing the component $x_i^{(p+1)}$, the previously updated components $x_1^{(p+1)}, x_2^{(p+1)}, \dots, x_{i-1}^{(p+1)}$ are immediately put to use, which yields the following iteration scheme [cf. formula (6.14)]:

$$x_i^{(p+1)} = -\sum_{j=1}^{i-1} \frac{a_{ij}}{a_{ii}} x_j^{(p+1)} - \sum_{j=i+1}^{n} \frac{a_{ij}}{a_{ii}} x_j^{(p)} + \frac{f_i}{a_{ii}}, \quad i = 1,2,\dots,n. \tag{6.15}$$

Using matrix notations, we can write the Gauss-Seidel method (6.15) as follows:

$$\boldsymbol{x}^{(p+1)} = -\boldsymbol{D}^{-1}\hat{\boldsymbol{L}}\boldsymbol{x}^{(p+1)} - \boldsymbol{D}^{-1}\hat{\boldsymbol{U}}\boldsymbol{x}^{(p)} + \boldsymbol{D}^{-1}\boldsymbol{f}, \tag{6.16}$$

where $\hat{\boldsymbol{L}} = \{\mathfrak{l}_{ij}\}$ is a lower triangular matrix with the entries: $\mathfrak{l}_{ij} = \begin{cases} a_{ij}, & j < i, \\ 0 & j \geq i, \end{cases}$ and $\hat{\boldsymbol{U}} = \{\mathfrak{u}_{ij}\}$ is an upper triangular matrix with the entries: $\mathfrak{u}_{ij} = \begin{cases} 0, & j \leq i, \\ a_{ij} & j > i, \end{cases}$ so that altogether $\boldsymbol{A} = \hat{\boldsymbol{L}} + \boldsymbol{D} + \hat{\boldsymbol{U}}$. By noticing that $(\boldsymbol{I}+\boldsymbol{D}^{-1}\hat{\boldsymbol{L}}) = \boldsymbol{D}^{-1}(\boldsymbol{D}+\hat{\boldsymbol{L}}) = \boldsymbol{D}^{-1}(\boldsymbol{A}-\hat{\boldsymbol{U}})$, we convert expression (6.16) to the form (6.4):

$$\boldsymbol{x}^{(p+1)} = -(\boldsymbol{A}-\hat{\boldsymbol{U}})^{-1}\hat{\boldsymbol{U}}\boldsymbol{x}^{(p)} + (\boldsymbol{A}-\hat{\boldsymbol{U}})^{-1}\boldsymbol{f} \tag{6.17}$$

and see that the iteration matrix $\boldsymbol{B}$ for the Gauss-Seidel method is given by: $\boldsymbol{B} = -(\boldsymbol{A}-\hat{\boldsymbol{U}})^{-1}\hat{\boldsymbol{U}}$. If the matrix $\boldsymbol{A}$ is diagonally dominant, see (6.12), then the Gauss-Seidel iterations (6.17) are known to converge to the solution $\boldsymbol{x}$ of system (6.1) for an arbitrary initial guess $\boldsymbol{x}^{(0)}$. We refer the reader to [Axe94] for the proof.

Example 3: The Over-Relaxation Methods

Note that the condition of diagonal dominance (6.12) of the matrix $\boldsymbol{A}$ that guarantees convergence of both the Jacobi method (6.14) and the Gauss-Seidel method (6.15) is only a sufficient and not a necessary condition of convergence. These methods may, in fact, converge for other types of matrices as well. Often, the convergence of an iteration is judged experimentally rather than studied theoretically. In this case, it may be beneficial to consider a broader family of algorithms that would provide more room for tuning the parameters in order to achieve a better convergence. A widely used generalization of both the Jacobi iteration and the Gauss-Seidel iteration is obtained by introducing the relaxation parameter γ and using a weighted update between the old and the new value, which leads to the Jacobi over-relaxation method (JOR):

$$x_i^{(p+1)} = \gamma\Bigg(-\sum_{\substack{j=1,\\ j\neq i}}^{n} \frac{a_{ij}}{a_{ii}} x_j^{(p)} + \frac{f_i}{a_{ii}}\Bigg) + (1-\gamma)x_i^{(p)}, \quad i = 1,2,\dots,n, \tag{6.18}$$

and to the successive over-relaxation method (SOR):

$$x_i^{(p+1)} = \gamma\Bigg(-\sum_{j=1}^{i-1} \frac{a_{ij}}{a_{ii}} x_j^{(p+1)} - \sum_{j=i+1}^{n} \frac{a_{ij}}{a_{ii}} x_j^{(p)} + \frac{f_i}{a_{ii}}\Bigg) + (1-\gamma)x_i^{(p)}, \quad i = 1,2,\dots,n. \tag{6.19}$$

Theoretical convergence results for the iterations (6.18) or (6.19) are only obtained for some particular cases. Again, we refer the reader to [Axe94] for detail. In general, the successive over-relaxation method (6.19) will not converge if $\gamma < 0$ or $\gamma > 2$. Otherwise, adjusting the value of γ can be used to speed up the convergence. Conversely, if it is observed in a numerical experiment that the convergence for a given case is "iffy," one may try and gain a better robustness by trading off the convergence rate, i.e., by assigning more weight to the old value and less weight to the new value, which means taking a small positive γ.

6.1.2 A Necessary and Sufficient Condition for Convergence

THEOREM 6.2

Let $\mathbb{L}$ be a complex n-dimensional linear space, and let $\boldsymbol{B}$ be an operator mapping this space onto itself $\boldsymbol{B} : \mathbb{L} \longmapsto \mathbb{L}$. A first order linear stationary iterative method (6.4):

$$\boldsymbol{x}^{(p+1)} = \boldsymbol{B}\boldsymbol{x}^{(p)} + \boldsymbol{\varphi}, \quad p = 0,1,2,\dots, \tag{6.20}$$

converges to the solution $\boldsymbol{x}$ of problem (6.1) in any norm and for an arbitrary initial guess $\boldsymbol{x}^{(0)} \in \mathbb{L}$ if and only if the spectral radius of the operator $\boldsymbol{B}$ is strictly less than one (here λ_j, $j = 1,\dots,n$, are the eigenvalues of $\boldsymbol{B}$):

$$\rho(\boldsymbol{B}) \stackrel{\text{def}}{=} \max_j |\lambda_j| < 1. \tag{6.21}$$

PROOF We will first prove the sufficiency, that is, if inequality (6.21) holds then the iterations (6.20) converge. Inequality (6.21) obviously implies that the number $\lambda = 1$ is not an eigenvalue of the operator $\boldsymbol{B}$. Consequently, the linear system $\boldsymbol{Bx} = 1 \cdot \boldsymbol{x}$ only has a trivial solution $\boldsymbol{x} = 0$ and as such, the system $\boldsymbol{x} = \boldsymbol{Bx} + \boldsymbol{\varphi}$ has a unique solution for any $\boldsymbol{\varphi} \in \mathbb{L}$.

As before, we define the error of the iterate $\boldsymbol{x}^{(p)}$ as follows: $\boldsymbol{\varepsilon}^{(p)} = \boldsymbol{x} - \boldsymbol{x}^{(p)}$. Let us first note that convergence of the sequence $\boldsymbol{x}^{(p)}$, $p = 0, 1, 2, \ldots$, for an arbitrary $\boldsymbol{x}^{(0)}$ is equivalent to convergence of the sequence $\boldsymbol{\varepsilon}^{(p)}$ to zero for an arbitrary $\boldsymbol{\varepsilon}^{(0)}$: $\boldsymbol{\varepsilon}^{(p)} \longrightarrow \boldsymbol{0}$ as $p \longrightarrow \infty$. Indeed, let us first assume that the sequence $\boldsymbol{x}^{(p)}$ converges. Then it can only converge to the solution $\boldsymbol{x}$, because by taking both sides of equality (6.20) to the limit $p \longrightarrow \infty$ we obtain that the limit $\lim\limits_{p\to\infty} \boldsymbol{x}^{(p)}$ furnishes a solution to the system $\boldsymbol{x} = \boldsymbol{Bx} + \boldsymbol{\varphi}$. Because of the uniqueness, $\lim\limits_{p\to\infty} \boldsymbol{x}^{(p)} = \boldsymbol{x}$, and consequently, $\lim\limits_{p\to\infty} \boldsymbol{\varepsilon}^{(p)} = \boldsymbol{0}$. Conversely, if $\boldsymbol{\varepsilon}^{(p)} \longrightarrow \boldsymbol{0}$ as $p \longrightarrow \infty$, then clearly $\boldsymbol{x}^{(p)} \longrightarrow \boldsymbol{x}$ as $p \longrightarrow \infty$.

Let us fix some $\boldsymbol{\varepsilon}^{(0)}$. For the error $\boldsymbol{\varepsilon}^{(p)}$ we can write:

$$\boldsymbol{\varepsilon}^{(p+1)} = \boldsymbol{B}\boldsymbol{\varepsilon}^{(p)}, \quad p = 0, 1, 2, \ldots,$$

which yields: $\|\boldsymbol{\varepsilon}^{(p)}\| \leq \|\boldsymbol{B}\|^p \|\boldsymbol{\varepsilon}^{(0)}\|$. Denote by $\boldsymbol{w}(\lambda)$, $\lambda \in \mathbb{C}$, the sum of the series of vector quantities:

$$\boldsymbol{w}(\lambda) = \sum_{p=0}^{\infty} \frac{\boldsymbol{\varepsilon}^{(p)}}{\lambda^p}. \tag{6.22}$$

This series converges uniformly (and absolutely) outside of any disk of radius $\|\boldsymbol{B}\| + \eta$, $\eta > 0$, centered at the origin on the complex plane of the variable λ. Indeed, $\forall \lambda \in \mathbb{C}$, $|\lambda| > \|\boldsymbol{B}\| + \eta$, series (6.22) is majorized (component-wise) by a convergent geometric series: $\|\boldsymbol{\varepsilon}^{(0)}\| \sum_{p=0}^{\infty} \frac{\|\boldsymbol{B}\|^p}{(\|\boldsymbol{B}\|+\eta)^p}$. According to the Weierstrass theorem proven in the courses of complex analysis, see, e.g., [Mar77, Chapter 3], the sum of a uniformly converging series of holomorphic functions is holomorphic. Therefore, in the region of convergence the function $\boldsymbol{w}(\lambda)$ is a holomorphic vector function of its argument λ, and the series (6.22) is its Laurent series. It is also easy to see that $\lambda \boldsymbol{w}(\lambda) - \lambda \boldsymbol{\varepsilon}^{(0)} = \boldsymbol{B}\boldsymbol{w}(\lambda)$, which, in turn, means: $\boldsymbol{w}(\lambda) = -\lambda(\boldsymbol{B} - \lambda \boldsymbol{I})^{-1}\boldsymbol{\varepsilon}^{(0)}$. Moreover, by multiplying the series (6.22) by λ^{p-1} and then integrating (counterclockwise) along the circle $|\lambda| = r$ on the complex plane, where the number r is to be chosen so that the contour of integration lie within the area of convergence, i.e., $r \geq \|\boldsymbol{B}\| + \eta$, we obtain:

$$\boldsymbol{\varepsilon}^{(p)} = \frac{1}{2\pi i} \int\limits_{|\lambda|=r} \lambda^{p-1} \boldsymbol{w}(\lambda) d\lambda = -\frac{1}{2\pi i} \int\limits_{|\lambda|=r} \lambda^p (\boldsymbol{B} - \lambda \boldsymbol{I})^{-1} \boldsymbol{\varepsilon}^{(0)} d\lambda. \tag{6.23}$$

Indeed, integrating the individual powers of λ on the complex plane, we have:

$$\int\limits_{|\lambda|=r} \lambda^k d\lambda = \begin{cases} 2\pi i, & k = -1, \\ 0, & k \neq -1. \end{cases}$$

In other words, formula (6.23) implies that $-\boldsymbol{\varepsilon}^{(p)}$ is the residue of the vector function $\lambda^{p-1}\boldsymbol{w}(\lambda)$ at infinity.

Next, according to inequality (6.21), all the eigenvalues of the operator $\boldsymbol{B}$ belong to the disk of radius $\rho < 1$ centered at the origin on the complex plane: $|\lambda_j| \leq \rho < 1$, $j = 1,2,\ldots,n$. Then the integrand in the second integral of formula (6.23) is an analytic vector function of λ outside of this disk, i.e., for $|\lambda| > \rho$, because the operator $(\boldsymbol{B} - \lambda \boldsymbol{I})^{-1}$ exists (i.e., is bounded) for all $\lambda: \ |\lambda| > \rho$. This function is the analytic continuation of the function $\lambda^{p-1}\boldsymbol{w}(\lambda)$, where $\boldsymbol{w}(\lambda)$ is originally defined by the series (6.22) that can only be proven to converge outside of a larger disk $|\lambda| \leq \|\boldsymbol{B}\| + \eta$. Consequently, the contour of integration in (6.23) can be altered, and instead of $r \geq \|\boldsymbol{B}\| + \eta$ one can take $r = \rho + \zeta$, where $\zeta > 0$ is arbitrary, without changing the value of the integral. Therefore, the error can be estimated as follows:

$$\begin{aligned} \|\boldsymbol{\varepsilon}^{(p)}\| &= \frac{1}{2\pi}\left\| \int_{|\lambda|=\rho+\zeta} \lambda^p (\boldsymbol{B} - \lambda \boldsymbol{I})^{-1} \boldsymbol{\varepsilon}^{(0)} d\lambda \right\| \\ &\leq (\rho+\zeta)^p \max_{|\lambda|=\rho+\zeta} \|(\boldsymbol{B} - \lambda \boldsymbol{I})^{-1}\| \|\boldsymbol{\varepsilon}^{(0)}\|. \end{aligned} \tag{6.24}$$

In formula (6.24), let us take $\zeta > 0$ sufficiently small so that $\rho + \zeta < 1$. Then, the right-hand side of inequality (6.24) vanishes as p increases, which implies the convergence: $\|\boldsymbol{\varepsilon}^{(p)}\| \longrightarrow 0$ when $p \longrightarrow \infty$. This completes the proof of sufficiency.

To prove the necessity, suppose that inequality (6.21) does not hold, i.e., that for some λ_k we have $|\lambda_k| \geq 1$. At the same time, contrary to the conclusion of the theorem, let us assume that the convergence still takes place for any choice of $\boldsymbol{x}^{(0)}$: $\boldsymbol{x}^{(p)} \longrightarrow \boldsymbol{x}$ as $p \longrightarrow \infty$. Then we can choose $\boldsymbol{x}^{(0)}$ so that $\boldsymbol{\varepsilon}^{(0)} = \boldsymbol{x} - \boldsymbol{x}^{(0)} = \boldsymbol{e}_k$, where $\boldsymbol{e}_k$ is the eigenvector of the operator $\boldsymbol{B}$ that corresponds to the eigenvalue λ_k. In this case, $\boldsymbol{\varepsilon}^{(p)} = \boldsymbol{B}^p \boldsymbol{\varepsilon}^{(0)} = \boldsymbol{B}^p \boldsymbol{e}_k = \lambda_k^p \boldsymbol{e}_k$. As $|\lambda_k| \geq 1$, the sequence $\lambda_k^p \boldsymbol{e}_k$ does not converge to $\boldsymbol{0}$ when p increases. The contradiction proves the necessity. $\square$

REMARK 6.2 Let us make an interesting and important observation of a situation that we encounter here for the first time. The problem of computing the limit $\boldsymbol{x} = \lim\limits_{p \to \infty} \boldsymbol{x}^{(p)}$ is ultimately well conditioned, because the result $\boldsymbol{x}$ does not depend on the initial data at all, i.e., it does not depend on the initial guess $\boldsymbol{x}^{(0)}$. Yet the algorithm for computing the sequence $\boldsymbol{x}^{(p)}$ that converges according to Theorem 6.2 may still appear computationally unstable. The instability may take place if along with the inequality $\max_j |\lambda_j\| = \rho < 1$ we have $\|\boldsymbol{B}\| > 1$. This situation is typical for non-self-adjoint (or non-normal) matrices $\boldsymbol{B}$ (opposite of Theorem 5.2).

Indeed, if $\|\boldsymbol{B}\| < 1$, then the norm of the error $\|\boldsymbol{\varepsilon}^{(p)}\| = \|\boldsymbol{B}^p \boldsymbol{\varepsilon}^{(0)}\|$ decreases monotonically, this is the result of Theorem 6.1. Otherwise, if $\|\boldsymbol{B}\| > 1$, then for some $\boldsymbol{\varepsilon}^{(0)}$ the norm $\|\boldsymbol{\varepsilon}^{(p)}\|$ will initially grow, and only then decrease. The behavior will be qualitatively similar to that shown in Figure 10.13 (see page 394) that pertains to the study of stability for finite-difference initial boundary value problems. In doing so, the height of the intermediate "hump" on the curve showing the dependence of $\|\boldsymbol{\varepsilon}^{(p)}\|$ on p may be arbitrarily high. A small relative error committed near the value of p that corresponds to the maximum of the "hump" will subsequently increase (i.e., its norm will increase) — this error will also evolve and undergo a maximum, etc. The resulting instability may appear so strong that the computation will become practically impossible already for moderate dimensions n and for the norms $\|\boldsymbol{B}\|$ that are only slightly larger than one. ▯

A rigorous definition of stability for the first order linear stationary iterative methods, as well as the classification of possible instabilities, the pertinent theorems, and examples can be found in work [Rya70].

6.1.3 The Richardson Method for $A = A^* > 0$

Consider equation (6.3): $\boldsymbol{x} = \boldsymbol{B}\boldsymbol{x} + \boldsymbol{\varphi}$, $\boldsymbol{x} \in \mathbb{L}$, assuming that $\mathbb{L}$ is an n-dimensional Euclidean space with the inner product $(\boldsymbol{x}, \boldsymbol{y})$ and the norm $\|\boldsymbol{x}\| = \sqrt{(\boldsymbol{x}, \boldsymbol{x})}$ (e.g., $\mathbb{L} = \mathbb{R}^n$). Also assume that $\boldsymbol{B} : \mathbb{L} \longmapsto \mathbb{L}$ is a self-adjoint operator: $\boldsymbol{B} = \boldsymbol{B}^*$, with respect to the chosen inner product. Let ν_j, $j = 1, 2, \ldots, n$, be the eigenvalues of $\boldsymbol{B}$ and let

$$\rho = \rho(\boldsymbol{B}) = \max_j |\nu_j|$$

be its spectral radius. Specify an arbitrary $\boldsymbol{x}^{(0)} \in \mathbb{L}$ and build a sequence of iterations:

$$\boldsymbol{x}^{(p+1)} = \boldsymbol{B}\boldsymbol{x}^{(p)} + \boldsymbol{\varphi}, \quad p = 0, 1, 2, \ldots. \tag{6.25}$$

LEMMA 6.1

1. *If $\rho < 1$ then the system $\boldsymbol{x} = \boldsymbol{B}\boldsymbol{x} + \boldsymbol{\varphi}$ has a unique solution $\boldsymbol{x} \in \mathbb{L}$; the iterates $\boldsymbol{x}^{(p)}$ of (6.25) converge to $\boldsymbol{x}$; and the Euclidean norm of the error $\|\boldsymbol{x} - \boldsymbol{x}^{(p)}\|$ satisfies the estimate:*

$$\|\boldsymbol{x} - \boldsymbol{x}^{(p)}\| \leq \rho^p \|\boldsymbol{x} - \boldsymbol{x}^{(0)}\|, \quad p = 0, 1, 2, \ldots. \tag{6.26}$$

   Moreover, there is a particular $\boldsymbol{x}^{(0)} \in \mathbb{L}$ for which inequality (6.26) transforms into a precise equality.

2. *Let the system $\boldsymbol{x} = \boldsymbol{B}\boldsymbol{x} + \boldsymbol{\varphi}$ have a solution $\boldsymbol{x} \in \mathbb{L}$ for a given $\boldsymbol{\varphi} \in \mathbb{L}$, and let $\rho \geq 1$. Then there is an initial guess $\boldsymbol{x}^{(0)} \in \mathbb{L}$ such that the corresponding sequence of iterations (6.25) does not converge to $\boldsymbol{x}$.*

PROOF According to Theorem 5.2, the Euclidean norm of a self-adjoint operator $\boldsymbol{B} = \boldsymbol{B}^*$ coincides with its spectral radius ρ. Therefore, the first conclusion of the lemma except its last statement holds by virtue of Theorem 6.1.

To find the initial guess that would turn (6.26) into an equality, we first introduce our standard notation $\boldsymbol{\varepsilon}^{(p)} = \boldsymbol{x} - \boldsymbol{x}^{(p)}$ for the error of the iterate $\boldsymbol{x}^{(p)}$, and subtract equation (6.25) from $\boldsymbol{x} = \boldsymbol{B}\boldsymbol{x} + \boldsymbol{\varphi}$, which yields: $\boldsymbol{\varepsilon}^{(p+1)} = \boldsymbol{B}\boldsymbol{\varepsilon}^{(p)}$, $p = 0,1,2,\ldots$. Next, suppose that $|\nu_k| = \max_j |\nu_j| = \rho$ and take $\boldsymbol{\varepsilon}^{(0)} = \boldsymbol{x} - \boldsymbol{x}^{(0)} = \boldsymbol{e}_k$, where $\boldsymbol{e}_k$ is the eigenvector of $\boldsymbol{B}$ that corresponds to the eigenvalue with maximum magnitude. Then we obtain: $\|\boldsymbol{\varepsilon}^{(p)}\| = |\nu_k|^p \|\boldsymbol{\varepsilon}^{(0)}\| = \rho^p \|\boldsymbol{\varepsilon}^{(0)}\|$.

To prove the second conclusion of the lemma, we take the particular eigenvalue ν_k that delivers the maximum: $|\nu_k| = \max_j |\nu_j| = \rho \geq 1$, and again select $\boldsymbol{\varepsilon}^{(0)} = \boldsymbol{x} - \boldsymbol{x}^{(0)} = \boldsymbol{e}_k$, where $\boldsymbol{e}_k$ is the corresponding eigenvector. In this case the error obviously does not vanish as $p \longrightarrow \infty$, because:

$$\boldsymbol{\varepsilon}^{(p)} = \boldsymbol{B}\boldsymbol{\varepsilon}^{(p-1)} = \ldots = \boldsymbol{B}^p \boldsymbol{\varepsilon}^{(0)} = \nu_k^p \boldsymbol{e}_k,$$

and consequently, $\|\boldsymbol{\varepsilon}^{(p)}\| = \rho^p \|\boldsymbol{e}_k\|$, where ρ^p will either stay bounded but will not vanish, or will increase when $p \longrightarrow \infty$. $\square$

Lemma 6.1 analyzes a special case $\boldsymbol{B} = \boldsymbol{B}^*$ and provides a simple illustration for the general conclusion of Theorem 6.2 that for the convergence of a first order linear stationary iteration it is necessary and sufficient that the spectral radius of the iteration matrix be strictly less than one. With the help of this lemma, we will now analyze the convergence of the stationary Richardson iteration (6.5) for the case $\boldsymbol{A} = \boldsymbol{A}^* > 0$.

THEOREM 6.3

Consider a system of linear algebraic equations:

$$\boldsymbol{A}\boldsymbol{x} = \boldsymbol{f}, \quad \boldsymbol{A} = \boldsymbol{A}^* > 0, \tag{6.27}$$

where $\boldsymbol{x}, \boldsymbol{f} \in \mathbb{L}$, and $\mathbb{L}$ is an n-dimensional Euclidean space (e.g., $\mathbb{L} = \mathbb{R}^n$). Let $\lambda_{\min}$ and $\lambda_{\max}$ be the smallest and the largest eigenvalues of the operator $\boldsymbol{A}$, respectively. Specify some $\tau \neq 0$ and recast system (6.27) in an equivalent form:

$$\boldsymbol{x} = (\boldsymbol{I} - \tau\boldsymbol{A})\boldsymbol{x} + \tau\boldsymbol{f}. \tag{6.28}$$

Given an arbitrary initial guess $\boldsymbol{x}^{(0)} \in \mathbb{L}$, consider a sequence of Richardson iterations:

$$\boldsymbol{x}^{(p+1)} = (\boldsymbol{I} - \tau\boldsymbol{A})\boldsymbol{x}^{(p)} + \tau\boldsymbol{f}, \quad p = 0,1,2,\ldots. \tag{6.29}$$

1. *If the parameter τ satisfies the inequalities:*

$$0 < \tau < \frac{2}{\lambda_{\max}}, \tag{6.30}$$

then the sequence $\boldsymbol{x}^{(p)}$ of (6.29) converges to the solution $\boldsymbol{x}$ of system (6.27). Moreover, the norm of the error $\|\boldsymbol{x} - \boldsymbol{x}^{(p)}\|$ is guaranteed to decrease when p increases with the rate given by the following estimate:

$$\|\boldsymbol{x} - \boldsymbol{x}^{(p)}\| \leq \rho^p \|\boldsymbol{x} - \boldsymbol{x}^{(0)}\|, \quad p = 0,1,2,\ldots. \tag{6.31}$$

The quantity ρ in formula (6.31) is defined as

$$\rho = \rho(\tau) = \max\{|1-\tau\lambda_{\min}|, |1-\tau\lambda_{\max}|\}. \quad (6.32)$$

This quantity is less than one, $\rho < 1$, as it is the maximum of two numbers, $|1-\tau\lambda_{\min}|$ and $|1-\tau\lambda_{\max}|$, neither of which may exceed one provided that inequalities (6.30) hold.

2. *Let the number τ satisfy (6.30). Then there is a special initial guess $\boldsymbol{x}^{(0)}$ for which estimate (6.31) cannot be improved, because for this $\boldsymbol{x}^{(0)}$ inequality (6.31) transforms into a precise equality.*

3. *If condition (6.30) is violated, so that either $\tau \geq 2/\lambda_{\max}$ or $\tau \leq 0$, then there is an initial guess $\boldsymbol{x}^{(0)}$ for which the sequence $\boldsymbol{x}^{(p)}$ of (6.29) does not converge to the solution $\boldsymbol{x}$ of system (6.27).*

4. *The number $\rho = \rho(\tau)$ given by formula (6.32) assumes its minimal (i.e., optimal) value $\rho_{\text{opt}} = \rho(\tau_{\text{opt}})$ when $\tau = \tau_{\text{opt}} = 2/(\lambda_{\min}+\lambda_{\max})$. In this case,*

$$\rho = \rho_{\text{opt}} = \frac{\lambda_{\max}-\lambda_{\min}}{\lambda_{\max}+\lambda_{\min}} = \frac{\mu(\boldsymbol{A})-1}{\mu(\boldsymbol{A})+1}, \quad (6.33)$$

where $\mu(\boldsymbol{A}) = \lambda_{\max}/\lambda_{\min}$ is the condition number of the operator $\boldsymbol{A}$ (see Theorem 5.3).

PROOF To prove Theorem 6.3, we will use Lemma 6.1. In this lemma, let us set $\boldsymbol{B} = \boldsymbol{I} - \tau\boldsymbol{A}$. Note that if $\boldsymbol{A} = \boldsymbol{A}^*$ then the operator $\boldsymbol{B} = \boldsymbol{I} - \tau\boldsymbol{A}$ is also self-adjoint, i.e., $\boldsymbol{B} = \boldsymbol{B}^*$:

$$\begin{aligned}(\boldsymbol{Bx},\boldsymbol{y}) &= ((\boldsymbol{I}-\tau\boldsymbol{A})\boldsymbol{x},\boldsymbol{y}) = (\boldsymbol{x},\boldsymbol{y}) - \tau(\boldsymbol{Ax},\boldsymbol{y}) \\ &= (\boldsymbol{x},\boldsymbol{y}) - \tau(\boldsymbol{x},\boldsymbol{Ay}) = (\boldsymbol{x},(\boldsymbol{I}-\tau\boldsymbol{A})\boldsymbol{y}) = (\boldsymbol{x},\boldsymbol{By}).\end{aligned}$$

Suppose that λ_j, $j = 1,2,\dots,n$, are the eigenvalues of the operator $\boldsymbol{A}$ arranged in the ascending order:

$$0 < \lambda_{\min} = \lambda_1 \leq \lambda_2 \leq \ldots \leq \lambda_n = \lambda_{\max}, \quad (6.34)$$

and $\{\boldsymbol{e}_1, \boldsymbol{e}_2, \dots, \boldsymbol{e}_n\}$ are the corresponding eigenvectors: $\boldsymbol{Ae}_j = \lambda_j\boldsymbol{e}_j$, $j = 1,2,\dots,n$, that form an orthonormal basis in the space $\mathbb{L}$. Then clearly, the same vectors $\boldsymbol{e}_j$, $j = 1,2,\dots,n$, are also eigenvectors of the operator $\boldsymbol{B}$, whereas the respective eigenvalues are given by:

$$\nu_j = \nu_j(\tau) = 1 - \tau\lambda_j, \quad j = 1,2,\dots,n. \quad (6.35)$$

Indeed,

$$\begin{gathered}\boldsymbol{Be}_j = (\boldsymbol{I}-\tau\boldsymbol{A})\boldsymbol{e}_j = \boldsymbol{e}_j - \tau\lambda_j\boldsymbol{e}_j = (1-\tau\lambda_j)\boldsymbol{e}_j = \nu_j\boldsymbol{e}_j, \\ j = 1,2,\dots,n.\end{gathered}$$

According to (6.34), if $\tau > 0$ then the eigenvalues ν_j given by formula (6.35) are arranged in the descending order, see Figure 6.1:

$$1 > \nu_1 \geq \nu_2 \geq \ldots \geq \nu_n.$$

From Figure 6.1 it is also easy to see that the largest among the absolute values $|\nu_j|$, $j = 1,2,\ldots,n$, may be either $|\nu_1| = |1 - \tau\lambda_1| \equiv |1 - \tau\lambda_{\min}|$ or $|\nu_n| = |1 - \tau\lambda_n| \equiv |1 - \tau\lambda_{\max}|$; the case $|\nu_n| = \max_j |\nu_j|$ is realized when $\nu_n = 1 - \tau\lambda_{\max} < 0$ and $|1 - \tau\lambda_{\max}| > |1 - \tau\lambda_{\min}|$. Consequently, the condition:

$$\rho = \max_j |\nu_j| < 1 \tag{6.36}$$

of Lemma 6.1 coincides with the condition [see formula (6.32)]:

$$\rho = \max\{|1 - \tau\lambda_{\min}|, |1 - \tau\lambda_{\max}|\} < 1.$$

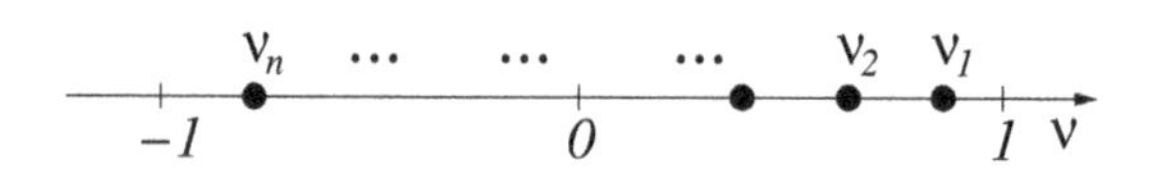

FIGURE 6.1: Eigenvalues of the matrix $\boldsymbol{B} = \boldsymbol{I} - \tau\boldsymbol{A}$.

Clearly, if $\tau > 0$ we can only guarantee $\rho < 1$ provided that the point ν_n on Figure 6.1 is located to the right of the point -1, i.e., if $\nu_n = 1 - \tau\lambda_{\max} > -1$. This means that along with $\tau > 0$ the second inequality of (6.30) also holds. Otherwise, if $\tau \geq 2/\lambda_{\max}$, then $\rho > 1$. If $\tau < 0$, then $\nu_j = 1 - \tau\lambda_j = 1 + |\tau|\lambda_j > 1$ for all $j = 1,2,\ldots,n$, and we will always have $\rho = \max_j |\nu_j| > 1$. Hence, condition (6.30) is equivalent to the requirement (6.36) of Lemma 6.1 for $\boldsymbol{B} = \boldsymbol{I} - \tau\boldsymbol{A}$ (or to requirement (6.21) of Theorem 6.2). We have thus proven the first three implications of Theorem 6.3.

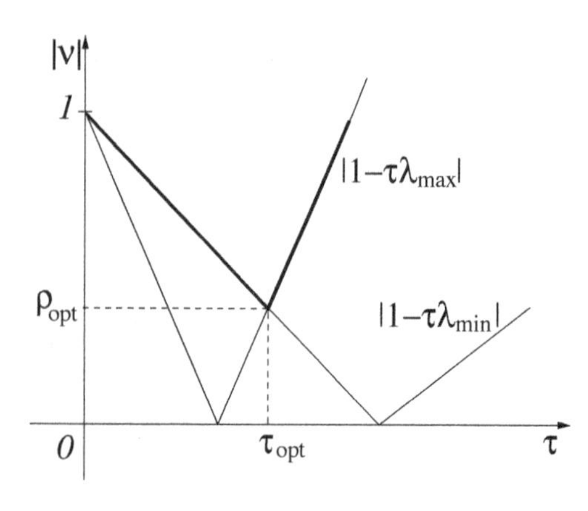

FIGURE 6.2: $|\nu_1|$ and $|\nu_n|$ as functions of τ.

To prove the remaining fourth implication, we need to analyze the behavior of the quantities $|\nu_1| = |1 - \tau\lambda_{\min}|$ and $|\nu_n| = |1 - \tau\lambda_{\max}|$ as functions of τ. We schematically show this behavior in Figure 6.2. From this figure, we determine that for smaller values of τ the quantity $|\nu_1|$ dominates, i.e., $|1 - \tau\lambda_{\min}| > |1 - \tau\lambda_{\max}|$, whereas for larger values of τ the quantity $|\nu_n|$ dominates, i.e., $|1 - \tau\lambda_{\max}| > |1 - \tau\lambda_{\min}|$. The value of $\rho(\tau) = \max\{|1 - \tau\lambda_{\min}|, |1 - \tau\lambda_{\max}|\}$ is shown by a bold polygonal line in Figure 6.2; it coincides with $|1-\tau\lambda_{\min}|$ before the intersection point, and after this point it coincides with $|1-\tau\lambda_{\max}|$. Consequently, the minimum value of $\rho=\rho_{\text{opt}}$ is achieved precisely at the intersection, i.e., at the value of $\tau=\tau_{\text{opt}}$ obtained from the following condition: $\nu_1(\tau)=|\nu_n(\tau)|=-\nu_n(\tau)$. This condition reads:

$$1-\tau\lambda_{\min}=\tau\lambda_{\max}-1,$$

which yields:

$$\tau_{\text{opt}}=\frac{2}{\lambda_{\min}+\lambda_{\max}}.$$

Consequently,

$$\rho_{\text{opt}}=\rho(\tau_{\text{opt}})=1-\tau_{\text{opt}}\lambda_{\min}=\frac{\lambda_{\max}-\lambda_{\min}}{\lambda_{\max}+\lambda_{\min}}=\frac{\mu(\boldsymbol{A})-1}{\mu(\boldsymbol{A})+1}.$$

This expression is identical to (6.33), which completes the proof. $\square$

Let us emphasize the following important consideration. Previously, we saw that the condition number of a matrix determines how sensitive the solution of the corresponding linear system will be to the perturbations of the input data (Section 5.3.2). The result of Theorem 6.3 provides the first evidence that the condition number also determines the rate of convergence of an iterative method. Indeed, from formula (6.33) it is clear that the closer the value of $\mu(\boldsymbol{A})$ to one, the closer the value of ρ_{opt} to zero, and consequently, the faster is the decay of the error according to estimate (6.31). When the condition number $\mu(\boldsymbol{A})$ increases, so does the quantity ρ_{opt} (while still remaining less than one) and the convergence slows down.

According to formulae (6.31) and (6.33), the optimal choice of the iteration parameter $\tau=\tau_{\text{opt}}$ enables the following error estimate:

$$\|\boldsymbol{\varepsilon}^{(p)}\|\leq\left(\frac{1-\xi}{1+\xi}\right)^p\|\boldsymbol{\varepsilon}^{(0)}\|,\quad\text{where}\quad\xi=\frac{\lambda_{\min}}{\lambda_{\max}}=\frac{1}{\mu(\boldsymbol{A})}.$$

Moreover, Lemma 6.1 implies that this estimate is actually attained for some particular $\boldsymbol{\varepsilon}^{(0)}$, i.e., that there is an initial guess for which the inequality transforms into a precise equality. Therefore, in order to guarantee that the initial error drops by a prescribed factor in the course of the iteration, i.e., in order to guarantee the estimate:

$$\|\boldsymbol{\varepsilon}^{(p)}\|\leq\sigma\|\boldsymbol{\varepsilon}^{(0)}\|, \tag{6.37}$$

where $\sigma>0$ is given, it is necessary and sufficient to select p that would satisfy:

$$\left(\frac{1-\xi}{1+\xi}\right)^p\leq\sigma,\quad\text{i.e.,}\quad p\geq-\frac{\ln\sigma}{\ln(1+\xi)-\ln(1-\xi)}.$$

A more practical estimate for the number p can also be obtained. Note that

$$\ln(1+\xi)-\ln(1-\xi)=2\xi\sum_{k=0}^{\infty}\frac{\xi^{2k}}{2k+1},$$

where

$$1 \leq \sum_{k=0}^{\infty} \frac{\xi^{2k}}{2k+1} \leq \frac{1}{1-\xi^2}.$$

Therefore, for the estimate (6.37) to hold it is sufficient that p satisfy:

$$p \geq -\frac{1}{2}\ln\sigma \cdot \mu(\boldsymbol{A}), \qquad \mu(\boldsymbol{A}) = \frac{1}{\xi}, \tag{6.38a}$$

and it is necessary that

$$p \geq -\frac{1}{2}\ln\sigma \cdot (1-\xi^2)\mu(\boldsymbol{A}). \tag{6.38b}$$

Altogether, the number of Richardson iterations required for reducing the initial error by a predetermined factor *is proportional to the condition number of the matrix.*

REMARK 6.3 In many cases, for example when approximating elliptic boundary value problems using finite differences (see, e.g., Section 5.1.3), the operator $\boldsymbol{A} : \mathbb{L} \longmapsto \mathbb{L}$ of the resulting linear system (typically, $\mathbb{L} = \mathbb{R}^n$) appears self-adjoint and positive definite ($\boldsymbol{A} = \boldsymbol{A}^* > 0$) in the sense of some natural inner product. However, most often one cannot find the precise minimum and maximum eigenvalues for such operators. Instead, only the estimates a and b for the boundaries of the spectrum may be available:

$$0 < a \leq \lambda_{\min} \leq \lambda_{\max} \leq b. \tag{6.39}$$

In this case, the Richardson iteration (6.29) can still be used for solving the system $\boldsymbol{Ax} = \boldsymbol{f}$. □

The key difference, though, between the more general case outlined in Remark 6.3 and the case of Theorem 6.3, for which the precise boundaries of the spectrum are known, is the way the iteration parameter τ is selected. If instead of $\lambda_{\min}$ and $\lambda_{\max}$ we only know a and b, see formula (6.39), then the best we can do is take $\tau' = 2/(a+b)$ instead of $\tau_{\text{opt}} = 2/(\lambda_{\min} + \lambda_{\max})$. Then, instead of ρ_{opt} given by formula (6.33):

$$\rho_{\text{opt}} = (\lambda_{\min} - \lambda_{\max})/(\lambda_{\min} + \lambda_{\max})$$

another quantity

$$\rho' = \max\{|1 - \tau'\lambda_{\min}|, |1 - \tau'\lambda_{\max}|\},$$

which is larger than ρ_{opt}, will appear in the guaranteed error estimate (6.31). As has been shown, for any value of τ within the limits (6.30), and for the respective value of $\rho = \rho(\tau)$ given by formula (6.32), there is always an initial guess $\boldsymbol{x}^{(0)}$ for which estimate (6.31) becomes a precise equality. Therefore, for $\tau = \tau' \neq \tau_{\text{opt}}$ we obtain an unimprovable estimate (6.31) with $\rho = \rho' > \rho_{\text{opt}}$. In doing so, the rougher the estimate for the boundaries of the spectrum, the slower the convergence.

Example

Let us apply the Richardson iterative method to solving the finite-difference Dirichlet problem for the Poisson equation: $-\Delta^{(h)}u^{(h)} = f^{(h)}$ that we introduced in Section 5.1.3. In this case, formula (6.29) becomes:

$$u^{(h,p+1)} = (\boldsymbol{I} + \tau\Delta^{(h)})u^{(h,p)} + \tau f^{(h)}, \quad p = 0, 1, 2, \ldots.$$

The eigenvalues of the operator $-\Delta^{(h)}$ are given by formula (5.109) of Section 5.7.2. In the same Section 5.7.2, we have shown that the operator $-\Delta^{(h)} : U^{(h)} \longmapsto U^{(h)}$ is self-adjoint with respect to the natural scalar product (5.20) on $U^{(h)}$:

$$(u^{(h)}, v^{(h)}) = h^2 \sum_{m_1,m_2=1}^{M-1} u_{m_1,m_2} v_{m_1,m_2}.$$

Finally, we have estimated its condition number: $\mu(-\Delta^{(h)}) = \mathcal{O}(h^{-2})$, see formula (5.115).

Therefore, according to formulae (6.37) and (6.38), when $\tau = \tau_{\text{opt}}$,[1] the number of iterations p required for reducing the initial error, say, by a factor of e is $p \approx \frac{1}{2}\mu(-\Delta^{(h)}) = \mathcal{O}(h^{-2})$. Every iteration requires $\mathcal{O}(h^{-2})$ arithmetic operations and consequently, the overall number of operations is $\mathcal{O}(h^{-4})$.

In Section 5.7.2, we represented the exact solution of the system $-\Delta^{(h)}u^{(h)} = f^{(h)}$ in the form of a finite Fourier series, see formula (5.110). However, in the case of a non-rectangular domain, or in the case of an equation with variable coefficients (as opposed to the Poisson equation):

$$\frac{\partial}{\partial x}\left(a\frac{\partial u}{\partial x}\right) + \frac{\partial}{\partial y}\left(b\frac{\partial u}{\partial y}\right) = f, \quad a = a(x,y) > 0, \quad b = b(x,y) > 0, \tag{6.40}$$

we typically do not know the eigenvalues and eigenvectors of the problem and as such, cannot use the discrete Fourier series. At the same time, an iterative algorithm, such as the Richardson method, can still be implemented in quite the same way as it was done previously. In doing so, we only need to make sure that the discrete operator is self-adjoint, and also obtain reasonable estimates for the boundaries of its spectrum.

In Section 6.2, we will analyze other iterative methods for the system $\boldsymbol{Ax} = \boldsymbol{f}$, where $\boldsymbol{A} = \boldsymbol{A}^* > 0$, and will show that even for an ill conditioned operator $\boldsymbol{A}$, say, with the condition number $\mu(\boldsymbol{A}) = \mathcal{O}(h^{-2})$, it is possible to build the methods that will be far more efficient than the Richardson iteration. A better efficiency will be achieved by obtaining a more favorable dependence of the number of required iterations p on the condition number $\mu(\boldsymbol{A})$. We will have $p \gtrsim \sqrt{\mu(\boldsymbol{A})}$ as opposed to $p \gtrsim \mu(\boldsymbol{A})$, which is guaranteed by formulae (6.38).

[1] In this case, $\tau_{\text{opt}} = 2/(\lambda_{11} + \lambda_{M-1,M-1}) \approx h^2/[4(1+\sin^2\frac{\pi h}{2})]$, see formulae (5.112) and (5.114).

6.1.4 Preconditioning

As has been mentioned, some alternative methods that are less expensive computationally than the Richardson iteration will be described in Section 6.2. They will have a slower than linear rate of increase of p as a function of $\mu(A)$. A complementary strategy for reducing the number of iterations p consists of modifying the system $Ax = f$ itself, to keep the solution intact and at the same time make the condition number μ smaller.

Let P be a non-singular square matrix of the same dimension as that of A. We can equivalently recast system (6.1) by multiplying it from the left by P^{-1}:

$$P^{-1}Ax = P^{-1}f. \tag{6.41}$$

The matrix P is known as *a preconditioner.* The solution x of system (6.41) is obviously the same as that of system (6.1).

Accordingly, instead of the standard stationary Richardson iteration (6.5) or (6.6), we now obtain its preconditioned version:

$$x^{(p+1)} = (I - \tau P^{-1}A)x^{(p)} + \tau P^{-1}f, \quad p = 0, 1, 2, \ldots, \tag{6.42}$$

or equivalently,

$$x^{(p+1)} = x^{(p)} - \tau P^{-1}r^{(p)}, \quad \text{where} \quad r^{(p)} = Ax^{(p)} - f^{(p)}. \tag{6.43}$$

As a matter of fact, we have already seen some examples of a preconditioned Richardson method. By comparing formulae (6.13) and (6.42), we conclude that the Jacobi method (Example 1 of Section 6.1.1) can be interpreted as a preconditioned Richardson iteration with $\tau = 1$ and $P = D = \text{diag}\{a_{ii}\}$. Similarly, by comparing formulae (6.17) and (6.42) and by noticing that

$$-(A - \hat{U})^{-1}\hat{U} = -(A - \hat{U})^{-1}(A - (A - \hat{U})) = I - (A - \hat{U})^{-1}A,$$

we conclude that the Gauss-Seidel method (Example 2 of Section 6.1.1) can be interpreted as a preconditioned Richardson iteration with $\tau = 1$ and $P = A - \hat{U}$.

Of course, we need to remember that the purpose of preconditioning is not to analyze the equivalent system (6.41) "for the sake of it," but rather to reduce the condition number, so that $\mu(P^{-1}A) < \mu(A)$ or ideally, $\mu(P^{-1}A) \ll \mu(A)$. Unfortunately, relatively little systematic theory is available in the literature for the design of efficient preconditioners. Different types of problems may require special individual tools for analysis, and we refer the reader, e.g., to [Axe94] for detail.

For our subsequent considerations, let us assume that the operator A is self-adjoint and positive definite, as in Section 6.1.3, so that we need to solve the system:

$$Ax = f, \quad A = A^* > 0, \quad x \in \mathbb{L}, \quad f \in \mathbb{L}. \tag{6.44}$$

Introduce an operator $P = P^* > 0$, which can be taken arbitrarily in the meantime, and multiply both sides of system (6.44) by P^{-1}, which yields an equivalent system:

$$Cx = g, \quad C = P^{-1}A, \quad g = P^{-1}f. \tag{6.45}$$

Note that the new operator C of (6.45) is, generally speaking, no longer self-adjoint.

Let us, however, introduce a new inner product on the space $\mathbb{L}$ by means of the operator $\boldsymbol{P}$: $[\boldsymbol{x},\boldsymbol{y}]_{\boldsymbol{P}} \stackrel{\text{def}}{=} (\boldsymbol{P}\boldsymbol{x},\boldsymbol{y})$. Then, the operator $\boldsymbol{C}$ of (6.45) appears self-adjoint and positive definite in the sense of this new inner product. Indeed, as the inverse of a self-adjoint operator is also self-adjoint, we can write:

$$\begin{aligned}
[\boldsymbol{Cx},\boldsymbol{y}]_{\boldsymbol{P}} &= (\boldsymbol{PCx},\boldsymbol{y}) = (\boldsymbol{PP}^{-1}\boldsymbol{Ax},\boldsymbol{y}) = (\boldsymbol{Ax},\boldsymbol{y}) = (\boldsymbol{x},\boldsymbol{Ay}) \\
&= (\boldsymbol{P}^{-1}\boldsymbol{Px},\boldsymbol{Ay}) = (\boldsymbol{Px},\boldsymbol{P}^{-1}\boldsymbol{Ay}) = [\boldsymbol{x},\boldsymbol{Cy}]_{\boldsymbol{P}}, \\
[\boldsymbol{Cx},\boldsymbol{x}]_{\boldsymbol{P}} &= (\boldsymbol{PCx},\boldsymbol{x}) = (\boldsymbol{PP}^{-1}\boldsymbol{Ax},\boldsymbol{x}) = (\boldsymbol{Ax},\boldsymbol{x}) > 0, \text{ if } \boldsymbol{x} \neq \boldsymbol{0}.
\end{aligned}$$

As of yet, the choice of the preconditioner $\boldsymbol{P}$ for system (6.45) was arbitrary. For example, we can choose $\boldsymbol{P} = \boldsymbol{A}$ and obtain $\boldsymbol{C} = \boldsymbol{A}^{-1}\boldsymbol{A} = \boldsymbol{I}$, which immediately yields the solution $\boldsymbol{x}$. As such, $\boldsymbol{P} = \boldsymbol{A}$ can be interpreted as the ideal preconditioner; it provides an indication of what the ultimate goal should be. However, in real life setting $\boldsymbol{P} = \boldsymbol{A}$ is totally impractical. Indeed, the application of the operator $\boldsymbol{P}^{-1} = \boldsymbol{A}^{-1}$ is equivalent to solving the system $\boldsymbol{Ax} = \boldsymbol{f}$ directly, which is precisely what we are trying to avoid by employing an iterative scheme. Recall, an iterative method only requires computing $\boldsymbol{Az}$ for a given $\boldsymbol{z} \in \mathbb{L}$, but does not require computing $\boldsymbol{A}^{-1}\boldsymbol{f}$. It therefore only makes sense to select the operator $\boldsymbol{P}$ among those for which the computation of $\boldsymbol{P}^{-1}\boldsymbol{z}$ for a given $\boldsymbol{z}$ is considerably easier than the computation of $\boldsymbol{A}^{-1}\boldsymbol{z}$. The other extreme, however, would be setting $\boldsymbol{P} = \boldsymbol{I}$, which does not require doing anything, but does not bring along any benefits either. In other words, the preconditioner $\boldsymbol{P}$ should be chosen so as to be easily invertible on one hand, and on the other hand, to "resemble" the operator $\boldsymbol{A}$. In this case we can expect that the operator $\boldsymbol{C} = \boldsymbol{P}^{-1}\boldsymbol{A}$ will "resemble" the unit operator $\boldsymbol{I}$, and the boundaries of its spectrum $\lambda_{\min}$ and $\lambda_{\max}$, as well as the condition number, will all be "closer" to one.

THEOREM 6.4

Let $\boldsymbol{P} = \boldsymbol{P}^* > 0$, let the two numbers $\gamma_1 > 0$ and $\gamma_2 > 0$ be fixed, and let the following inequalities hold:

$$\gamma_1(\boldsymbol{Px},\boldsymbol{x}) \leq (\boldsymbol{Ax},\boldsymbol{x}) \leq \gamma_2(\boldsymbol{Px},\boldsymbol{x}) \tag{6.46}$$

for all $\boldsymbol{x} \in \mathbb{L}$. Then the eigenvalues $\lambda_{\min}(\boldsymbol{C})$, $\lambda_{\max}(\boldsymbol{C})$ and the condition number $\mu_{\boldsymbol{P}}(\boldsymbol{C})$ of the operator $\boldsymbol{C} = \boldsymbol{P}^{-1}\boldsymbol{A}$ satisfy the inequalities:

$$\begin{aligned}
\gamma_1 \leq \lambda_{\min}(\boldsymbol{C}) \leq \lambda_{\max}(\boldsymbol{C}) \leq \gamma_2, \\
\mu_{\boldsymbol{P}}(\boldsymbol{C}) \leq \gamma_2/\gamma_1.
\end{aligned} \tag{6.47}$$

PROOF From the courses of linear algebra it is known that the eigenvalues of a self-adjoint operator can be obtained in the form of the Rayleigh-Ritz quotients (see, e.g., [HJ85, Section 4.2]):

$$\lambda_{\min}(C) = \min_{x\in\mathbb{L},\ x\neq 0} \frac{[Cx,x]_P}{[x,x]_P} = \min_{x\in\mathbb{L},\ x\neq 0} \frac{(Ax,x)}{(Px,x)},$$
$$\lambda_{\max}(C) = \max_{x\in\mathbb{L},\ x\neq 0} \frac{[Cx,x]_P}{[x,x]_P} = \max_{x\in\mathbb{L},\ x\neq 0} \frac{(Ax,x)}{(Px,x)}.$$

By virtue of (6.46), these relations immediately yield (6.47). □

The operators A and P that satisfy inequalities (6.46) are called equivalent by spectrum or equivalent by energy, with the equivalence constants γ_1 and γ_2.

Let us emphasize that the transition from system (6.44) to system (6.45) is only justified if

$$\mu_P(C) \leq \frac{\gamma_2}{\gamma_1} \ll \mu(A).$$

Indeed, since in this case the condition number of the transformed system becomes much smaller than that of the original system, the convergence of the iteration noticeably speeds up. In other words, the rate of decay of the error $\varepsilon^{(p)} = x - x^{(p)}$ in the norm $\|\cdot\|_P$ increases substantially compared to (6.31):

$$\|\varepsilon^{(p)}\|_P = \|x - x^{(p)}\|_P \leq \rho_P^p \|x - x^{(0)}\|_P = \rho_P^p \|\varepsilon^{(0)}\|_P,$$
$$\rho_P = \frac{\mu_P(C) - 1}{\mu_P(C) + 1}.$$

The mechanism of the increase is the drop $\rho_P \ll \rho$ that takes place because $\rho = \dfrac{\mu(A) - 1}{\mu(A) + 1}$ according to formula (6.33), and $\mu_P(C) \ll \mu(A)$. Consequently, for one and the same value of p we will have $\rho_P^p \ll \rho^p$.

A typical situation when preconditioners of type (6.46) prove efficient arises in the context of discrete approximations for elliptic boundary value problems. It was first identified and studied by D'yakonov in the beginning of the sixties; the key results have then been summarized in a later monograph [D'y96].

Consider a system of linear algebraic equations:

$$A_n x = f, \quad f \in \mathbb{R}^n, \quad x \in \mathbb{R}^n,$$

obtained as a discrete approximation of an elliptic boundary value problem. For example, it may be a system of finite-difference equations introduced in Section 5.1.3. In doing so, the better the operator A_n approximates the original elliptic differential operator, the higher the dimension n of the space $\mathbb{R}^n$ is. As such, we are effectively dealing with a sequence of approximating spaces $\mathbb{R}^n$, $n \longrightarrow \infty$, that we will assume Euclidean with the scalar product $(x,y)^{(n)}$.

Let $A_n : \mathbb{R}^n \longmapsto \mathbb{R}^n$ be a sequence of operators such that $A_n = A_n^* > 0$, and let $A_n x = f$, where $x, f \in \mathbb{R}^n$, be a sequence of systems to be solved in the respective spaces. Suppose that the condition number, $\mu(A_n)$, increases when the dimension

n increases so that $\mu(\boldsymbol{A}_n) \sim n^s$, where $s > 0$ is a constant. Then, according to formulae (6.38), it will take $\mathcal{O}(-n^s \ln \sigma)$ iterations to find the solution $\boldsymbol{x} \in \mathbb{R}^n$ with the guaranteed accuracy $\sigma > 0$.

Next, let $\boldsymbol{P}_n : \mathbb{R}^n \longmapsto \mathbb{R}^n$ be a sequence of operators equivalent to the respective operators $\boldsymbol{A}_n$ by energy, with the equivalence constants γ_1 and γ_2 *that do not depend on* n. Then, for $\boldsymbol{C}_n = \boldsymbol{P}_n^{-1}\boldsymbol{A}_n$ we obtain:

$$\mu_{\boldsymbol{P}_n}(\boldsymbol{C}_n) \leq \gamma_2/\gamma_1 = \text{const.} \tag{6.48}$$

Hence we can replace the original system (6.44): $\boldsymbol{A}_n\boldsymbol{x} = \boldsymbol{f}$ by its equivalent (6.45):

$$\boldsymbol{C}_n\boldsymbol{x} = \boldsymbol{g}, \quad \boldsymbol{C}_n = \boldsymbol{P}_n^{-1}\boldsymbol{A}_n, \quad \boldsymbol{g} = \boldsymbol{P}_n^{-1}\boldsymbol{f}. \tag{6.49}$$

In doing so, because of a uniform boundedness of the condition number with respect to n, see formula (6.48), the number of iterations required for reducing the $\|\cdot\|_{\boldsymbol{P}_n}$ norm of the initial error by a predetermined factor of σ:

$$\|\boldsymbol{x} - \boldsymbol{x}^{(p)}\|_{\boldsymbol{P}_n} \leq \sigma \|\boldsymbol{x} - \boldsymbol{x}^{(0)}\|_{\boldsymbol{P}_n}, \tag{6.50}$$

will not increase when the dimension n increases, and will remain $\mathcal{O}(\ln \sigma)$.

Furthermore, let the norms:

$$\|\boldsymbol{x}\| = \sqrt{(\boldsymbol{x},\boldsymbol{x})^{(n)}} \quad \text{and} \quad \|\boldsymbol{x}\|_{\boldsymbol{P}_n} = \sqrt{[\boldsymbol{x},\boldsymbol{x}]_{\boldsymbol{P}_n}} \equiv \sqrt{(\boldsymbol{P}_n\boldsymbol{x},\boldsymbol{x})^{(n)}}$$

be related to one another via the inequalities:

$$n^{-l}\|\boldsymbol{x}\|_{\boldsymbol{P}_n} \leq \|\boldsymbol{x}\| \leq n^l \|\boldsymbol{x}\|_{\boldsymbol{P}_n}, \quad \text{where} \quad l \geq 0, \ l = \text{const.}$$

Then, in order to guarantee the original error estimate (6.37):

$$\|\boldsymbol{x} - \boldsymbol{x}^{(p)}\| \leq \sigma \|\boldsymbol{x} - \boldsymbol{x}^{(0)}\| \tag{6.51}$$

when iterating system (6.49), it is sufficient that the following inequality hold:

$$\|\boldsymbol{x} - \boldsymbol{x}^{(p)}\|_{\boldsymbol{P}_n} \leq \frac{\sigma}{n^l} \|\boldsymbol{x} - \boldsymbol{x}^{(0)}\|_{\boldsymbol{P}_n}. \tag{6.52}$$

Inequality (6.52) is obtained from (6.50) by replacing σ with σn^{-l}, so that solving the preconditioned system (6.49) by the Richardson method will only require $\mathcal{O}(-\ln(\sigma n^{-l})) = \mathcal{O}(\ln n - \ln \sigma)$ iterations, as opposed to $\mathcal{O}(-n^s \ln \sigma)$ iterations required for solving the original non-preconditioned system (6.44).

Of course, the key question remains of how to design a preconditioner equivalent by spectrum to the operator $\boldsymbol{A}$, see (6.46). In the context of elliptic boundary value problems, good results can often be achieved when preconditioning a discretized operator with variable coefficients, such as the one from equation (6.40), with the discretized Laplace operator. On a regular grid, a preconditioner of this type $\boldsymbol{P} = -\Delta^{(h)}$, see Section 5.1.3, can be easily inverted with the help of the FFT, see Section 5.7.3.

Overall, the task of designing an efficient preconditioner is highly problem-dependent. One general approach is based on availability of some a priori knowledge of where the matrix $\boldsymbol{A}$ originates from. A typical example here is the aforementioned spectrally equivalent elliptic preconditioners. Another approach is purely algebraic and only uses the information contained in the structure of a given matrix $\boldsymbol{A}$. Examples include incomplete factorizations ($\boldsymbol{LU}$, Cholesky, modified unperturbed and perturbed incomplete $\boldsymbol{LU}$), polynomial preconditioners (e.g., truncated Neumann series), and various ordering strategies, foremost the multilevel recursive orderings that are conceptually close to the idea of multigrid (Section 6.4). For further detail, we refer the reader to specialized monographs [Axe94, Saa03, vdV03].

6.1.5 Scaling

One reason for a given matrix $\boldsymbol{A}$ to be poorly conditioned, i.e., to have a large condition number $\mu(\boldsymbol{A})$, may be large disparity in the magnitudes of its entries. If this is the case, then scaling the rows of the matrix so that the largest magnitude among the entries in each row becomes equal to one often helps improve the conditioning.

Let $\boldsymbol{D}$ be a non-singular diagonal matrix ($d_{ii} \neq 0$), and instead of the original system $\boldsymbol{Ax} = \boldsymbol{f}$ let us consider its equivalent:

$$\boldsymbol{DAx} = \boldsymbol{Df}. \tag{6.53}$$

The entries d_{ii}, $i = 1, 2, \dots, n$ of the matrix $\boldsymbol{D}$ are to be chosen so that the maximum absolute value of the entry in each row of the matrix $\boldsymbol{DA}$ be equal to one.

$$\max_j |d_{ii}a_{ij}| = 1, \quad i = 1, 2, \dots, n. \tag{6.54}$$

The transition from the matrix $\boldsymbol{A}$ to the matrix $\boldsymbol{DA}$ is known as *scaling* (of the rows of $\boldsymbol{A}$). By comparing equations (6.53) and (6.41) we conclude that it can be interpreted as a particular approach to preconditioning with $\boldsymbol{P}^{-1} = \boldsymbol{D}$. Note that different strategies of scaling can be employed; instead of (6.54) we can require, for example, that all diagonal entries of $\boldsymbol{DA}$ have the same magnitude.

Scaling typically reduces the condition number of a system: $\mu(\boldsymbol{DA}) < \mu(\boldsymbol{A})$. To solve the system $\boldsymbol{Ax} = \boldsymbol{f}$ by iterations, we can first transform it to an equivalent system $\boldsymbol{Cx} = \boldsymbol{g}$ with a self-adjoint positive definite matrix $\boldsymbol{C} = \boldsymbol{A}^*\boldsymbol{A}$ and the right-hand side $\boldsymbol{g} = \boldsymbol{A}^*\boldsymbol{f}$, and then apply the Richardson method. According to Theorem 6.3, the rate of convergence of the Richardson iteration will be determined by the condition number $\mu(\boldsymbol{C})$. It is possible to show that for the Euclidean condition numbers we have: $\mu(\boldsymbol{C}) = \mu^2(\boldsymbol{A})$ (see Exercise 7 after Section 5.3). If the matrix $\boldsymbol{A}$ is scaled ahead of time, see formula (6.53), then the convergence of the iterations will be faster, because $\mu((\boldsymbol{DA})^*(\boldsymbol{DA})) = \mu^2(\boldsymbol{DA}) < \mu^2(\boldsymbol{A})$.

Note that the transition from a given $\boldsymbol{A}$ to $\boldsymbol{C} = \boldsymbol{A}^*\boldsymbol{A}$ is almost never used in practice as a means of enabling the solution by iterations that require a self-adjoint matrix, because an additional matrix multiplication may eventually lead to large errors when computing with finite precision. Therefore, the foregoing example shall only be regarded as a simple theoretical illustration. However, scaling can also help when solving the system $\boldsymbol{Ax} = \boldsymbol{f}$ by a direct method rather than by iterations. For a matrix $\boldsymbol{A}$ with large disparity in the magnitudes of entries it may improve stability of the Gaussian elimination algorithm (Section 5.4). Besides, the system $\boldsymbol{Ax} = \boldsymbol{f}$ with a general matrix $\boldsymbol{A}$ can be solved by an iterative method that does not require a self-adjoint matrix, e.g., by a Krylov subspace iteration (see Section 6.3). In this case, scaling may be very helpful in reducing the condition number $\mu(\boldsymbol{A})$.

Exercises

1. Assume that the eigenvalues of the operator $\boldsymbol{A} : \mathbb{R}^{100} \longmapsto \mathbb{R}^{100}$ are known:

$$\lambda_k = k^2, \quad k = 1, 2, \ldots, 100. \tag{6.55}$$

   The system $\boldsymbol{Ax} = \boldsymbol{f}$ is to be solved by the non-stationary Richardson iterative method:

$$\boldsymbol{x}^{(p+1)} = (\boldsymbol{I} - \tau_p \boldsymbol{A})\boldsymbol{x}^{(p)} + \tau_p \boldsymbol{f}, \quad p = 0, 1, 2, \ldots, \tag{6.56}$$

   where τ_p, $p = 0, 1, 2, \ldots$, are some positive parameters.

   Find a particular set of parameters $\{\tau_0, \tau_1, \ldots, \tau_{99}\}$ that would guarantee $\boldsymbol{x}^{(100)} = \boldsymbol{x}$, where $\boldsymbol{x}$ is the exact solution of the system $\boldsymbol{Ax} = \boldsymbol{f}$.

   Hint. First make sure that $\boldsymbol{x} - \boldsymbol{x}^{(p+1)} \equiv \boldsymbol{\varepsilon}^{(p+1)} = (\boldsymbol{I} - \tau_p \boldsymbol{A})\boldsymbol{\varepsilon}^{(p)} \equiv (\boldsymbol{I} - \tau_p \boldsymbol{A})(\boldsymbol{x} - \boldsymbol{x}^{(p)})$, $p = 0, 1, 2, \ldots$ Then expand the initial error:

$$\boldsymbol{\varepsilon}^{(0)} = \varepsilon_1^{(0)} \boldsymbol{e}_1 + \varepsilon_2^{(0)} \boldsymbol{e}_2 + \ldots + \varepsilon_{100}^{(0)} \boldsymbol{e}_{100}, \tag{6.57}$$

   where $\boldsymbol{e}_1, \boldsymbol{e}_2, \ldots, \boldsymbol{e}_{100}$ are the eigenvectors of $\boldsymbol{A}$ that correspond to the eigenvalues (6.55). Finally, as the eigenvalues (6.55) are given explicitly, choose the iteration parameters $\{\tau_0, \tau_1, \ldots, \tau_{99}\}$ in such a way that each iteration will eliminate precisely one term from the expansion of the error (6.57).

2. Let the iteration parameters in Exercise 1 be chosen as follows:

$$\tau_p = \frac{1}{(p+1)^2}, \quad p = 0, 1, 2, \ldots, 99. \tag{6.58}$$

   a) Show that in this case $\boldsymbol{x}^{(100)} = \boldsymbol{x}$.

   b) Implementation of the algorithm (6.56) with iteration parameters (6.58) on a real computer encounters a critical obstacle. Very large numbers are generated in the course of computation that quickly supersede the largest number that can be represented in the machine. This makes the computation practically impossible. Explain the mechanism of the foregoing phenomenon.

      Hint. Take expansion (6.57) and operate on it with the matrices $(\boldsymbol{I} - \tau_p \boldsymbol{A})$, where τ_p are chosen according to (6.58). Show that before the components with large indexes (close to 100) get canceled, they may become excessively large.

3. Let the iteration parameters in Exercise 1 be chosen as follows:

$$\tau_p = \frac{1}{(100-p)^2}, \quad p = 0, 1, 2, \ldots, 99. \tag{6.59}$$

a) Show that in this case also $\boldsymbol{x}^{(100)} = \boldsymbol{x}$.

b) Implementation of the algorithm (6.56) with iteration parameters (6.59) on a real computer encounters another critical obstacle. Small round-off errors rapidly increase and destroy the overall accuracy. This, again, makes the computation practically impossible. Explain the mechanism of the aforementioned phenomenon.

Hint. When expansion (6.57) is operated on by the matrices $(\boldsymbol{I} - \tau_p \boldsymbol{A})$ with τ_p of (6.59), components of the error with large indexes are canceled first. The cancellation, however, is not exact, its accuracy is determined by the machine precision. Show that the corresponding round-off errors will subsequently grow.

6.2 Chebyshev Iterations and Conjugate Gradients

For the linear system:

$$\boldsymbol{A}\boldsymbol{x} = \boldsymbol{f}, \quad \boldsymbol{A} = \boldsymbol{A}^* > 0, \quad \boldsymbol{x} \in \mathbb{R}^n, \quad \boldsymbol{f} \in \mathbb{R}^n, \tag{6.60}$$

we will describe two iterative methods of solution that offer a better performance (faster convergence) compared to the Richardson method of Section 6.1. We will also discuss the conditions that may justify preferring one of these methods over the other. The two methods are known as the Chebyshev iterative method and the method of conjugate gradients, both are described in detail, e.g., in [SN89b].

As we require that $\boldsymbol{A} = \boldsymbol{A}^* > 0$, all eigenvalues λ_j, $j = 1, 2, \ldots, n$, of the operator $\boldsymbol{A}$ are strictly positive. With no loss of generality, we will assume that they are arranged in ascending order. We will also assume that two numbers $a > 0$ and $b > 0$ are known such that:

$$0 < a \leq \lambda_1 \leq \ldots \leq \lambda_n \leq b. \tag{6.61}$$

The two numbers a and b in formula (6.61) are called boundaries of the spectrum of the operator $\boldsymbol{A}$. If $a = \lambda_1$ and $b = \lambda_n$ these boundaries are referred to as sharp. As in Section 6.1.3, we will also introduce their ratio:

$$\xi = \frac{a}{b} < 1.$$

If the boundaries of the spectrum are sharp, then clearly $\xi = \mu(\boldsymbol{A})^{-1}$, where $\mu(\boldsymbol{A})$ is the Euclidean condition number of $\boldsymbol{A}$ (Theorem 5.3).

6.2.1 Chebyshev Iterations

Let us specify the initial guess $\boldsymbol{x}^{(0)} \in \mathbb{R}^n$ arbitrarily, and let us then compute the iterates $\boldsymbol{x}^{(p)}$, $p = 1, 2, \ldots$, according to the following formulae:

$$\begin{gathered} \boldsymbol{x}^{(1)} = (\boldsymbol{I} - \tau \boldsymbol{A})\boldsymbol{x}^{(0)} + \tau \boldsymbol{f}, \\ \boldsymbol{x}^{(p+1)} = \alpha_{p+1}(\boldsymbol{I} - \tau \boldsymbol{A})\boldsymbol{x}^{(p)} + (1 - \alpha_{p+1})\boldsymbol{x}^{(p-1)} + \tau \alpha_{p+1} \boldsymbol{f}, \\ p = 1, 2, \ldots, \end{gathered} \tag{6.62a}$$

where the parameters τ and α_p are given by:

$$\tau = \frac{2}{a+b}, \quad \alpha_1 = 2, \quad \alpha_{p+1} = \frac{4}{4-\rho_0^2\alpha_p}, \quad \rho_0 = \frac{1-\xi}{1+\xi}, \quad p = 1,2,\dots \tag{6.62b}$$

We see that the first iteration of (6.62a) coincides with that of the Richardson method, and altogether formulae (6.62) describe a second order linear non-stationary iteration scheme. It is possible to show that the error $\boldsymbol{\varepsilon}^{(p)} = \boldsymbol{x} - \boldsymbol{x}^{(p)}$ of the iterate $\boldsymbol{x}^{(p)}$ satisfies the estimate:

$$\|\boldsymbol{\varepsilon}^{(p)}\| \leq \frac{2\rho_1^p}{1+\rho_1^{2p}}\|\boldsymbol{\varepsilon}^{(0)}\|, \quad \rho_1 = \frac{1-\sqrt{\xi}}{1+\sqrt{\xi}}, \quad p = 1,2,\dots \tag{6.63}$$

where $\|\cdot\| = \sqrt{(\cdot,\cdot)}$ is a Euclidean norm on $\mathbb{R}^n$. Based on formula (6.63), the analysis very similar to the one performed in the end of Section 6.1.3 will lead us to the conclusion that in order to reduce the norm of the initial error by a predetermined factor σ, i.e., in order to guarantee the following estimate:

$$\|\boldsymbol{\varepsilon}^{(p)}\| \leq \sigma|\boldsymbol{\varepsilon}^{(0)}\|, \tag{6.64}$$

it is sufficient to choose the number of iterations p so that:

$$p \geq -\frac{1}{2}\left(\ln\frac{\sigma}{2}\right)\frac{1}{\sqrt{\xi}} = -\frac{1}{2}\left(\ln\frac{\sigma}{2}\right)\sqrt{\frac{b}{a}}. \tag{6.65}$$

This number is about $\sqrt{\frac{b}{a}}$ times smaller than the number of iterations required for achieving the same error estimate (6.64) using the Richardson method. In particular, when the sharp boundaries of the spectrum are available, we have:

$$\frac{b}{a} = \frac{1}{\xi} = \mu(\boldsymbol{A}).$$

Then, by comparing formulae (6.38a) and (6.65) we can see that whereas for the Richardson method the number of iterations p is proportional to the condition number $\mu(\boldsymbol{A})$ itself, for the new method (6.62) it is only proportional to the square root $\sqrt{\mu(\boldsymbol{A})}$. Therefore, the larger the condition number, the more substantial is the relative economy offered by the new approach. It is also important to mention that the iterative scheme (6.62) is computationally stable.

The construction of the iterative method (6.62), the proof of its error estimate (6.63), as well as the proof of its computational stability and other properties, are based on the analysis of Chebyshev polynomials (see Section 3.2.3) and of some related polynomials. We omit this analysis here and refer the reader, e.g., to [SN89b] for detail. We only mention that because of its relation to Chebyshev polynomials, the iterative method (6.62) is often referred to as the second order Chebyshev method in the literature.

Let us also note that along with the second order Chebyshev method (6.62), there is also a first order method of a similar design and with similar properties. It is known as the first order Chebyshev iterative method, see [SN89b]. Its key advantage is that for computing the iterate $\boldsymbol{x}^{(p+1)}$ this method only requires the knowledge of $\boldsymbol{x}^{(p)}$ so that the iterate $\boldsymbol{x}^{(p-1)}$ no longer needs to be stored in the computer memory. However, the first order Chebyshev method is prone to computational instabilities.

6.2.2 Conjugate Gradients

The method of conjugate gradients was introduced in Section 5.6 as a direct method for solving linear systems of type (6.60). Starting from an arbitrary initial guess $\boldsymbol{x}^{(0)} \in \mathbb{R}^n$, the method computes successive approximations $\boldsymbol{x}^{(p)}$, $p = 1,2,\ldots$, to the solution $\boldsymbol{x}$ of $\boldsymbol{A}\boldsymbol{x} = \boldsymbol{f}$ according to the formulae (see Exercise 2 after Section 5.6):

$$
\begin{aligned}
\boldsymbol{x}^{(1)} &= (\boldsymbol{I} - \tau_1 \boldsymbol{A})\boldsymbol{x}^{(0)} + \tau_1 \boldsymbol{f}, \\
\boldsymbol{x}^{(p+1)} &= \gamma_{p+1}(\boldsymbol{I} - \tau_{p+1}\boldsymbol{A})\boldsymbol{x}^{(p)} + (1 - \gamma_{p+1})\boldsymbol{x}^{(p-1)} + \gamma_{p+1}\tau_{p+1}\boldsymbol{f},
\end{aligned}
\tag{6.66a}
$$

where

$$
\begin{aligned}
\tau_{p+1} &= \frac{(\boldsymbol{r}^{(p)}, \boldsymbol{r}^{(p)})}{(\boldsymbol{A}\boldsymbol{r}^{(p)}, \boldsymbol{r}^{(p)})}, \quad \boldsymbol{r}^{(p)} = \boldsymbol{A}\boldsymbol{x}^{(p)} - \boldsymbol{f}, \quad p = 0,1,2,\ldots, \\
\gamma_1 = 1, \quad \gamma_{p+1} &= \left(1 - \frac{\tau_{p+1}}{\tau_p} \frac{(\boldsymbol{r}^{(p)}, \boldsymbol{r}^{(p)})}{(\boldsymbol{r}^{(p-1)}, \boldsymbol{r}^{(p-1)})} \frac{1}{\gamma_p}\right)^{-1}, \quad p = 1,2,\ldots.
\end{aligned}
\tag{6.66b}
$$

After at most n iterations, the method produces the exact solution $\boldsymbol{x}$ of system (6.60), and the computational complexity of obtaining this solution is $\mathcal{O}(n^3)$ arithmetic operations, i.e., asymptotically the same as that of the Gaussian elimination (Section 5.4).

However, in practice the method of conjugate gradients is never used as a direct method. The reason is that its convergence to the exact solution after a finite number of steps can only be guaranteed if the computations are conducted with infinite precision. On a real computer with finite precision the method is prone to instabilities, especially when the condition number $\mu(\boldsymbol{A})$ is large. Therefore, the method of conjugate gradients is normally used in the capacity of an iteration scheme. It is a second order linear non-stationary iterative method and a member of a larger family of iterative methods known as methods of conjugate directions, see [SN89b].

The rate of convergence of the method of conjugate gradients (6.66) is at least as fast as that of the Chebyshev method (6.62) when the latter is built based on the knowledge of the sharp boundaries a and b, see formula (6.61). In other words, the same error estimate (6.63) as pertains to the Chebyshev method is guaranteed by the method of conjugate gradients as well. In doing so, the quantity ξ in formula (6.63) needs to be interpreted as:

$$
\xi = \frac{a}{b} = \frac{\lambda_{\min}}{\lambda_{\max}} = \frac{1}{\mu(\boldsymbol{A})}.
$$

The key advantage of the method of conjugate gradients compared to the Chebyshev method is that the boundaries of the spectrum do not need to be known explicitly, yet the method guarantees that the number of iterations required to reduce the initial error by a prescribed factor will only be proportional to the square root of the condition number $\sqrt{\mu(\boldsymbol{A})}$, see formula (6.65).

The key shortcoming of the method of conjugate gradients is its computational instability that manifests itself stronger for larger condition numbers $\mu(\boldsymbol{A})$. The most favorable situation for applying the method of conjugate gradients is when it is known that the condition number $\mu(\boldsymbol{A})$ is not too large, while the boundaries of the spectrum are unknown, and the dimension n of the system is much higher than the number of iterations p needed for achieving a given accuracy. In practice, one can also use special stabilization procedures for the method of conjugate gradients, as well as implement restarts after every so many iterations. The latter, however, slow down the convergence.

Note also that both the Chebyshev iterative method and the method of conjugate gradients can be preconditioned, similarly to how the Richardson method was preconditioned in Section 6.1.4. Preconditioning helps reduce the condition number and as such, speeds up the convergence.

Exercises

1. Consider the same second order central difference scheme as the one built in Exercise 4 after Section 5.4 (page 156):

$$\frac{u_{j+1}-2u_j+u_{j-1}}{h^2}-u_j=f_j,\quad j=1,2,\dots,N-1,$$
$$h=\frac{1}{N},\quad u_0=0,\quad u_N=0.$$

   a) Write down this scheme as a system of linear algebraic equations with an $(N-1)\times(N-1)$ matrix $\boldsymbol{A}$. Use the apparatus of finite Fourier series (Section 5.7) to prove that the matrix $-\boldsymbol{A}$ is symmetric positive definite: $-\boldsymbol{A}=-\boldsymbol{A}^*>0$, find sharp boundaries of its spectrum $\lambda_{\min}$ and $\lambda_{\max}$, end estimate its Euclidean condition number $\mu(-\boldsymbol{A})$ as it depends on $h=1/N$ for large N.

   Hint. The eigenvectors and eigenvalues of $-\boldsymbol{A}$ are given by $\psi_j^{(k)}=\sqrt{2}\sin\frac{k\pi j}{N}$, $j,k=1,2,\dots,N$, and $\lambda_k=\left(\frac{4}{h^2}\sin^2\frac{k\pi}{2N}+1\right)$, $k=1,2,\dots,N-1$, respectively.

   b) Solve the system

$$-\boldsymbol{A}\boldsymbol{u}=-\boldsymbol{f}$$

   on the computer by iterations, with the initial guess $\boldsymbol{u}^{(0)}=\boldsymbol{0}$ and the same right-hand side $\boldsymbol{f}$ as in in Exercise 4 after Section 5.4: $f_j=f(x_j)$, $j=1,\dots,N-1$, where $f(x)=(-\pi^2\sin(\pi x)+2\pi\cos(\pi x))e^x$ and $x_j=j\cdot h$; the corresponding values of the exact solution are $u(x_j)$, $j=1,\dots,N-1$, where $u(x)=\sin(\pi x)e^x$.

   Implement three different iteration schemes: The non-preconditioned stationary Richardson method of Section 6.1.3, the Chebyshev method of Section 6.2.1, and the conjugate gradients method of Section 6.2.2. For each scheme, use four different grids: $N=32$, 64, 128, and 256. Evaluate the error $\boldsymbol{u}_{\text{exact}}-\boldsymbol{u}^{(p)}$ in the Euclidean norm based on the inner product (5.98): $(\boldsymbol{v},\boldsymbol{w})=h\sum\limits_{j=1}^{N-1}v_jw_j$ (recall, we always have $u_0 = u_N = 0$). In every case, stop the iteration once the norm of the initial error $\|\boldsymbol{\varepsilon}^{(0)}\| = \|\boldsymbol{u}_{\text{exact}} - \boldsymbol{u}^{(0)}\|$ has dropped by a factor of $\sigma = 10^{-3}$: $\|\boldsymbol{\varepsilon}^{(p)}\| \leq \sigma\|\boldsymbol{\varepsilon}^{(0)}\|$, and find the actual number of iterations p required to reach that level of error reduction. Determine the dependence of this number p on $\mu(-\boldsymbol{A})$, and as such, on the dimension N of the grid, for each iteration scheme.

Hint. For the Richardson method, $p \gtrsim \mu(-\boldsymbol{A})$ and consequently, p should scale quadratically with N. For the Chebyshev method and the method of conjugate gradients, $p \gtrsim \sqrt{\mu(-\boldsymbol{A})}$ and consequently, p should scale linearly with N.

6.3 Krylov Subspace Iterations

The key idea behind the Krylov subspace iterations is conceptually close to that behind the method of conjugate gradients. Recall, in the method of conjugate gradients (see Sections 5.6 and 6.2.2) we solve the system $\boldsymbol{Ax} = \boldsymbol{f}$, $\boldsymbol{A} = \boldsymbol{A}^* > 0$, where $\boldsymbol{x} \,\&\, \boldsymbol{f} \in \mathbb{R}^n$, by building a sequence of descent directions $\boldsymbol{d}^{(p)}$, $p = 0, 1, 2, \ldots$, and a sequence of iterates $\boldsymbol{x}^{(p)}$, $p = 0, 1, 2, \ldots$, such that every residual $\boldsymbol{r}^{(p+1)} = \boldsymbol{Ax}^{(p+1)} - \boldsymbol{f}$ is orthogonal to all the previous descent directions $\boldsymbol{d}^{(j)}$, $j = 0, 1, \ldots, p$. In their own turn, the descent directions $\boldsymbol{d}^{(j)}$ are chosen $\boldsymbol{A}$-conjugate, i.e., orthogonal in the sense of the inner product defined by $\boldsymbol{A}$, so that for every $p = 1, 2, \ldots$: $[\boldsymbol{d}^{(j)}, \boldsymbol{d}^{(p)}]_{\boldsymbol{A}} = 0$, $j = 0, 1, \ldots, p-1$. Then, given an arbitrary $\boldsymbol{d}^{(0)}$, the pool of all available linearly independent directions $\boldsymbol{d}^{(j)}$, $j = 0, 1, 2, \ldots, p$, in the space $\mathbb{R}^n$ gets exhausted at $p = n-1$, and because of the orthogonality: $\boldsymbol{r}^{(p+1)} \perp \operatorname{span}\{\boldsymbol{d}^{(0)}, \boldsymbol{d}^{(1)}, \ldots, \boldsymbol{d}^{(p)}\}$, the method converges to the exact solution $\boldsymbol{x}$ after at most n steps.

A key provision that facilitates implementation of the method of conjugate gradients is that a scalar product can be introduced in the space $\mathbb{R}^n$ by means of the matrix $\boldsymbol{A}$. To enable that, the matrix $\boldsymbol{A}$ must be symmetric positive definite. Then, a system of linearly independent directions $\boldsymbol{d}^{(j)}$ can be built in $\mathbb{R}^n$ using the notion of orthogonality in the sense of $[\cdot,\cdot]_{\boldsymbol{A}}$. Iterative methods based on Krylov subspaces also exploit the idea of building a sequence of linearly independent directions in the space $\mathbb{R}^n$. Then, some quantity related to the error is minimized on the subspaces that span those directions. However, the Krylov subspace iterations apply to more general systems $\boldsymbol{Ax} = \boldsymbol{f}$ rather than only $\boldsymbol{A} = \boldsymbol{A}^* > 0$. In doing so, the orthogonality in the sense of $[\cdot,\cdot]_{\boldsymbol{A}}$ is no longer available, and the problem of building the linearly independent directions for the iteration scheme becomes more challenging.

In this section, we only provide a very brief and introductory account of Krylov subspace iterative methods. This family of methods was developed fairly recently, and in many aspects it still represents an active research area. For further detail, we refer the reader, e.g., to [Axe94, Saa03, vdV03].

6.3.1 Definition of Krylov Subspaces

Consider a non-preconditioned non-stationary Richardson iteration (6.7):

$$\boldsymbol{x}^{(p+1)} = \boldsymbol{x}^{(p)} - \tau_p \boldsymbol{r}^{(p)}, \quad p = 0,1,2,\ldots. \tag{6.67}$$

For the residuals $\boldsymbol{r}^{(p)} = \boldsymbol{A}\boldsymbol{x}^{(p)} - \boldsymbol{f}$, formula (6.67) yields:

$$\boldsymbol{r}^{(p+1)} = \boldsymbol{r}^{(p)} - \tau_p \boldsymbol{A}\boldsymbol{r}^{(p)} = (\boldsymbol{I} - \tau_p \boldsymbol{A})\boldsymbol{r}^{(p)}, \quad p = 0,1,2,\ldots.$$

Consequently, we obtain:

$$\boldsymbol{r}^{(p)} = \left[\prod_{j=0}^{p-1} (\boldsymbol{I} - \tau_j \boldsymbol{A})\right] \boldsymbol{r}^{(0)}. \tag{6.68}$$

In other words, the residual $\boldsymbol{r}^{(p)}$ can be written in the form: $\boldsymbol{r}^{(p)} = Q_p(\boldsymbol{A})\boldsymbol{r}^{(0)}$, where $Q_p(\boldsymbol{A})$ is a matrix polynomial with respect to $\boldsymbol{A}$ of degree no greater than p.

DEFINITION 6.1 *For a given $n \times n$ matrix $\boldsymbol{A}$ and a given $\boldsymbol{u} \in \mathbb{R}^n$, the Krylov subspace of order m is defined as a linear span of the m vectors obtained by successively applying the powers of $\boldsymbol{A}$ to the vector $\boldsymbol{u}$:*

$$K_m(\boldsymbol{A}, \boldsymbol{u}) = \text{span}\{\boldsymbol{u}, \boldsymbol{A}\boldsymbol{u}, \ldots, \boldsymbol{A}^{m-1}\boldsymbol{u}\}. \tag{6.69}$$

Clearly, the Krylov subspace is a space of all n-dimensional vectors that can be represented in the form $Q_{m-1}(\boldsymbol{A})\boldsymbol{u}$, where Q_{m-1} is a polynomial of degree no greater than $m-1$. From formulae (6.68) and (6.69) it is clear that

$$\boldsymbol{r}^{(p)} \in K_{p+1}(\boldsymbol{A}, \boldsymbol{r}^{(0)}), \quad p = 0,1,2,\ldots.$$

As for the iterates $\boldsymbol{x}^{(p)}$ themselves, from formula (6.67) we find:

$$\boldsymbol{x}^{(p)} = \boldsymbol{x}^{(0)} - \sum_{j=0}^{p-1} \tau_j \boldsymbol{r}^{(j)}.$$

Therefore, we can write:

$$\boldsymbol{x}^{(p)} \in N_p \stackrel{\text{def}}{=} \left\{\boldsymbol{x} \in \mathbb{R}^n \,\middle|\, \boldsymbol{x} = \boldsymbol{x}^{(0)} + \boldsymbol{z},\ \boldsymbol{z} \in K_p(\boldsymbol{A}, \boldsymbol{r}^{(0)})\right\}, \tag{6.70}$$

which means that $\boldsymbol{x}^{(p)}$ belongs to an affine space N_p obtained as a translation of $K_p(\boldsymbol{A}, \boldsymbol{r}^{(0)})$ from Definition 6.1 by a fixed vector $\boldsymbol{x}^{(0)}$. Of course, a specific $\boldsymbol{x}^{(p)} \in N_p$ can be chosen in a variety of ways.

Different iterative methods from the Krylov subspace family differ precisely in how one selects a particular $\boldsymbol{x}^{(p)} \in N_p$, see (6.70). Normally, one employs some optimization criterion. Of course, the best thing would be to minimize the norm of the error $\|\boldsymbol{x} - \boldsymbol{z}\|$:

$$\boldsymbol{x}^{(p)} = \arg\min_{\boldsymbol{z} \in N_p} \|\boldsymbol{x} - \boldsymbol{z}\|,$$

where $\boldsymbol{x}$ is the solution to $\boldsymbol{Ax}=\boldsymbol{f}$. This, however, is obviously impossible in practice, because the solution $\boldsymbol{x}$ is not known. Therefore, alternative strategies must be used. For example, one can enforce the orthogonality $\boldsymbol{r}^{(p)} \perp K_p(\boldsymbol{A},\boldsymbol{r}^{(0)})$, i.e., require that $\forall \boldsymbol{u} \in K_p(\boldsymbol{A},\boldsymbol{r}^{(0)}):\ (\boldsymbol{u},\boldsymbol{Ax}^{(p)}-\boldsymbol{f})=0$. This leads to the Arnoldi method, which is also called the full orthogonalization method (FOM). This method is known to reduce to conjugate gradients (Section 5.6) if $\boldsymbol{A}=\boldsymbol{A}^*>0$. Alternatively, one can minimize the Euclidean norm of the residual $\boldsymbol{r}^{(p)}=\boldsymbol{Ax}^{(p)}-\boldsymbol{f}$:

$$\boldsymbol{x}^{(p)} = \arg\min_{z\in N_p} \|\boldsymbol{A}z-\boldsymbol{f}\|_2.$$

This strategy defines the so-called method of generalized minimal residuals (GMRES) that we describe in Section 6.3.2.

It is clear though that before FOM, GMRES, or any other Krylov subspace method can actually be implemented, we need to thoroughly describe the minimization search space N_p of (6.70). This is equivalent to describing the Krylov subspace $K_p(\boldsymbol{A},\boldsymbol{r}^{(0)})$. In other words, we need to construct a basis in the space $K_p(\boldsymbol{A},\boldsymbol{r}^{(0)})$ introduced by Definition 6.1 for $p=m$ and $\boldsymbol{r}^{(0)}=\boldsymbol{u}$.

The first result which is known is that in general the dimension of the space $K_m(\boldsymbol{A},\boldsymbol{u})$ is a non-decreasing function of m. This dimension, however, may actually be lower than m as it is not guaranteed ahead of time that all the vectors: $\boldsymbol{u}$, $\boldsymbol{Au}$, $\boldsymbol{A}^2\boldsymbol{u}$, ..., $\boldsymbol{A}^{m-1}\boldsymbol{u}$ are linearly independent.

For a given m one can obtain an orthonormal basis in the space $K_m(\boldsymbol{A},\boldsymbol{u})$ with the help of the so-called Arnoldi process, which is based on the well known Gram-Schmidt orthonormalization algorithm (all norms are Euclidean):

$$\begin{aligned}
\boldsymbol{u}_1 &= \frac{\boldsymbol{u}}{\|\boldsymbol{u}\|},\\
\boldsymbol{v}_1 &= \boldsymbol{Au}_1-(\boldsymbol{u}_1,\boldsymbol{Au}_1)\boldsymbol{u}_1, \quad \boldsymbol{u}_2=\frac{\boldsymbol{v}_1}{\|\boldsymbol{v}_1\|},\\
\boldsymbol{v}_2 &= \boldsymbol{Au}_2-(\boldsymbol{u}_1,\boldsymbol{Au}_2)\boldsymbol{u}_1-(\boldsymbol{u}_2,\boldsymbol{Au}_2)\boldsymbol{u}_2, \quad \boldsymbol{u}_3=\frac{\boldsymbol{v}_2}{\|\boldsymbol{v}_2\|},\\
&\cdots\cdots\cdots\cdots\cdots\cdots\cdots\cdots\cdots\cdots\\
\boldsymbol{v}_k &= \boldsymbol{Au}_k-\sum_{j=1}^{k}(\boldsymbol{u}_j,\boldsymbol{Au}_k)\boldsymbol{u}_j, \quad \boldsymbol{u}_{k+1}=\frac{\boldsymbol{v}_k}{\|\boldsymbol{v}_k\|},\\
&\cdots\cdots\cdots\cdots\cdots\cdots\cdots\cdots\cdots\cdots
\end{aligned} \tag{6.71}$$

Obviously, all the resulting vectors $\boldsymbol{u}_1,\boldsymbol{u}_2,\ldots,\boldsymbol{u}_k,\ldots$ are orthonormal. If the Arnoldi process terminates at step m, then the vectors $\{\boldsymbol{u}_1,\boldsymbol{u}_2,\ldots,\boldsymbol{u}_m\}$ will form an orthonormal basis in $K_m(\boldsymbol{A},\boldsymbol{u})$. The process can also terminate prematurely, i.e., yield $\boldsymbol{v}_k=\boldsymbol{0}$ at some $k<m$. This will indicate that the dimension of the corresponding Krylov subspace is lower than m. Note also that the classical Gram-Schmidt orthogonalization is prone to numerical instabilities. Therefore, in practice one often uses its stabilized version (see Remark 7.4 on page 219). The latter is not not completely fail proof either, yet it is more robust and somewhat more expensive computationally.

Let us now introduce the matrix $U_k = [u_1, u_2, \ldots, u_k]$ composed of the n-dimensional orthonormal vectors $u_1, u_2, \ldots, u_k$ as columns. From the last equation (6.71) is is clear that the vector Au_k is a linear combination of the vectors $u_{k+1}, u_k, \ldots, u_1$. Consequently, we can write:

$$AU_k = U_{k+1}H_k, \tag{6.72}$$

where H_k is a matrix in the upper Hessenberg form. This matrix has $k+1$ rows and k columns, and all its entries below the first sub-diagonal are equal to zero:

$$H_k = \{\mathfrak{h}_{i,j} \,|\, i = 1, 2, \ldots, k+1,\ j = 1, 2, \ldots, k,\ \mathfrak{h}_{i,j} = 0 \text{ for } i > j+1\}.$$

Having introduced the procedure for building a basis in a Krylov subspace, we can now analyze a particular iteration method — the GMRES.

6.3.2 GMRES

In GMRES, we are minimizing the Euclidean norm of the residual, $\|\cdot\| \equiv \|\cdot\|_2$:

$$\begin{gathered}\|r^{(p)}\| = \|Ax^{(p)} - f\| = \min_{z \in N_p} \|Az - f\|, \\ N_p = x^{(0)} + K_p(A, r^{(0)}).\end{gathered} \tag{6.73}$$

Let U_p be an $n \times p$ matrix with orthonormal columns such that these columns form a basis in the space $K_p(A, r^{(0)})$; this matrix is obtained by means of the Arnoldi process described in Section 6.3.1. Then we can write:

$$x^{(p)} = x^{(0)} + U_p w^{(p)},$$

where $w^{(p)}$ is a vector of dimension p that needs to be determined. Hence, for the residual $r^{(p)}$ we have according to formula (6.72):

$$r^{(p)} = r^{(0)} + AU_p w^{(p)} = r^{(0)} + U_{p+1}H_p w^{(p)} = U_{p+1}(q^{(p+1)} + H_p w^{(p)}), \tag{6.74}$$

where $q^{(p+1)}$ is a $p+1$-dimensional vector defined as follows:

$$q^{(p+1)} = [\|r^{(0)}\|, \underbrace{0, 0, \ldots, 0}_{p}]^T.$$

Indeed, as the first column of the matrix U_{p+1} generated by the Arnoldi process (6.71) is given by $r^{(0)}/\|r^{(0)}\|$, we clearly have: $U_{p+1}q^{(p+1)} = r^{(0)}$ in formula (6.74). Consequently, minimization in the sense of (6.73) reduces to:

$$\|r^{(p)}\| = \min_{w^{(p)} \in \mathbb{R}^p} \|q^{(p+1)} + H_p w^{(p)}\|. \tag{6.75}$$

Equality (6.75) holds because $U_{p+1}^T U_{p+1} = I_{p+1}$, which implies that for the Euclidean norm $\|\cdot\| \equiv \|\cdot\|_2$ we have: $\|U_{p+1}(q^{(p+1)} + H_p w^{(p)})\| = \|q^{(p+1)} + H_p w^{(p)}\|$.

Problem (6.75) is a typical problem of solving an overdetermined system of linear algebraic equations in the sense of the least squares. Indeed, the matrix $\boldsymbol{H}_p$ has $p+1$ rows and p columns, i.e., there are $p+1$ equations and only p unknowns. Solutions of such systems can, generally speaking, only be found in the weak sense, in particular, in the sense of the least squares. The concept of weak, or generalized, solutions, as well as the methods for their computation, are discussed in Chapter 7.

In the meantime, let us mention that if the Arnoldi orthogonalization process (6.71) does not terminate on or before $k=p$, then the minimization problem (6.75) has a unique solution. As shown in Section 7.2, this is an implication of the matrix $\boldsymbol{H}_p$ being full rank. The latter assertion, in turn, is true because according to formula (6.72), equation number k of (6.71) can be recast as follows:

$$\boldsymbol{A}\boldsymbol{u}_k = \boldsymbol{v}_k + \sum_{j=1}^{k} \mathfrak{h}_{jk}\boldsymbol{u}_j = \boldsymbol{u}_{k+1}\|\boldsymbol{v}_k\| + \sum_{j=1}^{k} \mathfrak{h}_{jk}\boldsymbol{u}_j = \sum_{j=1}^{k+1} \mathfrak{h}_{jk}\boldsymbol{u}_j,$$

which means that $\mathfrak{h}_{k+1,k} = \|\boldsymbol{v}_k\| \neq 0$. As such, all columns of the matrix $\boldsymbol{H}_p$ are linearly independent since every column has an additional non-zero entry $\mathfrak{h}_{k+1,k}$ compared to the previous column. Consequently, the vector $\boldsymbol{w}^{(p)}$ can be obtained as a solution to the linear system:

$$\boldsymbol{H}_p^T\boldsymbol{H}_p\boldsymbol{w}^{(p)} = \boldsymbol{H}_p^T\boldsymbol{q}^{(p+1)}. \tag{6.76}$$

The solution $\boldsymbol{w}^{(p)}$ of system (6.76) is unique because the matrix $\boldsymbol{H}_p^T\boldsymbol{H}_p$ is non-singular (Exercise 2). In practice, one does not normally reduce the least squares minimization problem (6.75) to linear system (6.76) since this reduction may lead to the introduction of large additional errors (amplification of round-off). Instead, problem (6.75) is solved using the $\boldsymbol{QR}$ factorization of the matrix $\boldsymbol{H}_p$, see Section 7.2.2.

Note that in the course of the previous analysis we assumed that the dimension of the Krylov subspaces $K_p(\boldsymbol{A},\boldsymbol{r}^{(0)})$ would increase monotonically as a function of p. Let us now see what happens if the alternative situation takes place, i.e., if the Arnoldi process terminates prematurely.

THEOREM 6.5

Let p be the smallest integer number for which the Arnoldi process (6.71) terminates:

$$\boldsymbol{A}\boldsymbol{u}_p - \sum_{j=1}^{p} (\boldsymbol{u}_j, \boldsymbol{A}\boldsymbol{u}_p)\boldsymbol{u}_j = \boldsymbol{0}.$$

Then the corresponding iterate yields the exact solution:

$$\boldsymbol{x}^{(p)} = \boldsymbol{x} = \boldsymbol{A}^{-1}\boldsymbol{f}.$$

PROOF By hypothesis of the theorem, $\boldsymbol{A}\boldsymbol{u}_p \in K_p$. Consequently, $\boldsymbol{A}K_p \subset K_p$. This implies [cf. formula (6.72)]:

$$\boldsymbol{A}\boldsymbol{U}_p = \boldsymbol{U}_p\tilde{\boldsymbol{H}}, \tag{6.77}$$

where $\tilde{\boldsymbol{H}}$ is a $p \times p$ matrix. This matrix is non-singular because otherwise it would have had linearly dependent rows. Then, according to formula (6.77), each column of $\boldsymbol{A}\boldsymbol{U}_p$ could be represented as a linear combination of only a subset of the columns from $\boldsymbol{U}_p$ rather than as a linear combination of all of its columns. This, in turn, means that the Arnoldi process terminates earlier than $k = p$, which contradicts the hypothesis of the theorem.

For the norm of the residual $\boldsymbol{r}^{(p)} = \boldsymbol{A}\boldsymbol{x}^{(p)} - \boldsymbol{f}$ we can write:

$$\|\boldsymbol{r}^{(p)}\| = \|\boldsymbol{A}\boldsymbol{x}^{(p)} - \boldsymbol{f}\| = \|\boldsymbol{A}(\boldsymbol{x}^{(p)} - \boldsymbol{x}^{(0)}) + \boldsymbol{r}^{(0)}\|. \tag{6.78}$$

Next, we notice that since $\boldsymbol{x}^{(p)} \in N_p$, then $\boldsymbol{x}^{(p)} - \boldsymbol{x}^{(0)} \in K_p(\boldsymbol{A}, \boldsymbol{r}^{(0)})$, and consequently, $\exists \boldsymbol{w} \in \mathbb{R}^p: \ \boldsymbol{x}^{(p)} - \boldsymbol{x}^{(0)} = \boldsymbol{U}_p \boldsymbol{w}$, because the columns of the matrix $\boldsymbol{U}_p$ provide a basis in the space $K_p(\boldsymbol{A}, \boldsymbol{r}^{(0)})$. Let us also introduce a p-dimensional vector $\boldsymbol{q}^{(p)}$ with real components:

$$\boldsymbol{q}^{(p)} = \big[\|\boldsymbol{r}^{(0)}\|, \underbrace{0, 0, \dots, 0}_{p-1}\big]^T,$$

so that $\boldsymbol{U}_p \boldsymbol{q}^{(p)} = \boldsymbol{r}^{(0)}$. Then, taking into account equality (6.77), as well as the orthonormality of the columns of the matrix $\boldsymbol{U}_p$: $\boldsymbol{U}_p^T \boldsymbol{U}_p = \boldsymbol{I}_p$, we obtain from formula (6.78):

$$\|\boldsymbol{r}^{(p)}\| = \|\boldsymbol{U}_p(\boldsymbol{q}^{(p)} + \tilde{\boldsymbol{H}}\boldsymbol{w})\| = \|\boldsymbol{q}^{(p)} + \tilde{\boldsymbol{H}}\boldsymbol{w}\|. \tag{6.79}$$

Finally, we recall that on every iteration of GMRES we minimize the norm of the residual: $\|\boldsymbol{r}^{(p)}\| \longrightarrow \min$. Then, we can simply set $\boldsymbol{w} = -\tilde{\boldsymbol{H}}^{-1}\boldsymbol{q}^{(p)}$ in formula (6.79), which immediately yields $\|\boldsymbol{r}^{(p)}\| = 0$. This is obviously a minimum of the norm, and it implies $\boldsymbol{r}^{(p)} = \boldsymbol{0}$, i.e., $\boldsymbol{A}\boldsymbol{x}^{(p)} = \boldsymbol{f} \implies \boldsymbol{x}^{(p)} = \boldsymbol{A}^{-1}\boldsymbol{f} = \boldsymbol{x}$. □

We can now summarize two possible scenarios of behavior of the GMRES iteration. If the Arnoldi process terminates prematurely at some $p < n$ (n is the dimension of the space), then, according to Theorem 6.5, $\boldsymbol{x}^{(p)}$ is the exact solution to $\boldsymbol{A}\boldsymbol{x} = \boldsymbol{f}$. Otherwise, the maximum number of iterations that the GMRES can perform is equal to n. Indeed, if the Arnoldi process does not terminate prematurely, then $\boldsymbol{U}_n$ will contain n linearly independent vectors of dimension n and consequently, $K_n(\boldsymbol{A}, \boldsymbol{r}^{(0)}) = \mathbb{R}^n$. As such, the last minimization of the residual in the sense of (6.73) will be performed over the entire space $\mathbb{R}^n$, which obviously yields the exact solution $\boldsymbol{x} = \boldsymbol{A}^{-1}\boldsymbol{f}$. Therefore, technically speaking, the GMRES can be regarded as a direct method for solving $\boldsymbol{A}\boldsymbol{x} = \boldsymbol{f}$, in much the same way as we regarded the method of conjugate gradients as a direct method (see Section 5.6).

In practice, however, the GMRES is never used in the capacity of a direct solver, it its only used as an iterative scheme. The reason is that for high dimensions n it is only feasible to perform very few iterations, and one should hope that the approximate solution obtained after these iterations will be sufficiently accurate in a given context. The limitations for the number of iterations come primarily from the large storage requirements for the Krylov subspace basis $\boldsymbol{U}_p$, as well as from the increasing computational costs associated with solving the sequence of the least squares problems (6.75) for $p = 1, 2, \ldots$. Note that the method of conjugate gradients does not entail this type of limitations because its descent directions are automatically $\boldsymbol{A}$-orthogonal. These additional constraints that characterize the GMRES are the "price to pay" for its broader applicability and ability to handle general matrices $\boldsymbol{A}$, as opposed to only symmetric positive definite matrices, for which the method of conjugate gradients works. However, another inherent limitation of the GMRES fully translates to the method of conjugate gradients (or the other way around). Indeed, the exact solution of $\boldsymbol{Ax} = \boldsymbol{f}$ can only be obtained by means of the GMRES if the computations are conducted with infinite precision. On a finite precision computer the method is prone to numerical instabilities. No universal cure is available for this problem; some partial remedies, such as restarts, are discussed, e.g., in [Saa03].

Exercises

1. Prove that the Arnoldi process (6.71) indeed yields an orthonormal system of vectors: $\boldsymbol{u}_1, \boldsymbol{u}_2, \ldots$.
2. Prove that the system matrix $\boldsymbol{H}_p^T \boldsymbol{H}_p$ in (6.76) is symmetric positive definite.

6.4 Multigrid Iterations

We have seen previously that in many cases numerical methods with superior performance can be developed at the expense of narrowing down the class of problems that they are designed to solve. In the framework of direct methods, examples include the tri-diagonal elimination (Section 5.4.2), as well as the methods that exploit the finite Fourier series and the FFT (Section 5.7). In the framework of iterative methods, a remarkable example of that kind is given by multigrid.

Multigrid methods have been originally developed for solving elliptic boundary value problems discretized by finite differences (Chapter 12). A key distinctive characteristic of these methods is that the number of iterations required for reducing the initial error by a prescribed factor does not depend on the dimension of the grid at all. Accordingly, the required number of arithmetic operations is directly proportional to the grid dimension N^n, where N is the number of grid nodes along one coordinate direction and n is the dimension of the space $\mathbb{R}^n$. This is clearly an asymptotically unimprovable behavior, because the overall number of quantities to be computed (solution values on the grid) is also directly proportional to the grid dimension. As the grid dimension determines the condition number of the corresponding matrix,[2]

[2]The latter is typically inversely proportional to the square of the grid size: $\mu = \mathcal{O}(h^{-2})$, see formula (5.115), i.e., $\mu = \mathcal{O}(N^2)$.

we conclude that the number of multigrid iterations needed for achieving a given accuracy does not depend on the condition number μ. In contradistinction to that, for the best iterative methods we have analyzed before, the Chebyshev method (Section 6.2.1) and the method of conjugate gradients (Section 6.2.2), the number of iterations is proportional to the square root of the condition number $\sqrt{\mu}$, see formula (6.65). Accordingly, the required number of arithmetic operations is $\mathcal{O}(N^{n+1})$.

Multigrid iterations will apply to basically the same range of elliptic finite-difference problems to which the Richardson iterations apply. An additional constraint is that of the "smoothness," or "regularity," of the first eigenfunctions of the corresponding operator (matrix). For elliptic problems, it normally holds.

A rigorous analysis of multigrid is quite involved. Therefore, we will restrict ourselves to a qualitative description of its key idea (Section 6.4.1) and of the simplest version of the actual numerical algorithm (Section 6.4.2). Further detail, general constructions, and proofs can be found in the literature quoted in Section 6.4.3.

6.4.1 Idea of the Method

Introduce a uniform Cartesian grid on the square $D = \{(x,y) | 0 \leq x \leq 1, 0 \leq y \leq 1\}$:

$$(x_{m_1}, y_{m_2}) = (m_1 h, m_2 h), \quad m_1, m_2 = 0, 1, \ldots, M, \quad h = M^{-1},$$

define the grid boundary Γ_h as the set of nodes that belong to $\Gamma = \partial D$:

$$\Gamma_h = \{(x_{m_1}, y_{m_2}) | m_1 = 0, M \ \& \ m_2 = 0, M\},$$

and consider the same homogeneous finite-difference Dirichlet problem for the Poisson equation as we analyzed in Section 5.1.3:

$$\begin{aligned} -\Delta_h u_{m_1,m_2} \equiv -\Bigg(& \frac{u_{m_1+1,m_2} - 2u_{m_1,m_2} + u_{m_1-1,m_2}}{h^2} \\ & + \frac{u_{m_1,m_2+1} - 2u_{m_1,m_2} + u_{m_1,m_2-1}}{h^2} \Bigg) = f_{m_1,m_2}, \\ & m_1, m_2 = 1, 2, \ldots, M-1, \\ u\big|_{\Gamma_h} = 0. & \end{aligned} \tag{6.80}$$

For solving problem (6.80), we will use the standard stationary Richardson iteration (Section 6.1) as our starting point:

$$\begin{aligned} u^{(p+1)}_{m_1,m_2} &= u^{(p)}_{m_1,m_2} + \tau \Delta_h u^{(p)}_{m_1,m_2} + \tau f_{m_1,m_2}, \\ m_1, m_2 &= 1, 2, \ldots, M-1, \quad p = 0, 1, 2, \ldots \\ u^{(p+1)}\big|_{\Gamma_h} &= 0, \quad u^{(0)}_{m_1,m_2} \text{ is given.} \end{aligned} \tag{6.81}$$

Iterations (6.81) are generally known to converge slowly. This slowness, however, is not uniform across the spectrum of the problem. To see that, let us introduce the error $\varepsilon^{(p)} = u - u^{(p)}$ of the iterate $u^{(p)}$ and represent it in the form of a finite Fourier series according to the methodology of Section 5.7:

$$\varepsilon^{(p)} = \sum_{r,s=1}^{M-1} [1 - \tau\lambda_{rs}]^p c_{rs}^{(0)} \psi^{(r,s)}. \tag{6.82}$$

In formula (6.82), $\psi^{(r,s)}$ are eigenfunctions of the discrete Laplacian $-\Delta_h$ given by (5.105):

$$\psi^{(r,s)} = \left\{2\sin\frac{r\pi m_1}{M}\sin\frac{s\pi m_2}{M}\right\}, \quad r,s = 1,2,\ldots,M-1,$$

and λ_{rs} are the corresponding eigenvalues given by (5.109):

$$\lambda_{rs} = \frac{4}{h^2}\left(\sin^2\frac{r\pi}{2M} + \sin^2\frac{s\pi}{2M}\right), \quad r,s = 1,2,\ldots,M-1.$$

The amplification factors $\nu_{rs}(\tau) \stackrel{\text{def}}{=} [1 - \tau\lambda_{rs}]$ in formula (6.82) belong to the interval: $\nu_{\min} \leq \nu_{rs} \leq \nu_{\max}$, where:

$$\nu_{\min} = 1 - \tau\lambda_{M-1,M-1} \approx 1 - 8\tau M^2 \quad \text{and} \quad \nu_{\max} = 1 - \tau\lambda_{1,1} \approx 1 - 2\tau\pi^2.$$

Let us specify the iteration parameter τ as follows:

$$\tau = \frac{1}{5M^2}. \tag{6.83}$$

This choice of τ guarantees that if at least one of the numbers r or s is greater than $M/2$, then

$$|\nu_{rs}| < \frac{3}{5}.$$

Therefore, the contribution of the high frequency harmonics $\psi^{(r,s)}$ (with $r \geq M/2$ or $s \geq M/2$) to the error (6.82) reduces by almost a factor of two on every iteration. As such, this contribution soon becomes small and after several iterations (6.81) the error $\varepsilon^{(p)}$ will be composed primarily of the smooth components on a given grid, i.e., of the low frequencies $\psi^{(r,s)}$ that correspond to $r < M/2$ and $s < M/2$. Indeed, the amplification factors $\nu_{rs}(\tau) = 1 - \tau\lambda_{rs}$ for low frequency harmonics $\psi^{(r,s)}$ are closer to 1. The slowest decaying harmonic is $\psi^{(1,1)}$, because for the parameter τ chosen by formula (6.83) the resulting amplification factor is

$$\nu_{1,1} = 1 - \tau\lambda_{1,1} \approx 1 - \frac{2\pi^2}{5M^2}, \tag{6.84}$$

which is clearly very close to 1 for large M. Hence we see that the high frequency error content on a given grid decays fast, whereas the low frequencies decay slowly. It will therefore be natural to consider a given problem on a sequence of grids with different fineness. In doing so, the key idea is to have a special grid for every part of the spectrum, such that the corresponding harmonics on this grid can be regarded as high frequencies. These high frequencies (i.e., short waves on the scale of the grid size) will decay fast in the course of the Richardson iteration (6.81).

Let $u^{(p)}$ be the approximate solution obtained by the iteration process (6.81). For simplicity, we will use an alternative notation $u^{(p)} = \mathfrak{U}$ hereafter. Let us also re-denote the error of the iterate $u^{(p)}$: $\varepsilon^{(p)} = u - u^{(p)} = u - \mathfrak{U} = \phi$. If we knew the error ϕ, then we would have immediately found the solution: $u = \mathfrak{U} + \phi$. We, however, do not know ϕ per se, we only know that it solves the boundary value problem:

$$-\Delta_h \phi = -\eta, \quad \phi\big|_{\Gamma_h} = 0, \tag{6.85}$$

where η is the residual of the iterate $u^{(p)}$ that it generates if substituted into (6.80):

$$\eta = -\Delta_h u^{(p)} - f \equiv -\Delta_h \mathfrak{U} - f.$$

Problem (6.85) with the correction ϕ as the unknown is only simpler than the original problem (6.80) in the sense that ϕ is known to be a smooth grid function ahead of time (with little or no high frequency content). Therefore, to approximately compute ϕ we can consider the same problem as (6.85) but on a twice as coarse grid. If M is even, then this new grid with size $2h$ instead of h will have $M/2+1$ nodes in each direction and will merely be a sub-grid of the original grid with every other node in every coordinate direction dropped out. The new problem can be written as

$$-\Delta_{2h} \tilde{\phi} = -\tilde{\eta}, \quad \tilde{\phi}\big|_{\Gamma_{2h}} = 0, \tag{6.86}$$

where the tildes denote the quantities on the coarser grid. The grid boundary is now given by:

$$\Gamma_{2h} = \{(2hm_1, 2hm_2) | \, m_1 = 0, M/2 \;\&\; m_2 = 0, M/2\}.$$

Note that the transition from a fine grid to a coarser grid is often called restriction. Problem (6.86) is to be solved by the Richardson iteration similar to (6.81):

$$\begin{gathered}
\tilde{\phi}^{(p+1)}_{m_1,m_2} = \tilde{\phi}^{(p)}_{m_1,m_2} + \tilde{\tau}\Delta_{2h}\tilde{\phi}^{(p)}_{m_1,m_2} - \tilde{\tau}\tilde{\eta}_{m_1,m_2}, \\
m_1, m_2 = 1, 2, \ldots, \tilde{M}-1, \quad p = 0, 1, 2, \ldots \\
\tilde{\phi}^{(p+1)}\big|_{\Gamma_{2h}} = 0, \quad \tilde{\phi}^{(0)}_{m_1,m_2} = 0,
\end{gathered} \tag{6.87}$$

where $\tilde{M} = M/2$ and $\tilde{\tau} = 4\tau$ [see formula (6.83)].

Each iteration (6.87) is four times less expensive than one iteration (6.81), because there are four times fewer computational nodes. Moreover, as $\tilde{\tau} = 4\tau$ the slowest decaying component of the error is still decreasing faster on the coarser grid than on the original grid. Indeed, according to (6.84) we have:

$$\tilde{\nu}_{1,1} = 1 - \tilde{\tau}\tilde{\lambda}_{1,1} \approx 1 - \frac{2\pi^2}{5\tilde{M}^2} = 1 - 4\frac{2\pi^2}{5M^2} < \nu_{1,1}. \tag{6.88}$$

Let us require that for a given $\sigma \in (0,1)$:

$$(\tilde{\nu}_{1,1})^p \approx \left(1 - \frac{2\pi^2}{5\tilde{M}^2}\right)^p \leq \sigma.$$

Then, assuming that the subtrahend in formula (6.88) is still small, $3\pi^2/8M^2 \ll 1$, i.e., that M is large, we can use the Taylor formula for $\ln(\,\cdot\,)$ and write:

$$p\ln\left(1-\frac{2\pi^2}{5\tilde{M}^2}\right) \le \ln\sigma \quad\Longrightarrow\quad -p\frac{2\pi^2}{5\tilde{M}^2} \le \ln\sigma$$
$$\Longrightarrow\quad p \ge -\frac{5\tilde{M}^2}{2\pi^2}\ln\sigma = -\frac{1}{4}\frac{5M^2}{2\pi^2}\ln\sigma.$$

As such, for reducing the contribution of $\tilde{\psi}^{(1,1)}$ into the error by a prescribed factor σ we will need approximately four times fewer iterations on the coarse grid with size $2h$ than for reducing the contribution of $\psi^{(1,1)}$ on the fine grid with size h.

Let us denote by $\tilde{\Phi}$ the grid function obtained as a result of the iteration process (6.87). This function is defined on the coarse grid with size $2h$. We will interpolate it (linearly) from this coarse grid onto the original fine grid with size h and obtain the function Φ. Note that the transition from the coarse grid to a finer grid is referred to as prolongation in the multigrid framework. In doing so, the smooth components will be obtained almost correctly on the fine grid. The corresponding interpolation error will be small relative to the smooth interpolated function. However, the Fourier expansion of the interpolation error will contain all harmonics, because the interpolation error itself has kinks at the interpolation nodes and cannot be regarded as a smooth function. Moreover, as the grid function $\tilde{\Phi}$ is obtained by iteration (6.87), it has a non-smooth component of its own. The latter has basically nothing to do with the correction ϕ that we are looking for. It will, however, yield an additional (random) contribution to the non-smooth part of the resulting interpolant Φ. Altogether, we conclude that the smooth component of the sum $\mathfrak{U}+\Phi$ on the fine grid will be close to the smooth component of the unknown solution $u = \mathfrak{U}+\phi$, whereas the non-smooth component may not necessarily be very small and will basically have a random nature. Therefore, after the prolongation it is necessary to perform a few more fine grid iterations (6.81) while choosing $\mathfrak{U}+\Phi$ as their initial guess. This will facilitate a rapid suppression of the non-smooth component of the error introduced by interpolation, because every iteration (6.81) damps the high frequency part of the spectrum by almost a factor of two.

6.4.2 Description of the Algorithm

The speedup of convergence achieved using a coarser grid with size $2h$ and the iteration process (6.87) may still be insufficient. If M is large, the complexity of the coarse grid problem (6.86) will nonetheless remain fairly high. Therefore, when solving this problem it may be advisable to coarsen the grid one more time and obtain yet another problem similar to (6.86), but on the grid of size $4h$. For simplicity, let us assume that the initial grid dimension is given by a power of two, $M = 2^k$. Then, a number of coarsening steps can be performed, and a sequence of embedded grids and respective problems of type (6.86) can be introduced and exploited.

On the initial fine grid, we first make several iterations (6.81) to smooth out the error, i.e., reduce its high frequency content. As the error itself is not known, we can monitor the residual $-\Delta_h u^{(p)} - f$ instead, because it also becomes smoother in the course of iteration (6.81). The result of these iterations $u^{(p)} = \mathfrak{U}$ is to be stored in the computer memory. Then we consider a coarser grid problem (6.86) for the correction ϕ, make several iterations (6.87) in order to smooth out the correction to the correction, and again store the result $\tilde{\Phi}$ in the computer memory (it requires four times less space than $\mathfrak{U}$). To actually compute the correction to $\tilde{\Phi}$, we consider yet another coarser grid problem, this time with size $4h$, perform several iterations with step $\tilde{\tau}^{(2)} = 4\tilde{\tau} = 16\tau$ and store the result $\tilde{\tilde{\Phi}}^{(2)}$. This nested process of computing corrections to corrections on twice as coarse embedded grids is run k times until the coarsest grid is reached and the corresponding correction $\tilde{\tilde{\Phi}}^{(k)}$ is obtained.

Then, we start the process of returning to the fine grid. First we interpolate the coarsest grid correction $\tilde{\tilde{\Phi}}^{(k)}$ to the second to last grid, which is twice as fine. On this grid, we add the correction $\Phi^{(k-1)}$ to the previously stored solution and also make several iterations to damp the interpolation error. The result of these iterations is interpolated to the next finer grid and then used on this grid for correcting the stored function $\tilde{\tilde{\Phi}}^{(k-2)}$. Subsequently, several iterations are conducted and yet another interpolation is made. On the second to last step, once the correction $\Phi^{(2)}$ is introduced and iterations performed on the grid $2h$, we obtain the last correction $\tilde{\Phi}$, interpolate it to the finest grid h, make several iterations (6.81) starting with the initial guess $\mathfrak{U} + \Phi$, and obtain the final result.

In the modern theory of multigrid methods, the algorithm we have just described is referred to as a V-cycle; it is schematically shown in Figure 6.3(a). In practice, several consecutive V-cycles may be required for obtaining a sufficiently accurate approximation to the solution u of problem (6.80). Alternatively, one can use the so-called W-cycles that are shown schematically in Figure 6.3(b). Each individual W-cycle may be more expensive computationally than the corresponding V-cycle. However, fewer W-cycles are normally required for reducing the initial error by a prescribed factor.

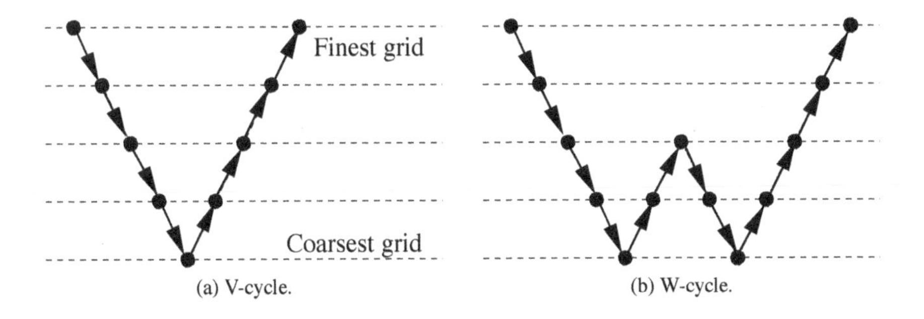

(a) V-cycle. (b) W-cycle.

FIGURE 6.3: Multigrid cycles.

6.4.3 Bibliography Comments

The concept of what has later become known as multigrid methods was first introduced in a 1961 paper by Fedorenko [Fed61]. The author called his iteration scheme a relaxation method. In a subsequent 1964 paper [Fed64], Fedorenko has also provided first estimates of the convergence rate of his relaxation method when it was applied to solving a finite-difference Dirichlet problem. Similar estimates for other problems were later obtained by Bakhvalov [Bak66] and by Astrakhantsev [Ast71]. A detailed summary of this early development of multigrid methods can be found in the review paper by Fedorenko [Fed73].

Subsequent years, starting from the mid-seventies, witnessed a rapid growth of attention to multigrid methods, and an "explosion" of work on their theoretical analysis, algorithmic implementation, and applications to a wide variety of problems far beyond simple elliptic discretizations. These methods have proven extremely successful and superior to other techniques even when their actual performance for difficult problems was not as good as predicted theoretically, say, for the Poisson equation. For example, multigrid methods have enabled a historical breakthrough in the performance of numerical solvers used in computational fluid dynamics for the quantitative analysis of aerodynamic configurations. This dramatic progress in the development of multigrid is associated with the names of many researchers; fundamental contributions were made by Brandt, Hackbusch, Jameson, and others. Advances in the area of multigrid methods are summarized in a number of papers and books, see, e.g., [Bra77], [Bra84, Hac85, Wes92, Bra93, BHM00, TOS01].

A separate research direction in this area is the so-called algebraic multigrid methods, when similar multilevel ideas are applied directly to a given matrix, without any regard to where this matrix originates. This approach, in particular, led to the development of the recursive ordering preconditioners.

Exercises

1. Redefine the notion of the high frequencies on the grid as those harmonics $\psi^{(r,s)}$ for which both $r > M/2$ and $s > M/2$. What value of the iteration parameter τ shall one choose instead of (6.83) so that to guarantee the best possible damping of the high frequencies by the iteration scheme (6.81)? What is the corresponding maximum value of the amplification factor: $\max_{r>\frac{M}{2},s>\frac{M}{2}} |v_{rs}|$?

   Hint. Use the condition $v_{\min}(\tau) \equiv v_{M-1,M-1}(\tau) = -v_{\frac{M}{2},\frac{M}{2}}(\tau)$. Explain why this choice of τ will guarantee the best damping.

Chapter 7

Overdetermined Linear Systems. The Method of Least Squares

7.1 Examples of Problems that Result in Overdetermined Systems

7.1.1 Processing of Experimental Data. Empirical Formulae

Assume that the quantity y is a function of the argument t, and suppose that a table of values $y_k = y(t_k)$, $k = 1,2,3,4$, was obtained for this function as a result of a series of measurements, see Figure 7.1. By merely looking at the resulting table, an experimentalist can guess that the dependence $y = y(t)$ can be at least approximately interpreted as linear:

$$y = x_1 t + x_2. \tag{7.1}$$

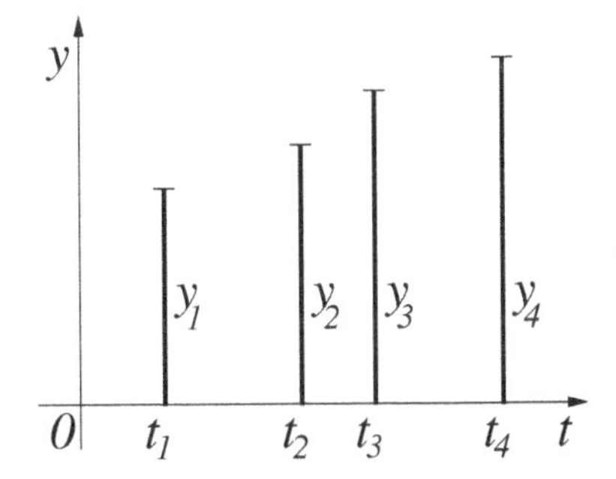

FIGURE 7.1: Experimental data.

Of course, the values of the parameters x_1 and x_2 in the empirical formula (7.1) should ideally be chosen so that to satisfy the actual experimental results for all $t = t_k$, $k = 1,2,3,4$, i.e., so that to satisfy the following equalities:

$$\begin{aligned} x_1 t_1 + x_2 &= y_1, \\ x_1 t_2 + x_2 &= y_2, \\ x_1 t_3 + x_2 &= y_3, \\ x_1 t_4 + x_2 &= y_4. \end{aligned} \tag{7.2}$$

System (7.2) is a system of four linear algebraic equations with respect to only two unknowns: x_1 and x_2. This system does not, generally speaking, have a classical solution, because there is no straight line that would cross precisely through all four experimental points, see Figure 7.1. Systems of this type are called overdetermined.

As there is no classical solution, one will need to adopt an alternative strategy for determining the values of the parameters x_1 and x_2 in formula (7.1). For example, one can choose x_1 and x_2 as a pair of numbers for which the residuals

$$r_k \stackrel{\text{def}}{=} [x_1 t_k + x_2 - y_k], \quad k = 1,2,3,4,$$

of system (7.2) will be minimal. Of course, the notion of minimizing the residuals r_k needs to be quantified. One can, for instance, define a function of two variables:

$$\Phi(x_1,x_2) = \sum_{k=1}^{4} [x_1 t_k + x_2 - y_k]^2 \tag{7.3}$$

as the sum of the squares of these residuals. Then, a generalized, or weak, solution of system (7.2) can be introduced as that particular pair of numbers x_1 and x_2, for which the function $\Phi = \Phi(x_1,x_2)$ of (7.3) assumes its minimal value. Obviously, the point (x_1,x_2) must then be a stationary point of the quadratic function $\Phi(x_1,x_2)$:

$$\frac{\partial \Phi}{\partial x_1}(x_1,x_2) = 0, \quad \frac{\partial \Phi}{\partial x_2}(x_1,x_2) = 0. \tag{7.4}$$

System (7.4) is, in fact, a system of two linear algebraic equations with two unknowns x_1 and x_2:

$$\begin{aligned} x_1\Big[\sum_{k=1}^{4} t_k^2\Big] + x_2\Big[\sum_{k=1}^{4} t_k\Big] &= \sum_{k=1}^{4} t_k y_k, \\ x_1\Big[\sum_{k=1}^{4} t_k\Big] + x_2\Big[\sum_{k=1}^{4} 1\Big] &= \sum_{k=1}^{4} y_k. \end{aligned} \tag{7.5}$$

It can be shown that system (7.5) is non-singular and has a unique classical solution for any y_k, $k = 1,2,3,4$, and any t_k, $k = 1,2,3,4$, as long as no two points t_k coincide. We will prove this result for a more general setting in Section 7.2 (a particular example given by (7.5) is analyzed in Exercise 2 after Section 7.2).

The function $\Phi(x_1,x_2)$ that defines the generalized solution (x_1,x_2) of the overdetermined system (7.2) can obviously be chosen in a variety of different ways, not necessarily according to formula (7.3). For example, individual measurements can be assigned different weights: $b_k > 0$, $k = 1,2,3,4$, if there is an application-related reason to emphasize some particular measurements and de-emphasize the others. In this case, the function $\Phi(x_1,x_2)$ of (7.3) is replaced by:

$$\Phi(x_1,x_2) = \sum_{k=1}^{4} b_k [x_1 t_k + x_2 - y_k]^2, \tag{7.6}$$

The stationary point of the new function $\Phi(x_1,x_2)$ is still defined by equalities (7.4); however, the resulting 2×2 linear system changes accordingly [it will no longer be equivalent to (7.5)], and so does the generalized solution (x_1,x_2) of system (7.2). In other words, one can give alternative definitions of the generalized solution, and the resulting values of the parameters (x_1,x_2) in formula (7.1) will naturally depend on the specific definition employed, e.g., on the choice of weights in formula (7.6).

In general, the definition of a weak solution of system (7.2) through the minimum of $\Phi(x_1,x_2)$ is an example of applying the method of least squares. The method draws its name from the squares of the residuals contained in $\Phi(x_1,x_2)$ [either non-weighted, see (7.3), or weighted, see (7.6)]. Accordingly, the resulting solutions (x_1,x_2) are known as weak, or generalized, solutions in the sense of the least squares.

One, of course, does not have to use a quadratic function $\Phi(x_1,x_2)$ at all. For example, moduli can be employed instead of the squares as measures of the residuals, which yields [cf. formula (7.6)]:

$$\Phi(x_1,x_2) = \sum_{k=1}^{4} b_k |x_1 t_k + x_2 - y_k|. \tag{7.7}$$

A generalized solution of system (7.2) obtained by minimizing the function $\Phi(x_1,x_2)$ of (7.7) is known as the generalized, or weak, solution in the sense of (a weighted) l_1. However, one cannot define a stationary point for the function $\Phi(x_1,x_2)$ of (7.7), because contrary to the functions (7.3) and (7.6), the function (7.7) is not differentiable. As such, the minimum value of $\Phi(x_1,x_2)$ can no longer be found with the help of equalities (7.4). Instead, it can be computed by the method of linear programming (see, e.g., [Van01]). The result (x_1,x_2) will, generally speaking, differ from that obtained by means of the least squares. The advantage of using the least squares is that the computation of the corresponding generalized solution is typically much simpler than, say, the computation of the generalized solution in the sense of l_1.

7.1.2 Improving the Accuracy of Experimental Results by Increasing the Number of Measurements

The problem analyzed in Section 7.1.1 can be given a slightly different interpretation. Suppose that the dependence of the quantity $y = y(t)$ on its argument t is known to be linear ahead of time. In other words, assume that formula (7.1) is not empirical (i.e., it is not "guessed" based on observations) but rather reflects the actual behavior of the observable function $y = y(t)$. The coefficients x_1 and x_2 of the linear law (7.1) are not known ahead of time, and the task is to calculate those coefficients, given the results of experimental measurements of the quantity $y(t)$ for several distinct values of t. As before, let us think that the measurements are taken at $t = t_k$, $k = 1,2,3,4$, and that their results are summarized in Figure 7.1.

Then, we will obtain the same overdetermined linear system (7.2) for the unknown values x_1 and x_2. This system will, generally speaking, be inconsistent. The reason why no straight line can actually cross through all four points on Figure 7.1, even though in theory there is one, is the inevitable experimental inaccuracies, i.e., errors of the measurements. As such, we again need to define a generalized solution. It can be introduced as a pair of numbers (x_1,x_2) that minimizes a quadratic function of type (7.6). This solution can be found by solving system (7.4).

Let us emphasize that if we only were to perform two measurements, then system (7.2) would not be overdetermined and it would have a classical solution (x_1,x_2). However, given the errors of the measurements, this solution would be a fairly inaccurate approximation of the true parameters x_1 and x_2 of the linear law (7.1). At the same time, it is known that performing some additional measurements, even if they also carry an error, still allows us to reduce the overall impact of individual experimental errors on the final result. In other words, we can expect that having four approximate measurements instead of two enables a more accurate reconstruction of the linear function (7.1).

In the framework of the probability theory, one can rigorously formulate and analyze the problem of improving the accuracy of the experimental results that entail a random error content. The improvement is achieved by means of increasing the number of measurements to be performed, as well as by processing the data using the method of least squares. Note also that overdetermined linear systems arise in many other applications, besides those mentioned in this section.

7.2 Weak Solutions of Full Rank Systems. *QR* Factorization

7.2.1 Existence and Uniqueness of Weak Solutions

An overdetermined linear system can be represented in the canonical form:

$$\begin{aligned} a_{11}x_1 + a_{12}x_2 + \ldots + a_{1n}x_n &= f_1, \\ \ldots\ldots\ldots\ldots\ldots\ldots\ldots\ldots\ldots\ldots\ldots&\ldots\ldots \\ a_{m1}x_1 + a_{m2}x_2 + \ldots + a_{mn}x_n &= f_m, \end{aligned} \tag{7.8}$$

where $m > n$, i.e., the number of equations is greater than the number of unknowns. Hereafter we will assume that $\boldsymbol{x} = \begin{bmatrix} x_1 \\ \ldots \\ x_n \end{bmatrix} \in \mathbb{R}^n$ and $\boldsymbol{f} = \begin{bmatrix} f_1 \\ \ldots \\ f_m \end{bmatrix} \in \mathbb{R}^m$. Then, we can introduce an $m \times n$ rectangular matrix of the coefficients of system (7.8):

$$\boldsymbol{A} = \begin{bmatrix} a_{11} & \ldots & a_{1n} \\ \ldots & \ldots & \ldots \\ a_{m1} & \ldots & a_{mn} \end{bmatrix}, \tag{7.9}$$

and recast system (7.8) as:

$$\boldsymbol{A}\boldsymbol{x} = \boldsymbol{f}, \quad \boldsymbol{x} \in \mathbb{R}^n, \quad \boldsymbol{f} \in \mathbb{R}^m. \tag{7.10}$$

Hereafter, we will also need a scalar product on the space $\mathbb{R}^m$. The simplest Euclidean product is given by:

$$(\boldsymbol{f}, \boldsymbol{g})^{(m)} = \sum_{k=1}^{m} f_k g_k. \tag{7.11}$$

As shown in Section 5.2.1, there are many alternative ways of introducing a scalar product on $\mathbb{R}^m$. Namely, any symmetric positive definite matrix $\boldsymbol{B} = \boldsymbol{B}^* > 0$ [recall, $\forall \boldsymbol{f} \in \mathbb{R}^m, \boldsymbol{f} \neq \boldsymbol{0}: \ (\boldsymbol{Bf},\boldsymbol{f}) > 0$] defines a scalar product as follows:

$$[\boldsymbol{f},\boldsymbol{g}]_{\boldsymbol{B}}^{(m)} \stackrel{\text{def}}{=} (\boldsymbol{Bf},\boldsymbol{g})^{(m)}, \quad \boldsymbol{f},\boldsymbol{g} \in \mathbb{R}^m. \tag{7.12}$$

It is also known that any scalar product on $\mathbb{R}^m$ (see the axioms given on page 126) can be represented by formula (7.12) with an appropriate choice of the matrix $\boldsymbol{B}$.

In general, system (7.8) [or, alternatively, system (7.10)] does not have a classical solution. In other words, no set of n numbers $\{x_1,x_2,\ldots,x_n\}$ will simultaneously turn every equation of (7.8) into an identity. We therefore define a solution of system (7.8) in the following weak sense.

DEFINITION 7.1 *Let $\boldsymbol{B}: \mathbb{R}^m \longmapsto \mathbb{R}^m$, $\boldsymbol{B} = \boldsymbol{B}^* > 0$, be fixed. Introduce a scalar function $\Phi = \Phi(\boldsymbol{x})$ of the vector argument $\boldsymbol{x} \in \mathbb{R}^n$:*

$$\Phi(\boldsymbol{x}) = [\boldsymbol{Ax}-\boldsymbol{f},\boldsymbol{Ax}-\boldsymbol{f}]_{\boldsymbol{B}}^{(m)}. \tag{7.13}$$

A generalized, or weak, solution $\boldsymbol{x_B}$ of system (7.8) is the vector $\boldsymbol{x_B} \in \mathbb{R}^n$ that delivers a minimum value to the quadratic function (7.13).

REMARK 7.1 There is a fair degree of flexibility in choosing the matrix $\boldsymbol{B}$. The meaning of this matrix is that of a "weight" matrix (see Section 7.1.1). It can be chosen based on some "external" considerations to emphasize certain components of the residual of system (7.10) and de-emphasize the others. □

REMARK 7.2 The functions $\Phi = \Phi(x_1,x_2)$ introduced by formulae (7.3) and (7.6) as particular examples of the method of least squares, can also be obtained according to the general formula (7.13). In this sense, Definition 7.1 can be regarded as the most general definition of the least squares weak solution for a full rank system. □

THEOREM 7.1

Let the rectangular matrix $\boldsymbol{A}$ of (7.9) have full rank equal to n, i.e., let it have linearly independent columns. Then, there exists one and only one generalized solution $\boldsymbol{x_B}$ of system (7.8) in the sense of Definition 7.1. This generalized solution coincides with the classical solution of the system:

$$\boldsymbol{A}^*\boldsymbol{BAx} = \boldsymbol{A}^*\boldsymbol{Bf} \tag{7.14}$$

that contains n linear algebraic equations with respect to precisely as many scalar unknowns: $x_1,x_2,\ldots,x_n$.

PROOF Denote the columns of the matrix $\boldsymbol{A}$ by $\boldsymbol{a}_k$, $\boldsymbol{a}_k \in \mathbb{R}^m$:

$$\boldsymbol{a}_k = \begin{bmatrix} a_{1k} \\ \vdots \\ a_{mk} \end{bmatrix}, \quad k = 1, 2, \dots, n.$$

The matrix $\boldsymbol{C} = \boldsymbol{A}^*\boldsymbol{B}\boldsymbol{A}$ of system (7.14) is a symmetric square matrix of dimension $n \times n$. Indeed, the entry c_{ij} of this matrix that is located at the intersection of row number i and column number j is given by:

$$c_{ij} = (\boldsymbol{a}_i, \boldsymbol{B}\boldsymbol{a}_j)^{(m)} = (\boldsymbol{B}\boldsymbol{a}_i, \boldsymbol{a}_j)^{(m)} = [\boldsymbol{a}_i, \boldsymbol{a}_j]_{\boldsymbol{B}}^{(m)},$$

and we conclude that $c_{ij} = c_{ji}$, i.e., $\boldsymbol{C} = \boldsymbol{C}^*$. Let us now show that the matrix $\boldsymbol{C}$ is non-singular and moreover, positive definite:[1]

$$(\boldsymbol{C}\boldsymbol{\xi}, \boldsymbol{\xi})^{(n)} > 0, \quad \boldsymbol{\xi} \in \mathbb{R}^n, \quad \boldsymbol{\xi} \neq \boldsymbol{0}. \tag{7.15}$$

To prove inequality (7.15), we first recall a well-known formula:

$$(\boldsymbol{f}, \boldsymbol{A}\boldsymbol{\xi})^{(m)} \equiv (\boldsymbol{A}^*\boldsymbol{f}, \boldsymbol{\xi})^{(n)}, \quad \boldsymbol{f} \in \mathbb{R}^m, \quad \boldsymbol{\xi} \in \mathbb{R}^n. \tag{7.16}$$

This formula can be verified directly, by writing the matrix-vector products and the inner products involved via the individual entries and components. Next, let $\boldsymbol{\xi} \in \mathbb{R}^n$, $\boldsymbol{\xi} = \begin{bmatrix} \xi_1 \\ \dots \\ \xi_n \end{bmatrix} \neq \begin{bmatrix} 0 \\ \dots \\ 0 \end{bmatrix}$. Then expression

$$\boldsymbol{A}\boldsymbol{\xi} = \xi_1 \boldsymbol{a}_1 + \xi_2 \boldsymbol{a}_2 + \dots + \xi_n \boldsymbol{a}_n$$

can be interpreted as a linear combination of the linearly independent vectors $\boldsymbol{a}_k$, $k = 1, 2, \dots, n$, with the coefficients ξ_k, $k = 1, 2, \dots, n$, that are not all simultaneously equal to zero. Consequently, $\mathbb{R}^m \ni \boldsymbol{A}\xi \neq \boldsymbol{0}$. This implies that the scalar product of the vector $\boldsymbol{A}\boldsymbol{\xi}$ with itself is positive, and using formula (7.16) we obtain:

$$0 < [\boldsymbol{A}\boldsymbol{\xi}, \boldsymbol{A}\boldsymbol{\xi}]_{\boldsymbol{B}}^{(m)} = (\boldsymbol{B}\boldsymbol{A}\boldsymbol{\xi}, \boldsymbol{A}\boldsymbol{\xi})^{(m)} = (\boldsymbol{A}^*\boldsymbol{B}\boldsymbol{A}\boldsymbol{\xi}, \boldsymbol{\xi})^{(n)} = (\boldsymbol{C}\boldsymbol{\xi}, \boldsymbol{\xi})^{(n)}$$

so that inequality (7.15) indeed holds. As the matrix $\boldsymbol{C} = \boldsymbol{A}^*\boldsymbol{B}\boldsymbol{A}$ is positive definite, it is non-singular. Therefore, system (7.14) has a unique solution $\boldsymbol{x}_{\boldsymbol{B}} \in \mathbb{R}^n$ for any given $\boldsymbol{f} \in \mathbb{R}^m$.

---

[1] The scalar product $(\,\cdot\,,\,\cdot\,)^{(n)}$ in formula (7.15) is defined by the same formula (7.11), only for the n-dimensional space $\mathbb{R}^n$ rather than for the m-dimensional space $\mathbb{R}^m$.

It only remains to prove that this $\boldsymbol{x_B}$ yields a unique weak solution of system (7.10) in the sense of Definition 7.1. In other words, we need to prove that for any $\boldsymbol{x} = \boldsymbol{x_B} + \boldsymbol{\delta}$, $\boldsymbol{\delta} \in \mathbb{R}^n$, $\boldsymbol{\delta} \neq \boldsymbol{0}$, the following strict inequality holds:

$$\Phi(\boldsymbol{x_B} + \boldsymbol{\delta}) > \Phi(\boldsymbol{x_B}), \tag{7.17}$$

where the function Φ is defined by formula (7.13). To prove this inequality, let us first notice that for any $\boldsymbol{\delta} \in \mathbb{R}^{(n)}$ we have:

$$[\boldsymbol{Ax_B} - \boldsymbol{f}, \boldsymbol{A\delta}]_{\boldsymbol{B}}^{(m)} = (\boldsymbol{B}(\boldsymbol{Ax_B} - \boldsymbol{f}), \boldsymbol{A\delta})^{(m)} = (\boldsymbol{A^*BAx_B} - \boldsymbol{A^*Bf}, \boldsymbol{\delta})^{(n)} = 0. \tag{7.18}$$

Equality (7.18) is true because $\boldsymbol{x_B}$ is the solution to system (7.14), i.e., $\boldsymbol{A^*BAx_B} = \boldsymbol{A^*Bf}$. We can now establish inequality (7.17) with the help of formulae (7.15) and (7.18):

$$\begin{aligned}
\Phi(\boldsymbol{x_B} + \boldsymbol{\delta}) &= [\boldsymbol{A}(\boldsymbol{x_B} + \boldsymbol{\delta}) - \boldsymbol{f}, \boldsymbol{A}(\boldsymbol{x_B} + \boldsymbol{\delta}) - \boldsymbol{f}]_{\boldsymbol{B}}^{(m)} \\
&= [\boldsymbol{Ax_B} - \boldsymbol{f}, \boldsymbol{Ax_B} - \boldsymbol{f}]_{\boldsymbol{B}}^{(m)} - 2[\boldsymbol{Ax_B} - \boldsymbol{f}, \boldsymbol{A\delta}]_{\boldsymbol{B}}^{(m)} + [\boldsymbol{A\delta}, \boldsymbol{A\delta}]_{\boldsymbol{B}}^{(m)} \\
&= \Phi(\boldsymbol{x_B}) - 2[\boldsymbol{Ax_B} - \boldsymbol{f}, \boldsymbol{A\delta}]_{\boldsymbol{B}}^{(m)} + (\boldsymbol{C\delta}, \boldsymbol{\delta})^{(n)} \\
&= \Phi(\boldsymbol{x_B}) + (\boldsymbol{C\delta}, \boldsymbol{\delta})^{(n)} > \Phi(\boldsymbol{x_B}).
\end{aligned}$$

This completes the proof of the theorem. □

REMARK 7.3 An example of the least squares minimization problem in the sense of Definition 7.1 was given in Section 6.3.2, when we built the GMRES iteration. The function Φ to be minimized in the context of GMRES is introduced by formula (6.75), it is the Euclidean norm of the residual, and $\boldsymbol{B} = \boldsymbol{I}$. Equation (7.14) for GMRES is written as (6.76). □

7.2.2 Computation of Weak Solutions. *QR* Factorization

The matrix $\boldsymbol{C} = \boldsymbol{A^*BA}$ of system (7.14) is symmetric positive definite. Therefore, in theory this system can be efficiently solved either by a direct method (e.g., Cholesky factorization, Section 5.4.5) or by iterations (e.g., conjugate gradients, Section 6.2.2). In practice, however, system (7.14) is almost never used for computing the weak solution of system (7.10) directly, unless the matrix $\boldsymbol{C} = \boldsymbol{A^*BA}$ appears particularly well conditioned, see, e.g., Exercise 8. The reason is that additional matrix multiplications may lead to a strong amplification of the round-off errors. Instead, the weak solution is typically obtained with the help of the $\boldsymbol{QR}$ factorization.

LEMMA 7.1
Let $\boldsymbol{A}$ *be a real* $m \times n$ *matrix of rank* n $(m \geq n)$. *There exists a unique representation of* $\boldsymbol{A}$ *as a product:*

$$\boldsymbol{A} = \boldsymbol{QR}, \tag{7.19}$$

where Q and R are real matrices of dimensions $m \times n$ and $n \times n$ respectively, Q has orthonormal columns, and R is upper triangular.

PROOF To build Q, we will apply the well-known Gram-Schmidt orthogonalization procedure to the linearly independent columns a_k, $k = 1,2,\ldots,n$, of the matrix A. The first column of the matrix Q is the normalized[2] first column of the matrix A:

$$q_1 = \frac{a_1}{\|a_1\|}. \tag{7.20a}$$

Then, for each subsequent column a_{k+1} of the matrix A, we take its orthogonal projection onto the linear span of the previously obtained columns of Q, subtract from a_{k+1} itself:

$$v_{k+1} = a_{k+1} - \sum_{i=1}^{k} (q_i, a_{k+1}) q_i, \quad k = 1,2,\ldots,n-1, \tag{7.20b}$$

and finally normalize:

$$q_{k+1} = \frac{v_{k+1}}{\|v_{k+1}\|}. \tag{7.20c}$$

As $\text{rank} A = n$, formulae (7.20) imply that the the column vectors q_k, $k = 1,2,\ldots,n$, of the matrix Q will be orthonormal. Formulae (7.20b), (7.20c) also show that every column a_k of the matrix A is a linear combination of only the first k columns $\{q_1,\ldots,q_k\}$ of the matrix Q. Therefore, the column number k of the $n \times n$ square matrix R in formula (7.19) may have only zero entries below the main diagonal: $r_{ik} = 0$ for $i > k$. In other words, the matrix R (7.19) is indeed upper triangular. □

COROLLARY 7.1

The matrix R in formula (7.19) coincides with the upper triangular Cholesky factor of the symmetric positive definite matrix A^*A, see formula (5.78).

PROOF Using orthonormality of the columns of Q, $Q^*Q = I$, we have:

$$A^*A = (QR)^*(QR) = R^*Q^*QR = R^*R,$$

and the desired result follows because the Cholesky factorization is unique (Section 5.4.5). □

The next theorem will show how the QR factorization can be employed for computing the generalized solutions of full rank overdetermined systems in the sense of Definition 7.1. For simplicity, we are considering the case $B = I$.

[2]All norms in this section are Euclidean: $\|\cdot\| \equiv \|\cdot\|_2 = \sqrt{(\cdot,\cdot)}$.

THEOREM 7.2
Let $\boldsymbol{A}$ be an $m \times n$ matrix with real entries, $m \geq n$, and let $\text{rank}\boldsymbol{A} = n$. *Then, there is a unique vector $\tilde{\boldsymbol{x}} \in \mathbb{R}^m$ that minimizes the quadratic function:*

$$\Phi(\boldsymbol{x}) = (\boldsymbol{A}\boldsymbol{x} - \boldsymbol{f}, \boldsymbol{A}\boldsymbol{x} - \boldsymbol{f})^{(m)}. \tag{7.21}$$

This vector is given by:

$$\tilde{\boldsymbol{x}} = \boldsymbol{R}^{-1}\boldsymbol{Q}^*\boldsymbol{f}, \tag{7.22}$$

where the matrices $\boldsymbol{Q}$ and $\boldsymbol{R}$ are introduced in Lemma 7.1.

PROOF Consider an arbitrary complement $\hat{\boldsymbol{Q}}$ of the matrix $\boldsymbol{Q}$ to an $m \times m$ orthogonal square matrix: $\hat{\boldsymbol{Q}}^*\hat{\boldsymbol{Q}} = \hat{\boldsymbol{Q}}\hat{\boldsymbol{Q}}^* = \boldsymbol{I}_m$. Consider also a rectangular $m \times n$ matrix $\hat{\boldsymbol{R}}$ with its first n rows equal to those of $\boldsymbol{R}$ and the remaining $m - n$ rows filled with zeros. Clearly, $\boldsymbol{Q}\boldsymbol{R} = \hat{\boldsymbol{Q}}\hat{\boldsymbol{R}}$. Then, using definition (7.21) of the function $\Phi(\boldsymbol{x})$ we can write:

$$\begin{aligned}
\Phi(\boldsymbol{x}) &= (\boldsymbol{A}\boldsymbol{x} - \boldsymbol{f}, \boldsymbol{A}\boldsymbol{x} - \boldsymbol{f})^{(m)} = (\boldsymbol{Q}\boldsymbol{R}\boldsymbol{x} - \boldsymbol{f}, \boldsymbol{Q}\boldsymbol{R}\boldsymbol{x} - \boldsymbol{f})^{(m)} \\
&= (\hat{\boldsymbol{Q}}\hat{\boldsymbol{R}}\boldsymbol{x} - \hat{\boldsymbol{Q}}\hat{\boldsymbol{Q}}^*\boldsymbol{f}, \hat{\boldsymbol{Q}}\hat{\boldsymbol{R}}\boldsymbol{x} - \hat{\boldsymbol{Q}}\hat{\boldsymbol{Q}}^*\boldsymbol{f})^{(m)} = (\hat{\boldsymbol{R}}\boldsymbol{x} - \hat{\boldsymbol{Q}}^*\boldsymbol{f}, \hat{\boldsymbol{Q}}^*\hat{\boldsymbol{Q}}\hat{\boldsymbol{R}}\boldsymbol{x} - \hat{\boldsymbol{Q}}^*\hat{\boldsymbol{Q}}\hat{\boldsymbol{Q}}^*\boldsymbol{f})^{(m)} \\
&= (\hat{\boldsymbol{R}}\boldsymbol{x} - \hat{\boldsymbol{Q}}^*\boldsymbol{f}, \hat{\boldsymbol{R}}\boldsymbol{x} - \hat{\boldsymbol{Q}}^*\boldsymbol{f})^{(m)} = (\boldsymbol{R}\boldsymbol{x} - \boldsymbol{Q}^*\boldsymbol{f}, \boldsymbol{R}\boldsymbol{x} - \boldsymbol{Q}^*\boldsymbol{f})^{(n)} + \sum_{k=n+1}^{m} ([\hat{\boldsymbol{Q}}^*\boldsymbol{f}]_k)^2.
\end{aligned}$$

Clearly, the minimum of the previous expression is attained when the first term in the sum is equal to zero, i.e., when $\boldsymbol{R}\boldsymbol{x} = \boldsymbol{Q}^*\boldsymbol{f}$. This implies (7.22). ▯

REMARK 7.4 In practice, the classical Gram-Schmidt orthogonalization (7.20) cannot be employed for computing the $\boldsymbol{Q}\boldsymbol{R}$ factorization of $\boldsymbol{A}$ because it is prone to numerical instabilities due to amplification of the round-off errors. Instead, one can use its stabilized version. In this version, one does not compute the orthogonal projection of $\boldsymbol{a}_{k+1}$ onto $\text{span}\{\boldsymbol{q}_1, \ldots, \boldsymbol{q}_k\}$ first and then subtract according to formula (7.20b), but rather does projections and subtractions successively for individual vectors $\{\boldsymbol{q}_1, \ldots, \boldsymbol{q}_k\}$:

$$\begin{aligned}
\boldsymbol{v}_{k+1}^{(1)} &= \boldsymbol{a}_{k+1} - (\boldsymbol{q}_1, \boldsymbol{a}_{k+1})\boldsymbol{q}_1, \\
\boldsymbol{v}_{k+1}^{(2)} &= \boldsymbol{v}_{k+1}^{(1)} - (\boldsymbol{q}_2, \boldsymbol{v}_{k+1}^{(1)})\boldsymbol{q}_2, \\
\boldsymbol{v}_{k+1}^{(3)} &= \boldsymbol{v}_{k+1}^{(2)} - (\boldsymbol{q}_3, \boldsymbol{v}_{k+1}^{(2)})\boldsymbol{q}_3, \\
&\cdots\cdots\cdots\cdots\cdots\cdots \\
\boldsymbol{v}_{k+1}^{(k)} &= \boldsymbol{v}_{k+1}^{(k-1)} - (\boldsymbol{q}_k, \boldsymbol{v}_{k+1}^{(k-1)})\boldsymbol{q}_k.
\end{aligned}$$

In the end, the normalization is performed as before, according to formula (7.20c): $\boldsymbol{q}_{k+1} = \boldsymbol{v}_{k+1}^{(k)} / \|\boldsymbol{v}_{k+1}^{(k)}\|$. This modified Gram-Schmidt algorithm is more robust than the original one. It also entails a higher computational cost. ▯

REMARK 7.5 Note that there is an alternative way of computing the $\boldsymbol{QR}$ factorization (7.19). It employs special types of transformations for matrices, known as the Householder transformations (orthogonal reflections, or symmetries, with respect to a given hyperplane in the vector space) and Givens transformations (two-dimensional rotations), see, e.g., [GVL89, Chapter 5]. ☐

7.2.3 Geometric Interpretation of the Method of Least Squares

Linear system (7.8) can be written as follows:

$$x_1\boldsymbol{a}_1 + x_2\boldsymbol{a}_2 + \ldots + x_n\boldsymbol{a}_n = \boldsymbol{f}, \tag{7.23}$$

where the vectors $\boldsymbol{a}_k = \begin{bmatrix} a_{1k} \\ \ldots \\ a_{mk} \end{bmatrix} \in \mathbb{R}^m$, $k = 1,2,\ldots,n$, are columns of the matrix $\boldsymbol{A}$ of (7.9), and the right-hand side is given: $\boldsymbol{f} = \begin{bmatrix} f_1 \\ \ldots \\ f_m \end{bmatrix} \in \mathbb{R}^m$. According to Definition 7.1, we need to find such coefficients $x_1, x_2, \ldots, x_n$ of the linear combination $x_1\boldsymbol{a}_1 + x_2\boldsymbol{a}_2 + \ldots + x_n\boldsymbol{a}_n$ that the deviation of this linear combination from the vector $\boldsymbol{f}$ be minimal:

$$\Big\|\boldsymbol{f} - \sum_{k=1}^{n} x_k\boldsymbol{a}_k\Big\|_{\boldsymbol{B}} \longmapsto \min.$$

The columns $\boldsymbol{a}_k$, $k = 1,2,\ldots,n$, of the matrix $\boldsymbol{A}$ are assumed linearly independent. Therefore, their linear span, $\operatorname{span}\{\boldsymbol{a}_1,\ldots,\boldsymbol{a}_n\}$, forms a subspace of dimension n in the space $\mathbb{R}^m$: $\mathbb{R}^n(\boldsymbol{A}) \subset \mathbb{R}^m$. If $\{x_1, x_2, \ldots, x_n\}$ is a weak solution of system (7.8) in the sense of Definition 7.1, then the linear combination $\sum_{k=1}^n x_k\boldsymbol{a}_k$ is the orthogonal projection of the vector $\boldsymbol{f} \in \mathbb{R}^m$ onto the subspace $\mathbb{R}^n(\boldsymbol{A})$, where the orthogonality is understood in the sense of the inner product $[\,\cdot\,,\cdot\,]_{\boldsymbol{B}}^{(m)}$ given by formula (7.12).

Indeed, according to formula (7.13) we have $\Phi(\boldsymbol{x}) = \|\boldsymbol{f} - \boldsymbol{A}\boldsymbol{x}\|_{\boldsymbol{B}}^2$. Therefore, the element $\sum x_k\boldsymbol{a}_k = \boldsymbol{A}\boldsymbol{x}$ of the space $\mathbb{R}^n(\boldsymbol{A})$ that has minimum deviation from the given $\boldsymbol{f} \in \mathbb{R}^m$ minimizes $\Phi(\boldsymbol{x})$, i.e., yields the generalized solution of system (7.10) in the sense of Definition 7.1. In other words, the minimum deviation is attained for $\boldsymbol{A}\boldsymbol{x}_{\boldsymbol{B}} \in \mathbb{R}^n(\boldsymbol{A})$, where $\boldsymbol{x}_{\boldsymbol{B}}$ is the solution to system (7.14). On the other hand, any vector from $\mathbb{R}^n(\boldsymbol{A})$ can be represented as $\boldsymbol{A}\boldsymbol{\delta} = \delta_1\boldsymbol{a}_1 + \delta_2\boldsymbol{a}_2 + \ldots + \delta_n\boldsymbol{a}_n$, where $\boldsymbol{\delta} \in \mathbb{R}^n$. Formula (7.18) then implies that $\boldsymbol{f} - \boldsymbol{A}\boldsymbol{x}_{\boldsymbol{B}}$ is orthogonal to any $\boldsymbol{A}\boldsymbol{\delta} \in \mathbb{R}^n(\boldsymbol{A})$ in the sense of $[\,\cdot\,,\cdot\,]_{\boldsymbol{B}}^{(m)}$. As such, $\boldsymbol{A}\boldsymbol{x}_{\boldsymbol{B}}$ is the orthogonal projection of $\boldsymbol{f}$ onto $\mathbb{R}^n(\boldsymbol{A})$.

If a different basis $\{\hat{\boldsymbol{a}}_1, \hat{\boldsymbol{a}}_2, \ldots, \hat{\boldsymbol{a}}_n\}$ is introduced in the space $\mathbb{R}^n(\boldsymbol{A})$ instead of the original basis $\{\boldsymbol{a}_1, \boldsymbol{a}_2, \ldots, \boldsymbol{a}_n\}$, then system (7.14) is replaced by

$$\hat{\boldsymbol{A}}^*\boldsymbol{B}\hat{\boldsymbol{A}}\boldsymbol{x} = \hat{\boldsymbol{A}}^*\boldsymbol{B}\boldsymbol{f}, \tag{7.24}$$

where $\hat{\boldsymbol{A}}$ is the matrix composed of the columns $\hat{\boldsymbol{a}}_k$, $k = 1,2,\ldots,n$. Instead of the solution $\boldsymbol{x}_{\boldsymbol{B}}$ of system (7.14) we will obtain a new solution $\hat{\boldsymbol{x}}_{\boldsymbol{B}}$ of system (7.24).

However, the projection of $\boldsymbol{f}$ onto $\mathbb{R}^n(\boldsymbol{A})$ will obviously remain the same, so that the following equality will hold:

$$\sum_{k=1}^{n} x_k \boldsymbol{a}_k = \sum_{k=1}^{n} \hat{x}_k \hat{\boldsymbol{a}}_k.$$

In general, if we need to find a projection of a given vector $\boldsymbol{f} \in \mathbb{R}^m$ onto a given subspace $\mathbb{R}^n \subset \mathbb{R}^m$, then it is beneficial to try and choose a basis $\{\hat{\boldsymbol{a}}_1, \hat{\boldsymbol{a}}_2, \ldots, \hat{\boldsymbol{a}}_n\}$ in this subspace that would be as close to orthonormal as possible (in the sense of a particular scalar product $[\cdot,\cdot]_{\boldsymbol{B}}^{(m)}$). On one hand, the projection itself does not depend on the choice of the basis. On the other hand, the closer this basis is to orthonormal, the better is the conditioning of the system matrix in (7.24). Indeed, the entries $\hat{c}_{ij}$ of the matrix $\hat{\boldsymbol{C}} = \hat{\boldsymbol{A}}^* \boldsymbol{B} \hat{\boldsymbol{A}}$ are given by: $\hat{c}_{ij} = [\hat{\boldsymbol{a}}_i, \hat{\boldsymbol{a}}_j]_{\boldsymbol{B}}^{(m)}$, so that for an orthonormal basis we obtain $\hat{\boldsymbol{C}} = \boldsymbol{I}$, which is an "ultimately" well conditioned matrix. (See Exercise 8 after the section for a specific example).

7.2.4 Overdetermined Systems in the Operator Form

Besides its canonical form (7.8), an overdetermined system can be specified in the operator form. Namely, let $\mathbb{R}^n$ and $\mathbb{R}^m$ be two real vector spaces, $n < m$, let $\boldsymbol{A} : \mathbb{R}^n \longmapsto \mathbb{R}^m$ be a linear operator, and let $\boldsymbol{f} \in \mathbb{R}^m$ be a fixed vector. One needs to find such a vector $\boldsymbol{x} \in \mathbb{R}^n$ that satisfies:

$$\boldsymbol{A}\boldsymbol{x} = \boldsymbol{f}, \quad \boldsymbol{x} \in \mathbb{R}^n, \quad \boldsymbol{f} \in \mathbb{R}^m. \tag{7.25}$$

The difference between this form and the previously analyzed form (7.10) is that in (7.25) we do not necessarily assume that the vectors $\boldsymbol{x}$ and $\boldsymbol{f}$ are specified by their components with respect to the corresponding bases, and we do not assume that the operator $\boldsymbol{A}$ is specified explicitly by its matrix either.

Generally speaking, system (7.25) does not have a classical solution, because for any $\boldsymbol{x} \in \mathbb{R}^n$ the vector $\boldsymbol{A}\boldsymbol{x} \in \mathbb{R}^m$ belongs to the n-dimensional subspace $\mathbb{R}^n(\boldsymbol{A}) \subset \mathbb{R}^m$, which is the image of $\mathbb{R}^n$ under the transformation $\boldsymbol{A} : \mathbb{R}^n \longmapsto \mathbb{R}^m$. For the classical solvability of (7.25) it is necessary that the right-hand side $\boldsymbol{f}$ belong to the same subspace $\mathbb{R}^n(\boldsymbol{A})$, which does not always have to be the case.

To define a generalized, or weak, solution of (7.25), we first introduce the simplest scalar product $(\boldsymbol{f}, \boldsymbol{g})^{(m)}$ on $\mathbb{R}^m$. All other scalar products have the form:

$$[\boldsymbol{f}, \boldsymbol{g}]_{\boldsymbol{B}}^{(m)} = (\boldsymbol{B}\boldsymbol{f}, \boldsymbol{g})^{(m)},$$

where $\boldsymbol{B} : \mathbb{R}^m \longmapsto \mathbb{R}^m$, $\boldsymbol{B} = \boldsymbol{B}^* > 0$. Next, we introduce a quadratic function $\Phi_{\boldsymbol{B}} = \Phi_{\boldsymbol{B}}(\boldsymbol{x})$ [cf. formula (7.13)]:

$$\Phi_{\boldsymbol{B}}(\boldsymbol{x}) = [\boldsymbol{A}\boldsymbol{x} - \boldsymbol{f}, \boldsymbol{A}\boldsymbol{x} - \boldsymbol{f}]_{\boldsymbol{B}}^{(m)},$$

and define the generalized, or weak, solution $\boldsymbol{x}_{\boldsymbol{B}}$ of system (7.25) as

$$\boldsymbol{x}_{\boldsymbol{B}} = \arg\min_{\boldsymbol{x} \in \mathbb{R}^n} \Phi_{\boldsymbol{B}}(\boldsymbol{x}).$$

Example

Consider two uniform Cartesian grids on the square $\{0 \leq x \leq 1, 0 \leq y \leq 1\}$, a coarse grid with the size $H = 10^{-1}$ and the nodes:

$$(X_m, Y_n) = (mH, nH), \quad m, n = 1, 2, \ldots, 10,$$

and a fine grid with the size $h = 10^{-2}$ and the nodes:

$$(x_h, y_h) = (mh, nh), \quad m, n = 1, 2, \ldots, 100.$$

Introduce two linear spaces: $U^{(H)}$ and $F^{(h)}$. The first space will contain all grid functions $u = \{u_{mn}\}$ defined on the coarse grid, and the second space will contain all grid functions $f = \{f_{mn}\}$ defined on the fine grid. Let $\boldsymbol{A} : U^{(H)} \longmapsto F^{(h)}$ be an operator that for a given $u \in U^{(H)}$ builds the corresponding $f \in F^{(h)}$ using linear interpolation from the nodes of the coarse grid to the nodes of the fine grid in both coordinate directions x and y. Let also a particular $f = \{f_{mn}\} \in F^{(h)}$ be specified; for example, this function can be obtained by the approximate measurements of some physical scalar field at the nodes of the fine grid.

Let us formulate a problem of solving the overdetermined system:

$$\boldsymbol{A}u = f, \quad u \in U^{(H)}, \quad f \in F^{(h)}, \tag{7.26}$$

in the weak sense, where the scalar product in the space $F^{(h)}$ is defined as:

$$(f, g) = h^2 \sum_{m,n=0}^{100} f_{mn} g_{mn}, \quad f, g \in F^{(h)}.$$

This problem arises when one needs to smooth out the experimental data and reduce the effect of the random errors that inevitably accompany all experimental measurements. Once the generalized solution u of system (7.26) has been obtained, one can use a regularized function $\tilde{f} = \boldsymbol{A}u$ instead of the actual experimental function f.

Let us emphasize that system (7.26) is given in the operator form rather than in the canonical form. Yet efficient numerical algorithms can be constructed for computing its weak solution without reducing the system to the canonical form. Of course, the foregoing reduction can be implemented, but this may often appear less efficient.

Exercises

1. Show how the functions $\Phi = \Phi(x_1, x_2)$ given by formulae (7.3) and (7.6) can be obtained according to the general formula (7.13).
2. Show that the 2×2 system (7.5) that yields the least squares weak solution for a particular example analyzed in Section 7.1.1 can, in fact, be interpreted as system (7.14) if the matrix $\boldsymbol{A}$ is taken as the system matrix of (7.2):

$$\boldsymbol{A} = \begin{bmatrix} t_1 & 1 \\ t_2 & 1 \\ t_3 & 1 \\ t_4 & 1 \end{bmatrix},$$

and $\boldsymbol{B}$ is a 4×4 identity matrix: $\boldsymbol{B} = \boldsymbol{I}$.

3. Solve the following overdetermined system in the sense of the least squares:

$$x+y=1, \quad x-y=2, \quad 2x+y=2.4.$$

4. A total of m approximate measurements are taken of the length l of some object, and the results are: $l=l_1, l=l_2, \ldots, l=l_m$. Solve the resulting system of m equations with respect to one unknown l in the sense of the least squares.

5. The electric resistivity R of a wire depends linearly on the temperature T, so that $R=a+bT$. Determine the values of a and b in the sense of the least squares from the following approximate measurements:

T	19.1	25.0	30.1	36.0
R	76.30	77.80	79.75	80.80

Find the resistivity of the wire at $T=21.28$.

6. The speed v of a ship is related to the power p of its engine by the empirical formula $p=a+bv^3$. Determine the values of a and b in the sense of the least squares from the following experimental data:

v	6	8	10	12
p	420	805	1370	2370

7. The measurements of the angles α, β, and γ of a triangle yield the following values: $\alpha=52°5'$, $\beta=50°1'$, and $\gamma=78°6'$. The sum is equal to $180°12'$, it contains a residual of $12'$ due to experimental inaccuracies. "Fix" the problem with the help of the method of least squares.

   Hint. One minute is $1/60$ of one degree.

8. Let the function $y=y(x)$ be specified by means of its table of values: $y_k=1+x_k+x_k^2+\sin x_k$ sampled at the nodes: $x_k=-1+10^{-2}k$, $k=0,1,2,\ldots,200$ (i.e., on a uniform grid with size $h=10^{-2}$ that spans the interval $[-1,1]$). With the help of a computer, one needs to construct a polynomial $Q_m(x)$ of degree $m\leq 50$ for which the sum:

$$\sum_{k=0}^{200} |Q_m(x_k)-y_k|^2$$

   assumes its minimum value. Analyze and compare the following three variants of the solution and compute the values of $Q_m(x)$ at the midpoints: $x_{k+1/2}=(x_k+x_{k+1})/2$, $k=0,1,2,\ldots,199$.

   Variant 1. The polynomial $Q_m(x)$ is to be sought for in the form of a sum $Q_m(x)=\sum_{s=0}^{50} c_s x^s$ with the undetermined coefficients c_s, $s=1,2,\ldots,50$.

   Variant 2. The polynomial $Q_m(x)$ is to be sought for in the form of a sum

$$Q_m(x)=\sum_{s=0}^{50} c_s P_s(x),$$

where c_s, $s = 1, 2, \ldots, 50$, are undetermined coefficients, and $P_s(x)$ are the Legendre polynomials of degree s. The Legendre polynomials are defined recursively:

$$P_0(x) = 1, \quad P_1(x) = x,$$
$$P_{l+1}(x) = \frac{2l}{l+1} x P_l(x) - \frac{l}{l+1} P_{l-1}(x), \quad l = 1, 2, \ldots.$$

It is known that these polynomials are orthogonal on the interval $-1 \le x \le 1$:

$$\int_{-1}^{1} P_k(x) P_l(x) dx = 0, \quad k \ne l,$$

and also that

$$\int_{-1}^{1} P_k^2(x) dx = \frac{2}{2k+1}.$$

Variant 3. The polynomial $Q_m(x)$ is to be sought for in the form of a sum

$$Q_m(x) = \sum_{s=0}^{50} c_s \tilde{P}_s(x),$$

where c_s, $s = 1, 2, \ldots, 50$, are undetermined coefficients, and $\tilde{P}_s(x)$ are the normalized Legendre polynomials of degree s:

$$\tilde{P}_s(x) = \sqrt{\frac{2s+1}{2}} P_s(x).$$

For all three variants, use the following algorithm of determining the coefficients c_s, $s = 1, 2, \ldots, 50$:

1. Compute the matrices $\boldsymbol{A}^*\boldsymbol{BA}$ and $\boldsymbol{A}^*\boldsymbol{B}$ that define the linear system:

$$\boldsymbol{A}^*\boldsymbol{BAc} = \boldsymbol{A}^*\boldsymbol{By}, \quad \boldsymbol{c} = \begin{bmatrix} c_0 \\ \ldots \\ c_{50} \end{bmatrix}, \quad \boldsymbol{y} = \begin{bmatrix} y_0 \\ \ldots \\ y_{200} \end{bmatrix},$$

of the method of least squares. Choose the weight matrix $\boldsymbol{B} = \boldsymbol{I}$.

2. Solve the resulting system on a computer using the method of conjugate gradients (Section 6.2.2) with the initial guess: $\boldsymbol{c}^{(0)} = \begin{bmatrix} 0 \\ \ldots \\ 0 \end{bmatrix}$ and the stopping criterion: $\max_s |c_s^{(p)} - c_s^{(p-1)}| \le 10^{-3}$, which is based on having a small difference between two consecutive iterates.

3. Explain why the condition number for Variant 1 is large, whereas for Variant 3 it is close to 1.

9.* Build a computer code for the numerical solution of the overdetermined operator equation introduced in the Example in the end of Section 7.2.4.

Hint. Use the following scalar product in the space $F^{(h)}$:

$$(f, g) = h^2 \sum_{m,n=0}^{100} f_{mn} g_{mn}, \quad f, g \in F^{(h)}.$$

Then, the method of least squares yields the operator equation $\boldsymbol{A}^*\boldsymbol{A}u = \boldsymbol{A}^*f$. Solve it numerically using the method of conjugate gradients (Section 6.2.2).

7.3 Rank Deficient Systems. Singular Value Decomposition

In this section, we introduce and study the concept of a weak solution for the linear system:

$$\boldsymbol{A}\boldsymbol{x} = \boldsymbol{f}, \quad \boldsymbol{x} \in \mathbb{R}^n, \quad \boldsymbol{f} \in \mathbb{R}^m, \tag{7.27}$$

where $\boldsymbol{A}$ is an $m \times n$ matrix with real entries, $n \leq m$, and with linearly dependent columns, i.e., $\text{rank}\boldsymbol{A} = r < n$. In this case, the matrix $\boldsymbol{A}$ is said to be rank deficient, and there are non-trivial linear combinations of the columns $\boldsymbol{a}_k$, $k = 1, 2, \dots, n$, of the matrix $\boldsymbol{A}$ that are equal to zero: $\xi_1 \boldsymbol{a}_1 + \xi_2 \boldsymbol{a}_2 + \dots + \xi_n \boldsymbol{a}_n = \boldsymbol{0}$, while $\sum_{k=1}^n |\xi_k|^2 \neq 0$. Alternatively, we say that the matrix $\boldsymbol{A}$ has a non-empty kernel:

$$\mathbb{R}^n \supset \text{Ker}\boldsymbol{A} \stackrel{\text{def}}{=} \{\boldsymbol{\xi} \in \mathbb{R}^n, \boldsymbol{\xi} \neq \boldsymbol{0} \,|\, \boldsymbol{A}\boldsymbol{\xi} = \boldsymbol{0}\} \neq \emptyset.$$

For a rank deficient matrix $\boldsymbol{A}$, the previous definition of a generalized solution needs to be refined. The reason is that even if we find a weak solution $\boldsymbol{x}_{\boldsymbol{B}}$ according to Definition 7.1, we will not be able to guarantee its uniqueness. Indeed, let $\mathbb{R}^n \ni \boldsymbol{x}_{\boldsymbol{B}} = \arg\min_{\boldsymbol{x} \in \mathbb{R}^n} \Phi(\boldsymbol{x})$, where the function $\Phi(\boldsymbol{x})$ is defined by formula (7.13). Then, clearly, $\forall \boldsymbol{\xi} \in \text{Ker}\boldsymbol{A} : \ \Phi(\boldsymbol{x}_{\boldsymbol{B}} + \boldsymbol{\xi}) = \Phi(\boldsymbol{x}_{\boldsymbol{B}})$. In terms of matrices, one can show that the matrix $\boldsymbol{C} = \boldsymbol{A}^*\boldsymbol{B}\boldsymbol{A}$ of system (7.14) will be singular in this case.

Recall that when we introduced the original definition of a weak solution there were many alternatives. We could use different weight matrices $\boldsymbol{B}$ for the function $\Phi(\boldsymbol{x})$ of (7.13), and we did not necessarily even have to choose the function Φ quadratic, see (7.7). The resulting generalized solution was obviously determined by the choice of Φ, and it changed when a different Φ was selected. Likewise, when we need to refine Definition 7.1 for the case of rank deficient matrices, there may be alternative strategies. A particular approach based on selecting the vector $\boldsymbol{x}$ with the minimum Euclidean length (i.e., norm) of its own is described in Section 7.3.2. There are other approaches that will yield different weak solutions. To implement the strategy based on minimizing the Euclidean length among all those vectors that minimize Φ, we need the apparatus of singular value decomposition, see Section 7.3.1.

7.3.1 Singular Value Decomposition and Moore-Penrose Pseudoinverse

In this section, we only provide a brief summary of the results from linear algebra; further detail can be found, e.g., in [HJ85, Chapter 7]. Let $\boldsymbol{A}$ be a rectangular $m \times n$ matrix with complex entries. There exist two unitary matrices, $\boldsymbol{U}$ of dimension $m \times m$ and $\boldsymbol{W}$ of dimension $n \times n$, such that

$$\begin{gathered} \boldsymbol{U}^*\boldsymbol{A}\boldsymbol{W} = \boldsymbol{\Sigma} \equiv \text{diag}_{m \times n}\{\sigma_1, \sigma_2, \dots, \sigma_k\}, \quad k = \min(m,n), \\ \sigma_1 \geq \sigma_2 \geq \dots \geq \sigma_k \geq 0. \end{gathered} \tag{7.28}$$

The matrix $\boldsymbol{\Sigma}$ has the same dimension $m \times n$ as that of the matrix $\boldsymbol{A}$. The only non-zero entries that $\boldsymbol{\Sigma}$ may have are the diagonal entries $\sigma_{ii} \equiv \sigma_i$, $i = 1, 2, \ldots, k$. Note that as $\boldsymbol{\Sigma}$ is, generally speaking, not a square matrix, the diagonal that originates at its upper left corner does not necessarily terminate at the lower right corner.

Representation (7.28) is known as the singular value decomposition (SVD) of the matrix $\boldsymbol{A}$. The numbers $\sigma_i = \sigma_i(\boldsymbol{A})$ are called singular values. Note that there may be both non-zero (positive) and zero singular values. Also note that if $\boldsymbol{A}$ happens to be an $m \times n$ matrix with real entries, then both $\boldsymbol{U}$ and $\boldsymbol{W}$ in formula (7.28) can be taken as real orthogonal matrices.

Hereafter, we will consider the case $n \leq m$. Then, the singular values of $\boldsymbol{A}$ can be obtained as $\sigma_i = \sqrt{\lambda_i}$, $i = 1, 2, \ldots, n$, where λ_i are eigenvalues of the matrix $\boldsymbol{A}^*\boldsymbol{A}$. Indeed, from formula (7.28) we derive $\boldsymbol{A} = \boldsymbol{U}\boldsymbol{\Sigma}\boldsymbol{W}^*$ and consequently, $\boldsymbol{A}^* = \boldsymbol{W}\boldsymbol{\Sigma}^*\boldsymbol{U}^*$. Clearly, $\boldsymbol{A}^*\boldsymbol{A}$ is a square $n \times n$ matrix, and there is similarity by means of the orthogonal matrix $\boldsymbol{W}$:

$$\boldsymbol{A}^*\boldsymbol{A} = \boldsymbol{W}\boldsymbol{\Sigma}^*\boldsymbol{\Sigma}\boldsymbol{W}^*, \tag{7.29}$$

where $\boldsymbol{\Sigma}^*\boldsymbol{\Sigma} = \operatorname{diag}_{n\times n}\{\sigma_1^2, \sigma_2^2, \ldots, \sigma_n^2\}$ is a square diagonal $n \times n$ matrix with real entries. Hence we conclude that $\sigma_i = \sqrt{\lambda_i}$, $i = 1, 2, \ldots, n$, and also that the columns of $\boldsymbol{W}$ are eigenvectors of $\boldsymbol{A}^*\boldsymbol{A}$. They are called the right singular vectors of $\boldsymbol{A}$.

Likewise, we can write:

$$\boldsymbol{A}\boldsymbol{A}^* = \boldsymbol{U}\boldsymbol{\Sigma}\boldsymbol{\Sigma}^*\boldsymbol{U}^*,$$

and conclude that the $m \times m$ square matrix $\boldsymbol{A}\boldsymbol{A}^*$ is unitarily similar to the diagonal square matrix $\boldsymbol{\Sigma}\boldsymbol{\Sigma}^* = \operatorname{diag}_{m\times m}\{\sigma_1^2, \sigma_2^2, \ldots, \sigma_n^2, \underbrace{0, \ldots, 0}_{m-n}\}$, so that the columns of $\boldsymbol{U}$ are eigenvectors of $\boldsymbol{A}\boldsymbol{A}^*$. They are called the left singular vectors of $\boldsymbol{A}$.

Suppose that only the first r singular values of $\boldsymbol{A}$ are non-zero ($r < n$):

$$\sigma_1 \geq \sigma_2 \geq \ldots \geq \sigma_r > \sigma_{r+1} = \ldots = \sigma_n = 0.$$

Then the last $n - r$ columns of the matrix $\boldsymbol{\Sigma}$ are zero, and recasting formula (7.28) as $\boldsymbol{A}\boldsymbol{W} = \boldsymbol{U}\boldsymbol{\Sigma}$ we see that the last $n - r$ columns of the matrix $\boldsymbol{A}\boldsymbol{W}$ are also zero. Since the columns $\boldsymbol{w}_i$, $i = 1, 2, \ldots, n$, of the matrix $\boldsymbol{W}$ are orthonormal, we conclude that

$$\operatorname{rank}\boldsymbol{A} = r \quad \text{and} \quad \operatorname{Ker}\boldsymbol{A} = \operatorname{span}\{\boldsymbol{w}_{r+1}, \ldots, \boldsymbol{w}_n\}.$$

With the help of SVD, we introduce an important notion of the Moore-Penrose pseudoinverse for a rectangular matrix $\boldsymbol{A}$.

DEFINITION 7.2 *Let $\boldsymbol{A}$ be an $m \times n$ matrix, $m \geq n$, and let $\operatorname{rank}\boldsymbol{A} = r$. Assume that there is a singular value decomposition of type (7.28) for $\boldsymbol{A}$: $\boldsymbol{U}^*\boldsymbol{A}\boldsymbol{W} = \boldsymbol{\Sigma} = \operatorname{diag}_{m\times n}\{\sigma_1, \sigma_2, \ldots, \sigma_n\} = \operatorname{diag}_{m\times n}\{\sigma_1, \sigma_2, \ldots, \sigma_r, \underbrace{0, \ldots, 0}_{n-r}\}$. Introduce an $n \times m$ matrix $\boldsymbol{\Sigma}^+$ obtained by transposing $\boldsymbol{\Sigma}$ and inverting its non-zero diagonal entries: $\boldsymbol{\Sigma}^+ = \operatorname{diag}_{n\times m}\{\frac{1}{\sigma_1}, \frac{1}{\sigma_2}, \ldots, \frac{1}{\sigma_r}, \underbrace{0, \ldots, 0}_{n-r}\}$. The $n \times m$ matrix*

$$\boldsymbol{A}^+ = \boldsymbol{W}\boldsymbol{\Sigma}^+\boldsymbol{U}^* \tag{7.30}$$

is called the Moore-Penrose pseudoinverse of $\boldsymbol{A}$.

This definition is quite general and applies to many cases when the conventional inverse does not exist or cannot even be defined. However, in the case when there is a conventional inverse, the Moore-Penrose pseudoinverse coincides with it. Indeed, let $m = n = r$. Then all the matrices involved are square matrices of the same dimension. Moreover, it is clear that $\boldsymbol{\Sigma}^+ = \text{diag}_{n\times n}\{\frac{1}{\sigma_i}\} = \boldsymbol{\Sigma}^{-1}$, and consequently:

$$\begin{aligned}\boldsymbol{A}^+\boldsymbol{A} &= (\boldsymbol{W}\boldsymbol{\Sigma}^{-1}\boldsymbol{U}^*)(\boldsymbol{U}\boldsymbol{\Sigma}\boldsymbol{W}^*) = \boldsymbol{W}\boldsymbol{\Sigma}^{-1}\boldsymbol{\Sigma}\boldsymbol{W}^* = \boldsymbol{W}\boldsymbol{W}^* = \boldsymbol{I},\\ \boldsymbol{A}\boldsymbol{A}^+ &= (\boldsymbol{U}\boldsymbol{\Sigma}\boldsymbol{W}^*)(\boldsymbol{W}\boldsymbol{\Sigma}^{-1}\boldsymbol{U}^*) = \boldsymbol{U}\boldsymbol{\Sigma}\boldsymbol{\Sigma}^{-1}\boldsymbol{U}^* = \boldsymbol{U}\boldsymbol{U}^* = \boldsymbol{I},\end{aligned}$$

which implies $\boldsymbol{A}^+ = \boldsymbol{A}^{-1}$. Furthermore, if $\boldsymbol{A}$ is a rectangular matrix, $n < m$, but still has full rank, $\text{rank}\boldsymbol{A} = r = n$, then

$$\boldsymbol{A}^+ = (\boldsymbol{A}^*\boldsymbol{A})^{-1}\boldsymbol{A}^*. \tag{7.31}$$

Indeed, we first notice that if $\text{rank}\boldsymbol{A} = n$, then $\boldsymbol{A}^*\boldsymbol{A}$ is a non-singular square matrix. Next, using formula (7.29) we can write: $\boldsymbol{A}^*\boldsymbol{A} = \boldsymbol{W}\boldsymbol{\Sigma}^*\boldsymbol{\Sigma}\boldsymbol{W}^* = \boldsymbol{W}\text{diag}_{n\times n}\{\sigma_i^2\}\boldsymbol{W}^*$ and consequently, $(\boldsymbol{A}^*\boldsymbol{A})^{-1} = \boldsymbol{W}\text{diag}_{n\times n}\{\frac{1}{\sigma_i^2}\}\boldsymbol{W}^*$. Finally, we have:

$$\begin{aligned}(\boldsymbol{A}^*\boldsymbol{A})^{-1}\boldsymbol{A}^* &= \left(\boldsymbol{W}\text{diag}_{n\times n}\{\tfrac{1}{\sigma_i^2}\}\boldsymbol{W}^*\right)\left(\boldsymbol{W}\boldsymbol{\Sigma}^*\boldsymbol{U}^*\right)\\ &= \boldsymbol{W}\text{diag}_{n\times n}\{\tfrac{1}{\sigma_i^2}\}\boldsymbol{\Sigma}^*\boldsymbol{U}^* = \boldsymbol{W}\boldsymbol{\Sigma}^+\boldsymbol{U}^* = \boldsymbol{A}^+.\end{aligned}$$

If, again, $\boldsymbol{A}$ is a non-singular square matrix, then the previous property yields: $\boldsymbol{A}^+ = (\boldsymbol{A}^*\boldsymbol{A})^{-1}\boldsymbol{A}^* = \boldsymbol{A}^{-1}(\boldsymbol{A}^*)^{-1}\boldsymbol{A}^* = \boldsymbol{A}^{-1}$. Thus, we see that the pseudoinverse matrix introduced according to Definition 7.2 resembles the conventional inverse in the sense that it reduces to conventional inverse when the latter exists and also reproduces some other properties that are characteristic of the conventional inverse, see formula (7.31), as well as Exercise 1 after the section.

The SVD of a matrix can be computed, for example, using the Golub-Kahan-Reinsch algorithm, see [GVL89, Chapter 5]. This algorithm consists of two stages, the first one employs Householder transformations and the second one uses an iterative scheme based on the $\boldsymbol{QR}$ factorization.

7.3.2 Minimum Norm Weak Solution

As we have seen, if the system matrix in (7.27) is rank deficient, we need an alternative definition of the weak solution because of non-uniqueness. As an additional criterion for uniqueness, we will use the minimum Euclidean norm (length).

DEFINITION 7.3 *Let* $\boldsymbol{A}$ *be an* $m\times n$ *matrix with real entries,* $m \geq n$, *and let* $\text{rank}\boldsymbol{A} = r < n$. *Introduce a scalar function* $\Phi = \Phi(\boldsymbol{x})$ *of the vector argument* $\boldsymbol{x} \in \mathbb{R}^n$:

$$\Phi(\boldsymbol{x}) = (\boldsymbol{A}\boldsymbol{x} - \boldsymbol{f}, \boldsymbol{A}\boldsymbol{x} - \boldsymbol{f})^{(m)} \equiv \|\boldsymbol{A}\boldsymbol{x} - \boldsymbol{f}\|_2^2. \tag{7.32}$$

A minimum norm weak (generalized) solution of the overdetermined system (7.27) is the vector $\hat{\boldsymbol{x}} \in \mathbb{R}^n$ *that minimizes* $\Phi(\boldsymbol{x})$, *i.e.*, $\forall \boldsymbol{x} \in \mathbb{R}^n$: $\Phi(\boldsymbol{x}) \geq \Phi(\hat{\boldsymbol{x}})$, *and also such that* $\forall \boldsymbol{x} \in \mathbb{R}^n$ & $\Phi(\boldsymbol{x}) = \Phi(\hat{\boldsymbol{x}})$: $\|\boldsymbol{x}\|_2 \geq \|\hat{\boldsymbol{x}}\|_2$.

Note that the minimum norm weak solution introduced according to Definition 7.3 may exhibit strong sensitivity to the perturbations of the matrix $\boldsymbol{A}$ in the case when these perturbations change the rank of the matrix, see the example given in Exercise 2 after the section.

REMARK 7.6 Definition 7.3 can also be applied to the case of a full rank matrix $\boldsymbol{A}$, $\text{rank}\boldsymbol{A} = n$. Then it reduces to Definition 7.1 (for $\boldsymbol{B} = \boldsymbol{I}$), because according to Theorem 7.2 a unique least squares weak solution exists for a full rank overdetermined system, and consequently, the Euclidean norm of this solution is minimum. □

THEOREM 7.3
Let $\boldsymbol{A}$ *be an* $m \times n$ *matrix with real entries,* $m \geq n$, *and let* $\text{rank}\boldsymbol{A} = r < n$. *There is a unique weak solution of system (7.27) in the sense of Definition 7.3. This solution is given by the formula:*

$$\hat{\boldsymbol{x}} = \boldsymbol{A}^+ \boldsymbol{f}, \tag{7.33}$$

where $\boldsymbol{A}^+$ *is the Moore-Penrose pseudoinverse of* $\boldsymbol{A}$ *introduced in Definition 7.2.*

PROOF Using singular value decomposition, represent the system matrix of (7.27) in the form: $\boldsymbol{A} = \boldsymbol{U}\boldsymbol{\Sigma}\boldsymbol{W}^*$. Also define $\boldsymbol{y} = \boldsymbol{W}^*\boldsymbol{x}$. Then, according to formula (7.32), we can write:

$$\begin{aligned}\Phi(\boldsymbol{x}) &= (\boldsymbol{U}\boldsymbol{\Sigma}\boldsymbol{W}^*\boldsymbol{x} - \boldsymbol{f}, \boldsymbol{U}\boldsymbol{\Sigma}\boldsymbol{W}^*\boldsymbol{x} - \boldsymbol{f})^{(m)} = (\boldsymbol{U}\boldsymbol{\Sigma}\boldsymbol{y} - \boldsymbol{f}, \boldsymbol{U}\boldsymbol{\Sigma}\boldsymbol{y} - \boldsymbol{f})^{(m)} \\ &= (\boldsymbol{\Sigma}\boldsymbol{y} - \boldsymbol{U}^*\boldsymbol{f}, \boldsymbol{\Sigma}\boldsymbol{y} - \boldsymbol{U}^*\boldsymbol{f})^{(m)} = \|\boldsymbol{\Sigma}\boldsymbol{y} - \boldsymbol{U}^*\boldsymbol{f}\|_2^2,\end{aligned}$$

and we need to find the vector $\hat{\boldsymbol{y}} \in \mathbb{R}^n$ such that $\forall \boldsymbol{y} \in \mathbb{R}^n$: $\|\boldsymbol{\Sigma}\hat{\boldsymbol{y}} - \boldsymbol{U}^*\boldsymbol{f}\|_2^2 \leq \|\boldsymbol{\Sigma}\boldsymbol{y} - \boldsymbol{U}^*\boldsymbol{f}\|_2^2$. This vector $\hat{\boldsymbol{y}}$ must also have a minimum Euclidean norm, because the matrix $\boldsymbol{W}$ is orthogonal and since $\boldsymbol{y} = \boldsymbol{W}^*\boldsymbol{x}$ we have $\|\boldsymbol{y}\|_2 = \|\boldsymbol{x}\|_2$.

Next, recall that as $\text{rank}\boldsymbol{A} = r$, the matrix $\boldsymbol{A}$ has precisely r non-zero singular values σ_i. Then we have:

$$\|\boldsymbol{\Sigma}\boldsymbol{y} - \boldsymbol{U}^*\boldsymbol{f}\|_2^2 = \sum_{i=1}^{r} [\sigma_i y_i - (\boldsymbol{U}^*\boldsymbol{f})_i]^2 + \sum_{i=r+1}^{m} [(\boldsymbol{U}^*\boldsymbol{f})_i]^2, \tag{7.34}$$

where $(\boldsymbol{U}^*\boldsymbol{f})_i$ denotes component number i of the m-dimensional vector $\boldsymbol{U}^*\boldsymbol{f}$. Expression (7.34) attains its minimum value when the first sum on the right-hand side is equal to zero, because the second sum simply does not depend on $\boldsymbol{y}$. This immediately yields the first r components of the vector $\hat{\boldsymbol{y}}$:

$$\hat{y}_i = \frac{(\boldsymbol{U}^*\boldsymbol{f})_i}{\sigma_i}, \quad i = 1, 2, \ldots, r. \tag{7.35a}$$

As the desired weak solution $\hat{\boldsymbol{y}}$ is supposed to have a minimum Euclidean norm, and its first r components are already specified, we must merely set the rest of the components to zero:

$$\hat{y}_i = 0, \quad i = r+1, \ldots, n. \tag{7.35b}$$

Combining formulae (7.35a) and (7.35b), we obtain:

$$\hat{\boldsymbol{y}} = \boldsymbol{\Sigma}^+ \boldsymbol{U}^* \boldsymbol{f},$$

where the matrix $\boldsymbol{\Sigma}^+$ is introduced in Definition 7.2. Accordingly,

$$\hat{\boldsymbol{x}} = \boldsymbol{W}\boldsymbol{\Sigma}^+ \boldsymbol{U}^* \boldsymbol{f},$$

which is equivalent to (7.33). □

REMARK 7.7 Even when system $\boldsymbol{Ax} = \boldsymbol{f}$ is not overdetermined and rather has a non-singular square matrix $\boldsymbol{A}$, the method of least squares can still be applied for computing its solution. Either its full rank version based on $\boldsymbol{QR}$ factorization (Section 7.2) or the current SVD based version can be employed as an alternative to the Gaussian elimination (Section 5.4). Both alternatives are more expensive computationally than the Gaussian elimination, yet they are more robust, especially for ill conditioned systems. □

Exercises

1. Prove the following properties of the Moore-Penrose pseudoinverse (7.30):

$$\boldsymbol{A}\boldsymbol{A}^+\boldsymbol{A} = \boldsymbol{A},$$
$$\boldsymbol{A}^+\boldsymbol{A}\boldsymbol{A}^+ = \boldsymbol{A}^+.$$

   Notice that if $\boldsymbol{A}$ is a non-singular square matrix so that $\boldsymbol{A}^+ = \boldsymbol{A}^{-1}$, then these properties become standard properties of the inverse: $\boldsymbol{A}\boldsymbol{A}^{-1}\boldsymbol{A} = \boldsymbol{A}$ and $\boldsymbol{A}^{-1}\boldsymbol{A}\boldsymbol{A}^{-1} = \boldsymbol{A}^{-1}$.

2. Let $m = n = 2$, and consider the following rank 1 system:

$$\boldsymbol{Ax} = \boldsymbol{f}, \quad \boldsymbol{A} = \begin{bmatrix} 0 & 0 \\ 1 & 0 \end{bmatrix}, \quad \boldsymbol{f} = \begin{bmatrix} 1 \\ 1 \end{bmatrix}.$$

   Consider also a perturbed version of this system that has rank 2, while the perturbation is assumed small, $\varepsilon \ll 1$:

$$\boldsymbol{A}_\varepsilon \boldsymbol{x} = \boldsymbol{f}, \quad \boldsymbol{A}_\varepsilon = \begin{bmatrix} 0 & \varepsilon \\ 1 & 0 \end{bmatrix}, \quad \boldsymbol{f} = \begin{bmatrix} 1 \\ 1 \end{bmatrix}.$$

Solve both systems in the sense of Definition 7.3. Characterize the sensitivity of the solution to the perturbation ε.

Hint. Notice that the second system has a classical solution $\boldsymbol{x}_\varepsilon = \begin{bmatrix} 1 \\ \varepsilon^{-1} \end{bmatrix}$, with which any weak solution must coincide. The SVD of $\boldsymbol{A}$ is given by $\boldsymbol{U}^* = \begin{bmatrix} 0 & 1 \\ 1 & 0 \end{bmatrix}$ (a permutation matrix) and $\boldsymbol{W} = \boldsymbol{I}$ so that $\boldsymbol{\Sigma} = \boldsymbol{U}^*\boldsymbol{A}\boldsymbol{W} = \begin{bmatrix} 1 & 0 \\ 0 & 0 \end{bmatrix}$. Accordingly, $\boldsymbol{\Sigma}^+ = \boldsymbol{\Sigma}$ and $\boldsymbol{A}^+ = \boldsymbol{W}\boldsymbol{\Sigma}^+\boldsymbol{U}^* = \begin{bmatrix} 0 & 1 \\ 0 & 0 \end{bmatrix}$. Consequently, $\hat{\boldsymbol{x}} = \boldsymbol{A}^+\boldsymbol{f} = \begin{bmatrix} 1 \\ 0 \end{bmatrix}$. For small ε the difference is large, which implies strong sensitivity to perturbations.

Chapter 8

Numerical Solution of Nonlinear Equations and Systems

Consider a scalar nonlinear equation:

$$F(x) = 0,$$

where $F(x)$ is a given function. Very often, equations of this type cannot be solved analytically. For example, no analytic solution is available when $F(x)$ is a high degree algebraic polynomial or when $F(x)$ is a transcendental function, such as $F(x) = \sin x - \frac{1}{2}x$ or $F(x) = e^{-x} + \cos x$. Then, a numerical method must be employed for computing the solution x approximately. Such methods are often referred to as the methods of rootfinding, because the number x that solves the equation $F(x) = 0$ is called its root.

Along with the scalar nonlinear equations, we will also consider systems of such equations:

$$\boldsymbol{F}(\boldsymbol{x}) = \boldsymbol{0},$$

where $\boldsymbol{F}(\boldsymbol{x})$ is a given vector-function of the vector argument $\boldsymbol{x} = (x_1, x_2, \ldots, x_n)$. For example, if

$$\boldsymbol{F}(\boldsymbol{x}) = \begin{bmatrix} F_1(x_1,x_2) \\ F_2(x_1,x_2) \end{bmatrix} = \begin{bmatrix} x_1^2 + x_2^2 - 25 \\ x_2 - x_1^2 \end{bmatrix},$$

then the nonlinear system $\boldsymbol{F}(\boldsymbol{x}) = \boldsymbol{0}$ can be written in components as follows:

$$\begin{aligned} x_1^2 + x_2^2 - 25 &= 0, \\ x_2 - x_1^2 &= 0. \end{aligned}$$

When solving numerically either scalar nonlinear equations $F(x) = 0$ or systems $\boldsymbol{F}(\boldsymbol{x}) = \boldsymbol{0}$, one typically needs to address two issues. First, the solutions (roots) need to be isolated, i.e., the appropriate domains of the independent variable(s) need to be identified that will only contain one solution each. Then, the solutions (roots) need to be "refined," i.e., actually computed with a prescribed accuracy.

There are no universal approaches to solving the problem of the isolation of roots. One can use graphs, look into the intervals of monotonicity of the function $F(x)$ on which it changes sign, and employ other special "tricks." There is only one substantial class of functions, namely, algebraic polynomials with real coefficients, for which this problem has been solved completely in the most general form. The solution is given by the Sturm theorem, see, e.g., [Dör82, Section 24].

As for the second problem, that of the roots' refinement, it is normally solved by iterations. In this respect, we would like to emphasize that for solving linear algebraic systems there always was an alternative (at least in principle) between using a direct method (Chapter 5) or an iterative method (Chapter 6). In doing so, the final choice was typically motivated by the considerations of efficiency. In contradistinction to that, for the nonlinear equations and systems iterative methods basically provide the only feasible choice of the solution methodology.

An iterative method builds a sequence of iterates $x^{(p)}$, $p = 0, 1, 2, \ldots$, that is supposed to converge to the solution $\hat{x}$:

$$\lim_{p \to \infty} x^{(p)} = \hat{x}.$$

Furthermore, the method is said to converge with the order $\kappa \geq 1$ if there is a constant $c > 0$ such that

$$|x^{(p+1)} - \hat{x}| \leq c|x^{(p)} - \hat{x}|^{\kappa}, \quad p = 0, 1, 2, \ldots. \tag{8.1}$$

Note that in the case of a first order convergence, $\kappa = 1$, it is necessary that $c < 1$. Then the rate of convergence will be that of a geometric sequence with the base c.

Let us also emphasize that unlike in the linear case, the convergence of iterations for nonlinear equations typically depends on the initial guess $x^{(0)}$. In other words, one and the same iteration scheme may converge for some particular initial guesses and diverge for other initial guesses.

Similarly to the linear case, the notion of conditioning can also be introduced for the problem of nonlinear rootfinding. As always, conditioning provides a quantitative measure of how sensitive the solution (root) is to the perturbations of the input data. Let $\hat{x} \in \mathbb{R}$ be the desired root of the equation $F(x) = 0$. Assume that this root has multiplicity $m + 1 \geq 1$, which means that

$$\frac{d^k F}{dx^k}(\hat{x}) = 0, \quad k = 0, 1, \ldots, m-1.$$

Instead of the original equation $F(x) = 0$ consider its perturbed version $F(x) = \eta$. The perturbation η of the right-hand side will obviously cause a perturbation in the solution that we denote δ so that $F(\hat{x} + \delta) = \eta$. Using the Taylor formula, we can write:

$$F(\hat{x} + \delta) = \frac{1}{m!} \frac{d^m F}{dx^m}(\hat{x}) \cdot \delta^m + o(\delta^m) = \eta.$$

Dropping the higher order term $o(\delta^m)$, we obtain for the perturbation δ itself and for the condition number μ:

$$\delta \approx \left(\frac{m!\eta}{F^{(m)}(\hat{x})} \right)^{\frac{1}{m}} \quad \Longrightarrow \quad \mu = \frac{|\delta|}{|\eta|} \approx \left| \frac{m!\eta}{F^{(m)}(\hat{x})} \right|^{\frac{1}{m}} \cdot \frac{1}{|\eta|}.$$

We therefore conclude that the smaller the value of the first non-zero derivative $F^{(m)}$ at the root $\hat{x}$, the more sensitive the solution $\hat{x} + \delta$ is to the perturbation η, i.e., the poorer the conditioning of the problem. In particular, when the root $\hat{x}$ is simple, i.e., when $m = 1$, we have:

$$\mu \approx \frac{1}{|F'(\hat{x})|},$$

so that the condition number appears inversely proportional to the magnitude of the first derivative of $F(x)$ at the root $\hat{x}$.

8.1 Commonly Used Methods of Rootfinding

In this section, we will describe several commonly used methods of rootfinding. We will assume that the function $F = F(x)$ is continuous on the interval $[a,b]$, and that $F(a) \cdot F(b) < 0$. In other words, the value of the function $F(x)$ at one endpoint of the interval is supposed to be positive, and at the other endpoint — negative. This guarantees that there is at least one root $\hat{x} \in (a,b): \; F(\hat{x}) = 0$.

8.1.1 The Bisection Method

Denote $a^{(0)} = a$, $b^{(0)} = b$, and take the initial guess $x^{(0)}$ at the midpoint of the original interval (a,b):

$$x^{(0)} = \frac{a^{(0)} + b^{(0)}}{2}.$$

Then, for all $p = 0, 1, 2, \ldots$ define:

$$\begin{aligned} a^{(p+1)} &= a^{(p)}, \; b^{(p+1)} = x^{(p)}, \quad \text{if} \quad F(a^{(p)}) \cdot F(x^{(p)}) < 0, \\ a^{(p+1)} &= x^{(p)}, \; b^{(p+1)} = b^{(p)}, \quad \text{if} \quad F(x^{(p)}) \cdot F(b^{(p)}) < 0, \end{aligned}$$

and take the new iterate $x^{(p+1)}$, again, at the midpoint:

$$x^{(p+1)} = \frac{a^{(p+1)} + b^{(p+1)}}{2}.$$

In other words, on every iteration the interval that contains the root is partitioned into two equal subintervals, see Figure 8.1. This partitioning is the source of the name "bisection" that the method bears. The next interval that will contain the root is determined according to the same criterion, namely, as that

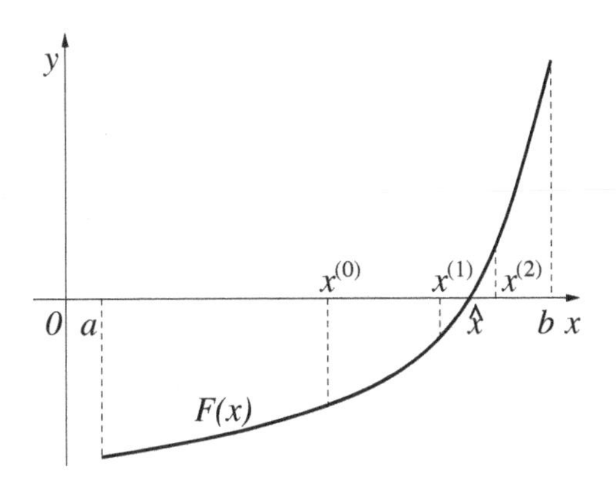

FIGURE 8.1: The method of bisection.

one of the two subintervals, for which the function $F(x)$ is positive at one endpoint and negative at the other.

The convergence of the method is robust yet fairly slow. If we introduce the error of the iterate $x^{(p)}$ as $\varepsilon^{(p)} = \hat{x} - x^{(p)}$, then, clearly, its magnitude will not exceed the size of the next subinterval that contains the root:

$$|\varepsilon^{(p)}| \leq \frac{b-a}{2^{p+1}}. \tag{8.2}$$

This estimate implies, in particular, that the method of bisection always converges provided that there is a root $a < \hat{x} < b$. However, if there are multiple roots on (a,b), then the method will converge to one of them and ignore the others.

If we require that the error be smaller than some initially prescribed quantity: $|\varepsilon^{(p)}| \leq \sigma$, then from formula (8.2) we can easily derive the estimate for the corresponding number of iterations:

$$p \geq \log_2\left(\frac{b-a}{\sigma}\right) - 1.$$

Note that the convergence of the bisection method is not necessarily monotonic. In other words, we cannot generally claim that it is even a first order method in the sense of estimate (8.1), i.e., that

$$|\varepsilon^{(p+1)}| \leq c|\varepsilon^{(p)}|, \quad c < 1.$$

For example, it is easy to see from Figure 8.1 that we can have $|\hat{x} - x^{(2)}| > |\hat{x} - x^{(1)}|$. In practice, the method of bisection is often used not completely on its own, but rather as a predictor for a faster method. The bisection would roughly approximate the root by confining it to a sufficiently small interval, after which a faster method is employed in order to efficiently compute an accurate approximation to the root.

8.1.2 The Chord Method

Several methods of rootfinding have been proposed that are faster than the method of bisection. If $\hat{x}$ is the root of $F(x) = 0$, we can take an arbitrary x and write:

$$F(\hat{x}) = F(x) + (\hat{x} - x)F'(\xi) = 0,$$

where ξ is a point between x and $\hat{x}$. Consequently, for the root $\hat{x}$ we have:

$$\hat{x} = x - \frac{F(x)}{F'(\xi)}.$$

Then, given some initial guess $x^{(0)}$, we can build the iteration sequence as follows:

$$x^{(p+1)} = x^{(p)} - \frac{F(x^{(p)})}{\eta^{(p)}}, \quad p = 0, 1, 2, \ldots,$$

where the quantity $\eta^{(p)}$ is supposed to approximate $F'(\xi)$ for ξ between $x^{(p)}$ and $x^{(p+1)}$. Different choices of $\eta^{(p)}$ yield different iteration schemes.

For the chord method, the quantity $\eta^{(p)}$ is chosen constant, i.e., independent of p:

$$\eta^{(p)} = \eta = \frac{F(b) - F(a)}{b - a}.$$

In other words, the value of the derivative $F'(\xi)$ anywhere on $[a,b]$ is approximated by the slope of the chord connecting the points $(a, F(a))$ and $(b, F(b))$ on the graph of the function $F(x)$, see Figure 8.2. In doing so, the iteration scheme is written as follows:

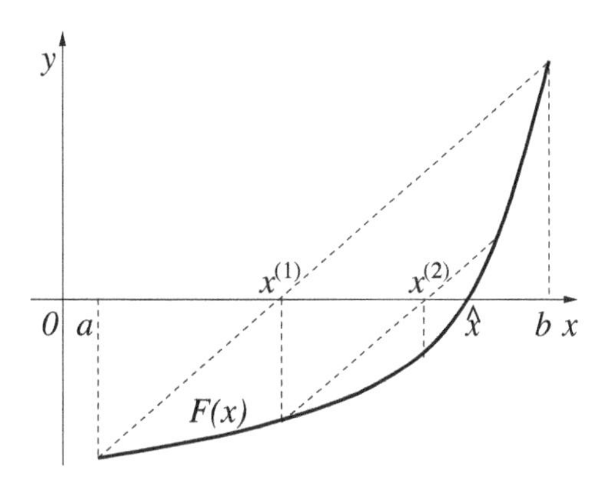

FIGURE 8.2: The chord method.

$$x^{(p+1)} = x^{(p)} - \frac{b - a}{F(b) - F(a)} F(x^{(p)}), \quad p = 0, 1, 2, \ldots,$$

and to initiate it, we can formally set, say, $x^{(0)} = b$. It is possible to show (Section 8.2.1) that the sequence of iterates $x^{(p)}$ generated by the chord method converges to the root $\hat{x}$ with order $\kappa = 1$ in the sense of formula (8.1). In practice, the convergence rate of the chord method may not always be faster than that of the bisection method, for which, however, order $\kappa = 1$ cannot be guaranteed (Section 8.1.1).

8.1.3 The Secant Method

In this method, the derivative $F'(\xi) \approx \eta^{(p)}$ is approximated by the slope of the secant on the previous interval, see Figure 8.3, i.e., by the slope of the straight line that connects the points $(x^{(p)}, F(x^{(p)}))$ and $(x^{(p-1)}, F(x^{(p-1)}))$ on the graph of $F(x)$:

$$\eta^{(p)} = \frac{F(x^{(p)}) - F(x^{(p-1)})}{x^{(p)} - x^{(p-1)}}.$$

This yields the following iteration scheme:

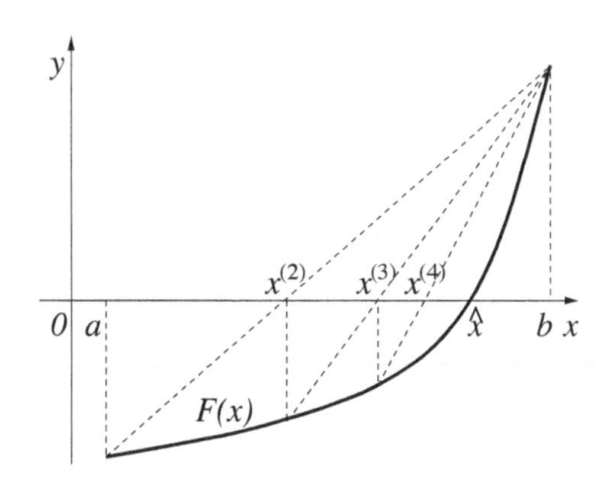

FIGURE 8.3: The secant method.

$$x^{(p+1)} = x^{(p)} - \frac{x^{(p)} - x^{(p-1)}}{F(x^{(p)}) - F(x^{(p-1)})}, \quad p = 1, 2, 3, \ldots,$$

which is started off by setting $x^{(1)} = b$ and $x^{(0)} = a$. Clearly, the first step of the secant method, $p = 1$, coincides with that of the chord method, $p = 0$, see Section 8.1.2.

The convergence of the secant method is faster than that of either the bisection method or the chord method. However, in order for the convergence to take place the two initial guesses, $x^{(0)} = a$ and $x^{(1)} = b$, must be sufficiently close to the root $\hat{x}$. In this case, it is possible to show that the method has order

$$\kappa = \frac{1+\sqrt{5}}{2} \approx 1.62$$

in the sense of estimate (8.1).

8.1.4 Newton's Method

In Newton's method, the derivative $F'(\xi) = \eta^{(p)}$ at the intermediate point ξ between $x^{(p)}$ and $x^{(p+1)}$ is approximated by the value of $F'(x^{(p)})$, see Figure 8.4. Let us assume that the function $F = F(x)$ is continuously differentiable, $F(x) \in C^1[a,b]$, and that $F'(\hat{x}) \neq 0$, where $\hat{x}$ is the root of $F(x) = 0$. Then, by continuity, we can also claim that $F'(x) \neq 0$ on some neighborhood of the root $\hat{x}$. Consequently, Newton's iteration can be carried through:

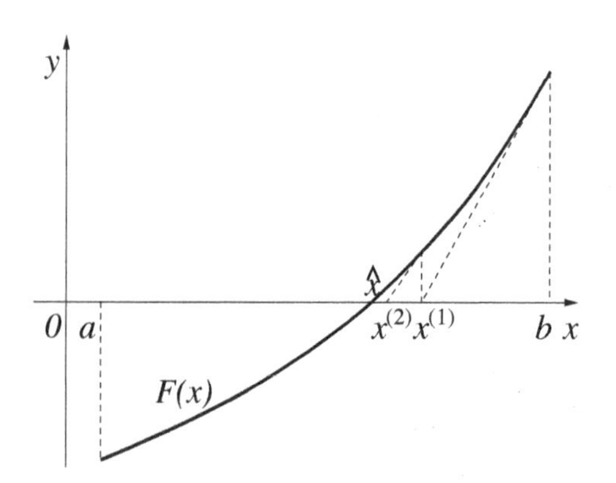

FIGURE 8.4: Newton's method.

$$x^{(p+1)} = x^{(p)} - \frac{F(x^{(p)})}{F'(x^{(p)})}, \quad p = 0,1,2,\dots,$$

because the denominator will differ from zero for all $x^{(p)}$ that are sufficiently close to $\hat{x}$. The initial guess for Newton's method can be taken, say, as $x^{(0)} = b$, see Figure 8.4. A remarkable property of Newton's method is its quadratic convergence, which means that we have $\kappa = 2$ in inequality (8.1). This result will be proved in Section 8.3. On the other hand, to ensure any convergence at all, the initial guess $x^{(0)}$ must be chosen sufficiently close to the (yet unknown) root $\hat{x}$.

It is easy to see that Newton's method is, in fact, based on linearization of the function $F(x)$, because the next iterate $x^{(p+1)}$ is obtained by solving the linear equation:

$$F(x^{(p)}) + F'(x^{(p)})(x^{(p+1)} - x^{(p)}) = 0.$$

On the left-hand side of this equation we have the first two terms of the Taylor expansion of $F(x)$, i.e., the linear part of the increment of $F(x)$ between $x^{(p)}$ and $x^{(p+1)}$.

Note also that compared to the previously analyzed methods (bisection, chord, and secant) Newton's method requires evaluation not only of $F(x^{(p)})$ but also of $F'(x^{(p)})$ for all $p = 0,1,2,\dots$. This implies an increased computational cost per iteration, which may, however, be compensated for by a faster convergence.

8.2 Fixed Point Iterations

8.2.1 The Case of One Scalar Equation

Suppose that we want to solve a nonlinear equation:

$$F(x) = 0, \tag{8.3}$$

for which the desired root $\hat{x}$ is known to belong to the interval (a,b). On this interval, equation (8.3) can be recast in an equivalent form:

$$x = f(x). \tag{8.4}$$

For example, this can be done by defining:

$$f(x) \stackrel{\text{def}}{=} x - \alpha(x)F(x),$$

where $\alpha(x)$ is an arbitrary non-zero function on $a < x < b$. The root $\hat{x} \in (a,b)$ of equation (8.4) is naturally called *a fixed point* of the function (mapping) $f = f(x)$. To find this root, let us specify some initial guess $x^{(0)}$ and then compute the iterates $x^{(1)}, x^{(2)}, \ldots$ with the help of the formula:

$$x^{(p+1)} = f(x^{(p)}), \quad p = 0, 1, 2, \ldots. \tag{8.5}$$

In doing so we assume that every iterate $x^{(p)}$ belongs to the domain of the function $f(x)$, so that the successive computation of $x^{(1)}, x^{(2)}, \ldots, x^{(p)}, \ldots$ is actually possible. Iteration scheme (8.5) is known as the fixed point iteration or Picard iteration.

THEOREM 8.1
Let the function $f(x)$ be continuous on $a < x < b$, let the initial guess $x^{(0)}$ be given, and let the sequence (8.5) converge to some $\hat{x} \in (a,b)$. Then, $\hat{x}$ is a root of equation (8.4), i.e., the following equality holds: $\hat{x} = f(\hat{x})$.

PROOF From formula (8.5) it immediately follows that $\lim\limits_{p\to\infty} x^{(p+1)} = \lim\limits_{p\to\infty} f(x^{(p)})$. At the same time, the continuity of $f(x)$ implies that $\lim\limits_{p\to\infty} f(x^{(p)}) = f(\lim\limits_{p\to\infty} x^{(p)})$. Consequently, $\lim\limits_{p\to\infty} x^{(p+1)} = f(\lim\limits_{p\to\infty} x^{(p)})$, which means $\hat{x} = f(\hat{x})$. $\square$

In practice, convergence of the iteration process (8.5) can be monitored experimentally, in the course of computing the iterates $x^{(p)}$, $p = 0, 1, 2, \ldots$. If it is observed that the difference between the successive iterates is steadily decreasing, then the sequence $x^{(p)}$ is likely to be convergent, and the iteration (8.5) can be terminated at some large number p, yielding a sufficiently accurate approximation to the root $\hat{x}$.

However, the sequence of iterates (8.5) does not always converge. In Figure 8.5, we are schematically showing two possible scenarios of its behavior. For the function $f(x)$ depicted in Figure 8.5(a) the iteration (8.5) converges, whereas for the function $f(x)$ depicted in Figure 8.5(b) it diverges.

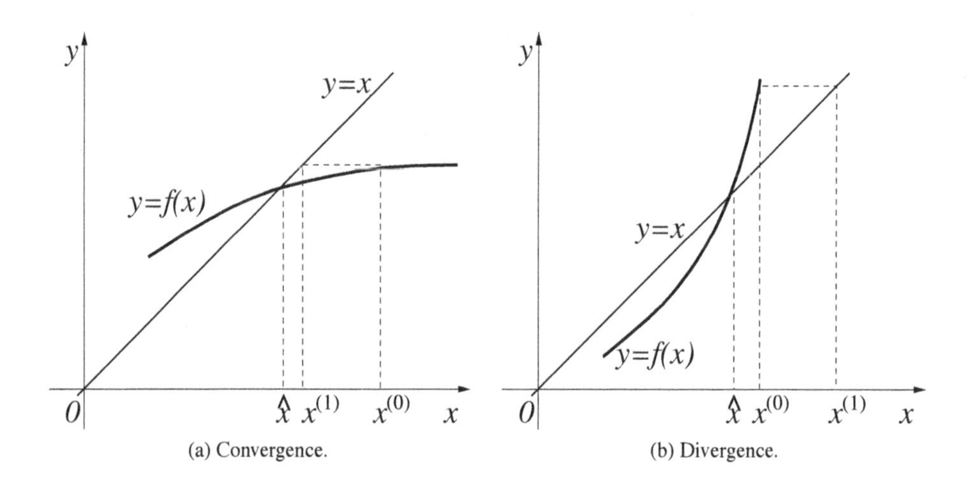

(a) Convergence. (b) Divergence.

FIGURE 8.5: Two possible scenarios of the behavior of iteration (8.5).

We can formulate a sufficient condition that will guarantee convergence of the iteration process (8.5). To do so, let us first introduce the following definition.

DEFINITION 8.1 *The function (mapping) $f = f(x)$ is called a contraction on the interval (open domain) Ω, if for any pair of arguments $x' \in \Omega$ and $x'' \in \Omega$ the following inequality holds:*

$$|f(x') - f(x'')| \leq q|x' - x''|, \tag{8.6}$$

where the constant q, $0 \leq q < 1$ is called the contraction coefficient.

THEOREM 8.2

Let the equation $x = f(x)$ have a solution $\hat{x} \in \Omega$, and let the mapping $f = f(x)$ be a contraction with the coefficient q, $0 \leq q < 1$, on the domain Ω. Then, the solution $\hat{x}$ is unique in Ω. Moreover, one can always take a sufficiently small number $R > 0$, such that if the initial guess $x^{(0)}$ is chosen according to $|\hat{x} - x^{(0)}| < R$, then all terms of the sequence (8.5) are well defined, and the following error estimate holds:

$$|\hat{x} - x^{(p)}| \leq q^p |\hat{x} - x^{(0)}|, \quad p = 0, 1, 2, \ldots. \tag{8.7}$$

In other words, if $|\hat{x}-x^{(0)}|<R$, then iteration (8.5) converges to the solution (fixed point) $\hat{x}$ with the rate of a geometric sequence with the base q.

PROOF Let us first prove uniqueness of the root $\hat{x}$. Assume the opposite, i.e., that there is another root $\hat{x}' \in \Omega$, $\hat{x}'=f(\hat{x}')$. Then, as $f(x)$ is a contraction, see formula (8.6), we can write: $|f(\hat{x}')-f(\hat{x})| \le q|\hat{x}'-\hat{x}|$. According to our assumption, we have both $\hat{x}=f(\hat{x})$ and $\hat{x}'=f(\hat{x}')$. Consequently, $|\hat{x}'-\hat{x}| \le q|\hat{x}'-\hat{x}| < |\hat{x}'-\hat{x}|$. The contradiction we obtained proves the uniqueness.

Next, introduce the set $\Omega(\hat{x},R) \stackrel{\text{def}}{=} \{x \mid |x-\hat{x}|<R\}$. Clearly, as long as $\hat{x}\in\Omega$ we can always chose the parameter $R>0$ sufficiently small so that $\Omega(\hat{x},R)\subset\Omega$. Let $x^{(0)}\in\Omega(\hat{x},R)$. Then,

$$|\hat{x}-x^{(1)}| = |f(\hat{x})-f(x^{(0)})| \le q|\hat{x}-x^{(0)}| \le qR < R.$$

Consequently, the next iterate $x^{(1)}$ also belongs to $\Omega(\hat{x},R)$, and by the same argument we can claim that it is true for all subsequent iterates as well, i.e., $x^{(p)}\in\Omega(\hat{x},R)$, $p=2,3,\ldots$. Regarding estimate (8.7), we can write:

$$\begin{aligned}
|\hat{x}-x^{(p)}| &= |f(\hat{x})-f(x^{(p-1)})| \le q|\hat{x}-x^{(p-1)}| \\
&= q|f(\hat{x})-f(x^{(p-2)})| \le q^2|\hat{x}-x^{(p-2)}| \\
&\cdots\cdots\cdots\cdots\cdots\cdots\cdots\cdots \\
&= q^{p-1}|f(\hat{x})-f(x^{(0)})| \le q^p|\hat{x}-x^{(0)}|.
\end{aligned}$$

This completes the proof. □

REMARK 8.1 Note that every contraction mapping on a complete space (the space on which every fundamental, or Cauchy, sequence converges) has a fixed point, i.e., a solution to the equation $x=f(x)$. This result is known as the contraction mapping principle, see [KF75, Section 8]. In fact, given the completeness of the real numbers (the Cauchy theorem, see [WW62, Secion 2.22]), only a slight modification of the formulation and proof of Theorem 8.2 would have allowed us to claim both existence and uniqueness of the root $\hat{x}$, rather than only its uniqueness. □

Theorem 8.2 does not provide a practical recipe of how to choose the parameter $R>0$ and the interval $\Omega(\hat{x},R)\subset\Omega$ that should contain the initial guess $x^{(0)}$. However, the very existence of such $\Omega(\hat{x},R)$ guaranteed by Theorem 8.2 already suggests that an algorithm for computing the root $\hat{x}$ can be successfully implemented on a computer. Indeed, let us take an arbitrary $x^{(0)}\in\Omega$. If the corresponding sequence (8.5) happens to converge, then it converges to the root $\hat{x}\in\Omega$ according to Theorem 8.1. Otherwise, another initial guess from Ω needs to be tried. If the iteration does not converge either, yet another one should be taken. In doing so, we will eventually generate a collection of initial guesses that would sufficiently "densely" cover Ω. Then, at some stage the next initial guess will inevitably fall into the small neighborhood of the root $\Omega(\hat{x}, R)$ that guarantees convergence of the iteration (8.5).

Let us also point out a sufficient condition for a given function to be a contraction.

THEOREM 8.3
Let the function $f = f(x)$ be differentiable on Ω, i.e., at every point of the interval $a < x < b$, and let its derivative be bounded for all $x \in (a,b)$: $|f'(x)| < q$, where $q = \text{const}$, $0 \leq q < 1$. Then, $f(x)$ is a contraction mapping on Ω with the contraction coefficient q.

PROOF According to the Lagrange theorem (the first mean value theorem, see [WW62, Section 5.41]), for any $x', x'' \in \Omega$ we can write:

$$f(x') - f(x'') = (x' - x'')f'(\xi),$$

where ξ is an intermediate point between x' and x''. Consequently,

$$|f(x') - f(x'')| = |x' - x''||f'(\xi)| \leq q|x' - x''|,$$

which means that $f(x)$ is indeed a contraction in the sense of Definition 8.1. ▯

Example

Consider the chord method of Section 8.1.2:

$$x^{(p+1)} = x^{(p)} - \frac{b-a}{F(b) - F(a)} F(x^{(p)}), \quad p = 0, 1, 2, \ldots.$$

Define the auxiliary function $f(x)$ as follows:

$$f(x) = x - \frac{b-a}{F(b) - F(a)} F(x).$$

Then, according to Theorem 8.2, the fixed point iteration: $x^{(p+1)} = f(x^{(p)})$, $p = 0, 1, 2, \ldots$, will converge provided that the mapping $f = f(x)$ is a contraction. This, in turn, is true if $|f'(x)| \leq q < 1$, see Theorem 8.3, i.e., if

$$\left| 1 - \frac{b-a}{F(b) - F(a)} F'(x) \right| \leq q < 1 \ \text{ for all } \ x \in (a,b).$$

The convergence will be linear, $\kappa = 1$, in the sense of estimate (8.1).

8.2.2 The Case of a System of Equations

Consider a system of nonlinear equations:

$$\boldsymbol{F}(\boldsymbol{x}) = \boldsymbol{0}, \tag{8.8}$$

where $\boldsymbol{F}(\boldsymbol{x})$ is an n-dimensional vector function of the argument $\boldsymbol{x} = \begin{bmatrix} x_1 \\ \dots \\ x_n \end{bmatrix}$ defined on some domain Ω of the space $\mathbb{R}^n$. Hereafter, we will assume that this system has a solution $\hat{\boldsymbol{x}}$ on Ω. To solve system (8.8) by iterations, we first recast it in an equivalent form:

$$\boldsymbol{x} = \boldsymbol{f}(\boldsymbol{x}), \tag{8.9}$$

where $\boldsymbol{f}(x)$ is another vector function of dimension n. There are many different ways of transitioning from system (8.8) to system (8.9). For example, we can define:

$$\boldsymbol{f}(\boldsymbol{x}) = \boldsymbol{x} - \boldsymbol{\alpha}(\boldsymbol{x})\boldsymbol{F}(\boldsymbol{x}),$$

where $\boldsymbol{\alpha}(\boldsymbol{x})$ is an $n \times n$ matrix function with non-zero determinant everywhere on Ω.

Next, we specify $\boldsymbol{x}^{(0)} \in \Omega$ and build the sequence of iterates according to the formula:

$$\boldsymbol{x}^{(p+1)} = \boldsymbol{f}(\boldsymbol{x}^{(p)}), \quad p = 0, 1, 2, \dots. \tag{8.10}$$

The space $\mathbb{R}^n$ will hereafter be assumed normed. Then, all the key formulations and results of Section 8.2.1 (Theorem 8.1, Definition 8.1, and Theorem 8.2) easily translate from the case of a scalar nonlinear equation to the case of a nonlinear system, provided that the absolute values of the real numbers involved are replaced by the norms of the corresponding vectors.

A sufficient condition for a given mapping to be a contraction, which was formulated by Theorem 8.3 in the case of a scalar function, needs to be refined for the case of a vector function. To do so, we first introduce the notion of a convex domain.

DEFINITION 8.2 *The domain $\Omega \subset \mathbb{R}^n$ is called convex, if along with any pair of points $\boldsymbol{a} \in \Omega$ and $\boldsymbol{b} \in \Omega$ the domain also contains the entire interval of the straight line between these two points:*

$$\boldsymbol{x} = t\boldsymbol{a} + (1-t)\boldsymbol{b} \in \Omega \quad \text{for any } 0 \le t \le 1.$$

THEOREM 8.4
Let the vector function

$$\boldsymbol{f}(\boldsymbol{x}) = \begin{bmatrix} f_1(x_1, \dots, x_n) \\ \vdots \\ f_n(x_1, \dots, x_n) \end{bmatrix}$$

be defined on a convex domain $\Omega \subset \mathbb{R}^n$ of the independent variable $\boldsymbol{x}$. Let the components $f_1, \dots, f_n$ of this vector function have uniformly continuous first partial derivatives everywhere on Ω. Consider the Jacobi matrix of the mapping $\boldsymbol{f} = \boldsymbol{f}(\boldsymbol{x})$:

$$\frac{\partial \boldsymbol{f}}{\partial \boldsymbol{x}} \stackrel{\text{def}}{=} \begin{bmatrix} \frac{\partial f_1}{\partial x_1} & \cdots & \frac{\partial f_1}{\partial x_n} \\ \dots & \dots & \dots \\ \frac{\partial f_n}{\partial x_1} & \cdots & \frac{\partial f_n}{\partial x_n} \end{bmatrix}.$$

If the norm of this matrix is bounded by some number q, $0 \leq q < 1$, uniformly for all $\boldsymbol{x} \in \Omega$, then the mapping $\boldsymbol{f} = \boldsymbol{f}(\boldsymbol{x})$ is a contraction on Ω, i.e., the following inequality holds:

$$\|\boldsymbol{f}(\boldsymbol{x}') - \boldsymbol{f}(\boldsymbol{x}'')\| \leq q\|\boldsymbol{x}' - \boldsymbol{x}''\|,$$

where $\boldsymbol{x}'$ and $\boldsymbol{x}''$ are two arbitrary points from Ω.

The proof of this theorem is the subject of Exercise 1.

To actually use the fixed point iterations (8.10) for computing an approximation to the solution $\hat{\boldsymbol{x}}$ of the equation $\boldsymbol{F}(\boldsymbol{x}) = \boldsymbol{0}$, one needs to exploit the flexibility that exists in choosing the auxiliary mapping $\boldsymbol{f}(\boldsymbol{x})$ so as to make it a contraction with the smallest possible coefficient q. This will guarantee the fastest convergence.

REMARK 8.2 In Section 6.1, we solved the linear system $\boldsymbol{Ax} = \boldsymbol{b}$ by first reducing it to an equivalent form $\boldsymbol{x} = \boldsymbol{Bx} + \boldsymbol{\varphi}$ and then employing the iteration $\boldsymbol{x}^{(p+1)} = \boldsymbol{Bx}^{(p)} + \boldsymbol{\varphi}$, $p = 0, 1, 2, \ldots$. In the context of this section, we can define $\boldsymbol{f}(\boldsymbol{x}) = \boldsymbol{Bx} + \boldsymbol{\varphi}$, in which case the Jacobi matrix $\frac{\partial \boldsymbol{f}}{\partial \boldsymbol{x}} = \boldsymbol{B}$. Then, according to Theorem 8.4, if $\|\boldsymbol{B}\| = q < 1$ then the mapping $\boldsymbol{f} = \boldsymbol{f}(\boldsymbol{x})$ is a contraction. This, in turn, guarantees convergence of the fixed point iteration (8.10). As such, we see that Theorem 6.1 (see page 175) that guarantees convergence of the linear iteration provided that $\|\boldsymbol{B}\| = q < 1$, can be considered a direct implication of the results of this section in the case of linear systems. □

Exercises

1.* Prove Theorem 8.4.

Hint. Represent the increment of the function $\boldsymbol{f}(\boldsymbol{x})$ as the integral of the derivative in the direction $\boldsymbol{x}' - \boldsymbol{x}''$.

8.3 Newton's Method

8.3.1 Newton's Linearization for One Scalar Equation

As shown in Section 8.1.4, Newton's method for finding the solution of the nonlinear equation $F(x) = 0$ is based on linearization of the function $F(x)$. Let $x^{(0)}$ be the initial guess, and let $x^{(p)}$ be the current iterate. For any given $p = 0, 1, 2, \ldots$ we can write the following approximate linear formula:

$$F(x) \approx F(x^{(p)}) + F'(x^{(p)})(x - x^{(p)}),$$

which is obtained by truncating the Taylor expansion after its second term. Then, instead of the original equation $F(x) = 0$, let us use the linear equation:

$$F(x^{(p)}) + F'(x^{(p)})(x - x^{(p)}) = 0.$$

The solution of this equation yields the next iterate $x^{(p+1)}$:

$$x^{(p+1)} = x^{(p)} - \frac{F(x^{(p)})}{F'(x^{(p)})}, \quad p = 0, 1, 2, \ldots. \tag{8.11}$$

Newton's method admits a straightforward geometric interpretation, see Figure 8.4 on page 236. The curvilinear graph of the function $F(x)$ is replaced by its tangent at the point $(x^{(p)}, F(x^{(p)})$. The next iterate $x^{(p+1)}$ is obtained as the intersection point of this tangent with the abscissa axis.

Newton's method can also be put into the framework of the Picard iterations of Section 8.2.1. Let us assume that $F'(x) \neq 0$ on some neighborhood of the desired root $\hat{x}$. Then, instead of the original equation $F(x) = 0$ we can consider its equivalent:

$$x = f(x),$$
$$f(x) = x - \left[F'(x)\right]^{-1} F(x).$$

The corresponding fixed point iteration will coincide with Newton's iteration (8.11). Besides, by direct differentiation we can obtain:

$$f'(x) = 1 - \frac{[F'(x)]^2 - F''(x)F(x)}{[F'(x)]^2} = \frac{F''(x)F(x)}{[F'(x)]^2},$$

which means that at the root $\hat{x}$, $F(\hat{x}) = 0$, we also have $f'(\hat{x}) = 0$. Let us additionally assume that the second derivative $F''(x)$ is continuous. Consequently, the function $f'(x)$ is continuous as well, and for any $0 < q < 1$ there is always a sufficiently small neighborhood of the root $\hat{x}$, on which $|f'(x)| \leq q$. By virtue of Theorem 8.3, the function $f = f(x)$ is then a contraction mapping with the contraction coefficient q, and the smaller the neighborhood, the stronger the contraction (i.e., the smaller the q). According to Theorem 8.2, this implies that the iterates of Newton's method (8.11) will converge to the root $\hat{x}$ provided that the initial guess $x^{(0)}$ is chosen sufficiently close to $\hat{x}$. Moreover, the rate of convergence for the error $|\hat{x} - x^{(p)}|$ will be faster than that of any geometric sequence q^p with the fixed base $0 < q < 1$.

THEOREM 8.5

Let $\hat{x}$ be a root of the nonlinear equation $F(x) = 0$. Let $F'(\hat{x}) \neq 0$, and let the second derivative $F''(x)$ be continuous and bounded on some neighborhood of this root. Then, Newton's iteration (8.11) will converge to $\hat{x}$ quadratically in the sense of estimate (8.1), i.e.,

$$|\hat{x} - x^{(p+1)}| \leq \text{const} \cdot |\hat{x} - x^{(p)}|^2,$$

provided that the initial guess $x^{(0)}$ is chosen sufficiently close to $\hat{x}$.

PROOF As always, we denote the error of the iterate $x^{(p)}$ by $\varepsilon^{(p)} = \hat{x} - x^{(p)}$. Then, we can write:

$$0 = F(\hat{x}) = F(x^{(p)} + \varepsilon^{(p)}) = F(x^{(p)}) + F'(x^{(p)})\varepsilon^{(p)} + \frac{1}{2}F''(\xi)(\varepsilon^{(p)})^2,$$

where ξ is some intermediate point between $\hat{x}$ and $x^{(p)}$. Consequently,

$$\varepsilon^{(p)} = \hat{x} - x^{(p)} = -\frac{1}{F'(x^{(p)})}\left[F(x^{(p)}) + \frac{1}{2}F''(\xi)(\varepsilon^{(p)})^2\right].$$

On the other hand, according to the definition of Newton's method, see formula (8.11), we have:

$$x^{(p+1)} - x^{(p)} = -\frac{F(x^{(p)})}{F'(x^{(p)})},$$

which, after the substitution into the previous formula, yields:

$$\varepsilon^{(p+1)} = \hat{x} - x^{(p+1)} = -\frac{1}{2}\frac{F''(\xi)}{F'(x^{(p)})}(\varepsilon^{(p)})^2.$$

By the hypothesis of the theorem, on some neighborhood of the root $\hat{x}$ we have:

$$\frac{1}{F'(x)} \leq C_1 \quad \text{and} \quad |F''(x)| \leq C_2,$$

where C_1 and C_2 are two constants. Therefore,

$$|\varepsilon^{(p+1)}| = |\hat{x} - x^{(p+1)}| \leq \frac{1}{2}C_1C_2|\varepsilon^{(p)}|^2 = \frac{1}{2}C_1C_2|\hat{x} - x^{(p)}|^2,$$

which implies quadratic convergence. ∎

8.3.2 Newton's Linearization for Systems

Similarly to the scalar case, Newton's method can also be applied to solving systems of nonlinear equations $\boldsymbol{F}(\boldsymbol{x}) = \boldsymbol{0}$, where $\boldsymbol{F}$ is a mapping, $\boldsymbol{F}: \mathbb{R}^n \longmapsto \mathbb{R}^n$. Hereafter, we will assume that $\boldsymbol{F}(\boldsymbol{x})$ is continuously differentiable on some domain $\Omega \subseteq \mathbb{R}^n$ that contains the desired solution $\hat{\boldsymbol{x}}$: $\boldsymbol{F}(\hat{\boldsymbol{x}}) = \boldsymbol{0}$. Then, for any $\boldsymbol{x}^{(p)}$, $p = 0, 1, 2, \ldots$, the Taylor-based linearization yields:

$$\boldsymbol{F}(\boldsymbol{x}) \approx \boldsymbol{F}(\boldsymbol{x}^{(p)}) + \frac{\partial \boldsymbol{F}(\boldsymbol{x}^{(p)})}{\partial \boldsymbol{x}}(\boldsymbol{x} - \boldsymbol{x}^{(p)}),$$

where $\frac{\partial \boldsymbol{F}}{\partial \boldsymbol{x}}$ is the Jacobi matrix, or Jacobian, of the mapping $\boldsymbol{F}$:

$$\frac{\partial \boldsymbol{F}}{\partial \boldsymbol{x}} \stackrel{\text{def}}{=} \begin{bmatrix} \frac{\partial F_1}{\partial x_1} & \cdots & \frac{\partial F_1}{\partial x_n} \\ \cdots & \cdots & \cdots \\ \frac{\partial F_n}{\partial x_1} & \cdots & \frac{\partial F_n}{\partial x_n} \end{bmatrix} \equiv \boldsymbol{J_F}.$$

Accordingly, instead of the original nonlinear system $\boldsymbol{F}(\boldsymbol{x}) = \boldsymbol{0}$ we will write the linear system:

$$\boldsymbol{F}(\boldsymbol{x}^{(p)}) + \boldsymbol{J_F}(\boldsymbol{x}^{(p)})(\boldsymbol{x} - \boldsymbol{x}^{(p)}) = \boldsymbol{0},$$

and obtain the next iterate $\boldsymbol{x}^{(p+1)}$ as its solution:

$$\boldsymbol{x}^{(p+1)} = \boldsymbol{x}^{(p)} - \boldsymbol{J}_{\boldsymbol{F}}^{-1}(\boldsymbol{x}^{(p)})\boldsymbol{F}(\boldsymbol{x}^{(p)}), \quad p = 0, 1, 2, \ldots. \tag{8.12}$$

Clearly, to be able to implement Newton's method we must require that the Jacobian be nonsingular at the solution, i.e., that the inverse $\boldsymbol{J}_{\boldsymbol{F}}^{-1}(\hat{\boldsymbol{x}})$ exist. Then, by continuity, the Jacobian will be non-singular on some neighborhood of $\hat{\boldsymbol{x}}$, and the iteration (8.12) can be carried through. The requirement of invertibility of $\boldsymbol{J_F}(\hat{\boldsymbol{x}})$ is similar to the condition $F'(\hat{x}) \neq 0$ that we introduced in the scalar case (Section 8.3.1).

Also as in the scalar case, Newton's method for systems can be interpreted as a fixed-point iteration, see Section 8.2.2. To do so, we recast the original equation $\boldsymbol{F}(\boldsymbol{x}) = \boldsymbol{0}$ in the equivalent form:

$$\boldsymbol{x} = \boldsymbol{f}(\boldsymbol{x}),$$
$$\boldsymbol{f}(\boldsymbol{x}) = \boldsymbol{x} - \boldsymbol{J}_{\boldsymbol{F}}^{-1}(\boldsymbol{x})\boldsymbol{F}(\boldsymbol{x}),$$

and then iterate: $\boldsymbol{x}^{(p+1)} = \boldsymbol{f}(\boldsymbol{x}^{(p)})$, $p = 0, 1, 2, \ldots$, starting from some initial guess $\boldsymbol{x}^{(0)}$. Recall, in Section 8.3.1 we obtained that in the scalar case $f'(\hat{x}) = 0$. Likewise, it is possible to show that $\frac{\partial \boldsymbol{f}}{\partial \boldsymbol{x}}$ is a zero matrix for $\boldsymbol{x} = \hat{\boldsymbol{x}}$. By continuity, we can then claim that $\|\frac{\partial \boldsymbol{f}(\boldsymbol{x})}{\partial \boldsymbol{x}}\|$ is small near the solution $\hat{\boldsymbol{x}}$. Therefore, according to Theorem 8.4 the mapping $\boldsymbol{f} = \boldsymbol{f}(\boldsymbol{x})$ will be a contraction on some convex neighborhood of $\hat{\boldsymbol{x}}$. (Clearly, uniform continuity of the derivatives can always be guaranteed on a smaller compact set.) Moreover, the closer we come to $\hat{\boldsymbol{x}}$ the smaller $\|\frac{\partial \boldsymbol{f}(\boldsymbol{x})}{\partial \boldsymbol{x}}\|$ will be, and as such, Newton's method (8.12) can be considered a fixed-point iteration with the contraction coefficient q that decreases as the iterates $\boldsymbol{x}^{(p)}$ approach the solution $\hat{\boldsymbol{x}}$. As in the scalar case, this indicates that the method will converge faster than linear. In fact, the convergence of Newton's method is quadratic:

$$\|\boldsymbol{x}^{(p+1)} - \hat{\boldsymbol{x}}\| \le \text{const} \cdot \|\boldsymbol{x}^{(p)} - \hat{\boldsymbol{x}}\|^2. \tag{8.13}$$

The proof of this result is very similar to that of Theorem 8.5 and we omit it here. We only mention that the assumptions needed to establish estimate (8.13) are continuous differentiability of $\boldsymbol{F}(\boldsymbol{x})$ on a convex domain $\Omega \subseteq \mathbb{R}^n$ that contains the solution, $\hat{\boldsymbol{x}} \in \Omega$, existence of $\boldsymbol{J}_{\boldsymbol{F}}^{-1}(\hat{\boldsymbol{x}})$, and Lipshitz-continuity of the Jacobian:

$$\|\boldsymbol{J_F}(\boldsymbol{x}') - \boldsymbol{J_F}(\boldsymbol{x}'')\| \le \text{const} \cdot \|\boldsymbol{x}' - \boldsymbol{x}''\|, \quad \boldsymbol{x}', \boldsymbol{x}'' \in \Omega.$$

The latter assumption is somewhat weaker than boundedness of the second derivative $F''(x)$ that we required in the scalar case. It is also important to note that convergence of Newton's method is only guaranteed if the initial guess $\boldsymbol{x}^{(0)}$ is chosen sufficiently close to the solution $\hat{\boldsymbol{x}}$.

To actually implement Newton's iteration (8.12), we need to solve an $n \times n$ linear system with the matrix $\boldsymbol{J_F}(\boldsymbol{x}^{(p)})$ for every $p = 0, 1, 2, \ldots$. To do so, we can employ the Gaussian elimination or some other method (iterative). However, for large dimensions, this may prove computationally expensive, especially if the Jacobian $\boldsymbol{J_F}(\boldsymbol{x}^{(p)})$ is ill-conditioned. To reduce the implementation cost of the original Newton's method, several modified versions of the algorithm have been proposed.

8.3.3 Modified Newton's Methods

Instead of inverting the Jacobian $\boldsymbol{J_F}(\boldsymbol{x}^{(p)})$ on every iteration, it may only be inverted once per several iterations. In the scalar context this means that we do not compute the new tangent to the graph of $F(x)$ for every $x^{(p)}$, $p = 0, 1, 2, \ldots$, as in the original method, see Figure 8.4, but rather "freeze" the same slope of the straight line for several iterations, and after that update. In the context of systems, this approach may considerably reduce the computational cost. Its downside, though, is a slower convergence of the method, which is the case for both scalar equations and systems.

Alternatively, when the linear system:

$$\boldsymbol{F}(\boldsymbol{x}^{(p)}) + \boldsymbol{J_F}(\boldsymbol{x}^{(p)})(\boldsymbol{x} - \boldsymbol{x}^{(p)}) = \boldsymbol{0}$$

is solved by iterations, instead of iterating until, say, a predetermined level of the residual reduction is reached, we can only run a fixed number of linear iterations chosen ahead of time. While reducing the cost, this, of course, only yields a fairly "crude" approximation of the solution to the linear system, and also slows down the overall convergence. Depending on what particular linear iteration is used (Chapter 6), there are Newton-Jacobi, Newton-SOR, Newton-Krylov, and other methods of this type.

In addition to the costs associated with inverting the Jacobian $\boldsymbol{J_F}(\boldsymbol{x}^{(p)})$, the cost of computing the matrix $\boldsymbol{J_F}(\boldsymbol{x}^{(p)})$ itself for every $p = 0, 1, 2, \ldots$ may be quite noticeable. To reduce it, instead of evaluating the Jacobian exactly, one often evaluates it approximately using the finite differences. Either forward differencing or central differencing can be employed; the analysis of the corresponding errors can be found in Chapter 9. To keep the quadratic convergence of the method when the Jacobian is evaluated with the help of the finite differences, one needs to satisfy certain constraints for the step size, otherwise the convergence rate drops to linear.

Finally, there is a family of the so-called quasi-Newton methods that can be built for both exact and finite-difference Jacobians [cf. formula (8.12)]:

$$\boldsymbol{x}^{(p+1)} = \boldsymbol{x}^{(p)} - \gamma_p \boldsymbol{J}_{\boldsymbol{F}}^{-1}(\boldsymbol{x}^{(p)})\boldsymbol{F}(\boldsymbol{x}^{(p)}), \quad p = 0, 1, 2, \ldots.$$

The quantities γ_p are known as the damping parameters.

Besides Newton's method, other methods are available for the solution of nonlinear systems. There is, for example, a multidimensional version of the secant method called the Broyden method. We do not discuss it in this book. We rather refer the reader to the specialized literature for further detail on the numerical solution of the nonlinear scalar equations and systems, see, e.g., the monographs [OR00, Kel95, Kel03].

Exercises

1.* Prove quadratic convergence of Newton's method for systems, i.e., establish estimate (8.13) given the conditions right after equation (8.13) on page 245 and also assuming that the initial guess $\boldsymbol{x}^{(0)}$ is sufficiently close to the solution $\hat{\boldsymbol{x}}$.

   Hint. Increment of $F(x)$ is to be obtained by integrating the corresponding directional derivative.

2. Build an algorithm for computing $\sqrt{5}$ with a given precision. Interpret $\sqrt{5}$ as a solution to the equation $x^2 - 5 = 0$ and employ Newton's method. Show that the iterations will converge for any initial guess $x^{(0)} > 0$.

3. Let $\phi = \phi(x)$ and $\psi = \psi(x)$ be two functions with bounded second derivatives. For solving the equation $F(x) \equiv \phi(x) - \psi(x) = 0$ one can use Newton's method.

   a) Let the graphs of the functions $\phi = \phi(x)$ and $\psi = \psi(x)$ be plotted on the Cartesian plane (x, y), and let their intersection point [the root of $F(x) = 0$] have the abscissa $x = \hat{x}$. Assume that the Newton iterate $x^{(p)}$ is already computed and provide a geometric interpretation of how one obtains the next iterate $x^{(p+1)}$. Use the following form of Newton's linearization:

   $$\phi(x^{(p)}) - \psi(x^{(p)}) + (\phi'(x^{(p)}) - \psi'(x^{(p)}))(x^{(p+1)} - x^{(p)}) = 0.$$

   b) Assume, for definiteness, that $\phi(x) > \psi(x)$ for $x > \hat{x}$, and $\phi'(\hat{x}) - \psi'(\hat{x}) > 0$. Let also the graph of $\phi(x)$ be convex, i.e., $\phi''(x) > 0$, and the graph of $\psi(x)$ be concave, i.e., $\psi''(x) < 0$. Show that Newton's method will converge for any $x^{(0)} > \hat{x}$.

4. Use Newton's method to compute real solutions to the following systems of nonlinear equations. Obtain the results with five significant digits:

   a) $\sin x - y = 1.30$; $\cos y - x = -0.84$.

   b) $x^2 + 4y^2 = 1$; $x^4 + y^4 = 0.5$.

5. A nonlinear boundary value problem:

   $$\frac{d^2u}{dx^2} - u^3 = x^2, \quad x \in [0,1],$$
   $$u(0) = 1, \quad u(1) = 3,$$

   with the unknown function $u = u(x)$ is approximated on the uniform grid $x_j = j \cdot h$, $j = 0, 1, \ldots, N$, $h = \frac{1}{N}$, using the following second-order central-difference scheme:

   $$\frac{u_{j+1} - 2u_j + u_{j-1}}{h^2} - u_j^3 = (jh)^2, \quad j = 1, 2, \ldots, N-1,$$
   $$u_0 = 1, \quad u_N = 3,$$

   where u_j, $j = 0, 1, 2, \ldots, N$, is the unknown discrete solution on the grid.
   Solve the foregoing discrete system of nonlinear equations with respect to u_j, $j = 0, 1, 2, \ldots, N$, using Newton's method. Introduce the following initial guess: $u_j^{(0)} = 1 + 2(jh)$, $j = 0, 1, 2, \ldots, N$. This grid function obviously satisfies the boundary conditions:

$u_0^{(0)} = 1$, $u_N^{(0)} = 3$. For every iterate $\left\{u_j^{(p)}\right\}$, $p \geq 0$, the next iterate $\left\{u_j^{(p+1)}\right\}$ is defined as

$$u_j^{(p+1)} = u_j^{(p)} + \delta u_j^{(p)}, \quad j = 0, 1, 2, \ldots, N,$$

where the increment $\left\{\delta u_j^{(p)}\right\}$ should satisfy a new system of equations obtained by Newton's linearization.

a) Show that this linear system of equations with respect to $\delta u_j^{(p)}$, $j = 0, 1, 2, \ldots, N$, can be written as follows:

$$\frac{\delta u_{j+1}^{(p)} - 2\delta u_j^{(p)} + \delta u_{j-1}^{(p)}}{h^2} - 3\left(u_j^{(p)}\right)^2 \delta u_j^{(p)} =$$

$$= (jh)^2 + \left(u_j^{(p)}\right)^3 - \frac{u_{j+1}^{(p)} - 2u_j^{(p)} + u_{j-1}^{(p)}}{h^2},$$

$$\text{for } \; j = 1, 2, \ldots, N-1,$$

$$\delta u_0^{(p)} = 0, \quad \delta u_N^{(p)} = 0.$$

Prove that it can be solved by the tri-diagonal elimination of Section 5.4.2.

b) Implement Newton's iteration on the computer. For every $p \geq 0$, use the tri-diagonal elimination algorithm (Section 5.4.2) to solve the linear system with respect to $\delta u_j^{(p)}$, $j = 1, 2, \ldots, N-1$. Consider a sequence of subsequently more fine grids: $N = 32, 64, \ldots$. On every grid, iterate till $\max_{1 \leq j \leq N-1} \left|u_j^{(p+1)} - u_j^{(p)}\right| < 0.25 \cdot 10^{-5}$. Denote the resulting approximate solution $u_j^{(p_N)}$, $j = 0, 1, 2, \ldots, N$. Then, go to the next grid with $2N$ cells and obtain $u_j^{(p_{2N})}$, $j = 0, 1, 2, \ldots, 2N$. Keep refining the grid till $\max_{1 \leq j \leq N-1} \left|u_{2j}^{(p_{2N})} - u_j^{(p_N)}\right| < 10^{-5}$. Then stop, plot the solution on the most fine grid, and also compare the resulting numbers p_N for all the grids that have been used.

Part III

The Method of Finite Differences for the Numerical Solution of Differential Equations

Perhaps, the most commonly used approach to the numerical solution of both ordinary and partial differential equations is based on the method of finite differences. This method has a large number of modifications and "flavors" tuned for different specific problems and/or classes of problems. In all versions of this method, one first introduces a special collection of discrete nodes in the domain of the solution. This set of nodes is called *the finite-difference grid*; and the solution of interest is subsequently sought on the grid. For the unknown grid function one constructs a system of scalar equations that it is supposed to satisfy. The solution of this system is interpreted as an approximate discrete table of values for the original continuous solution.

A most straightforward approach to obtaining the foregoing system of scalar equations, which is called *the finite-difference scheme,* is based on approximately replacing the derivatives contained in the original differential equation, as well as in the initial and/or boundary conditions, by the appropriate difference quotients. This is where the method of finite differences draws its name from.

The literature on numerical solution of differential equations, and in particular, on the method of finite differences, is very broad. Here, we only provide a few references to the books that discuss ordinary differential equations [HNW93, HW96, FM87, But03] and to the books that discuss partial differential equations [RM67, GR87, MM05, Str04, GKO95]. Other sources of reference, including those that cover more narrow subjects, are quoted later on in the text.

Chapter 9

Numerical Solution of Ordinary Differential Equations

Hereafter, we will assume that a given ordinary differential equation, or a system of ordinary differential equations, is to be solved on the interval $a \leq x \leq b$ of the independent variable x, subject to the additional initial or boundary conditions.

9.1 Examples of Finite-Difference Schemes. Convergence

For both initial and boundary value problems to be analyzed in this chapter, we will be using the same generic notation:

$$\boldsymbol{L}u = f. \tag{9.1}$$

Several simple examples of such problems are provided below.

An initial value (Cauchy) problem for the first order ordinary differential equation:

$$\frac{du}{dx} + \frac{x}{1+u^2} = \cos x, \quad 0 \leq x \leq 1, \quad u(0) = 3. \tag{9.2}$$

An initial value problem for the second order ordinary differential equation:

$$\begin{gathered} \frac{d^2u}{dx^2} = (1+x^2)u + \sqrt{x+1}, \quad 0 \leq x \leq 1, \\ u(0) = 2, \quad \left.\frac{du}{dx}\right|_{x=0} = 1. \end{gathered} \tag{9.3}$$

A boundary value problem for the second order ordinary differential equation, with the boundary conditions given at the endpoints $x = 0$ and $x = 1$ of the interval:

$$\begin{gathered} \frac{d^2u}{dx^2} = (1+x^2)u + \sqrt{x+1}, \quad 0 \leq x \leq 1, \\ u(0) = 2, \quad u(1) = 1. \end{gathered} \tag{9.4}$$

An initial value problem for the system of two first order ordinary differential equations:

$$\begin{aligned} \frac{dv}{dx} + xvw &= x^2 - 3x + 1, \quad 0 \le x \le 1, \quad v(0) = 1, \\ \frac{dw}{dx} + \frac{v+w}{1+x^2} &= \cos^2 x, \quad 0 \le x \le 1, \quad w(0) = -3. \end{aligned} \tag{9.5}$$

Solution of this system is a two-dimensional vector function $u = [v, w]^T$ with the components that satisfy the differential equations of (9.5).

Note that in all these examples — (9.2) through (9.5) — it is only for definiteness and not for any other reason that we are considering each problem on the specific interval $\bar{D} = \{0 \le x \le 1\}$ rather than on a general interval $\{a \le x \le b\}$.

9.1.1 Examples of Difference Schemes

We will always assume that the solution $u = u(x)$ of problem (9.1) does exist on the interval $0 \le x \le 1$. To compute this solution using the method of finite differences, one first needs to specify a finite set of (distinct) points on the interval $\bar{D}$ that will be called the grid and denoted D_h. Subsequently, instead of the continuous solution $u(x)$ of problem (9.1), a discrete table $[u]_h$ of its values on the grid D_h is to be considered as the unknown function of the problem. It is assumed that the grid D_h depends on the positive parameter h that can be arbitrarily small. This parameter is referred to as the grid size, so that the smaller the h the finer the grid. For example, one can consider $h = 1/N$, where N is a positive integer, and compose the grid D_h of the $N+1$ equally spaced points (also called nodes): $x_0 = 0$, $x_1 = h$, ..., $x_n = nh$, ..., $x_N = 1$, with the distance between each two neighboring points being equal to h. Finite-difference grids of this type are called uniform. The value of the unknown grid function $[u]_h$ at the node x_n of D_h will be denoted by u_n.

Consider, for example, problem (9.2). To approximately compute the trace $[u]_h \stackrel{\text{def}}{=} \{u(x_n)\}$ of its solution $u(x)$ on the grid D_h, one can use the system of equations:

$$\frac{u_{n+1} - u_n}{h} + \frac{x_n}{1 + u_n^2} = \cos x_n, \quad n = 0, 1, \dots, N-1, \quad u_0 = 3, \tag{9.6}$$

obtained via replacing the values of the derivative du/dx at the nodes x_n, $n = 0, 1, \dots, N-1$, by difference quotients according to the approximate formula:

$$\frac{du}{dx} \approx \frac{u(x+h) - u(x)}{h}.$$

Solution $u^{(h)} = \left\{u_0^{(h)}, u_1^{(h)}, \dots, u_N^{(h)}\right\}$ of system (9.6) is defined on the same grid D_h as the unknown grid function $[u]_h$ is. The values $u_1^{(h)}, u_2^{(h)}, \dots, u_N^{(h)}$ of this solution at the nodes $x_1, x_2, \dots, x_N$ are computed consecutively for $n = 0, 1, \dots, N-1$ from system (9.6), while $u_0^{(h)} = 3$ is determined from the initial condition. For simplicity, we are omitting the superscript "(h)" for all $u_n^{(h)}$ in equations (9.6). Henceforth, we will omit this superscript on similar occasions provided that it may not cause any confusion. The system of equations (9.6) provides an example of a finite-difference scheme (or difference scheme, or simply scheme) for the initial value problem (9.2).

In the case of problem (9.4), to obtain a grid function $u^{(h)}$ that would approximate the desired exact table of values $[u]_h$, one can employ the following scheme:

$$\begin{gathered} \frac{u_{n+1}-2u_n+u_{n-1}}{h^2}-(1+x_n^2)u_n=\sqrt{x_n+1}, \quad n=1,\dots,N-1, \\ u_0=2, \quad u_N=1. \end{gathered} \tag{9.7}$$

This scheme arises when the values of the second derivative d^2u/dx^2 from the differential equation of (9.4) are replaced at the nodes $x_1,x_2,\dots,x_{N-1}$ of the grid D_h according to the approximate formula:

$$\frac{d^2u}{dx^2}\approx\frac{u(x+h)-2u(x)+u(x-h)}{h^2}.$$

To actually compute the solution $u^{(h)}$ of problem (9.7), one can use the algorithm of tri-diagonal elimination described in Section 5.4.2 of Chapter 5.

Let us also write down a scheme that could be used for approximately computing the solution of problem (9.5):

$$\begin{aligned} &\frac{v_{n+1}-v_n}{h}+x_nv_nw_n=x_n^2-3x_n+1, && n=0,1,\dots,N-1, && v_0=1, \\ &\frac{w_{n+1}-w_n}{h}+\frac{v_n+w_n}{1+x_n^2}=\cos^2 x_n, && n=0,1,\dots,N-1, && w_0=-3. \end{aligned} \tag{9.8}$$

Here $u_0^{(h)}=\begin{bmatrix}v_0\\w_0\end{bmatrix}=\begin{bmatrix}1\\-3\end{bmatrix}$ is given. When $n=0$, one can find $u_1^{(h)}=\begin{bmatrix}v_1\\w_1\end{bmatrix}$ from equations (9.8). In general, if $u_n^{(h)}=\begin{bmatrix}v_n\\w_n\end{bmatrix}$ is known, one can obviously compute the discrete solution at the next grid node: $u_{n+1}^{(h)}=\begin{bmatrix}v_{n+1}\\w_{n+1}\end{bmatrix}$. In doing so, the solution can be obtained on the entire grid D_h consecutively, one node after another. This numerical procedure is known as *marching*.

In the previous examples the grid D_h was taken uniform. This, however, is not a necessity. Instead of having $N+1$ equally spaced nodes with $h=1/N$ on the interval $[0,1]$, we could have chosen D_h as follows:

$$x_0=0, \quad x_1=x_0+h_0, \quad x_2=x_1+h_1, \ \dots, \ x_{n+1}=x_n+h_n, \ \dots, \ x_N=1, \tag{9.9}$$

where the grid sizes $h_n>0$, $n=0,1,\dots,N-1$, do not have to be equal. We do require, though, that $\max_n h_n \longrightarrow 0$ when $N\longrightarrow\infty$. The grids of type (9.9) are called nonuniform. By appropriately choosing the node locations of a nonuniform grid for a given fixed N, one can make the desired table $[u]_h$ more detailed in those areas where the solution $u(x)$ undergoes stronger variations. Those areas can sometimes be identified ahead of time from the physics considerations, or from the preliminary low-accuracy computations. Besides, the information about the behavior of $u(x)$ also becomes available in the course of computation. In particular, the rate of change of $u(x)$ is readily obtained when the solution is marched: $u_1, u_2, \ldots, u_n, \ldots$. This information can be taken into account every time when choosing the location of the next grid node x_{n+1}. Altogether, the strategies of building special nonuniform grids that would accommodate particular features of the computed solution are known as *grid adaptation.*

We have provided several examples that illustrate the important concept of the grid and that of the grid function we are seeking — the trace, or projection, $[u]_h$ of the continuous solution $u(x)$ onto the grid. More examples will be given as we introduce and study properties of finite-difference schemes.

Note that we are interested in computing the grid function $u^{(h)}$ because when the grid is refined, i.e., when $h \longrightarrow 0$, we expect it to furnish an increasingly accurate table of values for the unknown solution $u(x)$ and as such, provide its more detailed description altogether. Once the discrete table of values is known, we could use interpolation and reconstruct the continuous solution $u(x)$ everywhere on the domain D with the accuracy that would improve as $h \longrightarrow 0$. It is clear that for a fixed dimension of the grid D_h, i.e., fixed number of nodes $N+1$, the actual reconstruction accuracy will depend on the location of those nodes, as well as on the additional information available about the continuous solution, such as estimates for its derivatives.

One can find a considerably more detailed discussion of the reconstruction-related issues in Chapters 2 and 3 devoted to the interpolation theory. In this chapter, we will rather restrict ourselves by the foregoing brief comments only, and will focus on the computation of the table $u^{(h)}$ instead. In this context, it will be natural to say that problem (9.1) has been solved exactly already if the grid function $[u]_h$ is obtained.[1] In general, however, we cannot compute the table $[u]_h$ exactly either. Instead, we compute its approximation $u^{(h)}$, which is expected to converge to $[u]_h$ as the grid is refined. It is for computing $u^{(h)}$ that one can employ the finite-difference equations.

9.1.2 Convergent Difference Schemes

Our primary objective hereafter will be to study the techniques for the construction and analysis of convergent schemes for ordinary differential equations. However, before we can even approach this agenda, we will need to attach an accurate quantitative interpretation to the very notion of convergence $u^{(h)} \longrightarrow [u]_h$, which is the key property that we will require of the schemes we build. To do so, let us consider a normed linear space U_h of all the functions defined on the grid D_h. As those grid functions are essentially vectors of dimension $N+1$, the definitions and examples that pertain to vector norms and related concepts, see Section 5.2 of Chapter 5, also apply here. Very briefly, the norm $\|u^{(h)}\|_{U_h}$ of the grid function $u^{(h)} \in U_h$ is a non-

[1] As opposed to the actual continuous solution $u(x)$ that requires reconstruction from the grid.

negative real number that can be interpreted as length of the corresponding vector and quantifies the extent of its deviation from the identical zero on the grid.

There are many different ways one can define an appropriate norm for the grid function. For example, the least upper bound of all its absolute values at the grid nodes is a norm that is called the maximum norm:

$$\|u^{(h)}\|_{U_h} = \sup_n |u_n| = \max_n |u_n|. \tag{9.10}$$

If $u^{(h)}$ is a vector function, say, $u^{(h)} = \begin{bmatrix} v_n \\ u_n \end{bmatrix}$, $n = 0, 1, \ldots, N$, as in scheme (9.8), then the norm similar to (9.10) can be defined as the least upper bound of the absolute values of both functions v_n and w_n on the corresponding grid.

If the grid D_h is uniform, i.e., if the grid functions $u^{(h)} \in U_h$ are defined at the equally spaced nodes $x_n = nh$, $h > 0$, $n = 0, 1, \ldots, N$, then the following Euclidean norm is often used:

$$\|u^{(h)}\|_{U_h} = \left[h \sum_{n=0}^{N} u_n^2 \right]^{1/2}.$$

This norm is analogous to the continuous L_2 norm for the square integrable functions:

$$\|u(x)\| = \left[\int_0^1 u^2(x) dx \right]^{1/2}$$

Henceforth, we will always assume (for simplicity) that the maximum norm (9.10) is used, unless explicitly stated otherwise.

Having introduced the normed space U_h, we can now quantify the discrepancy between any two functions in this space. Let $a^{(h)} \in U_h$ and $b^{(h)} \in U_h$ be a pair of arbitrary functions defined on the grid D_h. The measure of their deviation from one another is naturally given by the norm of their difference:

$$\|a^{(h)} - b^{(h)}\|_{U_h}.$$

The latter quantification finally enables us to give an accurate definition of convergence for finite-difference schemes.

Let us denote by

$$\boldsymbol{L}_h u^{(h)} = f^{(h)} \tag{9.11}$$

the system of scalar equations to be used for approximately computing the solution of problem (9.1). In other words, solution of system (9.11) is supposed to yield an approximation to $[u]_h$, which is the discrete table of values for the continuous solution $u(x)$ of problem (9.1). Specific examples of the systems of type (9.11) are given by the difference schemes (9.6), (9.7), and (9.8) built for problems (9.2), (9.4), and (9.5), respectively. To recast scheme (9.6) in the general form (9.11) on a uniform grid $x_n = nh$, $n = 0, 1, \dots, N$, one can define the operator $\boldsymbol{L}_h$ and the right-hand side $f^{(h)}$ as follows:

$$\boldsymbol{L}_h u^{(h)} = \begin{cases} \dfrac{u_{n+1} - u_n}{h} + \dfrac{nh}{1 + u_n^2}, & n = 0, 1, \dots, N-1, \\ u_0, \end{cases}$$

$$f^{(h)} = \begin{cases} \cos(nh), & n = 0, 1, \dots, N-1, \\ 3. \end{cases}$$

Scheme (9.7) can be written in the form (9.11) once we set:

$$\boldsymbol{L}_h u^{(h)} = \begin{cases} \dfrac{u_{n+1} - 2u_n + u_{n-1}}{h^2} - [1 + (nh)^2]u_n, & n = 1, 2, \dots, N-1, \\ u_0, \\ u_N, \end{cases}$$

$$f^{(h)} = \begin{cases} \sqrt{1 + nh}, & n = 1, 2, \dots, N-1, \\ 2, \\ 1. \end{cases}$$

Scheme (9.8) for a system of differential equations transforms into (9.11) if:

$$\boldsymbol{L}_h u^{(h)} = \boldsymbol{L}_h \begin{bmatrix} v^{(h)} \\ w^{(h)} \end{bmatrix} = \begin{cases} \dfrac{v_{n+1} - v_n}{h} + (nh) v_n w_n, & n = 0, 1, \dots, N-1, \\ \dfrac{w_{n+1} - w_n}{h} + \dfrac{v_n + w_n}{1 + (nh)^2}, & n = 0, 1, \dots, N-1, \\ v_0, \\ w_0, \end{cases}$$

$$f^{(h)} = \begin{cases} (nh)^2 - 3nh + 1, & n = 0, 1, \dots, N-1, \\ \cos^2(nh), & n = 0, 1, \dots, N-1, \\ 1, \\ -3. \end{cases}$$

As we can see, in general system (9.11) depends on the grid size h. Therefore, it shall be written for all those values of h, for which we introduce the grid D_h and the exact solution $[u]_h$. Consequently, the discrete problem (9.11) is to be interpreted as the entire family of systems of equations parameterized by the quantity h, rather than as one single system of equations. Henceforth, we will assume that for each sufficiently small h the corresponding system (9.11) has a unique solution $u^{(h)} \in U_h$.

We will say that solution $u^{(h)}$ to the finite-difference problem (9.11) *converges* to the solution $u(x)$ of the original differential problem (9.1) as the grid is refined, if

$$\|[u]_h - u^{(h)}\|_{U_h} \longrightarrow 0, \quad \text{when} \quad h \longrightarrow 0, \tag{9.12}$$

where $[u]_h = \{u(x_n) | n = 0, 1, \dots, N\}$. If, in addition, $k > 0$ happens to be the largest integer such that the following inequality holds for all sufficiently small h:

$$\|[u]_h - u^{(h)}\|_{U_h} \le ch^k, \quad c = \text{const}, \tag{9.13}$$

where c does not depend on h, then we say that *the convergence rate* of scheme (9.11) is $\mathcal{O}(h^k)$, or alternatively, that *the error* of the approximate solution, which is the quantity (rather, function) under the norm on the left-hand side of (9.13), has order k with respect to the grid size h. Sometimes we would also say that the order of convergence is equal to $k > 0$.

Convergence is a fundamental requirement that one imposes on the difference scheme (9.11) so that to make it an appropriate tool for the numerical solution of the original differential (also referred to as continuous) problem (9.1). If convergence does take place, then the solution $[u]_h$ can be computed using scheme (9.11) with any initially prescribed accuracy by simply choosing a sufficiently small grid size h.

Having rigorously defined the concept of convergence, we have now arrived at the central question of how to construct a convergent difference scheme (9.11) for computing the solution of problem (9.1). The examples of Section 9.1.1 suggest a simple initial idea for building the schemes: One should first generate a grid and subsequently replace the derivatives in the governing equation(s) by appropriate difference quotients. However, for the same continuous problem (9.1) one can obviously obtain a large variety of schemes (9.11) by choosing different grids D_h and different ways to approximate the derivatives by difference quotients. In doing so, it turns out that some of the schemes do converge, while the others do not.

9.1.3 Verification of Convergence for a Difference Scheme

Let us therefore reformulate our central question in a somewhat different way. Suppose that the finite-difference scheme $\boldsymbol{L}_h u^{(h)} = f^{(h)}$ has already been constructed, and we expect that it could be convergent:

$$\|[u]_h - u^{(h)}\|_{U_h} \longrightarrow 0, \quad \text{when} \quad h \longrightarrow 0.$$

How can we actually check whether it really converges or not?

Assume that problem (9.11) has a unique solution $u^{(h)}$, and let us substitute the grid function $[u]_h$ instead of $u^{(h)}$ into the left-hand side of (9.11). If equality (9.11) were to hold exactly upon this substitution, then uniqueness would have implied $u^{(h)} = [u]_h$, which is ideal for convergence. Indeed, it would have meant that solution $u^{(h)}$ to the discrete problem $\boldsymbol{L}_h u^{(h)} = f^{(h)}$ coincides with the grid function $[u]_h$ that is sought for and that we have agreed to interpret as the unknown exact solution.

However, most often one cannot construct system (9.11) so that the solution $[u]_h$ would satisfy it exactly. Instead, the substitution of $[u]_h$ into the left-hand side of (9.11) would typically generate a residual $\delta f^{(h)}$:

$$\boldsymbol{L}_h [u]_h = f^{(h)} + \delta f^{(h)} \tag{9.14}$$

also known as *the truncation error*. If the truncation error $\delta f^{(h)}$ tends to zero as $h \longrightarrow 0$, then we say that the finite-difference scheme $\boldsymbol{L}_h u^{(h)} = f^{(h)}$ is *consistent*, or alternatively, that it approximates the differential problem $\boldsymbol{L}u = f$ on the solution $u(x)$ of the latter. This notion indeed makes sense because smaller residuals for smaller grid sizes mean that the exact solution $[u]_h$ satisfies equation (9.11) with better and better accuracy as h vanishes.

When the scheme is consistent, one can think that equation (9.14) for $[u]_h$ is obtained from equation (9.11) for $u^{(h)}$ by adding a small perturbation $\delta f^{(h)}$ to the right-hand side $f^{(h)}$ ($\delta f^{(h)}$ is small provided that h is small). Consequently, if the solution $u^{(h)}$ of problem (9.11) happens to be only weakly sensitive to the perturbations of the right-hand side, or in other words, if small changes of the right-hand side $f^{(h)}$ may only induce small changes in the solution $u^{(h)}$, then the difference between the solution $u^{(h)}$ of problem (9.11) and the solution $[u]_h$ of problem (9.14) will be small. In other words, consistency $\delta f^{(h)} \longrightarrow 0$ will imply convergence:

$$u^{(h)} \longrightarrow [u]_h, \quad \text{as} \quad h \longrightarrow 0.$$

The aforementioned weak sensitivity of the finite-difference solution $u^{(h)}$ to perturbations of $f^{(h)}$ can, in fact, be defined using rigorous terms. This definition will lead us to the fundamental notion of *stability* for finite-difference schemes.

Altogether, we can now outline our approach to verifying the convergence (9.12). We basically suggest to split this difficult task into two potentially simpler tasks. First, we would need to see whether the scheme (9.11) is consistent. Then, we will need to find out whether the scheme (9.11) is stable. This approach also indicates how one might actually construct convergent schemes for the numerical solution of problem (9.1). Namely, one would first need to obtain a consistent scheme, and then among many such schemes select those that would also be stable.

The foregoing general approach to the analysis of convergence obviously requires that both consistency and stability be defined rigorously, so that a theorem can eventually be proven on convergence as an implication of consistency and stability. The previous definitions of consistency and stability are, however, vague. As far as consistency, we need to be more specific on what the truncation error (residual) $\delta f^{(h)}$ is in the general case, and how to properly define its magnitude. As far as stability, we need to assign a precise meaning to the words "small changes of the right-hand side $f^{(h)}$ may only induce small changes in the solution $u^{(h)}$" of problem (9.11). We will devote two separate sections to the rigorous definitions of consistency and stability.

9.2 Approximation of Continuous Problem by a Difference Scheme. Consistency

In this section, we will rigorously define the concept of consistency, and explain what it actually means when we say that the finite-difference scheme (9.11) approximates the original differential problem (9.1) on its solution $u = u(x)$.

9.2.1 Truncation Error $\delta f^{(h)}$

To do so, we will first need to elaborate on what the truncation error $\delta f^{(h)}$ is, and show how to introduce its magnitude in a meaningful way. According to formula (9.14), $\delta f^{(h)}$ is the residual that arises when the exact solution $[u]_h$ is substituted into the left-hand side of (9.11). The decay of the magnitude of $\delta f^{(h)}$ as $h \longrightarrow 0$ is precisely what we term as consistency of the finite-difference scheme (9.11).

We begin with analyzing an example of a difference scheme for solving the following second order initial value problem:

$$\begin{gathered} \frac{d^2u}{dx^2} + a(x)\frac{du}{dx} + b(x)u = \cos x, \quad 0 \le x \le 1, \\ u(0) = 1, \quad u'(0) = 2. \end{gathered} \tag{9.15}$$

The grid D_h on the interval $[0,1]$ will be uniform: $x_n = nh$, $n = 0,1,\dots,N$, $h = 1/N$. To obtain the scheme, we approximately replace all the derivatives in relations (9.15) by difference quotients:

$$\frac{d^2u}{dx^2} \approx \frac{u(x+h) - 2u(x) + u(x-h)}{h^2}, \tag{9.16a}$$

$$\frac{du}{dx} \approx \frac{u(x+h) - u(x-h)}{2h}, \tag{9.16b}$$

$$\left.\frac{du}{dx}\right|_{x=0} \approx \frac{u(h) - u(0)}{h}. \tag{9.16c}$$

Substituting expressions (9.16a)–(9.16c) into (9.15), we arrive at the system of equations that can be used for the approximate computation of $[u]_h$:

$$\begin{gathered} \frac{u_{n+1} - 2u_n + u_{n-1}}{h^2} + a(x_n)\frac{u_{n+1} - u_{n-1}}{2h} + b(x_n)u_n = \cos x_n, \\ n = 1,2,\dots,N-1, \\ u_0 = 1, \quad \frac{u_1 - u_0}{h} = 2. \end{gathered} \tag{9.17}$$

Scheme (9.17) can be easily converted to form (9.11) if we denote:

$$\begin{gathered} \boldsymbol{L}_h u^{(h)} = \begin{cases} \dfrac{u_{n+1} - 2u_n + u_{n-1}}{h^2} + a(x_n)\dfrac{u_{n+1} - u_{n-1}}{2h} + b(x_n)u_n, \\ \qquad\qquad n = 1,2,\dots,N-1, \\ u_0, \\ \dfrac{u_1 - u_0}{h}, \end{cases} \\ f^{(h)} = \begin{cases} \cos x_n, \quad n = 1,2,\dots,N-1, \\ 1, \\ 2. \end{cases} \end{gathered} \tag{9.18}$$

To estimate the magnitude of the truncation error $\delta f^{(h)}$ that arises when the grid function $[u]_h$ is substituted into the left-hand side of (9.11), we will need to "sharpen" the approximate equalities (9.16a)–(9.16c). This can be done by evaluating the quantities on the right-hand sides of (9.16a)–(9.16c) with the help of the Taylor formula.

For *the central difference* approximation of the second derivative (9.16a), we will need four terms of the Taylor expansion of u at the point x:

$$\begin{aligned} u(x+h) &= u(x) + hu'(x) + \frac{h^2}{2!}u''(x) + \frac{h^3}{3!}u'''(x) + \frac{h^4}{4!}u^{(4)}(\xi_1), \\ u(x-h) &= u(x) - hu'(x) + \frac{h^2}{2!}u''(x) - \frac{h^3}{3!}u'''(x) + \frac{h^4}{4!}u^{(4)}(\xi_2). \end{aligned} \tag{9.19a}$$

For *the central difference* approximation of the first derivative (9.16b), it will be sufficient to use three terms of the expansion:

$$\begin{aligned} u(x+h) &= u(x) + hu'(x) + \frac{h^2}{2!}u''(x) + \frac{h^3}{3!}u'''(\xi_3), \\ u(x-h) &= u(x) - hu'(x) + \frac{h^2}{2!}u''(x) - \frac{h^3}{3!}u'''(\xi_4), \end{aligned} \tag{9.19b}$$

and for *the forward difference* approximation of the first derivative (9.16c) one would only need two terms of the Taylor formula (to be used at $x = 0$):

$$u(x+h) = u(x) + hu'(x) + \frac{h^2}{2!}u''(\xi_5). \tag{9.19c}$$

Note that the last term in each of the expressions (9.19a)–(9.19c) is the remainder (or error) of the Taylor formula in the so-called Lagrange form, where $\xi_1, \xi_3, \xi_5 \in [x, x+h]$ and $\xi_2, \xi_4 \in [x-h, x]$. Substituting expressions (9.19a), (9.19b), and (9.19c) into formulae (9.16a), (9.16c), and (9.16c), respectively, we obtain:

$$\frac{u(x+h) - 2u(x) + u(x-h)}{h^2} = u''(x) + \frac{h^2}{24}\left(u^{(4)}(\xi_1) + u^{(4)}(\xi_2)\right), \tag{9.20a}$$

$$\frac{u(x+h) - u(x-h)}{2h} = u'(x) + \frac{h^2}{12}\left(u'''(\xi_3) + u'''(\xi_4)\right), \tag{9.20b}$$

$$\left.\frac{u(x+h) - u(x)}{h}\right|_{x=0} = \left.u'(x)\right|_{x=0} + \frac{h}{2}u''(\xi_5). \tag{9.20c}$$

9.2.2 Evaluation of the Truncation Error $\delta f^{(h)}$

Let us now assume that the solution $u = u(x)$ of problem (9.15) has bounded derivatives up to the fourth order everywhere on the interval $0 \le x \le 1$. Then, according to formulae (9.20a)–(9.20c) one can write:

$$\begin{aligned} &\frac{u(x+h) - 2u(x) + u(x-h)}{h^2} + a(x)\frac{u(x+h) - u(x-h)}{2h} + b(x)u(x) \\ &= \frac{d^2u}{dx^2} + a(x)\frac{du}{dx} + b(x)u \quad + \quad h^2\left[\frac{u^{(4)}(\xi_1) + u^{(4)}(\xi_2)}{24} + a(x)\frac{u'''(\xi_3) + u'''(\xi_4)}{12}\right], \end{aligned}$$

where $\xi_1, \xi_2, \xi_3, \xi_4 \in [x-h, x+h]$. Consequently, the expression for $\boldsymbol{L}_h[u]_h$ given by formula (9.18):

$$\boldsymbol{L}_h[u]_h = \begin{cases} \dfrac{u(x_n+h)-2u(x_n)+u(x_n-h)}{h^2} + a(x_n)\dfrac{u(x_n+h)-u(x_n-h)}{2h} \\ \qquad\qquad\qquad\qquad +b(x_n)u(x_n), \qquad n=1,2,\dots,N-1, \\ u(0), \\ \dfrac{u(h)-u(0)}{h}, \end{cases}$$

can be rewritten as follows:

$$\boldsymbol{L}_h[u]_h = \begin{cases} \cos x_n + h^2\left[\dfrac{u^{(4)}(\xi_1)+u^{(4)}(\xi_2)}{24} + a(x_n)\dfrac{u'''(\xi_3)+u'''(\xi_4)}{12}\right], \\ \qquad\qquad\qquad\qquad n=1,2,\dots,N-1, \\ 1+0, \\ 2+h\dfrac{u''(\xi_5)}{2}, \end{cases}$$

where $\xi_1, \xi_2, \xi_3, \xi_4 \in [x_n-h, x_n+h], n=1,2,\dots,N-1$, and $\xi_5 \in [0,h]$. Alternatively, we can write:

$$\boldsymbol{L}_h[u]_h = f^{(h)} + \delta f^{(h)},$$

where $f^{(h)}$ is also defined in (9.18), and

$$\delta f^{(h)} = \begin{cases} h^2\left[\dfrac{u^{(4)}(\xi_1)+u^{(4)}(\xi_2)}{24} + a(x_n)\dfrac{u'''(\xi_3)+u'''(\xi_4)}{12}\right], \\ \qquad\qquad\qquad\qquad n=1,2,\dots,N-1, \\ 0, \\ h\dfrac{u''(\xi_5)}{2}. \end{cases} \tag{9.21}$$

To quantify the truncation error $\delta f^{(h)}$, i.e., to introduce its magnitude, it will be convenient to assume that $f^{(h)}$ and $\delta f^{(h)}$ belong to a linear normed space F_h that consists of the following elements:

$$g^{(h)} = \begin{cases} \varphi_n, & n=1,2,\dots,N-1, \\ \psi_0, \\ \psi_1, \end{cases} \tag{9.22}$$

where $\varphi_1, \varphi_2, \dots, \varphi_{N-1}$, and ψ_0, ψ_1 is an arbitrary ordered system of numbers. One can think of $g^{(h)}$ given by formula (9.22) as of a combination of the grid function φ_n, $n=1,2,\dots,N-1$, and an ordered pair of numbers (ψ_0, ψ_1). Addition of the elements from the space F_h, as well as their multiplication by scalars, are performed component-wise. Hence, it is clear that F_h is a linear (vector) space of dimension

$N+1$. It can be supplemented with a variety of different norms. If the norm in F_h is introduced as the maximum absolute value of all the components of $g^{(h)}$:

$$\|g^{(h)}\|_{F_h} = \max\left\{|\psi_0|, |\psi_1|, \max_n |\varphi_n|\right\},$$

then according to (9.21) we obviously have for all sufficiently small grid sizes h:

$$\|\delta f^{(h)}\| \leq ch, \tag{9.23}$$

where c is a constant that may, generally speaking, depend on $u(x)$, but should not depend on h. Inequality (9.23) guarantees that the truncation error $\delta f^{(h)}$ does vanish when $h \longrightarrow 0$, and that the scheme (9.17) has first order accuracy.

If the difference equation (9.17) that we have considered as an example is represented in the form $\boldsymbol{L}_h u^{(h)} = f^{(h)}$, see formula (9.18), then one can interpret $\boldsymbol{L}_h$ as an operator, $\boldsymbol{L}_h : U_h \longmapsto F_h$. Given a grid function $v^{(h)} = \{v_n\}$, $n = 0, 1, \ldots, n$, that belongs to the linear space U_h of functions defined on the grid D_h, the operator $\boldsymbol{L}_h$ maps it onto some element $g^{(h)} \in F_h$ of type (9.22) according to the following formula:

$$\boldsymbol{L}_h v^{(h)} = \begin{cases} \dfrac{v_{n+1} - 2v_n + v_{n-1}}{h^2} + a(x_n)\dfrac{v_{n+1} - v_{n-1}}{2h} + b(x_n)v_n, \\ \qquad\qquad\qquad\qquad n = 1, 2, \ldots, N-1, \\ v_0, \\ \dfrac{v_1 - v_0}{h}. \end{cases}$$

The operator interpretation holds for the general finite-difference problem (9.11) as well. Assume that the right-hand sides of all individual equations contained in $\boldsymbol{L}_h u^{(h)} = f^{(h)}$ are components of the vector $f^{(h)}$ that belongs to some linear normed space F_h. Then, $\boldsymbol{L}_h$ of (9.11) becomes an operator that maps any given grid function $u^{(h)} \in U_h$ onto some element $f^{(h)} \in F_h$, i.e., $\boldsymbol{L}_h : U_h \longmapsto F_h$. Accordingly, expression $\boldsymbol{L}_h[u]_h$ stands for the result of application of the operator $\boldsymbol{L}_h$ to the function $[u]_h$ that is an element of U_h; as such, $\boldsymbol{L}_h[u]_h \in F_h$. Consequently, $\delta f^{(h)} \stackrel{\text{def}}{=} \boldsymbol{L}_h[u]_h - f^{(h)} \in F_h$ as a difference between two elements of the space F_h. The magnitude of the residual $\delta f^{(h)}$, i.e., the magnitude of the truncation error, is given by the norm $\|\delta f^{(h)}\|_{F_h}$.

9.2.3 Accuracy of Order h^k

DEFINITION 9.1 *We will say that the finite-difference scheme* $\boldsymbol{L}_h u^{(h)} = f^{(h)}$ *approximates the differential problem* $\boldsymbol{L}u = f$ *on its solution* $u = u(x)$ *if the truncation error vanishes when the grid is refined, i.e.,* $\|\delta f^{(h)}\|_{F_h} \longrightarrow 0$ *as* $h \longrightarrow 0$. *A scheme that approximates the original differential problem is called consistent. If, in addition,* $k > 0$ *happens to be the largest integer that guarantees the following estimate:*

$$\|\delta f^{(h)}\|_{F_h} \leq c_1 h^k, \qquad c_1 = \text{const},$$

where c_1 does not depend on the grid size h, then we say that the scheme has order of accuracy $\mathcal{O}(h^k)$ or that its accuracy is of order k with respect to h.[2]

Note that in Definition 9.1 the function $u = u(x)$ is considered a solution of problem $\boldsymbol{L}u = f$. This assumption can provide useful information about u, e.g., bounds for its derivatives, that can subsequently be exploited when constructing the scheme, as well as when verifying its consistency. This is the reason for incorporating a solution of problem (9.1) into Definition 9.1. Let us emphasize, however, that the notion of approximation of the problem $\boldsymbol{L}u = f$ by the scheme $\boldsymbol{L}_h u^{(h)} = f^{(h)}$ does not rely on the equality $\boldsymbol{L}u = f$ for the function u. The central requirement of Definition 9.1 is only the decay of the quantity $\|\boldsymbol{L}_h[u]_h - f^{(h)}\|_{F_h}$ when the grid size h vanishes.Therefore, if there is a function $u = u(x)$ that meets this requirement, then we could simply say that the truncation error of the scheme $\boldsymbol{L}_h u^{(h)} = f^{(h)}$ has some order $k > 0$ with respect to h on a given function u, without going into detail regarding the origin of this function. In this context, it may often be helpful to use a slightly different concept of approximation — that of the differential operator $\boldsymbol{L}$ by the difference operator $\boldsymbol{L}_h$ (see Section 10.2.2 of Chapter 10 for more detail).

Namely, we will say that the finite-difference operator $\boldsymbol{L}_h$ approximates the differential operator $\boldsymbol{L}$ on some function $u = u(x)$ if $\|\boldsymbol{L}_h[u]_h - [\boldsymbol{L}u]_h\|_{F_h} \longrightarrow 0$ as $h \longrightarrow 0$. In the previous expression, $[u]_h$ denotes the trace of the continuous function $u(x)$ on the grid as before, and likewise, $[\boldsymbol{L}u]_h$ denotes the trace of the continuous function $\boldsymbol{L}u$ on the grid. For example, equality (9.20a) can be interpreted as approximation of the differential operator $\boldsymbol{L}u \equiv u''$ by the central difference on the left-hand side of (9.20a), with the accuracy $\mathcal{O}(h^2)$. In so doing, $u(x)$ may be any function with the bounded fourth derivative. Consequently, the approximation can be considered on the class of all functions u with bounded derivatives up to the order four, rather than on a single function u. Similarly, equality (9.20b) can be interpreted as approximation of the differential operator $\boldsymbol{L}u \equiv u'$ by the central difference on the left-hand side of (9.20b), with the accuracy $\mathcal{O}(h^2)$, and this approximation holds on the class of all functions $u = u(x)$ that have bounded third derivatives.

9.2.4 Examples

Example 1

According to formula (9.21) and estimate (9.23), the finite-difference scheme (9.17) is consistent and has accuracy $\mathcal{O}(h)$, i.e., it approximates problem (9.15) with the first order with respect to h. In fact, scheme (9.17) can be easily improved so that it would gain second order accuracy. To achieve that, let us first notice that every component of the vector $\delta f^{(h)}$ except for the last one, see (9.21), decays with the rate $\mathcal{O}(h^2)$ when h decreases, and the second to last component is even equal to zero exactly. It is only the last component of the vector $\delta f^{(h)}$ that displays a slower rate of decay, namely, $\mathcal{O}(h)$. This last component is the residual generated by substituting

[2] Sometimes also referred to as the order of approximation of the continuous problem by the scheme.

$u = u(x)$ into the last equation $(u_1 - u_0)/h = 2$ of system (9.17). Fortunately, the slower rate of decay for this component, which hampers the overall accuracy of the scheme, can be easily sped up.

Using the Taylor formula with the remainder in the Lagrange form, we can write:

$$\frac{u(h)-u(0)}{h} = u'(0) + \frac{h}{2}u''(0) + \frac{h^2}{6}u'''(\xi), \quad \text{where } 0 \leq \xi \leq h.$$

At the same time, the original differential equation, along with its initial conditions, see (9.15), yield:

$$u''(0) = -a(0)u'(0) - b(0)u(0) + \cos 0 = -2a(0) - b(0) + 1.$$

Therefore, if we replace the last equality of (9.17) with the following:

$$\frac{u_1 - u_0}{h} = 2 + \frac{h}{2}u''(0) = 2 - \frac{h}{2}[2a(0) + b(0) - 1], \tag{9.24}$$

then we obtain a new expression for the right-hand side $f^{(h)}$, instead of the one given in formula (9.18):

$$f^{(h)} = \begin{cases} \cos x_n, & n = 1, 2, \ldots, N-1, \\ 1, \\ 2 - \frac{h}{2}[2a(0) + b(0) - 1]. \end{cases}$$

This, in turn, yields a new expression for the truncation error:

$$\delta f^{(h)} = \begin{cases} h^2 \left[\frac{u^{(4)}(\xi_1) + u^{(4)}(\xi_2)}{24} + a(x_n)\frac{u'''(\xi_3) + u'''(\xi_4)}{12} \right], \\ \qquad\qquad n = 1, 2, \ldots, N-1, \\ 0, \\ \frac{h^2}{6}u'''(\xi). \end{cases} \tag{9.25}$$

Formula (9.25) implies that under the previous assumption of boundedness of all the derivatives up to the order four, we have $\|\delta f^{(h)}\| \leq ch^2$, where the constant c does not depend on h. Thus, the accuracy of the scheme becomes $\mathcal{O}(h^2)$ instead of $\mathcal{O}(h)$.

Let us emphasize that in order to obtain the new difference initial condition (9.24) not only have we used the original continuous initial conditions of (9.15), but also the differential equation itself. It is possible to say that we have exploited the additional initial condition:

$$u''(x) + a(x)u'(x) + b(x)u(x)\big|_{x=0} = \cos x\big|_{x=0},$$

which is a direct implication of the given differential equation of (9.15) at $x = 0$.

Example 2

Let

$$L_h u^{(h)} = \begin{cases} \dfrac{u_{n+1} - u_{n-1}}{2h} + Au_n, & n = 1, 2, \ldots, N-1, \\ u_0, \\ u_1, \end{cases}$$
$$f^{(h)} = \begin{cases} 1 + x_n^2, & n = 1, 2, \ldots, N-1, \\ b, \\ b, \end{cases} \tag{9.26}$$

and let us determine the order of accuracy of the scheme:

$$L_h u^{(h)} = f^{(h)} \tag{9.27}$$

on the solution $u = u(x)$ of the initial value problem:

$$\frac{du}{dx} + Au = 1 + x^2, \quad u(0) = b. \tag{9.28}$$

For the exact solution $[u]_h$ we obtain:

$$L_h [u]_h = \begin{cases} \dfrac{u(x_n + h) - u(x_n - h)}{2h} + Au(x_n), & n = 1, 2, \ldots, N-1, \\ u(0), \\ u(h), \end{cases}$$

which, according to formula (9.20b), yields:

$$L_h [u]_h = \begin{cases} \left[\dfrac{du(x_n)}{dx} + Au(x_n) \right] + \dfrac{h^2}{12} (u'''(\xi_3) + u'''(\xi_4)), & n = 1, 2, \ldots, N-1, \\ u(0), \\ u(0) + hu'(\xi_0), \end{cases}$$

where $\xi_0 \in [0, h]$ and for each n: $\xi_3 \in [x_n, x_n + h]$ and $\xi_4 \in [x_n - h, x_n]$. As u is a solution to (9.28), we have:

$$\frac{du(x_n)}{dx} + Au(x_n) = 1 + x_n^2,$$

and for the truncation error $\delta f^{(h)}$ we obtain:

$$\delta f^{(h)} = \begin{cases} \dfrac{h^2}{12} (u'''(\xi_3) + u'''(\xi_4)), & n = 1, 2, \ldots, N-1, \\ 0, \\ hu'(\xi_0). \end{cases}$$

Consequently, scheme (9.26)–(9.27) altogether has first order accuracy with respect to h. It is interesting to note, however, that similarly to Example 1, different components of the truncation error have different orders with respect to the grid size. The system of difference equations:

$$\frac{u_{n+1}-u_{n-1}}{2h}+Au_n=1+x_n^2, \quad n=1,2,\dots,N-1,$$

is satisfied by the exact solution $[u]_h$ with the residual $\frac{h^2}{12}(u'''(\xi_3)+u'''(\xi_4))$ of order $\mathcal{O}(h^2)$. The first initial condition $u_0=b$ is satisfied exactly, and only the second initial condition $u_1=b$ is satisfied by $[u]_h$ with the residual $hu'(\xi_0)$ of order $\mathcal{O}(h)$.

Example 3

Finally, let us illustrate the comments we have made right after the definition of consistency, see Definition 9.1, in Section 9.2.3. For simplicity, we will be considering problems with no initial or boundary conditions, defined on the grid: $x_n=nh$, $h>0$, $n=0,\pm1,\pm2,\dots$. Let

$$\boldsymbol{L}_h u^{(h)}=\frac{u_{n+1}-2u_n+u_{n-1}}{h^2}+u_n.$$

Formula (9.20a) immediately implies that $\boldsymbol{L}_h$ approximates the differential operator:

$$\boldsymbol{L}u=u''+u$$

with second order accuracy on the class of functions with bounded fourth derivatives. Indeed, according to (9.20a) for any such function we can write:

$$\boldsymbol{L}_h[u]_h-[\boldsymbol{L}u]_h=\frac{h^2}{24}\left(u^{(4)}(\xi_1)+u^{(4)}(\xi_2)\right),$$

and consequently, $\|\boldsymbol{L}_h[u]_h-[\boldsymbol{L}u]_h\|\le ch^2$.

Let us now take $u(x)=\sin x$ and show that the homogeneous difference scheme $\boldsymbol{L}_h u^{(h)}=0$ is consistent on this function. Indeed,

$$\begin{aligned}\boldsymbol{L}_h[u]_h&=\frac{u(x+h)-2u(x)+u(x-h)}{h^2}+u(x)\\&=u''(x)+\frac{h^2}{24}\left(u^{(4)}(\xi_1)+u^{(4)}(\xi_2)\right)+u(x)\\&=-\sin x+\frac{h^2}{24}\left(u^{(4)}(\xi_1)+u^{(4)}(\xi_2)\right)+\sin x\\&=\frac{h^2}{24}\left(u^{(4)}(\xi_1)+u^{(4)}(\xi_2)\right),\end{aligned}$$

which means that $\|\boldsymbol{L}_h[u]_h-0\|\le \text{const}\cdot h^2/12$. Yet we note that in the previous argument we have used the consideration that $(\sin x)''+\sin x=0$, i.e., that $u(x)=\sin x$

is not just a function with bounded fourth derivatives, but a solution to the homogeneous differential equation $\boldsymbol{L}u \equiv u'' + u = 0$. In other words, while the definition of consistency per se does not explicitly require that $u(x)$ be a solution to the underlying differential problem, this fact may still be needed when actually evaluating the magnitude of the truncation error.

However, for the specific example that we are currently analyzing, consistency can also be established independently. We have:

$$\begin{aligned}
\boldsymbol{L}_h[u]_h &= \frac{\sin(x+h) - 2\sin(x) + \sin(x-h)}{h^2} + \sin(x) \\
&= \frac{\sin x \cos h + \cos x \sin h - 2\sin(x) + \sin x \cos h - \cos x \sin h}{h^2} + \sin(x) \\
&= \sin x \frac{2(\cos h - 1)}{h^2} + \sin x = \left(1 - \frac{4}{h^2}\sin^2\frac{h}{2}\right)\sin x.
\end{aligned}$$

Using the Taylor formula, one can easily show that for small grid sizes h the expression in the brackets above is $\mathcal{O}(h^2)$, and as $\sin x = u(x)$ is bounded, we, again, conclude that the scheme $\boldsymbol{L}_h u^{(h)} = 0$ has second order accuracy.

9.2.5 Replacement of Derivatives by Difference Quotients

In all our previous examples, in order to obtain a finite-difference scheme we would replace the derivatives in the original continuous problem (differential equation and initial conditions) by the appropriate difference quotients. This approach is quite universal, and for any differential problem with a sufficiently smooth solution $u(x)$ it enables construction of a scheme with any prescribed order of accuracy.

9.2.6 Other Approaches to Constructing Difference Schemes

However, the replacement of derivatives by difference quotients is by no means the only method of building the schemes, and often not the best one either. In Section 9.4, we describe a family of popular methods of approximation that lead to the widely used Runge-Kutta schemes. Here, we only provide some examples.

The simplest scheme:

$$\begin{gathered}
\frac{u_{n+1} - u_n}{h} - G(x_n, u_n) = 0, \quad n = 0, 1, \ldots, N-1, \\
u_0 = a, \quad h = 1/N,
\end{gathered} \tag{9.29}$$

known as *the forward Euler scheme* is consistent and has first order accuracy with respect to h on the solutions of the differential problem:

$$\frac{du}{dx} - G(x, u) = 0, \quad 0 \le x \le 1, \quad u(0) = a. \tag{9.30}$$

Solution of the finite-difference equation (9.29) can be found by marching, i.e., it can be computed consecutively one grid node after another using the formula:

$$u_{n+1} = u_n + hG(x_n, u_n). \tag{9.31}$$

This formula yields u_{n+1} in the closed form once the values of u_n at the previous nodes are available. That's why the forward Euler scheme is called *explicit.*

Let us now take the same formula (9.31) as in the forward Euler method but use it only as *a predictor* at the preliminary stage of computation. At this stage we obtain the intermediate quantity $\tilde{u}$:

$$\tilde{u} = u_n + hG(x_n, u_n).$$

Then, the actual value of the finite-difference solution at the next grid node u_{n+1} is obtained by applying *the corrector* at the final stage:

$$u_{n+1} = u_n + \frac{h}{2}[G(x_n, u_n) + G(x_{n+1}, \tilde{u})].$$

The overall resulting *predictor-corrector scheme*:

$$\begin{aligned} \tilde{u} &= u_n + hG(x_n, u_n), \\ \frac{u_{n+1} - u_n}{h} &= \frac{1}{2}[G(x_n, u_n) + G(x_{n+1}, \tilde{u})], \\ u_0 &= a, \end{aligned} \tag{9.32}$$

can be shown to have second order accuracy with respect to h on the solutions of problem (9.30). It is also explicit, and is, in fact, a member of the Runge-Kutta family of methods, see Section 9.4. In the literature, scheme (9.32) is sometimes also referred to as the Heun scheme.

Let us additionally note that the corrector stage of scheme (9.32) can actually be considered as an independent single stage finite-difference method of its own:

$$\begin{gathered} \frac{u_{n+1} - u_n}{h} = \frac{1}{2}[G(x_n, u_n) + G(x_{n+1}, u_{n+1})], \\ u_0 = a. \end{gathered} \tag{9.33}$$

Scheme (9.33) is known as the *Crank-Nicolson scheme.* It approximates problem (9.30) with second order of accuracy. Solution of the finite-difference equation (9.33) can also be found by marching. However, unlike in the forward Euler method (9.29), equation (9.33) contains u_{n+1} both on its left-hand side and on its right-hand side, as an argument of $G(x_{n+1}, u_{n+1})$. Therefore, to obtain u_{n+1} for each $n = 0, 1, 2, \ldots$, one basically needs to solve an algebraic equation. This may require special techniques, such as Newton's method of Section 8.3 of Chapter 8, when the function G happens to be nonlinear with respect to its argument u. Altogether, we see that for the Crank-Nicolson scheme (9.33) the value of u_{n+1} is not immediately available in the closed form as a function of the previous u_n. That's why the Crank-Nicolson method is called *implicit.*

Finally, perhaps the simplest implicit scheme for problem (9.30) is the so-called *backward Euler* scheme:

$$\frac{u_{n+1} - u_n}{h} - G(x_{n+1}, u_{n+1}) = 0, \quad u_0 = a, \tag{9.34}$$

that has first order accuracy with respect to h.

Nowadays, a number of well established universal and efficient methods for solving the Cauchy problem for ordinary differential equations and systems are available as standard implementations in numerical software libraries. There is also a large number of specialized methods developed for solving particular (often, narrow) classes of problems that may present difficulties of a specific type.

Exercises

1. Verify that the forward Euler scheme (9.29) has first order accuracy on a smooth solution $u = u(x)$ of problem (9.30).
2. Modify the second initial condition $u_1 = b$ in the scheme (9.26)-(9.27) so that to achieve the overall second order accuracy.
3.* Verify that the predictor-corrector scheme (9.32) has second order accuracy on a smooth solution $u = u(x)$ of problem (9.30).
   Hint. See the analysis in Section 9.4.1.
4. Verify that the Crank-Nicolson scheme (9.33) has second order accuracy on a smooth solution $u = u(x)$ of problem (9.30).
5. Verify that the backward Euler scheme (9.34) has first order accuracy on a smooth solution $u = u(x)$ of problem (9.30).

9.3 Stability of Finite-Difference Schemes

In the previous sections, we have constructed a number of consistent finite-difference schemes for ordinary differential equations. It is, in fact, possible to show that those schemes are also convergent, and that the convergence rate for each scheme coincides with its respective order of accuracy.

One can, however, build examples of consistent yet divergent (i.e., non-convergent) finite-difference schemes. For instance, it is easy to see that the scheme:

$$\begin{gathered} 4\frac{u_{n+1}-u_{n-1}}{2h} - 3\frac{u_{n+1}-u_n}{h} + Au_n = 0, \quad n = 1,2,\dots,N-1, \\ u_0 = b, \quad u_1 = be^{-Ah} \end{gathered} \tag{9.35}$$

approximates the Cauchy problem:

$$\begin{gathered} \frac{du}{dx} + Au = 0, \quad 0 \leq x \leq 1, \\ u(0) = b, \quad A = \text{const}, \end{gathered} \tag{9.36}$$

on its solution $u = u(x)$ with first order accuracy with respect to h. However, the solution $u^{(h)}$ obtained with the help of this scheme does not converge to $[u]_h$, and does not even remain bounded as $h \longrightarrow 0$.

Indeed, the general solution of the difference equation from (9.35) has the form:

$$u_n = c_1 q_1^n + c_2 q_2^n,$$

where c_1 and c_2 are arbitrary constants, and q_1 and q_2 are roots of the algebraic characteristic equation: $-q^2 + (3 + Ah)q - 2 = 0$. Choosing the constants c_1 and c_2 to satisfy the initial conditions, we obtain:

$$u_n = u_0 \left[\frac{q_2}{q_2 - q_1} q_1^n - \frac{q_1}{q_2 - q_1} q_2^n \right] + u_1 \left[-\frac{1}{q_2 - q_1} q_1^n + \frac{1}{q_2 - q_1} q_2^n \right]. \tag{9.37}$$

Analysis of formula (9.37) shows that $\max\limits_{0 \le nh \le 1} |u_n| \longrightarrow \infty$ as $h \longrightarrow 0$. Therefore, consistency alone is generally not sufficient for convergence. *Stability* is also required.

9.3.1 Definition of Stability

Consider an initial or boundary value problem

$$\boldsymbol{L}u = f, \tag{9.38}$$

and assume that the finite-difference scheme

$$\boldsymbol{L}_h u^{(h)} = f^{(h)} \tag{9.39}$$

is consistent and has accuracy $\mathscr{O}(h^k)$, $k > 0$, on the solution u of problem (9.38). This means that the truncation error $\delta f^{(h)}$ defined by the formula:

$$\boldsymbol{L}_h [u]_h = f^{(h)} + \delta f^{(h)}, \tag{9.40}$$

where $[u]_h$ is the projection of the exact solution $u(x)$ onto the grid D_h, satisfies the following inequality (cf. Definition 9.1):

$$\|\delta f^{(h)}\|_{F_h} \le c_1 h^k, \tag{9.41}$$

where c_1 is a constant that does not depend on h.

We will now introduce two alternative definitions of stability for scheme (9.39).

DEFINITION 9.2 *Scheme (9.39) will be called stable if one can choose two numbers $h_0 > 0$ and $\delta > 0$ such that for any $h < h_0$ and for any $\varepsilon^{(h)} \in F_h$, $\|\varepsilon^{(h)}\|_{F_h} < \delta$, the finite-difference problem:*

$$\boldsymbol{L}_h z^{(h)} = f^{(h)} + \varepsilon^{(h)} \tag{9.42}$$

obtained from (9.39) by adding the perturbation $\varepsilon^{(h)}$ to the right-hand side, has one and only one solution $z^{(h)}$ whose deviation from the solution $u^{(h)}$ of the unperturbed problem (9.39) satisfies the estimate:

$$\|z^{(h)} - u^{(h)}\|_{U_h} \le c_2 \|\varepsilon^{(h)}\|_{F_h}, \quad c_2 = \text{const}, \tag{9.43}$$

where c_2 does not depend on h, $\varepsilon^{(h)}$.

Inequality (9.43) implies that a small perturbation $\varepsilon^{(h)}$ of the right-hand side $f^{(h)}$ may only cause a small perturbation $z^{(h)} - u^{(h)}$ of the corresponding solution. In other words, it implies weak sensitivity of the solution to perturbations of the right-hand side that drives it. This is a setup that applies to both linear and nonlinear problems, because it discusses stability of individual solutions with respect to small perturbations of their respective source terms. Hence, Definition 9.2 can basically be referred to as the definition for nonlinear equations. Note that for a different solution $u^{(h)}$ that corresponds to another right-hand side $f^{(h)}$, the value of c_2 in inequality (9.43) may, generally speaking, change. The key point though is that c_2 does not increase when the grid size h becomes smaller.

Let us now assume that the operator $\boldsymbol{L}_h : U_h \longmapsto F_h$ is linear. Then, Definition 9.2 is equivalent to the following definition of stability for linear equations:

DEFINITION 9.3 *Scheme (9.39) with a linear operator $\boldsymbol{L}_h$ will be called stable if for any given $f^{(h)} \in F_h$ the equation $\boldsymbol{L}_h u^{(h)} = f^{(h)}$ has a unique solution $u^{(h)}$ such that*

$$\|u^{(h)}\|_{U_h} \leq c_2 \|f^{(h)}\|_{F_h}, \quad c_2 = \text{const}, \tag{9.44}$$

where c_2 does not depend on h, $f^{(h)}$.

We emphasize that the value of c_2 in inequality (9.44) is one and the same for all $f^{(h)} \in F_h$. The equivalence of Definitions 9.3 and 9.2 (stability in the nonlinear and in the linear sense) for the case of a linear operator $\boldsymbol{L}_h$ will be rigorously proven in the context of partial differential equations, see Lemma 10.1 of Chapter 10.

9.3.2 The Relation between Consistency, Stability, and Convergence

The following theorem is of central importance for the entire analysis.

THEOREM 9.1
Let the scheme $\boldsymbol{L}_h u^{(h)} = f^{(h)}$ be consistent with the order of accuracy $\mathcal{O}(h^k)$, $k > 0$, on the solution $u = u(x)$ of the problem $\boldsymbol{L}u = f$, and let this scheme also be stable. Then, the finite-difference solution $u^{(h)}$ converges to the exact solution $[u]_h$, and the following estimate holds:

$$\|[u]_h - u^{(h)}\|_{U_h} \leq c_1 c_2 h^k, \tag{9.45}$$

where c_1 and c_2 are the constants from estimates (9.41) and (9.43).

The result of Theorem 9.45 basically means that a consistent and stable scheme converges with the rate equal to its order of accuracy.

PROOF Let $\varepsilon^{(h)} = \delta f^{(h)}$ and $[u]_h = z^{(h)}$. Then, estimate (9.43) becomes:

$$\|[u]_h - u^{(h)}\|_{U_h} \leq c_2 \|\delta f^{(h)}\|_{F_h}.$$

Taking into account (9.41), we immediately obtain (9.45). ∎

Note that when proving Theorem 9.1, we interpreted stability in the nonlinear sense (Definition 9.2). Because of the equivalence of Definitions 9.2 and 9.3 for a linear operator $\boldsymbol{L}_h$, Theorem 9.1, of course, covers the linear case as well. There may be situations, however (mostly encountered for partial differential equations, see Chapter 10), when stability can be established for the linear case only, whereas stability for nonlinear problems may not necessarily lend itself to easy analysis. Then, the result of Theorem 9.1 will hold only for linear finite-difference schemes.

Let us now analyze some examples. First, we will prove stability of the forward Euler scheme:

$$\begin{gathered}\frac{u_{n+1} - u_n}{h} - G(x_n, u_n) = \varphi_n, \quad n = 0, 1, \ldots, N-1, \\ u_0 = \psi,\end{gathered} \tag{9.46}$$

where $x_n = nh$, $n = 0, 1, \ldots, N$, and $h = 1/N$. Scheme (9.46) can be employed for the numerical solution of the initial value problem:

$$\frac{du}{dx} - G(x, u) = \varphi(x), \quad 0 \leq x \leq 1, \quad u\big|_{x=0} = \psi. \tag{9.47}$$

Henceforth, we will assume that the functions $G = G(x, u)$ and $\varphi = \varphi(x)$ are such that problem (9.47) has a solution $u = u(x)$ with the bounded second derivative so that $|u''(x)| \leq \text{const}$. We will also assume that $G(x, u)$ has a bounded partial derivative with respect to u:

$$\left|\frac{\partial G}{\partial u}\right| \leq M = \text{const}. \tag{9.48}$$

Next, we will introduce the norms:

$$\begin{aligned}\|u^{(h)}\|_{U_h} &= \max_n |u_n|, \\ \|f^{(h)}\|_{F_h} &= \max\{|\psi|, \max_n |\varphi_n|\},\end{aligned}$$

and verify that the forward Euler scheme (9.46) is indeed stable. This scheme can be recast in the operator form (9.39) if we define:

$$\boldsymbol{L}_h u^{(h)} = \begin{cases} \dfrac{u_{n+1} - u_n}{h} - G(x_n, u_n), & n = 0, 1, 2, \ldots, N-1, \\ u_0, \end{cases}$$

$$f^{(h)} = \begin{cases} \varphi(x_n), & n = 0, 1, 2, \ldots, N-1, \\ \psi. \end{cases}$$

At the same time, the perturbed problem (9.42) in detail reads [cf. formula (9.46)]:

$$\frac{z_{n+1}-z_n}{h} - G(x_n, z_n) = \varphi(x_n) + \varepsilon_n, \quad n = 0, 1, \dots, N-1, \qquad z_0 = \psi + \varepsilon, \tag{9.49}$$

which means that in formula (9.42) we can take:

$$\varepsilon^{(h)} = \begin{cases} \varepsilon_n, & n = 0, 1, 2, \dots, N-1, \\ \varepsilon. \end{cases}$$

Let us now subtract equations (9.46) from the respective equations (9.49) termwise. In so doing, let us also denote $z_n - u_n = w_n$ and take into account that

$$G(x_n, z_n) - G(x_n, u_n) = \frac{\partial G(x_n, \xi_n)}{\partial u} w_n \equiv M_n^{(h)} w_n,$$

where ξ_n is some number in between z_n and u_n. We will thus obtain the following system of equations with the unknown $w^{(h)} = \{w_0, w_1, \dots, w_N\}$:

$$\frac{w_{n+1}-w_n}{h} - M_n^{(h)} w_n = \varepsilon_n, \quad n = 0, 1, \dots, N-1, \qquad w_0 = \varepsilon. \tag{9.50}$$

According to (9.48), we can claim that $\forall n$: $|M_n^{(h)}| \leq M$. Then, system (9.50) yields:

$$\begin{aligned} |w_{n+1}| = &|(1 + hM_n^{(h)})w_n + h\varepsilon_n| \leq (1 + Mh)|w_n| + h|\varepsilon_n| \\ \leq &(1 + Mh)^2 |w_{n-1}| + h(1 + Mh)|\varepsilon_{n-1}| + h|\varepsilon_n| \\ \leq &(1 + Mh)^2 |w_{n-1}| + 2h(1 + Mh)\|\varepsilon^{(h)}\|_{F_h} \\ \leq &(1 + Mh)^3 |w_{n-2}| + 3h(1 + Mh)^2 \|\varepsilon^{(h)}\|_{F_h} \\ &\dots\dots\dots\dots\dots\dots\dots\dots\dots\dots \\ \leq &(1 + Mh)^{n+1} |w_0| + (n+1)h(1 + Mh)^n \|\varepsilon^{(h)}\|_{F_h} \\ \leq &2(1 + Mh)^N \|\varepsilon^{(h)}\|_{F_h} \leq 2e^M \|\varepsilon^{(h)}\|_{F_h}, \end{aligned}$$

because $(n+1)h \leq Nh = 1$. Hence:

$$|w_{n+1}| \leq 2e^M \|\varepsilon^{(h)}\|_{F_h}.$$

This inequality obviously implies an estimate of type (9.43):

$$\|w^{(h)}\|_{U_h} \leq 2e^M \|\varepsilon^{(h)}\|_{F_h},$$

which is equivalent to stability with the constant $c_2 = 2e^M$ in the nonlinear sense (Definition 9.2). Let us also recall that scheme (9.46) has accuracy $\mathcal{O}(h)$ on the solution $u = u(x)$ of problem (9.47), see Exercise 1 of the previous Section 9.2.

Theorem 9.1 would therefore guarantee a first order convergence with respect to h of the forward Euler scheme (9.46) on the interval $0 \leq x \leq 1$.

In our next example, we will analyze convergence of the finite-difference scheme (9.7) for the second order boundary value problem (9.4). This scheme is consistent and guarantees second order accuracy due to formula (9.20a). However, stability of the scheme (9.7) yet remains to be verified.

Due to the linearity, it is sufficient to establish the unique solvability of the problem:

$$\begin{gathered} \frac{u_{n+1} - 2u_n + u_{n-1}}{h^2} - (1 + x_n^2)u_n = g_n, \quad n = 1, 2, \ldots, N-1, \\ u_0 = \alpha, \quad u_N = \beta, \end{gathered} \tag{9.51}$$

for arbitrary $\{g_n\}$, α, and β, and to obtain the estimate:

$$\max_n |u_n| \leq c \max\{|\alpha|, |\beta|, \max_n |g_n|\}. \tag{9.52}$$

We have, in fact, previously considered a problem of type (9.51) in the context of the tri-diagonal elimination, see Section 5.4.2 of Chapter 5. For the problem:

$$\begin{gathered} a_n u_{n-1} + b_n u_n + c_n u_{n-1} = g_n, \\ u_0 = \alpha, \quad u_N = \beta \end{gathered}$$

under the assumption:

$$\forall n: \ |b_n| > |a_n| + |c_n| + \delta, \quad \delta > 0,$$

we have proven its unique solvability and the estimate:

$$\max_n |u_n| \leq \max\left\{|\alpha|, |\beta|, \frac{2}{\delta} \max_m |g_m|\right\}. \tag{9.53}$$

In the case of problem (9.51), we have:

$$a_n = \frac{1}{h^2}, \quad c_n = \frac{1}{h^2}, \quad b_n = -\frac{2}{h^2} - 1 - x_n^2.$$

Consequently, $|b_n| > |a_n| + |c_n| + 1$, and estimate (9.53) therefore implies estimate (9.52) with $c = 2$. We have thus proven stability of the scheme (9.7). Along with the second order accuracy of the scheme, stability yields its convergence with the rate $\mathcal{O}(h^2)$. Note that the scheme (9.7) is linear, and its stability was proven in the sense of Definition 9.3. Because of the linearity, it is equivalent to Definition 9.2 that was used when proving convergence in Theorem 9.1.

Let us also emphasize that the approach to proving convergence of the scheme via independently verifying its consistency and stability is, in fact, quite general. Formula $\boldsymbol{L}u = f$ does not necessarily have to denote an initial or boundary value problem for an ordinary differential equation; it can actually be an operator equation from a rather broad class. We outline this general framework in Section 10.1.6 of Chapter 10, where we discuss the Kantorovich theorem. Here we just mention that it is not very important what type of problem the function u solves. The continuous operator equation $\boldsymbol{L}u = f$ is only needed in order to be able to construct its difference counterpart $\boldsymbol{L}_h u^{(h)} = f^{(h)}$. In Section 9.3.3, we elaborate on the latter consideration. Namely, we provide an example of a consistent, stable, and therefore convergent scheme for an integral rather than differential equation.

9.3.3 Convergent Scheme for an Integral Equation

We will construct and analyze a difference scheme for solving the equation:

$$\boldsymbol{L}u \equiv u(x) - \int_0^1 K(x,y)u(y)dy = f(x), \quad 0 \leq x \leq 1, \tag{9.54}$$

while assuming that the kernel in (9.54) is bounded: $|K(x,y)| \leq \rho < 1$.

Let us specify a positive integer N, set $h = 1/N$, and introduce a uniform grid D_h on $[0,1]$ as before: $x_n = nh$, $n = 0,1,2,\ldots,N$. As usual, our unknown grid function will be the table of values $[u]_h$ of the exact solution $u(x)$ on the grid D_h.

To obtain a finite-difference scheme for the approximate computation of $[u]_h$ we will replace the integral in each equality:

$$u(x_n) - \int_0^1 K(x_n,y)u(y)dy = f(x_n), \quad n = 0,1,2,\ldots,N, \tag{9.55}$$

by a finite sum using one of the numerical quadrature formulas discussed in Chapter 4, namely, the trapezoidal rule of Section 4.1.2. It says that for an arbitrary twice differentiable function $\varphi = \varphi(y)$ on the interval $0 \leq y \leq 1$, the following approximate equality holds:

$$\int_0^1 \varphi(y)dy \approx h\left(\frac{\varphi_0}{2} + \varphi_1 + \ldots + \varphi_{N-1} + \frac{\varphi_N}{2}\right), \quad h = \frac{1}{N},$$

and the corresponding approximation error is $\mathcal{O}(h^2)$. Having replaced the integral in every equality (9.55) with the aforementioned sum, we obtain:

$$\begin{aligned} u_n - h\Bigg(\frac{K(x_n,0)}{2}u_0 + K(x_n,h)u_1 + \ldots + K(x_n,(N-1)h)u_{N-1} \\ + \frac{K(x_n,1)}{2}u_N\Bigg) = f_n, \quad n = 0,1,2,\ldots,N. \end{aligned} \tag{9.56}$$

The system of equations (9.56) can be recast in the standardized form $\boldsymbol{L}_h u^{(h)} = f^{(h)}$ if we set:

$$\boldsymbol{L}_h u^{(h)} = \begin{cases} g_0, \\ g_1, \\ \vdots \\ g_N, \end{cases} \qquad f^{(h)} = \begin{cases} f(0), \\ f(h), \\ \vdots \\ f(Nh) \equiv f(1), \end{cases}$$

where

$$g_n = u_n - h\left(\frac{K(x_n,0)}{2}u_0 + K(x_n,h)u_1 + \ldots + \frac{K(x_n,1)}{2}u_N\right),$$
$$n = 0,1,2,\ldots,N.$$

The scheme $\boldsymbol{L}_h u^{(h)} = f^{(h)}$ we have just constructed is consistent and has second order accuracy on the solution $u = u(x)$ of the integral equation $\boldsymbol{L}u = f$ of (9.54), because the error of the trapezoidal rule is $\mathcal{O}(h^2)$.

Let us now determine whether the scheme (9.56) is also stable. Assume that $u^{(h)} = \{u_0, u_1, \ldots, u_N\}$ is a solution of system (9.56), and let u_s be its component with the maximum absolute value:

$$|u_s| \geq |u_n|, \quad n = 0,1,2,\ldots,N.$$

If there are more than one component with the same maximum absolute value, we will take one of those. Then, from the equation with $n = s$ of system (9.56) we find:

$$\begin{aligned}|f(x_s)| &= \left|u_s - h\left(\frac{K(x_s,0)}{2}u_0 + K(x_s,h)u_1 + \ldots + \frac{K(x_s,1)}{2}u_N\right)\right| \\ &\geq |u_s| - h\left(\frac{\rho}{2} + \underbrace{\rho + \ldots + \rho}_{N-1 \text{ times}} + \frac{\rho}{2}\right)|u_s| = (1 - Nh\rho)|u_s| = (1-\rho)|u_s|,\end{aligned}$$

because $|K(x,y)| \leq \rho < 1$. Consequently,

$$\|u^{(h)}\|_{U_h} = \max_n |u_n| = |u_s| \leq \frac{1}{1-\rho}|f(x_s)| \leq \frac{1}{1-\rho}\|f^{(h)}\|_{F_h}. \tag{9.57}$$

In particular, when $f(x_n) \equiv 0$, inequality (9.57) implies that the homogeneous counterpart of system (9.56) may only have a trivial solution. Therefore, system (9.56) is uniquely solvable for any given right-hand side $f_n \equiv f(x_n)$, $n = 0,1,\ldots,N$. For the solution $u^{(h)} = \{u_n\}$, inequality (9.57) implies stability in the sense of Definition 9.3, with the constant in estimate (9.44) equal to $c_2 = 1/(1-\rho)$. According to Theorem 9.1, solution $u^{(h)}$ then converges to $[u]_h$ with the second order:

$$\|[u]_h - u^{(h)}\|_{U_h} = \max_n |u(nh) - u_n| \leq ch^2, \qquad c = \text{const}.$$

9.3.4 The Effect of Rounding

A consistent and stable finite-difference scheme converges, and can therefore be used, at least in theory, for computing a sequence of increasingly more accurate approximations to the unknown solution (as the grid is refined). In practice, however, the original design of the scheme is never realized exactly, due to the round-off errors in specifying both its coefficients and its right-hand sides.

Convergence with the rate $\mathcal{O}(h^k)$, $k > 0$, as $h \longrightarrow 0$ means that we are expecting to obtain an answer with approximately $\ln(1/h)$ accurate decimal digits. Therefore, it is natural to require that when the grid size decreases, the coefficients of the scheme and its right-hand sides be specified with increasing accuracy as well. In doing so, one has to maintain a similar number of correct digits, on the order of $\ln(1/h)$. This requirement is not very stringent, because $\ln(1/h)$ is a slowly growing function of h.

Suppose that the accuracy of specifying the coefficients of a stable scheme is commensurate with the accuracy that the original unperturbed scheme would have on a given solution of the underlying differential problem [say, both are $\mathcal{O}(h^k)$]. Then it is possible to show that the actual perturbed scheme will remain stable and will also keep the same order of accuracy, $\mathcal{O}(h^k)$. Therefore, according to Theorem 9.1, its k-th order of convergence with respect to the grid size h will not deteriorate.

If one conducts the computations and refines the grid while not synchronously increasing the number of correct decimal digits in the definition of the scheme, then one should not, generally speaking, expect any increase of the overall accuracy. In practice, however, there is usually no need to continuously redefine and improve the coefficients as $h \longrightarrow 0$. The reason is that on actual computers the number of decimal digits available for the machine representation of numbers is fixed. Most often one uses either single precision given by 23 binary digits or equivalently, about 7 decimal digits, or double precision given by 52 binary digits or equivalently, about 16 decimal digits. If, for example, the coefficients of the scheme and its right-hand sides are specified with double precision, then this accuracy would typically far supersede the magnitude of the truncation error on any realistic grid.

A perhaps even more important issue is the round-off errors that inevitably and continuously appear in the course of computing the difference solution by a given scheme (as opposed to one-time errors in coefficients). These errors may accumulate, which, in turn, may necessitate running the computations with ever increasing number of correct decimal digits. If this number happens to grow "too fast" as a function of $1/h$, then the corresponding algorithm is deemed unstable and inappropriate for computations. Of course, had we only been able to conduct the computations with an infinite precision, we would not have had this problem at all.

For serious real-life problems, it may generally prove both difficult and cumbersome to analyze the effect of rounding errors on the performance of actual algorithms. Often, we can only make a judgment based on experimental observations. However, for a number of relatively simple cases we can basically see how many spare digits are needed using only the stability of the original scheme and the previous result that allows to specify this scheme approximately rather than exactly. The idea of the argument is that the round-off errors committed in the course of computation can sometimes be interpreted as errors in specifying the right-hand side of the scheme, up to a given factor of h^m, $m > 0$. Then the previous result implies that for a stable scheme these errors would not hamper the convergence, provided that the computations are conducted with the number of digits that slowly increases as $c\ln(1/h)$, $c = \text{const}$. In practice, again, the number of digits is fixed ahead of time at a value that would be a priori sufficient.

9.3.5 General Comments. A-stability

Consider the Cauchy problem (9.36) for a first order constant coefficient ordinary differential equation; solution of this problem is $u(x) = be^{-Ax}$. Problem (9.36) can be approximated by the forward Euler scheme (9.46) with first order accuracy, here $G(x_n, u_n) = -Au_n$, $\varphi_n = 0$, $n = 0, 1, \ldots, N-1$, and $\psi = b$. Solution of the corresponding finite-difference equation is then given by the formula:

$$u_n = b(1 - Ah)^n. \tag{9.58}$$

Let $A < 0$. Then we easily see from (9.58) that $\|u^{(h)}\|_{U_h} = \max\limits_{0 \le n \le N} |u_n| = |u_N| = |b|(1 + |A|h)^{\frac{1}{h}}$. We also note that in this simple homogeneous case the discrete right-hand side $f^{(h)}$ consists only of the initial value b. As such, $\|u^{(h)}\|_{U_h} = (1+|A|h)^{\frac{1}{h}}\|f^{(h)}\|_{F_h}$, so that in the stability inequality:

$$\|u^{(h)}\|_{U_h} \le c_2 \|f^{(h)}\|_{F_h}, \quad c_2 = \text{const}, \tag{9.59}$$

the constant c_2 has a lower bound: $c_2 \ge (1+|A|h)^{\frac{1}{h}} \longrightarrow e^{|A|} \equiv e^{-A}$, as $h \longrightarrow 0$. At the same time, in our general stability analysis for the forward Euler method, see page 275, we have obtained an upper bound for this constant: $c_2 \le 2e^M = 2e^{-A}$. We therefore see that not only the general upper bound for the stability constant in problem (9.36) with $A < 0$ has exponential behavior, but also that it is impossible to choose a stability constant that would grow slower than the exponential e^{-A}.

Clearly, for $|A| \gg 1$ the foregoing stability constant becomes very large. This situation is actually quite general. No convergent scheme that approximates problem (9.36) for $A < 0$, $|A| \gg 1$, may have a small stability constant. Indeed, for small grid sizes the finite-difference solution is supposed to be close to the continuous solution. However, for this continuous solution itself we have $\|u\| = \max\limits_{x \in [0,1]} |u(x)| = \max\limits_{x \in [0,1]} |be^{-Ax}| = |b|e^{-A}$, so that its maximum on the interval $[0,1]$ is a factor of e^{-A} greater than the magnitude of the initial value b, where e^{-A} is large.

In contradistinction to the previous case, when $A > 0$ stability constants do not grow. For the homogeneous equation (9.36), it immediately follows from formula (9.58). Indeed, let $h < 2/A$. Then, $\|u^{(h)}\|_{U_h} = \max\limits_{0 \le n \le N} |u_n| = \max\limits_{0 \le n \le N} |b||1 - Ah|^n = |b||1 - Ah|^0 = |b| = \|f^{(h)}\|_{F_h}$. Therefore, we can set $h_0 = 2/A$ in Definition 9.2 and consider $c_2 = 1$ in the stability inequality (9.59). For the general inhomogeneous equation, a straightforward modification of the argument given on page 275 will allow us to conclude that $c_2 \le 2$.

The foregoing drastic change of behavior for the stability constants is obviously accounted for by the change of behavior of the corresponding continuous solution $u(x) = be^{-Ax}$. If $A < 0$, we have $\max\limits_{0 \le x \le 1} |u(x)| = e^{|A|}$, which is large for $|A| \gg 1$, whereas if $A > 0$, then $\max\limits_{0 \le x \le 1} |u(x)| = u(0) = 1$ irrespective of A. We therefore see that on one hand, stability in the sense of Definitions 9.2 and 9.3 is an intrinsic property of the scheme concerned with its asymptotic behavior as the grid dimension N grows, or equivalently, as the grid size h vanishes, on a finite fixed interval of the independent variable x, say, $0 \leq x \leq 1$. On the other hand, for a stable scheme the constant c_2 that provides a quantitative measure of stability, see inequality (9.59), does depend on the properties of the continuous solution approximated by the scheme.

One should not think though that any stable scheme that approximates problem (9.36) in the "good" case $A > 0$ has a small stability constant. For instance, one can show that the central-difference scheme:

$$\begin{gathered} \frac{u_{n+1} - u_{n-1}}{2h} + Au_n = 0, \quad n = 1, 2, \ldots, N-1, \\ u_0 = b, \quad u_1 = b(1 - Ah), \end{gathered} \tag{9.60}$$

approximates problem (9.36) on its solution $u(x) = be^{-Ax}$ with second order accuracy and is stable, but the stability constant c_2 for $A \gg 1$ is large: $c_2 \geq e^A$.

For practical computations, not only the property of stability itself, but also the values of the stability constants are important — the smaller the constants the better error estimates one can expect to get. Indeed when proving Theorem 9.1, we have obtained inequality (9.45) for the error. Assume that the truncation error $\sim c_1 h^k$ is small. Then according to (9.45), to have the magnitude of the the solution error $\| [u]_h - u^{(h)} \|_{U_h}$ also small, one may not have a very large stability constant c_2.

Let us introduce *the relative error* $\frac{\| [u]_h - u^{(h)} \|_{U_h}}{\| [u]_h \|_{U_h}}$ in the solution. If the stability constant of the scheme happens to be of the same order of magnitude as the norm of the exact solution: $c_2 = \mathcal{O}\left(\| [u]_h \|_{U_h} \right)$, which is the case for the forward Euler scheme applied to problem (9.36) with $A < 0$, then the foregoing relative error basically appears bounded by the (small) value of the truncation error, and the scheme still remains appropriate for computations. If, however, we have a large stability constant while $\| [u]_h \|_{U_h} = \mathcal{O}(1)$ as in the case of scheme (9.60) for $A \gg 1$, then obtaining the solution with a prescribed accuracy may require an overly fine grid and consequently, an excessive and unjustified amount of computations.

On the other hand, we emphasize that scheme (9.60) with large A represents just one particular example, and we would like to caution the reader from "mechanically" generalizing it. Otherwise, the reader can get a distorted perception of all second order schemes. Note that even for the same scheme (9.60), but applied to a problem with $A = \mathcal{O}(1)$, one can easily show that it requires far fewer grid points for achieving a comparable accuracy than the "competing" first order scheme (9.46). Altogether, the advantages and disadvantages of using a given scheme are determined not only by the scheme itself, but also by the problem it is applied to.

Let us now elaborate on the case $A > 0$ just a little further. Every solution of the differential equation $u' + Au = 0$ remains bounded for all $x \geq 0$. It is natural to expect a similar behavior of the finite-difference solution as well. More precisely, consider the same model problem (9.36), but on the semi-infinite interval $x \geq 0$ rather than on $0 \leq x \leq 1$. Let $u^{(h)}$ be its finite-difference solution defined on the grid $x_n = nh$, $n = 0, 1, 2, \ldots$. The corresponding scheme is called asymptotically stable, or *A-stable*, if for any *fixed grid size* h the solution is bounded on the entire grid:

$$\sup_{0 \le n < \infty} |u_n| \le \text{const}.$$

According to formula (9.58), the explicit forward Euler scheme meets the requirement of A-stability if $|1 - Ah| \le 1$, i.e., when $h \le 2/A$. In other words, to guarantee A-stability, the grid size should be sufficiently small, which makes the forward Euler method conditionally A-stable.

The implicit backward Euler scheme:

$$\frac{u_{n+1} - u_n}{h} + Au_{n+1} = 0, \quad u_0 = b, \tag{9.61}$$

approximates the original Cauchy problem with the same first order accuracy. For this scheme, however, instead of formula (9.58) we obtain:

$$u_n = \frac{b}{(1 + Ah)^n}. \tag{9.62}$$

For $A > 0$, formula (9.62) guarantees boundedness of u_n with no constraints on the grid size. That's why the backward Euler scheme is unconditionally A-stable.

To conclude this section, we describe a convenient criterion for the stability analysis of a large class of methods known as linear. The scheme that approximates the Cauchy problem (9.30) is called linear if it can be represented in the form:

$$\sum_{k=0}^{m} \alpha_k u_{n+k} = h \sum_{k=0}^{m} \beta_k G(x_{n+k}, u_{n+k}), \tag{9.63}$$

where it is assumed that $\alpha_m \ne 0$ and $\alpha_0^2 + \beta_0^2 \ne 0$. The forward Euler scheme (9.29) is linear with $m = 1$, $\alpha_0 = -1$, $\alpha_1 = 1$, $\beta_0 = 1$, $\beta_1 = 0$. It is called a single-step method because $m = 1$. The backward Euler scheme (9.34) and the Crank-Nicolson scheme (9.33) are also linear single-step methods. If $m > 1$, the linear method is called multistep; for example, for the schemes (9.35) and (9.60) we have $m = 2$. The popular Runge-Kutta schemes described in Section 9.4 do not, generally speaking, belong to the class of linear methods. Note also that linearity of the scheme in the sense of (9.63) does not necessarily mean that if this scheme is written in the canonical form $\boldsymbol{L}_h u^{(h)} = f^{(h)}$, then the operator $\boldsymbol{L}_h$ will be linear. In other words, linearity with respect to $u^{(h)}$ is not presumed, as the function $G(x, u)$ may be nonlinear.

For an m-step method, by appropriately choosing the coefficients α_k and β_k, $k = 0, \dots, m$, one can always achieve the order of accuracy $\mathcal{O}(h^m)$. The corresponding choice of the coefficients is clearly not unique, as already the examples of the forward and backward Euler schemes demonstrate. If $m > 1$, achieving the overall accuracy $\mathcal{O}(h^m)$ may also require additional initial conditions for the scheme, beyond the condition $u_0 = u(0) = a$ given in (9.30), see Exercise 2 after Section 9.2.

Stability of a linear method can be characterized in terms of the roots of a certain polynomial associated with the scheme. The following Theorem proven, e.g., in

[Sch02], provides a necessary and sufficient condition for stability in the sense of Definition 9.2.

THEOREM 9.2

Assume that the function $G = G(x,u)$ has a bounded partial derivative with respect to u: $\left|\frac{\partial G}{\partial u}\right| \leq M = \text{const}$. Define the following polynomial of degree m:

$$p(t) = \sum_{k=0}^{m} \alpha_k t^k. \tag{9.64}$$

The scheme (9.63) is stable in the sense of Definition 9.2 if and only if all roots q_j, $j = 1,\dots,m$, of the algebraic equation $p(t) = 0$, where $p(t)$ is given by (9.64), belong to the unit disk on the complex plane: $|q_j| \leq 1$, $j = 1,\dots,m$, and no multiple root (if any) can belong to the unit circle (i.e., for multiple roots we may not have $|q_j| = 1$).

Note that the stability criterion presented by Theorem 9.2 is related to the so-called spectral stability theory, which is based on the analysis of the appropriately defined eigenvalues that characterize the scheme, see [GR87]. The spectral theory of stability can be developed for both ordinary and partial differential equations, and we discuss it in Chapter 10, see Section 10.3.

Exercises

1. Consider the finite-difference scheme (9.35).
   a) Verify that it has first order accuracy on the solutions of problem (9.36).
   b) Make sure that formula (9.37) yields solution of equations (9.35).
   c) Prove directly that for u_n given by (9.37): $\max\limits_{0 \leq nh \leq 1} |u_n| \longrightarrow \infty$ as $h \longrightarrow 0$.
   d) Alternatively, employ Theorem 9.2 to analyze stability of the scheme.
2. Analyze stability of the backward Euler scheme (9.34).
   Hint. Use the argument from page 275. Alternatively, employ Theorem 9.2.
3. Analyze stability of the Crank-Nicolson scheme (9.33).
   Hint. Use the argument from page 275. Alternatively, employ Theorem 9.2.
   Also obtain an accurate estimate for the stability constant in the case of a linear constant-coefficient equation $u' + Au = 0$. Make sure it is not large for $A \gg 1$.
4.* Prove that scheme (9.60) has second order accuracy and is stable.
   Hint. For the stability analysis, use the approach similar to the one outlined in the very beginning of Section 9.3. Show that if q_1 and q_2 are roots of the characteristic equation $q^2 + 2Ahq - 1 = 0$ that corresponds to scheme (9.60), then the quantities in front of u_0 and u_1 in formula (9.37) remain bounded for all $n = 0, 1, \dots, N = 1/h$ when $h \longrightarrow 0$. Alternatively, employ Theorem 9.2.
5. Show that the Crank-Nicolson scheme (9.33) for the equation $u' + Au = 0$, $A > 0$, is unconditionally A-stable.

9.4 The Runge-Kutta Methods

We will first discuss a class of widely used finite-difference schemes for solving the Cauchy problem for a scalar first order ordinary differential equation:

$$\frac{du}{dx} - G(x,u) = 0, \quad 0 \leq x \leq 1, \quad u(0) = a. \tag{9.65}$$

Then, we will analyze extensions to first order systems. This is the most general case as an equation or system of any order can be recast as a first order system.

Again, let the grid D_h be uniform on the interval $[0,1]$:

$$x_n = nh, \quad n = 0,1,2,\dots,N, \quad h = 1/N.$$

We will be building schemes for the approximate computation of the table $[u]_h = \{u(x_n)|n = 0,1,2,\dots,N\}$, where $u = u(x)$ is the exact solution of problem (9.65).

We have already described the simplest explicit scheme for solving problem (9.65), which is the forward Euler scheme:

$$\begin{gathered}\frac{u_{n+1} - u_n}{h} - G(x_n, u_n) = 0, \quad n = 0,1,\dots,N-1, \\ u_0 = a.\end{gathered} \tag{9.66}$$

This scheme has first order accuracy, it also converges with the rate $\mathscr{O}(h)$. Computations according to scheme (9.66) can be given a straightforward geometric interpretation. If u_n is already known, then computing u_{n+1} according to the marching formula: $u_{n+1} = u_n + hG(x_n, u_n)$ is basically equivalent to the translation from the point (x_n, u_n) to the point (x_{n+1}, u_{n+1}) on the (x,u) plane along the line tangent to the integral curve of the equation $\frac{du}{dx} - G(x,u) = 0$ that crosses through the point (x_n, u_n).

The Runge-Kutta methods yield explicit schemes capable of attaining higher orders of accuracy; they are most popular for solving ordinary differential equations.

9.4.1 The Runge-Kutta Schemes

Assume that the approximate solution u_n is already known at the node x_n. We need to find u_{n+1} at the next node $x_{n+1} = x_n + h$. To do so, we first choose a positive integer l and write down the following expressions:

$$\begin{aligned}
k_1 &= G(x_n, u_n), \\
k_2 &= G(x_n + \alpha_2 h, u_n + \alpha_2 h k_1), \\
k_3 &= G(x_n + \alpha_3 h, u_n + \alpha_3 h k_2), \\
&\dots\dots\dots\dots\dots\dots\dots\dots \\
k_l &= G(x_n + \alpha_l h, u_n + \alpha_l h k_{l-1}).
\end{aligned}$$

Then, we construct the scheme itself:

$$\frac{u_{n+1}-u_n}{h} - (p_1k_1 + p_2k_2 + \ldots + p_lk_l) = 0, \quad n = 0,1,\ldots,N-1,$$
$$u_0 = a.$$

In doing so, the coefficients $\alpha_2, \alpha_3, \ldots, \alpha_l$, $p_1, p_2, \ldots, p_l$ are selected to obtain the maximum possible order of accuracy for a given l. Once u_n is known and the coefficients have been determined, then one can first calculate $k_1, k_2, \ldots, k_l$, and subsequently obtain the solution $u_{n+1} = u_n + h(p_1k_1 + p_2k_2 + \ldots + p_lk_l)$. Clearly, the computations according to the Runge-Kutta scheme remain explicit.

The simplest Runge-Kutta scheme is the previously analyzed forward Euler scheme (9.66) that corresponds to $l = 1$ and $p_1 = 1$. The predictor-corrector scheme (9.32) is a Runge-Kutta scheme with $l = 2$, $p_1 = p_2 = 1/2$, and $\alpha_2 = 1$.

The Runge-Kutta scheme:

$$\frac{u_{n+1}-u_n}{h} - \frac{1}{6}(k_1 + 2k_2 + 2k_3 + k_4) = 0, \quad n = 0,1,\ldots,N-1, \qquad (9.67)$$
$$u_0 = a,$$

has fourth-order accuracy if the additional quantities are defined as:

$$k_1 = G(x_n, u_n), \qquad k_2 = G\left(x_n + \frac{h}{2}, u_n + \frac{k_1h}{2}\right),$$
$$k_3 = G\left(x_n + \frac{h}{2}, u_n + \frac{k_2h}{2}\right), \qquad k_4 = G(x_n + h, u_n + k_3h).$$

The Runge-Kutta scheme:

$$\frac{u_{n+1}-u_n}{h} - \left(\frac{2\alpha-1}{2\alpha}k_1 + \frac{1}{2\alpha}k_2\right) = 0, \quad n = 0,1,\ldots,N-1, \qquad (9.68)$$
$$u_0 = a,$$

where $k_1 = G(x_n, u_n)$ and $k_2 = G(x_n + \alpha h, u_n + \alpha hk_1)$ has second order accuracy for any fixed $\alpha \neq 0$. We will only analyze the accuracy of scheme (9.68); the analysis for scheme (9.67) can be conducted similarly, but it is more cumbersome.

Solution $u = u(x)$ of the differential equation $u' = G(x,u)$ satisfies the identities:

$$\frac{du}{dx} \equiv G(x, u(x)),$$
$$\frac{d^2u}{dx^2} \equiv \frac{d}{dx}G(x, u(x)) = \frac{\partial G}{\partial x} + \frac{\partial G}{\partial u}G.$$

Therefore, the Taylor formula:

$$\frac{u(x_n+h) - u(x_n)}{h} = u'(x_n) + \frac{h}{2}u''(x_n) + \mathcal{O}(h^2)$$

leads to the following equality for the solution $u(x)$:

$$\frac{u(x_n+h)-u(x_n)}{h} - \left[G+\frac{h}{2}\left(\frac{\partial G}{\partial x}+\frac{\partial G}{\partial u}G\right)\right]\Bigg|_{x=x_n,u=u(x_n)} = \mathcal{O}(h^2). \quad (9.69)$$

Next, we employ the Taylor formula for the function of two variables: $k_2 = G(x_n + \alpha h, u_n + \alpha h k_1)$, and by retaining the terms up to the first order, obtain:

$$\begin{aligned}\frac{2\alpha-1}{2\alpha}k_1 + \frac{1}{2\alpha}k_2 \\ &= \frac{2\alpha-1}{2\alpha}G + \frac{1}{2\alpha}\left[G+\frac{\partial G}{\partial x}\alpha h + \frac{\partial G}{\partial u}\alpha h G + \mathcal{O}(h^2)\right]\Bigg|_{x=x_n,u=u(x_n)} \\ &= \left[G+\frac{h}{2}\left(\frac{\partial G}{\partial x}+\frac{\partial G}{\partial u}G\right)\right]\Bigg|_{x=x_n,u=u(x_n)} + \mathcal{O}(h^2). \quad (9.70)\end{aligned}$$

Consequently, if we substitute $u(x_n)$ instead of u_n and $u(x_{n+1})$ instead of u_{n+1} into the left-hand side of the first equality of (9.68), where $u(x)$ is the solution to (9.65), the resulting expression would coincide with the left-hand side of (9.69) with accuracy $\mathcal{O}(h^2)$. As such, the left-hand side of the first equality of (9.68) has second order accuracy on the solution $u = u(x)$ of the equation $u' = G(x,u)$. Since the initial condition $u_0 = a$ is specified exactly, this completes the proof of the overall second order accuracy of scheme (9.68).

9.4.2 Extension to Systems

All the schemes that we have introduced in Section 9.4.1 for solving the Cauchy problem for a scalar first order ordinary differential equation (9.65) can be easily generalized to systems of such equations. To do so, in formula (9.65) we simply need to consider the vector functions $\boldsymbol{u} = \boldsymbol{u}(x)$ and $\boldsymbol{G} = \boldsymbol{G}(x,\boldsymbol{u})$ instead of $u(x)$ and $G(x,u)$ respectively, and a fixed vector $\boldsymbol{a}$ instead of the given scalar quantity a.[3] For example, the system of ordinary differential equations:

$$\begin{aligned}\frac{dv}{dx} - (x+v^2+\sin w) &= 0,\\ \frac{dw}{dx} + xvw &= 0,\\ v(0) = a_1, \quad w(0) &= a_2\end{aligned}$$

can be recast in the form:

$$\begin{aligned}\frac{d\boldsymbol{u}}{dx} - \boldsymbol{G}(x,\boldsymbol{u}) &= 0,\\ \boldsymbol{u}(0) &= \boldsymbol{a},\end{aligned}$$

[3]All vectors must be of the same dimension.

if we set:

$$\boldsymbol{u}(x) = \begin{bmatrix} v(x) \\ w(x) \end{bmatrix}, \quad \boldsymbol{G}(x, \boldsymbol{u}) = \begin{bmatrix} x + v^2 + \sin w \\ -xvw \end{bmatrix}, \quad \boldsymbol{a} = \begin{bmatrix} a_1 \\ a_2 \end{bmatrix}.$$

The formula for $\boldsymbol{u}_{n+1}$ in the forward Euler scheme for this system:

$$\boldsymbol{u}_{n+1} = \boldsymbol{u}_n + h\boldsymbol{G}(x_n, \boldsymbol{u}_n)$$

can be written in components as follows:

$$\begin{aligned} v_{n+1} &= v_n + h(x_n + v_n^2 + \sin w_n), \\ w_{n+1} &= w_n - hx_n v_n w_n. \end{aligned}$$

All the considerations regarding the order of accuracy for one scalar equation also translate to systems. In so doing, the derivative $\frac{\partial G}{\partial u}$ in formulae (9.69) and (9.70) is replaced by the Jacobi matrix:

$$\frac{\partial \boldsymbol{G}}{\partial \boldsymbol{u}} = \begin{bmatrix} \frac{\partial G_1}{\partial u_1} & \frac{\partial G_1}{\partial u_2} & \dots & \frac{\partial G_1}{\partial u_n} \\ \frac{\partial G_2}{\partial u_1} & \frac{\partial G_2}{\partial u_2} & \dots & \frac{\partial G_2}{\partial u_n} \\ \vdots & \vdots & \ddots & \vdots \\ \frac{\partial G_n}{\partial u_1} & \frac{\partial G_n}{\partial u_2} & \dots & \frac{\partial G_n}{\partial u_n} \end{bmatrix}.$$

An arbitrary system of ordinary differential equations resolved with respect to the highest-order derivatives can be reduced to a system of first order equations, $d\boldsymbol{u}/dx = \boldsymbol{G}(x, \boldsymbol{u})$, by introducing additional unknowns. We illustrate this technique by the following example. The Cauchy problem:

$$\begin{aligned} \frac{d^2v}{dx^2} + \sin\left(x\frac{dv}{dx} + v^2 + w\right) &= 0, \\ \frac{dw}{dx} + \sqrt{x^2 + v^2 + \left(\frac{dv}{dx}\right)^2} + w^2 &= 0, \\ v(0) = a, \quad \left.\frac{dv}{dx}\right|_{x=0} = b, \quad w(0) &= c \end{aligned}$$

is transformed into that for a first order system if we set: $u_1 = v$, $u_2 = \frac{dv}{dx}$, and $u_3 = w(x)$. Then we obtain:

$$\begin{aligned} \frac{du_1}{dx} - u_2 &= 0, \\ \frac{du_2}{dx} + \sin\left(xu_2 + u_1^2 + u_3\right) &= 0, \\ \frac{du_3}{dx} + \sqrt{x^2 + u_1^2 + u_2^2 + u_3^2} &= 0, \\ u_1(0) = a, \quad u_2(0) = b, \quad u_3(0) &= c. \end{aligned}$$

REMARK 9.1 Special Runge-Kutta type schemes have been developed that can solve second order ordinary differential equations directly, i.e., without initially reducing them to first order systems. □

Exercises

1.* Assume that the function $G = G(x,u)$ has a bounded partial derivative with respect to u, i.e., that inequality (9.48) holds. Prove that the second order Runge-Kutta scheme (9.68) for the Cauchy problem (9.65) is stable in the sense of Definition 9.2.

   Hint. Develop an argument similar to the one used in Section 9.3.2 when proving stability of the forward Euler method. Use the Taylor formula with the error term (remainder) of order $\mathcal{O}(h)$ in the Lagrange form.

2. Consider a Cauchy problem for the first order ordinary differential equation: $u' + u = 0$, $u(0) = 1$, $0 \le x \le 1$. To solve this problem numerically, implement on the computer the forward Euler scheme (9.66), the second order Runge-Kutta scheme (9.68) [more precisely, (9.32)], and the fourth order Runge-Kutta scheme (9.67). Verify experimentally the first order convergence, the second order convergence, and the fourth-order convergence, respectively, of the corresponding finite-difference solutions to the exact solution $u(x) = e^{-x}$ of the differential problem.

3. Consider a Cauchy problem for the second order ordinary differential equation: $u'' + u = 0$, $u(0) = 0\ u'(0) = 1$, $0 \le x \le \pi/2$. To solve this problem numerically, first reduce it to a system of two first order equations, and then follow the recipes of Exercise 2; the exact solution is $u(x) = \sin x$.

9.5 Solution of Boundary Value Problems

We will outline some techniques for the numerical solution of boundary value problems using the following simple example of a boundary value problem for the second order ordinary differential equation:

$$\begin{aligned} y'' &= f(x,y,y'), \quad 0 \le x \le 1, \\ y(0) &= Y_0, \quad y(1) = Y_1, \end{aligned} \tag{9.71}$$

where the boundary conditions are specified at both endpoints of the interval $[0,1]$. This is in contradistinction to the Cauchy, or initial value problem, for which the initial conditions are always specified only at one endpoint.

9.5.1 The Shooting Method

In Section 9.4, we have described a number of numerical methods for solving the Cauchy problem, i.e., a problem of the type:

$$y'' = f(x,y,y'), \quad 0 \le x \le 1,$$
$$y(0) = Y_0, \quad \left.\frac{dy}{dx}\right|_{x=0} = \tan\alpha, \tag{9.72}$$

where Y_0 is the ordinate of the point $(0,Y_0)$ on the (x,y) plane from which the integral curve of the differential equation $y'' = f(x,y,y')$ originates, and $\tan\alpha$ is the slope of the tangent to this integral curve with respect to the horizontal axis, see Figure 9.1.

For a fixed Y_0, the solution of problem (9.72) is a function of the independent variable x and of the parameter α: $y = y(x,\alpha)$. When $x = 1$, the solution $y = y(x,\alpha)$ depends only on α:

$$y(x,\alpha)\big|_{x=1} = y(1,\alpha).$$

Using this consideration, we can reformulate the boundary value problem (9.71) as follows: Find such an angle $\alpha = \alpha^*$, for which the integral curve of the equation $y'' = f(x,y,y')$ would hit exactly the point $(1,Y_1)$:

$$y(1,\alpha) = Y_1, \tag{9.73}$$

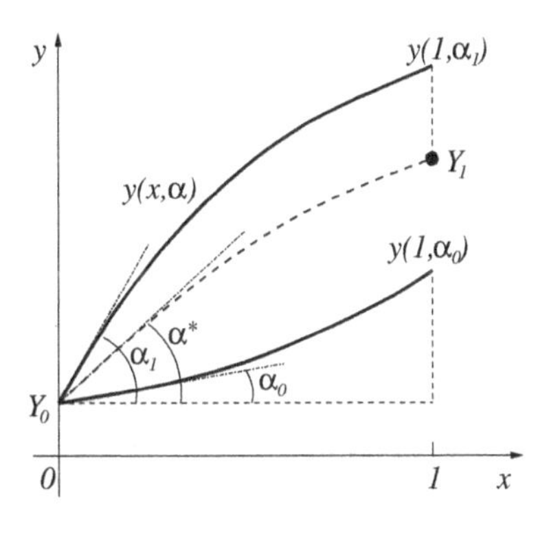

FIGURE 9.1: The shooting method.

provided that this curve originates at the point $(0,Y_0)$ and has slope $\tan\alpha$ with respect to the horizontal axis at its origin. For this particular $\alpha = \alpha^*$, the solution of the Cauchy problem (9.72) will obviously coincide with the solution of the boundary value problem (9.71) that we are seeking. As such, the overall problem reduces to solving equation (9.73).

Equation (9.73) is a functional equation of the general type $F(\alpha) = 0$, where $F(\alpha) = y(1,\alpha) - Y_1$. The only difference between this equation and the traditional equations is that the function $F = F(\alpha)$ is not specified by an analytic expression and is rather introduced through an algorithm for solving the Cauchy problem (9.72). The essence of the shooting method is precisely the foregoing reduction of the boundary value problem (9.71) to the initial value problem (9.72).

To solve equation (9.73), one can employ any of the conventional rootfinding techniques for scalar algebraic equations (Chapter 8). Perhaps the simplest approach employs the well-known idea of bisection. We start with specifying α_0 and α_1, the initial guesses, so that the differences $y(1,\alpha_0) - Y_1$ and $y(1,\alpha_1) - Y_1$ have opposite signs, see Figure 9.1. Then, we take the mid-point of the interval $[\alpha_0, \alpha_1]$:

$$\alpha_2 = \frac{\alpha_0 + \alpha_1}{2}$$

and evaluate $y(1,\alpha_2)$. The next iteration, α_3, is found according to one of the two formulae:

$$\alpha_3 = \frac{\alpha_0+\alpha_2}{2} \quad \text{or} \quad \alpha_3 = \frac{\alpha_2+\alpha_1}{2},$$

depending on whether the differences $y(1,\alpha_0)-Y_1$ and $y(1,\alpha_2)-Y_1$ have the opposite signs or the same sign. Then, the quantity $y(1,\alpha_3)$ is computed. The process is continued until the desired accuracy $|y(1,\alpha_n)-Y_1| < \varepsilon$ is reached.

If one chooses to use the secant method for solving equation (9.73), then, having specified the two initial values α_0 and α_1, one computes the subsequent α_n according to the recursion formula:

$$\alpha_{n+1} = \alpha_n - \frac{F(\alpha_n)}{F(\alpha_n)-F(\alpha_{n-1})}(\alpha_n-\alpha_{n-1}), \quad n=1,2,\ldots.$$

One can also employ Newton's method for solving equation (9.73), see Section 8.3 of Chapter 8.

In general, the shooting method that reduces the solution of the boundary value problem (9.71) to the repeated computation of the solution to the initial value problem (9.72), works well if the solution $y=y(x,\alpha)$ is not "overly sensitive" to the value of α. Otherwise, the method may become computationally unstable even if problem (9.71) is well conditioned.

Indeed, consider the following example of a linear boundary value problem:

$$\begin{aligned} &y''-a^2y=0, \quad 0\le x\le 1,\\ &y(0)=Y_0, \quad y(1)=Y_1, \end{aligned} \tag{9.74}$$

where $a^2=\text{const}$. The solution of problem (9.74) is given by the expression:

$$y(x) = \frac{e^{-ax}-e^{-a(2-x)}}{1-e^{-2a}}Y_0 + \frac{e^{-a(1-x)}-e^{-a(1+x)}}{1-e^{-2a}}Y_1. \tag{9.75}$$

Note that the coefficients in front of Y_0 and Y_1 in formula (9.75) remain bounded functions of their argument $x\in[0,1]$ uniformly with respect to $a>0$, namely, both coefficients never exceed one. Consequently, small perturbations of the values of Y_0 and Y_1 may only cause small perturbations in the solution $y(x)$ given by (9.75), and hence problem (9.74) is well conditioned.

Let us now analyze the Cauchy problem that corresponds to (9.74):

$$\begin{aligned} &y''-a^2y=0, \quad 0\le x\le 1,\\ &y(0)=Y_0, \quad \left.\frac{dy}{dx}\right|_{x=0}=\tan\alpha, \end{aligned} \tag{9.76}$$

The solution of problem (9.76) can be written as:

$$y(x) = \frac{aY_0+\tan\alpha}{2a}e^{ax} + \frac{aY_0-\tan\alpha}{2a}e^{-ax}. \tag{9.77}$$

If we have an error of magnitude ε when specifying $\tan\alpha$, then the value of the solution $y(x)$ of (9.77) at $x = 1$ gets perturbed by the quantity:

$$\Delta y(1) = \frac{\varepsilon}{2a}e^{a} - \frac{\varepsilon}{2a}e^{-a}. \tag{9.78}$$

If a is large, then the subtrahend on the right-hand side of formula (9.78) is negligibly small, however, the coefficient in front of ε in the minuend becomes large. Therefore, the quantity $y(1,\alpha)$ that we need to repeatedly compute in the course of shooting for different values of α may actually appear too sensitive to α itself. This will make the shooting method inappropriate for practical computation when the boundary value problem (9.74) needs to be solved numerically for a large a.

9.5.2 Tri-Diagonal Elimination

To compute the solution of the linear boundary value problem:

$$\begin{gathered} y'' - p(x)y = f(x), \quad 0 \le x \le 1, \\ y(0) = Y_0, \quad y(1) = Y_1, \end{gathered}$$

where $p(x) \ge \delta > 0$, one can employ the second order scheme:

$$\begin{gathered} \frac{y_{n+1} - 2y_n + y_{n-1}}{h^2} - p(x_n)y_n = f(x_n), \quad n = 1,2,\dots,N-1, \\ y_0 = Y_0, \quad y_N = Y_1, \end{gathered}$$

and subsequently use the tri-diagonal elimination for solving the resulting system of linear algebraic equations. One can easily see that the sufficient conditions for using tri-diagonal elimination, see Section 5.4 of Chapter 5, will be satisfied.

9.5.3 Newton's Method

As we have seen, the shooting method of Section 9.5.1 may become unstable even if the original boundary value problem is well conditioned. On the other hand, the disadvantage of the tri-diagonal elimination as it applies to solving boundary value problems (Section 9.5.2) is that it can only be used for the linear case.

An alternative is offered by the direct application of Newton's method. Recall, in Section 9.5.1 we have indicated that Newton's method can be used for finding roots of the scalar algebraic equation (9.73) in the context of shooting. This method, however, can also be used directly for solving systems of nonlinear algebraic and differential equations, see Section 8.3 of Chapter 8. Newton's method is based on linearization, i.e., on reducing the solution of a nonlinear problem to the solution of a sequence of linear problems.

Consider the same boundary value problem (9.71) and assume that there is a function $y = y_0(x)$ that satisfies the boundary conditions of (9.71) exactly, and also roughly approximates the unknown solution $y = y(x)$. We will use this function

$y = y_0(x)$ as the initial guess for Newton's iteration. Let

$$y(x) = y_0(x) + v(x), \tag{9.79}$$

where $v(x)$ is the correction to the initial guess $y_0(x)$. Substitute equality (9.79) into formula (9.71) and linearize the problem around y_0 using the following relations:

$$y'(x) = y_0'(x) + v'(x), \quad y''(x) = y_0''(x) + v''(x),$$

$$f(x, y_0 + v, y_0' + v') = f(x, y_0, y_0') + \frac{\partial f(x, y_0, y_0')}{\partial y} v + \frac{\partial f(x, y_0, y_0')}{\partial y'} v' + \mathcal{O}(v^2 + |v'|^2).$$

Disregarding the quadratic remainder $\mathcal{O}(v^2 + |v'|^2)$, we arrive at the linear problem for the correction $v(x)$:

$$v'' = pv' + qv + \varphi(x), \qquad v(0) = v(1) = 0, \tag{9.80}$$

where

$$p = p(x) = \frac{\partial f(x, y_0, y_0')}{\partial y'}, \quad q = q(x) = \frac{\partial f(x, y_0, y_0')}{\partial y},$$
$$\varphi(x) = f(x, y_0, y_0') - y_0''(x).$$

By solving the linear problem (9.80) either analytically or numerically, we determine the correction v and subsequently define the next Newton iteration as follows:

$$y_1 = y_0 + v.$$

Then, the procedure cyclically repeats itself, and a sequence of iterations $y_1, y_2, y_3, \ldots$ is computed until the difference between the two consecutive iterations becomes smaller than some initially prescribed tolerance, which means convergence of Newton's method.

Note that the foregoing procedure can also be applied directly to the nonlinear finite-difference problem that arises when the differential problem (9.71) is approximated on the grid, see Exercise 5 after Section 8.3 of Chapter 8.

Exercises

1. Show that the coefficients in front of Y_1 and Y_0 in formula (9.75) are indeed bounded by 1 for all $x \in [0, 1]$ and all $a > 0$.
2. Use the shooting method to solve the boundary value problem (9.74) for a "moderate" value of $a = 1$; also take $Y_0 = 1, Y_1 = -1$.

   a) Approximate the original differential equation with the second order central differences on a uniform grid: $x_n = nh, n = 0, 1, \ldots, N, h = 1/N$. Set $y(0) = y_0 = 0$, and use $y(h) = y_1 = \alpha$ as the parameter to be determined by shooting.

   b) Reduce the original second order differential equation to a system of two first order equations and employ the Runge-Kutta scheme (9.68). The unknown parameter for shooting will then be $y'(0) = \alpha$, exactly as in problem (9.76).

In both cases, conduct the computations on a sequence of consecutively more fine grids (reduce the size h by a factor of two several times). Verify experimentally that the numerical solution converges to the exact solution (9.75) with the second order with respect to h.

3.* Investigate applicability of the shooting method to solving the boundary value problem:

$$y'' + a^2 y = 0, \quad 0 \leq x \leq 1,$$
$$y(0) = Y_0, \quad y(1) = Y_1,$$

which has a "+" sign instead of the "−" in the governing differential equation, but otherwise is identical to problem (9.74).

9.6 Saturation of Finite-Difference Methods by Smoothness

Previously, we explored the saturation of numerical methods by smoothness in the context of algebraic interpolation (piecewise polynomials, see Section 2.2.5, and splines, see Section 2.3.2). Very briefly, the idea is to see whether or not a given method of approximation fully utilizes all the information available, and thus attains the optimal accuracy limited only by the threshold of the unavoidable error. When the method introduces its own error threshold, which may only be larger than that of the unavoidable error and shall be attributed to the specific design, we say that the phenomenon of saturation takes place. For example, we have seen that the interpolation by means of algebraic polynomials on uniform grids saturates, whereas the interpolation by means of trigonometric or Chebyshev polynomials does not, see Sections 3.1.3, 3.2.4, and 3.2.7. In the current section, we will use a number of very simple examples to demonstrate that the approximations by means of finite-difference schemes are, generally speaking, also prone to saturation by smoothness.

Before we continue, let us note that in the context of finite differences the term "saturation" may sometimes acquire an alternative meaning. Namely, saturation theorems are proven as to what maximum order of accuracy may a (stable) scheme have if it approximates a particular equation or class of equations on a given fixed stencil.[4] Results of this type are typically established for partial differential equations, see, e.g., [EO80, Ise82]. Some very simple conclusions can already be drawn based on the method of undetermined coefficients (see Section 10.2.2). Note that this alternative notion of saturation is similar to the one we have previously introduced in this book (see Section 2.2.5) in the sense that it also discusses certain accuracy limits. The difference is, however, that these accuracy limits pertain to a particular class of discretization methods (schemes on a given stencil), whereas previously

[4]The stencil of a difference scheme is a set of grid nodes, on which the finite-difference operator is built that approximates the original differential operator.

we discussed accuracy limits that are not related to specific methods and are rather accounted for by the loss of information in the course of discretization.

Let us now consider the following simple boundary value problem for a second order ordinary differential equation:

$$u'' = f(x), \quad 0 \le x \le 1, \quad u(0) = 0, \quad u(1) = 0, \tag{9.81}$$

where the right-hand side $f(x)$ is assumed given.

We introduce a uniform grid on the interval $[0,1]$:

$$x_n = nh, \quad n = 0, 1, \dots, N, \quad Nh = 1,$$

and approximate problem (9.81) using central differences:

$$\begin{gathered} \frac{u_{n+1} - 2u_n + u_{n-1}}{h^2} = f_n \equiv f(x_n), \quad n = 1, 2, \dots, N-1, \\ u_0 = 0, \quad u_N = 0. \end{gathered} \tag{9.82}$$

Provided that the solution $u = u(x)$ of problem (9.81) is sufficiently smooth, or more precisely, provided that its fourth derivative $u^{(4)}(x)$ is bounded for $0 \le x \le 1$, the approximation (9.82) is second-oder accurate, see formula (9.20a). In this case, if the scheme (9.82) is stable, then it will converge with the rate $\mathcal{O}(h^2)$.

However, for such a simple difference scheme as (9.82) one can easily study the convergence directly, i.e., without using Theorem 9.1. A study of that type will be particularly instrumental because on one hand the regularity of the solution may not always be sufficient to guarantee consistency, and on the other hand, it will allow one to see whether or not the convergence accelerates for the functions that are smoother than those minimally required for obtaining $\mathcal{O}(h^2)$.

Note that the degree of regularity of the solution $u(x)$ to problem (9.81) is immediately determined by that of the right-hand side $f(x)$. Namely, the solution $u(x)$ will always have two additional derivatives. It will therefore be convenient to use different right-hand sides $f(x)$ with different degree of regularity, and to investigate directly the convergence properties of scheme (9.82). In doing so, we will analyze both the case when the regularity is formally insufficient to guarantee the second order convergence, and the opposite case when the regularity is "excessive" for that purpose. In the latter case we will, in fact, see that the convergence still remains only second order with respect to h, which implies saturation.

Let us first consider a discontinuous right-hand side:

$$f(x) = \begin{cases} 0, & 0 \le x \le \frac{1}{2}, \\ 1, & \frac{1}{2} < x \le 1. \end{cases} \tag{9.83}$$

On each of the two sub-intervals: $[0, 1/2]$ and $[1/2, 1]$, the solution can be found as a combination of the general solution to the homogeneous equation and a particular solution to the inhomogeneous equation. The latter is equal to zero on $[0, 1/2]$ and on

$[1/2,1]$ it is easily obtained using undetermined coefficients. Therefore, the overall solution of problem (9.81), (9.83) can be found in the form:

$$u(x)=\begin{cases} c_1+c_2x, & 0\le x\le \frac{1}{2},\\ c_3+c_4x+\frac{1}{2}x^2, & \frac{1}{2}<x\le 1, \end{cases} \tag{9.84}$$

where the constants c_1, c_2, c_3, and c_4 are to be chosen so that to satisfy the boundary conditions $u(0)=u(1)=1$ and the continuity requirements:

$$u\left(\frac{1}{2}-0\right)=u\left(\frac{1}{2}+0\right),\quad u'\left(\frac{1}{2}-0\right)=u'\left(\frac{1}{2}+0\right). \tag{9.85}$$

Altogether this yields:

$$\begin{aligned} c_1&=0, & c_3+c_4+\frac{1}{2}&=0,\\ c_1+\frac{c_2}{2}-c_3-\frac{c_4}{2}-\frac{1}{8}&=0, & c_2-c_4-\frac{1}{2}&=0. \end{aligned} \tag{9.86}$$

Solving equations (9.86) we find:

$$c_1=0,\quad c_2=-\frac{1}{8},\quad c_3=\frac{1}{8},\quad c_4=-\frac{5}{8}, \tag{9.87}$$

so that

$$u(x)=\begin{cases} -\frac{1}{8}x, & 0\le x\le \frac{1}{2},\\ \frac{1}{8}-\frac{5}{8}x+\frac{1}{2}x^2, & \frac{1}{2}<x\le 1. \end{cases} \tag{9.88}$$

In the finite-difference case, instead of (9.83) we have:

$$f_n=\begin{cases} 0, & n=0,1,\dots,\frac{N}{2},\\ 1, & n=\frac{N}{2}+1,\frac{N}{2}+2,\dots,N. \end{cases} \tag{9.89}$$

Accordingly, the solution is to be sought for in the form:

$$u_n=\begin{cases} c_1+c_2(nh), & n=0,1,\dots,\frac{N}{2}+1,\\ c_3+c_4(nh)+\frac{1}{2}(nh)^2, & n=\frac{N}{2},\frac{N}{2}+1,\dots,N, \end{cases} \tag{9.90}$$

where on each sub-interval we have a combination of the general solution to the homogeneous difference equation and a particular solution of the inhomogeneous difference equation (obtained by the method of undetermined coefficients). Notice that unlike in the continuous case (9.84), the two grid sub-intervals in formula (9.90) overlap across the entire cell $[N/2,N/2+1]$ (we are assuming that N is even). Therefore, the constants c_1, c_2, c_3, and c_4 in (9.90) are to be determined from the boundary conditions at the endpoints of the interval $[0,1]$: $u_0=u_N=0$, and from the matching conditions in the middle that are given simply as [cf. formula (9.85)]:

$$\begin{aligned} c_1+c_2\left(\frac{N}{2}h\right)&=c_3+c_4\left(\frac{N}{2}h\right)+\frac{1}{2}\left(\frac{N}{2}h\right)^2,\\ c_1+c_2\left(\frac{N}{2}h+h\right)&=c_3+c_4\left(\frac{N}{2}h+h\right)+\frac{1}{2}\left(\frac{N}{2}h+h\right)^2. \end{aligned} \tag{9.91}$$

Altogether this yields:

$$\begin{aligned} c_1 &= 0, & c_3 + c_4 + \frac{1}{2} &= 0, \\ c_1 + \frac{c_2}{2} - c_3 - \frac{c_4}{2} - \frac{1}{8} &= 0, & c_2 - c_4 - \frac{1}{2} - \frac{h}{2} &= 0, \end{aligned} \tag{9.92}$$

where the last equation of system (9.92) was obtained by subtracting the first equation of (9.91) from the second equation of (9.91) and subsequently dividing by h.

Notice that system (9.92) which characterizes the finite-difference case is almost identical to system (9.86) which characterizes the continuous case, except that there is an $\mathcal{O}(h)$ discrepancy in the fourth equation. Accordingly, there is also an $\mathcal{O}(h)$ difference in the values of the constants [cf. formula (9.87)]:

$$c_1 = 0, \quad c_2 = -\frac{1}{8} + \frac{h}{4}, \quad c_3 = \frac{1}{8} + \frac{h}{4}, \quad c_4 = -\frac{5}{8} - \frac{h}{4},$$

so that the solution to problem (9.82), (9.89) is given by:

$$u_n = \begin{cases} -\frac{1}{8}(nh) + \frac{h}{4}(nh), & n = 0, 1, \dots, \frac{N}{2} + 1, \\ \frac{1}{8} - \frac{5}{8}(nh) + \frac{1}{2}(nh)^2 + \frac{h}{4}(1 - nh), & n = \frac{N}{2}, \frac{N}{2} + 1, \dots, N. \end{cases} \tag{9.93}$$

By comparing formulae (9.88) and (9.93), where $nh = x_n$, we conclude that

$$\|[u]_h - u^{(h)}\| = \max_n |u(x_n) - u_n| = \mathcal{O}(h),$$

i.e., that the solution of the finite-difference problem (9.82), (9.89) converges to the solution of the differential problem (9.81), (9.83) with the first order with respect to h. Note that scheme (9.82), (9.89) falls short of the second order convergence because the solution of the differential problem (9.82), (9.89) is not sufficiently smooth.

Instead of the discontinuous right-hand side (9.83) let us now consider a continuous function with discontinuous first derivative:

$$f(x) = \begin{cases} -x, & 0 \le x \le \frac{1}{2}, \\ x - 1, & \frac{1}{2} < x \le 1. \end{cases} \tag{9.94}$$

Solution to problem (9.81), (9.94) can be found in the form:

$$u(x) = \begin{cases} c_1 + c_2 x - \frac{1}{6}x^3, & 0 \le x \le \frac{1}{2}, \\ c_3 + c_4 x + \frac{1}{6}x^3 - \frac{1}{2}x^2, & \frac{1}{2} < x \le 1, \end{cases}$$

where the constants c_1, c_2, c_3, and c_4 are again to be chosen so that to satisfy the boundary conditions $u(0) = u(1) = 1$ and the continuity requirements (9.85):

$$\begin{aligned} c_1 &= 0, & c_3 + c_4 - \frac{1}{3} &= 0, \\ c_1 + \frac{c_2}{2} - \frac{1}{48} - c_3 - \frac{c_4}{2} + \frac{5}{48} &= 0, & c_2 - \frac{1}{8} - c_4 + \frac{3}{8} &= 0. \end{aligned} \tag{9.95}$$

Solving equations (9.95) we find:

$$c_1 = 0, \quad c_2 = \frac{1}{8}, \quad c_3 = -\frac{1}{24}, \quad c_4 = \frac{3}{8}, \tag{9.96}$$

so that

$$u(x) = \begin{cases} \frac{1}{8}x - \frac{1}{6}x^3, & 0 \le x \le \frac{1}{2}, \\ -\frac{1}{24} + \frac{3}{8}x + \frac{1}{6}x^3 - \frac{1}{2}x^2, & \frac{1}{2} < x \le 1. \end{cases} \tag{9.97}$$

In the discrete case, instead of (9.94) we write:

$$f_n = \begin{cases} -(nh), & n = 0, 1, \dots, \frac{N}{2}, \\ (nh) - 1, & n = \frac{N}{2} + 1, \frac{N}{2} + 2, \dots, N, \end{cases} \tag{9.98}$$

and then look for the solution u_n to problem (9.82), (9.98) in the form:

$$u_n = \begin{cases} c_1 + c_2(nh) - \frac{1}{6}(nh)^3, & n = 0, 1, \dots, \frac{N}{2} + 1, \\ c_3 + c_4(nh) + \frac{1}{6}(nh)^3 - \frac{1}{2}(nh)^2, & n = \frac{N}{2}, \frac{N}{2} + 1, \dots, N. \end{cases}$$

For the matching conditions in the middle we now have [cf. formulae (9.91)]:

$$\begin{aligned} c_1 + c_2\left(\frac{N}{2}h\right) - \frac{1}{6}\left(\frac{N}{2}h\right)^3 &= c_3 + c_4\left(\frac{N}{2}h\right) + \frac{1}{6}\left(\frac{N}{2}h\right)^3 - \frac{1}{2}\left(\frac{N}{2}h\right)^2, \\ c_1 + c_2\left(\frac{N}{2}h + h\right) - \frac{1}{6}\left(\frac{N}{2}h + h\right)^3 &= c_3 + c_4\left(\frac{N}{2}h + h\right) \\ &\quad + \frac{1}{6}\left(\frac{N}{2}h + h\right)^3 - \frac{1}{2}\left(\frac{N}{2}h + h\right)^2, \end{aligned} \tag{9.99}$$

and consequently:

$$\begin{aligned} c_1 &= 0, & c_3 + c_4 - \frac{1}{3} &= 0, \\ c_1 + \frac{c_2}{2} - \frac{1}{48} - c_3 - \frac{c_4}{2} + \frac{5}{48} &= 0, & c_2 - \frac{1}{8} - c_4 + \frac{3}{8} - \frac{h^2}{3} &= 0, \end{aligned} \tag{9.100}$$

where the last equation of (9.100) was obtained by subtracting the first equation of (9.99) from the second equation of (9.99) and subsequently dividing by h.

Solving equations (9.100) we obtain [cf. formula (9.96)]:

$$c_1 = 0, \quad c_2 = \frac{1}{8} + \frac{h^2}{6}, \quad c_3 = -\frac{1}{24} + \frac{h^2}{6}, \quad c_4 = \frac{3}{8} - \frac{h^2}{6},$$

and

$$u_n = \begin{cases} \frac{1}{8}(nh) - \frac{1}{6}(nh)^3 + \frac{h^2}{6}(nh), & n = 0, 1, \dots, \frac{N}{2} + 1, \\ -\frac{1}{24} + \frac{3}{8}(nh) + \frac{1}{6}(nh)^3 - \frac{1}{2}(nh)^2 & \\ \qquad + \frac{h^2}{6}(1 - nh), & n = \frac{N}{2}, \frac{N}{2} + 1, \dots, N. \end{cases} \tag{9.101}$$

It is clear that the error between the continuous solution (9.97) and the discrete solution (9.101) is estimated as

$$\|[u]_h - u^{(h)}\| = \max_n |u(x_n) - u_n| = \mathcal{O}(h^2),$$

which means that the solution of the finite-difference problem (9.82), (9.98) converges to the solution of the differential problem (9.81), (9.94) with the second order with respect to h. Note that second order convergence is attained here even though the degree of regularity of the solution — third derivative is discontinuous — is formally insufficient to guarantee second order accuracy (consistency).

In much the same way one can analyze the case when the right-hand side $f(x)$ has one continuous derivative (the so-called C^1 space of functions), for example:

$$f(x) = \begin{cases} -(x-\frac{1}{2})^2, & 0 \le x \le \frac{1}{2}, \\ (x-\frac{1}{2})^2, & \frac{1}{2} < x \le 1. \end{cases} \tag{9.102}$$

For problem (9.81), (9.102), it is also possible to prove the second order convergence, which is the subject of Exercise 1 at the end of the section.

The foregoing examples demonstrate that the rate of finite-difference convergence depends on the regularity of the solution to the underlying continuous problem. It is therefore interesting to see what happens when the regularity increases beyond C^1.

Consider the right-hand side in the form of a quadratic polynomial:

$$f(x) = x(x-1). \tag{9.103}$$

This function is obviously infinitely differentiable (C^∞ space), and so is the solution $u = u(x)$ of problem (9.81), (9.103), which is given by:

$$u(x) = \frac{1}{12}x + \frac{1}{12}x^4 - \frac{1}{6}x^3. \tag{9.104}$$

Scheme (9.82) with the right-hand side

$$f_n = nh(nh-1), \quad n = 0, 1, \dots, N, \tag{9.105}$$

approximates problem (9.81), (9.103) with second order accuracy. The solution of the finite-difference problem (9.82), (9.105) can be found in the form:

$$u_n = \underbrace{c_1 + c_2(nh)}_{u_n^{(g)}} + \underbrace{(nh)^2(A(nh)^2 + B(nh) + C)}_{u_n^{(p)}}, \tag{9.106}$$

where $u_n^{(g)}$ is the general solution to the homogeneous equation and $u_n^{(p)}$ is a particular solution to the inhomogeneous equation. The values of A, B, and C are to be found using the method of undetermined coefficients:

$$A\frac{((n+1)h)^4-2(nh)^4+((n-1)h)^4}{h^2} \\ +B\frac{((n+1)h)^3-2(nh)^3+((n-1)h)^3}{h^2} \\ +C\frac{((n+1)h)^2-2(nh)^2+((n-1)h)^2}{h^2}=(nh)^2-(nh),$$

which yields:

$$A(12(nh)^2+2h^2)+B(6nh)+2C=(nh)^2-(nh)$$

and accordingly,

$$A=\frac{1}{12}, \quad B=-\frac{1}{6}, \quad C=-Ah^2=-\frac{1}{12}h^2. \tag{9.107}$$

For the constants c_1 and c_2 we substitute the expression (9.106) and the already available coefficients A, B, and C of (9.107) into the boundary conditions of (9.82) and write:

$$u_0=c_1=0, \quad u_N=c_2+A+B+C=0.$$

Consequently,

$$c_1=0 \text{ and } c_2=\frac{1}{12}+\frac{1}{12}h^2,$$

so that for the overall solution u_n of problem (9.81), (9.105) we obtain:

$$u_n=\frac{1}{12}(nh)+\frac{1}{12}h^2(nh)+(nh)^2\left(\frac{1}{12}(nh)^2-\frac{1}{6}(nh)-\frac{1}{12}h^2\right). \tag{9.108}$$

Comparing the continuous solution $u(x)$ given by (9.104) with the discrete solution u_n given by (9.108) we conclude that

$$\|[u]_h-u^{(h)}\|=\max_n |u(x_n)-u_n|=\mathcal{O}(h^2),$$

which implies that notwithstanding the infinite smoothness of the right-hand side $f(x)$ of (9.103) and that of the solution $u(x)$, scheme (9.82), (9.105) still shows only second order convergence. This is a manifestation of the phenomenon of *saturation by smoothness*. The rate of decay of the approximation error is determined by the specific approximation method employed on a given grid, and does not reach the level of the pertinent unavoidable error.

To demonstrate that the previous observation is not accidental, let us consider another example of an infinitely differentiable (C^∞) right-hand side:

$$f(x)=\sin(\pi x). \tag{9.109}$$

The solution of problem (9.81), (9.109) is given by:

$$u(x) = -\frac{1}{\pi^2}\sin(\pi x). \tag{9.110}$$

The discrete right-hand side that corresponds to (9.109) is:

$$f_n = \sin(\pi nh), \quad n = 0, 1, \dots, N, \tag{9.111}$$

and the solution to the finite-difference problem (9.82), (9.111) is to be sought for in the form $u_n = A\sin(\pi nh) + B\cos(\pi nh)$ with the undetermined coefficients A and B, which eventually yields:

$$u_n = -\frac{h^2}{4\sin^2\frac{\pi h}{2}}\sin(\pi nh). \tag{9.112}$$

The error between the continuous solution given by (9.110) and the discrete solution given by (9.112) is easy to estimate provided that the grid size is small, $h \ll 1$:

$$\begin{aligned}
\|[u]_h - u^{(h)}\| = \max_n |u(x_n) - u_n| &= \left|\frac{1}{\pi^2} - \frac{h^2}{4\sin^2\frac{\pi h}{2}}\right| \\
&\approx \left|\frac{1}{\pi^2} - \frac{h^2}{4\left[\frac{\pi h}{2} - \frac{1}{6}\left(\frac{\pi h}{2}\right)^3\right]^2}\right| \approx \left|\frac{1}{\pi^2} - \frac{h^2}{4\left[\frac{\pi h}{2} - \frac{1}{3}\left(\frac{\pi h}{2}\right)^4\right]}\right| \\
&\approx \left|\frac{1}{\pi^2} - \frac{h^2}{4\left(\frac{\pi h}{2}\right)^2}\left[1 + \frac{1}{3}\left(\frac{\pi h}{2}\right)^2\right]\right| = \frac{1}{\pi^2}\frac{1}{3}\left(\frac{\pi h}{2}\right)^2 = \mathcal{O}(h^2).
\end{aligned}$$

This, again, corroborates the effect of saturation, as the convergence of the scheme (9.82), (9.111) is only second order in spite of the infinite smoothness of the data.

In general, all finite-difference methods are prone to saturation. This includes the methods for solving ordinary differential equations described in this chapter, as well as the methods for partial differential equations described in Chapter 10. There are, however, other methods for the numerical solution of differential equations. For example, the so-called spectral methods described briefly in Section 9.7 do not saturate and exhibit convergence rates that self-adjust to the regularity of the corresponding solution (similarly to how the error of the trigonometric interpolation adjusts to the smoothness of the interpolated function, see Theorem 3.5 on page 68). The literature on the subject of spectral methods is vast, and we can refer the reader, e.g., to the monographs [GO77, CHQZ88, CHQZ06], and textbooks [Boy01, HGG06].

Exercise

1. Consider scheme (9.82) with the right-hand side:

$$f_n = \begin{cases} -(nh - \frac{1}{2})^2, & n = 0, 1, 2, \dots, \frac{N}{2}, \\ (nh - \frac{1}{2})^2, & \frac{N}{2}, \frac{N}{2}+1, \dots, N. \end{cases}$$

This scheme approximates problem (9.81), (9.102). Obtain the finite-difference solution in closed form and prove second order convergence.

9.7 The Notion of Spectral Methods

In this section, we only provide one particular example of a spectral method. Namely, we solve a simple boundary value problem using a Fourier-based technique. Our specific goal is to demonstrate that alternative discrete approximations to differential equations can be obtained that, unlike finite-difference methods, will not suffer from the saturation by smoothness (see Section 9.6). A comprehensive account of spectral methods can be found, e.g., in the monographs [GO77, CHQZ88, CHQZ06], as well as in the textbooks [Boy01, HGG06]. The material of this section is based on the analysis of Chapter 3 and can be skipped during the first reading.

Consider the same boundary value problem as in Section 9.6:

$$u'' = f(x), \quad 0 \leq x \leq 1, \quad u(0) = 0, \quad u(1) = 0, \tag{9.113}$$

where the right-hand side $f(x)$ is assumed given. In this section, we will not approximate problem (9.113) on the grid using finite differences. We will rather look for an approximate solution to problem (9.113) in the form of a trigonometric polynomial.

Trigonometric polynomials were introduced and studied in Chapter 3. Let us formally extend both the unknown solution $u = u(x)$ and the right-hand side $f = f(x)$ to the interval $[-1,1]$ antisymmetrically, i.e., $u(-x) = -u(x)$ and $f(-x) = -f(x)$, so that the resulting functions are odd. We can then represent the solution $u(x)$ of problem (9.113) approximately as a trigonometric polynomial:

$$u^{(n)}(x) = \sum_{k=1}^{n+1} B_k \sin(\pi k x) \tag{9.114}$$

with the coefficients B_k to be determined. Note that according to Theorem 3.3 (see page 66), the polynomial (9.114), which is a linear combination of the sine functions only, is suited specifically for representing the odd functions. Note also that for any choice of the coefficients B_k the polynomial $u^{(n)}(x)$ of (9.114) satisfies the boundary conditions of problem (9.113) exactly.

Let us now introduce the same grid (of dimension $n+1$) as we used in Section 3.1:

$$x_m = \frac{1}{n+1}m + \frac{1}{2(n+1)}, \quad m = 0, 1, \ldots, n, \tag{9.115}$$

and interpolate the given function $f(x)$ on this grid by means of the trigonometric polynomial with $n+1$ terms:

$$f^{(n)}(x) = \sum_{k=1}^{n+1} b_k \sin(\pi k x). \tag{9.116}$$

The coefficients of the polynomial (9.116) are given by:

$$\begin{aligned} b_k &= \frac{2}{n+1}\sum_{m=0}^{n} f(x_m)\sin k\left(\frac{\pi}{n+1}m + \frac{\pi}{2(n+1)}\right), \quad k = 1,2,\dots,n, \\ b_{n+1} &= \frac{1}{n+1}\sum_{m=0}^{n} f(x_m)(-1)^m. \end{aligned} \tag{9.117}$$

To approximate the differential equation $u'' = f$ of (9.113), we require that the second derivative of the approximate solution $u^{(n)}(x)$:

$$\frac{d^2}{dx^2}u^{(n)}(x) = -\pi^2 \sum_{k=1}^{n+1} B_k k^2 \sin(\pi k x) \tag{9.118}$$

coincide with the interpolant of the right-hand side $f^{(n)}(x)$ at every node x_m of the grid (9.115):

$$\frac{d^2}{dx^2}u^{(n)}(x_m) = f^{(n)}(x_m), \quad m = 0,1,\dots,n. \tag{9.119}$$

Note that both the interpolant $f^{(n)}(x)$ given by formula (9.116) and the derivative $\frac{d^2}{dx^2}u^{(n)}(x)$ given by formula (9.118) are sine trigonometric polynomials of the same order $n+1$. According to formula (9.119), they coincide at x_m for all $m = 0,1,\dots,n$. Therefore, due to the uniqueness of the trigonometric interpolating polynomial (see Theorem 3.1 on page 62), these two polynomials are, in fact, the same everywhere on the interval $0 \le x \le 1$. Consequently, their coefficients are identically equal:

$$-\pi^2 k^2 B_k = b_k, \quad k = 1,2,\dots,n+1. \tag{9.120}$$

Equalities (9.120) allow one to find B_k provided that b_k are known.

Consider a particular example analyzed in the end of Section 9.6:

$$f(x) = \sin(\pi x). \tag{9.121}$$

The exact solution of problem (9.113), (9.121) is given by:

$$u(x) = -\frac{1}{\pi^2}\sin(\pi x). \tag{9.122}$$

According to formulae (9.117), the coefficients b_k that correspond to the right-hand side $f(x)$ given by (9.121) are:

$$b_1 = 1 \text{ and } b_k = 0, \ k = 2,3,\dots,n+1.$$

Consequently, relations (9.120) imply that

$$B_1 = -\frac{1}{\pi^2} \text{ and } B_k = 0, \ k = 2,3,\dots,n+1.$$

Therefore,

$$u^{(n)}(x) = -\frac{1}{\pi^2}\sin(\pi x). \tag{9.123}$$

By comparing formulae (9.123) and (9.122), we conclude that the approximate method based on enforcing the differential equation $u'' = f$ via the finite system of equalities (9.119) reconstructs the exact solution of problem (9.113), (9.121). The error is therefore equal to zero. Of course, one should not expect that this ideal behavior of the error will hold in general. The foregoing particular result only takes place because of the specific choice of the right-hand side (9.121). However, in a variety of other cases one can obtain a rapid decay of the error as n increases.

Consider the odd function $f(-x) = -f(x)$ obtained on the interval $[-1,1]$ by extending the right-hand side of problem (9.113) antisymmetrically from the interval $[0,1]$. Assume that this function can also be translated along the entire real axis:

$$\forall x \in [2l+1, 2(l+1)+1]: \; f(x) = f(x-2(l+1)), \quad l = 0, 1, \pm 2, \pm 3, \ldots$$

so that the resulting periodic function with the period $L = 2$ be smooth. More precisely, we require that the function $f(x)$ constructed this way possess continuous derivatives of order up to $r > 0$ everywhere, and a square integrable derivative of order $r+1$:

$$\int_{-1}^{1} \left[f^{(r+1)}(x) \right]^2 dx < \infty.$$

Clearly the function $f(x) = \sin(\pi x)$, see formula (9.121), satisfies these requirements. Another example which, unlike (9.121), leads to a full infinite Fourier expansion is $f(x) = \sin(\pi \sin(\pi x))$. Both functions are periodic with the period $L = 2$ and infinitely smooth everywhere ($r = \infty$).

Let us represent $f(x)$ as the sum of its sine Fourier series:

$$f(x) = \sum_{k=1}^{\infty} \beta_k \sin(k\pi x), \tag{9.124}$$

where the coefficients β_k are defined as:

$$\beta_k = 2 \int_0^1 f(x) \sin(k\pi x) dx. \tag{9.125}$$

The series (9.124) converges to the function $f(x)$ uniformly and absolutely. The rate of convergence was obtained when proving Theorem 3.5, see pages 68–71. Namely, if we define the partial sum $S_n(x)$ and the remainder $\delta S_n(x)$ of the series (9.124) as done in Section 3.1.3:

$$S_n(x) = \sum_{k=1}^{n+1} \beta_k \sin(k\pi x), \quad \delta S_n(x) = \sum_{k=n+2}^{\infty} \beta_k \sin(k\pi x), \tag{9.126}$$

then

$$|f(x) - S_n(x)| = |\delta S_n(x)| \leq \frac{\zeta_n}{n^{r+\frac{1}{2}}}, \tag{9.127}$$

where ζ_n is a numerical sequence such that $\zeta_n = o(1)$, i.e., $\lim_{n\to\infty} \zeta_n = 0$. Substituting the expressions:

$$f(x_m) = S_n(x_m) + \delta S_n(x_m), \quad m = 0, 1, \ldots, n,$$

into the definition (9.117) of the coefficients b_k we obtain:

$$
\begin{aligned}
b_k &= \underbrace{\frac{2}{n+1}\sum_{m=0}^{n} S_n(x_m)\sin(\pi k x_m)}_{\beta_k} + \underbrace{\frac{2}{n+1}\sum_{m=0}^{n} \delta S_n(x_m)\sin(\pi k x_m)}_{\delta\beta_k}, \quad k = 1,2,\dots,n, \\
b_{n+1} &= \underbrace{\frac{1}{n+1}\sum_{m=0}^{n} S_n(x_m)(-1)^m}_{\beta_{n+1}} + \underbrace{\frac{1}{n+1}\sum_{m=0}^{n} \delta S_n(x_m)(-1)^m}_{\delta\beta_{n+1}}.
\end{aligned}
\tag{9.128}
$$

The first sum on the right-hand side of each equality (9.128) is indeed equal to the genuine Fourier coefficient β_k of (9.125), $k = 1,2,\dots,n+1$, because the partial sum $S_n(x)$ given by (9.126) coincides with its own trigonometric interpolating polynomial[5] for all $0 \le x \le 1$. As for the "corrections" to the coefficients, $\delta\beta_k$, they come from the remainder $\delta S_n(x)$ and their magnitudes can be easily estimated using inequality (9.127) and formulae (9.117):

$$
|\delta\beta_k| \le 2\frac{\zeta_n}{n^{r+1/2}}, \quad k = 1,2,\dots,n+1. \tag{9.129}
$$

Let us now consider the exact solution $u = u(x)$ of problem (9.113). Given the assumptions made regarding the right-hand side $f = f(x)$, the solution u is also a smooth odd periodic function with the period $L = 2$. It can be represented as its own Fourier series:

$$
u(x) = \sum_{k=1}^{\infty} \gamma_k \sin(k\pi x), \tag{9.130}
$$

where the coefficients γ_k are given by:

$$
\gamma_k = 2\int_0^1 u(x)\sin(k\pi x)dx. \tag{9.131}
$$

Series (9.130) converges uniformly. Moreover, the same argument based on the periodicity and smoothness implies that the Fourier series for the derivatives $u'(x)$ and $u''(x)$ also converge uniformly.[6] Consequently, series (9.130) can be differentiated (at least) twice termwise:

$$
u''(x) = -\pi^2 \sum_{k=1}^{\infty} k^2 \gamma_k \sin(k\pi x). \tag{9.132}
$$

Recall, we must enforce the equality $u'' = f$. Then by comparing the expansions (9.132) and (9.124) and using the orthogonality of the trigonometric system, we have:

$$
\gamma_k = -\frac{1}{\pi^2 k^2}\beta_k, \quad k = 1,2,\dots. \tag{9.133}
$$

[5]Due to the uniqueness of the trigonometric interpolating polynomial, see Theorem 3.1.

[6]The first derivative $u'(x)$ will be an even function rather than odd.

Next, recall that the coefficients B_k of the approximate solution $u^{(n)}(x)$ defined by (9.114) are given by formula (9.120). Using the representation $b_k = \beta_k + \delta\beta_k$, see formula (9.128), and also employing relations (9.133), we obtain:

$$\begin{aligned} B_k &= -\frac{1}{\pi^2k^2}b_k \\ &= -\frac{1}{\pi^2k^2}\beta_k - \frac{1}{\pi^2k^2}\delta\beta_k \\ &= \gamma_k - \frac{1}{\pi^2k^2}\delta\beta_k, \quad k = 1,2,\ldots,n+1. \end{aligned} \tag{9.134}$$

Formula (9.134) will allow us to obtain an error estimate for the approximate solution $u^{(n)}(x)$. To do so, we first rewrite the Fourier series (9.130) for the exact solution $u(x)$ as its partial sum plus the remainder [cf. formula (9.126)]:

$$u(x) = \tilde{S}_n(x) + \delta\tilde{S}_n(x) = \sum_{k=1}^{n+1} \gamma_k \sin(k\pi x) + \sum_{k=n+2}^{\infty} \gamma_k \sin(k\pi x), \tag{9.135}$$

and obtain an estimate for the convergence rate [cf. formula (9.127)]:

$$|u(x) - \tilde{S}_n(x)| = |\delta\tilde{S}_n(x)| \le \frac{\eta_n}{n^{r+\frac{5}{2}}}, \tag{9.136}$$

where $\eta_n = o(1)$ as $n \longrightarrow \infty$. Note that according to the formulae (9.136) and (9.127), the series (9.130) converges faster than the series (9.124), with the rates $o\left(n^{-\left(r+\frac{5}{2}\right)}\right)$ and $o\left(n^{-\left(r+\frac{1}{2}\right)}\right)$, respectively. The reason is that if the right-hand side $f = f(x)$ of problem (9.113) has r continuous derivatives and a square integrable derivative of order $r+1$, then the solution $u = u(x)$ to this problem would normally have $r+2$ continuous derivatives and a square integrable derivative of order $r+3$.

Next, using equalities (9.114), (9.134), and (9.135) and estimates (9.127) and (9.136), we can write $\forall x \in [0,1]$:

$$\begin{aligned} |u(x) - u^{(n)}(x)| &= \left|\tilde{S}_n(x) + \delta\tilde{S}_n(x) - \sum_{k=1}^{n+1} B_k \sin(\pi k x)\right| \\ &= \left|\tilde{S}_n(x) + \delta\tilde{S}_n(x) - \sum_{k=1}^{n+1} \gamma_k \sin(\pi k x) + \sum_{k=1}^{n+1} \frac{\delta\beta_k}{\pi^2k^2}\sin(\pi k x)\right| \\ &= \left|\delta\tilde{S}_n(x) + \sum_{k=1}^{n+1} \frac{\delta\beta_k}{\pi^2k^2}\sin(\pi k x)\right| \le |\delta\tilde{S}_n(x)| + \sum_{k=1}^{n+1} |\delta\beta_k| \\ &\le \frac{\eta_n}{n^{r+\frac{5}{2}}} + \frac{\zeta_n}{n^{r-\frac{1}{2}}} \le \frac{\sigma_n}{n^{r-\frac{1}{2}}}, \end{aligned} \tag{9.137}$$

where σ_n is another infinitesimal sequence: $\sigma_n = o(1)$ as $n \longrightarrow \infty$. The key distinctive feature of error estimate (9.137) is that it provides for a more rapid convergence in the case when the right-hand side $f(x)$ that drives the problem has higher regularity. In other words, similarly to the original trigonometric interpolation (see Section 3.1.3), the foregoing method of obtaining an approximate solution to problem (9.113) *does not get saturated by smoothness.* Indeed, the approximation error self-adjusts to the regularity of the data without us having to change anything in the algorithm. Moreover, if the right-hand side $f(x)$ of problem (9.113) has continuous periodic derivatives of all orders, then according to estimate (9.137) the method *will converge with a spectral rate,* i.e., faster than any inverse power of n. For that reason, methods of this type are referred to as spectral methods in the literature.

Note that the simple Fourier-based spectral method that we have outlined in this section will only work for smooth periodic functions, i.e., for the functions that withstand smooth periodic extensions. There are many examples of smooth right-hand sides that do not satisfy this constraint, for example the quadratic function $f(x) = x(x-1)$ used in Section 9.6, see formula (9.103). However, a spectral method can be built for problem (9.113) with this right-hand side as well. In this case, it will be convenient to look for a solution as a linear combination of Chebyshev polynomials, rather than in the form of a trigonometric polynomial (9.114). This approach is similar to Chebyshev-based interpolations discussed in Section 3.2.

Note also that in this section we enforced the differential equation of (9.113) by requiring that the two trigonometric polynomials, $\frac{d^2}{dx^2}u^{(n)}(x)$ and $f^{(n)}(x)$, coincide at the nodes of the grid (9.115), see equalities (9.119). In the context of spectral methods, the points x_m given by (9.115) are often referred to as the collocation points, and the corresponding methods are known as the spectral collocation methods. Alternatively, one can use Galerkin approximations for building spectral methods. The Galerkin method is a very useful and general technique that has many applications in numerical analysis and beyond; we briefly describe it in Section 12.2.3 when discussing finite elements.

Similarly to any other method of approximation, one generally needs to analyze accuracy and stability when designing spectral methods. Over the recent years, a number of efficient spectral methods have been developed for solving a wide variety of initial and boundary value problems for ordinary and partial differential equations. For further detail, we refer the reader to [GO77,CHQZ88,Boy01,CHQZ06,HGG06].

Exercise

1. Solve problem (9.113) with the right-hand side $f(x) = \sin(\pi \sin(\pi x))$ on a computer using the Fourier collocation method described in this section. Alternatively, apply the second order difference method of Section 9.6. Demonstrate experimentally the difference in convergence rates.

Chapter 10

Finite-Difference Schemes for Partial Differential Equations

In Chapter 9, we defined the concepts of convergence, consistency, and stability in the context of finite-difference schemes for ordinary differential equations. We have also proved that if the scheme is consistent and stable, then the discrete solution converges to the corresponding differential solution as the grid is refined. This result provides a recipe for constructing convergent schemes. One should start with building consistent schemes and subsequently select stable schemes among them.

We emphasize that the definitions of convergence, consistency, and stability, as well as the theorem that establishes the relation between them, are quite general. In fact, these three concepts can be introduced and studied for arbitrary functional equations. In Chapter 9, we illustrated them with examples of the schemes for ordinary differential equations and for an integral equation. In the current chapter, we will discuss the construction of finite-difference schemes for partial differential equations, and consider approaches to the analysis of their stability. Moreover, we will prove the theorem that consistent and stable schemes converge.

In the context of partial differential equations, we will come across a number of important and essentially new circumstances compared to the case of ordinary differential equations. First and foremost, stability analysis becomes rather elaborate and non-trivial. It is typically restricted to the linear case, and even in the linear case, very considerable difficulties are encountered, e.g., when analyzing stability of initial-boundary value problems (as opposed to the Cauchy problem). At the same time, a consistent scheme picked arbitrarily most often turns out unstable. The "pool" of grids and approximation techniques that can be used is very wide; and often, special methods are required for solving the resulting system of finite-difference equations.

10.1 Key Definitions and Illustrating Examples

10.1.1 Definition of Convergence

Assume that a continuous (initial-)boundary value problem:

$$Lu = f \tag{10.1}$$

is to be solved on some domain D with the boundary $\Gamma = \partial D$. To approximately compute the solution u of problem (10.1) given the data f, one first needs to specify a discrete set of points $D_h \subset \{D \cup \Gamma\}$ that is called *the grid*. Then, one should introduce a linear normed space U_h of all discrete functions defined on the grid D_h, and subsequently identify the discrete exact solution $[u]_h$ of problem (10.1) in the space U_h. As $[u]_h$ will be a grid function and not a continuous function, it shall rather be regarded as a table of values for the continuous solution u. The most straightforward way to obtain this table is by merely sampling the values of u at the nodes D_h; in this case $[u]_h$ is said to be the trace (or projection) of the continuous solution u on the grid.[1]

Since, generally speaking, neither the continuous exact solution u nor its discrete counterpart $[u]_h$ are known, the key objective is to be able to compute $[u]_h$ approximately. For that purpose one needs to construct a system of equations

$$\boldsymbol{L}_h u^{(h)} = f^{(h)} \tag{10.2}$$

with respect to the unknown function $u^{(h)} \in U_h$, such that *the convergence* would take place of *the approximate solution* $u^{(h)}$ to *the exact solution* $[u]_h$ as the grid is refined:

$$\| [u]_h - u^{(h)} \|_{U_h} \longrightarrow 0, \quad \text{as} \quad h \longrightarrow 0. \tag{10.3}$$

As has been mentioned, the most intuitive way of building the discrete operator $\boldsymbol{L}_h$ in (10.2) consists of replacing the continuous derivatives contained in $\boldsymbol{L}$ by the appropriate difference quotients, see Section 10.2.1. In this case the discrete system (10.2) is referred to as *a finite-difference scheme*. Regarding the right-hand side $f^{(h)}$ of (10.2), again, the simplest way of obtaining it is to take the trace of the continuous right-hand side f of (10.1) on the grid D_h.

The notion of convergence (10.3) of the difference solution $u^{(h)}$ to the exact solution $[u]_h$ can be quantified. Namely, if $k > 0$ is the largest integer such that the following inequality holds for the solution $u^{(h)}$ of the discrete (initial-)boundary value problem (10.2):

$$\| [u]_h - u^{(h)} \|_{U_h} \leq ch^k, \qquad c = \text{const}, \tag{10.4}$$

then we say that *the convergence rate* is $\mathcal{O}(h^k)$ or alternatively, that the magnitude of the error, i.e., that of the discrepancy between the approximate solution and the exact solution, has order k with respect to the grid size h in the chosen norm $\| \cdot \|_{U_h}$. Note that the foregoing definitions of convergence (10.3) and its rate, or order (10.4), are basically the same as those for ordinary differential equations, see Section 9.1.2.

Let us also emphasize that the way we define convergence for finite-difference schemes, see formulae (10.3) and (10.4), differs from the traditional definition of convergence in a vector space. Namely, when $h \longrightarrow 0$ the overall number of nodes in the grid D_h will increase, and so will the dimension of the space U_h. As such, U_h shall, in fact, be interpreted as a sequence of spaces of increasing dimension parameterized by the grid size h, rather than a single vector space with a fixed dimension.

[1] There are many other ways to define $[u]_h$, e.g., those based on integration over the grid cells.

Accordingly, the limit in (10.3) shall be interpreted as a limit of the sequence of norms in vector spaces that have higher and higher dimension.

The problem of constructing a convergent scheme (10.2) is usually split into two sub-problems. First, one obtains a scheme that is *consistent,* i.e., that approximates problem (10.1) on its solution u in some specially defined sense. Then, one should verify that the chosen scheme (10.2) is *stable.*

10.1.2 Definition of Consistency

To assign a tangible meaning to the notion of consistency, one should first introduce a norm in the linear space F_h that contains the right-hand side $f^{(h)}$ of equation (10.2). Similarly to U_h, F_h should rather be interpreted as a sequence of spaces of increasing dimension parameterized by the grid size h, see Section 10.1.1. We say that the finite-difference scheme (10.2) is consistent, or in other words, that the discrete problem (10.2) approximates the differential problem (10.1) on the solution u of the latter, if the residual $\delta f^{(h)}$ that is defined by the following equality:

$$L_h[u]_h = f^{(h)} + \delta f^{(h)}, \tag{10.5}$$

i.e., that arises when the exact solution $[u]_h$ is substituted into the left-hand side of (10.2), vanishes as the grid is refined:

$$\|\delta f^{(h)}\|_{F_h} \longrightarrow 0, \quad \text{as} \quad h \longrightarrow 0. \tag{10.6}$$

If, in addition, $k > 0$ happens to be the largest integer such that

$$\|\delta f^{(h)}\|_{F_h} \leq c_1 h^k, \tag{10.7}$$

where c_1 is a constant that does not depend on h, then the approximation is said to have order k with respect to the grid size h. In the literature, the residual $\delta f^{(h)}$ of the exact solution $[u]_h$, see formula (10.5), is referred to as *the truncation error.* Accordingly, if inequality (10.7) holds, we say that the scheme (10.2) has *order of accuracy k with respect to h* (cf. Definition 9.1 of Chapter 9).

Let us, for example, construct a consistent finite-difference scheme for the following Cauchy problem:

$$\begin{aligned} &\frac{\partial u}{\partial t} - \frac{\partial u}{\partial x} = \varphi(x,t), \quad && -\infty < x < \infty, \quad 0 < t \leq T, \\ &u(x,0) = \psi(x), && -\infty < x < \infty. \end{aligned} \tag{10.8}$$

Problem (10.8) can be recast in the general form (10.1) if we define

$$Lu = \begin{cases} \dfrac{\partial u}{\partial t} - \dfrac{\partial u}{\partial x}, & -\infty < x < \infty, \quad 0 < t \leq T, \\ u(x,0), & -\infty < x < \infty, \end{cases}$$

$$f = \begin{cases} \varphi(x,t), & -\infty < x < \infty, \quad 0 < t \leq T, \\ \psi(x), & -\infty < x < \infty. \end{cases}$$

We will use a simple uniform Cartesian grid D_h with the nodes given by intersections of the two families of vertical and horizontal equally spaced straight lines:

$$x = x_m \equiv mh, \quad t = t_p \equiv p\tau, \quad m = 0, \pm 1, \pm 2, \dots, \quad p = 0, 1, \dots, [T/\tau], \tag{10.9}$$

where $h > 0$ and $\tau > 0$ are fixed real numbers, and $[T/\tau]$ denotes the integer part of T/τ. We will also assume that the temporal grid size τ is proportional to the spatial size h: $\tau = rh$, where $r = \text{const}$, so that in essence the grid D_h be parameterized by only one quantity h. The discrete exact solution $[u]_h$ on the grid D_h will be defined in the simplest possible way outlined in Section 10.1.1 — as a trace, i.e., by sampling the values of the continuous solution $u(x,t)$ of problem (10.8) at the nodes (10.9): $[u]_h = \{u(mh, p\tau)\}$.

We proceed now to building a consistent scheme (10.2) for problem (10.8). The value of the grid function $u^{(h)}$ at the node $(x_m, t_p) = (mh, p\tau)$ of the grid D_h will hereafter be denoted by u_m^p. To actually obtain the scheme, we replace the partial derivatives $\frac{\partial u}{\partial t}$ and $\frac{\partial u}{\partial x}$ by the first order difference quotients:

$$\left.\frac{\partial u}{\partial t}\right|_{(x,t)} \approx \frac{u(x, t+\tau) - u(x,t)}{\tau},$$

$$\left.\frac{\partial u}{\partial x}\right|_{(x,t)} \approx \frac{u(x+h, t) - u(x,t)}{h}.$$

Then, we can write down the following system of equations:

$$\begin{aligned} \frac{u_m^{p+1} - u_m^p}{\tau} - \frac{u_{m+1}^p - u_m^p}{h} &= \varphi_m^p, \quad m = 0, \pm 1, \pm 2, \dots, \; p = 0, 1, \dots, [T/\tau] - 1, \\ u_m^0 &= \psi_m, \quad m = 0, \pm 1, \pm 2, \dots, \end{aligned} \tag{10.10}$$

where the right-hand sides φ_m^p and ψ_m are obtained, again, by sampling the values of the continuous right-hand sides $\varphi(x,t)$ and $\psi(x)$ of (10.8) at the nodes of D_h:

$$\begin{aligned} \varphi_m^p &= \varphi(mh, p\tau), \quad m = 0, \pm 1, \pm 2, \dots, p = 0, 1, \dots, [T/\tau] - 1, \\ \psi_m &= \psi(mh), \qquad m = 0, \pm 1, \pm 2, \dots. \end{aligned} \tag{10.11}$$

The scheme (10.10) can be recast in the universal form (10.2) if the operator $\boldsymbol{L}_h$ and the right-hand side $f^{(h)}$ are defined as follows:

$$\boldsymbol{L}_h u^{(h)} = \begin{cases} \dfrac{u_m^{p+1} - u_m^p}{\tau} - \dfrac{u_{m+1}^p - u_m^p}{h}, & m = 0, \pm 1, \pm 2, \dots, \; p = 0, 1, \dots, [T/\tau] - 1, \\ u_m^0, & m = 0, \pm 1, \pm 2, \dots, \end{cases}$$

$$f^{(h)} = \begin{cases} \varphi_m^p, & m = 0, \pm 1, \pm 2, \dots, \; p = 0, 1, \dots, [T/\tau] - 1, \\ \psi_m, & m = 0, \pm 1, \pm 2, \dots, \end{cases}$$

where φ_m^p and ψ_m are given by (10.11). Thus, the right-hand side $f^{(h)}$ is basically a pair of grid functions φ_m^p and ψ_m such that the first one is defined on the two-dimensional grid (10.9) and the second one is defined on the one-dimensional grid:

$$(x_m, 0) = (mh, 0), \qquad m = 0, \pm 1, \pm 2, \dots.$$

The difference equation from (10.10) can be easily solved with respect to u_m^{p+1}:

$$u_m^{p+1} = (1-r)u_m^p + ru_{m+1}^p + \tau\varphi_m^p. \tag{10.12}$$

Thus, if we know the values u_m^p, $m = 0, \pm 1, \pm 2, \ldots$, of the approximate solution $u^{(h)}$ at the grid nodes that correspond to the time level $t = p\tau$, then we can use (10.12) and compute the values u_m^{p+1} at the grid nodes that correspond to the next time level $t = (p+1)\tau$. As the values u_m^0 at $t = 0$ are given by the equalities $u_m^0 = \psi_m$, see (10.10), then we can successively compute the discrete solution u_m^p one time level after another for $t = \tau$, $t = 2\tau$, $t = 3\tau$, etc., i.e., everywhere on the grid D_h. The schemes, for which solution on the upper time level can be obtained as a closed form expression that contains only the values of the solution on the lower time level(s), such as in formula (10.12), are called *explicit*. The foregoing process of computing the solution one time level after another is known as *the time marching*.

Let us now determine what order of accuracy the scheme (10.10) has. The linear space F_h will consist of all pairs of bounded grid functions $f^{(h)} = [\varphi_m^p, \psi_m]^T$ with the norm:[2]

$$\|f^{(h)}\|_{F_h} \stackrel{\text{def}}{=} \max_{m,p} |\varphi_m^p| + \max_m |\psi_m|. \tag{10.13}$$

We note that one can use various norms for the analysis of consistency, and the choice of the norm can, in fact, make a lot of difference. Hereafter in this section, we will only be using the maximum norm defined by (10.13).

Assume that the solution $u(x,t)$ of problem (10.8) has bounded second derivatives. Then, the Taylor formula yields:

$$\begin{aligned}
\frac{u(x_m+h,t_p)-u(x_m,t_p)}{h} &= \frac{\partial u(x_m,t_p)}{\partial x} + \frac{h}{2}\frac{\partial^2 u(x_m+\xi,t_p)}{\partial x^2} \\
\frac{u(x_m,t_p+\tau)-u(x_m,t_p)}{\tau} &= \frac{\partial u(x_m,t_p)}{\partial t} + \frac{\tau}{2}\frac{\partial^2 u(x_m,t_p+\eta)}{\partial t^2},
\end{aligned}$$

where ξ and η are some numbers that may depend on m, p, and h, and that satisfy the inequalities $0 \le \xi \le h$, $0 \le \eta \le \tau$. These formulae allow us to recast the expression

$$L_h[u]_h = \begin{cases} \dfrac{u(x_m,t_p+\tau)-u(x_m,t_p)}{\tau} - \dfrac{u(x_m+h,t_p)-u(x_m,t_p)}{h}, \\ u(x_m,0), \end{cases}$$

in the following form:

$$L_h[u]_h = \begin{cases} \left(\dfrac{\partial u}{\partial t} - \dfrac{\partial u}{\partial x}\right)\Big|_{(x_m,t_p)} + \dfrac{\tau}{2}\dfrac{\partial^2 u(x_m,t_p+\eta)}{\partial t^2} - \dfrac{h}{2}\dfrac{\partial^2 u(x_m+\xi,t_p)}{\partial x^2}, \\ u(x_m,0)+0, \end{cases}$$

[2]If either $\max|\varphi_m^p|$ or $\max|\psi_m|$ is not reached, then the least upper bound, i.e., supremum, $\sup|\varphi_m^p|$ or $\sup|\psi_m|$ should be used instead in formula (10.13).

or alternatively:

$$\boldsymbol{L}_h[u]_h = f^{(h)} + \delta f^{(h)},$$

where

$$\delta f^{(h)} = \begin{cases} \dfrac{\tau}{2}\dfrac{\partial^2 u(x_m, t_p+\eta)}{\partial t^2} - \dfrac{h}{2}\dfrac{\partial^2 u(x_m+\xi, t_p)}{\partial x^2}, \\ 0. \end{cases}$$

Consequently [see (10.13)],

$$\|\delta f^{(h)}\|_{F_h} \leq \frac{h}{2}\left(r \sup\left|\frac{\partial^2 u}{\partial t^2}\right| + \sup\left|\frac{\partial^2 u}{\partial x^2}\right|\right). \tag{10.14}$$

We can therefore conclude that the finite-difference scheme (10.10) renders *first order accuracy* with respect to h on the solution $u(x,t)$ of problem (10.8) that has bounded second partial derivatives.

10.1.3 Definition of Stability

DEFINITION 10.1 *The scheme (10.2) is said to be stable if one can find $\delta > 0$ and $h_0 > 0$ such that for any $h < h_0$ and any grid function $\varepsilon^{(h)} \in F_h$: $\|\varepsilon^{(h)}\| < \delta$, the finite-difference problem*

$$\boldsymbol{L}_h z^{(h)} = f^{(h)} + \varepsilon^{(h)} \tag{10.15}$$

is uniquely solvable, and its solution $z^{(h)}$ satisfies the following inequality

$$\|z^{(h)} - u^{(h)}\|_{U_h} \leq c_2 \|\varepsilon^{(h)}\|_{F_h}, \qquad c_2 = \text{const}, \tag{10.16}$$

where $u^{(h)}$ is the solution of (10.2), and c_2 does not depend on h or $\varepsilon^{(h)}$.

Definition 10.1 introduces stability for nonlinear problems, i.e., stability of each individual solution $u^{(h)}$ with respect to small perturbations of the right-hand side $f^{(h)}$ that drives it. This definition is similar to Definition 9.2 for ordinary differential equations. The constant c_2 in inequality (10.16) may, generally speaking, depend on $f^{(h)}$, but the most important consideration is that it does not depend on h.

Alternatively, one can introduce stability for linear problems; the following definition is similar to Definition 9.3 for ordinary differential equations.

DEFINITION 10.2 *The finite-difference problem (10.2) with a linear operator $\boldsymbol{L}_h$ is called stable if there is an $h_0 > 0$ such that for any $h < h_0$ and $f^{(h)} \in F_h$ it is uniquely solvable, and the solution $u^{(h)}$ satisfies*

$$\|u^{(h)}\|_{U_h} \leq c_2 \|f^{(h)}\|_{F_h}, \qquad c_2 = \text{const}, \tag{10.17}$$

where c_2 does not depend either on h or on $f^{(h)}$.

LEMMA 10.1
If the operator $\boldsymbol{L}_h$ is linear, then Definitions 10.1 and 10.2 are equivalent.

PROOF We first assume that the scheme (10.2) is stable in the sense of Definition 10.2. By subtracting equality (10.2) from equality (10.15) we obtain

$$\boldsymbol{L}_h\left(z^{(h)} - u^{(h)}\right) = \varepsilon^{(h)}.$$

Then, estimate (10.17) immediately yields inequality (10.16) for an arbitrary $\varepsilon^{(h)} \in F_h$, which implies stability in the sense of Definition 10.1. Moreover, the constant c_2 will be the same for all $f^{(h)} \in F_h$.

Conversely, let us now assume that the scheme (10.2) is stable in the sense of Definition 10.1. Then, for a given $f^{(h)}$, some $h_0 > 0$, $\delta > 0$, and for arbitrary $h < h_0$, $\varepsilon^{(h)} \in F_h$: $\|\varepsilon^{(h)}\| < \delta$, both equations

$$\boldsymbol{L}_h z^{(h)} = f^{(h)} + \varepsilon^{(h)} \qquad \text{and} \qquad \boldsymbol{L}_h u^{(h)} = f^{(h)}$$

have unique solutions. Now let $w^{(h)} = z^{(h)} - u^{(h)}$, and subtract the foregoing two equations from one another. This yields

$$\boldsymbol{L}_h w^{(h)} = \varepsilon^{(h)}, \tag{10.18}$$

whereas estimate (10.16) translates into

$$\|w^{(h)}\|_{U_h} \leq c_2 \|\varepsilon^{(h)}\|_{F_h}.$$

By merely changing the notations for the solution and the right-hand side of equation (10.18), one can reformulate the last result as follows. For arbitrary $h < h_0$ and any $f^{(h)} \in F_h$: $\|f^{(h)}\|_{F_h} < \delta$, problem (10.2) has a unique solution $u^{(h)}$. This solution satisfies estimate (10.17). To establish stability in the sense of Definition 10.2, we, however, need this result to hold not only for all $f^{(h)}$ that satisfy $\|f^{(h)}\| < \delta$, but for all other $f^{(h)} \in F_h$ as well.

To complete the proof, we invoke a simple argument based on scaling. Let $\|f^{(h)}\|_{F_h} > \delta$, and let us justify the unique solvability and estimate (10.17) in this case. Define the two new functions:

$$\tilde{u}^{(h)} \stackrel{\text{def}}{=} \frac{\delta}{2\|f^{(h)}\|_{F_h}} u^{(h)} \qquad \text{and} \qquad \tilde{f}^{(h)} \stackrel{\text{def}}{=} \frac{\delta}{2\|f^{(h)}\|_{F_h}} f^{(h)}. \tag{10.19}$$

Then, due to the linearity of the problem, we can write down the following equation for $\tilde{u}^{(h)}$:

$$\boldsymbol{L}_h \tilde{u}^{(h)} = \tilde{f}^{(h)}, \tag{10.20}$$

Moreover, because of (10.19) we have:

$$\|\tilde{f}^{(h)}\|_{F_h} = \frac{\delta}{2\|f^{(h)}\|_{F_h}} \|f^{(h)}\|_{F_h} = \frac{\delta}{2} < \delta,$$

and consequently, equation (10.20) is uniquely solvable, and the following estimate holds for its solution:

$$\|\tilde{u}^{(h)}\|_{U_h} \leq c_2 \|\tilde{f}^{(h)}\|_{F_h}.$$

Relations (10.19) then imply that problem (10.2) will have a unique solution for any $f^{(h)} \in F_h$ that will satisfy inequality (10.17), which means stability in the sense of Definition 10.2. ∎

The property of stability can be interpreted as uniform with respect to h sensitivity of the solution $u^{(h)}$ to the perturbations $\varepsilon^{(h)}$ of the right-hand side of problem (10.2). We emphasize that this (asymptotic) uniformity as $h \longrightarrow 0$ is of foremost importance, and that the concept of stability goes far beyond the simple well-posedness of a given problem (10.2) for a fixed h. Stability, in fact, requires that the entire family of these problems parameterized by the grid size be well-posed, and that the well-posedness constant c_2 in (10.17) be independent of h.

We emphasize that according to Definitions 10.1 and 10.2, stability is an intrinsic property of the scheme. The formulation of this property does not involve any direct relation to the original continuous problem. For example, even if the continuous problem is uniquely solvable and well-posed, one can still obtain both stable and unstable discretizations.[3] Moreover, the property of stability is formulated independently of either consistency or convergence. However, the following key theorem establishes a fundamental relation between consistency, stability, and convergence.

THEOREM 10.1

If the finite-difference problem (10.2) is consistent and stable, then the approximate solution $u^{(h)}$ converges to the exact solution $[u]_h$ [see formula (10.3)] as the grid is refined. Moreover, the rate of convergence according to the inequality $\|[u]_h - u^{(h)}\|_{U_h} \leq ch^k$ [see formula (10.4)] coincides with the order of accuracy of the scheme given by $\|\delta f^{(h)}\|_{F_h} \leq c_1 h^k$ [see formula (10.7)].

The conclusion of Theorem 10.1 constitutes one implication of *the Lax equivalence theorem* (1953), although the actual formulation we present follows a later work by Filippov (1955), see Section 10.1.8 for more detail.

PROOF In equality (10.15), let us set $z^{(h)} = [u]_h$, which obviously means $\varepsilon^{(h)} = \delta f^{(h)}$. Then, stability estimate (10.16) transforms into:

$$\|[u]_h - u^{(h)}\|_{U_h} \leq c_2 \|\delta f^{(h)}\|_{F_h},$$

and along with the consistency in the sense of (10.7), this yields:

$$\|[u]_h - u^{(h)}\|_{U_h} \leq c_2 c_1 h^k \equiv ch^k,$$

which is equivalent to convergence with the rate $\mathcal{O}(h^k)$, see formula (10.4). ∎

[3] If the original continuous problem is ill-posed, then chances for obtaining a stable discretization are slim.

Note that when proving Theorem 10.1, we interpreted stability in the nonlinear sense of Definition 10.1. This follows the approach of Chapter 9, see Theorem 9.1. We emphasize, however, that compared to the ordinary differential equations, the analysis of finite-difference stability for partial differential equations is considerably more involved. Most often, it is restricted to the linear case as there are basically no systematic means for studying stability of nonlinear finite-difference equations beyond the necessary conditions based on the principle of frozen coefficients (Section 10.4.1). Hence, the result of Theorem 10.1 for partial differential equations is typically constrained to the case of linear operators $\boldsymbol{L}_h$. The difficulties in applying Theorem 10.1 to the analysis of finite-difference schemes that approximate nonlinear partial differential equations are typically related to the loss of regularity of the solution, which is very common even when the initial and/or boundary data are smooth (see, e.g., Chapter 11). In certain cases when the solution is smooth though, direct proofs of convergence can be developed that do not rely on Theorem 10.1, see, e.g., work by Strang [Str64] and its account in [RM67, Chapter 5].

Let us now show that the finite-difference scheme (10.10) is stable for $r = \tau/h \leq 1$. To analyze the stability, we will define the norm $\|\cdot\|_{U_h}$ in the space U_h as follows:

$$\|u^{(h)}\|_{U_h} \stackrel{\text{def}}{=} \max_p \sup_m |u_m^p|, \tag{10.21}$$

and the norm $\|\cdot\|_{F_h}$ in the space F_h as follows:

$$\|f^{(h)}\|_{F_h} \stackrel{\text{def}}{=} \max_p \sup_m |\varphi_m^p| + \sup_m |\psi_m|. \tag{10.22}$$

In the finite-difference problem (10.10), the right-hand sides φ_m^p and ψ_m shall now be interpreted as arbitrary grid functions rather than traces of the continuous right-hand sides $\varphi(x,t)$ and $\psi(x)$ of (10.8) defined in the sense of (10.11). Of course, with these arbitrary right-hand sides, the scheme (10.10) can still be recast in the form (10.12):

$$u_m^{p+1} = (1-r)u_m^p + ru_{m+1}^p + \tau\varphi_m^p, \qquad u_m^0 = \psi_m. \tag{10.23}$$

As $r \leq 1$, then $1 - r \geq 0$. Consequently, the following estimate holds:

$$|(1-r)u_m^p + ru_{m+1}^p| \leq |(1-r)+r|\max\{|u_m^p|, |u_{m+1}^p|\} = \max\{|u_m^p|, |u_{m+1}^p|\} \leq \sup_m |u_m^p|.$$

Using this estimate along with representation (10.23), we have

$$|u_m^{p+1}| \leq \sup_m |u_m^p| + \tau \sup_m |\varphi_m^p| \leq \sup_m |u_m^p| + \tau \max_p \sup_m |\varphi_m^p|. \tag{10.24}$$

Note that if $\varphi_m^p \equiv 0$, then inequality (10.24) implies that $\sup_m |u_m^p|$ does not increase as p increases. This property of the finite-difference scheme is known as *the maximum principle*. For simplicity, we will sometimes apply this term to the entire inequality (10.24).

The right-hand side of inequality (10.24) does not depend on m. Therefore, on its left-hand side we can replace $|u_m^{p+1}|$ with $\sup_m |u_m^{p+1}|$ and thus obtain

$$\sup_m |u_m^{p+1}| \leq \sup_m |u_m^p| + \tau \max_p \sup_m |\varphi_m^p|.$$

Similarly, we can obtain the inequalities:

$$\sup_m |u_m^p| \leq \sup_m |u_m^{p-1}| + \tau \max_p \sup_m |\varphi_m^p|,$$

$$\dots \quad \dots \quad \dots \quad \dots \quad \dots \quad \dots$$

$$\sup_m |u_m^1| \leq \sup_m |u_m^0| + \tau \max_p \sup_m |\varphi_m^p|.$$

By combining all these inequalities, we finally arrive at:

$$\sup_m |u_m^{p+1}| \leq \sup_m |u_m^0| + \tau(p+1) \max_p \sup_m |\varphi_m^p|,$$

which immediately yields:

$$\sup_m |u_m^{p+1}| \leq \sup_m |\psi_m| + T \max_p \sup_m |\varphi_m^p| \leq (1+T)\|f^{(h)}\|_{F_h}.$$

The previous inequality does hold for any p, therefore it will still hold if on its left-hand side instead of $\sup_m |u_m^{p+1}|$ we write $\max_p \sup_m |u_m^{p+1}| = \|u^{(h)}\|_{U_h}$:

$$\|u^{(h)}\|_{U_h} \leq (1+T)\|f^{(h)}\|_{F_h}. \tag{10.25}$$

Estimate (10.25) implies stability of the finite-difference scheme (10.10) (in the sense of Definition 10.2), because problem (10.23) is obviously uniquely solvable for arbitrary bounded φ_m^p and ψ_m. In so doing, the quantity $1+T$ plays the role of the constant c_2 in inequality (10.17).

In the previous example, stability (10.25) of the scheme (10.10) along with its consistency (10.14) are sufficient for convergence according to Theorem 10.1. However, consistency of the scheme (10.2) alone, see formula (10.7), is not, generally speaking, sufficient for convergence in the sense of (10.3). Indeed, in the beginning of Section 9.3, we constructed an example of consistent yet divergent scheme for an ordinary differential equation. For partial differential equations, instability (and as such, inappropriateness) of an arbitrarily selected consistent scheme shall, in fact, be regarded as a general situation, whereas construction of a stable (and, consequently, convergent) scheme appears to be one of the key tasks of a numerical analyst.

Recall, for example, that we have proven stability of the finite-difference scheme (10.10) under the assumption that $r \leq 1$. In the case $r > 1$ one can easily see that the scheme (10.10) will remain consistent, i.e., it will still provide the $\mathcal{O}(h)$ accuracy on the solution $u(x,t)$ of the differential problem (10.8) that has bounded second derivatives. However, the foregoing stability proof will no longer work. Let us show that for $r > 1$ the solution $u^{(h)}$ of the finite-difference problem (10.10) does not, in fact, converge to the exact solution $[u]_h$ (trace of the solution $u(x,t)$ to problem (10.8) on the grid D_h of (10.9)). Therefore, there may be no stability either, as otherwise consistency and stability would have implied convergence by Theorem 10.1.

10.1.4 The Courant, Friedrichs, and Lewy Condition

In this section, we will prove that if $r = \tau/h > 1$, then the finite-difference scheme (10.10) will not converge for a general function $\psi(x)$. For simplicity, and with no loss of generality (as far as the no convergence argument), we will assume that $\varphi(x,t) \equiv 0$, which implies $\varphi_m^p = \varphi(mh, p\tau) \equiv 0$. Let us also set $T = 1$, and choose the spatial grid size h so that the point $(0,1)$ on the (x,t) plane be a grid node, or in other words, so that the number

$$N = \frac{1}{\tau} = \frac{1}{rh}$$

be integer, see Figure 10.1. Using the difference equation in the form (10.12):

$$u_m^{p+1} = (1-r)u_m^p + ru_{m+1}^p,$$

one can easily see that the value u_0^{p+1} of the discrete solution $u^{(h)}$ at the grid node $(0,1)$ (in this case $p+1 = N$) can be expressed through the values of u_0^p and u_1^p of the solution at the nodes $(0, 1-\tau)$ and $(h, 1-\tau)$, respectively, see Figure 10.1.

Likewise, the values of u_0^p and u_1^p are expressed through the values of u_0^{p-1}, u_1^{p-1}, and u_2^{p-1} at the three grid nodes: $(0, 1-2\tau)$, $(h, 1-2\tau)$, and $(2h, 1-2\tau)$. Similarly, the values of u_0^{p-1}, u_1^{p-1}, and u_2^{p-1} are, in turn, expressed through the values of the solution at the four grid nodes: $(0, 1-3\tau)$, $(h, 1-3\tau)$, $(2h, 1-3\tau)$, and $(3h, 1-2\tau)$. Eventually we obtain that u_0^{p+1} is expressed through the values of the solution $u_m^0 = \psi_m$ at the nodes: $(0,0)$, $(h,0)$, $(2h,0)$, ..., $(hT/\tau, 0)$. All these nodes belong to the interval:

$$0 \le x \le \frac{hT}{\tau} = \frac{h}{\tau} = \frac{1}{r}$$

of the horizontal axis $t = 0$, see Figure 10.1, where the initial condition

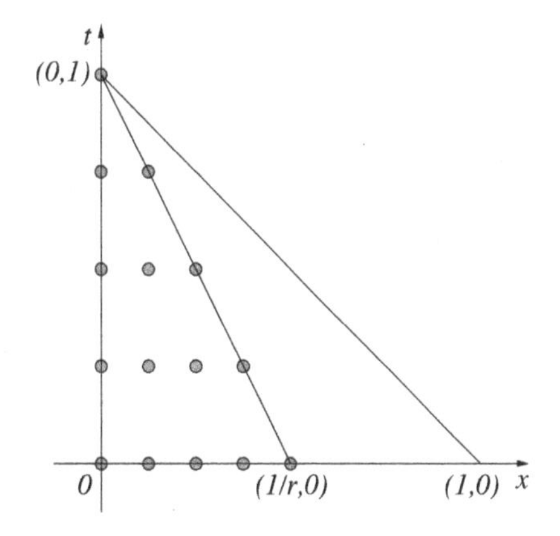

FIGURE 10.1: Continuous and discrete domain of dependence.

$$u(x,0) = \psi(x)$$

for the Cauchy problem (10.8) is specified.

Thus, solution u_m^p to the finite-difference problem (10.10) at the point $(x,t) = (0,1) \Leftrightarrow (m,p) = (0,N)$ does not depend on the values of the function $\psi(x)$ at the points x that do not belong to the interval $0 \le x \le 1/r$. It is, however, easy to see that solution to the original continuous initial value problem:

$$\begin{aligned} \frac{\partial u}{\partial t} - \frac{\partial u}{\partial x} &= 0, & -\infty < x < \infty, \quad 0 < t \le T, \\ u(x,0) &= \psi(x), & -\infty < x < \infty, \end{aligned}$$

is given by the function

$$u(x,t) = \psi(x+t).$$

This function is constant along every characteristic $x+t = \text{const}$ of the differential equation $\frac{\partial u}{\partial t} - \frac{\partial u}{\partial x} = 0$. In particular, it maintains a constant value along the straight line $x+y=1$, which crosses through the points $(0,1)$ and $(1,0)$ on the plane (x,t), see Figure 10.1. As such, $u(x,t)\big|_{(x,t)=(0,1)} = \psi(1)$. Therefore, if $r>1$, one should not, generally speaking, expect convergence of the scheme (10.10).

Indeed, in this case the interval $0 \le x \le 1/r < 1$ of the horizontal axis does not contain the point $(1,0)$. Let us assume for a moment that for some particular $\psi(x)$ the convergence, by accident, does take place. Then, while keeping the values of $\psi(x)$ on the interval $0 \le x \le 1/r$ unaltered, we can disrupt this convergence by modifying $\psi(x)$ at and around the point $x=1$. This is easy to see, because the latter modification will obviously affect the value $u(0,1)$ of the continuous solution, whereas the value u_0^N of the discrete solution at $(0,1)$ will remain unchanged, as it is fully determined by $\psi(x)$ on the interval $0 \le x \le 1/r$. Moreover, one can modify the function $\psi(x)$ at the point $x=1$ and in its vicinity so that it will stay sufficiently smooth (twice differentiable). Then, the solution $u(x,t) = \psi(x+t)$ will inherit this smoothness, and the scheme (10.10) will therefore remain consistent, see (10.14). Under these conditions, stability of the scheme (10.10) would have implied convergence. As, however, there is no convergence for $r>1$, there may be no stability either.

The argument we have used to prove that the scheme (10.10) is divergent (and thus unstable) when $r = \tau/h > 1$ is, in fact, quite general. It can be presented as follows.

Assume that some function ψ is involved in the formulation of the original problem, i.e., that it provides all or part of the required input data. Let P be an arbitrary point from the domain of the solution u. Let also the value $u(P)$ depend on the values of the function ψ at the points of some region $G_\psi = G_\psi(P)$ that belongs to the domain of ψ. This means that by modifying ψ in a small neighborhood of any point $Q \in G_\psi(P)$ one can basically alter the solution $u(P)$. The set $G_\psi(P)$ is referred to as the continuous domain of dependence for the solution u at the point P. In the previous example, $G_\psi(P)\big|_{P=(0,1)} = (1,0)$, see Figure 10.1.

Suppose now that the solution u is computed by means of a scheme that we denote $L_h u^{(h)} = f^{(h)}$, and that in so doing the value of the discrete solution $u^{(h)}$ at the point $P^{(h)}$ of the grid closest to P is fully determined by the values of the function ψ on some region $G_\psi^{(h)}(P)$. This region is called the discrete domain of dependence, and in the previous example: $G_\psi^{(h)}(P)\big|_{P=(0,1)} = \{x \,|\, 0 \le x \le 1/r\}$, see Figure 10.1.

The Courant, Friedrichs, and Lewy condition says that *for the convergence $u^{(h)} \longrightarrow u$ to take place as $h \longrightarrow 0$, the scheme should necessarily be designed in such a way that in any neighborhood of an arbitrary point from $G_\psi(P)$ there always be a point from $G_\psi^{(h)}(P)$, provided that h is chosen sufficiently small.*

In the literature, this condition is commonly referred to as *the CFL condition*. The number $r = \tau/h$ that in the previous example distinguishes between the convergence and divergence as $r \lessgtr 1$, is known as the CFL number or the Courant number.

Let us explain why in general there is no convergence if the foregoing CFL condition does not hold. Assume that it is not met so that no matter how small the grid size h is, there are no points from the set $G_{\psi}^{(h)}(P)$ in some fixed neighborhood of a particular point $Q \in G_{\psi}(P)$. Even if for some ψ the convergence $u^{(h)} \longrightarrow u$ does incidentally take place, we can modify ψ inside the aforementioned neighborhood of Q, while keeping it unchanged elsewhere. This modification will obviously cause a change in the value of $u(P)$. At the same time, for all sufficiently small h the values of $u^{(h)}$ at the respective grid nodes $P^{(h)}$ closest to P will remain the same because the function ψ has not changed at the points of the sets $G_{\psi}^{(h)}(P)$. Therefore, for the new function ψ the scheme may no longer converge.

Note that the CFL condition can be formulated as a theorem, while the foregoing arguments can be transformed into its proof. We also re-emphasize that for a consistent scheme the CFL condition is necessary not only for its convergence, but also for stability. If this condition does not hold, then there may be no stability, because otherwise consistency and stability would have implied convergence by Theorem 10.1.

10.1.5 The Mechanism of Instability

The proof of instability for the scheme (10.10) given in Section 10.1.4 for the case $r = \tau/h > 1$ is based on using the CFL condition that is necessary for convergence and for stability. As such, this proof is non-constructive in nature. It would, however, be very helpful to actually see how the instability of the scheme (10.10) manifests itself when $r > 1$, i.e., how it affects the sensitivity of the solution $u^{(h)}$ to the perturbations of the data $f^{(h)}$. Recall that according to Section 10.1.3 the scheme is called stable if the finite-difference solution is weakly sensitive to the errors in $f^{(h)}$, and the sensitivity is uniform with respect to h.

Assume for simplicity that for all h the right-hand sides of equations (10.10) are identically equal to zero: $\varphi_m^p \equiv 0$ and $\psi_m \equiv 0$, so that

$$f^{(h)} = \begin{bmatrix} \varphi_m^p \\ \psi_m \end{bmatrix} = 0,$$

and consequently, solution $u^{(h)} = \{u_m^p\}$ of problem (10.10) is also identically equal to zero: $u_m^p \equiv 0$. Suppose now that an error has been committed in the initial data, and instead of $\psi_m = 0$ a different function $\tilde{\psi}_m = (-1)^m \varepsilon$ ($\varepsilon = \text{const}$) has been specified, which yields:

$$\tilde{f}^{(h)} = \begin{bmatrix} 0 \\ \tilde{\psi}_m \end{bmatrix}, \qquad \|\tilde{f}^{(h)}\|_{F_h} = \varepsilon,$$

on the right-hand side of (10.10) instead of $f^{(h)} = 0$. Let us denote the corresponding solution by $\tilde{u}^{(h)}$. In accordance with the finite-difference equation:

$$\tilde{u}_m^{p+1} = (1-r)\tilde{u}_m^p + r\tilde{u}_{m+1}^p, \qquad u_m^0 = (-1)^m \varepsilon,$$

for $\tilde{u}_m^1$ we obtain: $\tilde{u}_m^1 = (1-r)\tilde{u}_m^0 + r\tilde{u}_{m+1}^0 = (1-2r)\tilde{u}_m^0$. We thus see that the error committed at $p = 0$ has been multiplied by the quantity $1-2r$ when advancing to

$p = 1$. For yet another time level $p = 2$ we have:

$$\tilde{u}_m^2 = (1-2r)^2 \tilde{u}_m^0,$$

and in general,

$$\tilde{u}_m^p = (1-2r)^p \tilde{u}_m^0.$$

When $r > 1$ we obviously have $1 - 2r < -1$, so that each time step, i.e., each transition between the levels p and $p+1$, implies yet another multiplication of the initial error $\tilde{u}_m^0 = (-1)^m \varepsilon$ by a negative quantity with the absolute value greater than one. For $p = [T/\tau]$ we thus obtain:

$$|\tilde{u}_m^p| = |1-2r|^{[T/\tau]} |\tilde{u}_m^0|.$$

Consequently,

$$\|\tilde{u}^{(h)}\|_{U_h} = \max_p \sup_m |\tilde{u}_m^p| = |1-2r|^{[T/(rh)]} \sup_m \|\tilde{\psi}_m\| = |1-2r|^{[T/(rh)]} \|\tilde{f}^{(h)}\|_{F_h}.$$

In other words, we see that for a given fixed T the error $(-1)^m \varepsilon$ originally committed when specifying the initial data for the finite-difference problem, grows at a rapid exponential rate $|1-2r|^{[T/(rh)]}$ as $h \longrightarrow 0$. This is a manifestation of the instability for scheme (10.10) when $r > 1$.

10.1.6 The Kantorovich Theorem

For a wide class of linear operator equations, the theory of their approximate solution (not necessarily numerical) can be developed, and the foregoing concepts of consistency, stability, and convergence can be introduced and studied in a unified general framework. In doing so, the exact equation and the approximating equation should basically be considered in different spaces of functions — the original space and the approximating space (like, for example, the space of continuous functions and the space of grid functions).[4] However, it will also be very helpful to establish a fundamental relation between consistency, stability, and convergence for the most basic setup, when all the operators involved are assumed to act in the same space of functions. The corresponding result is known as the Kantorovich theorem; and a number of more specific results, such as that of Theorem 10.1, can be interpreted as its particular realizations. Of course, each individual realization will still require some non-trivial constructive steps pertaining primarily to the proper definition of the spaces and operators involved. A detailed description of the corresponding developments can be found in [KA82], including a comprehensive analysis of the case when the approximating equation is formulated on a different space (subspace) of functions. Monograph [RM67] provides for an account oriented more toward computational methods. The material of this section is more advanced, and can be skipped during the first reading.

[4]The approximating space often appears isomorphic to a subspace of the original space.

Let U and F be two Banach spaces, and let $\boldsymbol{L}$ be a linear operator: $\boldsymbol{L}: U \longmapsto F$ that has a bounded inverse, $\boldsymbol{L}^{-1}: F \longmapsto U$, $\|\boldsymbol{L}^{-1}\| < \infty$. In other words, we assume that the problem

$$\boldsymbol{L}u = f \tag{10.26}$$

is uniquely solvable for every $f \in F$ and well-posed.

Let $\boldsymbol{L}_h: U \longmapsto F$ be a family of operators parameterized by some h (for example, we may have $h = 1/n$, $n = 1,2,3,\ldots$). Along with the original problem (10.26), we introduce a series of its "discrete" counterparts:

$$\boldsymbol{L}_h u^{(h)} = f, \tag{10.27}$$

where $u^{(h)} \in U$ and each $\boldsymbol{L}_h$ is also assumed to have a bounded inverse, $\boldsymbol{L}_h^{-1}: F \longmapsto U$, $\|\boldsymbol{L}_h^{-1}\| < \infty$. The operators $\boldsymbol{L}_h$ are referred to as approximating operators.

We say that problem (10.27) is consistent, or in other words, that the operators $\boldsymbol{L}_h$ of (10.27) approximate the operator $\boldsymbol{L}$ of (10.26), if for any $u \in U$ we have

$$\|\boldsymbol{L}_h u - \boldsymbol{L}u\|_F \longrightarrow 0, \quad \text{as} \quad h \longrightarrow 0. \tag{10.28}$$

Note that any given $u \in U$ can be interpreted as solution to problem (10.26) with the right-hand side defined as $F \ni f \stackrel{\text{def}}{=} \boldsymbol{L}u$. Then, the general notion of consistency (10.28) becomes similar to the concept of approximation on a solution introduced in Section 10.1.2, see formula (10.6).

Problem (10.27) is said to be stable if all the inverse operators are bounded uniformly:

$$\|\boldsymbol{L}_h^{-1}\| \leq C = \text{const}, \tag{10.29}$$

which means that C does not depend on h. This is obviously a stricter condition than simply having each $\boldsymbol{L}_h^{-1}$ bounded; it is, again, similar to Definition 10.2 of Section 10.1.3.

THEOREM 10.2 (Kantorovich)

Provided that properties (10.28) and (10.29) hold, the solution $u^{(h)}$ of the approximating problem (10.27) converges to the solution u of the original problem (10.26):

$$\|u - u^{(h)}\|_U \longrightarrow 0, \quad \text{as} \quad h \longrightarrow 0. \tag{10.30}$$

PROOF Given (10.28) and (10.29), we have

$$\begin{aligned}\|u - u^{(h)}\|_U &\leq \|\boldsymbol{L}_h^{-1}\boldsymbol{L}_h u - \boldsymbol{L}_h^{-1} f\|_F \leq \|\boldsymbol{L}_h^{-1}\| \|\boldsymbol{L}_h u - f\|_F \\ &\leq C\|\boldsymbol{L}_h u - f\|_F = \leq C\|\boldsymbol{L}_h u - \boldsymbol{L}u + \boldsymbol{L}u - f\|_F \\ &= C\|\boldsymbol{L}_h u - \boldsymbol{L}u\|_F \longrightarrow 0, \quad \text{as} \quad h \longrightarrow 0,\end{aligned}$$

because $\boldsymbol{L}u = f$ and $\boldsymbol{L}_h u^{(h)} = f$. ∎

Theorem 10.2 actually asserts only one implication of the complete Kantorovich theorem, namely, that consistency and stability imply convergence, cf. Theorem 10.1. The other implication says that if the approximate problem is consistent and convergent, then it is also stable. Altogether it means, that *under the assumption of consistency, the notions of stability and convergence are equivalent.* The proof of this second implication is based on the additional result known as the principle of uniform boundedness for families of operators, see [RM67]. This proof is generally not difficult, but we omit it here because otherwise it would require introducing additional notions and concepts from functional analysis.

10.1.7 On the Efficacy of Finite-Difference Schemes

Let us now briefly comment on the approach we have previously adopted for assessing the quality of approximation by a finite-difference scheme. In Section 10.1.2, it is characterized by the norm of the truncation error $\|\delta f^{(h)}\|_{F_h}$, measured as a power of the grid size h, see formula (10.7). As we have seen in Theorem 10.1, for stable schemes the foregoing characterization, known as the order of accuracy, coincides with the (asymptotic) order of the error $\|[u]_h - u^{(h)}\|_{U_h}$, also referred to as the convergence rate, see formula (10.4). As the latter basically tells us how good the approximate solution is, it is natural to characterize the scheme by the amount of computations required for achieving a given order of error. In its own turn, the amount of computations is generally proportional to the overall number of grid nodes N. For ordinary differential equations N is inversely proportional to the grid size h. Therefore, when we say that ε is the norm of the error, where $\varepsilon = \mathcal{O}(h^k)$, we actually imply that $\varepsilon = \mathcal{O}(N^{-k})$, i.e., that driving the error down by a factor of two would require increasing the computational effort by roughly a factor of $\sqrt[k]{2}$. Therefore, in the case of ordinary differential equations, the order of accuracy with respect to h adequately characterizes the required amount of computations.

For partial differential equations, however, this is no longer the case. In the previously analyzed example of a problem with two independent variables x and t, the grid is defined by the two sizes: h and τ. The number of nodes N that belong to a given bounded region of the plane (x,t) is obviously of order $1/(\tau h)$. Let $\tau = rh$. In this case $N = \mathcal{O}(h^{-2})$, and saying that $\varepsilon = \mathcal{O}(h^k)$ is equivalent to saying that $\varepsilon = \mathcal{O}(N^{-k/2})$. If $\tau = rh^2$, then $N = \mathcal{O}(h^{-3})$, and saying that $\varepsilon = \mathcal{O}(h^k)$ is equivalent to saying that $\varepsilon = \mathcal{O}(N^{-k/3})$.

We see that in the case of partial differential equations the magnitude of the error may be more natural to quantify using the powers of N^{-1} rather than those of h. Indeed, this would allow one to immediately determine the amount of additional work (assumed proportional to N) needed to reduce the error by a prescribed factor. Hereafter, we will nonetheless adhere to the previous way of characterizing the accuracy by means of the powers of h, because it is more convenient for derivations, and it also has some "historical" reasons in the literature. The reader, however, should always keep in mind the foregoing considerations when comparing properties of the schemes and determining their suitability for solving a given problem.

We should also note that our previous statement on the proportionality of the computational work to the number of grid nodes N does not always hold either. There are, in fact, examples of finite-difference schemes that require $\mathcal{O}(N^{1+\alpha})$ arithmetic operations for finding the discrete solution, where α may be equal to 1/2, 1, or even 2. One may encounter situations like that when solving finite-difference problems that approximate elliptic equations, or when solving the problems with three or more independent variables [e.g., $u(x,y,t)$].

In the context of real computations, one normally takes the code execution time as a natural measure of the algorithm quality and uses it as a key criterion for the comparative assessment of numerical methods. However, the execution time is not necessarily proportional to the number of floating point operations. It may also be affected by the integer, symbolic, or logical operations. There are other factors that may play a role, for example, the so-called memory bandwidth that determines the rate of data exchange between the computer memory and the CPU. For multiprocessor computer platforms with distributed memory, the overall algorithm efficacy is to a large extent determined by how efficiently the data can be exchanged between different blocks of the computer memory. All these considerations need to be taken into account when choosing the method and when subsequently interpreting its results.

10.1.8 Bibliography Comments

The notion of stability for finite-difference schemes was first introduced by von Neumann and Richtmyer in a 1950 paper [VNR50] that discusses the computation of gasdynamic shocks. In that paper, stability was treated as sensitivity of finite-difference solutions to perturbations of the input data, in the sense of whether or not the initial errors will get amplified as the time elapses. No attempt was made at proving convergence of the corresponding approximation.

In work [OHK51], O'Brien, Hyman, and Kaplan studied finite-difference schemes for the heat equation. They used the apparatus of finite Fourier series (see Section 5.7) to analyze the sensitivity of their finite-difference solutions to perturbations of the input data. In modern terminology, the analysis of [OHK51] can be qualified as a spectral study of stability for a particular case, see Section 10.3.

The first comprehensive systems of definitions of stability and consistency that enabled obtaining convergence as their implication, were proposed by Ryaben'kii in 1952, see [Rya52], and by Lax in 1953, see [RM67, Chapter 3].

Ryaben'kii in work [Rya52] analyzed the Cauchy problem for linear partial differential equations with coefficients that may only depend on t. He derived necessary and sufficient conditions for stability of finite-difference approximations in the sense of the definition that he introduced in the same paper. He also built constructive examples of stable finite-difference schemes for the systems that are hyperbolic or p-parabolic in the sense of Petrowsky.

Lax, in his 1953 system of definitions, see [RM67, Chapter 3], considered finite-difference schemes for time-dependent operator equations. His assumption was that these schemes operate in the same Banach spaces of functions as the original differential operators do. The central result of the Lax theory is known as *the equivalence theorem.* It says that if the original continuous problem is uniquely solvable and well-posed, and if its finite-difference approximation is consistent, then stability of the approximation is necessary and sufficient for its convergence.

The system of basic definitions adopted in this book, as well as the form of the key theorem that stability and consistency imply convergence, are close to those proposed by Filippov in 1955, see [Fil55] and also [RF56]. The most important difference is that we use a more universal definition of consistency compared to the one by Filippov, [Fil55]. We emphasize that, in the approach by Filippov, the analysis is conducted entirely in the space of discrete functions. It allows for all types of equations, not necessarily time-dependent, see [RF56]. In general, the framework of [Fil55], [RF56] can be termed as somewhat more universal and as such, less constructive, than the previous more narrow definitions by Lax and by Ryaben'kii.

Continuing along the same lines, the Kantorovich theorem of Section 10.1.6 can, perhaps, be interpreted as the "ultimate" generalization of all the results of this type.

Let us also note that in the 1928 paper by Courant, Friedrichs, and Lewy [CFL28], as well as in many other papers that employ the method of finite differences to prove existence of solutions to differential equations, the authors establish inequalities that could nowadays be interpreted as stability with respect to particular norms. However, the specific notion of *finite-difference stability* has rather been introduced in the context of using the schemes for the approximate computation of solutions, under the assumption that those solutions already exist. Therefore, stability in the framework of approximate computations is usually studied in weaker norms than those that would have been needed for the existence proofs.

In the literature, the method of finite differences has apparently been first used for proving existence of solutions to partial differential equations by Lyusternik in 1924, see his later publication [Lyu40]. He analyzed the Laplace equation.

Exercises

1. For the Cauchy problem (10.8), analyze the following finite-difference scheme:

$$\frac{u_m^{p+1}-u_m^p}{\tau}-\frac{u_m^p-u_{m-1}^p}{h}=\varphi_m^p, \quad m=0,\pm1,\pm2,\dots, \quad p=0,1,\dots,[T/\tau]-1,$$
$$u_m^0=\psi_m, \quad m=0,\pm1,\pm2,\dots,$$

where $\tau=rh$ and $r=\text{const}$.

a) Explicitly write down the operator $\boldsymbol{L}_h$ and the right-hand side $f^{(h)}$ that arise when recasting the foregoing scheme in the form $\boldsymbol{L}_h u^{(h)}=f^{(h)}$.

b) Depict graphically the stencil of the scheme, i.e., locations of the three grid nodes, for which the difference equation connects the values of the discrete solution $u^{(h)}$.

c) Show that the finite-difference scheme is consistent and has first order accuracy with respect to h on a solution $u(x,t)$ of the original differential problem that has bounded second derivatives.

d) Find out whether the scheme is stable for any choice of r.

2. For the Cauchy problem:

$$u_t + u_x = \varphi(x,t), \quad -\infty < x < \infty,$$
$$u(x,0) = \psi(x), \quad 0 \le t \le T,$$

analyze the following two difference schemes according to the plan of Exercise 1:

$$\frac{u_m^{p+1} - u_m^p}{\tau} + \frac{u_m^p - u_{m-1}^p}{h} = \varphi_m^p, \quad u_m^0 = \psi_m,$$
$$\frac{u_m^{p+1} - u_m^p}{\tau} + \frac{u_{m+1}^p - u_m^p}{h} = \varphi_m^p, \quad u_m^0 = \psi_m.$$

3. Consider a Cauchy problem for the heat equation:

$$\begin{aligned} \frac{\partial u}{\partial t} - \frac{\partial^2 u}{\partial x^2} &= \varphi(x,t), & -\infty < x < \infty,\ 0 \le t \le T, \\ u(x,0) &= \psi(x), & -\infty < x < \infty, \end{aligned} \tag{10.31}$$

and introduce the following finite-difference scheme:

$$\begin{gathered} \frac{u_m^{p+1} - u_m^p}{\tau} - \frac{u_{m+1}^p - 2u_m^p + u_{m-1}^p}{h^2} = \varphi_m^p, \\ m = 0, \pm 1, \pm 2, \ldots, \quad p = 0, 1, \ldots, [T/\tau] - 1, \\ u_m^0 = \psi_m, \quad m = 0, \pm 1, \pm 2, \ldots. \end{gathered} \tag{10.32}$$

Assume that norms in the spaces U_h and F_h are defined according to (10.21) and (10.22), respectively.

a) Verify that when $\tau/h^2 = r = \text{const}$, scheme (10.32) approximates problem (10.31) with accuracy $\mathcal{O}(h^2)$.

b)* Show that when $r \le 1/2$ and $\varphi(x,t) \equiv 0$ the following maximum principle holds for the finite-difference solution u_m^p:

$$\sup_m |u_m^{p+1}| \le \sup_m |u_m^p|.$$

Hint. See the analysis in Section 10.6.1.

c)* Using the same argument as in Question b)*, prove that the finite-difference scheme (10.32) is stable for $r \le 1/2$, including the general case $\varphi(x,t) \not\equiv 0$.

4. Solution of the Cauchy problem (10.31) for the one-dimensional homogeneous heat equation, $\varphi(x,t) \equiv 0$, is given by the Poisson integral:

$$u(x,t) = \frac{1}{2\sqrt{\pi t}} \int_{-\infty}^{\infty} \psi(\xi) e^{-\frac{(x-\xi)^2}{4t}} d\xi.$$

See whether or not it is possible that a consistent explicit scheme (10.32) would also be convergent and stable if $\tau = h$, $h \longrightarrow 0$.

Hint. Compare the domains of dependence for the continuous and discrete problems, and use the Courant, Friedrichs, and Lewy condition.

5. The acoustics system of equations:

$$\frac{\partial v}{\partial t} = \frac{\partial w}{\partial x}, \ \frac{\partial w}{\partial t} = \frac{\partial v}{\partial x}, \quad -\infty < x < \infty, \ 0 \le t \le T,$$
$$v(x,0) = \varphi(x), \ \ w(x,0) = \psi(x), \quad -\infty < x < \infty,$$

has solutions of the type:

$$v(x,t) = \frac{\varphi(x-t) - \psi(x-t) + \varphi(x+t) + \psi(x+t)}{2},$$
$$w(x,t) = \frac{-\varphi(x-t) + \psi(x-t) + \varphi(x+t) + \psi(x+t)}{2}.$$

Can the following finite-difference scheme converge?

$$\frac{v_m^{p+1} - v_m^p}{\tau} - \frac{w_{m+1}^p - w_m^p}{h} = 0, \quad \frac{w_m^{p+1} - w_m^p}{\tau} - \frac{v_{m+1}^p - v_m^p}{h} = 0,$$
$$m = 0, \pm 1, \pm 2, \dots, \ \ p = 0, 1, \dots, [T/\tau] - 1,$$
$$v_m^0 = \varphi_m, \ \ w_m^0 = \psi_m, \ \ m = 0, \pm 1, \pm 2, \dots.$$

Hint. Compare the domains of dependence for the continuous and discrete problems.

6.* The Cauchy problem:

$$\frac{\partial u}{\partial t} - \frac{\partial u}{\partial x} = 0, \ \ -\infty < x < \infty, \ \ t > 0,$$
$$u(x,0) = e^{i\alpha x}, \ \ 0 < \alpha < 2\pi, \ \ \alpha = \text{const},$$

has the following solution:

$$u(x,t) = e^{i\alpha t} e^{i\alpha x}.$$

Solution of the corresponding finite-difference scheme:

$$\frac{u_m^{p+1} - u_m^p}{\tau} - \frac{u_{m+1}^p - u_m^p}{h} = 0, \qquad m = 0, \pm 1, \pm 2, \dots, \ \ p = 0, 1, \dots,$$
$$u_m^0 = e^{i\alpha(mh)}, \quad m = 0, \pm 1, \pm 2, \dots,$$

is given by (verify that):

$$u_m^p = (1 - r + re^{i\alpha h})^p e^{i\alpha(mh)}.$$

For $p = t/\tau$ and $m = x/h$, this discrete solution converges to the foregoing continuous solution as $h \longrightarrow 0$ for any fixed value of $r = \tau/h$. At the same time, when $r > 1$ the difference scheme does not satisfy the Courant, Friedrichs, and Lewy condition, which is necessary for convergence. Explain the apparent paradox.

10.2 Construction of Consistent Difference Schemes

10.2.1 Replacement of Derivatives by Difference Quotients

Perhaps the simplest approach to constructing finite-difference schemes that would approximate the corresponding differential problems is based on replacing the derivatives in the original equations by appropriate difference quotients. Below, we illustrate this approach with several examples. In these examples, we use the following approximation formulae, assuming that h is small:

$$\begin{aligned} \frac{df(x)}{dx} &\approx \frac{f(x+h)-f(x)}{h}, \quad \frac{df(x)}{dx} \approx \frac{f(x)-f(x-h)}{h}, \\ \frac{df(x)}{dx} &\approx \frac{f(x+h)-f(x-h)}{2h}, \\ \frac{d^2f(x)}{dx^2} &\approx \frac{f(x+h)-2f(x)+f(x-h)}{h^2}. \end{aligned} \tag{10.33}$$

Suppose that the function $f=f(x)$ has sufficiently many bounded derivatives. Then, the approximation error for equalities (10.33) can be estimated similarly to how it has been done in Section 9.2.1 of Chapter 9. Using the Taylor formula, we can write:

$$\begin{aligned} f(x+h) &= f(x)+hf'(x)+\frac{h^2}{2!}f''(x)+\frac{h^3}{3!}f'''(x)+\frac{h^4}{4!}f^{(4)}(x)+o(h^4), \\ f(x-h) &= f(x)-hf'(x)+\frac{h^2}{2!}f''(x)-\frac{h^3}{3!}f'''(x)+\frac{h^4}{4!}f^{(4)}(x)+o(h^4). \end{aligned} \tag{10.34}$$

where $o(h^\alpha)$ is a conventional generic notation for any quantity that satisfies: $\lim_{h\longrightarrow 0}\{o(h^\alpha)/h^\alpha\}=0$. Substituting expansions (10.34) into the right-hand sides of approximate equalities (10.33), one can obtain the expressions for the error:

$$\begin{aligned} \frac{f(x+h)-f(x)}{h} &= f'(x)+\left[\frac{h}{2}f''(x)+o(h)\right], \\ \frac{f(x)-f(x-h)}{h} &= f'(x)+\left[-\frac{h}{2}f''(x)+o(h)\right], \\ \frac{f(x+h)-f(x-h)}{2h} &= f'(x)+\left[\frac{h^2}{6}f'''(x)+o(h^2)\right], \\ \frac{f(x+h)-2f(x)+f(x-h)}{h^2} &= f''(x)+\left[\frac{h^2}{12}f^{(4)}(x)+o(h^2)\right]. \end{aligned} \tag{10.35}$$

The error of each given approximation formula in (10.33) is the term in rectangular brackets on the right-hand side of the corresponding equality (10.35).

It is clear that formulae of type (10.33) and (10.35) can also be used for approximating partial derivatives with difference quotients. For example:

$$\frac{\partial u(x,t)}{\partial t} \approx \frac{u(x,t+\tau)-u(x,t)}{\tau},$$

where

$$\frac{u(x,t+\tau)-u(x,t)}{\tau} = \frac{\partial u(x,t)}{\partial t} + \left[\frac{\tau}{2}\frac{\partial^2 u(x,t)}{\partial t^2} + o(\tau)\right].$$

Similarly, the following formula holds:

$$\frac{\partial u(x,t)}{\partial x} \approx \frac{u(x+h,t)-u(x,t)}{h},$$

and

$$\frac{u(x+h,t)-u(x,t)}{h} = \frac{\partial u(x,t)}{\partial x} + \left[\frac{h}{2}\frac{\partial^2 u(x,t)}{\partial t^2} + o(h)\right].$$

Example 1

Let us consider again the Cauchy problem (10.8) of Section 10.1.2:

$$\begin{aligned} \frac{\partial u}{\partial t} - \frac{\partial u}{\partial x} &= \varphi(x,t), \quad -\infty < x < \infty, \quad 0 < t \leq T, \\ u(x,0) &= \psi(x), \quad -\infty < x < \infty. \end{aligned}$$

We will build and analyze three finite-difference schemes that approximate this problem. For all these schemes, we will use a uniform Cartesian grid D_h on the plane (x,t). The grid is defined by formula (10.9), and its nodes are obtained as intersections of equally spaced horizontal and vertical straight lines. In so doing, the temporal grid size τ will be assumed proportional to the spatial grid size h: $\tau = rh$, where r is a positive constant.

The simplest scheme has, in fact, already been constructed in Section 10.1.2. It is defined by formulae (10.10), (10.11):

$$\begin{aligned} \frac{u_m^{p+1}-u_m^p}{\tau} - \frac{u_{m+1}^p - u_m^p}{h} &= \varphi_m^p, \quad m = 0, \pm 1, \pm 2, \ldots, \quad p = 0, 1, \ldots, [T/\tau]-1, \\ u_m^0 &= \psi_m, \quad m = 0, \pm 1, \pm 2, \ldots, \end{aligned}$$

and has been obtained by using the following difference quotients in place of the derivatives:

$$\frac{\partial u(x,t)}{\partial t} \approx \frac{u(x,t+\tau)-u(x,t)}{\tau}, \qquad \frac{\partial u(x,t)}{\partial x} \approx \frac{u(x+h,t)-u(x,t)}{h}.$$

The truncation error $\delta f^{(h)}$ of scheme (10.10), i.e.,the residual of the exact solution $[u]_h$ once it has been substituted into the finite-difference equation: $L_h[u]_h = f^{(h)} + \delta f^{(h)}$, is given by the expression (see the analysis on pages 311–312):

$$\delta f^{(h)} = \begin{cases} \dfrac{\tau}{2}\dfrac{\partial^2 u(x_m,t_p)}{\partial t^2} - \dfrac{h}{2}\dfrac{\partial^2 u(x_m,t_p)}{\partial x^2} + \mathcal{O}(\tau^2+h^2), \\ 0. \end{cases}$$

Let us use the same maximum norm (10.22) for the elements $f^{(h)}$ of the space F_h, as we have used previously in Section 10.1. Then, it is easy to see that

$$\|\delta f^{(h)}\|_{F_h} = \mathscr{O}(\tau + h) = \mathscr{O}(rh + h) = \mathscr{O}(h),$$

and consequently the scheme has first order accuracy. In the literature, scheme (10.10) is often referred to as the first order upwind, because the direction of the spatial differencing forward $\frac{u(x+h,t)-u(x,t)}{h}$ is opposite of the advection governed by the differential equation $u'_t - u'_x = 0$.

The second scheme differs from the first one precisely in this respect — instead of the spatial differencing forward we use differencing backward:

$$\frac{\partial u(x,t)}{\partial x} \approx \frac{u(x,t) - u(x-h,t)}{h}.$$

The overall scheme then becomes (see Exercise 1 of Section 10.1):

$$\begin{aligned} \frac{u_m^{p+1} - u_m^p}{\tau} - \frac{u_m^p - u_{m-1}^p}{h} &= \varphi_m^p, \quad m = 0, \pm 1, \pm 2, \ldots, \ p = 0, 1, \ldots, [T/\tau] - 1, \\ u_m^0 &= \psi_m, \quad m = 0, \pm 1, \pm 2, \ldots, \end{aligned}$$

and it also has first order accuracy. Naturally, it is sometimes referred to as the first order downwind scheme.

At a first glance, the difference between the upwind and the downwind schemes is minute. However, it turns out that the downwind scheme cannot be used for computations at all, because it violates the Courant, Friedrichs, and Lewy condition of Section 10.1.4 for any $r = \tau/h$.

Note also that the notions of upwind and downwind pertain to the particular operator $\boldsymbol{L}u = u'_t - u'_x$. If we were to change the sign and consider $\boldsymbol{L}u = u'_t + u'_x$ instead, then the upwind and downwind schemes would accordingly change places.

The third scheme is known as the leap-frog scheme:

$$\begin{aligned} \frac{u_m^{p+1} - u_m^{p-1}}{2\tau} - \frac{u_{m+1}^p - u_{m-1}^p}{2h} &= \varphi_m^p, \quad m = 0, \pm 1, \pm 2, \ldots, \quad p = 1, 2, \ldots, [T/\tau] - 1, \\ u_m^0 = \psi_m, \quad u_m^1 &= \psi_m + \left[\psi'(x_m) + \varphi_m^0\right]\tau, \quad m = 0, \pm 1, \pm 2, \ldots. \end{aligned} \tag{10.36}$$

It has overall second order accuracy, see Exercise 1. Note that as we are using differences of the second order with respect to time in (10.36), the scheme requires one more initial condition, i.e., specification of u_m^1, $m = 0, \pm 1, \pm 2, \ldots$, in addition to the specification of u_m^0. Of course, a fair degree of flexibility exists in how we can actually specify u_m^1. In doing so, second order accuracy needs to be maintained, and we therefore employ two terms of the Taylor expansion along with the original differential equation and its initial condition:

$$\begin{aligned} u(x,\tau) &= u(x,0) + u'_t(x,0)\tau + \mathscr{O}(\tau^2) \\ &= u(x,0) + \left[u'_x(x,0) + \varphi(x,0)\right]\tau + \mathscr{O}(\tau^2) \\ &= \psi(x) + \left[\psi'(x) + \varphi(x,0)\right]\tau + \mathscr{O}(\tau^2). \end{aligned}$$

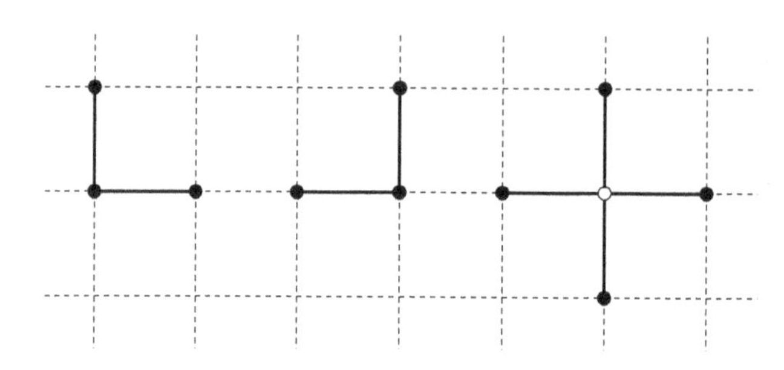

FIGURE 10.2: Stencils of thc upwind, downwind, and the leap-frog schemes.

For a given scheme, the subset of the grid nodes connected by the corresponding finite-difference equation for any fixed m and p is called *the stencil.* It is often convenient to depict the stencils graphically; for the three schemes we have just considered this is done in Figure 10.2.

Example 2

Next, we will describe two finite-difference schemes that approximate the Cauchy problem (10.31) for the heat equation (see Exercise 3 of Section 10.1):

$$\begin{aligned} \frac{\partial u}{\partial t} - \frac{\partial^2 u}{\partial x^2} &= \varphi(x,t), \qquad -\infty < x < \infty, \;\; 0 \le t \le T, \\ u(x,0) &= \psi(x), \qquad -\infty < x < \infty. \end{aligned}$$

The simplest scheme is defined by formula (10.32), and in the operator form $\boldsymbol{L}_h^{(1)} u^{(h)} = f^{(h)}$ we have:

$$\boldsymbol{L}_h^{(1)} u^{(h)} = \begin{cases} \dfrac{u_m^{p+1} - u_m^p}{\tau} - \dfrac{u_{m+1}^p - 2u_m^p + u_{m-1}^p}{h^2}, & m = 0, \pm 1, \pm 2, \dots, \\ & p = 0, 1, \dots, [T/\tau] - 1, \\ u_m^0, & m = 0, \pm 1, \pm 2, \dots, \end{cases}$$
$$f^{(h)} = \begin{cases} \varphi_m^p \equiv \varphi(mh, p\tau), \quad m = 0, \pm 1, \pm 2, \dots, & \\ & p = 0, 1, \dots, [T/\tau] - 1, \\ \psi_m \equiv \psi(mh), \quad m = 0, \pm 1, \pm 2, \dots . & \end{cases} \tag{10.37}$$

Scheme (10.37) is obtained with the help of the approximate formulae:

$$\begin{aligned} \frac{\partial u(x,t)}{\partial t} &\approx \frac{u(x,t+\tau) - u(x,t)}{\tau}, \\ \frac{\partial^2 u(x,t)}{\partial x^2} &\approx \frac{u(x+h,t) - 2u(x,t) + u(x-h,t)}{h^2}. \end{aligned}$$

Alternatively, we could write:

$$\begin{aligned} \frac{\partial u(x,t+\tau)}{\partial t} &\approx \frac{u(x,t+\tau) - u(x,t)}{\tau}, \\ \frac{\partial^2 u(x,t+\tau)}{\partial x^2} &\approx \frac{u(x+h,t+\tau) - 2u(x,t+\tau) + u(x-h,t+\tau)}{h^2}. \end{aligned}$$

These formulae lead to a new scheme $\boldsymbol{L}^{(2)}u^{(h)} = f^{(h)}$ for the same Cauchy problem (10.31):

$$\boldsymbol{L}_h^{(2)}u^{(h)} = \begin{cases} \dfrac{u_m^{p+1}-u_m^p}{\tau} - \dfrac{u_{m+1}^{p+1}-2u_m^{p+1}+u_{m-1}^{p+1}}{h^2}, & m=0,\pm 1,\pm 2,\ldots, \\ & p=0,1,\ldots,[T/\tau]-1, \\ u_m^0, & m=0,\pm 1,\pm 2,\ldots, \end{cases}$$

$$f^{(h)} = \begin{cases} \varphi_m^{p+1} \equiv \varphi(mh,(p+1)\tau), & m=0,\pm 1,\pm 2,\ldots, \\ & p=0,1,\ldots,[T/\tau]-1, \\ \psi_m \equiv \psi(mh), & m=0,\pm 1,\pm 2,\ldots. \end{cases} \quad (10.38)$$

At a first glance the two schemes, (10.37) and (10.38), are not that much different from one another. In practice, however, they differ substantially.

The solution $u^{(h)}$ of the finite-difference system $\boldsymbol{L}^{(1)}u^{(h)} = f^{(h)}$ defined by (10.37) can be easily found one time level after another with the help of the explicit formula:

$$u_m^{p+1} = (1-2r)u_m^p + r(u_{m+1}^p + u_{m-1}^p) + \tau\varphi_m^p$$

that constitutes marching in time; here $r = \tau/h^2$. This formula is obtained by resolving the finite-difference system $\boldsymbol{L}^{(1)}u^{(h)} = f^{(h)}$ with respect to u_m^{p+1}. Once the solution u_m^p, $m=0,\pm 1,\pm 2,\ldots$, is known at $t=t_p=p\tau$, one can immediately compute its values u_m^{p+1}, $m=0,\pm 1,\pm 2,\ldots$, at the next level $t=t_{p+1}=(p+1)\tau$, etc. For this reason, scheme (10.37) is referred to as explicit.

Solution $u^{(h)}$ of the second finite-difference system $\boldsymbol{L}^{(2)}u^{(h)} = f^{(h)}$, which is defined by (10.38), can also be found by time marching, i.e., level t_{p+1} after level t_p. However, this second scheme does not possess the foregoing convenient property of the first one. Namely, the finite-difference system $\boldsymbol{L}^{(2)}u^{(h)} = f^{(h)}$ cannot be easily resolved with respect to u_m^{p+1}, and there is no explicit formula that would express the value of u_m^{p+1} for some fixed m and p through the known values of u_{m-1}^p, u_m^p, and u_{m+1}^p, or perhaps some other known values from the previous time level $t=t_p$. This is the reason why the scheme $\boldsymbol{L}^{(2)}u^{(h)} = f^{(h)}$ of (10.38) is referred to as implicit.

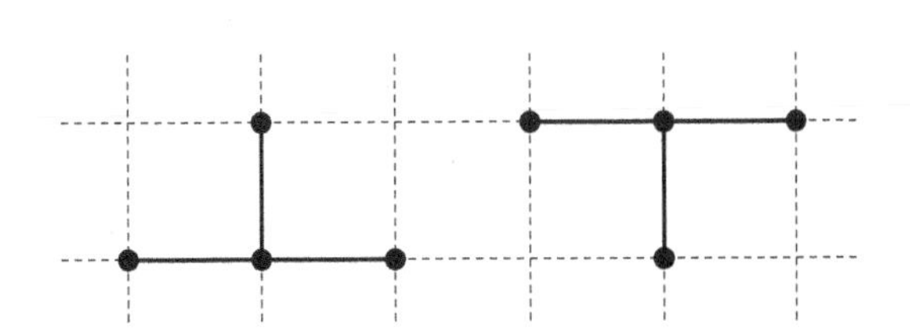

FIGURE 10.3: Stencils of the explicit and implicit schemes for the heat equation.

The system of finite-difference equations $\boldsymbol{L}^{(2)}u^{(h)} = f^{(h)}$ cannot be explicitly resolved with respect to u_m^{p+1} because for a given pair of indices m and p, the corresponding equation contains not only the unknown u_m^{p+1}, but also two other unknowns: u_{m-1}^{p+1} and u_{m+1}^{p+1}. Therefore, to determine u_m^{p+1}, $m=0,\pm 1,\pm 2,\ldots$, for each p, one would need to solve the entire system of equations with respect to the grid function u_m^{p+1} of the argument m. Nonetheless, it will later be shown (see Examples 6 and 7 of Section 10.3 and also Section 10.6) that the implicit scheme $\boldsymbol{L}^{(2)}u^{(h)} = f^{(h)}$ is often more convenient for practical computations compared to the explicit scheme $\boldsymbol{L}^{(1)}u^{(h)} = f^{(h)}$. The stencils of both schemes are schematically depicted in Figure 10.3.

When $\tau = rh^2$, $r = \text{const}$, both schemes have second order of accuracy. For the explicit scheme (10.37), this result is a part of Exercise 3 of Section 10.1. Let us therefore prove second order accuracy for the implicit scheme (10.38).

Using formulae (10.35), we can write:

$$\boldsymbol{L}_h^{(2)}[u]_h = \begin{cases} \dfrac{\partial u(x_m,t_{p+1})}{\partial t} - \dfrac{\partial^2 u(x_m,t_{p+1})}{\partial x^2} \\ \qquad\qquad - \dfrac{\tau}{2}\dfrac{\partial^2 u(x_m,t_{p+1})}{\partial t^2} - \dfrac{h^2}{12}\dfrac{\partial^4 u(x_m,t_{p+1})}{\partial x^4} + o(\tau+h^2), \\ u(x_m,0). \end{cases}$$

Consequently, considering that $\tau = rh^2$, we have:

$$\boldsymbol{L}_h^{(2)}[u]_h = \begin{cases} \varphi(mh,(p+1)\tau) + \mathcal{O}(h^2), \\ \psi(mh) + 0. \end{cases}$$

Therefore, for the truncation error $\delta f^{(h)} = \boldsymbol{L}_h^{(2)}[u]_h - f^{(h)}$ we obtain:

$$\delta f^{(h)} = \begin{cases} \mathcal{O}(h^2), \\ 0, \end{cases}$$

which obviously implies that

$$\|\delta f^{(h)}\|_{F_h} = \mathcal{O}(h^2).$$

Example 3

Let us now consider a simple finite-difference scheme that would approximate the inhomogeneous Dirichlet problem of type (5.7) for the Poisson equation on the square $D = \{(x,y)|0 < x < 1,\ 0 < y < 1\}$ with the boundary $\Gamma = \partial D$:

$$\begin{aligned} \frac{\partial^2 u}{\partial x^2} + \frac{\partial^2 u}{\partial y^2} &= \varphi(x,y), \qquad (x,y) \in D, \\ u\big|_\Gamma &= \psi(x,y), \qquad (x,y) \in \Gamma. \end{aligned}$$

As we did previously in Section 5.1.3, we will use a uniform Cartesian grid in D with size h in either coordinate direction; and we will assume that $M = 1/h$ is an integer:

$$D_h = \{(x_m,y_n)|x_m = mh,\ y_n = nh,\ m = 0,1,\dots,M,\ n = 0,1,\dots,M\}.$$

The scheme $L_h u^{(h)} = f^{(h)}$ is obtained by replacing the derivatives in the Poisson equation by the second order central differences at every interior node of the grid D_h, i.e., at the nodes with $m = 1, 2, \ldots, M-1$ and $n = 1, 2, \ldots, M-1$, see Figure 5.1(a) on page 122. This requires the use of a five-node symmetric stencil shown in Figure 10.4. As for those nodes of the grid D_h that happen to be on the boundary Γ, we rather impose there the Dirichlet boundary condition, which altogether yields:

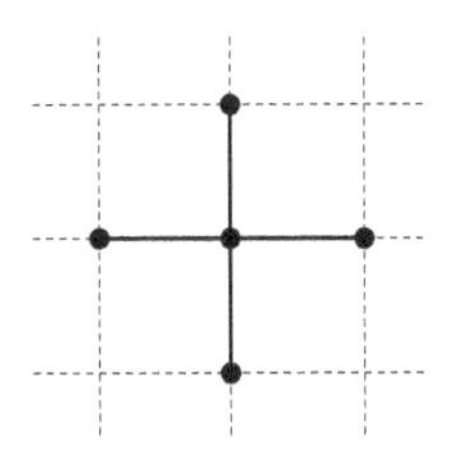

FIGURE 10.4: Five-node stencil for the central-difference scheme.

$$L_h u^{(h)} = \begin{cases} \dfrac{u_{m+1,n} - 2u_{m,n} + u_{m-1,n}}{h^2} + \dfrac{u_{m,n+1} - 2u_{m,n} + u_{m,n-1}}{h^2}, \\ \qquad\qquad (mh, nh) \in D_h,\ (mh, nh) \notin \Gamma, \\ u_{m,n}, \qquad (mh, nh) \in \Gamma, \end{cases}$$

$$f^{(h)} = \begin{cases} \varphi_{m,n} \equiv \varphi(mh, nh), & (mh, nh) \in D_h,\ (mh, nh) \notin \Gamma, \\ \psi_{m,n} \equiv \psi(mh, nh), & (mh, nh) \in \Gamma. \end{cases}$$

Then, according to the last formula of (10.35), for the truncation error $\delta f^{(h)} = L_h[u]_h - f^{(h)}$ we obtain:

$$f^{(h)} = \begin{cases} \dfrac{h^2}{12}\left(\dfrac{\partial^4 u}{\partial x^4} + \dfrac{\partial^4 u}{\partial y^4}\right) + o(h^2), \\ 0. \end{cases}$$

Therefore, we can conclude that the proposed central-difference scheme for the Poisson equation $L_h u^{(h)} = f^{(h)}$ has overall second order accuracy on those solutions of this equation that possess bounded fourth derivatives.

10.2.2 The Method of Undetermined Coefficients

The schemes that we have introduced and analyzed in the previous Section 10.2.1 have all been obtained by replacing each individual derivative in the corresponding differential equation by an appropriate difference quotient. A more general approach for constructing difference schemes is based on approximating the entire differential operator rather than the individual derivatives. We will illustrate this approach using several examples of the schemes for the Cauchy problem (10.8). Henceforth in this section, we will keep using the notations L and L_h for the full differential and difference operators, respectively, i.e., the operators that include the initial and/or boundary conditions. We will also introduce the notations Λ and Λ_h for the corresponding equations by themselves only. As before, the grid D_h on the region $\{(x,t) | -\infty < x < \infty,\ 0 \le t \le T\}$ will be assumed Cartesian and uniform:

$$D_h = \{(x_m, t_p) | x_m = mh,\ m = 0, \pm 1, \pm 2, \ldots;\ t_p = p\tau,\ p = 0, 1, \ldots, [T/\tau]\}.$$

Let us first consider the first order upwind scheme (10.10), (10.11) on the three-node stencil shown in the left part of Figure 10.2. The finite-difference equation:

$$\mathbf{\Lambda}_h u^{(h)} = \frac{u_m^{p+1} - u_m^p}{\tau} - \frac{u_{m+1}^p - u_m^p}{h} = \varphi_m^p \equiv \varphi(x_m, t_p)$$

employed in this scheme can be rewritten with generic notations for the coefficients:

$$\mathbf{\Lambda}_h u^{(h)} = a^0 u_m^{p+1} + a_0 u_m^p + a_1 u_{m+1}^p = \varphi_m^p.$$

For the moment, we will disregard that we already know the actual scheme (10.10), for which we have:

$$a^0 = \frac{1}{\tau}, \quad a_0 = \frac{1}{h} - \frac{1}{\tau}, \quad a_1 = -\frac{1}{h}.$$

Instead, we will treat the quantities a^0, a_0, and a^1 as undetermined coefficients, and will try and choose them to achieve first order accuracy, i.e., make sure that the following equality holds:

$$\mathbf{\Lambda}_h [u]_h \Big|_{(x_m,t_p)} = \left(\frac{\partial u}{\partial t} - \frac{\partial u}{\partial x} \right) \Big|_{(x_m,t_p)} + \mathcal{O}(h).$$

This equality can obviously be rewritten as:

$$\mathbf{\Lambda}_h [u]_h \Big|_{(x_m,t_p)} = \mathbf{\Lambda} u \Big|_{(x_m,t_p)} + \mathcal{O}(h), \tag{10.39}$$

where

$$\mathbf{\Lambda} u = \frac{\partial u}{\partial t} - \frac{\partial u}{\partial x}. \tag{10.40}$$

According to the Taylor formula, we have:

$$u(x_m, t_p + \tau) = u(x_m, t_p) + \tau \frac{\partial u(x_m, t_p)}{\partial t} + \mathcal{O}(\tau^2),$$
$$u(x_m + h, t_p) = u(x_m, t_p) + h \frac{\partial u(x_m, t_p)}{\partial x} + \mathcal{O}(h^2).$$

Substituting the previous expressions into the right-hand side of the equality:

$$\mathbf{\Lambda}_h [u]_h \Big|_{(x_m,t_p)} = a^0 u(x_m, t_p + \tau) + a_0 u(x_m, t_p) + a_1 u(x_m + h, t_p),$$

we obtain:

$$\begin{aligned} \mathbf{\Lambda}_h [u]_h \Big|_{(x_m,t_p)} &= (a^0 + a_0 + a_1) u(x_m, t_p) \\ &+ a^0 \tau \frac{\partial u(x_m, t_p)}{\partial t} + a_1 h \frac{\partial u(x_m, t_p)}{\partial x} + \mathcal{O}(a^0 \tau^2 + a_1 h^2). \end{aligned} \tag{10.41}$$

Since our objective is to facilitate approximation in the sense of (10.39) by selecting the appropriate coefficients a^0, a_0, and a_1, it will be natural to try and regroup the terms on the right-hand side of equality (10.41) in order to single out the term (10.40). Then, the rest of the terms will altogether yield the remainder of the approximation (i.e., the residual) that is supposed to be small. To single out the term $\mathbf{\Lambda}u$, one can replace either of the derivatives: $\frac{\partial u}{\partial t}$ or $\frac{\partial u}{\partial x}$, on the right-hand side of (10.41) using the respective identity:

$$\frac{\partial u}{\partial t} \equiv \mathbf{\Lambda}u + \frac{\partial u}{\partial x} \qquad \text{or} \qquad \frac{\partial u}{\partial x} \equiv \frac{\partial u}{\partial t} - \mathbf{\Lambda}u. \tag{10.42}$$

For definiteness, we will exploit the first formula (10.42).

In addition, we will impose the previously used constraint $\tau = rh$ on the grid sizes τ and h, where r is a positive constant. Then, equality (10.41) transforms into:

$$\begin{aligned} \mathbf{\Lambda}_h[u]_h\Big|_{(x_m,t_p)} &= a^0 rh\mathbf{\Lambda}u\Big|_{(x_m,t_p)} + (a^0 + a_0 + a_1)u(x_m,t_p) \\ &+ (a^0 r + a_1)h\frac{\partial u(x_m,t_p)}{\partial x} + \mathcal{O}(a^0 r^2 h^2) + \mathcal{O}(a_1 h^2). \end{aligned} \tag{10.43}$$

Next, we notice that among all smooth functions $u = u(x,t)$ one can obviously find those, for which the three quantities u, $\frac{\partial u}{\partial t}$, and $\frac{\partial u}{\partial x}$ will assume arbitrary independent values at any given fixed point (x,t). Consequently, the quantities:

$$u, \quad \frac{\partial u}{\partial x}, \quad \text{and } \mathbf{\Lambda}u \equiv \frac{\partial u}{\partial t} - \frac{\partial u}{\partial x} = \varphi(x,t)$$

can also be treated as independent entities. Therefore, formula (10.43) implies that in order to meet the consistency condition (10.39) for an arbitrary right-hand side $\varphi(x,t)$ of problem (10.8), the following equalities must hold:

$$a^0 rh = 1 + \mathcal{O}_1(h), \quad a^0 + a_0 + a_1 = 0 + \mathcal{O}_2(h), \quad (a^0 r + a_1)h = 0 + \mathcal{O}_3(h),$$

where $\mathcal{O}_1(h)$, $\mathcal{O}_2(h)$, and $\mathcal{O}_3(h)$ are some arbitrary quantities of order at most $\mathcal{O}(h)$.

For simplicity, let us first set $\mathcal{O}_1(h) = \mathcal{O}_2(h) = \mathcal{O}_3(h) = 0$. Then, the resulting system of equations:

$$a^0 rh = 1, \quad a^0 + a_0 + a_1 = 0, \quad a^0 r + a_1 = 0$$

has a unique solution:

$$a^0 = \frac{1}{rh} = \frac{1}{\tau}, \quad a_0 = \frac{r-1}{rh} = \frac{1}{h} - \frac{1}{\tau}, \quad a_1 = -\frac{1}{h} \tag{10.44}$$

that yields precisely the first order upwind scheme (10.10).

As such, we have corroborated one more time that the upwind scheme (10.10) has first order accuracy on the solution $u = u(x,t)$ of problem (10.8). Moreover, one can, in fact, claim that it is *the only* consistent scheme for the differential Cauchy problem (10.8) in the class of all schemes $\boldsymbol{L}_h u^{(h)} = f^{(h)}$, for which

$$\boldsymbol{L}_h u^{(h)} = \begin{cases} a^0 u_m^{p+1} + a_0 u_m^p + a_1 u_{m+1}^p, \\ u_m^0, \end{cases} \quad \text{and} \quad f^{(h)} = \begin{cases} \varphi(x_m, t_p), \\ \psi(x_m). \end{cases} \tag{10.45}$$

The aforementioned uniqueness of the consistent scheme (10.10) shall be understood here in the following sense. When obtaining the coefficients (10.44), we have disregarded the arbitrariness due to the freedom of choosing the vanishing quantities $\mathcal{O}_1(h)$, $\mathcal{O}_2(h)$, and $\mathcal{O}_3(h)$ (as $h \longrightarrow 0$). If we were to retain these quantities, then instead of the coefficients (10.44) we would have obtained:

$$\begin{aligned} \tilde{a}^0 &= \frac{1}{h}\left[\frac{1}{r} + \frac{1}{r} \cdot \mathcal{O}_1(h)\right], \\ \tilde{a}_0 &= \frac{1}{h}\left[\frac{r-1}{r} + \frac{r-1}{r}\mathcal{O}_1(h) - \mathcal{O}_3(h) + h \cdot \mathcal{O}_2(h)\right], \\ \tilde{a}_1 &= \frac{1}{h}\left[-1 - \mathcal{O}_1(h) + \mathcal{O}_3(h)\right]. \end{aligned} \tag{10.46}$$

By construction, the new coefficients $\tilde{a}^0$, $\tilde{a}_0$, and $\tilde{a}_1$ of (10.46) still guarantee first order accuracy with respect to h for any $\mathcal{O}_1$, $\mathcal{O}_2$, and $\mathcal{O}_3$. At the same time, the relative difference between the coefficients defined by formulae (10.46) and those given by formulae (10.44) becomes negligibly small as the grid size vanishes:

$$\lim_{h \longrightarrow 0} \frac{a^0 - \tilde{a}^0}{a^0} = 0, \quad \lim_{h \longrightarrow 0} \frac{a_0 - \tilde{a}_0}{a_0} = 0, \quad \lim_{h \longrightarrow 0} \frac{a_1 - \tilde{a}_1}{a_1} = 0.$$

Therefore, we can basically consider the entire collection of schemes (10.45) with the coefficients $\tilde{a}^0$, $\tilde{a}_0$, and $\tilde{a}_1$ given by (10.46), as being equivalent to one and the same first order upwind scheme (10.10) characterized by a fixed choice of the coefficients according to (10.44). Indeed, any scheme from this collection is consistent with order $\mathcal{O}(h)$, and the smaller the grid size the closer will any such scheme be to the "core" scheme (10.10). Moreover, one can show that the result quoted in Section 9.3.4 of Chapter 9 applies. The differences between the respective coefficients given by formulae (10.44) and (10.46) can be treated as small perturbations that do not ruin the stability and as such, do not affect the convergence.

Hereafter, we will always disregard the foregoing arbitrariness, and instead of considering individual schemes will rather operate with their equivalence classes, making no distinction within the class when all the variation is due to the quantities of type $\mathcal{O}_1(h)$, $\mathcal{O}_2(h)$, and $\mathcal{O}_3(h)$ in the definition of the coefficients, see, e.g., (10.46). In doing so, we may not always need to introduce those quantities explicitly, and will set them instead to zero from the very beginning.

Let us now see how one can construct a more general class of schemes $\boldsymbol{L}_h u^{(h)} = f^{(h)}$ for the same advection problem (10.8). Namely, we will be building explicit schemes on the four-node stencil from Figure 10.3(left) that we have otherwise used for the heat equation. The generic notation for the operator $\boldsymbol{L}_h$ will now involve four undetermined coefficients [as opposed to three coefficients in formula (10.45)]:

$$\boldsymbol{L}_h u^{(h)} = \begin{cases} a^0 u_m^{p+1} + a_{-1} u_{m-1}^p + a_0 u_m^p + a_1 u_{m+1}^p, \\ u_m^0, \end{cases} \quad f^{(h)} = \begin{cases} \varphi(x_m, t_p), \\ \psi(x_m), \end{cases} \tag{10.47}$$

whereas the right-hand side $f^{(h)}$ will still be the same as defined by formula (10.45). As before, we will assume that $\tau = rh$, $r = \text{const}$, and will also introduce the notation:

$$\boldsymbol{\Lambda}_h u^{(h)} = a^0 u_m^{p+1} + a_{-1} u_{m-1}^p + a_0 u_m^p + a_1 u_{m+1}^p. \tag{10.48}$$

Then, using the Taylor formula, we can write for any sufficiently smooth function $u = u(x,t)$:

$$\begin{aligned} \boldsymbol{\Lambda}_h [u]_h \Big|_{(x_m,t_p)} &= (a^0 + a_{-1} + a_0 + a_1) u(x_m,t_p) + a^0 rh \frac{\partial u(x_m,t_p)}{\partial t} \\ &+ (a_1 - a_{-1}) h \frac{\partial u(x_m,t_p)}{\partial x} + \frac{1}{2} a^0 r^2 h^2 \frac{\partial^2 u(x_m,t_p)}{\partial t^2} \\ &+ \frac{1}{2}(a_1 + a_{-1}) h^2 \frac{\partial^2 u(x_m,t_p)}{\partial x^2} + \mathcal{O}(a^0 r^3 h^3, a_1 h^3, a_{-1} h^3). \end{aligned}$$

On the right-hand side of the previous equality, we will single out the term $\boldsymbol{\Lambda} u$ of (10.40) with the help of the first identity of (10.42). This yields:

$$\begin{aligned} \boldsymbol{\Lambda}_h [u]_h \Big|_{(x_m,t_p)} &= a^0 rh \boldsymbol{\Lambda} u \Big|_{(x_m,t_p)} + (a^0 + a_{-1} + a_0 + a_1) u(x_m,t_p) \\ &+ (a^0 r + a_1 - a_{-1}) h \frac{\partial u(x_m,t_p)}{\partial x} + \frac{1}{2} a^0 r^2 h^2 \frac{\partial^2 u(x_m,t_p)}{\partial t^2} \\ &+ \frac{1}{2}(a_1 + a_{-1}) h^2 \frac{\partial^2 u(x_m,t_p)}{\partial x^2} + \mathcal{O}(a^0 r^3 h^3, a_1 h^3, a_{-1} h^3). \end{aligned} \tag{10.49}$$

Then, assuming that the quantity $\mathcal{O}(a^0 r^3 h^3, a_1 h^3, a_{-1} h^3)$ is small (this assumption will later be corroborated), we see that in order for the consistency condition (10.39) to hold for the operator $\boldsymbol{\Lambda}_h$ of (10.48), the four coefficients a^0, a_{-1}, a_0, and a_1 must satisfy the following three equations:

$$\begin{aligned} a^0 rh &= 1, \\ a^0 + a_{-1} + a_0 + a_1 &= 0, \\ h(a^0 r + a_1 - a_{-1}) &= 0. \end{aligned} \tag{10.50}$$

The system of linear algebraic equations (10.50) has multiple solutions. One of these solutions:

$$a^0 = \frac{1}{rh}, \quad a_{-1} = 0, \quad a_0 = \frac{r-1}{rh}, \quad a_1 = -\frac{1}{h}$$

leads to the previously analyzed upwind scheme (10.10). Another solution

$$a^0 = \frac{1}{rh} = \frac{1}{\tau}, \quad a_{-1} = \frac{1}{2h}, \quad a_0 = \frac{-1}{rh} = \frac{1}{h} - \frac{1}{\tau}, \quad a_1 = -\frac{1}{2h}$$

yields the scheme $\boldsymbol{L}_h u^{(h)} = f^{(h)}$ with the operator and the right-hand side defined as:

$$\boldsymbol{L}_h u^{(h)} = \begin{cases} \dfrac{u_m^{p+1} - u_m^p}{\tau} - \dfrac{u_{m+1}^p - u_{m-1}^p}{2h}, \\ u_m^0, \end{cases} \quad \text{and} \quad f^{(h)} = \begin{cases} \varphi(x_m, t_p), \\ \psi(x_m). \end{cases} \tag{10.51}$$

Given any solution of system (10.50), one, of course, needs to substitute it into the remainder of formula (10.49) and make sure that it is indeed small. For the previous two solutions that lead to the schemes (10.10) and (10.51), respectively, this substitution yields the remainder:

$$\frac{a^0 r^2 h^2}{2} \frac{\partial^2 u}{\partial t^2} + \frac{a_1 + a_{-1}}{2} h^2 \frac{\partial^2 u}{\partial x^2} + \mathcal{O}(a^0 r^3 h^3, a_1 h^3, a_{-1} h^3) \tag{10.52}$$

that has order $\mathcal{O}(h)$. Indeed, as the expressions for the coefficients a^0, a_{-1}, a_0, and a_1 all contain the grid size h in the denominator, we conclude that the first two terms in the previous sum have order $\mathcal{O}(h)$, while the quantity $\mathcal{O}(a^0 r^3 h^3, a_1 h^3, a_{-1} h^3)$ is, in fact, of order $\mathcal{O}(h^2)$.

In general, among the smooth functions $u = u(x,t)$ there are obviously polynomials of the second degree, for which the derivatives $\frac{\partial^2 u}{\partial t^2}$ and $\frac{\partial^2 u}{\partial x^2}$ can assume arbitrary independent values at any fixed point (x,t). In so doing, the term $\mathcal{O}(a^0 r^3 h^3, a_1 h^3, a_{-1} h^3)$ that contains the third derivatives of the polynomials vanishes. Therefore, in order to guarantee that the order of the residual (10.52) [remainder of the approximation (10.49)] be at least $\mathcal{O}(h)$, we must require that the coefficients in front of $\frac{\partial^2 u}{\partial t^2}$ and $\frac{\partial^2 u}{\partial x^2}$ both be of order h, independently of one another. On the other hand, from the first equation of (10.50) we always have $a^0 = 1/rh$, and consequently, the coefficient in front of $\frac{\partial^2 u}{\partial t^2}$ in the sum (10.52) is equal to $rh/2$. As such, the order of the residual with respect to the grid size may never exceed $\mathcal{O}(h)$.

Thus, we have established that one cannot construct a consistent scheme $\boldsymbol{L}_h u^{(h)} = f^{(h)}$ of type (10.47) that would approximate the Cauchy problem (10.8) with the accuracy better than $\mathcal{O}(h)$. To achieve, for example the accuracy of $\mathcal{O}(h^2)$, one would have to use a larger stencil, i.e., employ more grid nodes for building the difference operator $\boldsymbol{L}_h$.

However, under the additional assumption of $\varphi(x,t) \equiv 0$, there exists one and only one scheme $\boldsymbol{L}_h u^{(h)} = f^{(h)}$ of type (10.47) that approximates problem (10.8) with second order of accuracy with respect to h. Let us actually construct this scheme and in doing so also make sure that it is unique.[5] For that purpose, we first notice that the

[5]Uniqueness is to be understood in the same sense as on page 336.

identity $\frac{\partial u}{\partial t} - \frac{\partial u}{\partial x} = 0$ also implies $\frac{\partial^2 u}{\partial t^2} - \frac{\partial^2 u}{\partial x^2} = 0$. Indeed,

$$\frac{\partial^2 u}{\partial t^2} = \frac{\partial}{\partial t}\left(\frac{\partial u}{\partial t}\right) = \frac{\partial}{\partial t}\left(\frac{\partial u}{\partial x}\right) = \frac{\partial}{\partial x}\left(\frac{\partial u}{\partial t}\right) = \frac{\partial}{\partial x}\left(\frac{\partial u}{\partial x}\right) = \frac{\partial^2 u}{\partial x^2}.$$

Therefore, once the necessary conditions (10.50) for achieving first order accuracy hold, formula (10.49) transforms into:

$$\begin{aligned}\mathbf{\Lambda}_h[u]_h\Big|_{(x_m,t_p)} &= a^0 rh\mathbf{\Lambda} u\Big|_{(x_m,t_p)} \\ &+ \frac{1}{2}(a^0r^2 + a_1 + a_{-1})h^2\frac{\partial^2 u(x_m,t_p)}{\partial x^2} + \mathcal{O}(a^0r^3h^3, a_1h^3, a_{-1}h^3).\end{aligned} \tag{10.53}$$

To obtain a second order scheme while taking into account that $a^0 = 1/rh$, we need to require that $h^2(a^0r^2 + a_1 + a_{-1}) = \mathcal{O}(h^2)$. As has been discussed, the most straightforward way to satisfy this requirement is to take:

$$h^2(a^0r^2 + a_1 + a_{-1}) = 0. \tag{10.54}$$

The system of four linear equations (10.50), (10.54) has a unique solution:

$$a^0 = \frac{1}{rh}, \quad a_0 = -\frac{1}{rh} + \frac{r}{h}, \quad a_{-1} = \frac{1-r}{2h}, \quad a_1 = -\frac{1+r}{2h}. \tag{10.55}$$

If we were to exercise the existing flexibility and retain the vanishing quantities of order $\mathcal{O}(h^2)$ on the right-hand side of equations (10.50) and (10.54), then we would have only obtained small corrections to the coefficients (10.55).

As the third term on the right-hand side of (10.53) is $\mathcal{O}(h^2)$, we conclude that the scheme $\boldsymbol{L}_h u^{(h)} = f^{(h)}$ of (10.47), (10.55) is the only scheme built on the four-node stencil form Figure 10.3(left) that approximates problem (10.8) with $\varphi(x,t) = 0$ with second order accuracy. The corresponding finite-difference equation has the form:

$$\frac{u_m^{p+1} - u_m^p}{\tau} - \frac{u_{m+1}^p - u_{m-1}^p}{2h} - \frac{r}{2h}(u_{m+1}^p - 2u_m^p + u_{m-1}^p) = 0. \tag{10.56}$$

In the literature, scheme (10.56) is known as the Lax-Wendroff scheme. In Section 10.2.4, it will be discussed further from a somewhat different perspective. In the meantime, we only notice that according to the Lax-Wendroff example, one does not necessarily have to consider the second order differences in time, as in the leap-frog scheme (10.36), in order to achieve the overall second order of accuracy. On the other hand, it may still seem a little odd that second order accuracy can only be achieved for the homogeneous equation, i.e., when $\varphi(x,t) = 0$. This issue is addressed in Exercise 8 of this section.

The methods that have been described can also be applied when constructing the schemes for a wider range of formulations than only constant coefficients and uniform grids. In addition, differential equations with variable coefficients, classes of nonlinear equations, and nonuniform grids can be analyzed. For example, to obtain a difference scheme for the Poisson equation $u_{xx} + u_{yy} = \varphi(x,y)$ on a nonuniform Cartesian grid, one can use the following approximation formulae for the derivatives:

$$\left.\frac{\partial^2 u}{\partial x^2}\right|_{(x_m,y_n)} \approx \frac{2}{\Delta x_m + \Delta x_{m-1}} \cdot \left[\frac{u(x_{m+1},y_n) - u(x_m,y_n)}{\Delta x_m} - \frac{u(x_m,y_n) - u(x_{m-1},y_n)}{\Delta x_{m-1}}\right],$$

$$\left.\frac{\partial^2 u}{\partial y^2}\right|_{(x_m,y_n)} \approx \frac{2}{\Delta y_n + \Delta y_{n-1}} \cdot \left[\frac{u(x_m,y_{n+1}) - u(x_m,y_n)}{\Delta y_n} - \frac{u(x_m,y_n) - u(x_m,y_{n-1})}{\Delta y_{n-1}}\right].$$

Using the method of undetermined coefficients, one can actually make sure that these formulae are unique. More precisely, disregarding the non-essential lower order arbitrariness in the coefficients (see page 336), one can show that there exists one and only one set of coefficients a_{-1}, a_0, and a_1 that would guarantee first order accuracy for the approximation of the second derivative:

$$\left.\frac{\partial^2 u}{\partial x^2}\right|_{(x_m,y_n)} = a_{-1}u(x_{m-1},y_n) + a_0 u(x_m,y_n) + a_1 u(x_{m+1},y_n) + \mathcal{O}(\max\{\Delta x_m, \Delta x_{m-1}\}).$$

One can also show that unless $\Delta x_m = \Delta x_{m-1}$, one cannot construct a similar three-node approximation of the second derivative that would have second order of accuracy with respect to $\max\{\Delta x_m, \Delta x_{m-1}\}$. To achieve that, one would generally need to incorporate more grid points into the stencil.

10.2.3 Other Methods. Phase Error

For a given differential problem, consider the class of finite-difference schemes built on some fixed stencil. Let us require that the desired scheme satisfy two types of constraints — primary and secondary. These constraints may, generally speaking, be incompatible. The objective is to search through all the schemes on the stencil that meet the primary constraint(s), and find the specific scheme (or schemes) among them that would minimize the degree of violation of the secondary constraint(s).

To illustrate the idea, consider the class of explicit schemes:

$$u_m^{p+1} = \sum_{j=-j_{\text{left}}}^{j_{\text{right}}} b_j u_{m+j}^p, \tag{10.57}$$

where $j_{\text{left}} \geq 0$ and $j_{\text{right}} \geq 0$ are two given fixed integers, and b_j, $j = -j_{\text{left}}, \dots, j_{\text{right}}$, are the coefficients of the scheme. Many schemes that we have studied previously fall into the category (10.57). For example, under the assumption $\varphi(x,t) = 0$ the scheme (10.47) transforms into (10.57) if we take $j_{\text{left}} = j_{\text{right}} = 1$:

$$u_m^{p+1} = b_{-1}u_{m-1}^p + b_0 u_m^p + b_1 u_{m+1}^p,$$

and set:

$$b_{-1} = -\frac{a_{-1}}{a^0}, \quad b_0 = -\frac{a_0}{a^0}, \quad b_1 = -\frac{a_1}{a^0}.$$

Next, with every scheme of type (10.57) we identify the point $\boldsymbol{b} = \{b_{-j_{\text{left}}}, b_{-j_{\text{left}}+1}, \ldots, b_{j_{\text{right}}}\}$ in the J-dimensional linear (vector) space, where $J = j_{\text{left}} + j_{\text{right}} + 1$. Denote by M_{pr} the subset of those schemes (10.57), i.e., points in the J-dimensional space, that satisfy the given primary constraint(s). Often, the primary constraints are related to the approximation properties of the scheme. For instance, we may require that the scheme have a prescribed order of accuracy on smooth solutions of the corresponding differential problem, e.g., the Cauchy problem:

$$\frac{\partial u}{\partial t} - c\frac{\partial u}{\partial x} = 0, \quad u(x,0) = \psi(x).$$

Of course, the set M_{pr} may appear empty. For example, we have seen in Section 10.2.2 that no scheme built on the three-node upwind stencil from Figure 10.2(left) may have accuracy better than $\mathcal{O}(h)$. Thus, if we were to require second order accuracy on this stencil, then the set M_{pr} would have contained no points. Alternatively, the set M_{pr} may consist of a single point that we denote $\boldsymbol{b}_0$, which implies uniqueness of the scheme that satisfies the given primary constraint. For example, the Lax-Wendroff scheme (10.56) is the only second order scheme on the four-node stencil from Figure 10.3(left). Finally, the set M_{pr} may contain many points. For example, there are multiple first order schemes on the aforementioned four-node stencil from Figure 10.3(left).

Clearly, the first two scenarios provide no further room for "maneuvering." In either case: $M_{\text{pr}} = \emptyset$ or $M_{\text{pr}} = \boldsymbol{b}_0$, regardless of the secondary constraint that we may impose, no improvement can be made unless this constraint is satisfied ahead of time. However, the third scenario opens the possibility of quantifying the extent to which for the secondary constraint is violated, and subsequently minimizing the discrepancy. If we equip the J-dimensional vector space of schemes with some norm $\|\cdot\|$, then we can formulate the problem of finding $\boldsymbol{b} \in M_{\text{pr}}$ that would minimize the distance between the two sets M_{pr} and M_{sec}, where $\text{dist}(\boldsymbol{b}, M_{\text{sec}}) \stackrel{\text{def}}{=} \inf_{\boldsymbol{x} \in M_{\text{sec}}} \|\boldsymbol{b} - \boldsymbol{x}\|$. Depending on the particular situation, the problem of finding $\boldsymbol{b}$ that minimizes $\text{dist}(\boldsymbol{b}, M_{\text{sec}})$ can be solved using various numerical optimization techniques, such as linear programming, quadratic optimization, etc. This method of constructing the schemes has been first introduced by Kholodov. A detailed description of the corresponding developments, along with the analysis of numerous applications, can be found in the monograph [MK88].

In doing so, the nature of the constraints that can be taken into account may, of course, vary. Often, one employs the criteria related to the notions of the phase error and dispersion. Consider the differential equation $u_t - cu_x = 0$ with $c > 0$; the quantity $-c$ has the meaning of the propagation speed. The evolution of the initial data e^{ikx} by this equation is as follows: $u(x,t) = e^{ik(x+ct)} = e^{i(\omega t + kx)}$, where $\omega = kc$. Solutions of the type $e^{i(\omega t + kx)}$ are referred to as plane wave solutions; the quantity ω is known as frequency, and k is the wavenumber. The relation $\omega = kc$ is called the dispersion relation. In the theory of wave propagation, one commonly defines the phase speed of the waves v_{ph} as the speed with which the constant phase surfaces advance. For our current one-dimensional example, the surfaces of constant phase are actually point-wise locations. We see that their speed $v_{\text{ph}} = -\omega/k = -c$, i.e., the phase speed coincides with the constant propagation speed $-c$. It does not depend on either ω or k, in which case the propagation of waves is said to be dispersionless.

We will now see how the same initial data evolve in the framework of a first order upwind scheme of type (10.10). Let $u_m^p = \lambda^p e^{ikx_m}$, $\lambda \in \mathbb{C}$; substituting this expression into the homogeneous difference equation $u_m^{p+1} = u_m^p + cr(u_{m+1}^p - u_m^p)$ we obtain:

$$\lambda = 1 - cr + cre^{ikh} \stackrel{\text{def}}{=} |\lambda| e^{i\phi},$$

where $r = \tau/h = \text{const}$. Hence, for a given wave e^{ikx} both the amplitude $|\lambda|$ and the phase ϕ of thc amplification factor λ generally speaking depend on the wavenumber k. The study of the amplitude will be in the focus of the spectral stability analysis (see Section 10.3). Here, we will rather investigate the phase. We can write:

$$\tan\phi = \frac{\text{Im}\lambda}{\text{Re}\lambda} = \frac{cr\sin(kh)}{1 - cr + cr\cos(kh)}.$$

If $cr = 1$, then $\phi = kh = k\tau/r = k\tau c = \omega\tau$, and we conclude that there is no dispersion as in the continuous case. This is not surprising, because when the Courant number is equal to one, the scheme reconstructs the exact solution. Otherwise, assuming that $kh \ll 1$, which means that the wavelength $2\pi/k$ is much greater than the grid size h, we use the Taylor expansion and obtain:

$$\begin{aligned} \phi &= \arctan\left[\frac{cr\sin(kh)}{1 - cr + cr\cos(kh)}\right] \\ &= ck\tau\left[1 + (kh)^2\left(\frac{cr}{2} - \frac{1}{6} - \frac{(cr)^2}{3}\right) + \mathscr{O}\left((kh)^4\right)\right], \end{aligned} \tag{10.58}$$

because $rh = \tau$. Let us now interpret $\phi = \omega\tau$ as before. Then, formula (10.58) yields the frequency ω as a function of the wavenumber k. In other words, formula (10.58) provides a dispersion relation for the first order upwind scheme. We see that it is not as simple as $\omega = ck$. The expression for the phase speed is derived from (10.58):

$$v_{\text{ph}} = -\frac{\omega}{k} = -c\left[1 + (kh)^2\left(\frac{cr}{2} - \frac{1}{6} - \frac{(cr)^2}{3}\right) + \mathscr{O}\left((kh)^4\right)\right]. \tag{10.59}$$

Formula (10.59) indicates that the waves having different wavenumbers k propagate with different speeds (assuming $cr < 1$). This phenomenon is known as dispersion. The onset of dispersion is accounted for by the fact that the waves are governed by the difference equation as opposed to the original differential equation.

It is clear that as the phase speed for the scheme $v_{\text{ph}} = v_{\text{ph}}(k)$ differs from the phase speed for the differential equation $-c$, the discrepancy between the phase of the continuous solution and that of the discrete solution (for each individual k) will accumulate as the time elapses. Writing the solution in the form $u(x,t) = |u(x,t)|e^{ik(x - v_{\text{ph}}t)}$, we conclude that the accumulated phase difference is given by:

$$\delta\phi = \delta\phi(k,t) = (c - v_{\text{ph}}(k))t.$$

The quantity $\delta\phi$ is often referred to as the phase error.

In many applications related to the propagation of waves (for example, in computational aeroacoustics) it may be highly desirable to minimize the phase error of the scheme. Therefore, we can formulate the following problem: Among the schemes (10.57) that have a prescribed order of accuracy on a fixed stencil (primary constraint that defines the set M_{pr}), find the one that would have the minimum phase error on the largest possible subinterval $0 < k \leq k_0$ of the overall range of wavenumbers $0 < k \leq 2\pi/h$. Solving this problem would imply analyzing the dispersion relations similar to (10.58), but obtained for the general class of schemes (10.57).

In other words, the secondary constraint in this case can be introduced as a requirement that the phase error be zero. It can, however, be shown that among the schemes (10.57) only the simplest first order upwind scheme with the Courant number $cr = 1$ possesses this property. As such, the set M_{sec} that corresponds to the secondary constraint in the J-dimensional space of schemes will only contain one point. Consequently, for the schemes with accuracy higher than $\mathcal{O}(h)$, the primary and secondary constraints will indeed be incompatible, i.e., $M_{\text{pr}} \cap M_{\text{sec}} = \emptyset$. Then, finding the scheme with the minimum phase error on M_{pr} may require solving a fairly non-trivial optimization problem. Some particular formulations of this type have been studied in [TW93].

To provide another example of the phase error analysis, let us consider the Lax-Wendroff scheme [cf. formula (10.56)]:

$$\frac{u_m^{p+1} - u_m^p}{\tau} - c\frac{u_{m+1}^p - u_{m-1}^p}{2h} - c^2\frac{\tau}{2}\frac{u_{m+1}^p - 2u_m^p + u_{m-1}^p}{h^2} = 0.$$

Denoting $r = \tau/h = \text{const}$ as before, we obtain the amplification factor:

$$\lambda = 1 + icr\sin(kh) - 2(cr)^2\sin^2\frac{kh}{2} = |\lambda|e^{i\phi}.$$

Then, for the phase ϕ we have:

$$\tan\phi = \frac{cr\sin(kh)}{1 - 2(cr)^2\sin^2(kh/2)},$$

and similarly to the case of the first order upwind scheme we see that if $cr = 1$, then $\phi = kh$ and there is no dispersion. Otherwise, if $cr < 1$ we use the Taylor formula for the long waves $kh \ll 1$ and find the dispersion relation [cf. formula (10.58)]:

$$\begin{aligned}\phi &= \arctan\left[\frac{cr\sin(kh)}{1 - 2(cr)^2\sin^2(kh/2)}\right]\\ &= ck\tau\left[1 + (kh)^2\left(\frac{(cr)^2}{6} - \frac{1}{6}\right) + \mathcal{O}\left((kh)^4\right)\right].\end{aligned}$$

Hence, the phase speed for the Lax-Wendroff scheme is [cf. formula (10.59)]:

$$v_{\text{ph}} = -c\left[1 + (kh)^2\left(\frac{(cr)^2}{6} - \frac{1}{6}\right) + \mathcal{O}\left((kh)^4\right)\right].$$

We again conclude that the waves having different wavenumbers travel with different speeds, i.e., the propagation is accompanied by dispersion. Moreover, the long waves $kh \ll 1$ happen to travel slower than with the speed $-c$. The phase speed of these waves may only approach $-c$ as $kh \longrightarrow 0$.

10.2.4 Predictor-Corrector Schemes

When constructing difference schemes for time-dependent partial differential equations, one can exploit the same key idea that provides the foundation of Runge-Kutta schemes for ordinary differential equations. This is the idea of introducing the intermediate stages of computation, or equivalently, of employing the predictor-corrector strategy, see Sections 9.2.6 and 9.4 of Chapter 9. This strategy allows one to increase the order of accuracy that one would have obtained if only the original scheme were to be used by itself, i.e., with no intermediate stages. Besides, in the case of quasi-linear differential equations this strategy facilitates design of the so-called conservative finite-difference schemes that will be discussed in Chapter 11.

Let us recall the idea of the predictor-corrector approach using one of the simplest Runge-Kutta schemes as an example; this scheme will be applied to solving the Cauchy problem for a first order ordinary differential equation:

$$\frac{dy}{dt} = G(t,y), \quad y(0) = \psi, \quad 0 \leq t \leq T. \tag{10.60}$$

If the value of the solution y_p at the moment of time $t_p = p\tau$ is already computed, then in order to compute y_{p+1} we first find the auxiliary quantity $\tilde{y}_{p+1/2}$ using the standard forward Euler scheme in the capacity of a predictor:

$$\frac{\tilde{y}_{p+1/2} - y_p}{\tau/2} = G(t_p, y_p). \tag{10.61}$$

Subsequently, we apply the corrector scheme to compute y_{p+1}:

$$\frac{y_{p+1} - y_p}{\tau} = G(t_{p+1/2}, \tilde{y}_{p+1/2}). \tag{10.62}$$

The auxiliary quantity $\tilde{y}_{p+1/2}$ obtained by scheme (10.61) with first order accuracy helps us approximately evaluate the slope of the integral curve at the midpoint of the interval $[t_p, t_{p+1}]$ and thus obtain y_{p+1} by formula (10.62) with accuracy higher than that of the Euler scheme (10.61).

We have already mentioned in Section 9.4.2 of Chapter 9 that all these considerations will remain valid if y_p, $\tilde{y}_{p+1/2}$, and y_{p+1} were to be interpreted as finite-dimensional vectors and G, accordingly, was to be thought of as a vector function. However, one can go even further and consider y_p, $\tilde{y}_{p+1/2}$, and y_{p+1} as elements of some functional space, and G as an operator acting in this space.

For instance, the Cauchy problem:

$$\begin{aligned} &\frac{\partial u}{\partial t} + c\frac{\partial u}{\partial x} = 0, && -\infty < x < \infty, \quad 0 < t \leq T, \\ &u(x,0) = \psi(x), && -\infty < x < \infty. \end{aligned} \tag{10.63}$$

where $c = \text{const}$, can be interpreted as a problem of type (10.60) if we set $y(t) = u(x,t)$ so that for any given t the quantity y appears to be a function of the argument x, and the operation G stands for the differential operator $-c\frac{\partial}{\partial x}$. Let us now construct an example of a predictor-corrector scheme for problem (10.63).

Example

Assume that the grid function $u^p = \{u_m^p\}$, $m = 0, \pm 1, \pm 2, \ldots$, has already been computed for some value of p. Introduce the auxiliary grid function $\tilde{u}_{m+1/2}^{p+1/2}$, $m = 0, \pm 1, \pm 2, \ldots$, defined for the semi-integer time level $t_{p+1/2} = (p+1/2)\tau$ and at the semi-integer spatial points $x_{m+1/2} = (m+1/2)h$ (cell midpoints). This auxiliary grid function is to be computed using the following first order scheme:

$$\frac{\tilde{u}_{m+1/2}^{p+1/2} - (u_{m+1}^p + u_m^p)/2}{\tau/2} + c\frac{u_{m+1}^p - u_m^p}{h} = 0, \quad m = 0, \pm 1, \pm 2, \ldots. \tag{10.64}$$

known as the Lax-Friedrichs scheme. Then, we employ the leap-frog scheme (10.36) in the capacity of a corrector to obtain u^{p+1}:

$$\frac{u_m^{p+1} - u_m^p}{\tau} + c\frac{\tilde{u}_{m+1/2}^{p+1/2} - \tilde{u}_{m-1/2}^{p+1/2}}{h} = 0, \quad m = 0, \pm 1, \pm 2, \ldots. \tag{10.65}$$

Eliminating the intermediate quantities $\tilde{u}_{m+1/2}^{p+1/2}$ from equations (10.64) and (10.65) we arrive at the following scheme:

$$\begin{gathered} \frac{u_m^{p+1} - u_m^p}{\tau} + c\frac{u_{m+1}^p - u_{m-1}^p}{2h} - c^2\frac{\tau}{2}\frac{u_{m+1}^p - 2u_m^p + u_{m-1}^p}{h^2} = 0, \\ u_m^0 = \psi(mh), \quad m = 0, \pm 1, \pm 2, \ldots, \quad p = 0, 1, \ldots, [T/\tau] - 1. \end{gathered} \tag{10.66}$$

This is the Lax-Wendroff scheme that we have previously obtained in Section 10.2.2 for the case $c = -1$, see formula (10.56). Scheme (10.66) has second order accuracy with respect to h, provided that $\tau = rh$, $r = \text{const}$. Consequently, its "parent" predictor-corrector scheme (10.64)-(10.65) also has second order accuracy.

Another notable example is the predictor-corrector scheme of MacCormack, see formula (11.21) of Chapter 11; it is popular for fluid flow computations. For the case of a linear homogeneous equation, it reduces to the Lax-Wendroff scheme (10.66).

Exercises

1. Prove that the leap-frog scheme (10.36) has second order accuracy on a smooth solution $u = u(x,t)$ of the Cauchy problem (10.8).

2. Prove uniqueness of the Lax-Wendroff scheme (10.56) in the sense outlined on page 336, as the only second order scheme in its class (10.47) that approximates problem (10.8) when $\varphi(x,t) = 0$.

3. For the Cauchy problem in two space dimensions:

$$\frac{\partial u}{\partial t} - \left(\frac{\partial u}{\partial x} + \frac{\partial u}{\partial y}\right) = \varphi(x,y,t), \qquad -\infty < x,\, y < \infty,\;\; 0 < t \le T,$$
$$u(x,y,0) = \psi(x,y), \qquad -\infty < x,\, y < \infty,$$

use a uniform Cartesian grid: $x_m = mh$, $y_n = nh$, $t_p = p\tau$, and construct a consistent finite-difference scheme that would approximate this problem on its smooth solutions.

4. Consider a Cauchy problem for the homogeneous heat equation:

$$\begin{aligned} \frac{\partial u}{\partial t} - \frac{\partial^2 u}{\partial x^2} &= 0, \qquad -\infty < x < \infty,\;\; 0 < t \le T, \\ u(x,0) &= \psi(x), \qquad -\infty < x < \infty, \end{aligned} \tag{10.67}$$

and approximate it on a uniform rectangular grid with the help of the scheme:

$$\frac{u_m^{p+1} - u_m^p}{\tau} = \sigma \frac{u_{m+1}^{p+1} - 2u_m^{p+1} + u_{m-1}^{p+1}}{h^2} + (1-\sigma)\frac{u_{m+1}^p - 2u_m^p + u_{m-1}^p}{h^2},$$
$$u_m^0 = \psi(mh),$$

where σ is a real parameter, $0 \le \sigma \le 1$.

a) Show that for any σ the scheme is consistent and has the order of accuracy $\mathcal{O}(\tau + h^2)$ on a smooth solution $u = u(x,t)$.

b) Find the value of σ, for which the order of accuracy becomes $\mathcal{O}(\tau^2 + h^2)$.

c)* Assuming that $\tau/h^2 = r = \text{const}$, find the value of σ that yields approximation of order $\mathcal{O}(h^4)$.

d)* For $\sigma = 0$, find the value of $r = \tau/h^2$ that enables fourth order accuracy: $\mathcal{O}(h^4)$. **Hint.** In items c) and d) use: $u_{tt} = (u_{xx})_t = (u_t)_{xx} = u_{xxxx}$.

e)* For a fixed $r = \tau/h^2$, can one choose the value of σ that would yield higher than fourth order of accuracy on any smooth solution of problem (10.67)?

5. Consider a Cauchy problem for the heat equation with variable coefficients:

$$\frac{\partial u}{\partial t} - \frac{\partial}{\partial x}\left[a(x,t)\frac{\partial u}{\partial x}\right] = 0, \qquad -\infty < x < \infty,\;\; 0 < t \le T,$$
$$u(x,0) = \psi(x), \qquad -\infty < x < \infty.$$

Use a uniform rectangular grid: $x_m = mh$, $t_p = p\tau$, and construct a consistent finite-difference scheme that would approximate this problem on its smooth solutions.

6. Consider a Cauchy problem for the nonlinear heat equation:

$$\frac{\partial u}{\partial t} - \frac{\partial}{\partial x}\left[a(u)\frac{\partial u}{\partial x}\right] = 0, \qquad -\infty < x < \infty,\;\; 0 < t \le T,$$
$$u(x,0) = \psi(x), \qquad -\infty < x < \infty.$$

Use a uniform rectangular grid: $x_m = mh$, $t_p = p\tau$, and construct a consistent finite-difference scheme that would approximate this problem on its smooth solutions.

7. Consider Cauchy problem (10.67) for the heat equation, and approximate it with the predictor-corrector scheme designed as follows. The auxiliary grid function $\tilde{u}_m^{p+1/2}$ is to be computed with accuracy $\mathcal{O}(\tau + h^2)$ by the implicit method:

$$\frac{\tilde{u}_m^{p+1/2} - u_m^p}{\tau/2} - \frac{\tilde{u}_{m+1}^{p+1/2} - 2\tilde{u}_m^{p+1/2} + \tilde{u}_{m-1}^{p+1/2}}{h^2} = 0, \quad m = 0, \pm 1, \pm 2, \dots,$$

and the actual solution u_m^{p+1} at $t = t_{p+1}$ is to be computed by the scheme:

$$\frac{u_m^{p+1} - u_m^p}{\tau} - \frac{\tilde{u}_{m+1}^{p+1/2} - 2\tilde{u}_m^{p+1/2} + \tilde{u}_{m-1}^{p+1/2}}{h^2} = 0, \quad u_m^0 = \psi(mh).$$

Prove that the overall predictor-corrector scheme has accuracy $\mathcal{O}(\tau^2 + h^2)$ on the smooth solution $u = u(x,t)$.

8. Consider a modified scheme $\boldsymbol{L}_h u^{(h)} = f^{(h)}$ of type (10.47) (modification of the right-hand side only):

$$\boldsymbol{L}_h u^{(h)} = \begin{cases} a^0 u_m^{p+1} + a_{-1} u_{m-1}^p + a_0 u_m^p + a_1 u_{m+1}^p, \\ u_m^0, \end{cases}$$

$$f^{(h)} = \begin{cases} \varphi(x_m, t_p) + \dfrac{rh}{2} (\varphi_t + \varphi_x)\Big|_{(x_m, t_p)}, \\ \psi(x_m), \end{cases}$$

and define the coefficients of the operator $\boldsymbol{L}_h$ according to formula (10.55) that corresponds to the Lax-Wendroff method. Show that the resulting scheme approximates problem (10.8) with second order accuracy for an arbitrary sufficiently smooth right-hand side $\varphi(x,t)$ (not necessarily zero).

9. Represent scheme (10.47) with $\varphi(x,t) = 0$ in the form:

$$u_m^{p+1} = b_{-1} u_{m-1}^p + b_0 u_m^p + b_1 u_{m+1}^p, \tag{10.68}$$

where $b_{-1} = -a_{-1}/a^0$, $b_0 = -a_0/a^0$, and $b_1 = -a_1/a^0$, see Section 10.2.3. Scheme (10.68) is said to be *monotone* if $b_j \geq 0$, $j = -1, 0, 1$. Adopting the terminology of Section 10.2.3, let the primary constraint be the order of accuracy of at least $\mathcal{O}(h)$, and the secondary constraint be monotonicity of the scheme. In the three-dimensional space of vectors $\{b_{-1}, b_0, b_1\}$, describe the set $M_0 = M_{\text{pr}} \cap M_{\text{sec}}$ (in this case, the distance between the sets M_{pr} and M_{sec} is zero).

Answer. If $r = \tau/h > 1$, then $M_0 = \emptyset$. If $r = 1$, then $M_0 = \{(0,0,1)\}$. If $0 < r < 1$, then M_0 is the interval with the endpoints: $(\frac{1-r}{2}, 0, \frac{1+r}{2})$ and $(0, 1-r, r)$.

10. Prove that the monotone schemes defined the same way as in Exercise 9 for $\varphi = 0$ and $j = -j_{\text{left}}, \dots, j_{\text{right}}$ satisfy the maximum principle from page 315, i.e., that the maximum of the corresponding difference solution will not increase as the time elapses.

11.* **(Godunov theorem)** Prove that no one-step explicit monotone scheme may have accuracy higher than $\mathcal{O}(h)$ on smooth solutions of the differential equation $u_t + cu_x = 0$.
Hint. One version of the proof, due to Harten, Hyman, and Lax, can be found, e.g., in [Str04, pages 71–72]. Note also that there are exceptions, but they are all trivial. For example, the first order upwind scheme with $r = 1$ has zero error on the exact solution $u = u(x-t)$ of $u_t + u_x = 0$.

12. Consider a Cauchy problem for the one-dimensional wave (d'Alembert) equation:

$$\frac{\partial^2 u}{\partial t^2} - \frac{\partial^2 u}{\partial x^2} = \varphi(x,t), \quad -\infty < x < \infty, \quad 0 < t \leq T,$$
$$u(x,0) = \psi^{(0)}(x), \quad \frac{\partial u(x,0)}{\partial t} = \psi^{(1)}(x), \quad -\infty < x < \infty.$$

Analyze the approximation properties of the scheme $\boldsymbol{L}_h u^{(h)} = f^{(h)}$ on a smooth solution $u = u(x,t)$, where:

$$\boldsymbol{L}_h u^{(h)} = \begin{cases} \dfrac{u_m^{p+1} - 2u_m^p + u_m^{p-1}}{\tau^2} - \dfrac{u_{m+1}^p - 2u_m^p + u_{m-1}^p}{h^2}, \\ u_m^0, \\ \dfrac{u_m^1 - u_m^0}{\tau}, \end{cases}$$

$$f^{(h)} = \begin{cases} \varphi(x_m, t_p) \equiv \varphi_m^p, \\ \psi^{(0)}(x_m) \equiv \psi_m^{(0)}, \\ \psi^{(1)}(x_m) \equiv \psi_m^{(0)}. \end{cases}$$

Define the norm of $f^{(h)}$ as $\|f^{(h)}\|_{F_h} = \max\left\{\max_p \sup_m |\varphi_m^p|, \sup_m |\psi_m^{(0)}|, \sup_m |\psi_m^{(1)}|\right\}$ and show that when $\tau = rh$, $r = \text{const}$, the accuracy of the scheme is $\mathscr{O}(h)$.

Find an alternative way of specifying $\psi_m^{(1)}$ [instead of $\psi^{(1)}(x_m)$] so that to improve the order of accuracy and make it $\mathscr{O}(h^2)$? Note that $\varphi(x,t)$, $\psi^{(0)}(x)$, and $\psi^{(1)}(x)$ are given.

13. Consider the process of heat transfer on a finite interval, as opposed to the infinite line. It is governed by the heat equation subject to both initial and boundary conditions, which altogether yield the initial boundary value problem:

$$\begin{aligned} \frac{\partial u}{\partial t} - \frac{\partial^2 u}{\partial x^2} &= \varphi(x,t), & 0 < x < 1, \quad 0 < t \leq T, \\ u(x,0) &= \psi^{(0)}(x), & 0 < x < 1, \\ u(0,t) &= \psi^{(1)}(t), & 0 < t \leq T, \\ u(1,t) &= \psi^{(2)}(t), & 0 < t \leq T. \end{aligned}$$

Use the grid: $x_m = mh$, $t_p = p\tau$, and approximate this problem with the scheme:

$$\frac{u_m^{p+1} - u_m^p}{\tau} - \frac{u_{m+1}^p - 2u_m^p + u_{m-1}^p}{h^2} = \varphi(x_m, t_p),$$
$$m = 1, 2, \ldots, M-1, \quad h = 1/M, \quad p = 0, 1, \ldots, [T/\tau] - 1,$$
$$u_m^0 = \psi^{(0)}(x_m), \quad m = 0, 1, \ldots, M,$$
$$u_0^p = \psi^{(1)}(t_p), \quad p = 0, 1, \ldots, [T/\tau],$$
$$u_M^p = \psi^{(2)}(t_p), \quad p = 0, 1, \ldots, [T/\tau].$$

Define the norm in the space F_h as:

$$\|f^{(h)}\|_{F_h} = \max\left\{\max_{m,p} |\varphi(x_m, t_p)|, \max_m |\psi^{(0)}(x_m)|, \max_p |\psi^{(1)}(t_p)|, \max_p |\psi^{(2)}(t_p)|\right\},$$

and show that when $\tau = rh^2$, $r = \text{const}$, the accuracy of the scheme is $\mathscr{O}(h^2)$.

10.3 Spectral Stability Criterion for Finite-Difference Cauchy Problems

Perhaps the most widely used approach to the analysis of stability for finite-difference Cauchy problems has been proposed by von Neumann. In this section, we will introduce and illustrate it using several examples of difference equations with constant coefficients. The case of variable coefficients will be addressed in Section 10.4 and the case of initial boundary value problems (as opposed to pure initial value problems) will be explored in Section 10.5.

10.3.1 Stability with Respect to Initial Data

The simplest example of a finite-difference Cauchy problem is the first order up-wind scheme:

$$\boldsymbol{L}_h u^{(h)} = f^{(h)}, \tag{10.69}$$

for which the operator $\boldsymbol{L}_h$ and the right-hand side $f^{(h)}$ are given by:

$$\boldsymbol{L}_h u^{(h)} = \begin{cases} \dfrac{u_m^{p+1} - u_m^p}{\tau} - \dfrac{u_{m+1}^p - u_m^p}{h}, & m = 0, \pm 1, \pm 2, \dots, \\ & p = 0, 1, \dots, [T/\tau] - 1, \\ u_m^0, & m = 0, \pm 1, \pm 2, \dots, \end{cases} \tag{10.70}$$

$$f^{(h)} = \begin{cases} \varphi_m^p, & m = 0, \pm 1, \pm 2, \dots, \quad p = 0, 1, \dots, [T/\tau] - 1, \\ \psi_m, & m = 0, \pm 1, \pm 2, \dots. \end{cases}$$

We have already encountered this scheme on many occasions. Let us define the norms $\|u^{(h)}\|_{U_h}$ and $\|f^{(h)}\|_{F_h}$ as maximum norms (alternatively called l_∞ or C):

$$\|u^{(h)}\|_{U_h} \stackrel{\text{def}}{=} \max_p \sup_m |u_m^p|, \quad \|f^{(h)}\|_{F_h} \stackrel{\text{def}}{=} \max_p \sup_m |\varphi_m^p| + \sup_m |\psi_m|. \tag{10.71}$$

Then for the scheme (10.69)–(10.70), the stability condition (10.17):

$$\|u^{(h)}\|_{U_h} \leq c \|f^{(h)}\|_{F_h}$$

given by Definition 10.2 transforms into:

$$\sup_m |u_m^p| \leq c \left[\max_p \sup_m |\varphi_m^p| + \sup_m |\psi_m| \right], \quad p = 0, 1, \dots, [T/\tau], \tag{10.72}$$

where the constant c is not supposed to depend on h (or on $\tau = rh$, $r = \text{const}$).

Condition (10.72) must hold for any arbitrary $\{\psi_m\}$ and $\{\varphi_m^p\}$. In particular, it should obviously hold for an arbitrary $\{\psi_m\}$ and $\{\varphi_m^p\} \equiv 0$. In other words, for stability it is necessary that solution u_m^p to the problem:

$$\frac{u_m^{p+1} - u_m^p}{\tau} - \frac{u_{m+1}^p - u_m^p}{h} = 0, \quad m = 0, \pm 1, \pm 2, \dots, \quad p = 0, 1, \dots, [T/\tau] - 1,$$
$$u_m^0 = \psi_m, \quad m = 0, \pm 1, \pm 2, \dots, \tag{10.73}$$

satisfy the inequality:

$$\sup_m |u_m^p| \leq c \sup_m |\psi_m|, \quad p = 0, 1, \dots, [T/\tau], \tag{10.74}$$

for any bounded grid function ψ_m.

Property (10.74), which is *necessary* for the finite-difference scheme (10.69)–(10.70) to be stable in the sense (10.72), is called *stability with respect to perturbations of the initial data,* or simply stability with respect to the initial data. It means that if a perturbation is introduced into the initial data ψ_m of problem (10.73), then the corresponding perturbation of the solution will be no more than c times greater in magnitude than the original perturbation of the data, where the constant c does not depend on the grid size h.

10.3.2 A Necessary Spectral Condition for Stability

For the Cauchy problem (10.69)–(10.70) to be stable with respect to the initial data it is necessary, in particular, that inequality (10.74) hold for ψ_m being equal to a harmonic function:

$$u_m^0 = \psi_m = e^{i\alpha m}, \quad m = 0, \pm 1, \pm 2, \dots, \tag{10.75}$$

where α is a real parameter. One can easily solve problem (10.73) with the initial condition (10.75); the solution u_m^p can be found in the form:

$$u_m^p = \lambda^p e^{i\alpha m}, \quad m = 0, \pm 1, \pm 2, \dots, \quad p = 0, 1, \dots, [T/\tau], \tag{10.76}$$

where the quantity $\lambda = \lambda(\alpha)$ is determined by substitution of expression (10.76) into the homogeneous finite-difference equation of problem (10.73):

$$\lambda(\alpha) = 1 - r + re^{i\alpha}, \quad r = \tau/h = \text{const.} \tag{10.77}$$

Solution (10.76) satisfies the equality:

$$\sup_m |u_m^p| = |\lambda(\alpha)|^p \sup_m |u_m^0| = |\lambda(\alpha)|^p \sup_m |\psi_m|.$$

Therefore, for the stability condition (10.74) to be true it is necessary that the following inequality hold for all real α:

$$|\lambda(\alpha)|^p \leq c, \quad p = 0, 1, \dots, [T/\tau].$$

Equivalently, we can require that

$$|\lambda(\alpha)| \leq 1 + c_1\tau, \tag{10.78}$$

where c_1 is a constant that does not depend either on α or on τ. Inequality (10.78) represents the necessary spectral condition for stability due to von Neumann. It is called spectral because of the following reason.

Existence of the solution in the form (10.76) shows that the harmonic $e^{i\alpha m}$ is an eigenfunction of the transition operator from time level t_p to time level t_{p+1}:

$$u_m^{p+1} = (1-r)u_m^p + ru_{m+1}^p, \quad m = 0, \pm 1, \pm 2, \ldots.$$

According to the finite-difference equation (10.73), this operator maps the grid function $\{u_m^p\}$, $m = 0, \pm 1, \pm 2, \ldots$, defined for $t = t_p$ onto the grid function $\{u_m^{p+1}\}$, $m = 0, \pm 1, \pm 2, \ldots$, defined for $t = t_{p+1}$. The quantity $\lambda(\alpha)$ given by formula (10.77) is therefore an eigenvalue of the transition operator that corresponds to the eigenfunction $\{e^{i\alpha m}\}$. In the literature, $\lambda(\alpha)$ is sometimes also referred to as the amplification factor, we have encountered this concept in Section 10.2.3. The set of all complex numbers $\lambda = \lambda(\alpha)$ obtained when the parameter α sweeps through the real axis forms a curve on the complex plane. This curve is called the spectrum of the transition operator.

Consequently, the necessary stability condition (10.78) can be re-formulated as follows: The spectrum of the transition operator that corresponds to the difference equation of problem (10.73) must belong to the disk of radius $1 + c_1\tau$ centered at the origin on the complex plane. In our particular example, the spectrum (10.77) does not depend on τ at all. Therefore, condition (10.78) is equivalent to the requirement that the spectrum $\lambda = \lambda(\alpha)$ belong to the unit disk:

$$|\lambda(\alpha)| \leq 1. \tag{10.79}$$

Let us now use the criterion that we have formulated, and actually analyze stability of problem (10.69)–(10.70). The spectrum (10.77) forms a circle of radius r centered at the point $(1-r, 0)$ on the complex plane. When $r < 1$, this circle lies inside inside the unit disk, being tangent to the unit circle at the point $\lambda = 1$. When $r = 1$ the spectrum coincides with the unit circle. Lastly, when $r > 1$ the spectrum lies outside the unit disk, except one point $\lambda = 1$, see Figure 10.5. Accordingly, the necessary stability condition (10.79) is satisfied for $r \leq 1$ and is violated for $r > 1$. Let us now recall that in Section 10.1.3 we have studied the same difference problem and have proven that it is stable when $r \leq 1$ and is unstable when $r > 1$. Therefore, in this particular case the von Neumann necessary stability condition appears sufficiently sensitive to distinguish between the actual stability and instability.

In the case of general Cauchy problems for finite-difference equations and systems, we give the following

DEFINITION 10.3 *The spectrum of a finite-difference problem is given by the set of all those and only those $\lambda = \lambda(\alpha, h)$, for which the corresponding*

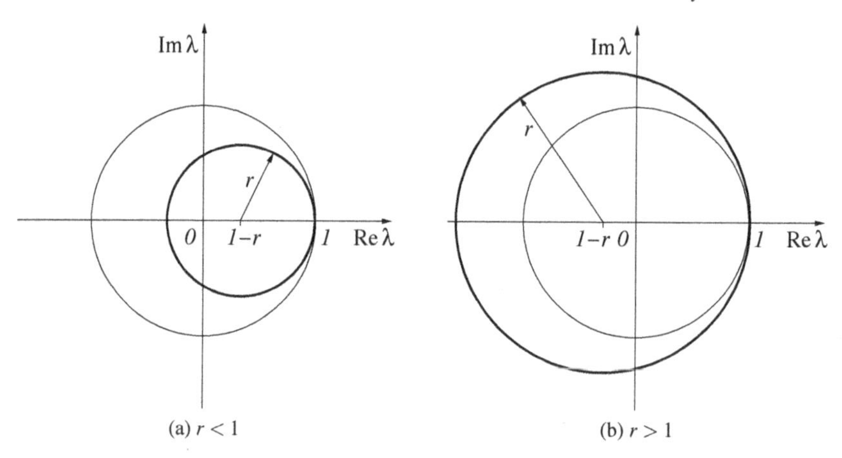

FIGURE 10.5: Spectra of the transition operators for the upwind scheme.

homogeneous finite-difference equation or system has a solution of the form:

$$\boldsymbol{u}_m^p = \lambda^p \left[\boldsymbol{u}^0 e^{i\alpha m}\right], \quad m = 0, \pm 1, \pm 2, \dots, \quad p = 0, 1, \dots, [T/\tau], \tag{10.80}$$

where $\boldsymbol{u}^0$ is either a fixed number in the case of one scalar difference equation (this number may be taken equal to one with no loss of generality), or a constant finite-dimensional vector in the case of a system of difference equations.

Then, the von Neumann necessary stability condition says that given an arbitrarily small number $\varepsilon > 0$, for all sufficiently small grid sizes h the spectrum $\lambda = \lambda(\alpha, h)$ of the difference problem has to lie inside the following disk on the complex plane:

$$|\lambda| \leq 1 + \varepsilon. \tag{10.81}$$

Note that if for a particular problem the spectrum appears to be independent of the grid size h (and τ), then condition (10.81) becomes equivalent to the requirement that the spectrum $\lambda(\alpha, h) = \lambda(\alpha)$ belong to the unit disk, see (10.79).

If the von Neumann necessary condition (10.81) is not satisfied, then one should not expect stability for any reasonable choice of norms. If, on the other hand, this condition is met, then one may hope that for a certain appropriate choice of norms the scheme will turn out stable.

10.3.3 Examples

We will exploit the von Neumann spectral condition to analyze stability of a number of interesting finite-difference problems. First, we will consider the schemes that approximate the Cauchy problem:

$$\begin{aligned} \frac{\partial u}{\partial t} - \frac{\partial u}{\partial x} &= \varphi(x,t), \quad -\infty < x < \infty, \ \ 0 < t \leq T, \\ u(x,0) &= \psi(x), \quad -\infty < x < \infty. \end{aligned} \tag{10.82}$$

Example 1

Consider the first order downwind scheme:

$$\frac{u_m^{p+1} - u_m^p}{\tau} - \frac{u_m^p - u_{m-1}^p}{h} = \varphi_m^p, \quad m = 0, \pm 1, \pm 2, \ldots, \; p = 0, 1, \ldots, [T/\tau] - 1,$$
$$u_m^0 = \psi_m, \quad m = 0, \pm 1, \pm 2, \ldots.$$

Substituting the solution of type (10.76) into the corresponding homogeneous finite-difference equation, we obtain:

$$\lambda(\alpha) = 1 + r - re^{-i\alpha}.$$

Therefore, the spectrum is a circle of radius r centered at the point $(1+r, 0)$ on the complex plane, see Figure 10.6. This spectrum does not depend on h. It is also clear that for no value of r does it belong to the unit circle. Consequently, the stability condition (10.79) may never be satisfied. This conclusion is expected, because the downwind scheme obviously violates the Courant, Friedrichs, and Lewy condition for any $r = \tau/h$ (see Section 10.1.4).

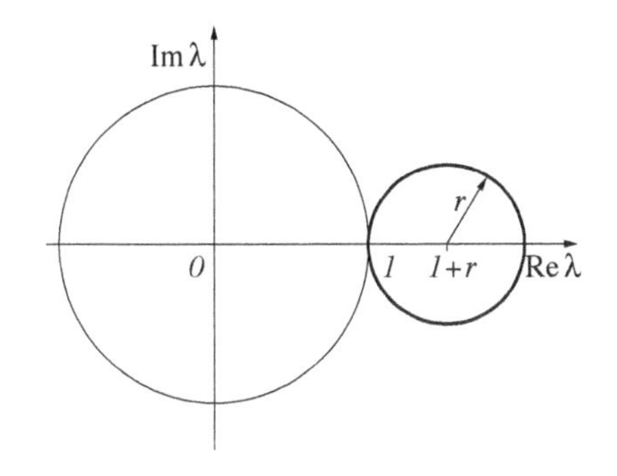

FIGURE 10.6: Spectrum for the downwind scheme.

Example 2

Next, consider the Lax-Wendroff scheme:

$$\frac{u_m^{p+1} - u_m^p}{\tau} - \frac{u_{m+1}^p - u_{m-1}^p}{2h} - \frac{\tau}{2h^2}(u_{m+1}^p - 2u_m^p + u_{m-1}^p) = \varphi_m^p, \qquad (10.83)$$
$$u_m^0 = \psi_m,$$

that approximates problem (10.82) with second order accuracy if $\varphi \equiv 0$, and with first order accuracy otherwise. For this scheme, the spectrum $\lambda = \lambda(\alpha, h)$ is determined from the equation:

$$\frac{\lambda - 1}{\tau} - \frac{e^{i\alpha} - e^{-i\alpha}}{2h} - \frac{\tau}{2h^2}(e^{i\alpha} - 2 + e^{-i\alpha}) = 0.$$

Let us denote $r = \tau/h$ as before, and notice that

$$\frac{e^{i\alpha} - e^{-i\alpha}}{2i} = \sin\alpha,$$

$$\frac{e^{i\alpha} - 2 + e^{-i\alpha}}{4} = -\left(\frac{e^{i\alpha/2} - e^{-i\alpha/2}}{2i}\right)^2 = -\sin^2\frac{\alpha}{2}.$$

Then,

$$\lambda(\alpha) = 1 + ir\sin\alpha - 2r^2\sin^2\frac{\alpha}{2},$$
$$|\lambda(\alpha)|^2 = \left(1 - 2r^2\sin^2\frac{\alpha}{2}\right)^2 + r^2\sin^2\alpha.$$

The spectrum does not depend on h, and from the previous equality we easily find that

$$1 - |\lambda|^2 = 4r^2(1-r^2)\sin^4\frac{\alpha}{2}.$$

The von Neumann condition is satisfied when the right-hand side of this equality is non-negative, which means $r \leq 1$; it is violated when $r > 1$.

Example 3

Consider the explicit central-difference scheme:

$$\begin{aligned} \frac{u_m^{p+1} - u_m^p}{\tau} - \frac{u_{m+1}^p - u_{m-1}^p}{2h} &= \varphi_m^p, \quad m = 0, \pm 1, \pm 2, \ldots, \quad p = 0, 1, \ldots, [T/\tau] - 1, \\ u_m^0 &= \psi_m, \quad m = 0, \pm 1, \pm 2, \ldots, \end{aligned} \tag{10.84}$$

for the same Cauchy problem (10.82).

Substituting expression (10.76) into the homogeneous counterpart of the difference equation from (10.84), we have:

$$\frac{\lambda - 1}{\tau} - \frac{e^{i\alpha} - e^{-i\alpha}}{2h} = 0,$$

which yields:

$$\lambda(\alpha) = 1 + i\left(\frac{\tau}{h}\sin\alpha\right).$$

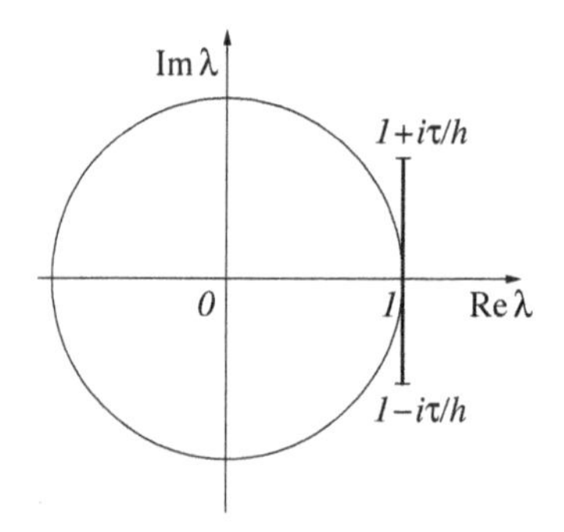

FIGURE 10.7: Spectrum for scheme (10.84).

The spectrum $\lambda = \lambda(\alpha)$ fills the vertical interval of length $2\tau/h$ that crosses through the point $(1,0)$ on the complex plane, see Figure 10.7.

If $\tau/h = r = \text{const}$, then the spectrum can be said to be independent of h (and of τ). This spectrum lies outside of the unit disk, the von Neumann condition (10.79) is not met, and the scheme is unstable. However, if we require that the temporal size τ be proportional to h^2 as $h \longrightarrow 0$, so that $r = \tau/h^2 = \text{const}$, then the point of the spectrum, which is most distant from the origin, will be $\lambda(\pi/2)$ [and also $\lambda(-\pi/2)$]. For this point we have:

$$|\lambda(\pi/2)| = \sqrt{1 + \left(\frac{\tau}{h}\right)^2} = \sqrt{1 + r\tau} < 1 + \frac{r}{2}\tau.$$

Then, the von Neumann condition (10.81) in the form $\lambda(\alpha) \leq 1 + c_1\tau$ is satisfied for $c_1 = r/2$. It is clear that the requirement $\tau = rh^2$ puts a considerably stricter constraint on the rate of decay of the temporal grid size τ as $h \longrightarrow 0$ than the previous requirement $\tau = rh$ does. Still, that previous requirement was sufficient for the von Neumann condition to hold for the difference schemes (10.69)–(10.70) and (10.83) that approximate the same Cauchy problem (10.82).

We also note that the Courant, Friedrichs, and Lewy condition of Section 10.1.4 allows us to claim that the scheme (10.84) is unstable only for $\tau/h = r > 1$, but does not allow us to judge the stability for $\tau/h = r \leq 1$. As such, it appears weaker than the von Neumann condition.

Example 4

The instability of scheme (10.84) for $\tau/h = r = \text{const}$ can be fixed by changing the way the time derivative is approximated. Instead of (10.84), consider the scheme:

$$\begin{aligned} \frac{u_m^{p+1} - (u_{m-1}^p + u_{m+1}^p)/2}{\tau} - \frac{u_{m+1}^p - u_{m-1}^p}{2h} &= \varphi_m^p, \\ u_m^0 &= \psi_m, \end{aligned} \tag{10.85}$$

obtained by replacing u_m^p with $(u_{m-1}^p + u_{m+1}^p)/2$. This is, in fact, a general approach attributed to Friedrichs, and the scheme (10.85) is known as the Lax-Friedrichs scheme; we have first introduced it in Section 10.2.4. The equation to determine the spectrum for the scheme (10.85) reads:

$$\frac{\lambda - (e^{i\alpha} + e^{-i\alpha})/2}{\tau} - \frac{e^{i\alpha} - e^{-i\alpha}}{2h} = 0,$$

which yields:

$$\frac{\lambda - \cos\alpha}{\tau} - \frac{i\sin\alpha}{h} = 0$$

and

$$\lambda(\alpha) = \cos\alpha + ir\sin\alpha,$$

where $r = \tau/h = \text{const}$. Consequently,

$$|\lambda(\alpha)|^2 = \cos^2\alpha + r^2\sin^2\alpha.$$

Clearly, the von Neumann condition (10.79) is satisfied for $r \leq 1$, because then $|\lambda|^2 \leq \cos^2\alpha + \sin^2\alpha = 1$. For $r > 1$, the von Neumann condition is violated.

Example 5

Finally, consider the leap-frog scheme (10.36) for problem (10.82). The corresponding homogeneous finite-difference equation is written as:

$$\frac{u_m^{p+1} - u_m^{p-1}}{2\tau} - \frac{u_{m+1}^p - u_{m-1}^p}{2h} = 0, \tag{10.86}$$

and for the spectrum we obtain:

$$\frac{\lambda - \lambda^{-1}}{2\tau} - \frac{e^{i\alpha} - e^{-i\alpha}}{2h} = 0.$$

This is a quadratic equation with respect to λ:

$$\lambda^2 - i2r\lambda \sin\alpha - 1 = 0,$$

where $r = \tau/h = \text{const}$. The roots of this equation are given by:

$$\lambda_{1,2} = ir\sin\alpha \pm \sqrt{1 - r^2\sin^2\alpha}.$$

We notice that when $r \le 1$, then $|\lambda_{1,2}|^2 = r^2\sin^2\alpha + (1 - r^2\sin^2\alpha) = 1$, so that the entire spectrum lies precisely on the unit circle and the von Neumann condition (10.79) is satisfied. Otherwise, when $r > 1$, we again take $\alpha = \pi/2$ and obtain: $|\lambda_1| = |ir + i\sqrt{r^2 - 1}| = r + \sqrt{r^2 - 1} > 1$, which means that the von Neumann condition is not met. This example illustrates how the von Neumann criterion can be applied to a finite-difference equation, such as equation (10.86), that connects the values of the solution on three, rather than two, consecutive time levels of the grid.

Next, we will consider two schemes that approximate the following Cauchy problem for the heat equation:

$$\begin{aligned} \frac{\partial u}{\partial t} - a^2\frac{\partial^2 u}{\partial x^2} &= \varphi(x,t), \qquad -\infty < x < \infty, \ 0 < t \le T, \\ u(x,0) &= \psi(x), \qquad -\infty < x < \infty. \end{aligned} \tag{10.87}$$

Example 6

The first scheme is explicit:

$$\begin{gathered} \frac{u_m^{p+1} - u_m^p}{\tau} - a^2\frac{u_{m+1}^p - 2u_m^p + u_{m-1}^p}{h^2} = \varphi_m^p, \\ u_m^0 = \psi_m, \quad m = 0, \pm 1, \pm 2, \dots, \quad p = 0, 1, \dots, [T/\tau] - 1. \end{gathered}$$

It allows us to compute $\{u_m^{p+1}\}$ in the closed form via $\{u_m^p\}$:

$$u_m^{p+1} = (1 - 2ra^2)u_m^p + ra^2(u_{m+1}^p + u_{m-1}^p) + \tau\varphi_m^p, \quad p = 0, 1, \dots, [T/\tau] - 1,$$

where $r = \tau/h^2 = \text{const}$. Substitution of $u_m^p = \lambda^p e^{i\alpha m}$ into the corresponding homogeneous difference equation yields:

$$\frac{\lambda - 1}{\tau} - a^2\frac{e^{-i\alpha} - 2 + e^{i\alpha}}{h^2} = 0.$$

By noticing that

$$\frac{e^{-i\alpha} - 2 + e^{i\alpha}}{4} = -\sin^2\frac{\alpha}{2},$$

we obtain:

$$\lambda(\alpha) = 1 - 4ra^2 \sin^2 \frac{\alpha}{2}, \quad r = \frac{\tau}{h^2}.$$

When α varies between 0 and 2π, the point $\lambda(\alpha)$ sweeps the interval $1 - 4ra^2 \leq \lambda \leq 1$ of the real axis, see Figure 10.8.

For stability, it is necessary that the left endpoint of this interval still be inside the unit circle (Figure 10.8), i.e., that $1 - 4ra^2 \geq -1$. This requirement translates into:

$$r \leq \frac{1}{2a^2}. \qquad (10.88)$$

Inequality (10.88) guarantees that the von Neumann stability condition will hold. Conversely, if we have $r > 1/(2a^2)$, then the point $\lambda(\alpha) = 1 - 4ra^2 \sin^2(\alpha/2)$ that corresponds to $\alpha = \pi$ will be located to the left of the point -1, i.e., outside the unit circle. In this case, the harmonic $e^{i\pi m} = (-1)^m$ generates the solution:

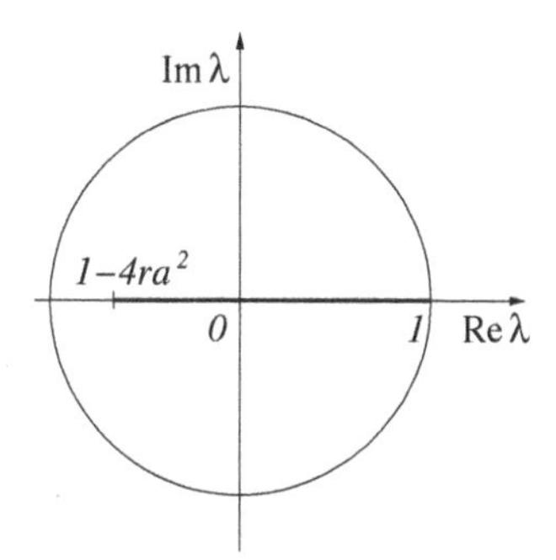

FIGURE 10.8: Spectrum of the explicit scheme for the heat equation.

$$u_m^p = (1 - 4ra^2)^p (-1)^m$$

that does not satisfy inequality (10.74) for any value of c.

Example 7

The second scheme is implicit:

$$\frac{u_m^{p+1} - u_m^p}{\tau} - a^2 \frac{u_{m+1}^{p+1} - 2u_m^{p+1} + u_{m-1}^{p+1}}{h^2} = \varphi_m^{p+1},$$
$$u_m^0 = \psi_m, \quad m = 0, \pm 1, \pm 2, \ldots, \quad p = 0, 1, \ldots, [T/\tau] - 1.$$

For this scheme, $\{u_m^{p+1}\}$ cannot be obtained from $\{u_m^p\}$ by an explicit formula because for a given m the difference equation contains three, rather than one, unknowns: u_{m+1}^{p+1}, u_m^{p+1}, and u_{m-1}^{p+1}. As in the previous case, we substitute $u_m^p = \lambda^p e^{i\alpha m}$ into the homogeneous equation and obtain:

$$\lambda(\alpha) = \frac{1}{1 + 4ra^2 \sin^2(\alpha/2)}, \quad r = \frac{\tau}{h^2}.$$

Consequently, the spectrum of the scheme fills the interval:

$$\left(1 + 4ra^2\right)^{-1} \leq \lambda \leq 1$$

of the real axis and the von Neumann condition $|\lambda| \leq 1$ is met for any r.

Example 8

The von Neumann analysis also applies when studying stability of a scheme in the case of more than one spatial variable. Consider, for instance, a Cauchy problem for the heat equation on the (x,y) plane:

$$\frac{\partial u}{\partial t} = \frac{\partial^2 u}{\partial x^2} + \frac{\partial^2 u}{\partial y^2} \qquad -\infty < x, y < \infty, \; 0 < t \leq T,$$
$$u(x,y,0) = \psi(x,y), \qquad -\infty < x, y < \infty.$$

We approximate this problem on the uniform Cartesian grid: $(x_m, y_n, t_p) = (mh, nh, p\tau)$. Replacing the derivatives with difference quotients we obtain:

$$\frac{u^{p+1}_{m,n} - u^p_{m,n}}{\tau} = \frac{u^p_{m+1,n} - 2u^p_{m,n} + u^p_{m-1,n}}{h^2} + \frac{u^p_{m,n+1} - 2u^p_{m,n} + u^p_{m,n-1}}{h^2}$$
$$u^0_{m,n} = \psi_{m,n}, \quad m,n = 0, \pm 1, \pm 2, \ldots, \quad p = 0, 1, \ldots, [T/\tau] - 1.$$

The resulting scheme is explicit. To analyze its stability, we specify $u^0_{m,n}$ in the form of a two-dimensional harmonic $e^{i(\alpha m + \beta n)}$ determined by two real parameters α and β, and generate a solution in the form:

$$u^p_{m,n} = \lambda^p(\alpha, \beta) e^{i(\alpha m + \beta n)}.$$

Substituting this expression into the homogeneous difference equation, we find after some easy equivalence transformations:

$$\lambda(\alpha, \beta) = 1 - 4r\left(\sin^2\frac{\alpha}{2} + \sin^2\frac{\beta}{2}\right).$$

When the real quantities α and β vary between 0 and 2π, the point $\lambda = \lambda(\alpha, \beta)$ sweeps the interval $1 - 8r \leq \lambda \leq 1$ of the real axis. The von Neumann stability condition is satisfied if $1 - 8r \geq -1$, i.e., when $r \leq 1/4$.

Example 9

In addition to the previous Example 5, let us now consider another example of the scheme that connects the values of the difference solution on the three, rather than two, consecutive time levels of the grid.

A Cauchy problem for the one-dimensional homogeneous d'Alembert (wave) equation:

$$\frac{\partial^2 u}{\partial t^2} - \frac{\partial^2 u}{\partial x^2} = 0, \quad -\infty < x < \infty, \quad 0 < t \leq T,$$
$$u(x,0) = \psi^{(0)}(x), \qquad \frac{\partial u(x,0)}{\partial t} = \psi^{(1)}(x), \quad -\infty < x < \infty,$$

can be approximated on a uniform Cartesian grid by the following scheme:

$$\frac{u_m^{p+1} - 2u_m^p + u_m^{p-1}}{\tau^2} - \frac{u_{m+1}^p - 2u_m^p + u_{m-1}^p}{h^2} = 0,$$

$$u_m^0 = \psi_m^{(0)}, \quad \frac{u_m^1 - u_m^0}{h} = \psi_m^{(1)},$$

$$m = 0, \pm 1, \pm 2, \ldots, \quad p = 1, 2, \ldots, [T/\tau] - 1.$$

Substituting a solution of type (10.76) into the foregoing finite-difference equation, we obtain the following quadratic equation for determining $\lambda = \lambda(\alpha)$:

$$\lambda^2 - 2\left(1 - 2r^2 \sin^2 \frac{\alpha}{2}\right)\lambda + 1 = 0, \quad r = \frac{\tau}{h}.$$

The product of the two roots of this equation is equal to one. If its discriminant:

$$D(\alpha) = 4r^2 \sin^2 \frac{\alpha}{2} \left(r^2 \sin^2 \frac{\alpha}{2} - 1\right)$$

is negative, then the roots $\lambda_1(\alpha)$ and $\lambda_2(\alpha)$ are complex conjugate and both have a unit modulus. If $r < 1$, the discriminant $D(\alpha)$ remains negative for all $\alpha \in [0, 2\pi)$.

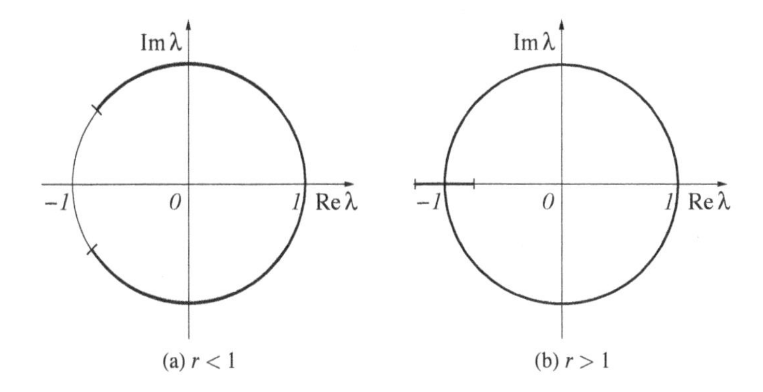

FIGURE 10.9: Spectrum of the scheme for the wave equation.

The spectrum for this case is shown in Figure 10.9(a); it fills an arc of the unit circle. If $r = 1$, the spectrum fills exactly the entire unit circle. When $r > 1$, the discriminant $D(\alpha)$ may be either negative or positive depending on the value of α. In this case, once the argument α increases from 0 to π the roots $\lambda_1(\alpha)$ and $\lambda_2(\alpha)$ depart from the point $\lambda = 1$ and move along the unit circle: One root moves clockwise and the other root counterclockwise, and then they merge at $\lambda = -1$. After that one root moves away from this point along the real axis to the right, and the other one to the left, because they are both real and their product $\lambda_1 \lambda_2 = 1$, see Figure 10.9(b). The von Neumann stability condition is met for $r \leq 1$.

Consider a Cauchy problem for the following first order hyperbolic system of equations that describes the propagation of acoustic waves:

$$\begin{gathered}\frac{\partial v}{\partial t}=\frac{\partial w}{\partial x}, \quad \frac{\partial w}{\partial t}=\frac{\partial v}{\partial x}, \\ -\infty<x<\infty, \quad 0<t\leq T, \\ v(x,0)=\psi^{(1)}(x), \quad w(x,0)=\psi^{(2)}(x).\end{gathered} \tag{10.89a}$$

Let us set:

$$\boldsymbol{u}(x,t)=\begin{bmatrix}v(x,t)\\ w(x,t)\end{bmatrix}, \quad \boldsymbol{\psi}(x)=\begin{bmatrix}\psi^{(1)}(x)\\ \psi^{(2)}(x)\end{bmatrix}.$$

Then, problem (10.89a) can be recast in the matrix form:

$$\begin{gathered}\frac{\partial \boldsymbol{u}}{\partial t}-\boldsymbol{A}\frac{\partial \boldsymbol{u}}{\partial x}=\boldsymbol{0}, \quad -\infty<x<\infty, \quad 0<t\leq T, \\ \boldsymbol{u}(x,0)=\boldsymbol{\psi}(x), \quad -\infty<x<\infty,\end{gathered} \tag{10.89b}$$

where

$$\boldsymbol{A}=\begin{bmatrix}0&1\\1&0\end{bmatrix}.$$

We will now analyze two difference schemes that approximate problem (10.89b).

Example 10

Consider the scheme:

$$\frac{\boldsymbol{u}_m^{p+1}-\boldsymbol{u}_m^p}{\tau}-\boldsymbol{A}\frac{\boldsymbol{u}_{m+1}^p-\boldsymbol{u}_m^p}{h}=0, \quad \boldsymbol{u}_m^0=\boldsymbol{\psi}_m. \tag{10.90}$$

We will be looking for a solution to the vector finite-difference equation (10.90) in the form (10.80):

$$\boldsymbol{u}_m^p=\lambda^p\begin{bmatrix}v^0\\ w^0\end{bmatrix}e^{i\alpha m}.$$

Substituting this expression into equation (10.90) we obtain:

$$\frac{\lambda-1}{\tau}\boldsymbol{u}^0-\boldsymbol{A}\frac{e^{i\alpha}-1}{h}\boldsymbol{u}^0=\boldsymbol{0},$$

or alternatively,

$$(\lambda-1)\boldsymbol{u}^0-r(e^{i\alpha}-1)\boldsymbol{A}\boldsymbol{u}^0=\boldsymbol{0}, \quad r=\tau/h. \tag{10.91}$$

Equality (10.91) can be considered as a vector form of the system of linear algebraic equations with respect to the components of the vector $\boldsymbol{u}^0$. System (10.91) can be written as:

$$\begin{bmatrix}\lambda-1 & -r(e^{i\alpha}-1)\\ -r(e^{i\alpha}-1) & \lambda-1\end{bmatrix}\begin{bmatrix}v^0\\ w^0\end{bmatrix}=\begin{bmatrix}0\\0\end{bmatrix}. \tag{10.92}$$

System (10.92) may only have a nontrivial solution $\boldsymbol{u}^0 = \begin{bmatrix} v^0 \\ w^0 \end{bmatrix}$ if its determinant turns into zero, which yields the following equation for $\lambda = \lambda(\alpha)$:

$$(\lambda - 1)^2 = r^2(e^{i\alpha} - 1)^2.$$

Consequently,

$$\lambda_1(\alpha) = 1 - r + re^{i\alpha},$$
$$\lambda_2(\alpha) = 1 + r - re^{i\alpha}.$$

When the parameter α varies between 0 and 2π, the roots $\lambda_1(\alpha)$ and $\lambda_1(\alpha)$ sweep two circles of radius r centered at the locations $1-r$ and $1+r$, respectively. Therefore, the von Neumann stability conditions may never be satisfied; irrespective of any particular value of r.

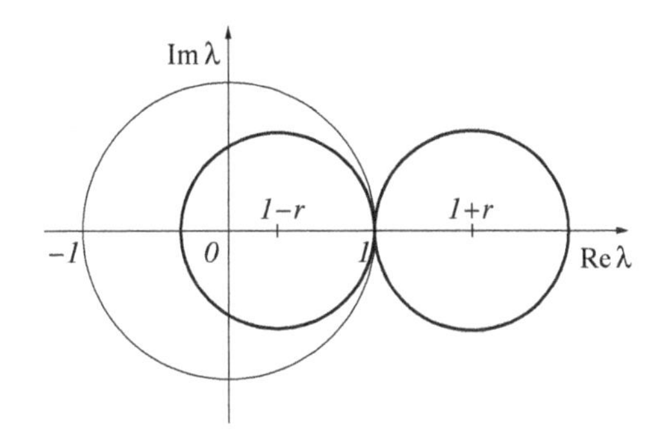

FIGURE 10.10: Spectrum of scheme (10.90).

Example 11

Consider the vector Lax-Wendroff scheme:

$$\begin{gathered}\frac{\boldsymbol{u}_m^{p+1} - \boldsymbol{u}_m^p}{\tau} - \boldsymbol{A}\frac{u_{m+1}^p - u_{m-1}^p}{2h} - \frac{\tau}{2h^2}\boldsymbol{A}^2(\boldsymbol{u}_{m+1}^p - 2\boldsymbol{u}_m^p + \boldsymbol{u}_{m-1}^p) = \boldsymbol{0}, \\ \boldsymbol{u}_m^0 = \boldsymbol{\psi}_m, \quad m = 0, \pm 1, \pm 2, \ldots, \quad p = 0, 1, \ldots, [T/\tau] - 1,\end{gathered} \tag{10.93}$$

that approximates problem (10.89b) on its smooth solutions with second order accuracy and that is analogous to the scalar scheme (10.83) for the Cauchy problem (10.82). As in Example 10, the finite-difference equation of (10.93) may only have a non-trivial solution of type (10.80) if the determinant of the corresponding linear system for finding $\boldsymbol{u}^0 = \begin{bmatrix} v^0 \\ w^0 \end{bmatrix}$ turns into zero.

Writing down this determinant and requiring that it be equal to zero, we obtain a quadratic equation with respect to $\lambda = \lambda(\alpha)$. Its roots can be easily found:

$$\lambda_1(\alpha) = 1 + ir\sin\alpha - 2r^2\sin^2\frac{\alpha}{2},$$
$$\lambda_2(\alpha) = 1 - ir\sin\alpha - 2r^2\sin^2\frac{\alpha}{2}.$$

These formulae are analogous to those obtained for the scalar Example 2, and similarly to that example we have:

$$1 - |\lambda_{1,2}(\alpha)|^2 = 4r^2\sin^4\frac{\alpha}{2}(1 - r^2).$$

We therefore see that the spectrum $\{\lambda_1(\alpha), \lambda_2(\alpha)\}$ belongs to the unit disk if $r \le 1$.

10.3.4 Stability in C

Let us emphasize that the type of stability we have analyzed in Sections 10.3.1–10.3.3 was stability in the sense of the maximum norm (10.71). Alternatively, it is referred to as stability in (the space) C. This space contains all bounded numerical sequences. The von Neumann spectral condition (10.78) is *necessary* for the scheme to be stable in C. As far as the sufficient conditions, in some simple cases stability in C can be proved directly, for example, using maximum principle, as done in Section 10.1.3 for the first order explicit upwind scheme and in Section 10.6.1 for an explicit scheme for the heat equation. Otherwise, sufficient conditions for stability in C turn out to be delicate and may require rather sophisticated arguments. The analysis of a general case even for one scalar constant coefficient difference equation goes beyond the scope of the current book, and we refer the reader to the original work by Fedoryuk [Fed67] (see also his monograph [Fed77, Chapter V, § 4]). In addition, in [RM67, Chapter 5] the reader can find an account of the work by Strang and by Thomee on the subject.

10.3.5 Sufficiency of the Spectral Stability Condition in l_2

However, *a sufficient condition for stability* may sometimes be easier to find if we were to use a different norm instead of the maximum norm (10.71). Let, for example,

$$
\begin{gathered}
\|u^p\|^2 = \sum_{m=-\infty}^{\infty} |u_m^p|^2, \quad \|\varphi^p\|^2 = \sum_{m=-\infty}^{\infty} |\varphi_m^p|^2, \quad \|\psi\|^2 = \sum_{m=-\infty}^{\infty} |\psi_m|^2, \\
\|u^{(h)}\|_{U_h} = \max_p \|u^p\|, \quad \|f^{(h)}\|_{F_h} = \left\| \begin{matrix} \varphi^p \\ \psi \end{matrix} \right\|_{F_h} = \max\{\|\psi\|, \max_p \|\varphi^p\|\}.
\end{gathered}
\tag{10.94}
$$

Relations (10.94) define Euclidean (i.e., l_2) norms for u^p, φ^p, and ψ. Accordingly, stability in the sense of the norms given by (10.94) is referred to as stability in (the space) l_2. We recall that the space l_2 is a Hilbert space of all numerical sequences, for which the sum of squares of absolute values of all their terms is bounded.

Consider a general constant coefficient finite-difference Cauchy problem:

$$
\begin{gathered}
\sum_{j=-j_{\text{left}}}^{j_{\text{right}}} b_j u_{m+j}^{p+1} - \sum_{j=-j_{\text{left}}}^{j_{\text{right}}} a_j u_{m+j}^p = \varphi_m^p, \\
u_m^0 = \psi_m, \quad m = 0, \pm 1, \pm 2, \dots, \quad p = 0, 1, \dots, [T/\tau] - 1,
\end{gathered}
\tag{10.95}
$$

under the assumption that

$$
\sum_{j=-j_{\text{left}}}^{j_{\text{right}}} b_j e^{i\alpha j} \neq 0, \quad 0 \le \alpha < 2\pi.
$$

Note that all spatially one-dimensional schemes from Section 10.3.3 fit into the category (10.95), even for a relatively narrow range: $j_{\text{left}} = j_{\text{right}} = 1$.

THEOREM 10.3

For the scheme (10.95) to be stable in l_2 with respect to the initial data, i.e., for the following inequality to hold:

$$\|u^p\| \leq c\|\psi\|, \quad p = 0, 1, \ldots, [T/\tau], \tag{10.96}$$

where the constant c does not depend on h [or on $\tau = \tau(h)$], is is necessary and sufficient that the von Neumann condition (10.78) be satisfied, i.e., that the spectrum of the scheme $\lambda = \lambda(\alpha)$ belong to the disk:

$$|\lambda(\alpha)| \leq 1 + c_1\tau, \tag{10.97}$$

where c_1 is another constant that does not depend either on α or on τ.

PROOF We will first prove the sufficiency. By hypotheses of the theorem, the number series $\sum\limits_{m=-\infty}^{\infty} |\psi_m|^2$ converges. Then, the function series of the independent variable α:

$$\sum_{m=-\infty}^{\infty} \psi_m e^{-i\alpha m}$$

also converges in the space $L_2[0, 2\pi]$, and its sum that we denote $\Psi(\alpha)$, $0 \leq \alpha \leq 2\pi$, is a function that has ψ_m as the Fourier coefficients:

$$\psi_m = \frac{1}{2\pi}\int_0^{2\pi} \Psi(\alpha) e^{i\alpha m} d\alpha, \quad m = 0, \pm 1, \pm 2, \ldots, \tag{10.98}$$

(a realization of the Riesz-Fischer theorem, see, e.g., [KF75, Section 16]). Moreover, the following relation holds:

$$\|\psi\|^2 = \sum_{m=-\infty}^{\infty} |\psi_m|^2 = \frac{1}{2\pi}\int_0^{2\pi} |\Psi(\alpha)|^2 d\alpha = \frac{1}{2\pi}\|\Psi\|_2^2$$

known as the Parseval equality.

Consider a homogeneous counterpart to the difference equation (10.95):

$$\sum_{j=-j_{\text{left}}}^{j_{\text{right}}} b_j u_{m+j}^{p+1} - \sum_{j=-j_{\text{left}}}^{j_{\text{right}}} a_j u_{m+j}^{p} = 0,$$
$$m = 0, \pm 1, \pm 2, \ldots, \quad p = 0, 1, \ldots, [T/\tau] - 1,$$

For any $\alpha \in [0, 2\pi)$ this equation obviously has a solution of the type:

$$u_m^p(\alpha) = \lambda^p(\alpha) e^{i\alpha m} \tag{10.99}$$

for some particular $\lambda = \lambda(\alpha)$ that can be determined by substitution:

$$\lambda(\alpha) = \left(\sum_{j=-j_{\text{left}}}^{j_{\text{right}}} a_j e^{i\alpha j}\right)\left(\sum_{j=-j_{\text{left}}}^{j_{\text{right}}} b_j e^{i\alpha j}\right)^{-1}.$$

Then, the grid function:

$$u_m^p = \frac{1}{2\pi}\int_0^{2\pi} \Psi(\alpha)\lambda^p(\alpha)e^{i\alpha m}d\alpha, \quad m = 0, \pm 1, \pm 2, \dots, \tag{10.100}$$

provides solution to the Cauchy problem (10.95) for the case $\varphi_m^p = 0$ because it is a linear combination of solutions $u_m^p(\alpha)$ of (10.99), and coincides with ψ_m for $p = 0$, see (10.98). Note that the integral on the right-hand side of formula (10.100) converges by virtue of the Parseval equality, because $\int_0^{2\pi}|\Psi(\alpha)|^2 d\alpha < \infty \Longrightarrow \int_0^{2\pi}|\Psi(\alpha)|d\alpha < \infty$.

If the von Neumann spectral condition (10.97) is satisfied, then

$$|\lambda(\alpha)|^p \le |1 + c_1\tau|^{T/\tau} \le e^{c_1 T}. \tag{10.101}$$

Consequently, using representation (10.100) along with the Parseval equality and inequality (10.101), we can obtain:

$$\|u^p\|^2 = \frac{1}{2\pi}\int_0^{2\pi}|\lambda^p(\alpha)\Psi(\alpha)|^2 d\alpha \le e^{c_1 T}\frac{1}{2\pi}\int_0^{2\pi}|\Psi(\alpha)|^2 d\alpha = e^{c_1 T}\|\psi\|^2,$$

which clearly implies stability with respect to the initial data: $\|u^p\| \le c\|\psi\|$.

To prove the necessity, we will need to show that if (10.97) holds for no fixed c_1, then the scheme is unstable. We should emphasize that to demonstrate the instability for the chosen norm (10.94) we may not exploit the unboundedness of the solution $u_m^p = \lambda^p(\alpha)e^{i\alpha m}$ that takes place in this case, because the grid function $\{e^{i\alpha m}\}$ does not belong to l_2.

Rather, let us take a particular $\Psi(\alpha) \in L_2[0, 2\pi]$ such that

$$\frac{1}{2\pi}\int_0^{2\pi}|\lambda(\alpha)|^{2p}|\Psi(\alpha)|^2 d\alpha \ge \max_\alpha\left(|\lambda(\alpha)|^{2p} - \varepsilon\right)\frac{1}{2\pi}\int_0^{2\pi}|\Psi(\alpha)|^2 d\alpha, \tag{10.102}$$

where $\varepsilon > 0$ is given. For an arbitrary ε, estimate (10.102) can always be guaranteed by selecting:

$$\Psi(\alpha) = \begin{cases} 1, & \text{if } \alpha \in [\alpha^* - \delta, \alpha^* + \delta], \\ 0, & \text{if } \alpha \notin [\alpha^* - \delta, \alpha^* + \delta], \end{cases}$$

where $\alpha^* = \arg\max_\alpha |\lambda(\alpha)|$ and $\delta > 0$. Indeed, as the function $|\lambda(\alpha)|^{2p}$ is continuous, inequality (10.102) will hold for a sufficiently small $\delta = \delta(\varepsilon)$.

If estimate (10.101) does not take place, then we can find a sequence h_k, $k = 0,1,2,3,\dots$, and the corresponding sequence $\tau_k = \tau(h_k)$ such that

$$c_k = \left(\max_{\alpha} |\lambda(\alpha, h_k)|\right)^{[T/\tau_k]} \longrightarrow \infty \quad \text{as} \quad k \longrightarrow \infty.$$

Let us set $\varepsilon = 1$ and choose $\Psi(\alpha)$ to satisfy (10.102). Define ψ_m as Fourier coefficients of the function $\Psi(\alpha)$, according to formula (10.98). Then, inequality (10.102) for $p_k = [T/\tau_k]$ transforms into:

$$\|u^{p_k}\|^2 \geq (c_k^2 - 1)\|\psi\|^2 \Longrightarrow \|u^{p_k}\| \geq (c_k - 1)\|\psi\|,$$
$$c_k \longrightarrow \infty \quad \text{as} \quad k \longrightarrow \infty,$$

i.e., there is indeed no stability (10.96) with respect to the initial data. □

Theorem 10.3 establishes *equivalence* between the von Neumann spectral condition and the l_2 stability of scheme (10.95) with respect to the initial data. In fact, one can go even further and prove that the von Neumann spectral condition is *necessary and sufficient* for the full-fledged l_2 stability of the scheme (10.95) as well, i.e., when the right-hand side φ_m^p is not disregarded. One implication, the necessity, immediately follows from Theorem 10.3, because if the von Neumann condition does not hold, then the scheme is unstable even with respect to the initial data. The proof of the other implication, the sufficiency, can be found in [GR87, § 25]. This proof is based on using the discrete Green's functions. In general, once stability with respect to the initial data has been established, stability of the full inhomogeneous problem can be derived using the Duhamel principle. This principle basically says that the solution to the inhomogeneous problem can be obtained as linear superposition of the solutions to some specially chosen homogeneous problems. Consequently, a stability estimate for the inhomogeneous problem can be obtained on the basis of stability estimates for a series of homogeneous problems, see [Str04, Chapter 9].

10.3.6 Scalar Equations vs. Systems

As of yet, our analysis of finite-difference stability has focused primarily on scalar equations; we have only considered a 2×2 system in Examples 10 and 11 of Section 10.3.3. Besides, in Examples 5 and 9 of Section 10.3.3 we have considered scalar difference equations that connect the values of the solution on more than two consecutive time levels of the grid; they can be reduced to systems of finite-difference equations on two consecutive time levels.

In general, a constant coefficient finite-difference Cauchy problem with vector unknowns (i.e., a system) can be written in the form similar to (10.95):

$$\sum_{j=-j_{\text{left}}}^{j_{\text{right}}} \boldsymbol{B}_j \boldsymbol{u}_{m+j}^{p+1} - \sum_{j=-j_{\text{left}}}^{j_{\text{right}}} \boldsymbol{A}_j \boldsymbol{u}_{m+j}^{p} = \boldsymbol{\varphi}_m^p, \tag{10.103}$$
$$\boldsymbol{u}_m^0 = \boldsymbol{\psi}_m, \quad m = 0, \pm 1, \pm 2, \dots, \quad p = 0, 1, \dots, [T/\tau] - 1,$$

under the assumption that the matrices

$$\sum_{j=-j_{\text{left}}}^{j_{\text{right}}} \boldsymbol{B}_j e^{i\alpha j}, \quad 0 \le \alpha < 2\pi,$$

are non-singular. In formula (10.103), $\boldsymbol{u}_m^p$, $\boldsymbol{\varphi}_m^p$, and $\boldsymbol{\psi}_m$ are grid vector functions of the same fixed dimension, and $\boldsymbol{A}_j = \boldsymbol{A}_j(h)$, $\boldsymbol{B}_j = \boldsymbol{B}_j(h)$, $j = -j_{\text{left}}, \dots, j_{\text{right}}$, are given square matrices of matching dimension.

Solution of the homogeneous counterpart to the finite-difference equation of (10.103) can be sought for in the form (10.80), where $\boldsymbol{u}^0 = \boldsymbol{u}^0(\alpha,h)$ and $\lambda = \lambda(\alpha,h)$ are the eigenvectors and eigenvalues, respectively, of the amplification matrix:

$$\boldsymbol{\Lambda} = \boldsymbol{\Lambda}(\alpha,h) = \left(\sum_{j=-j_{\text{left}}}^{j_{\text{right}}} \boldsymbol{B}_j e^{i\alpha j}\right)^{-1} \left(\sum_{j=-j_{\text{left}}}^{j_{\text{right}}} \boldsymbol{A}_j e^{i\alpha j}\right) \tag{10.104}$$

that corresponds to scheme (10.103).

The von Neumann spectral condition (10.97) is clearly necessary for stability of finite-difference systems in all norms. Indeed, if it is not met, then estimate (10.101) will not hold, and the scheme will develop a catastrophic exponential instability that cannot be fixed by any reasonable choice of norms.

Yet the von Neumann condition remains *only a necessary* stability condition for systems in either C or l_2. Regarding C, we have indicated previously (Section 10.3.4) that the analysis of sufficient conditions becomes cumbersome already for the scalar case. However, even in the case of l_2 that is supposedly more amenable for investigation, obtaining sufficient conditions for systems proves quite difficult.

Qualitatively, the aforementioned difficulties stem from the fact that the amplification matrix (10.104) may have multiple eigenvalues and as a consequence, may not necessarily have a full set of eigenvectors. If a multiple eigenvalue occurs exactly on the unit circle or just outside the unit disk, this may still cause instability even when all the eigenvalues satisfy the von Neumann constraint (10.97). These considerations are, in fact, similar to those that constitute the result of Theorem 9.2.

Of course, if the amplification matrix appears normal (a matrix that commutes with its adjoint) and therefore unitarily diagonalizable, then none of the aforementioned difficulties is present, and the von Neumann condition becomes not only necessary but also sufficient for stability of the vector scheme (10.103) in l_2.

Otherwise, the question of stability for scheme (10.103) can be *equivalently reformulated* using the new concept of stability for families of matrices. A family of square matrices (of a given fixed dimension) is said to be stable if there is a constant $K > 0$ such that for any particular matrix $\boldsymbol{\Lambda}$ from the family, and any positive integer p, the following estimate holds: $\|\boldsymbol{\Lambda}^p\| \le K$. Scheme (10.103) is stable in l_2 if and only if the family of amplification matrices $\boldsymbol{\Lambda} = \boldsymbol{\Lambda}(\alpha,h)$ given by (10.104) is stable in the sense of the previous definition (this family is parameterized by $\alpha \in [0,2\pi)$ and $h > 0$). The following theorem, known as the Kreiss matrix theorem, provides some necessary and sufficient conditions for a family of matrices to be stable.

THEOREM 10.4 (Kreiss)
Stability of a family of matrices $\mathbf{\Lambda}$ *is equivalent to any of the following conditions:*

1. *There is a constant* $C_1 > 0$ *such that for any matrix* $\mathbf{\Lambda}$ *from the given family, and any complex number* z, $|z| > 1$, *there is a resolvent* $(\mathbf{\Lambda} - z\mathbf{I})^{-1}$ *bounded as:*
$$\left\|(\mathbf{\Lambda} - z\mathbf{I})^{-1}\right\| \leq \frac{C_1}{|z| - 1}.$$
2. *There are constants* $C_2 > 0$ *and* $C_3 > 0$, *and for any matrix* $\mathbf{\Lambda}$ *from the given family there is a non-singular matrix* $\mathbf{M}$ *such that* $\|\mathbf{M}\| \leq C_2$, $\|\mathbf{M}^{-1}\| \leq C_2$, *and the matrix* $\mathbf{D} \stackrel{\text{def}}{=} \mathbf{M}\mathbf{\Lambda}\mathbf{M}^{-1}$ *is upper triangular, with the off-diagonal entries that satisfy:*
$$|d_{ij}| \leq C_3 \min\{1 - \kappa_i, 1 - \kappa_j\},$$
where $\kappa_i = d_{ii}$ *and* $\kappa_j = d_{jj}$ *are the corresponding diagonal entries of* $\mathbf{D}$, *i.e., the eigenvalues of* $\mathbf{\Lambda}$.
3. *There is a constant* $C_4 > 0$, *and for any matrix* $\mathbf{\Lambda}$ *from the given family there is a Hermitian positive definite matrix* $\mathbf{H}$, *such that*
$$C_4^{-1}\mathbf{I} \leq \mathbf{H} \leq C_4\mathbf{I} \quad \text{and} \quad \mathbf{\Lambda}^*\mathbf{H}\mathbf{\Lambda} \leq \mathbf{H}.$$

The proof of Theorem 10.4 can be found, e.g., in [RM67, Chapter 4] or [Str04, Chapter 9].

Exercises

1. Consider the so-called weighted scheme for the heat equation:
$$\frac{u_m^{p+1} - u_m^p}{\tau} = \sigma\frac{u_{m+1}^{p+1} - 2u_m^{p+1} + u_{m-1}^{p+1}}{h^2} + (1-\sigma)\frac{u_{m+1}^p - 2u_m^p + u_{m-1}^p}{h^2},$$
$$u_m^0 = \psi_m, \quad m = 0, \pm 1, \pm 2, \ldots, \quad p = 0, 1, \ldots, [T/\tau] - 1,$$
where the real parameter $\sigma \in [0, 1]$ is called the weight (between the fully explicit scheme, $\sigma = 0$, and fully implicit scheme, $\sigma = 1$). What values of σ guarantee that the scheme will meet the von Neumann stability condition for any $r = \tau/h^2 = \text{const}$?
2. Consider the Cauchy problem (10.87) for the heat equation. The scheme:
$$\frac{u_m^{p+1} - u_m^{p-1}}{2\tau} - a^2\frac{u_{m+1}^p - 2u_m^p + u_{m-1}^p}{h^2} = \varphi_m^p,$$
$$u_m^0 = \psi_m, \quad u_m^1 = \tilde{\psi}_m, \quad m = 0, \pm 1, \pm 2, \ldots, \quad p = 0, 1, \ldots, [T/\tau] - 1,$$
where we assume that $\varphi(x, 0) \equiv 0$ and define:
$$\tilde{\psi}_m = u(x,0) + \tau\left.\frac{\partial u(x,0)}{\partial t}\right|_{x=x_m} = u(x,0) + \tau a^2\left.\frac{\partial^2 u(x,0)}{\partial x^2}\right|_{x=x_m} = \psi_m + \tau a^2\psi''(x_m),$$
approximates problem (10.87) on its smooth solutions with the accuracy $\mathcal{O}(\tau^2 + h^2)$. Does this scheme satisfy the von Neumann spectral stability condition?

3. For the two-dimensional Cauchy problem:

$$\frac{\partial u}{\partial t} - \frac{\partial u}{\partial x} - \frac{\partial u}{\partial y} = \varphi(x,y,t), \quad -\infty < x,\, y < \infty, \quad 0 < t \leq T,$$
$$u(x,y,0) = \psi(x,y), \quad -\infty < x,\, y < \infty,$$

investigate the von Neumann spectral stability of:

a) The first order explicit scheme:

$$\frac{u_{m,n}^{p+1} - u_{m,n}^{p}}{\tau} - \frac{u_{m+1,n}^{p} - u_{m,n}^{p}}{h} - \frac{u_{m,n+1}^{p} - u_{m,n}^{p}}{h} = \varphi_m^p,$$
$$u_{m,n}^0 = \psi_{m,n}, \quad m,\, n = 0, \pm 1, \pm 2, \ldots, \quad p = 0, 1, \ldots, [T/\tau] - 1;$$

b) The second order explicit scheme:

$$\frac{u_{m,n}^{p+1} - u_{m,n}^{p-1}}{2\tau} - \frac{u_{m+1,n}^{p} - u_{m-1,n}^{p}}{2h} - \frac{u_{m,n+1}^{p} - u_{m,n-1}^{p}}{2h} = \varphi_m^p,$$
$$u_{m,n}^0 = \psi_{m,n}, \quad u_{m,n}^1 = \psi_{m,n} + \tau[\psi_x'(x_m, y_n) + \psi_y'(x_m, y_n) + \varphi(x_m, y_n, 0)],$$
$$m,\, n = 0, \pm 1, \pm 2, \ldots, \quad p = 0, 1, \ldots, [T/\tau] - 1.$$

4. Investigate the von Neumann spectral stability of the implicit two-dimensional scheme for the homogeneous heat equation:

$$\frac{u_{m,n}^{p+1} - u_{m,n}^{p}}{\tau} = \frac{u_{m+1,n}^{p+1} - 2u_{m,n}^{p+1} + u_{m-1,n}^{p+1}}{h^2} + \frac{u_{m,n+1}^{p+1} - 2u_{m,n}^{p+1} + u_{m,n-1}^{p+1}}{h^2}$$
$$u_{m,n}^0 = \psi_{m,n}, \quad m, n = 0, \pm 1, \pm 2, \ldots, \quad p = 0, 1, \ldots, [T/\tau] - 1.$$

5. Investigate the von Neumann stability of the implicit upwind scheme for the Cauchy problem (10.82):

$$\frac{u_m^{p+1} - u_m^{p}}{\tau} - \frac{u_{m+1}^{p+1} - u_m^{p+1}}{h} = \varphi_m^p, \tag{10.105}$$
$$u_m^0 = \psi_m, \quad m = 0, \pm 1, \pm 2, \ldots, \quad p = 0, 1, \ldots, [T/\tau] - 1.$$

6. Investigate the von Neumann stability of the implicit downwind scheme for the Cauchy problem (10.82):

$$\frac{u_m^{p+1} - u_m^{p}}{\tau} - \frac{u_m^{p+1} - u_{m-1}^{p+1}}{h} = \varphi_m^p, \tag{10.106}$$
$$u_m^0 = \psi_m, \quad m = 0, \pm 1, \pm 2, \ldots, \quad p = 0, 1, \ldots, [T/\tau] - 1.$$

7. Investigate the von Neumann stability of the implicit central scheme for the Cauchy problem (10.82):

$$\frac{u_m^{p+1} - u_m^{p}}{\tau} - \frac{u_{m+1}^{p+1} - u_{m-1}^{p+1}}{2h} = \varphi_m^p, \tag{10.107}$$
$$u_m^0 = \psi_m, \quad m = 0, \pm 1, \pm 2, \ldots, \quad p = 0, 1, \ldots, [T/\tau] - 1.$$

8.* Transform the leap-frog scheme (10.86) of Example 5, Section 10.3.3, and the central-difference scheme for the d'Alembert equation of Example 9, Section 10.3.3, to the schemes written for finite-difference systems, as opposed to scalar equations, but connecting only two, as opposed to three, consecutive time levels of the grid. Investigate the von Neumann stability by calculating spectra of the corresponding amplification matrices (10.104).

Hint. Use the difference $\{u_m^{p+1} - u_m^p\}$ as the second unknown grid function.

10.4 Stability for Problems with Variable Coefficients

The von Neumann necessary condition that we have introduced in Section 10.3 to analyze stability of linear finite-difference Cauchy problems with constant coefficients can, in fact, be applied to a wider class of formulations. A simple extension that we describe in this section allows one to exploit the von Neumann condition to analyze stability of problems with variable coefficients (continuous, but not necessarily constant) and even some nonlinear problems.

10.4.1 The Principle of Frozen Coefficients

Introduce a uniform Cartesian grid: $x_m = mh$, $m = 0, \pm 1, \pm 2, \dots$, $t_p = p\tau$, $p = 0, 1, 2, \dots$, and consider a finite-difference Cauchy problem for the homogeneous heat equation with the variable coefficient of heat conduction $a = a(x,t)$:

$$\begin{gathered} \frac{u_m^{p+1} - u_m^p}{\tau} - a(x_m, t_p)\frac{u_{m+1}^p - 2u_m^p + u_{m-1}^p}{h^2} = 0, \\ u_m^0 = \psi(x_m), \quad m = 0, \pm 1, \pm 2, \dots, \quad p \geq 0. \end{gathered} \tag{10.108}$$

Next, take an arbitrary point $(\tilde{x}, \tilde{t})$ in the domain of problem (10.108) and "freeze" the coefficients of problem (10.108) at this point. Then, we arrive at the constant-coefficient finite-difference equation:

$$\begin{gathered} \frac{u_m^{p+1} - u_m^p}{\tau} - a(\tilde{x}, \tilde{t})\frac{u_{m+1}^p - 2u_m^p + u_{m-1}^p}{h^2} = 0, \\ m = 0, \pm 1, \pm 2, \dots, \quad p \geq 0. \end{gathered} \tag{10.109}$$

Having obtained equation (10.109), we can formulate the following principle of frozen coefficients. *For the original variable-coefficient Cauchy problem (10.108) to be stable it is necessary that the constant-coefficient Cauchy problem for the difference equation (10.109) satisfy the von Neumann spectral stability condition.*

To justify the principle of frozen coefficients, we will provide an heuristic argument rather than a proof. When the grid is refined, the variation of the coefficient $a(x,t)$ in a neighborhood of the point $(\tilde{x}, \tilde{t})$ becomes smaller if measured over any finite fixed number of grid cells that have size h in space and size τ in time. This is true because of the continuity of the function $a = a(x,t)$. In other words, the finer the grid, the closer is $a(x,t)$ to $a(\tilde{x},\tilde{t})$ as long as (x,t) is no more than so many cells away from $(\tilde{x},\tilde{t})$. Consequently, if we were to perturb the solution of problem (10.108) on a fine grid at the moment of time $t = \tilde{t}$ near the space location $x = \tilde{x}$, then over short time intervals these perturbations would have evolved pretty much the as if they were perturbations of the solution to the constant-coefficient equation (10.109).

It is clear that the previous argument is quite general. It is not affected by the number of space dimensions, the number of unknown functions, or the specific type of the finite-difference equation or system.

In Section 10.3.3, we analyzed a Cauchy problem for the equation of type (10.109), see Example 6, and found that for the von Neumann stability condition to hold the ratio $r = \tau/h^2$ must satisfy the inequality:

$$r \leq \frac{1}{2a(\tilde{x},\tilde{t})}. \tag{10.110}$$

According to the principle of frozen coefficients, stability of scheme (10.108) requires that condition (10.110) be met for any $(\tilde{x},\tilde{t})$. Therefore, altogether the ratio $r = \tau/h^2$ must satisfy the inequality:

$$r \leq \frac{1}{2\max\limits_{x,t} a(x,t)}. \tag{10.111}$$

The principle of frozen coefficients can also provide an heuristic argument for the analysis of stability of nonlinear difference equations. We illustrate this using an example of a Cauchy problem for the nonlinear heat equation:

$$\begin{gathered}
\frac{\partial u}{\partial t} - (1+u^2)\frac{\partial^2 u}{\partial x^2} = 0, \quad -\infty < x < \infty, \quad 0 < t \leq T, \\
u(x,0) = \psi(x), \quad -\infty < x < \infty.
\end{gathered}$$

We approximate this problem by means of an explicit scheme:

$$\begin{gathered}
\frac{u_m^{p+1} - u_m^p}{\tau_p} - \left(1 + |u_m^p|^2\right) \frac{u_{m+1}^p - 2u_m^p + u_{m-1}^p}{h^2} = 0, \\
u_m^0 = \psi(x_m), \quad m = 0, \pm 1, \pm 2, \ldots, \quad p = 0, 1, 2, \ldots.
\end{gathered} \tag{10.112}$$

The scheme is built on a uniform spatial grid: $x_m = mh$, $m = 0, \pm 1, \pm 2, \ldots$, but with a non-uniform temporal grid, such that the size $\tau_p = t_{p+1} - t_p$ may vary from one time level to another. The finite-difference solution can still be obtained by marching.

Assume that we have already marched all the way up the time level t_p and computed the solution u_m^p, $m = 0, \pm 1, \pm 2, \ldots$. To continue marching, we first need to select the next grid size τ_p. This can be done by interpreting the finite-difference equation to be solved at $t = t_p$ with respect to u_m^{p+1} as a linear equation:

$$\frac{u_m^{p+1} - u_m^p}{\tau_p} - a_m^p \frac{u_{m+1}^p - 2u_m^p + u_{m-1}^p}{h^2} = 0 \tag{10.113}$$

with the given variable coefficient of heat conduction: $a_m^p \equiv \left(1+|u_m^p|^2\right)$. Indeed, it is natural to think that the values of the grid function u_m^p are close to the values $u(x_m,t_p)$ of the continuous solution $u(x,t)$. Then, the discrete heat conduction coefficient a_m^p will be close to the projection $a(x_m,t_p)$ of the continuous function $a(x,t) = 1 + u(x,t)^2$ onto the grid. This function may vary only slightly over a few temporal steps.

By applying the principle of frozen coefficients to equation (10.113), we arrive at the constraint (10.111) for the grid sizes that is necessary for stability:

$$\frac{\tau_p}{h^2} = r_p \leq \frac{1}{2\max\limits_m a_m^p} = \frac{1}{2\max\limits_m(1+|u_m^p|^2)}.$$

Consequently, when marching equation (10.112) one should select the temporal grid size τ_p for each $p = 0,1,2,\ldots$ based on the inequality:

$$\tau_p \leq \frac{h^2}{2\max\limits_m(1+|u_m^p|^2)}.$$

Numerical experiments corroborate correctness of these heuristic arguments.

If stability condition obtained by considering the Cauchy problem with frozen coefficients (at an arbitrary point of the domain) is violated, then we expect that there will be no stability. We re-emphasize though that our justification for the principle of frozen coefficients was heuristic rather than rigorous. There are, in fact, counter-examples, when the problem with variable coefficients is stable, whereas the problems with frozen coefficients are unstable. Those counter-examples, however, are fairly "exotic" in nature, like the one given by Strang in [Str66] that involves a second order differential equation with complex coefficients: $u_t = i[(\sin x)u_x]_x$. At the same time, the analysis of [Str66] shows that for some important classes of differential equations/systems that include, in particular, all first order systems (e.g., hyperbolic), the principle of frozen coefficients holds in the sense of l_2 for explicit finite-difference schemes. The same result is also known to be true for parabolic equations, see [RM67, Chapter 5]. Therefore, for all practical purposes hereafter, we will still regard the principle of frozen coefficients as a necessary condition for stability. Moreover, in Section 10.6.1 we will use the maximum principle and show that the explicit finite-difference scheme for a variable coefficient heat equation is actually stable if it satisfies the condition based on the principle of frozen coefficients.

We should also mention that the principle of frozen coefficients as formulated in this section applies only to Cauchy problems. For an initial boundary value problem formulated on a finite interval, the foregoing analysis is not sufficient. Even if the necessary stability condition based on the principle of frozen coefficients holds, the overall problem on the finite interval can still be either stable or unstable depending on the choice of the boundary conditions at the endpoints of the interval. In Section 10.5.1, we discuss the Babenko-Gelfand stability criterion that takes into account the effect of boundary conditions in the case of a problem on an interval.

10.4.2 Dissipation of Finite-Difference Schemes

A very useful concept that can lead to a sufficient condition of stability for some problems with variable coefficients is that of *dissipation*. We will first illustrate it using scalar finite-difference equations with constant coefficients. We will again use the notion of the spectrum $\lambda = \lambda(\alpha, h)$ of the transition operator introduced in Section 10.3.2 (see page 351) and subsequently studied in Section 10.3.3. For convenience, in this section we will assume that the frequency range is $-\pi < \alpha \leq \pi$, rather than $0 \leq \alpha < 2\pi$ that we have used previously. Both choices are clearly equivalent, because the harmonics $e^{i\alpha m}$ are 2π-periodic.

DEFINITION 10.4 *A scheme is said to be dissipative of order $2d$ if d is the smallest positive integer for which one can find a constant $\delta > 0$ that would not depend on h (and on τ either), such that for all $|\alpha| \leq \pi$ and all sufficiently small grid sizes h, the following inequality will hold:*

$$|\lambda(\alpha, h)| \leq 1 - \delta |\alpha|^{2d}. \tag{10.114}$$

Recall that in the case of constant coefficients we analyze solutions in the form $\lambda^p e^{i\alpha m}$, see formulae (10.76) and (10.80), and therefore, for a given α, the amplification factor $\lambda(\alpha, h)$ indicates how rapidly the harmonic $e^{i\alpha m}$ decays as the number of time steps p increases. From this perspective, by introducing the concept of dissipation, see inequality (10.114), we quantify the rate of this decay for individual harmonics in terms of a positive constant δ and some positive power $2d$ of the frequency α. Alternatively, if instead of inequality (10.114) we were to have

$$|\lambda(\alpha, h)| = 1,$$

i.e., if the spectrum were to lie entirely on the unit circle, then there would be no decay of the amplitude for any harmonic $e^{i\alpha m}$ (equivalently, no dissipation), and the scheme would be called *non-dissipative*, accordingly.

Let us now consider some typical examples.

Example 1

The first order upwind scheme (see Section 10.3.1):

$$\begin{gathered} \frac{u_m^{p+1} - u_m^p}{\tau} - \frac{u_{m+1}^p - u_m^p}{h} = 0, \\ u_m^0 = \psi(x_m), \quad m = 0, \pm 1, \pm 2, \ldots, \quad p = 0, 1, 2, \ldots, [T/\tau] - 1, \end{gathered} \tag{10.115}$$

for the Cauchy problem:

$$\begin{gathered} \frac{\partial u}{\partial t} - \frac{\partial u}{\partial x} = 0, \quad -\infty < x < \infty, \quad 0 < t \leq T, \\ u(x, 0) = \psi(x), \quad -\infty < x < \infty, \end{gathered} \tag{10.116}$$

has its spectrum defined by formula (10.77):

$$\lambda(\alpha) = 1 - r + re^{i\alpha}, \quad r = \tau/h = \text{const},$$

and consequently,

$$|\lambda(\alpha)|^2 = 1 - 4r(1-r)\sin^2\frac{\alpha}{2}.$$

For $r > 1$, this spectrum is outside the unit disk, see Figure 10.5(b), and it makes no sense to discuss dissipation according to Definition 10.4. If $r = 1$, then $|\lambda(\alpha)| = 1$, and the scheme is non-dissipative. If $r < 1$, we can use the inequality $|\alpha|/4 \leq |\sin(\alpha/2)|$ for $\alpha \in [-\pi, \pi]$, and by setting $\delta = r(1-r)/8 > 0$ we obtain:

$$\begin{aligned} |\lambda(\alpha)|^2 &= 1 - 4r(1-r)\sin^2\frac{\alpha}{2} \leq 1 - 4r(1-r)\frac{|\alpha|^2}{16} \\ &= 1 - 2\delta|\alpha|^2 \leq 1 - 2\delta|\alpha|^2 + \delta^2|\alpha|^4 = \left(1 - \delta|\alpha|^2\right)^2. \end{aligned}$$

Therefore, the first order upwind scheme for $r < 1$ is dissipative of order $2d = 2$.

Example 2

For the Lax-Wendroff scheme [see formula (10.83)]:

$$\begin{gathered} \frac{u_m^{p+1} - u_m^p}{\tau} - \frac{u_{m+1}^p - u_{m-1}^p}{2h} - \frac{\tau}{2h^2}(u_{m+1}^p - 2u_m^p + u_{m-1}^p) = 0, \\ u_m^0 = \psi_m, \end{gathered} \tag{10.117}$$

we determined in Example 2 of Section 10.3.3 that

$$|\lambda(\alpha)|^2 = 1 - 4r^2(1-r^2)\sin^4\frac{\alpha}{2}, \quad r = \frac{\tau}{h} = \text{const}.$$

The scheme is, again, non-dissipative when $r = 1$. For $r < 1$, it is easy to obtain:

$$|\lambda(\alpha)| \leq 1 - \delta|\alpha|^4,$$

where $\delta = r^2(1-r^2)/128$, i.e., the scheme appears dissipative of order $2d = 4$.

Example 3

In Section 9.2.6 of Chapter 9 we introduced the Crank-Nicolson scheme for ordinary differential equations, see formula (9.33). A very similar scheme can be designed and used in the case of partial differential equations. For the Cauchy problem (10.116) we write:

$$\begin{gathered} \frac{u_m^{p+1} - u_m^p}{\tau} - \frac{1}{2}\left[\frac{u_{m+1}^{p+1} - u_{m-1}^{p+1}}{2h} + \frac{u_{m+1}^p - u_{m-1}^p}{2h}\right] = 0, \\ u_m^0 = \psi(x_m), \quad m = 0, \pm 1, \pm 2, \ldots, \quad p = 0, 1, 2, \ldots, [T/\tau] - 1. \end{gathered} \tag{10.118}$$

Unlike the schemes discussed in Examples 1 & 2, the Crank-Nicolson scheme (10.118) is implicit. It approximates problem (10.116) with second order accuracy with respect to h, provided that $r = \tau/h = \text{const}$.

For the spectrum of scheme (10.118) we can easily find:

$$\frac{\lambda - 1}{\tau} - \frac{1}{2}\left[\lambda \frac{e^{i\alpha} - e^{-i\alpha}}{2h} + \frac{e^{i\alpha} - e^{-i\alpha}}{2h}\right] = 0,$$

which yields:

$$\lambda(\alpha) = \left(1 + \frac{ir}{2}\sin\alpha\right) \cdot \left(1 - \frac{ir}{2}\sin\alpha\right)^{-1}.$$

Consequently,

$$|\lambda(\alpha)| = 1$$

irrespective of the specific value of r. Therefore, the Crank-Nicolson scheme (10.118) is non-dissipative. We also see that it satisfies the von Neumann stability condition (10.78).

Example 4

Finally, consider a fully implicit first order upwind scheme (10.105):

$$\begin{gathered}\frac{u_m^{p+1} - u_m^p}{\tau} - \frac{u_{m+1}^{p+1} - u_m^{p+1}}{h} = 0, \\ u_m^0 = \psi(x_m), \quad m = 0, \pm 1, \pm 2, \dots, \quad p = 0, 1, 2, \dots, [T/\tau] - 1,\end{gathered} \tag{10.119}$$

for the same Cauchy problem (10.116). Substituting the solution in the form $u_m^p = c\lambda^p e^{i\alpha m}$, $-\pi \le \alpha \le \pi$, into the difference equation of (10.119), we obtain:

$$\frac{\lambda - 1}{\tau} - \lambda \frac{e^{i\alpha} - 1}{h} = 0,$$

which immediately yields the spectrum of the scheme (10.119):

$$\lambda(\alpha) = \frac{1}{1 + r - re^{i\alpha}}, \quad r = \tau/h = \text{const}.$$

Using the inequality: $|\alpha|/4 \le |\sin(\alpha/2)|$, $\alpha \in [-\pi, \pi]$, from Example 1, we have:

$$\begin{aligned}|\lambda|^2 &= \frac{1}{(1 + r - r\cos\alpha)^2 + r^2\sin^2\alpha} = \frac{1}{1 + 4r(1+r)\sin^2\frac{\alpha}{2}} \\ &= 1 - \frac{4r(1+r)\sin^2\frac{\alpha}{2}}{1 + 4r(1+r)\sin^2\frac{\alpha}{2}} \le 1 - \frac{4r(1+r)\sin^2\frac{\alpha}{2}}{1 + 4r(1+r)} \le 1 - \frac{4r(1+r)}{1 + 4r(1+r)}\frac{|\alpha|^2}{16}.\end{aligned}$$

Consequently, if we introduce $\delta > 0$ according to:

$$2\delta = \frac{1}{16} \cdot \frac{4r(1+r)}{1 + 4r(1+r)},$$

then we can write:

$$|\lambda|^2 \leq 1-2\delta|\alpha|^2 \leq 1-2\delta|\alpha|^2+\delta^2|\alpha^4| = \left(1-\delta|\alpha|^2\right)^2.$$

In other words,

$$|\lambda(\alpha)| \leq 1-\delta|\alpha|^2, \quad |\alpha| \leq \pi,$$

which means that scheme (10.119) is dissipative of order $2d = 2$ regardless of the value of $r = \tau/h$.

The concept of dissipation introduced in Definition 10.4 for scalar constant-coefficient difference equations can be extended to the case of variable coefficients and to the case of systems, including systems in multiple space dimensions [RM67, Chapter 5]. For the reason of simplicity, in this book we will only discuss one-dimensional systems. Instead of the scalar problem (10.116) consider a Cauchy problem for the hyperbolic system:

$$\begin{gathered}\frac{\partial \boldsymbol{u}}{\partial t} - \boldsymbol{A}(x)\frac{\partial \boldsymbol{u}}{\partial x} = \boldsymbol{0}, \quad -\infty < x < \infty, \quad 0 < t \leq T, \\ \boldsymbol{u}(x,0) = \boldsymbol{\psi}(x), \quad -\infty < x < \infty,\end{gathered} \tag{10.120}$$

where $\boldsymbol{u} = \boldsymbol{u}(x,t)$ is the unknown vector function of a given fixed dimension K. The $K \times K$ square matrix $\boldsymbol{A} = \boldsymbol{A}(x)$ is assumed Hermitian for all x.

We will approximate problem (10.120) by an explicit finite-difference scheme of type (10.103):

$$\begin{gathered}\boldsymbol{u}_m^{p+1} - \sum_{j=-j_{\text{left}}}^{j_{\text{right}}} \boldsymbol{A}_j(x,h)\boldsymbol{u}_{m+j}^p = \boldsymbol{0}, \\ \boldsymbol{u}_m^0 = \boldsymbol{\psi}_m, \quad m = 0, \pm 1, \pm 2, \ldots, \quad p = 0, 1, \ldots, [T/\tau]-1,\end{gathered} \tag{10.121}$$

where $\boldsymbol{A}_j(x,h)$, $j = -j_{\text{left}}, \ldots, j_{\text{right}}$, are also assumed Hermitian. The amplification matrix (10.104) for scheme (10.121) is obtained by exploiting the same idea as in Section 10.4.1, i.e., by freezing the coefficients $\boldsymbol{A}_j$ for each particular location x:

$$\boldsymbol{\Lambda} = \boldsymbol{\Lambda}(x,\alpha,h) = \sum_{j=-j_{\text{left}}}^{j_{\text{right}}} \boldsymbol{A}_j(x,h)e^{i\alpha j}. \tag{10.122}$$

DEFINITION 10.5 *Scheme (10.121) is said to be dissipative of order $2d$ if d is the smallest positive integer for which there is a constant $\delta > 0$ that does not depend on either h or x, such that for all $|\alpha| \leq \pi$, all $x \in (-\infty,\infty)$, and all sufficiently small grid sizes h, the following inequalities hold:*

$$|\lambda_k(x,\alpha,h)| \leq 1-\delta|\alpha|^{2d}, \quad k = 1, 2, \ldots, K, \tag{10.123}$$

where $\lambda_k = \lambda_k(x,\alpha,h)$ are eigenvalues of the amplification matrix $\boldsymbol{\Lambda}$ of (10.122).

The following theorem due to Kreiss provides a sufficient condition of stability for dissipative finite-difference schemes (see [RM67, Chapter 5] for the proof):

THEOREM 10.5 (Kreiss)

Let the matrix $\mathbf{A}(x)$ in equation (10.120) and the matrices $\mathbf{A}_j(x,h)$, $j = -j_{\text{left}}, \dots, j_{\text{right}}$, in equation (10.121) be Hermitian, uniformly bounded, and uniformly Lipshitz-continuous with respect to x. Then, if scheme (10.121) is dissipative of order $2d$ in the sense of Definition 10.5, and has accuracy of order $2d-1$ for some positive integer d, then this scheme is stable in l_2.

Having formulated the Kreiss Theorem 10.5, we can revisit our previously analyzed examples. The first order upwind scheme (10.115) of Example 1 is dissipative of order $2d = 2$ when $r = \tau/h < 1$. Its accuracy is $\mathscr{O}(h)$, and therefore we can set $d = 1$, apply Theorem 10.5, and conclude that this scheme is stable. The same conclusion can obviously be made regarding the implicit scheme (10.119) analyzed in Example 4. Moreover, in this case the dissipation of the scheme puts no constraints on the value of r, and therefore the stability is unconditional.

The Lax-Wendroff scheme (10.117) of Example 2 is dissipative of order $2d = 4$ when $r < 1$, i.e., we need to consider $d = 2$. However, this scheme is only second order accurate, while $2d - 1 = 3$. Therefore, Theorem 10.5 does not allow us to immediately judge the stability of scheme (10.117).[6]

The Crank-Nicolson scheme (10.118) of Example 3 is non-dissipative, and therefore Theorem 10.5 does not apply.

On the other hand, let us note that according to Theorem 10.3 (proven in Section 10.3.5), all three foregoing schemes are, in fact, stable in the sense of l_2 when $r \leq 1$, i.e., when they satisfy the von Neumann condition. Theorem 10.3 says that the von Neumann condition is necessary and sufficient for stability of a scalar constant-coefficient finite-difference scheme in l_2.

As such, we see that for the case of scalar constant-coefficient finite-difference equations, the result of Theorem 10.5 may, in fact, be superseded by that of Theorem 10.3. We know, however, that for systems the von Neumann criterion alone provides only a necessary condition for stability, see Section 10.3.6. Moreover, for the case of variable coefficients, the principle of frozen coefficients, see Section 10.4.1, also provides a necessary condition for stability only. These are the cases when the sufficient condition given by Theorem 10.5 is most helpful.

Note also that there is a special large group of methods used in particular for the computation of fluid flows, when dissipation (viscosity) is artificially added to the scheme to improve its stability characteristics, see Section 11.2.1.

---

[6]The Theorem can still be applied though, but only after a change of variables, see [RM67, Chapter 5].

Exercises

1. Prove that the Crank-Nicolson scheme (10.118) has accuracy $\mathcal{O}(h^2)$ provided that $r = \tau/h = \text{const}$.
2. Show that the Lax-Friedrichs scheme (10.85) is dissipative of order $2d = 2$, although not strictly in the sense of Definition 10.4. Prove that it rather satisfies inequality (10.114) for all $|\alpha| \le \pi - \varepsilon$, where $\varepsilon > 0$ can be arbitrary.
3. Use Theorem 10.5 to study stability of the scheme:

$$\frac{u_m^{p+1} - u_m^p}{\tau} - a(x_m)\frac{u_{m+1}^p - u_m^p}{h} = 0,$$
$$u_m^0 = \psi(x_m), \quad m = 0, \pm 1, \pm 2, \ldots, \quad p = 0, 1, 2, \ldots, [T/\tau] - 1,$$

   for the Cauchy problem:

$$\frac{\partial u}{\partial t} - a(x)\frac{\partial u}{\partial x} = 0, \quad -\infty < x < \infty, \quad 0 < t \le T,$$
$$u(x,0) = \psi(x), \quad -\infty < x < \infty,$$

   where $a(x)$ is a smooth function and $a_1 \ge a(x) \ge a_0 > 0$.
4. Show that the implicit downwind scheme (10.106) for the Cauchy problem (10.116) is dissipative of order $2d = 2$ when $r > 1$.
5. Show that the implicit central scheme (10.107) for the Cauchy problem (10.116) is dissipative of order $2d = 2$ in the same non-strict sense as outlined in Exercise 2.

10.5 Stability for Initial Boundary Value Problems

Instead of the Cauchy problem (10.108), let us now consider an initial boundary value problem for the heat equation formulated on the finite interval $0 \le x \le 1$:

$$\begin{gathered}\frac{u_m^{p+1} - u_m^p}{\tau} - a(x_m, t_p)\frac{u_{m+1}^p - 2u_m^p + u_{m-1}^p}{h^2} = 0,\\ u_m^0 = \psi(x_m), \quad l_1 u_0^{p+1} = 0, \quad l_2 u_M^{p+1} = 0,\\ m = 0, 1, 2, \ldots, M, \quad p \ge 0.\end{gathered} \tag{10.124}$$

In formula (10.124), we assume that the grid is uniform: $x_m = mh$, $m = 0, 1, 2, \ldots, M$, $M = 1/h$, $t_p = p\tau$, $p = 0, 1, 2, \ldots$, and denote by $\boldsymbol{l}_1$ and $\boldsymbol{l}_2$ the operators of the boundary conditions at the left and right endpoints of the interval, respectively.

10.5.1 The Babenko-Gelfand Criterion

To analyze stability of the difference problem (10.124), we will first develop a heuristic argument based on freezing the coefficients; this argument will further extend the previous considerations of Section 10.4.1. In Section 10.4.1, we have noticed that because of the continuity of the coefficient $a = a(x,t)$, its variation within a fixed number of cells around any given point $(\tilde{x},\tilde{t})$ becomes smaller when the grid is refined. In the context of the initial boundary value problem (10.124), as opposed to the initial value problem (10.108), we supplement this consideration by another obvious observation. If the point $(\tilde{x},\tilde{t})$ lies inside the domain, then the distance from this point to either of the endpoints, $x=0$ or $x=1$, measured in the number of grid cells (of size h) will increase with no bound when $h \longrightarrow 0$. In other words, on fine grids the point $(\tilde{x},\tilde{t})$ can be considered to be located far away from the boundaries. Consequently, we can still claim that for small h the perturbations superimposed on the solution of problem (10.124) at the moment of time $t=\tilde{t}$ near any interior space location $x=\tilde{x}$ will evolve similarly to how the perturbations of the solution to the same "old" constant-coefficient equation (10.109) would have evolved. This, in turn, implies that stability of the scheme (10.109) for every $(\tilde{x},\tilde{t})$ inside the domain is still necessary for the overall stability of scheme (10.124).

The foregoing heuristic argument, however, becomes far less convincing if the point $(\tilde{x},\tilde{t})$ happens to lie precisely on one of the lateral boundaries: $x=0$ or $x=1$. For example, when we let $\tilde{x}=0$, the distance from $(\tilde{x},\tilde{t})$ to any fixed location $x>0$ (and in particular, to the right endpoint $x=1$) measured in the number of grid cells will again increase with no bound as $h \longrightarrow 0$. Yet the number of grid cells to the left endpoint $x=0$ will not change and will remain equal to zero. In other words, the point $(\tilde{x},\tilde{t})$ will never be far from the left boundary, no matter how fine the grid may be. Consequently, we can no longer expect that perturbations of the solution to problem (10.124) near $\tilde{x}=0$ will behave similarly to perturbations of the solution to equation (10.109), as the latter is formulated on the grid infinite in both directions.

Instead, we shall rather expect that over the short periods of time the perturbations of the solution to problem (10.124) near the left endpoint $x=0$ will develop analogously to perturbations of the solution to the following constant-coefficient problem:

$$\begin{gathered}\frac{u_m^{p+1}-u_m^p}{\tau}-a(0,\tilde{t})\frac{u_{m+1}^p-2u_m^p+u_{m-1}^p}{h^2}=0, \\ l_1 u_0^{p+1}=0, \quad m=0,1,2,\ldots, \quad p\geq 0.\end{gathered} \tag{10.125}$$

Problem (10.125) is formulated on the semi-infinite grid: $m=0,1,2,\ldots$ (i.e., semi-infinite line $x\geq 0$). It is obtained from the original problem (10.124) by freezing the coefficient $a(x,t)$ at the left endpoint of the interval $0\leq x\leq 1$ and by simultaneously "pushing" the right boundary off all the way to $+\infty$. Problem (10.125) shall be analyzed only for those grid functions $u^p=\{u_0^p,u_1^p,\ldots\}$ that satisfy:

$$u_m^p \longrightarrow 0, \quad \text{as} \quad m \longrightarrow +\infty. \tag{10.126}$$

Indeed, only in this case will the perturbation be concentrated near the left boundary $x=0$, and only for the perturbations of this type will the problems (10.124) and (10.125) be similar to one another in the vicinity of $x=0$.

Likewise, the behavior of perturbations to the solution of problem (10.124) near the right endpoint $x = 1$ should resemble that for the problem:

$$\begin{aligned} &\frac{u_m^{p+1} - u_m^p}{\tau} - a(1,\tilde{t})\frac{u_{m+1}^p - 2u_m^p + u_{m-1}^p}{h^2} = 0, \\ &l_2 u_M^{p+1} = 0, \quad m = \dots, -2, -1, 0, 1, 2, \dots, M, \quad p \geq 0, \end{aligned} \tag{10.127}$$

that has only one boundary at $m = M$. Problem (10.127) is derived from problem (10.124) by freezing the coefficient $a(x,t)$ at the right endpoint of the interval $0 \leq x \leq 1$ and by simultaneously pushing the left boundary off all the way to $-\infty$. It should be considered only for the grid functions $u^p = \{\dots, u_{-1}^p, u_0^p, u_1^p, \dots, u_M^p\}$ that satisfy:

$$u_m^p \longrightarrow 0, \quad \text{as} \quad m \longrightarrow -\infty. \tag{10.128}$$

The three problems: (10.109), (10.125), and (10.127), are easier to investigate than the original problem (10.124), because they are all h independent provided that $r = \tau/h^2 = \text{const}$, and they all have constant coefficients.

Thus, the issue of studying stability for the scheme (10.124), with the effect of the boundary conditions taken into account, can be addressed as follows. One needs to formulate three auxiliary problems: (10.109), (10.125), and (10.127). For each of these three h independent problems, one needs to find all those numbers λ (eigenvalues of the transition operator from u^p to u^{p+1}), for which solutions of the type

$$u_m^p = \lambda^p u_m^0 \tag{10.129}$$

exist. In doing so, for problem (10.109), the function $u^0 = \{u_m^0\}$, $m = 0, \pm 1, \pm 2, \dots$, has to be bounded on the grid. For problem (10.125), the grid function $u^0 = \{u_m^0\}$, $m \geq 0$, has to satisfy: $u_m^0 \longrightarrow 0$ as $m \longrightarrow +\infty$, and for problem (10.125), the grid function $u^0 = \{u_m^0\}$, $m \leq M$, has to satisfy: $u_m^0 \longrightarrow 0$ as $m \longrightarrow -\infty$. For scheme (10.124) to be stable, it is necessary that the overall spectrum of the difference initial boundary value problem, i.e., all eigenvalues of all three problems: (10.109), (10.125), and (10.127), belong to the unit disk: $|\lambda| \leq 1$, on the complex plane. This is the Babenko-Gelfand stability criterion. Note that problem (10.109) has to be considered for every fixed $\tilde{x} \in (0,1)$ and all $\tilde{t}$.

REMARK 10.1 Before we continue to study problem (10.124), let us present an important intermediate conclusion that can already be drawn based on the foregoing qualitative analysis. For stability of the pure Cauchy problem (10.108) that has no boundary conditions it is necessary that finite-difference equations (10.109) be stable in the von Neumann sense $\forall(\tilde{x},\tilde{t})$. This requirement remains necessary for stability of the initial boundary value problem (10.124) as well. Moreover, when boundary conditions are present, two more auxiliary problems: (10.125) and (10.127), have to be stable in a similar sense. Therefore, adding boundary conditions to a finite-difference Cauchy problem *will not, generally speaking, improve its stability.* Boundary conditions may either remain neutral or hamper the overall stability if, for example, problem

(10.109) appears stable but one of the problems (10.125) or (10.127) happens to be unstable. Later on, we will discuss this phenomenon in more detail. □

Let us now assume for simplicity that $a(x,t) \equiv 1$ in problem (10.124), and let us calculate the spectra of the three auxiliary problems (10.109), (10.125), and (10.127) for various boundary conditions $l_1 u_0^{p+1} = 0$ and $l_2 u_M^{p+1} = 0$.

Substituting the solution in the form $u_m^p = \lambda^p u_m$ into the finite-difference equation (10.109), we obtain:

$$(\lambda - 1)u_m - r(u_{m+1} - 2u_m + u_{m-1}) = 0, \quad r = \tau/h^2,$$

which immediately yields:

$$u_{m+1} - \frac{\lambda - 1 + 2r}{r} u_m + u_{m-1} = 0. \tag{10.130}$$

This is a second order homogeneous ordinary difference equation. To find the general solution of equation (10.130) we write down its algebraic characteristic equation:

$$q^2 - \frac{\lambda - 1 + 2r}{r} q + 1 = 0. \tag{10.131}$$

If q is a root of the quadratic equation (10.131), then the grid function

$$u_m^p = \lambda^p q^m$$

solves the homogeneous finite-difference equation:

$$\frac{u_m^{p+1} - u_m^p}{\tau} - \frac{u_{m+1}^p - 2u_m^p + u_{m-1}^p}{h^2} = 0. \tag{10.132}$$

If $|q| = 1$, i.e., if $q = e^{i\alpha}$, then the grid function

$$u_m^p = \lambda^p e^{i\alpha m},$$

which is obviously bounded for $m \longrightarrow +\infty$ and $m \longrightarrow -\infty$, yields the solution of equation (10.132), provided that

$$\lambda = 1 - 4r\sin^2\frac{\alpha}{2}, \quad 0 \le \alpha < 2\pi,$$

see Example 6 of Section 10.3.3. These $\lambda = \lambda(\alpha)$ fill the interval $1 - 4r \le \lambda \le 1$ of the real axis, see Figure 10.8 on page 357. Therefore, interval $1 - 4r \le \lambda \le 1$ is the spectrum of problem (10.109) for $a(\tilde{x},\tilde{t}) = 1$, i.e., of problem (10.132). This problem has no eigenvalues that lie outside of the interval $1 - 4r \le \lambda \le 1$, because if the characteristic equation (10.131) does not have a root q with $|q| = 1$, then equation (10.130) may have no solution bounded for $m \longrightarrow \pm\infty$.

If λ does not belong to the interval $1 - 4r \le \lambda \le 1$, then the absolute values of both roots of the characteristic equation (10.131) differ from one. Their product, however, is still equal to one, see equation (10.131). Consequently, the absolute value of the first root of equation (10.131) will be greater than one, while that of the second root will be less than one. Let us denote $|q_1(\lambda)| < 1$ and $|q_2(\lambda)| > 1$. The general solution of equation (10.130) has the form:

$$u_m = c_1 q_1^m + c_2 q_2^m,$$

where c_1 and c_2 are arbitrary constants. Accordingly, the general solution that satisfies additional constraint (10.126), i.e., that decays as $m \longrightarrow +\infty$, is written as

$$u_m = c_1 q_1^m, \quad |q_1| = |q_1(\lambda)| < 1,$$

and the general solution that satisfies additional constraint (10.128), i.e., that decays as $m \longrightarrow -\infty$, is given by

$$u_m = c_2 q_2^m, \quad |q_2| = |q_2(\lambda)| > 1.$$

To calculate the eigenvalues of problem (10.125), one needs to substitute $u_m = c_1 q_1^m$ into the left boundary condition $l_1 u_0 = 0$ and find those q_1 and λ, for which it is satisfied. The corresponding λ will yield the spectrum of problem (10.125). If, for example, $l_1 u_0 \equiv u_0 = 0$, then equality $c_1 q_1^0 = 0$ will not hold for any $c_1 \neq 0$, so that problem (10.125) has no eigenvalues. (Recall, the eigenfunction must be nontrivial.) If $l_1 u_0 \equiv u_1 - u_0 = 0$, then condition $c_1(q_1 - q_1^0) = c_1(q_1 - 1) = 0$ again yields $c_1 = 0$ because $q_1 \neq 1$, so that there are no eigenvalues either. If, however, $l_1 u_0 \equiv 2u_1 - u_0 = 0$, then condition $c_1(2q_1 - q_1^0) = c_1(2q_1 - 1) = 0$ is satisfied for $c_1 \neq 0$ and $q_1 = 1/2 < 1$. Substituting $q_1 = 1/2$ into the characteristic equation (10.131) we find that

$$\lambda = 1 + r\left(q_1 - 2 + \frac{1}{q_1}\right) = 1 + \frac{r}{2}.$$

This is the only eigenvalue of problem (10.125). It does not belong to the unit disk on the complex plane, and therefore the necessary stability condition is violated.

The eigenvalues of the auxiliary problem (10.127) are calculated analogously. They are found from the equation $l_2 u_M = 0$ when

$$u_m = c_2 q_2^m, \quad |q_2| = |q_2(\lambda)| > 1, \quad m = M, M-1, M-2, \ldots.$$

For stability, it is necessary that they all belong to the unit disk on the complex plane.

We can now provide more specific comments following Remark 10.1. When boundary condition $l_1 u_0 \equiv 2u_1 - u_0 = 0$ is employed in problem (10.125) then the solution that satisfies condition (10.126) is found in the form $u_m^p = \lambda^p q_1^m$, where $q_1 = 1/2$ and $\lambda = 1 + r/2 > 1$. This solution is only defined for $m \geq 0$. If, however, we were to extend it to the region $m < 0$, we would have obtained an unbounded function: $u_m^p \longrightarrow \infty$ as $m \longrightarrow -\infty$. In other words, the function $u_m^p = \lambda^p q_1^m$ cannot be used in the framework of the standard von Neumann analysis of problem (10.109).

This consideration leads to a very simple explanation of the mechanism of instability. The introduction of a boundary condition merely expands the pool of candidate functions, on which the instability may develop. In the pure von Neumann case, with no boundary conditions, we have only been monitoring the behavior of the harmonics $e^{i\alpha m}$ that are bounded on the entire grid $m = 0, \pm 1, \pm 2, \ldots$. With the boundary conditions present, we may need to include additional functions that are bounded on the semi-infinite grid, but unbounded if extended to the entire grid. These functions do not belong to the von Neumann category. If any of them brings along an unstable eigenvalue $|\lambda| > 1$, such as $\lambda = 1 + r/2$, then the overall scheme becomes unstable as well. We therefore re-iterate that if the scheme that approximates some Cauchy problem is supplemented by boundary conditions and thus transformed into an initial boundary value problem, then its stability will not be improved. In other words, if the Cauchy problem was stable, then the initial boundary value problem may either remain stable or become unstable. If, however, the Cauchy problem is unstable, then the initial boundary value problem will not become stable.

Our next example will be the familiar first order upwind scheme, but built on a finite grid: $x_m = mh$, $m = 0, 1, 2, \ldots, M$, $Mh = 1$, rather than on the infinite grid $m = 0, \pm 1, \pm 2, \ldots$:

$$
\begin{gathered}
\frac{u_m^{p+1} - u_m^p}{\tau} - \frac{u_{m+1}^p - u_m^p}{h} = 0, \\
m = 0, 1, 2, \ldots, M-1, \quad p = 0, 1, 2, \ldots, [T/\tau] - 1, \\
u_m^0 = \psi(x_m), \quad u_M^{p+1} = 0.
\end{gathered}
\tag{10.133}
$$

Scheme (10.133) approximates the following first order hyperbolic initial boundary value problem:

$$
\begin{gathered}
\frac{\partial u}{\partial t} - \frac{\partial u}{\partial x} = 0, \quad 0 \le x \le 1, \quad 0 < t \le T, \\
u(x, 0) = \psi(x), \quad u(1, t) = 0,
\end{gathered}
$$

on the interval $0 \le x \le 1$. To investigate stability of scheme (10.133), we will employ the Babenko-Gelfand criterion. In other words, we will need to analyze three auxiliary problems: A problem with no lateral boundaries:

$$
\begin{gathered}
\frac{u_m^{p+1} - u_m^p}{\tau} - \frac{u_{m+1}^p - u_m^p}{h} = 0, \\
m = 0, \pm 1, \pm 2, \ldots,
\end{gathered}
\tag{10.134}
$$

a problem with only the left boundary:

$$
\begin{gathered}
\frac{u_m^{p+1} - u_m^p}{\tau} - \frac{u_{m+1}^p - u_m^p}{h} = 0, \\
m = 0, 1, 2, \ldots,
\end{gathered}
\tag{10.135}
$$

and a problem with only the right boundary:

$$
\begin{gathered}
\frac{u_m^{p+1} - u_m^p}{\tau} - \frac{u_{m+1}^p - u_m^p}{h} = 0, \\
m = M-1, M-2, \ldots, 1, 0, -1, \ldots, \\
u_M^{p+1} = 0.
\end{gathered}
\tag{10.136}
$$

Note that we do not set any boundary condition at the left boundary in problem (10.135) as we did not have any in the original problem (10.133) either.

We will need to find spectra of the three transition operators from u^p to u^{p+1} that correspond to the three auxiliary problems (10.134), (10.135), and (10.136), respectively, and determine under what conditions will all the eigenvalues belong to the unit disk $|\lambda| \leq 1$ on the complex plane.

Substituting a solution of the type:

$$u_m^p = \lambda^p u_m$$

into the finite-difference equation:

$$u_m^{p+1} = (1-r)u_m^p + ru_{m+1}^p, \quad r = \tau/h,$$

that corresponds to all three problems (10.134), (10.135), and (10.136), we obtain the following first order ordinary difference equation for the eigenfunction $\{u_m\}$:

$$(\lambda - 1 + r)u_m - ru_{m+1} = 0. \tag{10.137}$$

Its characteristic equation:

$$\lambda - 1 + r - rq = 0 \tag{10.138}$$

yields the relation between λ and q, so that the general solution of equation (10.137) can be written as

$$u_m = cq^m = c\left(\frac{\lambda - 1 + r}{r}\right)^m, \quad m = 0, \pm 1, \pm 2, \ldots, \quad c = \text{const.}$$

When $|q| = 1$, i.e., when $q = e^{i\alpha}, 0 \leq \alpha < 2\pi$, we have:

$$\lambda = 1 - r + re^{i\alpha}.$$

The point $\lambda = \lambda(\alpha)$ sweeps the circle of radius r centered at the point $(1-r, 0)$ on the complex plane. This circle gives the spectrum, i.e., the full set of eigenvalues, of the first auxiliary problem (10.134), see Figure 10.11(a). It is clearly the same spectrum as we have discussed in Section 10.3.2, see formula (10.77) and Figure 10.5(a).

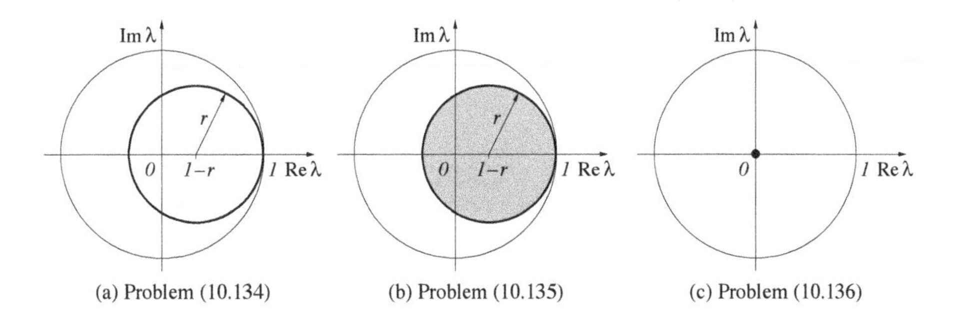

FIGURE 10.11: Spectra of auxiliary problems for the upwind scheme (10.133).

As far as the second auxiliary problem (10.135), we need to look for its non-trivial solutions that would decrease as $m \longrightarrow +\infty$, see formula (10.126). Such a solution, $u_m = c\lambda^p q^m$, obviously exists for any q: $|q| < 1$. The corresponding eigenvalues $\lambda = \lambda(q) = 1 - r + rq$ fill the interior of the disk bounded by the circle $\lambda = 1 - r + re^{i\alpha}$ on the complex plane, see Figure 10.11(b).

Solutions of the third auxiliary problem (10.136) that would satisfy (10.128), i.e., that would decay as $m \longrightarrow -\infty$, must obviously have the form: $u_m^p = c\lambda^p q^m$, where $|q| > 1$ and the relation between λ and q is, again, given by formula (10.138). The homogeneous boundary condition $u_M^{p+1} = 0$ of (10.136) implies that a non-trivial eigenfunction $u_m = cq^m$ may only exist when $\lambda = \lambda(q) = 0$, i.e., when $q = (r-1)/r$. The quantity q given by this expression may have its absolute value greater than one if either of the two inequalities holds:

$$\frac{r-1}{r} > 1 \quad \text{or} \quad \frac{r-1}{r} < -1.$$

The first inequality has no solutions. The solution to the second inequality is $r < 1/2$. Consequently, when $r < 1/2$, problem (10.136) has the eigenvalue $\lambda = 0$, see Figure 10.11(c).

In Figure 10.12, we are schematically showing the combined sets of all eigenvalues, i.e., combined spectra, for problems (10.134), (10.135), and (10.136) for the three different cases: $r < 1/2$, $1/2 < r < 1$, and $r > 1$.

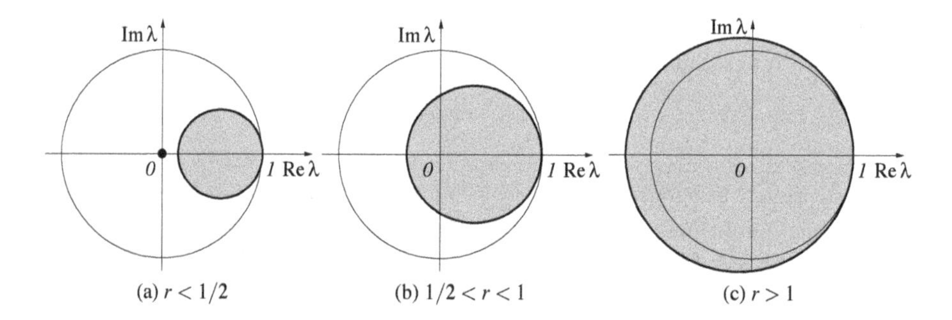

FIGURE 10.12: Combined spectra of auxiliary problems for scheme (10.133).

It is clear that the combined eigenvalues of all three auxiliary problems may only belong to the unit disk $|\lambda| \leq 1$ on the complex plane if $r \leq 1$. Therefore, condition $r \leq 1$ is necessary for stability of the difference initial boundary value problem (10.133).

Compared to the von Neumann stability condition of Section 10.3, the key distinction of the Babenko-Gelfand criterion is that it takes into account the boundary conditions for unsteady finite-difference equations on finite intervals. This criterion can also be generalized to systems of such equations. In this case, a scheme that may look perfectly natural and "benign" at a first glance, and that may, in particular, satisfy the von Neumann stability criterion, could still be unstable because of a poor approximation of the boundary conditions. Consequently, it is important to be able to build schemes that are free of this shortcoming.

In [GR87], the spectral criterion of Babenko and Gelfand is discussed from a more general standpoint, using a special new concept of *the spectrum of a family of operators* introduced by Godunov and Ryaben'kii. In this framework, one can rigorously prove that the Babenko-Gelfand criterion is necessary for stability, and also that when it holds, stability cannot be disrupted too severely. In the next section, we reproduce key elements of this analysis, while referring the reader to [GR87, Chapters 13 & 14] and [RM67, § 6.6 & 6.7] for further detail.

10.5.2 Spectra of the Families of Operators. The Godunov-Ryaben'kii Criterion

In this section, we briefly describe a rigorous approach, due to Godunov and Ryaben'kii, for studying stability of evolution-type finite-difference schemes on finite intervals. In other words, we study stability of the discrete approximations to initial boundary value problems for hyperbolic and parabolic partial differential equations. This material is more advanced, and can be skipped during the first reading.

As we have seen previously, for evolution finite-difference schemes the discrete solution $u^{(h)} = \{u_m^p\}$, which is defined on a two-dimensional space-time grid:

$$(x_m, t_p) \equiv (mh, p\tau), \quad m = 0, 1, \dots, M, \quad p = 0, 1, \dots, [T/\tau],$$

gets naturally split or "stratified" into a collection of one-dimensional grid functions $\{u^p\}$ defined for individual time layers t_p, $p = 0, 1, \dots, [T/\tau]$. For example, the first order upwind scheme:

$$\begin{gathered}
\frac{u_m^{p+1} - u_m^p}{\tau} - \frac{u_{m+1}^p - u_m^p}{h} = \varphi_m^p, \\
m = 0, 1, 2, \dots, M-1, \quad p = 0, 1, 2, \dots, [T/\tau] - 1, \\
u_m^0 = \psi_m, \quad u_M^{p+1} = \chi^{p+1},
\end{gathered} \tag{10.139}$$

for the initial boundary value problem:

$$\begin{gathered}
\frac{\partial u}{\partial t} - \frac{\partial u}{\partial x} = \varphi(x,t), \quad 0 \le x \le 1, \quad 0 < t \le T, \\
u(x,0) = \psi(x), \quad u(1,t) = \chi(t),
\end{gathered}$$

can be written as:

$$\begin{gathered}
u_m^{p+1} = \left[(1-r)u_m^p + r u_{m+1}^p\right] + \tau\varphi_m^p, \quad m = 0, 1, \dots, M-1, \\
u_M^{p+1} = \chi^{p+1}, \quad u_m^0 = \psi_m, \quad m = 0, 1, \dots, M,
\end{gathered} \tag{10.140}$$

where $r = \tau/h$. Form (10.140) suggests that the marching procedure for scheme (10.139) can be interpreted as consecutive computation of the grid functions:

$$u^0, u^1, \dots, u^p, \dots, u^{[T/\tau]},$$

defined on identical one-dimensional grids $m = 0, 1, \dots, M$ that can all be identified with one and the same grid. Accordingly, the functions u^p, $p = 0, 1, \dots, [T/\tau]$, can be considered elements of the linear space U'_h of functions $u = \{u_0, u_1, \dots, u_M\}$ defined on the grid $m = 0, 1, \dots, M$. We will equip this linear space with the norm, e.g.,

$$\|u\|_{U'_h} = \max_{0 \le m \le M} |u_m| \qquad \text{or} \qquad \|u\|_{U'_h} = \left[h \sum_m^M |u_m|^2 \right]^{\frac{1}{2}}.$$

We also recall that in the definitions of stability (Section 10.1.3) and convergence (Section 10.1.1) we employ the norm $\|u^{(h)}\|_{U_h}$ of the finite-difference solution $u^{(h)}$ on the entire two-dimensional grid. Hereafter, we will only be using norms that explicitly take into account the layered structure of the solution, namely, those that satisfy the equality:

$$\|u^{(h)}\|_{U_h} = \max_{0 \le p \le [T/\tau]} \|u^p\|_{U'_h}.$$

Having introduced the linear normed space U'_h, we can represent any evolution scheme, in particular, scheme (10.139), in *the canonical form:*

$$\begin{aligned} &u^{p+1} = \boldsymbol{R}_h u^p + \tau \rho^p, \\ &u^0 \ \text{is given.} \end{aligned} \tag{10.141}$$

In formula (10.141), $\boldsymbol{R}_h : U'_h \longmapsto U'_h$ is the transition operator between the consecutive time levels, and $\rho^p \in U'_h$. If we denote $v^{p+1} = \boldsymbol{R}_h u^p$, then formula (10.140) yields:

$$v_m^{p+1} = (1-r)u_m^p + r u_{m+1}^p, \quad m = 0, 1, \dots M-1. \tag{10.142a}$$

As far as the last component $m = M$ of the vector v^{p+1}, a certain flexibility exists in the definition of the operator $\boldsymbol{R}_h$ for scheme (10.139). For example, we can set:

$$v_M^{p+1} = u_M^p, \tag{10.142b}$$

which would also imply:

$$\rho_m^p = \varphi_m^p, \quad m = 0, 1, \dots M-1, \quad \text{and} \quad \rho_M^p = \frac{\chi^{p+1} - \chi^p}{\tau}, \tag{10.142c}$$

in order to satisfy the first equality of (10.141).

In general, the canonical form (10.141) for a given evolution scheme is not unique. For scheme (10.139), we could have chosen $v_M^{p+1} = 0$ instead of $v_M^{p+1} = u_M^p$ in formula (10.142b), which would have also implied $\rho_M^p = \frac{\chi^{p+1}}{\tau}$ in formula (10.142c). However, when building the operator $\boldsymbol{R}_h$, we need to make sure that the following rather natural conditions hold that require certain correlation between the norms in the spaces U_h' and F_h:

$$\begin{aligned} \|\rho^p\|_{U_h'} \le K_1 \|f^{(h)}\|_{F_h}, \quad p = 0, 1, \ldots, [T/\tau], \\ \|u^0\|_{U_h'} \le K_2 \|f^{(h)}\|_{F_h}. \end{aligned} \tag{10.143}$$

The constants K_1 and K_2 in inequalities (10.143) do not depend on h or on $f^{(h)}$. For scheme (10.139), if we define the norm in the space F_h as:

$$\|f^{(h)}\|_{F_h} = \max_{m,p} |\varphi_m^p| + \max_m |\psi_m| + \max_p \left| \frac{\chi^{p+1} - \chi^p}{\tau} \right|$$

and the norms of ρ^p and u_0 as $\|\rho^p\|_{U_h'} = \max_m |\rho_m^p|$ and $\|u^0\|_{U_h'} = \max_m |u_m^0|$, respectively, then conditions (10.143) obviously hold for the operator $\boldsymbol{R}_h$ and the source term ρ^p defined by formulae (10.142a)–(10.142c).

Let us now take an arbitrary $\hat{u}^0 \in U_h'$ and obtain $\hat{u}^1, \hat{u}^2, \ldots, \hat{u}^{[T/\tau]} \in U_h'$ using the recurrence formula $\hat{u}^{p+1} = \boldsymbol{R}_h \hat{u}^p$. Denote $\hat{u}^{(h)} = \{\hat{u}^p\}_{p=0}^{[T/\tau]}$ and evaluate $\hat{f}^{(h)} \stackrel{\text{def}}{=} \boldsymbol{L}_h \hat{u}^{(h)}$. Along with conditions (10.143), we will also require that

$$\|\hat{f}^{(h)}\|_{F_h} \le K_3 \|\hat{u}^0\|_{U_h'}, \tag{10.144}$$

where the constant K_3 does not depend on $\hat{u}^0 \in U_h'$ or on h.

In practice, inequalities (10.143) and (10.144) prove relatively non-restrictive.[7] These inequalities allow one to establish the following important theorem that provides a necessary and sufficient condition for stability in terms of the uniform boundedness of the powers of $\boldsymbol{R}_h$ with respect to the grid size h.

THEOREM 10.6

Assume that when reducing a given evolution scheme to the canonical form (10.141) the additional conditions (10.143) are satisfied. Then, for stability of the scheme in the linear sense (Definition 10.2) it is sufficient that

$$\|\boldsymbol{R}_h^p\| \le K, \quad p = 0, 1, \ldots, [T/\tau], \tag{10.145}$$

where the constant K in formula (10.145) does not depend on h. If the third additional condition (10.144) is met as well, then estimates (10.145) are also necessary for stability.

Theorem 10.6 is proven in [GR87, § 41].

For scheme (10.139), estimates (10.145) can be established directly, provided that $r \le 1$. Indeed, according to formula (10.142a), we have for $m = 0, 1, \ldots, M-1$:

$$|v_m^{p+1}| = |(1-r)u_m^p + ru_{m+1}^p| \le (1 - r + r) \max_m |u_m^p| = \|u^p\|_{U_h'},$$

[7]The first condition of (10.143) can, in fact, be further relaxed, see [GR87, § 42].

and according to formula (10.142b), we have for $m = M$:

$$|v_M^{p+1}| = |u_M^p| \leq \max_m |u_m^p| = \|u^p\|_{U_h'}.$$

Consequently,

$$\|\boldsymbol{R}_h u^p\|_{U_h'} = \|v^{p+1}\|_{U_h'} = \max_m |v_m^{p+1}| \leq \max_m |u_m^p| = \|u^p\|_{U_h'},$$

which means that $\|\boldsymbol{R}_h\| \leq 1$. Therefore, $\|\boldsymbol{R}_h^p\| \leq \|\boldsymbol{R}_h\|^p \leq 1$, and according to Theorem 10.6, scheme (10.139) is stable.

REMARK 10.2 We have already seen previously that the notion of stability for a finite-difference scheme can be reformulated as boundedness of powers for a family of matrices. Namely, in Section 10.3.6 we discussed stability of finite-difference Cauchy problems for systems of equations with constant coefficients (as opposed to scalar equations). We saw that finite-difference stability (Definition 10.2) was equivalent to stability of the corresponding family of amplification matrices. The latter, in turn, is defined as boundedness of their powers, and the Kreiss matrix theorem (Theorem 10.4) provides necessary and sufficient conditions for this property to hold.

Transition operators $\boldsymbol{R}_h$ can also be interpreted as matrices that operate on vectors from the space U_h'. In this perspective, inequality (10.145) implies uniform boundedness of all powers or stability of this family of operators (matrices). There is, however, a fundamental difference between the considerations of this section and those of Section 10.3.6. The amplification matrices that appear in the context of the Kreiss matrix theorem (Theorem 10.4) are parameterized by the frequency α and possibly the grid size h. Yet the dimension of all these matrices remains fixed and equal to the dimension of the original system, regardless of the grid size. In contradistinction to that, the dimension of the matrices $\boldsymbol{R}_h$ is inversely proportional to the grid size h, i.e., it grows with no bound as $h \longrightarrow 0$. Therefore, estimate (10.145) actually goes beyond the notion of stability for families of matrices of a fixed dimension (Section 10.3.6), as it implies stability (uniform bound on powers) for a family of matrices of increasing dimension. □

As condition (10.145) is equivalent to stability according to Theorem 10.6, then to investigate stability we need to see whether inequalities (10.145) hold. Let λ_h be an eigenvalue of the operator $\boldsymbol{R}_h$, and let $v^{(h)}$ be the corresponding eigenvector so that $\boldsymbol{R}_h v^{(h)} = \lambda_h v^{(h)}$. Then,

$$\|\boldsymbol{R}_h^p\| \|v^{(h)}\| \geq \|\boldsymbol{R}_h^p v^{(h)}\| = |\lambda_h|^p \|v^{(h)}\|$$

and consequently $\|\boldsymbol{R}_h^p\| \geq |\lambda_h|^p$. Since λ_h is an arbitrary eigenvalue, we have:

$$\|\boldsymbol{R}_h^p\| \geq [\max |\lambda_h|]^p, \quad p = 0, 1, \ldots [T/\tau],$$

where $[\max|\lambda_h|]$ is the largest eigenvalue of $\boldsymbol{R}_h$ by modulus. Hence, for the estimate (10.145) to hold, it is necessary that all eigenvalues λ of the transition operator $\boldsymbol{R}_h$ belong to the following disk on the complex plane:

$$|\lambda| \leq 1 + c_1\tau, \tag{10.146}$$

where the constant c_1 does not depend on the grid size h (or τ). It means that inequality (10.146) must hold *with one and the same constant* c_1 for any given transition operator from the family $\{\boldsymbol{R}_h\}$ parameterized by h.

Inequality (10.146) is known as *the spectral necessary condition* for the uniform boundedness of the powers $\|\boldsymbol{R}_h^p\|$. It is called spectral because as long as the operators $\boldsymbol{R}_h$ can be identified with matrices of finite dimension, the eigenvalues of those matrices yield the spectra of the operators. This spectral condition is also closely related to the von Neumann spectral stability criterion for finite-difference Cauchy problems on infinite grids that we have studied in Section 10.3, see formula (10.81).

Indeed, instead of the finite-difference initial boundary value problem (10.139), consider a Cauchy problem on the grid that is infinite in space:

$$\begin{gathered}\frac{u_m^{p+1} - u_m^p}{\tau} - \frac{u_{m+1}^p - u_m^p}{h} = \varphi_m^p, \\ u_m^0 = \psi_m, \\ m = 0, \pm 1, \pm 2, \ldots, \quad p = 0, 1, 2, \ldots, [T/\tau] - 1.\end{gathered} \tag{10.147}$$

The von Neumann analysis of Section 10.3.2 has shown that for stability it is necessary that $r = \tau/h \leq 1$. To apply the spectral criterion (10.146), we first reduce scheme (10.147) to the canonical form (10.141). The operator $\boldsymbol{R}_h : U_h' \longmapsto U_h'$, $\boldsymbol{R}_h u^p = v^{p+1}$, and the source term ρ^p are then given by [cf. formulae (10.142)]:

$$\begin{gathered}v_m^{p+1} = (1-r)u_m^p + ru_{m+1}^p, \quad \rho_m^p = \varphi_m^p, \\ m = 0, \pm 1, \pm 2, \ldots.\end{gathered}$$

The space U_h' contains infinite sequences $u = \{\ldots, u_{-m}, \ldots, u_{-1}, u_0, u_1, \ldots, u_m, \ldots\}$. We can supplement this space with the C norm: $\|u\| = \sup_m |u_m|$. The grid functions $u = \{u_m\} = \{e^{i\alpha m}\}$ then belong to the space U_h' for all $\alpha \in [0, 2\pi)$ and provide eigenfunctions of the transition operator:

$$\boldsymbol{R}_h u = (1-r)e^{i\alpha m} + re^{i\alpha(m+1)} = [(1-r) + re^{i\alpha}]e^{i\alpha m} = \lambda(\alpha)u,$$

where the eigenvalues are given by:

$$\lambda(\alpha) = (1-r) + re^{i\alpha}. \tag{10.148}$$

According to the spectral condition of stability (10.146), all eigenvalues must satisfy the inequality: $|\lambda(\alpha)| \leq 1 + c_1\tau$, which is the same as the von Neumann condition (10.78). As the eigenvalues (10.148) do not explicitly depend on the grid size, the spectral condition (10.146) reduces here to $|\lambda(\alpha)| \leq 1$, cf. formula (10.79).

Let us also recall that as shown in Section 10.3.5, the von Neumann condition is not only necessary, but also sufficient for the l_2 stability of the two-layer (one-step) scalar finite-difference Cauchy problems, see formula (10.95). If, however, the space U_h' is equipped with the l_2 norm: $\|u\| = \left[h\sum_{m=-\infty}^{\infty}|u_m|^2\right]^{1/2}$ (as opposed to the C norm), then the functions $\{e^{i\alpha m}\}$ no longer belong to this space, and therefore may no longer be the eigenfunctions of $\boldsymbol{R}_h$. Nonetheless, we can show that the points $\lambda(\alpha)$ of (10.148) still belong to the spectrum of the operator $\boldsymbol{R}_h$, provided that the latter is defined as traditionally done in functional analysis, see Definition 10.7 on page 393.[8] Consequently, if we interpret λ in formula (10.146) as all points of the spectrum rather than just the eigenvalues of the operator $\boldsymbol{R}_h$, then the spectral condition (10.146) also becomes sufficient for the l_2 stability of the Cauchy problems (10.95) on an infinite grid $m = 0, \pm 1, \pm 2, \ldots$.

Returning now to the difference equations on finite intervals and grids (as opposed to Cauchy problems), we first notice that one can most easily verify estimates (10.145) when the matrices of all operators $\boldsymbol{R}_h$ happen to be normal: $\boldsymbol{R}_h\boldsymbol{R}_h^* = \boldsymbol{R}_h^*\boldsymbol{R}_h$. Indeed, in this case there is an orthonormal basis in the space U_h' composed of the eigenvectors of the matrix $\boldsymbol{R}_h$, see, e.g., [HJ85, Chapter 2]. Using expansion with respect to this basis, one can show that the spectral condition (10.146) is necessary and sufficient for the l_2 stability of an evolution scheme with normal operators $\boldsymbol{R}_h$ on a finite interval. More precisely, the following theorem holds.

THEOREM 10.7

Let the operators $\boldsymbol{R}_h$ in the canonical form (10.141) be normal, and let them all be uniformly bounded with respect to the grid: $\|R_h\| \le c_2$, where c_2 does not depend on h. Let also all norms be chosen in the sense of l_2. Then, for the estimates (10.145) to hold, it is necessary and sufficient that the inequalities be satisfied:

$$\max_n |\lambda_n| \le 1 + c_1\tau, \quad c_1 = \text{const}, \tag{10.149}$$

where $\lambda_1, \lambda_2, \ldots, \lambda_N$ are eigenvalues of the matrix $\boldsymbol{R}_h$ and the constant c_1 in formula (10.149) does not depend on h.

One implication of Theorem 10.7, the necessity, coincides with the previous necessary spectral condition for stability that we have justified on page 389. The other implication, the sufficiency, is to be proven in Exercise 5 of this section. A full proof of Theorem 10.7 can be found, e.g., in [GR87, §43].

Unfortunately, in many practical situations the operators (matrices) $\boldsymbol{R}_h$ in the canonical form (10.141) are not normal. Then, the spectral condition (10.146) still remains necessary for stability. Moreover, we have just seen that in the special case of two-layer scalar constant-coefficient Cauchy problems it is also sufficient for stability and that sufficiency takes place regardless of whether or not $\boldsymbol{R}_h$ has a full sys-

[8] In general, the points $\lambda = \lambda(\alpha, h)$ given by Definition 10.3 on page 351 will be a part of the spectrum in the sense of its classical definition, see Definition 10.7 on page 393.

tem of orthonormal eigenfunctions. However, for general finite-difference problems on finite intervals the spectral condition (10.146) becomes pretty far detached from sufficiency and provides no adequate criterion for uniform boundedness of $\|\boldsymbol{R}_h^p\|$.

For instance, the matrix of the transition operator $\boldsymbol{R}_h$ defined by formulae (10.142a) and (10.142b) is given by:

$$\boldsymbol{R}_h = \begin{bmatrix} 1-r & r & 0 \cdots & 0 & 0 \\ 0 & 1-r & r \cdots & 0 & 0 \\ \vdots & \vdots & \vdots \ddots & \vdots & \vdots \\ 0 & 0 & 0 \cdots & 1-r & r \\ 0 & 0 & 0 \cdots & 0 & 1 \end{bmatrix}. \tag{10.150}$$

Its spectrum consists of the eigenvalues $\lambda = 1$ and $\lambda = 1 - r$ and as such, does not depend on h (or on τ). Consequently, for any $h > 0$ the spectrum of the operator $\boldsymbol{R}_h$ consists of only these two numbers: $\lambda = 1$ and $\lambda = 1 - r$. This spectrum belongs to the unit disk $|\lambda| \leq 1$ when $0 \leq r \leq 2$. However, for $1 < r \leq 2$, scheme (10.139) violates the Courant, Friedrichs, and Lewy condition necessary for stability, and hence, there may be no stability $\|\boldsymbol{R}_h^p\| \leq K$ for any reasonable choice of norms.

Thus, we have seen that the spectral condition (10.146) that employs the eigenvalues of the operators $\boldsymbol{R}_h$ and that is necessary for the uniform boundedness $\|\boldsymbol{R}_h^p\| \leq K$ appears too rough in the case of non-normal matrices. For example, it fails to detect the instability of scheme (10.139) for $1 < r \leq 2$.

To refine the spectral condition we will introduce a new concept. Assume, as before, that the operator $\boldsymbol{R}_h$ is defined on a normed linear space U_h'. We will denote by $\{\boldsymbol{R}_h\}$ the entire family of operators $\boldsymbol{R}_h$ for all legitimate values of the parameter h that characterizes the grid.[9]

DEFINITION 10.6 *A complex number λ is said to belong to* the spectrum of the family of operators $\{\boldsymbol{R}_h\}$ *if for any $h_0 > 0$ and $\varepsilon > 0$ one can always find such a value of h, $h < h_0$, that the inequality*

$$\|\boldsymbol{R}_h u - \lambda u\|_{U_h'} < \varepsilon \|u\|_{U_h'}$$

will have a solution $u \in U_h'$. The set of all such λ will be called the spectrum of the family of operators $\{\boldsymbol{R}_h\}$.

The following theorem employs the concept of the spectrum of a family of operators from Definition 10.6 and provides *a key necessary condition for stability.*

THEOREM 10.8 (Godunov-Ryaben'kii)
If even one point λ_0 of the spectrum of the family of operators $\{\boldsymbol{R}_h\}$ lies outside the unit disk on the complex plane, i.e., $|\lambda_0| > 1$, then there is no

[9]By the very nature of finite-difference schemes, h may assume arbitrarily small positive values.

common constant K such that the inequality

$$\|\boldsymbol{R}_h^p\| \leq K$$

will hold for all $h > 0$ and all integer values of p from 0 till some $p = p_0(h)$, where $p_0(h) \longrightarrow \infty$ as $h \longrightarrow 0$.

PROOF Let us first assume that no such numbers $h_0 > 0$ and $c > 0$ exist that for all $h < h_0$ the following estimate holds:

$$\|\boldsymbol{R}_h\| \leq c. \tag{10.151}$$

This assumption means that there is no uniform bound on the operators $\boldsymbol{R}_h$ themselves. As such, there may be no bound on the powers $\boldsymbol{R}_h^p$ either. Consequently, we only need to consider the case when there are such $h_0 > 0$ and $c > 0$ that for all $h < h_0$ inequality (10.151) is satisfied.

Let $|\lambda_0| = 1 + \delta$, where λ_0 is the point of the spectrum for which $|\lambda_0| > 1$. Take an arbitrary $K > 0$ and choose p and ε so that:

$$(1+\delta)^p > 2K,$$
$$1 - (1 + c + c^2 + \ldots + c^{p-1})\varepsilon > \frac{1}{2}.$$

According to Definition 10.6, one can find arbitrarily small h, for which there is a vector $u \in U_h'$ that solves the inequality:

$$\|\boldsymbol{R}_h u - \lambda_0 u\|_{U_h'} < \varepsilon \|u\|_{U_h'}.$$

Let u be the solution, and denote:

$$\boldsymbol{R}_h u = \lambda_0 u + z.$$

It is clear that $\|z\| < \varepsilon \|u\|$. Moreover, it is easy to see that

$$\boldsymbol{R}_h^p u = \lambda_0^p u + (\lambda_0^{p-1} z + \lambda_0^{p-2} \boldsymbol{R}_h z + \ldots + \boldsymbol{R}_h^{p-1} z).$$

As $|\lambda_0| > 1$, we have:

$$\|\lambda_0^{p-1} z + \lambda_0^{p-2} \boldsymbol{R}_h z + \ldots + \boldsymbol{R}_h^{p-1} z\| < |\lambda_0|^p (1 + \|\boldsymbol{R}_h\| + \|\boldsymbol{R}_h^2\| + \ldots + \|\boldsymbol{R}_h^{p-1}\|)\varepsilon \|u\|,$$

and consequently,

$$\begin{aligned}\|\boldsymbol{R}_h^p u\| &> |\lambda_0|^p [1 - \varepsilon(1 + c + c^2 + \ldots + c^{p-1})] \|u\| \\ &> (1+\delta)^p \frac{1}{2} \|u\| > 2K \frac{1}{2} \|u\| = K \|u\|.\end{aligned}$$

In doing so, the value of h can always be taken sufficiently small so that to ensure $p < p_0(h)$.

Since the value of K has been chosen arbitrarily, we have essentially proven that *for the estimate* $\|\boldsymbol{R}_h^p\| < K$ *to hold, it is necessary that all points of the spectrum of the family* $\{\boldsymbol{R}_h\}$ *belong to the unit disk* $|\lambda| \leq 1$ *on the complex plane.* □

Next recall that the following definition of the spectrum of an operator $\boldsymbol{R}_h$ (for a fixed h) is given in functional analysis.

DEFINITION 10.7 *A complex number* λ *is said to belong to the spectrum of the operator* $\boldsymbol{R}_h : U_h' \longmapsto U_h'$ *if for any* $\varepsilon > 0$ *the inequality*

$$\|\boldsymbol{R}_h u - \lambda u\|_{U_h'} < \varepsilon \|u\|_{U_h'}$$

has a solution $u \in U_h'$. *The set of all such* λ *is called the spectrum* $\boldsymbol{R}_h$.

At first glance, comparison of the Definitions 10.6 and 10.7 may lead one to thinking that the spectrum of the family of operators $\{\boldsymbol{R}_h\}$ consists of all those and only those points on the complex plane that are obtained by passing to the limit $h \longrightarrow 0$ from the points of the spectrum of $\boldsymbol{R}_h$, when h approaches zero along all possible sub-sequences. However, this assumption is, generally speaking, not correct.

Consider, for example, the operator $\boldsymbol{R}_h : U_h' \longmapsto U_h'$ defined by formulae (10.142a) and (10.142b). It is described by the matrix (10.150) and operates in the $M+1$-dimensional linear space U_h', where $M = 1/h$. The spectrum of a matrix consists of its eigenvalues, and the eigenvalues of the matrix (10.150) are $\lambda = 1$ and $\lambda = 1 - r$. These eigenvalues do not depend on h (or on τ) and consequently, the spectrum of the operator $\boldsymbol{R}_h$ consists of only two points, $\lambda = 1$ and $\lambda = 1 - r$, for any $h > 0$. As, however, we are going to see (pages 397-402), the spectrum of the family of operators $\{\boldsymbol{R}_h\}$ contains not only these two points, but also all points of the disk $|\lambda - 1 + r| \leq r$ of radius r centered at the point $(1 - r, 0)$ on the complex plane, see Figure 10.12 on page 384. When $r \leq 1$, the spectrum of the family of operators $\{\boldsymbol{R}_h\}$ belongs to the unit disk $|\lambda| \leq 1$, see Figures 10.12(a) and 10.12(b). However, when $r > 1$, this necessary spectral condition of stability does not hold, see Figure 10.12(c), and the inequality $\|\boldsymbol{R}_h^p\| \leq K$ can not be satisfied uniformly with respect to h.

Before we accurately compute the spectrum of the family of operators $\{\boldsymbol{R}_h\}$ given by formula (10.150), let us qualitatively analyze the behavior of the powers $\|\boldsymbol{R}_h^p\|$ for $r > 1$ and also show that the necessary stability criterion given by Theorem 10.8 is, in fact, rather close to sufficient.

We first notice that for any $h > 0$ there is only one eigenvalue of the matrix $\boldsymbol{R}_h$ that has unit modulus: $\lambda = 1$, and that the similarity transformation $\boldsymbol{S}_h^{-1}\boldsymbol{R}_h\boldsymbol{S}_h$, where

$$\boldsymbol{S}_h = \begin{bmatrix} 1 & 0 & 0 & \cdots & 0 & 1 \\ 0 & 1 & 0 & \cdots & 0 & 1 \\ \vdots & \vdots & \vdots & \ddots & \vdots & \vdots \\ 0 & 0 & 0 & \cdots & 1 & 1 \\ 0 & 0 & 0 & \cdots & 0 & 1 \end{bmatrix} \quad \text{and} \quad \boldsymbol{S}_h^{-1} = \begin{bmatrix} 1 & 0 & 0 & \cdots & 0 & -1 \\ 0 & 1 & 0 & \cdots & 0 & -1 \\ \vdots & \vdots & \vdots & \ddots & \vdots & \vdots \\ 0 & 0 & 0 & \cdots & 1 & -1 \\ 0 & 0 & 0 & \cdots & 0 & 1 \end{bmatrix},$$

reduces this matrix to the block-diagonal form:

$$S_h^{-1} R_h S_h = \begin{bmatrix} 1-r & r & 0 & \cdots & 0 & 0 \\ 0 & 1-r & r & \cdots & 0 & 0 \\ \vdots & \vdots & \vdots & \ddots & \vdots & \vdots \\ 0 & 0 & 0 & \cdots & 1-r & 0 \\ 0 & 0 & 0 & \cdots & 0 & 1 \end{bmatrix} \equiv B_h.$$

When $1 < r < 2$, we have $|1-r| < 1$ for the magnitude of the diagonal entry $1-r$. Then, it is possible to prove that $B_h^p \longrightarrow \operatorname{diag}\{0,0,\dots,0,1\}$ as $p \longrightarrow \infty$ (see Theorem 6.2 on page 178). In other words, the limiting matrix of B_h^p for the powers p approaching infinity has only one non-zero entry equal to one at the lower right corner. Consequently,

$$\lim_{p\to\infty} R_h^p = \begin{bmatrix} 0 & 0 & 0 & \cdots & 0 & 1 \\ 0 & 0 & 0 & \cdots & 0 & 1 \\ \vdots & \vdots & \vdots & \ddots & \vdots & \vdots \\ 0 & 0 & 0 & \cdots & 0 & 1 \\ 0 & 0 & 0 & \cdots & 0 & 1 \end{bmatrix},$$

and as such, $\lim_{p\to\infty} \|R_h^p\| = 1$. We can therefore see that regardless of the value of h, the norms of the powers of the transition operator R_h approach one and the same finite limit. In other words, we can write $\lim_{p\tau\to\infty} \|R_h^p\| = 1$, and this "benign" asymptotic behavior of $\|R_h^p\|$ for large $p\tau$ is indeed determined by the eigenvalues $\lambda = 1-r$ and $\lambda = 1$ that belong to the unit disk.

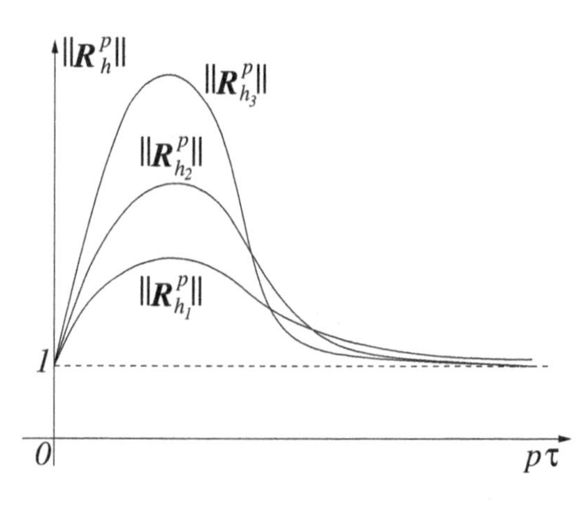

FIGURE 10.13: Schematic behavior of the powers $\|R_h^p\|$ for $1 < r < 2$ and $h_3 > h_2 > h_1$.

The fact that the spectrum of the family of operators $\{R_h\}$ does not belong to the unit disk for $r > 1$ manifests itself in the behavior of $\|R_h^p\|$ for $h \longrightarrow 0$ and for moderate (not so large) values of $p\tau$. The maximum value of $\|R_h^p\|$ on the interval $0 < p\tau < T$, where T is an arbitrary positive constant, will rapidly grow as h decreases, see Figure 10.13. This is precisely what leads to the instability, whereas the behavior of $\|R_h^p\|$ as $p\tau \longrightarrow \infty$, which is related to the spectrum of each individual operator R_h, is not important from the standpoint of stability.

Let us also emphasize that even though from a technical point of view Theorem 10.8 only provides a necessary condition for stability, this condition, *is, in fact, not so distant from sufficient*. More precisely, the following theorem holds.

THEOREM 10.9
Let the operators $\boldsymbol{R}_h$ be defined on a linear normed space U'_h for each $h > 0$, and assume that they are uniformly bounded with respect to h:

$$\|\boldsymbol{R}_h\| \leq c. \tag{10.152}$$

Let also the spectrum of the family of operators $\{\boldsymbol{R}_h\}$ belong to the unit disk on the complex plane: $|\lambda| \leq 1$.

Then for any $\eta > 0$, the norms of the powers of operators $\boldsymbol{R}_h$ satisfy the estimate:

$$\|R_h^p\| \leq A(\eta)(1+\eta)^p, \tag{10.153}$$

where $A = A(\eta)$ may depend on η, but does not depend on the grid size h.

Theorem 10.9 means that having the spectrum of the family of operators $\{\boldsymbol{R}_h\}$ lie inside the unit disk is *not only necessary for stability, but it also guarantees us from a catastrophic instability.* Indeed, if the conditions of Theorem 10.9 hold, then the quantity $\max\limits_{1 \leq p \leq [T/\tau]} \|\boldsymbol{R}_h^p\|$ either remains bounded as $h \longrightarrow 0$ or increases, but *slower than any exponential function,* i.e., slower than any $(1+\eta)^{[T/\tau]}$, where $\eta > 0$ may be arbitrarily small.

PROOF Let us first show that if the spectrum of the family of operators $\{\boldsymbol{R}_h\}$ belongs to the disk $|\lambda| \leq \rho$, then for any given λ that satisfies the inequality $|\lambda| \geq \rho + \eta$, $\eta > 0$, there are the numbers $A = A(\eta)$ and $h_0 > 0$ such that $\forall h < h_0$ and $\forall u \in U'_h$, $u \neq 0$, the following estimate holds:

$$\|\boldsymbol{R}_h u - \lambda u\|_{U'_h} > \frac{\rho+\eta}{A(\eta)} \|u\|_{U'_h}. \tag{10.154}$$

Assume the opposite. Then there exist: $\eta > 0$; a sequence of real numbers $h_k > 0$, $h_k \longrightarrow 0$ as $k \longrightarrow \infty$; a sequence of complex numbers λ_k, $|\lambda_k| > \rho + \eta$; and a sequence of vectors $u_{h_k} \in U'_{h_k}$ such that:

$$\|\boldsymbol{R}_{h_k} u_{h_k} - \lambda_k u_{h_k}\|_{U'_{h_k}} < \frac{\rho+\eta}{k} \|u_{h_k}\|_{U'_{h_k}}. \tag{10.155}$$

For sufficiently large values of k, for which $\frac{\rho+\eta}{k} < 1$, the numbers λ_k will not lie outside the disk $|\lambda| \leq c+1$ by virtue of estimate (10.152), because outside this disk we have:

$$\|\boldsymbol{R}_{h_k} u_{h_k} - \lambda u_{h_k}\|_{U'_{h_k}} \geq \left(|\lambda| - \|\boldsymbol{R}_{h_k}\|\right) \|u_{h_k}\|_{U'_{h_k}} \geq \|u_{h_k}\|_{U'_{h_k}}.$$

Therefore, the sequence of complex numbers λ_k is bounded and as such, has a limit point $\tilde{\lambda}$, $|\tilde{\lambda}| \geq \rho + \eta$. Using the triangle inequality, we can write: $\|\boldsymbol{R}_{h_k} u_{h_k} - \lambda_k u_{h_k}\|_{U'_{h_k}} \geq \|\boldsymbol{R}_{h_k} u_{h_k} - \tilde{\lambda} u_{h_k}\|_{U'_{h_k}} - |\lambda_k - \tilde{\lambda}| \|u_{h_k}\|_{U'_{h_k}}$. Substituting into inequality (10.155), we obtain:

$$\|\boldsymbol{R}_{h_k} u_{h_k} - \tilde{\lambda} u_{h_k}\|_{U'_{h_k}} < \underbrace{\left[\frac{\rho+\eta}{k} + |\lambda_k - \tilde{\lambda}|\right]}_{\varepsilon} \|u\|_{U'_{h_k}}.$$

Therefore, according to Definition 10.6 the point $\tilde{\lambda}$ belongs to the spectrum of the family of operators $\{\boldsymbol{R}_h\}$. This contradicts the previous assumption that the spectrum belongs to the disk $|\lambda| \leq \rho$.

Now let $\boldsymbol{R}$ be a linear operator on a finite-dimensional normed space U, $\boldsymbol{R} : U \longmapsto U$. Assume that for any complex λ, $|\lambda| \geq \gamma > 0$, and any $u \in U$ the following inequality holds for some $a = \text{const} > 0$:

$$\|\boldsymbol{R}u - \lambda u\| \geq a\|u\|. \tag{10.156}$$

Then,

$$\|\boldsymbol{R}^p\| \leq \frac{\gamma^{p+1}}{a}, \quad p = 1, 2, \ldots. \tag{10.157}$$

Inequality (10.157) follows from the relation:

$$\boldsymbol{R}^p = -\frac{1}{2\pi i} \oint\limits_{|\lambda|=\gamma} \lambda^p (\boldsymbol{R} - \lambda \boldsymbol{I})^{-1} d\lambda \tag{10.158}$$

combined with estimate (10.156), because the latter implies that $\|(\boldsymbol{R} - \lambda \boldsymbol{I})^{-1}\| \leq \frac{1}{a}$. To prove estimate (10.153), we set $\boldsymbol{R} = \boldsymbol{R}_h$, $\rho = 1$ so that $|\lambda| \geq 1 + \eta = \gamma$, and use (10.154) instead of (10.156). Then estimate (10.157) coincides with (10.153).

It only remains to justify equality (10.158). For that purpose, we will use an argument similar to that used when proving Theorem 6.2. Define

$$u^{p+1} = \boldsymbol{R}u^p \quad \text{and} \quad w(\lambda) = \sum_{p=0}^{\infty} \frac{u^p}{\lambda^p},$$

where the series converges uniformly at least for all $\lambda \in \mathbb{C}$, $|\lambda| > c$, see formula (10.152). Multiply the equality $u^{p+1} = \boldsymbol{R}u^p$ by λ^{-p} and take the sum with respect to p from $p = 0$ to $p = \infty$. This yields:

$$\lambda w(\lambda) - \lambda u^0 = \boldsymbol{R}w(\lambda),$$

or alternatively,

$$(\boldsymbol{R} - \lambda \boldsymbol{I}) w(\lambda) = -\lambda u^0, \qquad w(\lambda) = -\lambda (\boldsymbol{R} - \lambda \boldsymbol{I})^{-1} u^0.$$

From the definition of $w(\lambda)$ it is easy to see that $-u^p$ is the residue of the vector-function $\lambda^{p-1}w(\lambda)$i at infinity:

$$u^p = \frac{1}{2\pi i}\oint\limits_{|\lambda|=\gamma} \lambda^{p-1}w(\lambda)d\lambda = -\frac{1}{2\pi i}\oint\limits_{|\lambda|=\gamma} \lambda^{p}(\boldsymbol{R}-\lambda\boldsymbol{I})^{-1}u^0 d\lambda.$$

As $u^p = \boldsymbol{R}^p u^0$, the last equality is equivalent to (10.158). □

Altogether, we have seen that the question of stability for evolution finite-difference schemes on finite intervals reduces to studying the spectra of the families of the corresponding transition operators $\{\boldsymbol{R}_h\}$. More precisely, we need to find out whether the spectrum for a given family of operators $\{\boldsymbol{R}_h\}$ belongs to the unit disk $|\lambda| \le 1$. *If it does, then the scheme is either stable or, in the worst case scenario, it may only develop a mild instability.*

Let us now show how we can actually calculate the spectrum of a family of operators. To demonstrate the approach, we will exploit the previously introduced example (10.142a), (10.142b). *It turns out that the algorithm for computing the spectrum of the family of operators $\{\boldsymbol{R}_h\}$ coincides with the Babenko-Gelfand procedure described in Section 10.5.1.* Namely, we need to introduce three auxiliary operators: $\overleftrightarrow{\boldsymbol{R}}$, $\overrightarrow{\boldsymbol{R}}$, and $\overleftarrow{\boldsymbol{R}}$. The operator $\overleftrightarrow{\boldsymbol{R}}$, $v = \overleftrightarrow{\boldsymbol{R}}u$, is defined on the linear space of bounded grid functions $u = \{\dots,u_{-1},u_0,u_1,\dots\}$ according to the formula:

$$v_m = (1-r)u_m + ru_{m+1}, \quad m = 0,\pm1,\pm2,\dots, \tag{10.159}$$

which is obtained from (10.142a), (10.142b) by removing both boundaries. The operator $\overrightarrow{\boldsymbol{R}}$ is defined on the linear space of functions $u = \{u_0,u_1,\dots,u_m,\dots\}$ that vanish at infinity: $|u_m| \longrightarrow 0$ as $m \longrightarrow +\infty$. It is given by the formula:

$$v_m = (1-r)u_m + ru_{m+1}, \quad m = 0,1,2,\dots, \tag{10.160}$$

which is obtained from (10.142a), (10.142b) by removing the right boundary. Finally, the operator $\overleftarrow{\boldsymbol{R}}$ is defined on the linear space of functions $\{\dots,u_m,\dots,u_0,\dots,u_{M-1},u_M\}$ that satisfy: $|u_m| \longrightarrow 0$ as $m \longrightarrow -\infty$. It is given by the formula:

$$\begin{aligned} v_m &= (1-r)u_m + ru_{m+1}, \quad m = \dots,-1,0,1,\dots,M-1, \\ v_M &= u_M, \end{aligned} \tag{10.161}$$

which is obtained from (10.142a), (10.142b) by removing the left boundary. Note that the spaces of functions for the operators $\overrightarrow{\boldsymbol{R}}$ and $\overleftarrow{\boldsymbol{R}}$ are defined on semi-infinite grids $m = 0,1,2,\dots$ and $m = \dots,-1,0,1,\dots,M$, respectively.

None of the operators $\overleftrightarrow{\boldsymbol{R}}$, $\overrightarrow{\boldsymbol{R}}$, or $\overleftarrow{\boldsymbol{R}}$ depend on h. We will show that *the combination of all eigenvalues of these three auxiliary operators yields the spectrum of the family of operators $\{\boldsymbol{R}_h\}$*. In Section 10.5.1, we have, in fact, already computed the eigenvalues of the operators $\overleftrightarrow{\boldsymbol{R}}$ and $\overrightarrow{\boldsymbol{R}}$. For the operator $\overleftrightarrow{\boldsymbol{R}}$, the eigenvalues are all those and only those complex numbers λ, for which the equation $\overleftrightarrow{\boldsymbol{R}}u = \lambda u$ has a bounded solution $u = \{u_m\}$, $m = 0, \pm 1, \pm 2, \dots$. According to (10.159), this equation can be written as:

$$(1-r-\lambda)u_m + ru_{m+1} = 0, \quad m = 0, \pm 1, \pm 2, \dots,$$

and its general solution is $u_m = cq^m$, where q is a root of the characteristic equation: $(1-r-\lambda)+rq = 0$. This solution is bounded as $|m| \longrightarrow \infty$ if and only if $|q| = 1$, i.e., $q = e^{i\alpha}$, $\alpha \in [0, 2\pi)$. The corresponding eigenvalues are given by:

$$\lambda = 1 - r + rq = 1 - r + re^{i\alpha}, \quad \alpha \in [0, 2\pi).$$

The curve $\lambda = \lambda(\alpha)$ is a circle of radius r on the complex plane centered at the point $(1-r, 0)$, see Figure 10.11(a). We will denote this circle by $\overleftrightarrow{\Lambda}$.

The eigenvalues of the operator $\overrightarrow{\boldsymbol{R}}$ are all those and only those complex numbers λ, for which the equation $\overrightarrow{\boldsymbol{R}}u = \lambda u$ has a solution $u = \{u_0, u_1, \dots, u_m, \dots\}$ that satisfies $\lim\limits_{m \longrightarrow +\infty} |u_m| = 0$. Recasting this equation with the help of formula (10.160), we have:

$$(1-r-\lambda)u_m + ru_{m+1} = 0, \quad m = 0, 1, 2, \dots.$$

The solution $u_m = cq^m$ may only be bounded as $m \longrightarrow +\infty$ if $|q| < 1$. The corresponding eigenvalues $\lambda = 1 - r + rq$ completely fill the interior of the disk of radius r centered at the point $(1-r, 0)$, see Figure 10.11(b). We will denote this set by $\overrightarrow{\Lambda}$.

The eigenvalues of the operator $\overleftarrow{\boldsymbol{R}}$ are computed similarly. Using formula (10.161), we can write equation $\overleftarrow{\boldsymbol{R}}u = \lambda u$ as follows:

$$\begin{aligned} (1-r-\lambda)u_m + ru_{m+1} &= 0, \quad m = \dots, -1, 0, 1, \dots, M-1, \\ (1-\lambda)u_M &= 0. \end{aligned}$$

The general solution of the first equation from this pair is $u_m = cq^m$, and the relation between λ and q is $\lambda = 1 - r + rq$. The solution $u_m = cq^m$ may only vanish as $m \longrightarrow -\infty$ if $|q| > 1$. The second equation provides an additional constraint $(1-\lambda)q^M = 0$ so that $\lambda = 1$. However, for this particular λ we also have $q = 1$, which implies no decay as $m \longrightarrow -\infty$. We therefore conclude that the equation $\overleftarrow{\boldsymbol{R}}u = \lambda u$ has no solutions $u = \{u_m\}$ that satisfy $\lim\limits_{m \longrightarrow -\infty} |u_m| = 0$, i.e., there are no eigenvalues: $\overleftarrow{\Lambda} = \emptyset$.

The combination of all eigenvalues $\Lambda = \overleftrightarrow{\Lambda} \cup \overrightarrow{\Lambda} \cup \overleftarrow{\Lambda}$ is the disk $|\lambda - (1-r)| \leq r$ on the complex plane; it is centered at $(1-r, 0)$ and has radius r. We will now show that the spectrum of the family of operators $\{\boldsymbol{R}_h\}$ coincides with the set Λ. This is equivalent to showing that every point $\lambda_0 \in \Lambda$ belongs to the spectrum of $\{\boldsymbol{R}_h\}$ and that this spectrum contains no other points.

According to Definition 10.6, to prove the first implication it is sufficient to demonstrate that for any $\varepsilon > 0$ the inequality

$$\|\boldsymbol{R}_h u - \lambda_0 u\|_{U'_h} < \varepsilon \|u\|_{U'_h} \tag{10.162}$$

has a solution $u \in U'_h$ for all sufficiently small $h > 0$. As $\lambda_0 \in \Lambda$, then $\lambda_0 \in \overleftrightarrow{\Lambda}$ or $\lambda_0 \in \overrightarrow{\Lambda}$, because $\overleftarrow{\Lambda} = \emptyset$. Note that when ε is small one may call the solution u of inequality (10.162) "almost an eigenvector" of the operator $\boldsymbol{R}_h$, since a solution to the equation $\boldsymbol{R}_h u - \lambda_0 u = 0$ is its genuine eigenvector.

Let us first assume that $\lambda_0 \in \overleftrightarrow{\Lambda}$. To construct a solution u of inequality (10.162), we recall that by definition of the set $\overleftrightarrow{\Lambda}$ there exists q_0: $|q_0| = 1$, such that $\lambda_0 = 1 - r + rq_0$ and the equation $(1 - r - \lambda_0)v_m + rv_{m+1} = 0$, $m = 0, \pm 1, \pm 2, \ldots$, has a bounded solution $v_m = q_0^m$, $m = 0, \pm 1, \pm 2, \ldots$. We will consider this solution only for $m = 0, 1, 2, \ldots, M$, while keeping the same notation v. It turns out that the vector:

$$v = [v_0, v_1, v_2, \ldots, v_M] = \left[1, q_0, q_0^2, \ldots, q_0^M\right]$$

almost satisfies the operator equation $\boldsymbol{R}_h v - \lambda_0 v = 0$ that we write as:

$$\begin{aligned} (1 - r - \lambda_0)v_m + rv_{m+1} &= 0, \quad m = 0, 1, 2, \ldots, M - 1, \\ (1 - \lambda_0)v_M &= 0. \end{aligned}$$

The vector v would have completely satisfied the previous equation, which is an even stronger constraint than inequality (10.162), if it did not violate the last relation $(1 - \lambda_0)v_M = 0$.[10] This relation can be interpreted as a boundary condition for the difference equation:

$$(1 - r - \lambda_0)u_m + ru_{m+1} = 0, \quad m = 0, 1, 2, \ldots, M - 1.$$

The boundary condition is specified at $m = M$, i.e., at the right endpoint of the interval $0 \le x \le 1$. To satisfy this boundary condition, let us "correct" the vector $v = \left[1, q_0, q_0^2, \ldots, q_0^M\right]$ by multiplying each of its components v_m by the respective factor $(M - m)h$. The resulting vector will be denoted $u = [u_0, u_1, \ldots, u_M]$, $u_m = (M - m)hq_0^m$. Obviously, the vector u has unit norm:

$$\|u\|_{U'_h} = \max_m |u_m| = \max_m |(M - m)hq_0^m| = Mh = 1.$$

We will now show that this vector u furnishes a desired solution to the inequality (10.162). Define the vector $w \overset{\text{def}}{=} \boldsymbol{R}_h u - \lambda_0 u$, $w = [w_0, w_1, \ldots, w_M]$. We need to estimate its norm. For the individual components of w, we have:

$$\begin{aligned} |w_m| &= |(1 - r - \lambda_0)(M - m)hq_0^m + r(M - m - 1)hq_0^{m+1}| \\ &= |[(1 - r - \lambda_0) + rq_0](M - m)hq_0^m - rhq_0^{m+1}| \\ &= |0 \cdot (M - m)hq_0^m - rhq_0^{m+1}| = rh, \quad m = 0, 1, \ldots, M - 1, \\ |w_M| &= |u_M - \lambda_0 u_M| = |0 - \lambda_0 \cdot 0| = 0. \end{aligned}$$

Consequently, $\|w\|_{U'_h} = rh$, and for $h < \varepsilon/r$ we obtain: $\|w\|_{U'_h} = \|\boldsymbol{R}_h u - \lambda_0 u\|_{U'_h} < \varepsilon = \varepsilon\|u\|_{U'_h}$, i.e., inequality (10.162) is satisfied. Thus, we have shown that if $\lambda_0 \in \overleftrightarrow{\Lambda}$, then this point also belongs to the spectrum of the family of operators $\{\boldsymbol{R}_h\}$.

[10] Relation $(1 - \lambda_0)v_M = 0$ is violated unless $\lambda_0 = q_0 = 1$.

Next, let us assume that $\lambda_0 \in \overrightarrow{\Lambda}$ and show that in this case λ_0 also belongs to the spectrum of the family of operators $\{\boldsymbol{R}_h\}$. According to (10.160), equation $\overrightarrow{\boldsymbol{R}}v - \lambda_0 v = 0$ is written as:

$$(1 - r - \lambda_0)v_m + rv_{m+1} = 0, \quad m = 0, 1, 2, \ldots.$$

Since $\lambda_0 \in \overrightarrow{\Lambda}$, this equation has a solution $v_m = q_0^m$, $m = 0, 1, 2, \ldots$, where $|q_0| < 1$. We will consider this solution only for $m = 0, 1, 2, \ldots, M$:

$$u = [u_0, u_1, u_2, \ldots, u_M] = \left[1, q_0, q_0^2, \ldots, q_0^M\right], \qquad \|u\|_{U_h'} = 1.$$

As before, define $w \stackrel{\text{def}}{=} \boldsymbol{R}_h u - \lambda_0 u$. For the components of the vector w we have:

$$|w_m| = |(1 - r - \lambda_0)q_0^m + rq_0^{m+1}| = 0, \quad m = 0, 1, \ldots, M - 1,$$
$$|w_M| = |1 - \lambda_0| \cdot |q_0^M|.$$

Consequently, $\|w\|_{U_h'} = |1 - \lambda_0| \cdot |q_0|^M = |1 - \lambda_0| \cdot |q_0|^{1/h}$. Since $|q_0| < 1$, then for any $\varepsilon > 0$ we can always choose a sufficiently small h so that $|1 - \lambda_0| \cdot |q_0|^{1/h} < \varepsilon$. Then, $\|w\|_{U_h'} = \|\boldsymbol{R}_h u - \lambda_0 u\|_{U_h'} < \varepsilon = \varepsilon \|u\|_{U_h'}$ and the inequality (10.162) is satisfied.

Note that if the set $\overleftarrow{\Lambda}$ were not empty, then proving that each of its elements belongs to the spectrum of the family of operators $\{\boldsymbol{R}_h\}$ would have been similar. Altogether, we have thus shown that in our specific example given by equations (10.142) every $\lambda_0 \in \{\overleftrightarrow{\Lambda} \cup \overleftarrow{\Lambda} \cup \overrightarrow{\Lambda}\}$ is also an element of the spectrum of $\{\boldsymbol{R}_h\}$.

Now we need to prove that if $\lambda_0 \notin \{\overleftrightarrow{\Lambda} \cup \overleftarrow{\Lambda} \cup \overrightarrow{\Lambda}\}$ then it does not belong to the spectrum of the family of operators $\{\boldsymbol{R}_h\}$ either. To that end, it will be sufficient to show that there is an h-independent constant A, such that for any $u = [u_0, u_1, \ldots, u_M]$ the following inequality holds:

$$\|\boldsymbol{R}_h u - \lambda_0 u\|_{U_h'} \geq A\|u\|_{U_h'}. \tag{10.163}$$

Then, for $\varepsilon < A$, inequality (10.162) will have no solutions, and therefore the point λ_0 will not belong to the spectrum. Denote $f = \boldsymbol{R}_h u - \lambda_0 u$, then inequality (10.163) reduces to:

$$\|f\|_{U_h'} \geq A\|u\|_{U_h'}. \tag{10.164}$$

Our objective is to justify estimate (10.164). Rewrite the equation $\boldsymbol{R}_h u - \lambda_0 u = f$ as:

$$(1 - r - \lambda_0)u_m + ru_{m+1} = f_m, \quad m = 0, 1, \ldots, M - 1,$$
$$(1 - \lambda_0)u_M = f_M,$$

and interpret these relations as an equation with respect to the unknown $u = \{u_m\}$, whereas the right-hand side $f = \{f_m\}$ is assumed given. Let

$$u_m = v_m + w_m, \quad m = 0, 1, \ldots, M, \tag{10.165}$$

where v_m are components of the bounded solution $v = \{v_m\}$, $m = 0, \pm 1, \pm 2, \ldots$, to the following equation:

$$(1 - r - \lambda_0)v_m + rv_{m+1} = \hat{f}_m \stackrel{\text{def}}{=} \begin{cases} 0, & \text{if } m < 0, \\ f_m, & \text{if } m = 0, 1, \ldots, M-1, \\ 0, & \text{if } m \geq M. \end{cases} \tag{10.166}$$

Then because of the linearity, the grid function $w = \{w_m\}$ introduced by formula (10.165) solves the equation:

$$\begin{aligned} (1 - r - \lambda_0)w_m + rw_{m+1} &= 0, \qquad m = 0, 1, \ldots, M-1, \\ (1 - \lambda_0)w_M &= f_M - (1 - \lambda_0)v_M. \end{aligned} \tag{10.167}$$

Let us now recast estimate (10.164) as $|u_m| \leq A^{-1} \max_m |f_m|$. According to (10.165), to prove this estimate it is sufficient to establish individual inequalities:

$$|v_m| \leq A_1 \max_m |f_m|, \tag{10.168a}$$

$$|w_m| \leq A_2 \max_m |f_m|, \tag{10.168b}$$

where A_1 and A_2 are constants. We begin with inequality (10.168a). Notice that equation (10.166) is a first order constant-coefficient ordinary difference equation:

$$av_m + bv_{m+1} = \hat{f}_m, \quad m = 0, \pm 1, \pm 2, \ldots,$$

where $a = 1 - r - \lambda_0$, $b = r$. Its bounded fundamental solution is given by

$$G_m = \begin{cases} \dfrac{1}{a}\left(-\dfrac{a}{b}\right)^m, & m \leq 0, \\ 0, & m \geq 1, \end{cases}$$

because $\lambda_0 \notin \{\overleftrightarrow{\Lambda} \cup \overleftarrow{\Lambda} \cup \overrightarrow{\Lambda}\}$, i.e., $|\lambda_0 - (1-r)| > r$, and consequently $|a/b| > 1$. Representing the solution v_m in the form of a convolution: $v_m = \sum_{k=-\infty}^{\infty} G_{m-k}\hat{f}_k$ and summing up the geometric sequence we arrive at the estimate:

$$|v_m| \leq \frac{\max_m |\hat{f}_m|}{|a| - |b|} = \frac{\max_m |f_m|}{|a| - |b|}.$$

Introducing the distance δ_0 between the point λ_0 and the set $\{\overleftrightarrow{\Lambda} \cup \overleftarrow{\Lambda} \cup \overrightarrow{\Lambda}\}$, we can obviously claim that $|a| - |b| > \delta_0/2$, which makes the previous estimate equivalent to (10.168a). Estimate (10.168b) can be obtained by representing the solution of equation (10.167) in the form:

$$w_m = \frac{f_M - (1 - \lambda_0)v_M}{1 - \lambda_0} q_0^{m-M}, \tag{10.169}$$

where q_0 is determined by the relation $(1-r-\lambda_0)+rq_0=0$. Our assumption is that $\lambda_0 \notin \{\overset{\leftrightarrow}{\Lambda} \cup \overset{\leftarrow}{\Lambda} \cup \overset{\rightarrow}{\Lambda}\}$, i.e., that λ_0 lies outside of the disk of radius r on the complex plane centered at $(1-r,0)$. In this case $|q_0|>1$. Moreover, we can say that $|1-\lambda_0|=\delta_1>0$, because if $\lambda_0=1$, then λ_0 would have belonged to the set $\{\overset{\leftrightarrow}{\Lambda} \cup \overset{\leftarrow}{\Lambda} \cup \overset{\rightarrow}{\Lambda}\}$. As such, using formula (10.169) and taking into account estimate (10.168a) that we have already proved, we obtain the desired estimate (10.168b):

$$\begin{aligned}|w_m| &= \left|\frac{f_M-(1-\lambda_0)v_M}{1-\lambda_0}\right| \cdot |q_0^{m-M}| \leq \frac{|f_M|}{|1-\lambda_0|}+|v_M| \\ &\leq \frac{\max\limits_m |f_m|}{\delta_1} + A_1 \max_m |f_m| = A_2 \max_m |f_m|.\end{aligned}$$

We have thus proven that the spectrum of the family of operators $\{\boldsymbol{R}_h\}$ defined by formulae (10.142) coincides with the set $\{\overset{\leftrightarrow}{\Lambda} \cup \overset{\leftarrow}{\Lambda} \cup \overset{\rightarrow}{\Lambda}\}$ on the complex plane.

The foregoing algorithm for computing the spectrum of the family of operators $\{\boldsymbol{R}_h\}$ is, in fact, quite general. We have illustrated it using a particular example of the operators defined by formulae (10.142). However, for all other scalar finite-difference schemes the spectrum of the family of operators $\{\boldsymbol{R}_h\}$ can be obtained by performing the same Babenko-Gelfand analysis of Section 10.5.1. *The key idea is to take into account other candidate modes that may be prone to developing the instability, besides the eigenmodes $\{e^{i\alpha m}\}$ of the pure Cauchy problem that are accounted for by the von Neumann analysis.*

When systems of finite-difference equations need to be addressed as opposed to the scalar equations, the technical side of the procedure becomes more elaborate. In this case, the computation of the spectrum of a family of operators can be reduced to studying uniform bounds for the solutions of certain ordinary difference equations with matrix coefficients. A necessary and sufficient condition has been obtained in [Rya64] for the existence of such uniform bounds. This condition is given in terms of the roots of the corresponding characteristic equation and also involves the analysis of some determinants originating from the matrix coefficients of the system. For further detail, we refer the reader to [GR87, § 4 & § 45] and [RM67, § 6.6 & § 6.7], as well as to the original journal publication by Ryaben'kii [Rya64].

10.5.3 The Energy Method

For some evolution finite-difference problems, one can obtain the l_2 estimates of the solution directly, i.e., without employing any special stability criteria, such as spectral. The corresponding technique is known as the method of energy estimates. It is useful for deriving *sufficient conditions of stability,* in particular, because it can often be applied to problems with variable coefficients on finite intervals. We illustrate the energy method with several examples.

In the beginning, let us analyze the continuous case. Consider an initial boundary value problem for the first order constant-coefficient hyperbolic equation:

$$\begin{gathered}\frac{\partial u}{\partial t}-\frac{\partial u}{\partial x}=0, \quad 0 \leq x \leq 1, \quad 0<t \leq T, \\ u(x,0)=\psi(x), \quad u(1,t)=0.\end{gathered} \tag{10.170}$$

Note that both the differential equation and the boundary condition at $x=1$ in problem (10.170) are homogeneous. Multiply the differential equation of (10.170) by $u=u(x,t)$ and integrate over the entire interval $0 \leq x \leq 1$:

$$\begin{aligned}\frac{d}{dt}\int_0^1 \frac{u^2(x,t)}{2}dx - \int_0^1 \frac{\partial}{\partial x}\frac{u^2(x,t)}{2}dx \\ = \frac{d}{dt}\frac{\|u(\cdot,t)\|_2^2}{2} - \frac{u^2(1,t)}{2} + \frac{u^2(0,t)}{2} = 0,\end{aligned}$$

where $\|u(\cdot,t)\|_2 \stackrel{\text{def}}{=} \left(\int_0^1 u^2(x,t)dx\right)^{1/2}$ is the L_2 norm of the solution in space for a given moment of time t. According to formula (10.170), the solution at $x=1$ vanishes: $u(1,t)=0$, and we conclude that $\frac{d}{dt}\|u(\cdot,t)\|_2^2 \leq 0$, which means that $\|u(\cdot,t)\|_2$ is a non-increasing function of time. Consequently, we see that the L_2 norm of the solution will never exceed that of the initial data:

$$\|u(\cdot,t)\|_2 \leq \|\psi\|_2, \quad t \geq 0. \tag{10.171}$$

Inequality (10.171) is the simplest energy estimate. It draws its name from the fact that the quantities that are quadratic with respect to the solution are often interpreted as energy in the context of physics. Note that estimate (10.171) holds for all $t \geq 0$ rather than only $0 \leq t \leq T$.

Next, we consider a somewhat more general formulation compared to (10.170), namely, an initial boundary value problem for the hyperbolic equation with a variable coefficient:

$$\begin{gathered}\frac{\partial u}{\partial t}-a(x,t)\frac{\partial u}{\partial x}=0, \quad 0 \leq x \leq 1, \quad 0<t \leq T, \\ u(x,0)=\psi(x), \quad u(1,t)=0.\end{gathered} \tag{10.172}$$

We are assuming that $\forall x \in [0,1]$ and $\forall t \geq 0$: $a(x,t) \geq a_0 > 0$, so that the characteristic speed is negative across the entire domain. Then, the differential equation renders transport from the right to the left. Consequently, setting the boundary condition $u(1,t)=0$ at the right endpoint of the interval $0 \leq x \leq 1$ is legitimate.

Let us now multiply the differential equation of (10.172) by $u=u(x,t)$ and integrate over the entire interval $0 \leq x \leq 1$, while also applying integration by parts to the spatial term:

$$\frac{d}{dt}\int_0^1 \frac{u^2(x,t)}{2}dx - \int_0^1 a(x,t)\frac{\partial}{\partial x}\frac{u^2(x,t)}{2}dx$$

$$= \frac{d}{dt}\frac{\|u(\cdot,t)\|_2^2}{2} - a(1,t)\frac{u^2(1,t)}{2} + a(0,t)\frac{u^2(0,t)}{2} + \int_0^1 a'_x(x,t)\frac{u^2(x,t)}{2}dx = 0.$$

Using the boundary condition $u(1,t) = 0$, we find:

$$\frac{d}{dt}\frac{\|u(\cdot,t)\|_2^2}{2} = -a(0,t)\frac{u^2(0,t)}{2} - \int_0^1 a'_x(x,t)\frac{u^2(x,t)}{2}dx = 0.$$

The first term on the right-hand side of the previous equality is always non-positive. As far as the second term, let us denote $A = \sup_{(x,t)}[-a'_x(x,t)]$. Then we have:

$$\frac{d}{dt}\|u(\cdot,t)\|_2^2 \leq A\|u(\cdot,t)\|_2^2,$$

which immediately yields:

$$\|u(\cdot,t)\|_2 \leq e^{At/2}\|\psi\|_2, \quad t \geq 0.$$

If $A < 0$, the previous inequality implies that the L_2 norm of the solution decays as $t \longrightarrow +\infty$. If $A = 0$, then the norm of the solution stays bounded by the norm of the initial data. To obtain an overall uniform estimate of $\|u(\cdot,t)\|_2$ for $A \leq 0$ and all $t \geq 0$, we need to select the maximum value of the constant: $\max_t e^{At/2} = 1$, and then the desired inequality will coincide with (10.171). For $A > 0$, a uniform estimate can only be obtained for a given fixed interval $0 \leq t \leq T$, so that altogether we can write:

$$\|u(\cdot,t)\|_2 \leq \begin{cases} \|\psi\|_2, & \text{if } A \leq 0, \ t \geq 0, \\ e^{AT/2}\|\psi\|_2, & \text{if } A > 0, \ 0 \leq t \leq T. \end{cases} \tag{10.173}$$

Similarly to inequality (10.171), the energy estimate (10.173) also provides a bound for the L_2 norm of the solution in terms of the L_2 norm of the initial data. However, when $A > 0$ the constant in front of $\|\psi\|_2$ is no longer equal to one. Instead, $e^{AT/2}$ grows exponentially as the maximum time T elapses, and therefore estimate (10.173) for $A > 0$ may only be considered on a finite interval $0 \leq t \leq T$ rather than for $t \geq 0$.

In problems (10.170) and (10.172) the boundary condition at $x = 1$ was homogeneous. Let us now introduce yet another generalization and analyze the problem:

$$\begin{gathered} \frac{\partial u}{\partial t} - a(x,t)\frac{\partial u}{\partial x} = 0, \quad 0 \leq x \leq 1, \quad 0 < t \leq T, \\ u(x,0) = \psi(x), \quad u(1,t) = \chi(t) \end{gathered} \tag{10.174}$$

that differs from (10.172) by its inhomogeneous boundary condition: $u(1,t) = \chi(t)$. Otherwise everything is the same; in particular, we still assume that $\forall x \in [0,1]$ and $\forall t \geq 0$: $a(x,t) \geq a_0 > 0$ and denote $A = \sup_{(x,t)}[-a'_x(x,t)]$. Multiplying the differential equation of (10.174) by $u(x,t)$ and integrating by parts, we obtain:

$$\frac{d}{dt}\frac{\|u(\cdot,t)\|_2^2}{2} = -a(0,t)\frac{u^2(0,t)}{2} + a(1,t)\frac{\chi^2(t)}{2} - \int_0^1 a'_x(x,t)\frac{u^2(x,t)}{2}dx = 0.$$

Consequently,

$$\frac{d}{dt}\|u(\cdot,t)\|_2^2 \leq A\|u(\cdot,t)\|_2^2 + a(1,t)\chi^2(t).$$

Multiplying the previous inequality by e^{-At}, we have:

$$\frac{d}{dt}\left[e^{-At}\|u(\cdot,t)\|_2^2\right] \leq e^{-At}a(1,t)\chi^2(t),$$

which, after integrating over the time interval $0 \leq \theta \leq t$, yields:

$$\|u(\cdot,t)\|_2^2 \leq e^{At}\|\psi\|_2^2 + e^{At}\int_0^t e^{-A\theta}a(1,\theta)\chi^2(\theta)d\theta.$$

As in the case of a homogeneous boundary condition, we would like to obtain a uniform energy estimate for a given interval of time. This can be done if we again distinguish between $A \leq 0$ and $A > 0$. When $A \leq 0$ we can consider all $t \geq 0$ and when $A > 0$ we can only have the estimate on some fixed interval $0 \leq t \leq T$:

$$\|u(\cdot,t)\|_2^2 \leq \begin{cases} \|\psi\|_2^2 + \int_0^\infty a(1,\theta)\chi^2(\theta)d\theta, & A \leq 0, \ t \geq 0, \\ e^{AT}\left[\|\psi\|_2^2 + \int_0^T a(1,\theta)\chi^2(\theta)d\theta\right], & A > 0, \ 0 \leq t \leq T. \end{cases} \tag{10.175}$$

When deriving inequalities (10.175), we obviously need to assume that the integrals on the right-hand side of (10.175) are bounded. These integrals can be interpreted as weighted L_2 norms of the boundary data $\chi(t)$. Clearly, energy estimate (10.175) includes the previous estimate (10.173) as a particular case for $\chi(t) \equiv 0$.

All three estimates (10.171), (10.173), and (10.175) indicate that the corresponding initial boundary value problem is *well-posed* in the sense of L_2. Qualitatively, well-posedness means that the solution to a given problem is only weakly sensitive to perturbations of the input data (such as initial and/or boundary data). In the case of linear evolution problems, well-posedness can, for example, be quantified by means of the energy estimates. These estimates provide a bound for the norm of the solution in terms of the norms of the input data. In the finite-difference context, similar estimates would have implied stability in the sense of l_2, provided that the corresponding constants on the right-hand side of each inequality can be chosen independent of the grid. We will now proceed to demonstrate how energy estimates can be obtained for finite-difference schemes.

Consider the first order upwind scheme for problem (10.170):

$$\begin{gathered}\frac{u_m^{p+1}-u_m^p}{\tau}-\frac{u_{m+1}^p-u_m^p}{h}=0,\\ m=0,1,\dots,M-1,\quad p=0,1,\dots,[T/\tau]-1,\\ u_m^0=\psi_m,\quad u_M^p=0.\end{gathered}\tag{10.176}$$

To obtain an energy estimate for scheme (10.176), let us first consider two arbitrary functions $\{u_m\}$ and $\{v_m\}$ on the grid $m=0,1,2,\dots,M$. We will derive a formula that can be interpreted as a discrete analogue of the classical continuous integration by parts. In the literature, it is sometimes referred to as the summation by parts:

$$\begin{aligned}\sum_{m=0}^{M-1} u_m(v_{m+1}-v_m) &= \sum_{m=0}^{M-1} u_m v_{m+1} - \sum_{m=0}^{M-1} u_m v_m = \sum_{m=1}^{M} u_{m-1}v_m - \sum_{m=0}^{M-1} u_m v_m\\ &= -\sum_{m=1}^{M}(u_m-u_{m-1})v_m + u_M v_M - u_0 v_0\\ &= -\sum_{m=0}^{M-1}(u_{m+1}-u_m)v_{m+1} + u_M v_M - u_0 v_0\\ &= -\sum_{m=0}^{M-1}(u_{m+1}-u_m)v_m - \sum_{m=0}^{M-1}(u_{m+1}-u_m)(v_{m+1}-v_m)\\ &\quad + u_M v_M - u_0 v_0.\end{aligned}\tag{10.177}$$

Next, we rewrite the difference equation of (10.176) as

$$u_m^{p+1}=u_m^p+r(u_{m+1}^p-u_m^p),\quad r=\frac{\tau}{h}=\text{const},\quad m=0,1,\dots,M-1,$$

square both sides, and take the sum from $m=0$ to $m=M-1$. This yields:

$$\sum_{m=0}^{M-1}\left(u_m^{p+1}\right)^2=\sum_{m=0}^{M-1}(u_m^p)^2+r^2\sum_{m=0}^{M-1}\left(u_{m+1}^p-u_m^p\right)^2+2r\sum_{m=0}^{M-1}u_m^p\left(u_{m+1}^p-u_m^p\right).$$

To transform the last term on the right-hand side of the previous equality, we apply formula (10.177):

$$\begin{aligned}\sum_{m=0}^{M-1}\left(u_m^{p+1}\right)^2 &= \sum_{m=0}^{M-1}(u_m^p)^2+r^2\sum_{m=0}^{M-1}\left(u_{m+1}^p-u_m^p\right)^2+r\sum_{m=0}^{M-1}u_m^p\left(u_{m+1}^p-u_m^p\right)\\ &\quad + r\left[-\sum_{m=0}^{M-1}\left(u_{m+1}^p-u_m^p\right)u_m^p-\sum_{m=0}^{M-1}\left(u_{m+1}^p-u_m^p\right)^2+\left(u_M^p\right)^2-\left(u_0^p\right)^2\right]\\ &= \sum_{m=0}^{M-1}(u_m^p)^2+r(r-1)\sum_{m=0}^{M-1}\left(u_{m+1}^p-u_m^p\right)^2+r\left(u_M^p\right)^2-r\left(u_0^p\right)^2.\end{aligned}$$

Let us now assume that $r \leq 1$. Then, using the conventional definition of the l_2 norm: $\|u\|_2 = \left[h\sum_0^M |u_m|^2\right]^{1/2}$ and employing the homogeneous boundary condition $u_M^p = 0$ of (10.176), we obtain the inequality:

$$\|u^{p+1}\|_2 \leq \|u^p\|_2, \quad p = 0,1,2,\ldots,$$

which clearly implies the energy estimate:

$$\|u^p\|_2 \leq \|\psi\|_2, \quad p = 0,1,2,\ldots. \tag{10.178}$$

The discrete estimate (10.178) is analogous to the continuous estimate (10.171).

To approximate the variable-coefficient problem (10.172), we use the scheme:

$$\begin{gathered}\frac{u_m^{p+1} - u_m^p}{\tau} - a_m^p \frac{u_{m+1}^p - u_m^p}{h} = 0, \\ m = 0,1,\ldots,M-1, \quad p = 0,1,\ldots,[T/\tau]-1, \\ u_m^0 = \psi_m, \quad u_M^p = 0,\end{gathered} \tag{10.179}$$

where $a_m^p \equiv a(x_m, t_p)$. Applying a similar approach, we obtain:

$$\begin{aligned}\sum_{m=0}^{M-1} \left(u_m^{p+1}\right)^2 = \sum_{m=0}^{M-1} (u_m^p)^2 + r^2 \sum_{m=0}^{M-1} (a_m^p)^2 \left(u_{m+1}^p - u_m^p\right)^2 + r \sum_{m=0}^{M-1} a_m^p u_m^p \left(u_{m+1}^p - u_m^p\right) \\ + r\Bigg[-\sum_{m=0}^{M-1} \left(a_{m+1}^p u_{m+1}^p - a_m^p u_m^p\right) u_m^p \\ - \sum_{m=0}^{M-1} \left(a_{m+1}^p u_{m+1}^p - a_m^p u_m^p\right)\left(u_{m+1}^p - u_m^p\right) + a_M^p \left(u_M^p\right)^2 - a_0^p \left(u_0^p\right)^2\Bigg].\end{aligned}$$

Next, we notice that $a_{m+1}u_{m+1} = a_m u_{m+1} + (a_{m+1} - a_m)u_{m+1}$. Substituting this expression into the previous formula, we again perform summation by parts, which is analogous to the continuous integration by parts, and which yields:

$$\begin{aligned}\sum_{m=0}^{M-1} \left(u_m^{p+1}\right)^2 &= \sum_{m=0}^{M-1} (u_m^p)^2 + r^2 \sum_{m=0}^{M-1} (a_m^p)^2 \left(u_{m+1}^p - u_m^p\right)^2 + r \sum_{m=0}^{M-1} a_m^p u_m^p \left(u_{m+1}^p - u_m^p\right) \\ &\quad + r\Bigg[-\sum_{m=0}^{M-1} \left(a_m^p \left(u_{m+1}^p - u_m^p\right) + \left(a_{m+1}^p - a_m^p\right) u_{m+1}^p\right) u_m^p \\ &\quad - \sum_{m=0}^{M-1} \left(a_m^p \left(u_{m+1}^p - u_m^p\right) + \left(a_{m+1}^p - a_m^p\right) u_{m+1}^p\right)\left(u_{m+1}^p - u_m^p\right) \\ &\quad + a_M^p \left(u_M^p\right)^2 - a_0^p \left(u_0^p\right)^2\Bigg] \\ &= \sum_{m=0}^{M-1} (u_m^p)^2 + \sum_{m=0}^{M-1} \left(r^2 (a_m^p)^2 - r a_m^p\right)\left(u_{m+1}^p - u_m^p\right)^2\end{aligned}$$

$$-r\sum_{m=0}^{M-1}\left(a^p_{m+1}-a^p_m\right)\left(u^p_{m+1}\right)^2+ra^p_M\left(u^p_M\right)^2-ra^p_0\left(u^p_0\right)^2.$$

Let us now assume that for all m and p we have: $ra^p_m \le 1$. Equivalently, we can require that $r \le [\sup_{(x,t)} a(x,t)]^{-1}$. Let us also introduce:

$$A=\sup_{(m,p)}\left\{-\frac{a^p_{m+1}-a^p_m}{h}\right\}.$$

Then, using the homogeneous boundary condition $u^p_M=0$ and dropping the a priori non-positive term $\sum_{m=0}^{M-1} ra^p_m\left(ra^p_m-1\right)\left(u^p_{m+1}-u^p_m\right)^2$, we obtain:

$$\sum_{m=0}^{M-1}\left(u^{p+1}_m\right)^2 \le \sum_{m=0}^{M-1}(u^p_m)^2+rA\sum_{m=0}^{M}h\,(u^p_m)^2-ra^p_0\left(u^p_0\right)^2-rAh\left(u^p_0\right)^2.$$

If $A>0$, then for the last two terms on the right-hand side of the previous inequality we clearly have: $-r\left(u^p_0\right)^2(a^p_0+Ah)<0$. Even if $A\le 0$ we can still claim that $-r\left(u^p_0\right)^2(a^p_0+Ah)<0$ for sufficiently small h. Consequently, on fine grids the following inequality holds:

$$\|u^{p+1}\|_2^2 \le \|u^p\|_2^2+\tau A\|u^p\|_2^2=(1+A\tau)\|u^p\|_2^2,$$

which immediately implies:

$$\|u^p\|_2^2 \le (1+A\tau)^p\|\psi\|_2^2, \quad p=1,2,\dots.$$

If $A\le 0$, the norm of the discrete solution will either decay or remain bounded as p increases. If $A>0$, a uniform estimate of $\|u^p\|_2$ can only be obtained for $p=0,1,\dots,[T/\tau]$. Altogether, the solution $u^p=\{u^p_m\}$ to the finite-difference problem (10.179) satisfies the following energy estimate:

$$\|u^p\|_2 \le \begin{cases}\|\psi\|_2, & A\le 0,\ \ p=0,1,2,\dots,\\ e^{AT/2}\|\psi\|_2, & A>0,\ \ p=0,1,2,\dots,[T/\tau].\end{cases} \tag{10.180}$$

The discrete estimate (10.180) is analogous to the continuous estimate (10.173).

Finally, for problem (10.174) we use the scheme:

$$\begin{gathered}\frac{u^{p+1}_m-u^p_m}{\tau}-a^p_m\frac{u^p_{m+1}-u^p_m}{h}=0,\\ m=0,1,\dots,M-1, \quad p=0,1,\dots,[T/\tau]-1,\\ u^0_m=\psi_m, \quad u^p_M=\chi^p.\end{gathered} \tag{10.181}$$

Under the same assumptions that we introduced when deriving estimate (10.180) for scheme (10.179), we can now write for scheme (10.181):

$$\sum_{m=0}^{M-1}\left(u^{p+1}_m\right)^2 \le \sum_{m=0}^{M-1}(u^p_m)^2+rA\sum_{m=0}^{M}h\,(u^p_m)^2+ra^p_M(\chi^p)^2.$$

Denote $[\|u\|_2']^2 = h\sum_{m=0}^{M-1}|u_m|^2 = \|u\|_2^2 - hu_M^2$. Then the previous inequality implies:

$$[\|u^{p+1}\|_2']^2 \le (1+A\tau)[\|u^p\|_2']^2 + \tau(a_M^p + Ah)\,(\chi^p)^2, \quad p=1,2,\ldots,$$

and consequently:

$$[\|u^p\|_2']^2 \le (1+A\tau)^p[\|\psi\|_2']^2 + \sum_{k=1}^{p}(1+A\tau)^{p-k}\tau(a_M^{k-1}+Ah)(\chi^{k-1})^2, \quad p=1,2,\ldots.$$

We again need to distinguish between the cases $A \le 0$, $p=0,1,2,\ldots$, and $A>0$, $p=0,1,2,\ldots,[T/\tau]$:

$$[\|u^p\|_2']^2 \le \begin{cases} [\|\psi\|_2']^2 + \sum\limits_{k=1}^{\infty}\tau a_M^{k-1}(\chi^{k-1})^2, & A \le 0, \\ e^{AT}\left([\|\psi\|_2']^2 + \sum\limits_{k=1}^{[T/\tau]}\tau(a_M^{k-1}+Ah)(\chi^{k-1})^2\right), & A>0. \end{cases} \tag{10.182}$$

The discrete estimate (10.182) is analogous to the continuous estimate (10.175). To use the norms $\|\cdot\|_2$ instead of $\|\cdot\|_2'$ in (10.182), we only need to add a bounded quantity $h(\chi^0)^2 + h(\chi^p)^2$ on the right-hand side.

Energy estimates (10.178), (10.180), and (10.182) imply the l_2 stability of the schemes (10.176), (10.179), and (10.181), respectively, in the sense of the Definition 10.2 from page 312. Note that since the foregoing schemes are explicit, stability is not unconditional, and the Courant number has to satisfy $r \le 1$ for scheme (10.176) and $r \le [\sup_{(x,t)} a(x,t)]^{-1}$ for schemes (10.179) and (10.181).

In general, direct energy estimates appear helpful for studying stability of finite-difference schemes. Indeed, they may provide sufficient conditions for those difficult cases that involve variable coefficients, boundary conditions, and even multiple space dimensions. In addition to the scalar hyperbolic equations, energy estimates can be obtained for some hyperbolic systems, as well as for the parabolic equations. For detail, we refer the reader to [GKO95, Chapters 9 & 11], and to some fairly recent journal publications [Str94, Ols95a, Ols95b]. However, there is a key non-trivial step in proving energy estimates for finite-difference initial boundary value problems, namely, obtaining the discrete summation by parts rules appropriate for a given discretization [see the example given by formula (10.177)]. Sometimes, this step may not be obvious at all; otherwise, it may require using the alternative norms based on specially chosen inner products.

10.5.4 A Necessary and Sufficient Condition of Stability. The Kreiss Criterion

In Section 10.5.2, we have shown that for stability of a finite-difference initial boundary value problem it is necessary that the spectrum of the family of transition operators $\boldsymbol{R}_h$ belongs to the unit disk on the complex plane. We have also shown, see Theorem 10.9, that this condition is, in fact, not very far from a sufficient one, as it guarantees the scheme from developing a catastrophic exponential instability. However, it is not a fully sufficient condition, and there are examples of the schemes that satisfy the Godunov and Ryaben'kii criterion of Section 10.5.2, i.e., that have their spectrum of $\{\boldsymbol{R}_h\}$ inside the unit disk, yet they are unstable.

A comprehensive analysis of necessary and sufficient conditions for the stability of evolution-type schemes on finite intervals is rather involved. In the literature, the corresponding series of results is commonly referred to as the Gustafsson, Kreiss, and Sundström (GKS) theory, and we refer the reader to the monograph [GKO95, Part II] for detail. A concise narrative of this theory can also be found in [Str04, Chapter 11]. As often is the case when analyzing sufficient conditions, all results of the GKS theory are formulated in terms of the l_2 norm. An important tool used for obtaining stability estimates is the Laplace transform in time.

Although a full account of (and even a self-contained introduction to) the GKS theory is beyond the scope of this text, its key ideas are easy to understand on the qualitative level and easy to illustrate with examples. The following material is essentially based on that of Section 10.5.2 and can be skipped during the first reading.

Let us consider an initial boundary value problem for the first order constant coefficient hyperbolic equation:

$$\begin{gathered}\frac{\partial u}{\partial t}-\frac{\partial u}{\partial x}=0, \quad 0\leq x\leq 1, \quad 0<t\leq T,\\ u(x,0)=\psi(x), \quad u(1,t)=0.\end{gathered} \tag{10.183}$$

We introduce a uniform grid: $x_m = mh$, $m = 0,1,\ldots,M$, $h = 1/M$; $t_p = p\tau$, $p = 0,1,2,\ldots$, and approximate problem (10.183) with the leap-frog scheme:

$$\begin{gathered}\frac{u_m^{p+1}-u_m^{p-1}}{2\tau}-\frac{u_{m+1}^p-u_{m-1}^p}{2h}=0,\\ m=1,2,\ldots,M-1, \quad p=1,2,\ldots,[T/\tau]-1,\\ u_m^0=\psi(x_m), \quad u_m^1=\psi(x_m+\tau), \quad m=0,1,\ldots,M,\\ lu_0^{p+1}=0, \quad u_M^{p+1}=0, \quad p=1,2,\ldots,[T/\tau]-1.\end{gathered} \tag{10.184}$$

Notice that scheme (10.184) requires two initial conditions, and for simplicity we use the exact solution, which is readily available in this case, to specify u_m^1 for $m = 0,1,\ldots,M-1$. Also notice that the differential problem (10.183) does not require any boundary conditions at the "outflow" boundary $x = 0$, but the discrete problem (10.184) requires an additional boundary condition that we symbolically denote $lu_0^{p+1} = 0$. We will investigate two different outflow conditions for scheme (10.184):

$$u_0^{p+1}=u_1^{p+1} \tag{10.185a}$$

and

$$u_0^{p+1}=u_0^p+r(u_1^p-u_0^p), \tag{10.185b}$$

where we have used our standard notation $r = \frac{\tau}{h} = \text{const}$.

Let us first note that scheme (10.184) is not a one-step scheme. Therefore, to reduce it to the canonical form (10.141) so that to be able to investigate the spectrum of the family of operators $\{\boldsymbol{R}_h\}$, we would formally need to introduce additional variables (i.e., transform a scalar equation into a system) and then consider a one-step finite-difference equation, but with vector unknowns. It is, however, possible to show that even after a reduction of that type has been implemented, the Babenko-Gelfand procedure of Section 10.5.1 applied to the resulting vector scheme for calculating the spectrum of $\{\boldsymbol{R}_h\}$ will be equivalent to the Babenko-Gelfand procedure applied directly to the multi-step scheme (10.184). As such, we will skip the formal reduction of scheme (10.184) to the canonical form (10.141) and proceed immediately to computing the spectrum of the corresponding family of transition operators.

We need to analyze three model problems that follow from (10.184): A problem with no lateral boundaries:

$$\begin{gathered}\frac{u_m^{p+1} - u_m^{p-1}}{2\tau} - \frac{u_{m+1}^p - u_{m-1}^p}{2h} = 0, \\ m = 0, \pm 1, \pm 2, \ldots,\end{gathered} \tag{10.186}$$

a problem with only the left boundary:

$$\begin{gathered}\frac{u_m^{p+1} - u_m^{p-1}}{2\tau} - \frac{u_{m+1}^p - u_{m-1}^p}{2h} = 0, \\ m = 1, 2, \ldots, \\ lu_0^{p+1} = 0,\end{gathered} \tag{10.187}$$

and a problem with only the right boundary:

$$\begin{gathered}\frac{u_m^{p+1} - u_m^{p-1}}{2\tau} - \frac{u_{m+1}^p - u_{m-1}^p}{2h} = 0, \\ m = M-1, M-2, \ldots, 1, 0, -1, \ldots, \\ u_M^{p+1} = 0.\end{gathered} \tag{10.188}$$

Substituting a solution of the type:

$$u_m^p = \lambda^p u_m$$

into the finite-difference equation:

$$u_m^{p+1} - u_m^{p-1} = r(u_{m+1}^p - u_{m-1}^p), \quad r = \tau/h,$$

that corresponds to all three problems (10.186), (10.187), and (10.188), we obtain the following second order ordinary difference equation for the eigenfunction $\{u_m\}$:

$$(\lambda - \lambda^{-1})u_m - r(u_{m+1} - u_{m-1}) = 0. \tag{10.189}$$

Its characteristic equation:

$$(\lambda - \lambda^{-1}) - r(q - q^{-1}) = 0 \tag{10.190a}$$

has two roots: $q_1 = q_1(\lambda)$ and $q_2 = q_2(\lambda)$, so that the general solution of equation (10.189) can be written as

$$u_m = c_1 q_1^m + c_2 q_2^m, \quad m = 0, \pm 1, \pm 2, \ldots, \quad c_1 = \text{const}, \quad c_2 = \text{const}.$$

It will also be convenient to recast the characteristic equation (10.190a) in an equivalent form:

$$q^2 - \frac{\lambda - \lambda^{-1}}{r} q - 1 = 0. \tag{10.190b}$$

From equation (10.190b) one can easily see that $q_1 q_2 = -1$ and consequently, unless both roots have unit magnitude, we always have $|q_1(\lambda)| < 1$ and $|q_2(\lambda)| > 1$.

The solution of problem (10.186) must be bounded: $|u_m| \leq \text{const}$ for $m = 0, \pm 1, \pm 2, \ldots$. We therefore require that for this problem $|q_1| = |q_2| = 1$, which means $q_1 = e^{i\alpha}$, $0 \leq \alpha < 2\pi$, and $q_2 = -e^{-i\alpha}$. The spectrum of this problem was calculated in Example 5 of Section 10.3.3:

$$\overleftrightarrow{\Lambda} = \left\{ \lambda(\alpha) = ir\sin\alpha \pm \sqrt{1 - r^2 \sin^2\alpha} \,\middle|\, 0 \leq \alpha < 2\pi \right\}. \tag{10.191}$$

Provided that $r \leq 1$, the spectrum $\overleftrightarrow{\Lambda}$ given by formula (10.191) belongs to the unit circle on the complex plane.

For problem (10.188), we must have $u_m \longrightarrow 0$ as $m \longrightarrow -\infty$. Consequently, its general solution is given by:

$$u_m^p = c_2 \lambda^p q_2^m, \quad m = M, M-1, \ldots, 1, 0, -1, \ldots.$$

The homogeneous boundary condition $u_M^{p+1} = 0$ of (10.184) implies that a nontrivial eigenfunction $u_m = c_2 q_2^m$ may only exist if $\lambda = 0$. From the characteristic equation (10.190a) in yet another equivalent form $(\lambda^2 - 1)q - r\lambda(q^2 - 1) = 0$, we conclude that if $\lambda = 0$ then $q = 0$, which means that problem (10.188) has no eigenvalues:

$$\overleftarrow{\Lambda} = \emptyset. \tag{10.192}$$

To study problem (10.187), we first consider boundary condition (10.185a), known as the extrapolation boundary condition. The solution of problem (10.187) must satisfy $u_m \longrightarrow 0$ as $m \longrightarrow \infty$. Consequently, its general form is:

$$u_m^p = c_1 \lambda^p q_1^m, \quad m = 0, 1, 2, \ldots.$$

The extrapolation condition (10.185a) implies that a nontrivial eigenfunction $u_m = c_1 q_1^m$ may only exist if either $\lambda = 0$ or $c_1(1 - q_1) = 0$. However, we must have $|q_1| < 1$ for problem (10.187), and as such, we see that this problem has no eigenvalues either:

$$\overrightarrow{\Lambda} = \emptyset. \tag{10.193}$$

Combining formulae (10.191), (10.192), and (10.193), we obtain the spectrum of the family of operators:

$$\Lambda = \overleftrightarrow{\Lambda} \cup \overleftarrow{\Lambda} \cup \overrightarrow{\Lambda} = \overleftrightarrow{\Lambda}.$$

We therefore see that according to formula (10.191), the necessary condition for stability (Theorem 10.8) of scheme (10.184), (10.185a) is satisfied when $r \leq 1$.

Moreover, according to Theorem 10.9, even if there is no uniform bound on the powers of the transition operators for scheme (10.184), (10.185a), their rate of growth will be slower than any exponential function. Unfortunately, this is precisely what happens. Even though the necessary condition of stability holds for scheme (10.184), (10.185a), it still turns out unstable. The actual proof of instability can be found in [GKO95, Section 13.1] or in [Str04, Section 11.2]. We omit it here and only corroborate the instability with a numerical demonstration. In Figure 10.14 we show the results of numerical integration of problem (10.183) using scheme (10.184), (10.185a) with $r = 0.95$. We specify the exact solution of the problem as $u(x,t) = \cos 2\pi(x+t)$ and as such, supplement problem (10.183) with an inhomogeneous boundary condition $u(1,t) = \cos 2\pi(1+t)$.

In order to analyze what may have caused the instability of scheme (10.184), (10.185a), let us return to the proof of Theorem 10.9. If we were able to claim that the entire spectrum of the family of operators $\{\boldsymbol{R}_h\}$ lies strictly inside the unit disk, then a straightforward modification of that proof would immediately yield a uniform bound on the powers $\boldsymbol{R}_h^p$. This situation, however, is generally impossible. Indeed, in all our previous examples, the spectrum has always contained at least one point on the unit circle: $\lambda = 1$. It is therefore natural to assume that since the points λ inside the unit disk present no danger of instability according to Theorem 10.9, then the potential "culprits" should be sought on the unit circle.

Let us now get back to problem (10.187). We have shown that this problem has no nontrivial eigenfunctions in the class $u_m \longrightarrow 0$ as $m \longrightarrow \infty$ and accordingly, it has no eigenvalues either, see formula (10.193). As such, it does not contribute to the overall spectrum of the family of operators. However, even though the boundary condition (10.185a) in the form $c_1(1-q_1) = 0$ is not satisfied by any function $u_m = c_1 q_1^m$, where $|q_1| < 1$, we see that it is "almost satisfied" if the root q_1 is close to one. Therefore, the function $u_m = c_1 q_1^m$ is "almost an eigenfunction" of problem (10.187), and the smaller the quantity $|1-q_1|$, the more of a genuine eigenfunction it becomes.

To investigate stability, we need to determine whether or not the foregoing "almost an eigenfunction" can bring along an unstable eigenvalue, or rather "almost an eigenvalue," $|\lambda| > 1$. By passing to the limit $q_1 \longrightarrow 1$, we find from equation (10.190a) that $\lambda = 1$ or $\lambda = -1$. We should therefore analyze the behavior of the quantities λ and q in a neighborhood of each of these two values of λ, when the relation between λ and q is given by equation (10.190a).

First recall that according to formula (10.191), if $|q| = 1$, then $|\lambda| = 1$ (provided that $r \leq 1$). Consequently, if $|\lambda| > 1$, then $|q| \neq 1$, i.e., there are two distinct roots: $|q_1| < 1$ and $|q_2| > 1$. In particular, when λ is near $(1,0)$, there are still two roots: One with the magnitude greater than one and the other with the magnitude less than one. When $|\lambda - 1| \longrightarrow 0$ we will clearly have $|q_1| \longrightarrow 1$ and $|q_2| \longrightarrow 1$. We, however,

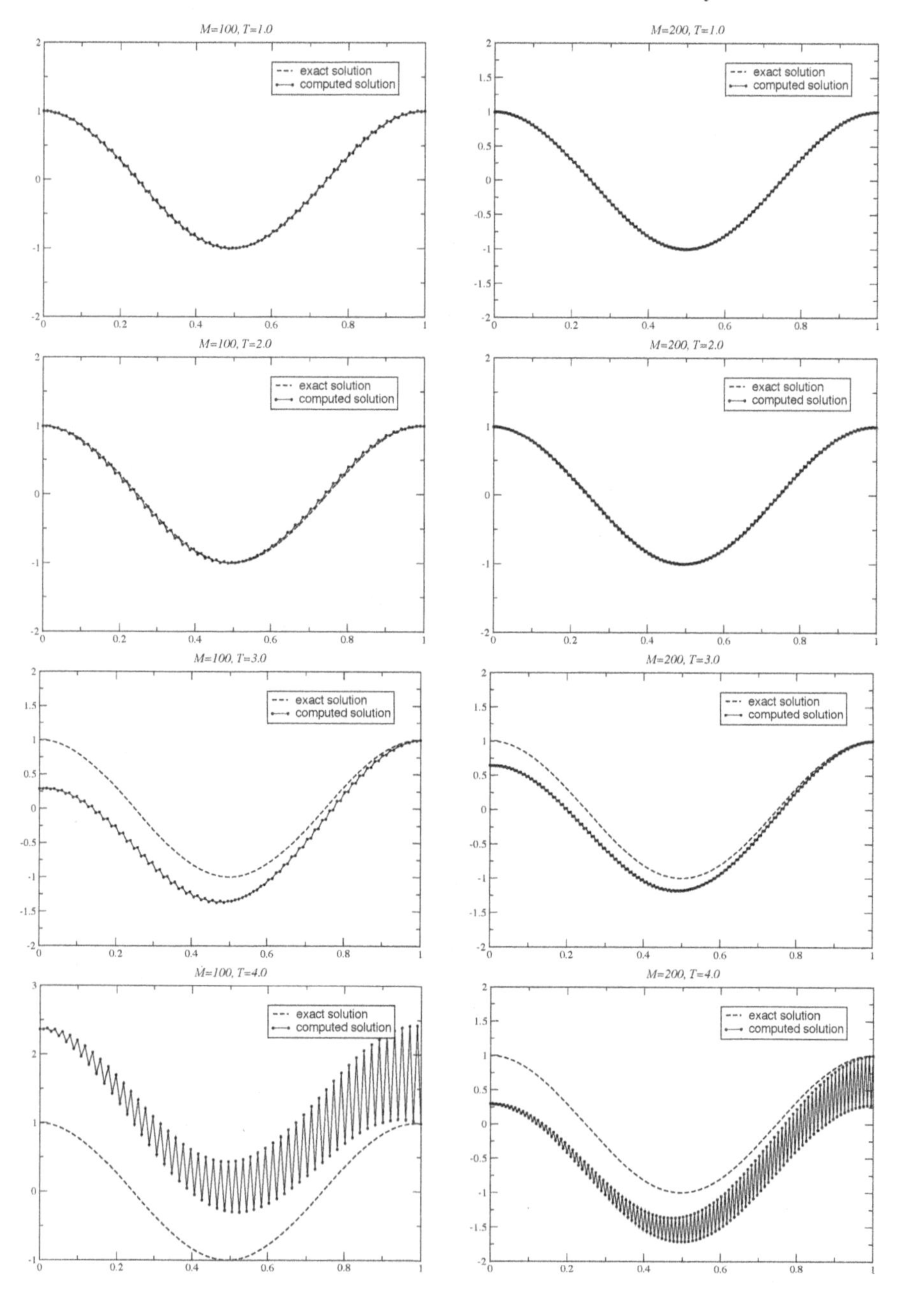

FIGURE 10.14: Solution of problem (10.183) with scheme (10.184), (10.185a).

don't know ahead of time which of the two possible scenarios actually takes place:

$$\lim_{|\lambda|>1,\,\lambda\to 1} q_1(\lambda) = 1, \qquad \lim_{|\lambda|>1,\,\lambda\to 1} q_2(\lambda) = -1 \tag{10.194a}$$

or

$$\lim_{|\lambda|>1,\,\lambda\to 1} q_1(\lambda) = -1, \qquad \lim_{|\lambda|>1,\,\lambda\to 1} q_2(\lambda) = 1. \tag{10.194b}$$

To find this out, let us notice that the roots $q_1(\lambda)$ and $q_2(\lambda)$ are continuous (in fact, analytic) functions of λ. Consequently, if we take λ in the form $\lambda = 1+\eta$, where $|\eta| \ll 1$, and if we want to investigate the root q that is close to one, then we can say that $q(\lambda) = 1+\zeta$, where $|\zeta| \ll 1$. From equation (10.190a) we then obtain:

$$2\eta + \mathcal{O}(\eta^2) = 2r\zeta + \mathcal{O}(\zeta^2). \tag{10.195}$$

Consider a special case of real $\eta > 0$, then ζ must obviously be real as well. From the previous equality we find that $\zeta > 0$ (because $r > 0$), i.e., $|q| > 1$. As such, we see that if $|\lambda| > 1$ and $\lambda \longrightarrow 1$, then

$$\{q = q(\lambda) \longrightarrow 1\} \Longrightarrow \{|q| > 1\}.$$

Indeed, for real η and ζ, we have $|q| = 1+\zeta > 1$; for other η and ζ the same result follows by continuity. Consequently, it is the root q_2 that approaches $(1,0)$ when $\lambda \longrightarrow 1$, and the true scenario is given by (10.194b) rather than by (10.194a).

We therefore see that when a potentially "dangerous" unstable eigenvalue $|\lambda| > 1$ approaches the unit circle at $(1,0)$: $\lambda \longrightarrow 1$, it is the grid function $u_m = c_2 q_2^m$, $|q_2| > 1$, that will almost satisfy the boundary condition (10.185a), because $c_2(1-q_2) \longrightarrow 0$. This grid function, however, does not satisfy the requirement $u_m \longrightarrow 0$ as $m \longrightarrow \infty$, i.e., it does not belong to the class of functions admitted by problem (10.187). On the other hand, the function $u_m = c_1 q_1^m$, $|q_1| > 1$, that satisfies $u_m \longrightarrow 0$ as $m \longrightarrow \infty$, will be very far from satisfying the boundary condition (10.185a) because $q_1 \longrightarrow -1$.

Next, recall that we actually need to investigate what happens when $q_1 \longrightarrow 1$, i.e., when $c_1 q_1^m$ is almost an eigenfunction. This situation appears opposite to the one we have analyzed. Consequently, when $q_1 \longrightarrow 1$ we will not have such a $\lambda(q_1) \longrightarrow 1$ where $|\lambda(q_1)| > 1$. Qualitatively, this indicates that there is no instability associated with "almost an eigenfunction" $u_m = c_1 q_1^m$, $|q_1| > 1$, of problem (10.187). In the framework of the GKS theory, this assertion can be proven rigorously.

Let us now consider the second case: $\lambda \longrightarrow -1$ while $|\lambda| > 1$. We need to determine which of the two scenarios holds:

$$\lim_{|\lambda|>1,\,\lambda\to -1} q_1(\lambda) = 1, \qquad \lim_{|\lambda|>1,\,\lambda\to -1} q_2(\lambda) = -1 \tag{10.196a}$$

or

$$\lim_{|\lambda|>1,\,\lambda\to -1} q_1(\lambda) = -1, \qquad \lim_{|\lambda|>1,\,\lambda\to -1} q_2(\lambda) = 1. \tag{10.196b}$$

Similarly to the previous analysis, let $\lambda = -1+\eta$, where $|\eta| \ll 1$, then also $q(\lambda) = 1+\zeta$, where $|\zeta| \ll 1$ (recall, we are still interested in $q \longrightarrow 1$). Consider a particular case of real $\eta < 0$, then equation (10.195) yields $\zeta < 0$, i.e., $|q| < 1$. Consequently, if $|\lambda| > 1$ and $\lambda \longrightarrow -1$, then

$$\{q = q(\lambda) \longrightarrow 1\} \Longrightarrow \{|q| < 1\}.$$

In other words, this time it is the root q_1 that approaches $(1,0)$ as $\lambda \longrightarrow -1$, and the scenario that gets realized is (10.196a) rather than (10.196b). In contradistinction to the previous case, this presents a potential for instability. Indeed, the pair (λ, q_1), where $|q_1| < 1$ and $|\lambda| > 1$, would have implied the instability in the sense of Section 10.5.2 if $c_1 q_1^m$ were a genuine eigenfunction of problem (10.187) and λ if were the corresponding genuine eigenvalue. As we know, this is not the case. However, according to the first formula of (10.196a), the actual setup appears to be a limit of the admissible yet unstable situation. In other words, the combination of "almost an eigenfunction" $u_m = c_1 q_1^m$, $|q_1| < 1$, that satisfies $u_m \longrightarrow 0$ as $m \longrightarrow \infty$ with "almost an eigenvalue" $\lambda = \lambda(q_1)$, $|\lambda| > 1$, is unstable. While remaining unstable, this combination becomes more of a genuine eigenpair of problem (10.187) as $\lambda \longrightarrow -1$. Again, a rigorous proof of the instability is given in the framework of the GKS theory using the technique based on the Laplace transform.

Thus, we have seen that two scenarios are possible when λ approaches the unit circle from the outside. In one case, there may be an admissible root q of the characteristic equation that almost satisfies the boundary condition, see formula (10.196a), and this situation is prone to instability. Otherwise, see formula (10.194b), there is no admissible root q that would ultimately satisfy the boundary condition, and as such, no instability will be associated with this λ.

In the unstable case exemplified by formula (10.196a), the corresponding limit value of λ is called *the generalized eigenvalue,* see [GKO95, Chapter 13]. In particular, $\lambda = -1$ is a generalized eigenvalue of problem (10.187). We re-emphasize that it is not a genuine eigenvalue of problem (10.187), because when $\lambda = -1$ then $q_1 = 1$ and the eigenfunction $u_m = c q_1^m$ does not belong to the admissible class: $u_m \longrightarrow 0$ as $m \longrightarrow \infty$. In fact, it is easy to see that $\|u\|_2 = \infty$. However, it is precisely the generalized eigenvalues that cause the instability of a scheme, even in the case when the entire spectrum of the family of operators $\{\boldsymbol{R}_h\}$ belongs to the unit disk.

Accordingly, *the Kreiss necessary and sufficient condition of stability* requires that the spectrum of the family of operators be confined to the unit disk as before, and additionally, that the scheme have no generalized eigenvalues $|\lambda| = 1$. Scheme (10.184), (10.185a) violates the Kreiss condition at $\lambda = -1$ and as such, is unstable. The instability manifests itself via the computations presented in Figure 10.14.

As, however, this instability is only due to a generalized eigenvalue with $|\lambda| = 1$, it is relatively mild, as expected. On the other hand, if we were to replace the marginally unstable boundary condition (10.185a) with a truly unstable one in the sense of Section 10.5.2, then the effect on the stability of the scheme would have been much more drastic. Instead of (10.185a), consider, for example:

$$u_0^{p+1} = 1.05 \cdot u_1^{p+1}. \tag{10.197}$$

This boundary condition generates an eigenfunction $u_m = c_1 q_1^m$ of problem (10.187) with $q_1 = \frac{1}{1.05} < 1$. The corresponding eigenvalues are given by:

$$\lambda(q_1) = \frac{r}{2}\left(q_1 - \frac{1}{q_1}\right) \pm \sqrt{1 + \frac{r^2}{4}\left(q_1 - \frac{1}{q_1}\right)^2},$$

and for one of these eigenvalues we obviously have $|\lambda| > 1$. Therefore, the scheme is unstable according to Theorem 10.8.

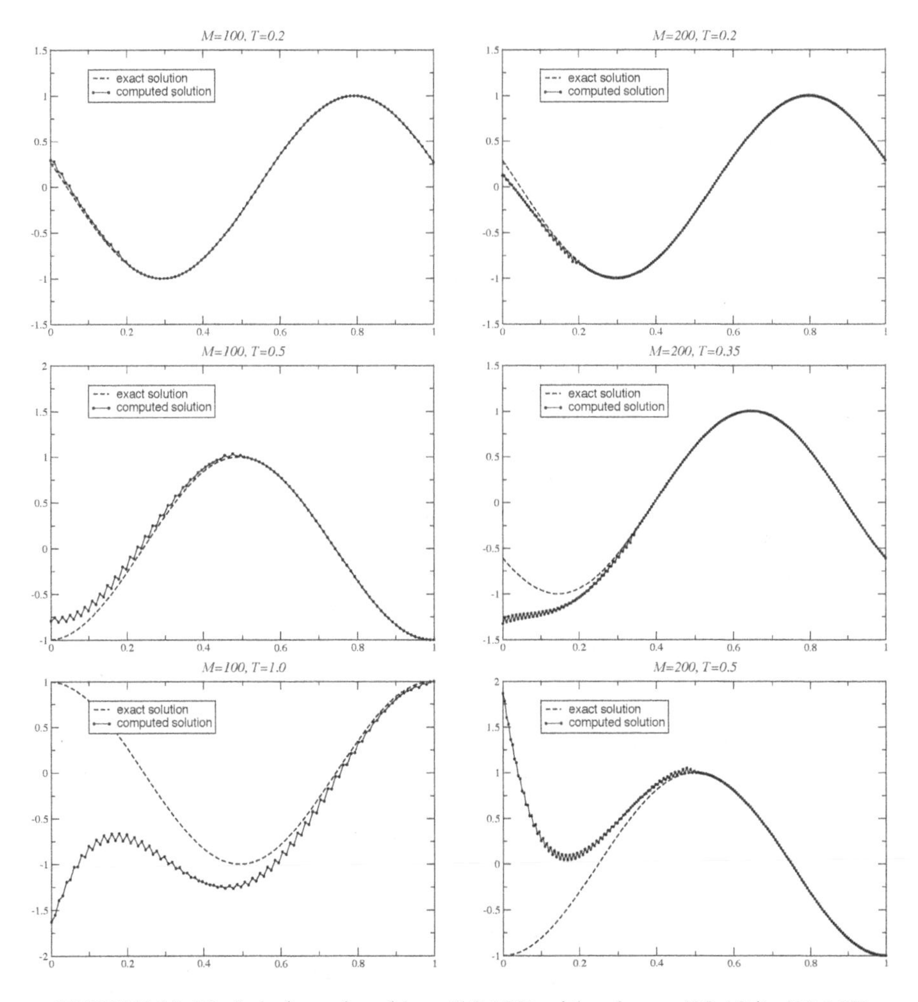

FIGURE 10.15: Solution of problem (10.183) with scheme (10.184), (10.197).

In Figure 10.15, we are showing the results of the numerical solution of problem (10.183) using the unstable scheme (10.184), (10.197). Comparing the plots in Figure 10.15 with those in Figure 10.14, we see that in the case of boundary condition

(10.197) the instability develops much more rapidly in time. Moreover, comparing the left column in Figure 10.15 that corresponds to the grid with $M = 100$ cells with the right column in the same figure that corresponds to $M = 200$, we see that the instability develops more rapidly on a finer grid, which is characteristic of an exponential instability.

Let now now analyze the second outflow boundary condition (10.185b):

$$u_0^{p+1} = u_0^p + r(u_1^p - u_0^p).$$

Unlike the extrapolation-type boundary condition (10.185a), which to some extent is arbitrary, boundary condition (10.185b) merely coincides with the first order upwind approximation of the differential equation itself that we have encountered previously on multiple occasions. To study stability, we again need to investigate three model problems: (10.186), (10.187), and (10.188). Obviously, only problem (10.187) changes due to the new boundary condition, where the other two stay the same. To find the corresponding λ and q we need to solve the characteristic equation (10.190a) along with a similar equation that stems from the boundary condition (10.185b):

$$\lambda = 1 - r + rq. \tag{10.198}$$

Consider first the case $r = 1$, then from equation (10.190a) we find that $\lambda = q$ or $\lambda = -1/q$, and equation (10.198) says that $\lambda = q$. Consequently, any solution $u_m^p = c_1 \lambda^p q_1^m$, $m = 0, 1, 2, \ldots$, of problem (10.187) will have $|\lambda| < 1$ as long as $|q| < 1$, i.e., there are only stable eigenfunctions/eigenvalues in the class $u_m \longrightarrow 0$ as $m \longrightarrow \infty$. Moreover, as $|\lambda| > 1$ is always accompanied by $|q| > 1$, we conclude that there are no generalized eigenvalues with $|\lambda| = 1$ either.

In the case $r < 1$, substituting λ from equation (10.198) into equation (10.190a) and subsequently solving for q, we find that there is only one solution: $q = 1$. For the corresponding λ, we then have from equation (10.198): $\lambda = 1$. Consequently, for $r < 1$, problem (10.187) also has no proper eigenfunctions/eigenvalues, which means that we again have $\overrightarrow{\Lambda} = \emptyset$. As far as the generalized eigenvalues, we only need to check one value of λ: $\lambda = 1$ (because $\lambda = -1$ does not satisfy equation (10.198) for $q = 1$). Let $\lambda = 1 + \eta$ and $q = 1 + \zeta$, where $|\eta| \ll 1$ and $|\zeta| \ll 1$. We then arrive at the same equation (10.195) that we obtained in the context of the previous analysis and conclude that $\lambda = 1$ does not violate the Kreiss condition, because $|\lambda| > 1$ implies $|q| > 1$. As such, the scheme (10.184), (10.185b) is stable.

Exercises

1. For the scalar Lax-Wendroff scheme [cf. formula (10.83)]:

$$\frac{u_m^{p+1} - u_m^p}{\tau} - \frac{u_{m+1}^p - u_{m-1}^p}{2h} - \frac{\tau}{2} \frac{u_{m+1}^p - 2u_m^p + u_{m-1}^p}{h^2} = 0,$$
$$p = 0, 1, \ldots, [T/\tau] - 1, \quad m = 1, 2, \ldots, M-1, \quad Mh = 1,$$
$$u_m^0 = \psi(x_m), \quad m = 0, 1, 2, \ldots, M,$$
$$\frac{u_0^{p+1} - u_0^p}{\tau} - \frac{u_1^p - u_0^p}{h} = 0, \quad u_M^{p+1} = 0, \quad p = 0, 1, \ldots, [T/\tau] - 1,$$

that approximates the initial boundary value problem:

$$\frac{\partial u}{\partial t} - \frac{\partial u}{\partial x} = 0, \quad 0 \le x \le 1, \quad 0 < t \le T,$$
$$u(x,0) = \psi(x), \quad u(1,t) = 0,$$

on the uniform rectangular grid: $x_m = mh$, $m = 0, 1, \dots, M$, $Mh = 1$, $t_p = p\tau$, $p = 0, 1, \dots, [T/\tau]$, find out when the Babenko-Gelfand stability criterion holds.

Answer. $r = \tau/h \le 1$.

scheme for the acoustics system, with the additional boundary conditions set using extrapolation of Riemann invariants along the characteristics. I could not solve it, stumbled across something in the algebra, and as such, removed the problem altogether. It is in the Russian edition of the book.

2.* Prove Theorem 10.6.

a) Prove the sufficiency part.

b) Prove the necessity part.

3.* Approximate the acoustics Cauchy problem:

$$\frac{\partial \boldsymbol{u}}{\partial t} - \boldsymbol{A}\frac{\partial \boldsymbol{u}}{\partial x} = \boldsymbol{\varphi}(x,t), \quad -\infty \le x \le \infty, \quad 0 < t \le T,$$
$$\boldsymbol{u}(x,0) = \boldsymbol{\psi}(x), \quad -\infty \le x \le \infty,$$
$$\boldsymbol{u}(x,t) = \begin{bmatrix} v(x,t) \\ w(x,t) \end{bmatrix}, \quad \boldsymbol{\varphi}(x) = \begin{bmatrix} \varphi^{(1)}(x) \\ \varphi^{(2)}(x) \end{bmatrix}, \quad \boldsymbol{\psi}(x) = \begin{bmatrix} \psi^{(1)}(x) \\ \psi^{(2)}(x) \end{bmatrix}, \quad \boldsymbol{A} = \begin{bmatrix} 0 & 1 \\ 1 & 0 \end{bmatrix},$$

with the Lax-Wendroff scheme:

$$\frac{\boldsymbol{u}_m^{p+1} - \boldsymbol{u}_m^p}{\tau} - \boldsymbol{A}\frac{\boldsymbol{u}_{m+1}^p - \boldsymbol{u}_{m-1}^p}{2h} - \frac{\tau}{2}\boldsymbol{A}^2\frac{\boldsymbol{u}_{m+1}^p - 2\boldsymbol{u}_m^p + \boldsymbol{u}_{m-1}^p}{h^2} = \boldsymbol{\varphi}_m^p,$$
$$p = 0, 1, \dots, [T/\tau] - 1, \quad m = 0, \pm 1, \pm 2, \dots,$$
$$\boldsymbol{u}_m^0 = \boldsymbol{\psi}(x_m), \quad m = 0, \pm 1, \pm 2, \dots.$$

Define $\boldsymbol{u}^p = \{\boldsymbol{u}_m^p\}$ and $\boldsymbol{\varphi}^p = \{\boldsymbol{\varphi}_m^p\}$, and introduce the norms as follows:

$$\|\boldsymbol{u}^{(h)}\|_{U_h} = \max_p \|\boldsymbol{u}^p\|, \quad \|\boldsymbol{f}^{(h)}\|_{F_h} = \max\left[\|\boldsymbol{\psi}\|, \max_p \|\boldsymbol{\varphi}^p\|\right],$$

where

$$\|\boldsymbol{u}^p\|^2 = \sum_m \left(|v_m^p|^2 + |w_m^p|^2\right), \quad \|\boldsymbol{\psi}\|^2 = \sum_m \left(|\psi^{(1)}(x_m)|^2 + |\psi^{(2)}(x_m)|^2\right),$$
$$\|\boldsymbol{\varphi}^p\|^2 = \sum_m \left(|\varphi^{(1)}(x_m,t_p)|^2 + |\varphi^{(2)}(x_m,t_p)|^2\right).$$

a) Show that when reducing the Lax-Wendroff scheme to the canonical form (10.141), inequalities (10.143) and (10.144) hold.

b) Prove that when $r = \frac{\tau}{h} \le 1$ the scheme is l_2 stable, and when $r > 1$ it is unstable.

Hint. To prove estimate (10.145) for the norms $\|\boldsymbol{R}_h^p\|$, first introduce the new unknown variables (called the Riemann invariants):

$$I_m^{(1)} = v_m + w_m \text{ and } I_m^{(2)} = v_m - w_m,$$

and transform the discrete system accordingly, and then employ the spectral criterion of Section 10.3.

4. Let the norm in the space U_h' be defined in the sense of l_2: $\|u\|_2 = \left[h \sum_{m=-\infty}^{\infty} |u_m|^2\right]^{1/2}$.

   Prove that in this case all complex numbers $\lambda(\alpha) = 1 - r + re^{i\alpha}$, $0 \leq \alpha < 2\pi$ [see formula (10.148)], belong to the spectrum of the transition operator $\boldsymbol{R}_h$ that corresponds to the difference Cauchy problem (10.147), where the spectrum is defined according to Definition 10.7.

   Hint. Construct the solution $u = \{u_m\}$, $m = 0, \pm 1, \pm 2, \ldots$, to the inequality that appears in Definition 10.7 in the form: $u_m = \begin{cases} q_1^m, & m \geq 0, \\ q_2^{-m}, & m < 0, \end{cases}$ where $q_1 = (1-\delta)e^{i\alpha}$, $q_2 = (1-\delta)e^{-i\alpha}$, and $\delta > 0$ is a small quantity.

5. Prove sufficiency in Theorem 10.7.

   Hint. Use expansion with respect to an orthonormal basis in U' composed of the eigenvectors of $\boldsymbol{R}_h$.

6. Compute the spectrum of the family of operators $\{\boldsymbol{R}_h\}$, $v = \boldsymbol{R}_h u$, given by the formulae:

$$\begin{aligned} v_m &= (1-r)u_m + ru_{m+1}, \quad m = 0, 1, \ldots, M-1, \\ v_M &= 0. \end{aligned}$$

   Assume that the norm is the maximum norm.

7. Prove that the spectrum of the family of operators $\{\boldsymbol{R}_h\}$, $v = \boldsymbol{R}_h u$, defined as:

$$\begin{aligned} v_m &= (1 - r + \gamma h)u_m + ru_{m+1}, \quad m = 0, 1, \ldots, M-1, \\ v_M &= u_M, \end{aligned}$$

   does not depend on the value of γ and coincides with the spectrum computed in Section 10.5.2 for the case $\gamma = 0$. Assume that the norm is the maximum norm.

   Hint. Notice that this operator is obtained by adding $\gamma h \boldsymbol{I}'$ to the operator $\boldsymbol{R}_h$ defined by formulae (10.142a) & (10.142b), and then use Definition 10.6 directly. Here $\boldsymbol{I}'$ is a modification of the identity operator that leaves all components of the vector u intact except the last component u_M that is set to zero.

8. Compute the spectrum of the family of operators $\{\boldsymbol{R}_h\}$, $v = \boldsymbol{R}_h u$, given by the formulae:

$$\begin{aligned} v_m &= (1-r)u_m + r(u_{m-1} + u_{m+1})/2, \quad m = 1, 2, \ldots, M-1, \\ v_M &= 0, \quad av_0 + bv_1 = 0, \end{aligned}$$

   where $a \in \mathbb{R}$ and $b \in \mathbb{R}$ are known and fixed. Consider the cases $|a| > |b|$ and $|a| < |b|$.

9.* Prove that the spectrum of the family of operators $\{\boldsymbol{R}_h\}$, $v = \boldsymbol{R}_h u$, defined by formulae (10.142a) & (10.142b) and analyzed in Section 10.5.2:

$$\begin{aligned} v_m &= (1-r)u_m + ru_{m+1}, \quad m = 0, 1, \ldots, M-1, \\ v_M &= u_M, \end{aligned}$$

will not change if the C norm: $\|u\| = \max_m |u_m|$ is replaced by the l_2 norm: $\|u\| = \left[h\sum_m u_m^2\right]^{1/2}$.

10. For the first order ordinary difference equation:

$$av_m + bv_{m+1} = f_m, \quad m = 0, \pm 1, \pm 2, \ldots,$$

the fundamental solution G_m is defined as *a bounded solution* of the equation:

$$aG_m + bG_{m+1} = \delta_m \equiv \begin{cases} 1, & m = 0, \\ 0, & m \neq 0. \end{cases}$$

a) Prove that if $|a/b| < 1$, then $G_m = \begin{cases} 0, & m \leq 0, \\ -\frac{1}{a}\left(-\frac{a}{b}\right)^m, & m \geq 1. \end{cases}$

b) Prove that if $|a/b| > 1$, then $G_m = \begin{cases} \frac{1}{a}\left(-\frac{a}{b}\right)^m, & m \leq 0, \\ 0, & m \geq 1. \end{cases}$

c) Prove that $v_m = \sum\limits_{k=-\infty}^{\infty} G_{m-k} f_k$.

11. Obtain energy estimates for the implicit first order upwind schemes that approximate problems (10.170), (10.172)*, and (10.174)*.

12.* Approximate problem (10.170) with the Crank-Nicolson scheme supplemented by one-sided differences at the left boundary $x = 0$:

$$\begin{gathered}
\frac{u_m^{p+1} - u_m^p}{\tau} - \frac{1}{2}\left[\frac{u_{m+1}^{p+1} - u_{m-1}^{p+1}}{2h} + \frac{u_{m+1}^p - u_{m-1}^p}{2h}\right] = 0, \\
m = 1, 2, \ldots, M-1, \quad p = 0, 1, \ldots, [T/\tau] - 1, \\
\frac{u_0^{p+1} - u_0^p}{\tau} - \frac{1}{2}\left[\frac{u_1^{p+1} - u_0^{p+1}}{h} + \frac{u_1^p - u_0^p}{h}\right] = 0, \quad u_M^p = 0, \\
p = 0, 1, \ldots, [T/\tau] - 1, \\
u_m^0 = \psi_m, \quad m = 0, 1, 2, \ldots, M.
\end{gathered} \tag{10.199}$$

a) Use an alternative definition of the l_2 norm: $\|u\|_2^2 = \frac{h}{2}(u_0^2 + u_M^2) + h\sum\limits_{m=1}^{M-1} u_m^2$ and develop an energy estimate for scheme (10.199).

Hint. Multiply the equation by $u_m^{p+1} + u_m^p$ and sum over the entire range of m.

b) Construct the schemes similar to (10.199) for the variable-coefficient problems (10.172) and (10.174) and obtain energy estimates.

13. Using the Kreiss condition, show that the leap-frog scheme (10.184) with the boundary condition:

$$u_0^{p+1} = u_1^p \tag{10.200a}$$

is stable, whereas with the boundary condition:

$$u_0^{p+1} = u_0^{p-1} + 2r(u_1^p - u_0^p) \tag{10.200b}$$

it is unstable.

14. Reproduce on the computer the results shown in Figures 10.14 and 10.15. In addition, conduct the computations using the leap-frog scheme with the boundary conditions (10.185b), (10.200a), and (10.200b), and demonstrate experimentally the stability and instability in the respective cases.
15.* Using the Kreiss condition, investigate stability of the Crank-Nicolson scheme applied to solving problem (10.183) and supplemented either with the boundary condition (10.185b) or with the boundary condition (10.200a).

10.6 Maximum Principle for the Heat Equation

Consider the following initial boundary value problem for a variable-coefficient heat equation, $a(x,t) > 0$:

$$\begin{gathered}\frac{\partial u}{\partial t} - a(x,t)\frac{\partial^2 u}{\partial x^2} = \varphi(x,t), \quad 0 \le x \le 1, \quad 0 \le t \le T,\\ u(x,0) = \psi(x), \quad 0 \le x \le 1,\\ u(0,t) = \vartheta(t), \quad u(1,t) = \chi(t), \quad 0 \le t \le T.\end{gathered} \tag{10.201}$$

To solve problem (10.201) numerically, we can use either an explicit or an implicit finite-difference scheme. We will analyze and compare both schemes. In doing so, we will see that quite often the implicit scheme has certain advantages compared to the explicit scheme, even though the algorithm of computing the solution with the help of an explicit scheme is simpler than that for the implicit scheme. The advantages of using an implicit scheme stem from its unconditional stability, i.e., stability that holds for any ratio between the spatial and temporal grid sizes.

10.6.1 An Explicit Scheme

We introduce a uniform grid on the interval $[0,1]$: $x_m = mh, \; m = 0,1,\dots,M,$ $Mh = 1,$ and build the scheme on the four-node stencil shown in Figure 10.3(left) (see page 331):

$$\begin{gathered}\frac{u_m^{p+1} - u_m^p}{\tau_p} - a(x_m,t_p)\frac{u_{m+1}^p - 2u_m^p + u_{m-1}^p}{h^2} = \varphi(x_m,t_p),\\ m = 1,2,\dots,M-1,\\ u_m^0 = \psi(x_m) \equiv \psi_m, \quad m = 0,1,\dots,M,\\ u_0^{p+1} = \vartheta(t_{p+1}), \quad u_M^{p+1} = \chi(t_{p+1}), \quad p \ge 0,\\ t_0 = 0, \quad t_p = \tau_0 + \tau_1 + \ldots + \tau_{p-1}, \quad p = 1,2,\dots.\end{gathered} \tag{10.202}$$

If the solution u_m^k, $m = 0,1,\dots,M,$ is already known for $k = 0,1,\dots,p,$ then, by virtue of (10.202), the values of u_m^{p+1} at the next time level $t = t_{p+1} = t_p + \tau_p$ can be computed with the help of an explicit formula:

$$u_m^{p+1} = u_m^p + \frac{\tau_p}{h^2} a(x_m, t_p)(u_{m+1}^p - 2u_m^p + u_{m-1}^p) + \tau_p \varphi(x_m, t_p), \\ m = 1, 2, \dots, M-1. \tag{10.203}$$

This explains why the scheme (10.202) is called *explicit*. Formula (10.203) allows one to easily march the discrete solution u_m^p from one time level to another.

According to the the principle of frozen coefficients (see Sections 10.4.1 and 10.5.1), when marching the solution by formula (10.203), one may only expect stability if the time step τ_p satisfies:

$$\tau_p \leq \frac{h^2}{2 \max_m a(x_m, t_p)}. \tag{10.204}$$

Let us prove that in this case scheme (10.202) is indeed stable, provided that the norms are defined as follows:

$$\|u^{(h)}\|_{U_h} = \max_p \max_m |u_m^p|,$$
$$\|f^{(h)}\|_{F_h} = \max\{\max_m |\psi(x_m)|, \max_p |\vartheta(t_p)|, \max_p |\chi(t_p)|, \max_{m,p} |\varphi(x_m, t_p)|\},$$

where $u^{(h)}$ and $f^{(h)}$ are the unknown solution and the right-hand side in the canonical form $\boldsymbol{L}_h u^{(h)} = f^{(h)}$ of the scheme (10.202). First, we will demonstrate that the following inequality holds known as the maximum principle:

$$\max_m |u_m^{p+1}| \leq \max\{\max_p |\vartheta(t_p)|, \max_p |\chi(t_p)|, \\ \max_m |u_m^p| + \tau_p \max_{m,p} |\varphi(x_m, t_p)|\}. \tag{10.205}$$

From formula (10.203) we have:

$$u_m^{p+1} = \left(1 - 2\frac{\tau_p}{h^2} a(x_m, t_p)\right) u_m^p + \frac{\tau_p}{h^2} a(x_m, t_p)(u_{m+1}^p + u_{m-1}^p) + \tau_p \varphi(x_m, t_p), \\ m = 1, 2, \dots, M-1.$$

If the necessary stability condition (10.204) is satisfied, then $1 - 2\frac{\tau_p}{h^2} a(x_m, t_p) \geq 0$, and consequently,

$$\begin{aligned} |u_m^{p+1}| &\leq \left(1 - 2\frac{\tau_p}{h^2} a(x_m, t_p)\right) \max_k |u_k^p| \\ &+ \frac{\tau_p}{h^2} a(x_m, t_p) \left(\max_k |u_k^p| + \max_k |u_k^p|\right) + \tau_p \max_{k,p} |\varphi(x_k, t_p)| \\ &= \max_k |u_k^p| + \tau_p \max_{k,p} |\varphi(x_k, t_p)|, \\ &m = 1, 2, \dots, M-1. \end{aligned}$$

Then, taking into account that $u_0^{p+1} = \vartheta(t_{p+1})$ and $u_M^{p+1} = \chi(t_{p+1})$, we obtain the maximum principle (10.205).

Let us now split the solution $u^{(h)}$ of the problem $L_h u^{(h)} = f^{(h)}$ into two components: $u^{(h)} = v^{(h)} + w^{(h)}$, where $v^{(h)}$ and $w^{(h)}$ satisfy the equations:

$$\boldsymbol{L}_h v^{(h)} = \begin{cases} 0 \\ \psi(x_m) \\ \vartheta(t_{p+1}) \\ \chi(t_{p+1}) \end{cases} \quad \text{and} \quad \boldsymbol{L}_h w^{(h)} = \begin{cases} \varphi(x_m, t_p) \\ 0 \\ 0 \\ 0 \end{cases}.$$

For the solution of the first sub-problem, the maximum principle (10.205) yields:

$$\begin{aligned} \max_m |v_m^{p+1}| &\le \max\{\max_p |\vartheta(t_p)|, \max_p |\chi(t_p)|, \max_m |v_m^p|\}, \\ \max_m |v_m^p| &\le \max\{\max_p |\vartheta(t_p)|, \max_p |\chi(t_p)|, \max_m |v_m^{p-1}|\}, \\ &\cdots\cdots\cdots\cdots\cdots\cdots\cdots\cdots \\ \max_m |v_m^1| &\le \max\{\max_p |\vartheta(t_p)|, \max_p |\chi(t_p)|, \max_m |\psi(x_m)|\}. \end{aligned}$$

For the solution of the second sub-problem, we obtain by virtue of the same estimate (10.205):

$$\begin{aligned} \max_m |w_m^{p+1}| &\le \max_m |w_m^p| + \tau_p \max_{m,p} |\varphi(x_m, t_p) \\ &\le \max_m |w_m^{p-1}| + (\tau_p + \tau_{p-1}) \max_{m,p} |\varphi(x_m, t_p) \\ &\cdots\cdots\cdots\cdots\cdots\cdots\cdots\cdots \\ &\le \underbrace{\max_m |w_m^0|}_{=0} + (\tau_p + \tau_{p-1} + \ldots + \tau_0) \max_{m,p} |\varphi(x_m, t_p) \\ &\le T \max_{m,p} |\varphi(x_m, t_p), \end{aligned}$$

where T is the terminal time, see formula (10.201). From the individual estimates established for v_m^{p+1} and w_m^{p+1} we derive:

$$\begin{aligned} \max_m |u_m^{p+1}| &= \max_m |v_m^{p+1} + w_m^{p+1}| \le \max_m |v_m^{p+1}| + \max_m |w_m^{p+1}| \\ &\le \max\{\max_p |\vartheta(t_p)|, \max_p |\chi(t_p)|, \max_m |\psi(x_m)|\} \\ &+ T \max_{m,p} |\varphi(x_m, t_p) \le c \|f^{(h)}\|_{F_h}, \end{aligned} \tag{10.206}$$

where $c = 2\max\{1, T\}$. Inequality (10.206) holds for all p. Therefore, we can write:

$$\|u^{(h)}\| \le c \|f^{(h)}\|_{F_h}, \tag{10.207}$$

which implies stability of scheme (10.202) in the sense of Definition 10.2. As such, we have shown that the necessary condition of stability (10.204) given by the principle of frozen coefficients is also sufficient for stability of the scheme (10.202).

Inequality (10.202) indicates that if the heat conduction coefficient $a(x_m,t_p)$ assumes large values near some point $(\tilde{x},\tilde{t})$, then computing the solution at time level $t=t_{p+1}$ will necessitate taking a very small time step $\tau=\tau_p$. Therefore, advancing the solution until a prescribed value of $t=T$ is reached may require an excessively large number of steps, which will make the computation impractical.

Let us also note that the foregoing restriction on the time step is of a purely numerical nature and has nothing to do with the physics behind problem (10.201). Indeed, this problem models the propagation of heat in a spatially one-dimensional structure, e.g., a rod, for which the heat conduction coefficient $a=a(x,t)$ may vary along the rod and also in time. Large values of $a(x,t)$ in a neighborhood of some point $(\tilde{x},\tilde{t})$ merely imply that this neighborhood can be removed, i.e., "cut off," from the rod without changing the overall pattern of heat propagation. In other words, we may think that this part of the rod consists of a material with zero heat capacity.

10.6.2 An Implicit Scheme

Instead of scheme (10.202), we can use the same grid and build the scheme on the stencil shown in Figure 10.3(right) (see page 331):

$$\begin{gathered}\frac{u_m^{p+1}-u_m^p}{\tau_p}-a(x_m,t_p)\frac{u_{m+1}^{p+1}-2u_m^{p+1}+u_{m-1}^{p+1}}{h^2}=\varphi(x_m,t_{p+1}),\\ m=1,2,\ldots,M-1,\\ u_m^0=\psi_m,\quad m=0,1,\ldots,M,\\ u_0^{p+1}=\vartheta(t_{p+1}),\quad u_M^{p+1}=\chi(t_{p+1}),\quad p\geq 0,\\ t_0=0,\quad t_p=\tau_0+\tau_1+\ldots+\tau_{p-1},\quad p=1,2,\ldots.\end{gathered}\tag{10.208}$$

Assume that the solution u_m^p, $m=0,1,\ldots,M$, at the time level $t=t_p$ is already known. According to formula (10.208), in order to compute the values of u_m^{p+1}, $m=0,1,\ldots,M$, at the next time level $t=t_{p+1}=t_p+\tau_p$ we need to solve the following system of linear algebraic equations with respect to $u_m\equiv u_m^{p+1}$:

$$\begin{gathered}u_0=\vartheta(t_{p+1}),\\ \alpha_m u_{m-1}+\beta_m u_m+\gamma_m u_{m+1}=f_m,\quad m=1,2,\ldots,M-1,\\ u_M=\chi(t_{p+1}),\end{gathered}\tag{10.209}$$

where

$$\begin{gathered}\alpha_m=\gamma_m=-\frac{\tau_p}{h^2}a(x_m,t_p),\quad \beta_m=1+2\frac{\tau_p}{h^2}a(x_m,t_p),\quad f_m=u_m^p+\tau_p\varphi(x_m,t_{p+1}),\\ m=1,2,\ldots,M-1,\\ \gamma_0=\alpha_M=0,\quad \beta_0=\beta_M=1.\end{gathered}$$

It is therefore clear that

$$\begin{gathered}|\beta_m|=|\alpha_m|+|\gamma_m|+\delta,\quad \delta=1>0,\\ m=0,1,2,\ldots,M,\end{gathered}$$

and because of the diagonal dominance, system (10.209) can be solved by the algorithm of tri-diagonal elimination described in Section 5.4.2. Note that in the case of scheme (10.208) there are no explicit formulae, i.e., closed form expressions such as formula (10.203), that would allow one to obtain the solution u_m^{p+1} at the upper time level given the solution u_m^p at the lower time level. Instead, when marching the solution in time one needs to solve systems (10.209) repeatedly, i.e., on every step, and that is why the scheme (10.208) is called *implicit*.

In Section 10.3.3 (see Example 7), we analyzed an implicit finite-difference scheme for the constant-coefficient heat equation and demonstrated that the von Neumann spectral stability condition holds for this scheme for any value of the ratio $r = \tau/h^2$. By virtue of the principle of frozen coefficients (see Section 10.4.1), the spectral stability condition will not impose any constraints on the time step τ even when the heat conduction coefficient $a(x,t)$ varies. This makes implicit scheme (10.208) unconditionally stable. It can be used efficiently even when the coefficient $a(x,t)$ assumes large values for some $(\tilde{x},\tilde{t})$. For convenience, when computing the solution of problem (10.201) with the help of scheme (10.208), one can choose a constant, rather than variable, time step $\tau_p = \tau$.

To conclude this section, let us note that unconditional stability of the implicit scheme (10.208) can be established rigorously. Namely, one can prove (see [GR87, § 28]) that the solution u_m^p of system (10.208) satisfies the same maximum principle (10.205) as holds for the explicit scheme (10.202). Then, estimate (10.207) for scheme (10.208) can be derived the same way as in Section 10.6.1.

Exercise

1. Let the heat conduction coefficient in problem (10.201) be defined as $a = 1 + u^2$, so that problem (10.201) becomes nonlinear.

   a) Introduce an explicit scheme and an implicit scheme for this new problem.

   b) Consider the following explicit scheme:

$$\frac{u_m^{p+1} - u_m^p}{\tau_p} - [1 + (u_m^p)^2]\frac{u_{m+1}^p - 2u_m^p + u_{m-1}^p}{h^2} = 0,$$

$$m = 1, 2, \dots, M-1,$$

$$u_m^0 = \psi(x_m) \equiv \psi_m, \quad m = 0, 1, \dots, M,$$

$$u_0^{p+1} = \vartheta(t_{p+1}), \quad u_M^{p+1} = \chi(t_{p+1}), \quad p \geq 0,$$

$$t_0 = 0, \quad t_p = \tau_0 + \tau_1 + \dots + \tau_{p-1}, \quad p = 1, 2, \dots.$$

   How should one choose τ_p, given the values of the solution u_m^p at the level p?

   c) Consider an implicit scheme based on the following finite-difference equation:

$$\frac{u_m^{p+1} - u_m^p}{\tau_p} - [1 + (u_m^{p+1})^2]\frac{u_{m+1}^{p+1} - 2u_m^{p+1} + u_{m-1}^{p+1}}{h^2} = 0.$$

   How should one modify this equation in order to employ the tri-diagonal elimination of Section 5.4 for the transition from u_m^p, $m = 0, 1, \dots, M$, to u_m^{p+1}, $m = 0, 1, \dots, M$?

Chapter 11

Discontinuous Solutions and Methods of Their Computation

A number of frequently encountered partial differential equations, such as the equations of hydrodynamics, elasticity, diffusion, and others, are in fact derived from the conservation laws of certain physical quantities (e.g., mass, momentum, energy, etc.). The conservation laws are typically written in an integral form, and the foregoing differential equations are only equivalent to the corresponding integral relations if the unknown physical fields (i.e., the unknown functions) are sufficiently smooth.

In every example considered previously (Chapters 9 and 10) we assumed that the differential initial (boundary) value problems had regular solutions. Accordingly, our key approach to building finite-difference schemes was based on approximating the derivatives using difference quotients. Subsequently, we could establish consistency with the help of Taylor expansions. However, the class of differentiable functions often appears too narrow for adequately describing many important physical phenomena and processes. For example, physical experiments demonstrate that the fields of density, pressure, and velocity in a supersonic flow of inviscid gas are described by functions with discontinuities (known as either shocks or contact discontinuities in the context of gas dynamics). It is important to emphasize that shocks may appear in the solution as time elapses, even in the case of smooth initial data.

From the standpoint of physical conservation laws, their integral form typically makes sense for both continuous and discontinuous solutions, because the corresponding discontinuous functions can typically be integrated. However, in the discontinuous case one can no longer obtain an equivalent differential formulation of the problem, because discontinuous functions cannot be differentiated. As a consequence, one can no longer enjoy the convenience of studying the properties of solutions to the differential equations, and can no longer use those for building finite-difference schemes. Therefore, prior to constructing the algorithms for computing discontinuous solutions of integral conservation laws, one will need to generalize the concept of solution to a differential initial (boundary) value problem. The objective is to make it meaningful and equivalent to the original integral conservation law even in the case of discontinuous solutions.

We will use a simple example below to illustrate the transition from the problem formulated in terms of an integral conservation law to an equivalent differential problem. The latter will only have a generalized (weak) solution, which is discontinuous. We will also demonstrate some approaches to computing this solution.

Assume that in the region (strip) $0 \le t \le T$ we need to find the function $u = u(x,t)$

that satisfies the integral equation:

$$\int_{\Gamma} \frac{u^k}{k} dx - \frac{u^{k+1}}{k+1} dt = 0 \tag{11.1a}$$

for an arbitrary closed contour Γ. The quantity k in formula (11.1a) is a fixed positive integer. We also require that $u = u(x,t)$ satisfies the initial condition:

$$u(x,0) = \psi(x), \quad -\infty < x < \infty. \tag{11.1b}$$

The left-hand side of equation (11.1a) can be interpreted as the flux of the vector field:

$$\phi(x,t) = \begin{bmatrix} u^k/k \\ u^{k+1}/(k+1) \end{bmatrix}$$

through the contour Γ. The requirement that the flux of this vector field through an arbitrary contour Γ be equal to zero can be thought of as a conservation law written in an integral form.

Problem (11.1a), (11.1b) provides the simplest formulation that leads to the formation of discontinuities albeit smooth initial data. It can serve as a model for understanding the methods of solving similar problems in the context of fluid dynamics.

11.1 Differential Form of an Integral Conservation Law

11.1.1 Differential Equation in the Case of Smooth Solutions

Let us first assume that the solution $u = u(x,t)$ to problem (11.1a), (11.1b) is continuously differentiable everywhere on the strip $0 \le t \le T$. We will then show that problem (11.1a), (11.1b) is equivalent to the following Cauchy problem:

$$\begin{aligned} &\frac{\partial u}{\partial t} + u\frac{\partial u}{\partial x} = 0, \quad 0 < t < T, \quad -\infty < x < \infty, \\ &u(x,0) = \psi(x), \quad -\infty < x < \infty. \end{aligned} \tag{11.2}$$

In the literature, the differential equation of (11.2) is known as the Burgers equation.

To establish the equivalence of problem (11.1a), (11.1b) and problem (11.2), we recall Green's formula. Let Ω be an arbitrary domain on the (x,t) plane, let $\Gamma = \partial\Omega$ be its boundary, and let the functions $\phi_1(x,t)$ and $\phi_2(x,t)$ have partial derivatives with respect to x and t on the domain Ω that are continuous everywhere up to the boundary Γ. Then, the following Green's formula holds:

$$\iint\limits_{\Omega} \left(\frac{\partial \phi_1}{\partial t} + \frac{\partial \phi_2}{\partial x} \right) dx dt = \int\limits_{\Gamma} \phi_1 dx - \phi_2 dt. \tag{11.3}$$

Identity (11.3) means that the integral of the divergence $\frac{\partial \phi_1}{\partial t} + \frac{\partial \phi_2}{\partial x}$ of the vector field $\phi = [\phi_1, \phi_2]^T$ over the domain Ω is equal to the flux of this vector field through the boundary $\Gamma = \partial\Omega$.

Using formula (11.3), we can write:

$$\int_\Gamma \frac{u^k}{k} dx - \frac{u^{k+1}}{k+1} dt = \iint_\Omega \left[\frac{\partial}{\partial t}\left(\frac{u^k}{k}\right) + \frac{\partial}{\partial x}\left(\frac{u^{k+1}}{k+1}\right) \right] dxdt. \tag{11.4}$$

Equality (11.4) implies that if a smooth function $u = u(x,t)$ satisfies the Burgers equation, see formula (11.2), then equation (11.1a) also holds. Indeed, if the Burgers equation is satisfied, then we also have:

$$u^{k-1}\left(\frac{\partial u}{\partial t} + u\frac{\partial u}{\partial x}\right) \equiv \frac{\partial}{\partial t}\left(\frac{u^k}{k}\right) + \frac{\partial}{\partial x}\left(\frac{u^{k+1}}{k+1}\right) = 0. \tag{11.5}$$

Consequently, the right-hand side of equality (11.4) becomes zero. The converse is also true: If a smooth function $u = u(x,t)$ satisfies the integral conservation law (11.1a), then at every point $(\tilde{x}, \tilde{t})$ of the strip $0 < t < T$ equation (11.5) holds, and hence equation (11.2) is true as well. To justify that, let us assume the opposite, and let us, for definiteness, take some point $(\tilde{x}, \tilde{t})$ for which:

$$\left.\frac{\partial}{\partial t}\left(\frac{u^k}{k}\right) + \frac{\partial}{\partial x}\left(\frac{u^{k+1}}{k+1}\right)\right|_{(\tilde{x},\tilde{t})} > 0.$$

Then, by continuity, we can always find a sufficiently small disk $\Omega \subset \{(x,t) | 0 < t < T\}$ centered at $(\tilde{x}, \tilde{t})$ such that

$$\left.\frac{\partial}{\partial t}\left(\frac{u^k}{k}\right) + \frac{\partial}{\partial x}\left(\frac{u^{k+1}}{k+1}\right)\right|_{(x,t)\in\Omega} > 0.$$

Hence, combining equations (11.1a) and (11.4) we obtain (recall, $\Gamma = \partial\Omega$):

$$0 = \int_\Gamma \frac{u^k}{k} dx - \frac{u^{k+1}}{k+1} dt = \iint_\Omega \left[\frac{\partial}{\partial t}\left(\frac{u^k}{k}\right) + \frac{\partial}{\partial x}\left(\frac{u^{k+1}}{k+1}\right) \right] dxdt > 0.$$

The contradiction we have just arrived at, $0 > 0$, proves that for smooth functions $u = u(x,t)$, problem (11.1a), (11.1b) and the Cauchy problem (11.2) are equivalent.

11.1.2 The Mechanism of Formation of Discontinuities

Let us first suppose that the solution $u = u(x,t)$ of the Cauchy problem (11.2) is smooth. Then we introduce the curves $x = x(t)$ defined by the following ordinary differential equation:

$$\frac{dx}{dt} = u(x,t). \tag{11.6}$$

These curves are known as characteristics of the Burgers equation: $u_t + uu_x = 0$.

Along every characteristic $x = x(t)$, the solution $u = u(x,t)$ can be considered as a function of the independent variable t only:

$$u(x,t) = u(x(t),t) = u(t).$$

Therefore, using (11.6) and (11.2), we can write:

$$\frac{du}{dt} = \frac{\partial u}{\partial t} + \frac{\partial u}{\partial x}\frac{dx}{dt} = 0.$$

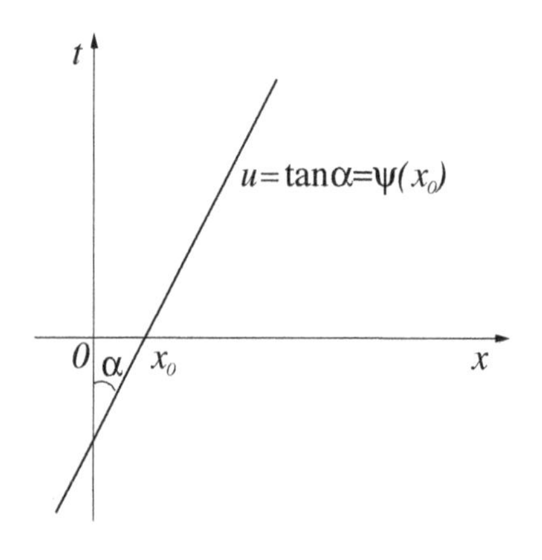

FIGURE 11.1: Characteristic (11.7).

Consequently, the solution is constant along every characteristic $x = x(t)$ defined by equation (11.6): $u(x,t)\big|_{x=x(t)} = \text{const}$. In turn, this implies that the characteristics of the Burgers equation are straight lines, because if $u = \text{const}$, then equation (11.6) yields:

$$x = ut + x_0. \tag{11.7}$$

In formula (11.7), x_0 denotes the abscissa of the point $(x_0, 0)$ on the (x,t) plane from which the characteristic originates, and $u = \psi(x_0) = \tan\alpha$ denotes its slope with respect to the vertical axis t, see Figure 11.1. Thus, we see that by specifying the initial function $u(x,0) = \psi(x)$ in the definition of problem (11.2), we fully determine the pattern of characteristics, as well as the values of the solution $u = u(x,t)$, at every point of the semi-plane $t > 0$.

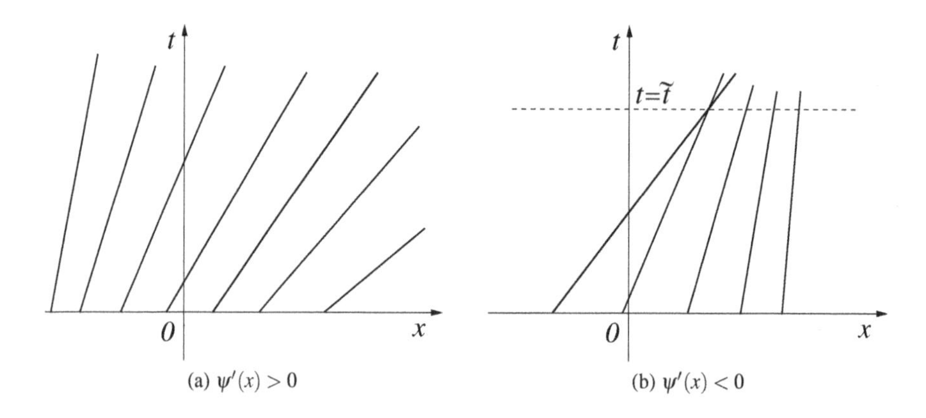

FIGURE 11.2: Schematic behavior of characteristics of the Burgers equation.

Note that under the assumption of a smooth solution, the characteristics of the Burgers equation cannot intersect. Otherwise, each characteristic would bring its own value of the solution into the intersection point and accordingly, the solution would not be a single-valued function. If the function $\psi = \psi(x)$ monotonically increases, i.e., if $\psi'(x) > 0$, then the angle α (see Figure 11.1) also increases as a function of x_0, and the characteristics do not intersect, see Figure 11.2(a). However, if $\psi = \psi(x)$ happens to be a monotonically decreasing function, i.e., if $\psi'(x) < 0$, then the characteristics converge toward one another as time elapses, and their intersections are unavoidable regardless of what regularity the function $\psi(x)$ has, see Figure 11.2(b). In this case, the smooth solution of problem (11.2) ceases to exist and a discontinuity forms, starting at the moment of time $t = \tilde{t}$ when at least two characteristics intersect, see Figure 11.2(b). The corresponding graphs of the function $u = u(x,t)$ at the moments of time $t = 0$, $t = \tilde{t}/2$, and $t = \tilde{t}$ are schematically shown in Figure 11.3.

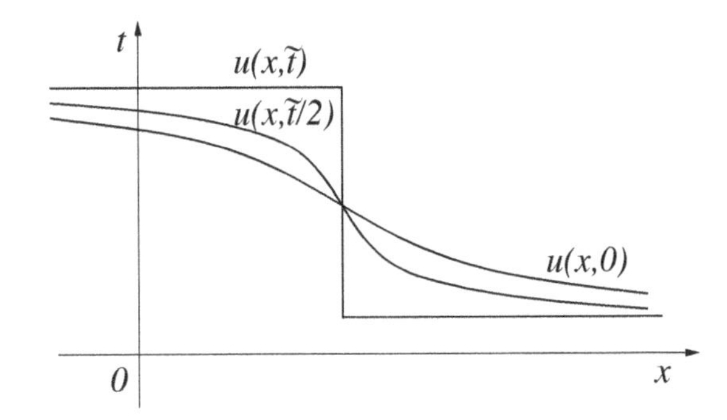

FIGURE 11.3: Schematic behavior of the solution to the Burgers equation (11.2) in the case $\psi'(x) < 0$.

11.1.3 Condition at the Discontinuity

Assume that there is a curve $L = \{(x,t)|x = x(t)\}$ inside the domain of the solution $u = u(x,t)$ of problem (11.1a), (11.1b), on which this solution undergoes a discontinuity of the first kind, alternatively called the jump. Assume also that when approaching this curve from the left and from the right we obtain the limit values:

$$u(x,t) = u_{\text{left}}(x,t)$$

and

$$u(x,t) = u_{\text{right}}(x,t),$$

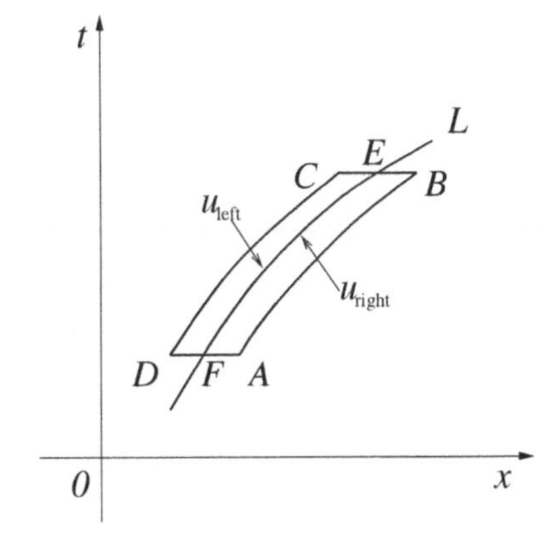

FIGURE 11.4: Contour of integration.

respectively, see Figure 11.4. It turns out that the values of $u_{\text{left}}(x,t)$ and $u_{\text{right}}(x,t)$ are related to the velocity of the jump $\dot{x} = dx/dt$ in a particular way, and altogether these quantities are not independent.

Let us introduce a contour $ABCD$ that straddles a part of the trajectory of the jump L, see Figure 11.4. The integral conservation law (11.1a) holds for any closed contour Γ and in particular for the contour $ABCDA$:

$$\int\limits_{ABCDA} \frac{u^k}{k} dx - \frac{u^{k+1}}{k+1} dt = 0. \tag{11.8}$$

Next, we start contracting this contour toward the curve L, i.e., start making it narrower. In doing so, the intervals BC and DA will shrink toward the points E and F, respectively, and the corresponding contributions to the integral (11.8) will obviously approach zero, so that in the limit we obtain:

$$\int\limits_{L'} \left[\frac{u^k}{k}\right] dx - \left[\frac{u^{k+1}}{k+1}\right] dt = 0,$$

or alternatively:

$$\int\limits_{L'} \left(\left[\frac{u^k}{k}\right] \frac{dx}{dt} - \left[\frac{u^{k+1}}{k+1}\right] \right) dt = 0.$$

Here the rectangular brackets: $[z] \stackrel{\text{def}}{=} z_{\text{right}} - z_{\text{left}}$ denote the magnitude of the jump of a given quantity z across the discontinuity, and L' denotes an arbitrary stretch of the jump trajectory.

Since L' is arbitrary, the integrand in the previous equality must be equal to zero at every point:

$$\left(\left[\frac{u^k}{k}\right] \frac{dx}{dt} - \left[\frac{u^{k+1}}{k+1}\right] \right) \Bigg|_{(x,t)\in L} = 0,$$

and consequently,

$$\frac{dx}{dt} = \left[\frac{u^{k+1}}{k+1}\right] \cdot \left[\frac{u^k}{k}\right]^{-1}. \tag{11.9}$$

Formula (11.9) indicates that for different values of k we can obtain different conditions at the trajectory of discontinuity L. For example, if $k = 1$ we have:

$$\frac{dx}{dt} = \frac{u_{\text{left}} + u_{\text{right}}}{2}, \tag{11.10}$$

and if $k = 2$ we can write:

$$\frac{dx}{dt} = \frac{2}{3} \frac{u_{\text{left}}^2 + u_{\text{left}} u_{\text{right}} + u_{\text{right}}^2}{u_{\text{left}} + u_{\text{right}}}.$$

We therefore conclude that the conditions that any discontinuous solution of problem (11.1a), (11.1b) must satisfy at the jump trajectory L depend on k.

11.1.4 Generalized Solution of a Differential Problem

Let us define a generalized solution to problem (11.2). This solution can be discontinuous, and we simply identify it with the solution to the integral conservation law (11.1a), (11.1b). Often the generalized solution is also called a weak solution.

In the case of a solution that has continuous derivatives everywhere, we have seen (Section 11.1.1) that the weak solution, i.e., solution to problem (11.1a), (11.1b), does not depend on k and coincides with the classical solution to the Cauchy problem (11.2). In other words, the solution in this case is a differentiable function $u = u(x,t)$ that turns the Burgers equation $u_t + uu_x = 0$ into an identity and also satisfies the initial condition $u(x,0) = \psi(x)$. We have also seen that even in the continuous case it is very helpful to consider both the integral formulation (11.1a), (11.1b) and the differential Cauchy problem (11.2). By staying only within the framework of problem (11.1a), (11.1b), we would make it more difficult to reveal the mechanism of the formation of discontinuity, as done in Section 11.1.2.

In the discontinuous case, the definition of a weak solution to problem (11.2) that we have just introduced does not enhance the formulation of problem (11.1a), (11.1b) yet; it merely renames it. Let us therefore provide an alternative definition of a generalized solution to problem (11.2). In doing so, we will only consider bounded solutions to problem (11.1a), (11.1b) that have continuous first partial derivatives everywhere on the strip $0 < t < T$, except perhaps for a set of smooth curves $x = x(t)$ along which the solution may undergo discontinuities of the first kind (jumps).

DEFINITION 11.1 *The function $u = u(x,t)$ is called a generalized (weak) solution to the Cauchy problem (11.2) that corresponds to the integral conservation law (11.1a) if:*

1. *The function $u = u(x,t)$ satisfies the Burgers equation [see formula (11.2)] at every point of the strip $0 < t < T$ that does not belong to the curves $x = x(t)$ which define the jump trajectories.*
2. *Condition (11.9) holds at the jump trajectory.*
3. *For every x for which the initial function $\psi = \psi(x)$ is continuous, the solution $u = u(x,t)$ is continuous at the point $(x,0)$ and satisfies the initial condition $u(x,0) = \psi(x)$.*

The proof of the equivalence of Definition 11.1 and the definition of a generalized solution given in the beginning of this section is the subject of Exercise 1.

Let us emphasize that in the discontinuous case, the generalized solution of the Cauchy problem (11.2) is not determined solely by equalities (11.2) themselves. It also requires that a particular conservation law, i.e., particular value of k, be specified that would relate the jump velocity with the magnitude of the jump across the discontinuity, see formula (11.9). Note that the general formulation of problem (11.1a), (11.1b) we have adopted provides no motivation for selecting any preferred value of k. However, in the problems that originate from real-world scientific applications, the integral conservation laws analogous to (11.1a) would normally express the conservation of some actual physical quantities. These conservation laws are, of course, well defined. In our subsequent considerations, we will assume for definiteness that the integral conservation law (11.1a) that corresponds to the value of $k = 1$ holds. Accordingly, condition (11.10) is satisfied at the jump trajectory.

In the literature, pioneering work on weak solutions was done by Lax [Lax54].

11.1.5 The Riemann Problem

Having defined the weak solutions of problem (11.2), see Definition 11.1, we will now see how a given initial discontinuity evolves when governed by the Burgers equation. In the literature, the problem of evolution of a discontinuity specified in the initial data is known as the Riemann problem.

Consider problem (11.2) with the following discontinuous initial function:

$$\psi(x) = \begin{cases} 2, & x < 0, \\ 1, & x > 0. \end{cases}$$

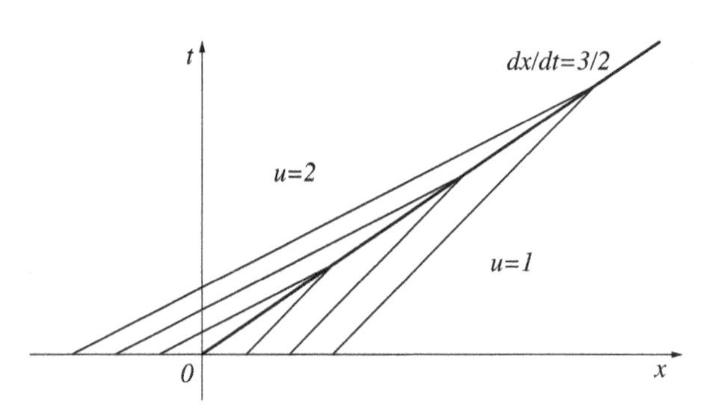

FIGURE 11.5: Shock.

The corresponding solution is shown in Figure 11.5. The evolution of the initial discontinuity consists of its propagation with the speed $\dot{x} = (2+1)/2 = 3/2$. This speed, which determines the slope of the jump trajectory in Figure 11.5, is obtained according to formula (11.10) as the arithmetic mean of the slopes of characteristics to the left and to the right of the shock. As can be seen, the characteristics on either side of the discontinuity impinge on it. In this case the discontinuity is called a shock; it is similar to the shock waves in the flows of ideal compressible fluid. One can show that the solution from Figure 11.5 is stable with respect to the small perturbations of initial data.

Next consider a different type of initial discontinuity:

$$\psi(x) = \begin{cases} 1, & x < 0, \\ 2, & x > 0. \end{cases} \tag{11.11}$$

One can obtain two alternative solutions for the initial data (11.11). The solution shown in Figure 11.6(a) has no discontinuities for $t > 0$. It consists of two regions with $u = 1$ and $u = 2$ bounded by the straight lines $\dot{x} = 1$ and $\dot{x} = 2$, respectively, that originate from $(0,0)$. These lines do not correspond to the trajectories of discontinuities, as the solution is continuous across both of them. The region in between these two lines is characterized by a family of characteristics that all originate at the

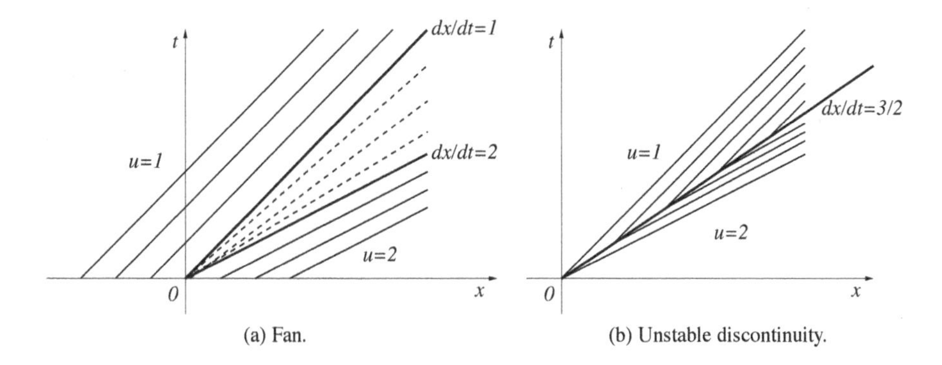

FIGURE 11.6: Solutions of the Burgers equation with initial data (11.11).

same point $(0,0)$. This structure is often referred to as a fan (of characteristics); in the context of gas dynamics it is known as the rarefaction wave.

The solution shown in Figure 11.6(b) is discontinuous; it consists of two regions $u=1$ and $u=2$ separated by the discontinuity with the same trajectory $\dot{x}=(1+2)/2=3/2$ as shown in Figure 11.5. However, unlike in the case of Figure 11.5, the characteristics in Figure 11.6(b) emanate from the discontinuity and veer away as the time elapses rather than impinge on it; this discontinuity is not a shock.

To find out which of the two solutions is actually realized, we need to incorporate additional considerations. Let us perturb the initial function $\psi(x)$ of (11.11) and consider:

$$\psi(x)=\begin{cases}1, & x<0,\\ 1+x/\varepsilon, & 0\le x\le\varepsilon,\\ 2, & x>\varepsilon.\end{cases} \tag{11.12}$$

The function $\psi(x)$ of (11.12) is continuous, and the corresponding solution $u(x,t)$ of problem (11.2) is determined uniquely. It is shown in Figure 11.7. When ε tends to zero, this solution approaches the continuous fan-type solution of problem (11.2), (11.11) shown in Figure 11.6(a). At the same time, the discontinuous solution of problem (11.2), (11.11) shown in Figure 11.6(b) appears unstable with respect to the small perturbations of initial data. Hence, it is the continuous solution with the fan that should

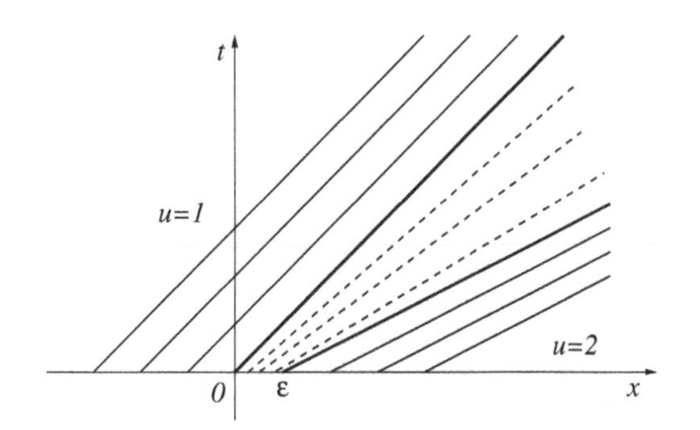

FIGURE 11.7: Solution of the Burgers equation with initial data (11.12).

be selected as the true solution of problem (11.2), (11.11), see Figure 11.6(a). As for the discontinuous solution from Figure 11.6(b), its exclusion due to the instability is similar to the exclusion of the so-called rarefaction shocks that appear as mathematical objects when analyzing the flows of ideal compressible fluid. Unstable solutions of this type are prohibited by the so-called entropy conditions that are introduced and analyzed in the theory of quasi-linear hyperbolic equations, see, e.g., [RJ83].

Exercises

1. Prove that the definition of a weak solution to problem (11.2) given in the very beginning of Section 11.1.4 is equivalent to Definition 11.1.

2.* Consider the following auxiliary problem:

$$\begin{gathered} \frac{\partial u}{\partial t} + u\frac{\partial u}{\partial x} = \mu\frac{\partial^2 u}{\partial x^2}, \quad 0 < t < T, \quad -\infty < x < \infty, \\ u(x,0) = \psi(x), \quad -\infty < x < \infty, \end{gathered} \tag{11.13}$$

where $\mu > 0$ is a parameter which is similar to viscosity in the context of fluid dynamics. The differential equation of (11.13) is parabolic rather than hyperbolic. It is known to have a smooth solution for any smooth initial function $\psi(x)$. If the initial function is discontinuous, the solution is also known to become smoother as time elapses.

Let $\psi(x) = 2$ for $x < 0$ and $\psi(x) = 1$ for $x > 0$. Prove that when $\mu \longrightarrow 0$, the solution of problem (11.13) approaches the generalized solution of problem (11.2) (see Definition 11.1) that corresponds to the conservation law (11.1a) with $k = 1$.

Hint. The solution $u = u(x,t)$ of problem (11.13) can be calculated explicitly with the help of the Hopf formula:

$$u(x,t) = \frac{\int\limits_{-\infty}^{\infty} (x-\xi)e^{-\frac{\lambda(x,\xi,t)}{2\mu}}d\xi}{\int\limits_{-\infty}^{\infty} te^{-\frac{\lambda(x,\xi,t)}{2\mu}}d\xi}, \quad \text{where} \quad \lambda(x,\xi,t) = \frac{(x-\xi)^2}{2t} + \int\limits_0^{\xi} \psi(\eta)d\eta.$$

More detail can be found in the monograph [RJ83], as well as in the original research papers [Hop50] and [Col51].

11.2 Construction of Difference Schemes

In this section, we will provide examples of finite-difference schemes for computing the generalized solution (see Definition 11.1) of problem (11.2):

$$\begin{gathered} \frac{\partial u}{\partial t} + u\frac{\partial u}{\partial x} = 0, \quad 0 < t < T, \quad -\infty < x < \infty, \\ u(x,0) = \psi(x), \quad -\infty < x < \infty, \end{gathered}$$

that corresponds to the integral conservation law (11.1a) with $k = 1$.

Let us assume for definiteness that $\psi(x) > 0$. Then, $u(x,t) > 0$ for all x and all $t > 0$. Naturally, the first idea is to consider the simple upwind scheme:

$$\begin{gathered} \frac{u_m^{p+1} - u_m^p}{\tau} + u_m^p \frac{u_m^p - u_{m-1}^p}{h} = 0, \\ m = 0, \pm 1, \pm 2, \dots, \quad p = 0, 1, \dots, \\ u_m^0 = \psi(mh). \end{gathered} \tag{11.14}$$

By freezing the coefficient u_m^p at a given grid location, we conclude that the resulting constant coefficient difference equation satisfies the maximum principle (see the analysis on page 315) when going from time level t_p to time level t_{p+1}, provided that the time step $\tau_p = t_{p+1} - t_p$ is chosen to satisfy the inequality:

$$r_p = \frac{\tau_p}{h} \leq \frac{1}{\sup_m |u_m^p|}.$$

Then, stability of scheme (11.14) can be expected. Furthermore, if the solution of problem (11.2) is smooth, then scheme (11.14) is clearly consistent. Therefore, it should converge, and indeed, numerical computations of the smooth test solutions corroborate the convergence.

However, if the solution of problem (11.2) is discontinuous, then no convergence of scheme (11.14) to the generalized solution of problem (11.2) is expected. The reason is that no information is built into the scheme (11.14) as to what specific conservation law is used for defining the generalized solution. The Courant, Friedrichs, and Lewy condition is violated here in the sense that the generalized solution given by Definition 11.1 depends on the specific value of k that determines the conservation law (11.1a), while the finite-difference solution does not depend on k.

Therefore, we need to use special techniques for the computation of generalized solutions. One approach is based on the equation with artificial viscosity $\mu > 0$:

$$\frac{\partial u}{\partial t} + u \frac{\partial u}{\partial x} = \mu \frac{\partial^2 u}{\partial x^2}.$$

As indicated previously (see Exercise 2 of Section 11.1), when $\mu \longrightarrow 0$ this equation renders a correct selection of the generalized solution to problem (11.2), i.e., it selects the solution that corresponds to the conservation law (11.1a) with $k = 1$. Moreover, it is also known to automatically filter out the unstable solutions of the type shown in Figure 11.6(b). Alternatively, we can employ the appropriate conservation law explicitly. In Sections 11.2.2 and 11.2.3 we will describe two different techniques based on this approach. The main difference between the two is that one technique introduces special treatment of the discontinuities, as opposed to all other areas in the domain of the solution, whereas the other technique employs uniform formulae for computation at all grid nodes.

11.2.1 Artificial Viscosity

Consider the following finite-difference scheme that approximates problem (11.13) and that is characterized by the artificially introduced small viscosity $\mu > 0$

[which is not present in the otherwise similar scheme (11.14)]:

$$\begin{gathered}\frac{u_m^{p+1}-u_m^p}{\tau}+u_m^p\frac{u_m^p-u_{m-1}^p}{h}=\mu\frac{u_{m+1}^p-2u_m^p+u_{m-1}^p}{h^2},\\ m=0,\pm1,\pm2,\dots,\quad p=0,1,\dots,\\ u_m^0=\psi(mh).\end{gathered}\tag{11.15}$$

Assume that $h \longrightarrow 0$, and also that the sufficiently small time step $\tau = \tau(h,\mu)$ is chosen accordingly to ensure stability. Then the solution $u^{(h)} = \{u_m^p\}$ of problem (11.15) converges to the generalized solution of problem (11.2), (11.10). The convergence is uniform in space and takes place everywhere except on the arbitrarily small neighborhoods of the discontinuities of the generalized solution. To ensure the convergence, the viscosity $\mu = \mu(h)$ must vanish as $h \longrightarrow 0$ with a certain (sufficiently slow) rate. Various techniques based on the idea of artificial dissipation (artificial viscosity) have been successfully implemented for the computation of compressible fluid flows, see, e.g., [RM67, Chapters 12 & 13] or [Tho95, Chapter 7]. Their common shortcoming is that they tend to smooth out the shocks. As an alternative, one can explicitly build the desired conservation law into the structure of the scheme used for computing the generalized solutions to problem (11.2).

11.2.2 The Method of Characteristics

In this method, we use special formulae to describe the evolution of discontinuities that appear in the process of computation, i.e., as the time elapses. These formulae are based on the condition (11.10) that must hold at the location of discontinuity. At the same time, in the regions of smoothness we use the differential form of the conservation law, i.e., the Burgers equation itself: $u_t + uu_x = 0$.

The key components of the method of characteristics are the following. For simplicity, consider a uniform spatial grid $x_m = mh$, $m = 0, \pm1, \pm2, \dots$. Suppose that the function $\psi(x)$ from the initial condition $u(x,0) = \psi(x)$ is smooth. From every point $(x_m, 0)$, we will "launch" a characteristic of the differential equation $u_t + uu_x = 0$. In doing so, we will assume that for the given function $\psi(x)$, we can always choose a sufficiently small τ such that on any time interval of duration τ, every characteristic intersects with no more than one neighboring characteristic. Take this τ and draw the horizontal grid lines $t = t_p = p\tau$, $p = 0, 1, 2, \dots$. Consider the intersection points of all the characteristics that emanate from the nodes $(x_m, 0)$ with the straight line $t = \tau$, and transport the respective values of the solution $u(x_m, 0) = \psi(x_m)$ along the characteristics from the time level $t = 0$ to these intersection points.

If no two characteristics intersect on the time interval $0 \le t \le \tau$, then we perform the next step, i.e., extend all the characteristics until the time level $t = 2\tau$ and again transport the values of the solution along the characteristics to the points of their intersection with the straight line $t = 2\tau$. If there are still no intersections between the characteristics for $\tau \le t \le 2\tau$, then we perform yet another step and continue this way until on some interval $t_p \le t \le t_{p+1}$ we find two characteristics that intersect.

Next assume that the two intersecting characteristics emanate from the nodes $(x_m, 0)$ and $(x_{m+1}, 0)$, see Figure 11.8. Then, we consider the midpoint of the interval $Q_{m+1}^{p+1}Q_m^{p+1}$ as the point that the incipient shock originates from. Subsequently, we replace the two different points Q_m^{p+1} and Q_{m+1}^{p+1} by one point Q (the midpoint), and assign two different values of the solution u_{left} and u_{right} to this point:

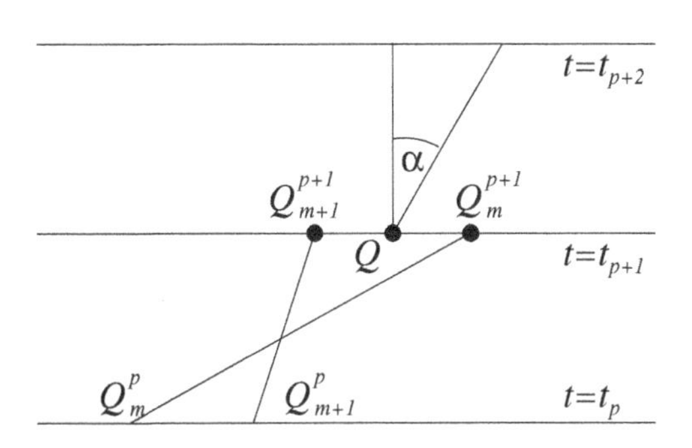

FIGURE 11.8: The method of characteristics.

$$u_{\text{left}}(Q) = u(Q_m^p), \quad u_{\text{right}}(Q) = u(Q_{m+1}^p).$$

From the point Q, we start up the trajectory of the shock until it intersects with the horizontal line $t = t_{p+2}$, see Figure 11.8. The slope of this trajectory with respect to the time axis is determined from the condition at the discontinuity (11.10):

$$\tan\alpha = \frac{u_{\text{left}} + u_{\text{right}}}{2}.$$

From the intersection point of the shock trajectory with the straight line $t = t_{p+2}$, we draw two characteristics backwards until they intersect with the straight line $t = t_{p+1}$. The slopes of these two characteristics are u_{left} and u_{right} from the previous time level, i.e., from the time level $t = t_{p+1}$. With the help of interpolation, we find the values of the solution u at the intersection points of the foregoing two characteristics with the line $t = t_{p+1}$. Subsequently, we use these values as the left and right values of the solution, respectively, at the location of the shock on the next time level $t = t_{p+2}$. This enables us to evaluate the new slope of the shock as the arithmetic mean of the new left and right values. Subsequently, the trajectory of the shock is extended one more time step τ, and the procedure repeats itself.

The advantage of the method of characteristics is that it allows us to track all of the discontinuities and to accurately compute the shocks starting from their inception. However, new shocks continually form in the course of the computation. In particular, the non-essential (low-intensity) shocks may start intersecting so that the overall solution pattern will become rather elaborate. Then, the computational logic becomes more complicated as well, and the requirements of the computer time and memory increase. This is a key disadvantage of the method of characteristics that singles out the discontinuities and computes those in a non-standard way.

11.2.3 Conservative Schemes. The Godunov Scheme

Finite-difference schemes that neither introduce the artificial dissipation (Section 11.2.1), nor explicitly use the condition at the discontinuity (11.10) (Section 11.2.2), must rely on the integral conservation law itself.

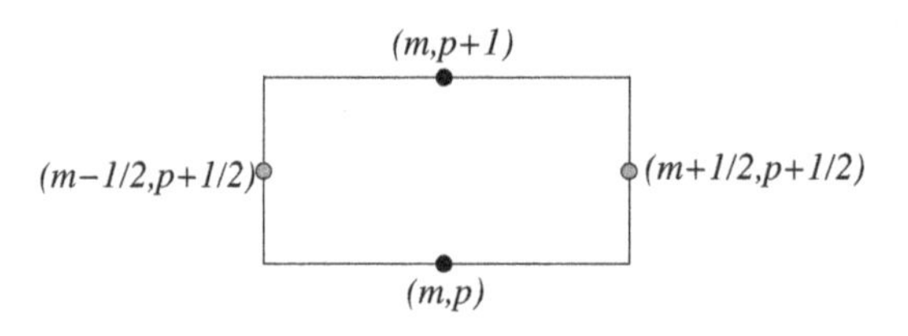

FIGURE 11.9: Grid cell.

Consider two families of straight lines on the plane (x,t): the horizontal lines $t = p\tau$, $p = 0,1,2,\ldots$, and the vertical lines $x = (m+1/2)h$, $m = 0,\pm 1,\pm 2,\ldots$. These lines partition the plane into rectangular cells. On the sides of each cell we will mark the respective midpoints, see Figure 11.9, and compose the overall grid D_h of the resulting nodes (we are not showing the coordinate axes in Figure 11.9).

The unknown function $[u]_h$ will be defined on the grid D_h. Unlike in many previous examples, when $[u]_h$ was introduced as a mere trace of the continuous exact solution $u(x,t)$ on the grid, here we define $[u]_h$ by averaging the solution $u(x,t)$ over the side of the grid cell (see Figure 11.9) that the given node belongs to:

$$[u]_h\Big|_{(x_m,t_p)} \stackrel{\text{def}}{=} \tilde{u}_m^p = \frac{1}{h}\int\limits_{x_{m-1/2}}^{x_{m+1/2}} u(x,t_p)dx,$$

$$[u]_h\Big|_{(x_{m+1/2},t_{p+1/2})} \stackrel{\text{def}}{=} \tilde{U}_{m+1/2}^{p+1/2} = \frac{1}{\tau}\int\limits_{t_p}^{t_{p+1}} u(x_{m+1/2},t)dt.$$

The approximate solution $u^{(h)}$ of our problem will be defined on the same grid D_h. The values of $u^{(h)}$ at the nodes (x_m,t_p) of the grid that belong to the horizontal sides of the rectangles, see Figure 11.9, will be denoted by u_m^p, and the values of the solution at the nodes $(x_{m+1/2},t_{p+1/2})$ that belong to the vertical sides of the rectangles will be denoted by $U_{m+1/2}^{p+1/2}$.

Instead of the discrete function $u^{(h)}$ defined only at the grid nodes (m,p) and $(m+1/2,p+1/2)$, let us consider its extension to the family piecewise constant functions of a continuous argument defined on the horizontal and vertical lines of the grid. In other words, we will think of the value u_m^p as associated with the entire horizontal side $\{(x,t)|x_{m-1/2} < x < x_{m+1/2},\ t = t_p\}$ of the grid cell that the node (x_m,t_p) belongs to, see Figure 11.9. Likewise, the value $U_{m+1/2}^{p+1/2}$ will be defined on the entire vertical grid interval $\{(x,t)|x = x_{m+1/2},\ t_p < t < t_{p+1}\}$. The relation between the quantities u_m^p and $U_{m+1/2}^{p+1/2}$, where $m = 0,\pm 1,\pm 2,\ldots$ and $p = 0,1,2,\ldots$, will be established based on the integral conservation law (11.1a) for $k = 1$:

$$\oint\limits_{\Gamma} udx - \frac{u^2}{2}dt = 0.$$

Let us consider the boundary of the grid cell from Figure 11.9 as the contour Γ:

$$-\oint_{\Gamma} u^{(h)}dx - \frac{[u^{(h)}]^2}{2}dt = 0. \tag{11.16}$$

Using the actual values of the foregoing piecewise constant function $u^{(h)}$, we can rewrite equality (11.16) as follows:

$$h[u_m^{p+1} - u_m^p] + \frac{\tau}{2}\left[\left(U_{m+1/2}^{p+1/2}\right)^2 - \left(U_{m-1/2}^{p+1/2}\right)^2\right] = 0. \tag{11.17}$$

Formula (11.17) implies that if there were a certain rule for the evaluation of the quantities $\left(U_{m+1/2}^{p+1/2}\right)^2$, $m = 0, \pm1, \pm2, \ldots$, given the quantities u_m^p, $m = 0, \pm1, \pm2, \ldots$, then we could have advanced one time step and obtained u_m^{p+1}, $m = 0, \pm1, \pm2, \ldots$. In other words, formula (11.17) would have enabled a marching algorithm. Note that the quantities $\left(U_{m+1/2}^{p+1/2}\right)^2$ are commonly referred to as fluxes. The reason is that the Burgers equation can be equivalently recast in the divergence form:

$$\frac{\partial u}{\partial t} + u\frac{\partial u}{\partial x} = \frac{\partial u}{\partial t} + \frac{\partial}{\partial x}\left(\frac{u^2}{2}\right) = \frac{\partial u}{\partial t} + \frac{\partial F(u)}{\partial x} = 0,$$

where $F(u)$ is known as the flux function for the general equation: $u_t + F_x(u) = 0$.

The fluxes $\left(U_{m+1/2}^{p+1/2}\right)^2$ can be computed using various approaches. However, regardless of the specific approach, the finite difference scheme (11.17) always appears conservative. This important characterization means the following.

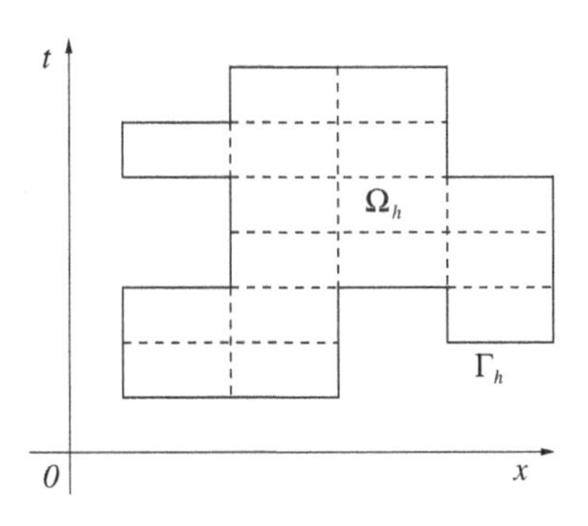

FIGURE 11.10: Grid domain.

Let us draw an arbitrary non-self-intersecting closed contour in the upper semi-plane $t > 0$ that would be completely composed of the grid segments, see Figure 11.10. Accordingly, this contour Γ_h encloses some domain Ω_h composed of the grid cells. Next, let us perform termwise summation of all the equations (11.17) that correspond to the grid cells from the domain Ω_h. Since equations (11.17) and (11.16) are equivalent and the only difference is in the notations, we may think that the summation is performed on equations (11.16). This immediately yields:

$$-\oint_{\Gamma_h} u^{(h)}dx - \frac{[u^{(h)}]^2}{2}dt = 0. \tag{11.18}$$

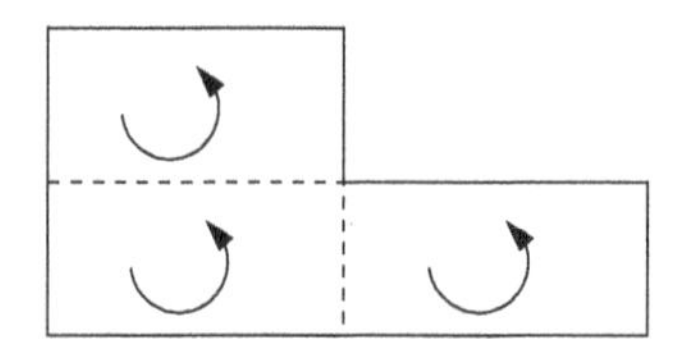

FIGURE 11.11: Directions of integration.

Formula (11.18) is easy to justify. The integrals along those sides of the grid rectangles that do not belong to the boundary Γ_h of the domain Ω_h, see Figure 11.10, mutually cancel out. Indeed, each of these interior grid segments belongs to two neighboring cells. Consequently, the integration of the function $u^{(h)}$ along each of those appears twice in the sum (11.18) and is conducted in the opposite directions, see Figure 11.11. Hence only the contributions due to the exterior boundary Γ_h do not cancel, and we arrive at equality (11.18).

Scheme (11.17) provides an example of what is known as conservative finite-difference schemes. In general, given a scheme, if we perform termwise summation of its finite-difference equations over the nodes of the grid domain Ω_h, and only those contributions to the sum remain that correspond to the boundary Γ_h, then the scheme is called conservative. Conservative schemes are analogous to the differential equations of divergence type, for example:

$$\operatorname{div}\phi = \frac{\partial \phi_1}{\partial t} + \frac{\partial \phi_2}{\partial x} = 0.$$

Once integrated over a two-dimensional domain Ω, these equations give rise to a contour integral along $\Gamma = \partial\Omega$, see formula (11.3). Finite-difference scheme (11.14) is not conservative, whereas scheme (11.17) is conservative.

REMARK 11.1 Let the grid function $u^{(h)}$ that satisfies equation (11.17) for $m = 0, \pm 1, \pm 2, \ldots$ and $p = 0, 1, 2, \ldots$, converge to a piecewise continuous function $u(x,t)$ when $h \longrightarrow 0$ uniformly on any closed region of space that does not contain the discontinuities. Also let $u^{(h)}$ be uniformly bounded with respect to h. Then, $u(x,t)$ satisfies the integral conservation law:

$$\oint_{\Gamma} u dx - \frac{u^2}{2} dt = 0,$$

where Γ is an arbitrary piecewise smooth contour. In other words, $u^{(h)}$ converges to the generalized solution of problem (11.2). This immediately follows from the possibility to approximate Γ by Γ_h, from formula (11.18), and from the convergence that we have just assumed. ☐

For the difference scheme (11.17) to make sense, we still need to define a procedure for evaluating the fluxes $\left(U_{m+1/2}^{p+1/2}\right)^2$ given the quantities u_m^p. To do that, we can exploit the solution to a special Riemann problem. This approach leads to one of the most popular and successful conservative schemes known as the Godunov scheme.

Assume that at the initial moment of time $t = 0$ the solution is specified as follows:

$$u(x,0) = \begin{cases} u_{\text{left}}, & x < 0, \\ u_{\text{right}}, & x > 0, \end{cases} \tag{11.19}$$

where $u_{\text{left}} = \text{const}$ and $u_{\text{right}} = \text{const}$. We can then obtain the corresponding generalized solution. In Section 11.1.5, we have analyzed specific examples for $u_{\text{left}} = 2$, $u_{\text{right}} = 1$ and for $u_{\text{left}} = 1$, $u_{\text{right}} = 2$. In the general case, the weak solution is obtained similarly. To compute the fluxes for the Godunov scheme, we need to know the value $U = u(0,t)$ of the generalized solution $u(x,t)$ at $x = 0$.

In fact, it is easy to see that for $x = 0$ the solution to the Riemann problem with the data (11.19) may be equal to u_{left}, u_{right}, or 0, depending on the specific values of u_{left} and u_{right}. In particular, if $u_{\text{left}} > 0$ and $u_{\text{right}} > 0$, then $u(0,t) = u_{\text{left}}$; in doing so, we will have the situation shown in Figure 11.5 if $u_{\text{left}} > u_{\text{right}}$, and the situation shown in Figure 11.6(a) if $u_{\text{left}} < u_{\text{right}}$. Likewise, if $u_{\text{left}} < 0$ and $u_{\text{right}} < 0$, then $u(0,t) = u_{\text{right}}$; again, we will have a shock if $|u_{\text{left}}| < |u_{\text{right}}|$ and a fan (rarefaction wave) if $|u_{\text{left}}| > |u_{\text{right}}|$. If $u_{\text{left}} > 0 > u_{\text{right}}$, then there will be a shock, and depending on whether $(u_{\text{left}} + u_{\text{right}})/2 > 0$ or $(u_{\text{left}} + u_{\text{right}})/2 < 0$ we will have either $u(0,t) = u_{\text{left}}$ or $u(0,t) = u_{\text{right}}$, respectively. If, conversely, $u_{\text{left}} < 0 < u_{\text{right}}$, then the vertical axis $x = 0$ will always be contained inside the fan, and $u(0,t) = 0$.

Accordingly, the quantity $U^{p+1/2}_{m+1/2} = U$ for scheme (11.17) will be determined by solving the Riemann problem for the initial discontinuity at the location $x = x_{m+1/2}$, when the constant values are specified to the left and to the right of the discontinuity: $u_{\text{left}} = u^p_m$ and $u_{\text{right}} = u^p_{m+1}$, respectively. For example, if $u^p_m > 0$, $m = 0, \pm 1, \pm 2, \ldots$, then $U^{p+1/2}_{m+1/2} = u_{\text{left}} = u^p_m$ for all $m = 0, \pm 1, \pm 2, \ldots$, and scheme (11.17) becomes:

$$\frac{u^{p+1}_m - u^p_m}{\tau} + \frac{1}{h}\left[\frac{(u^p_m)^2}{2} - \frac{(u^p_{m-1})^2}{2}\right] = 0,$$

$$u^0_m = \frac{1}{h}\int_{x_{m-1/2}}^{x_{m+1/2}} \psi(x)dx,$$

or alternatively:

$$\frac{u^{p+1}_m - u^p_m}{\tau} + \frac{u^p_{m-1} + u^p_m}{2} \cdot \frac{u^p_m - u^p_{m-1}}{h} = 0.$$

It is easy to see that when

$$r = \frac{\tau}{h} \leq \frac{1}{\sup_m |u^p_m|},$$

the maximum principle holds:

$$\sup_m |u^{p+1}_m| \leq \sup_m |u^p_m| \leq \ldots \leq \sup_m |u^0_m| \leq \sup_x |\psi(x)|. \tag{11.20}$$

Consequently, if $\tau \leq h[\sup_x |\psi(x)|]^{-1}$, we can expect that the resulting finite-difference scheme will be stable for some reasonable choice of norms. Numerical experiments corroborate that when the grid is refined, the solution $u^{(h)}$ of problem (11.17) with piecewise monotone and piecewise smooth initial data $\psi(x)$ converges to some function $u(x,t)$ that has a finite number of discontinuities. In doing so, the convergence is uniform outside of any neighborhood of the shocks.

Of course, the Godunov scheme, for which the fluxes are computed based on the solution to a specially chosen Riemann problem, is not the only conservative scheme for the problem (11.2), (11.10). Conservative schemes can be obtained using various approaches, in particular, one based on the predictor-corrector idea. For example, the well-known MacCormack scheme is conservative:

$$\begin{gathered}
\frac{\tilde{u}_m - u_m^p}{\tau} + \frac{1}{h}\left[\frac{(u_{m+1}^p)^2}{2} - \frac{(u_m^p)^2}{2}\right] = 0, \\
\frac{u_m^{p+1} - (u_m^p + \tilde{u}_m)/2}{\tau/2} + \frac{1}{h}\left[\frac{(\tilde{u}_m)^2}{2} - \frac{(\tilde{u}_{m-1})^2}{2}\right] = 0.
\end{gathered} \tag{11.21}$$

Proving this property is the subject of Exercise 2. Let us note, however, that even though the MacCormack scheme is consistent and conservative, its solutions do not always converge to the correct generalized solution of problem (11.2). Under certain conditions, the solution obtained by scheme (11.21) may contain a non-physical unstable discontinuity of the type shown in Figure 11.6(b), see [Tho99, Section 9.5].

Yet another example of a conservative predictor-corrector scheme consists of the implicit non-conservative predictor stage:

$$\frac{\tilde{u}_m - u_m^p}{\tau/2} + u_m^p \frac{\tilde{u}_m - \tilde{u}_{m-1}}{h} = 0 \tag{11.22a}$$

followed by the corrector stage rendered via scheme (11.17), where

$$U_{m+1/2}^{p+1/2} = \frac{1}{2}(\tilde{u}_m + \tilde{u}_{m+1}). \tag{11.22b}$$

Scheme (11.22a), (11.22b), (11.17) is second order accurate on smooth solutions of the Burgers equation. Its other properties are to be studied in Exercise 3.

A detailed discussion on numerical solution of conservation laws can be found, e.g., in [Tho99, Chapter 9], as well as in [LeV02].

Exercises

1. Prove the maximum principle (11.20).
2. Prove that the MacCormack scheme (11.21) is conservative.

   Hint. Define $\left(U_{m+1/2}^{p+1/2}\right)^2 = \frac{(u_{m+1}^p)^2}{2} + \frac{(\tilde{u}_m)^2}{2}$.
3. Analysis of scheme (11.22a), (11.22b), (11.17).
   a) Show that scheme (11.22a), (11.22b), (11.17) is conservative.
   b) For the linearized equation with the constant propagation speed: $u_t + au_x = 0$, show that the scheme is stable in the von Neumann sense for any $r = \tau/h$.

Chapter 12

Discrete Methods for Elliptic Problems

The simplest example of an elliptic partial differential equation is the Poisson equation in two space dimensions:

$$\Delta u \equiv \frac{\partial^2 u}{\partial x^2} + \frac{\partial^2 u}{\partial y^2} = \varphi(x,y). \tag{12.1}$$

Consider a domain $\Omega \in \mathbb{R}^2$ with the boundary $\Gamma = \partial\Omega$ and suppose that the following boundary condition supplements equation (12.1):

$$\left(\alpha u + \beta \frac{\partial u}{\partial n} \right) \bigg|_{\Gamma} = \psi(s), \tag{12.2}$$

where n denotes the direction of differentiation along the outward normal to Γ, $\alpha \geq 0$ and $\beta \geq 0$ are two fixed numbers, $\alpha^2 + \beta^2 = 1$, and s is the arc length along Γ. The functions $\varphi = \varphi(x,y)$ and $\psi = \psi(s)$ are assumed given. The combination of equations (12.1) and (12.2) is called a boundary value problem. If $\alpha = 1$ and $\beta = 0$, then it is a boundary value problem of the first kind, or Dirichlet boundary value problem. If $\alpha = 0$ and $\beta = 1$, then it is a boundary value problem of the second kind, or Neumann boundary value problem. If $\alpha > 0$ and $\beta > 0$, then it is a boundary value problem of the third kind, or Robin boundary value problem.

The foregoing three boundary value problems are the most frequently encountered problems for the Poisson equation, although other problems for this equation are also analyzed in the literature. Along with the Poisson equation, one often considers elliptic equations with variable coefficients, such as:

$$\frac{\partial}{\partial x}\left(a \frac{\partial u}{\partial x} \right) + \frac{\partial}{\partial y}\left(b \frac{\partial u}{\partial y} \right) = \varphi(x,y),$$

where $a = a(x,y) > 0$ and $b = b(x,y) > 0$. For this equation, typical formulations of boundary value problems also involve boundary conditions (12.2).

Elliptic equations and systems of elliptic equations typically appear when modeling various steady-state (i.e., time-independent) processes and phenomena. For example, the Poisson equation (12.1) may govern an electrostatic potential, the velocity potential for a steady-state flow of inviscid incompressible fluid, a steady-state temperature field in a homogeneous and isotropic material, the distribution of concentration for a substance that undergoes steady-state diffusion, etc.

Closed form analytic solutions for elliptic boundary value problems are only available on rare special occasions. Therefore, for most practical purposes those problems need to be solved numerically. In simple cases, when the solution across the entire domain is expected to vary only gradually, and the domain itself has a regular shape, one can use structured (often uniform) grids and construct finite-difference schemes by replacing the derivatives in the equation with appropriate difference quotients.

Otherwise, a structured grid (uniform or even non-uniform) may become unsuitable, because the regions of rapid variation of the solution, as well as, for example, the "bottleneck" parts of the domain (see, e.g., Figure 2.1 on page 59), will necessitate using a very fine grid. On the other hand, a structured grid, such as Cartesian, may appear difficult to refine locally in a multi-dimensional setting. An alternative is provided by unstructured grids, often obtained by triangulation. However, it is impossible to discretize differential equations on such grids by replacing the derivatives with difference quotients. Instead, one often builds discretizations that employ the variational formulation of the problem. The corresponding large group of numerical methods is commonly referred to as the method of finite elements.

In this chapter, we will address two questions. First, in Section 12.1 we will show that a simple central-difference scheme for the Dirichlet problem on a rectangular domain (we encountered this scheme previously, e.g., in Section 5.1.3) is consistent and stable, and as such, converges when the grid is refined. Then, in Section 12.2 we will provide a very brief and introductory account of the method of finite elements, including variational formulations of boundary value problems, the Ritz and Galerkin approximations, and basic concepts related to convergence.

Prior to actually starting the discussion on the subject, let us note that the discretization of linear elliptic boundary value problems, whether by finite differences or by finite elements, typically leads to large (i.e., high-order) systems of linear algebraic equations. Systems of this type can be solved either by direct methods analyzed in Chapter 5 of the book (Gaussian elimination including its tri-diagonal/banded versions, Cholesky factorization, separation of variables and FFT), or by iterative methods analyzed in Chapter 6 of the book (Richardson, Chebyshev, conjugate gradients, Krylov subspace methods — all normally with preconditioning, multigrid).

12.1 A Simple Finite-Difference Scheme. The Maximum Principle

Consider a Dirichlet problem for the Poisson equation on the square $\Omega = \{(x,y) \,|\, 0 < x < 1,\ 0 < y < 1\}$ with the boundary $\Gamma = \partial\Omega$:

$$\begin{aligned} \frac{\partial^2 u}{\partial x^2} + \frac{\partial^2 u}{\partial y^2} &= \varphi(x,y), \quad (x,y) \in \Omega, \\ u\Big|_{\Gamma} &= \psi(s). \end{aligned} \tag{12.3}$$

Introduce a uniform Cartesian grid on $\bar{\Omega} = \Omega \cup \Gamma$:

$$\Omega_h = \{(x_m, y_n) \,|\, (x_m, y_n) = (mh, nh),\ m, n = 0, 1, \ldots, M,\ h = 1/M\}, \tag{12.4}$$

where M is a fixed positive integer. The nodes of the grid Ω_h that belong strictly to the interior of the square Ω: $0 < x_m < 1$, $0 < y_n < 1$, will be called interior nodes, and the corresponding grid subset will be denoted Ω_h^0. Clearly,

$$\Omega_h^0 = \{(x_m, y_n) \,|\, (mh, nh),\ m, n = 1, 2, \ldots, M-1\}.$$

The nodes of the grid Ω_h that belong strictly to the boundary Γ of the square Ω will be called boundary nodes, and the corresponding grid subset will be denoted Γ_h. Likewise,

$$\Gamma_h = \{(x_m, y_n) \,|\, (mh, nh),\ m = 0, M,\ n = 0, M\}.$$

For problem (12.3), we use central differences and build the following scheme on the grid Ω_h of (12.4):

$$\boldsymbol{L}_h u^{(h)} = f^{(h)}, \tag{12.5}$$

where

$$\boldsymbol{L}_h u^{(h)} = \begin{cases} \dfrac{u_{m+1,n} - 2u_{mn} + u_{m-1,n}}{h^2} + \dfrac{u_{m,n+1} - 2u_{mn} + u_{m,n-1}}{h^2}, & (x_m, y_n) \in \Omega_h^0, \\ u_{mn}, & (x_m, y_n) \in \Gamma_h, \end{cases}$$

$$f^{(h)} = \begin{cases} \varphi(x_m, y_n), & (x_m, y_n) \in \Omega_h^0, \\ \psi(s_{mn}), & (x_m, y_n) \in \Gamma_h, \end{cases} \tag{12.6}$$

and $\psi(s_{mn})$ is the value of the function $\psi(s)$ at the node $(x_m, y_n) \in \Gamma_h$.

12.1.1 Consistency

Assume that the solution $u = u(x, y)$ of problem (12.3) has bounded partial derivatives up to the order four on the domain $\bar{\Omega}$. Then, we can use the Taylor formula with the remainder in the Lagrange form and write:

$$\begin{aligned} &\frac{u(x+h,y) - 2u(x,y) + u(x-h,y)}{h^2} + \frac{u(x,y+h) - 2u(x,y) + u(x,y+h)}{h^2} \\ &\qquad = \frac{\partial^2 u}{\partial x^2} + \frac{\partial^2 u}{\partial y^2} + \frac{h^2}{24}\left(\frac{\partial^4 u(x+\xi, y)}{\partial x^4} + \frac{\partial^4 u(x, y+\eta)}{\partial y^4}\right), \end{aligned} \tag{12.7}$$

where $0 \le \xi \le h$, $0 \le \eta \le h$. Consequently, when substituting the exact solution[1] $[u]_h$ into the left-hand side of scheme (12.5) we obtain:

$$\boldsymbol{L}_h [u]_h = \begin{cases} \varphi(x_m, y_n) + \mathcal{O}(h^2), & (x_m, y_n) \in \Omega_h^0, \\ \psi(s_{mn}) + 0, & (x_m, y_n) \in \Gamma_h. \end{cases}$$

[1] As before, $[u]_h$ denotes the trace of the solution $u(x, y)$ of problem (12.3) on the grid Ω_h.

Therefore, the corresponding residual (truncation error) $\delta f^{(h)}$ is given by:

$$\delta f^{(h)} = \begin{cases} \mathcal{O}(h^2), & (x_m, y_n) \in \Omega_h^0, \\ 0, & (x_m, y_n) \in \Gamma_h. \end{cases}$$

Next, we introduce the space of grid functions (discrete right-hand sides):

$$f^{(h)} = \begin{cases} \varphi_{mn}, & (x_m, y_n) \in \Omega_h^0, \\ \psi_{mn}, & (x_m, y_n) \in \Gamma_h, \end{cases}$$

and equip it with a maximum norm:

$$\|f^{(h)}\|_{F_h} \stackrel{\text{def}}{=} \max_{(mh,nh)\in\Omega_h^0} |\varphi_{mn}| + \max_{(mh,nh)\in\Gamma_h} |\psi_{mn}|.$$

Then, clearly:

$$\|\delta f^{(h)}\|_{F_h} = \mathcal{O}(h^2).$$

In other words, the finite-difference scheme (12.5), (12.6) is second order accurate, i.e., it approximates the Dirichlet problem (12.3) on its smooth solutions with the accuracy $\mathcal{O}(h^2)$, where h is the grid size.

12.1.2 Maximum Principle and Stability

Let U_h be the space of functions $u^{(h)} = \{u_{mn}\}$ defined on the grid Ω_h of (12.4). Similarly to F_h, we also supplement this space with a maximum norm:

$$\|u^{(h)}\|_{U_h} \stackrel{\text{def}}{=} \max_{(mh,nh)\in\Omega_h} |u_{mn}|.$$

To prove stability of the finite-difference scheme (12.5), (12.6) in the sense of Definition 10.2, we first need to show that problem (12.5) is uniquely solvable for any right-hand side $f^{(h)} \in F_h$ (this property is obviously not affected by the choice of the norm), and then establish the estimate:

$$\|u^{(h)}\|_{U_h} \leq c\|f^{(h)}\|_{F_h}, \tag{12.8}$$

where the constant c does not depend on either h or $f^{(h)}$.

LEMMA 12.1

Let the function $v^{(h)} = \{v_{mn}\}$ be defined on the grid Ω_h, and let it satisfy the inequality:

$$\mathbf{\Lambda}_h v^{(h)}\Big|_{(x_m,y_n)} \geq 0, \quad (x_m, y_n) \in \Omega_h^0, \tag{12.9}$$

at the interior nodes $(x_m, y_n) = (mh, nh)$ of the grid, where

$$\mathbf{\Lambda}_h v^{(h)}\Big|_{(x_m,y_n)} \stackrel{\text{def}}{=} \frac{v_{m+1,n} - 2v_{mn} + v_{m-1,n}}{h^2} + \frac{v_{m,n+1} - 2v_{mn} + v_{m,n-1}}{h^2}$$

is the finite-difference Laplace operator. Then, the maximum value of $v^{(h)}$ *on the grid* Ω_h *is attained at least at one point of the boundary* Γ_h.

PROOF Assume the opposite. Among all the nodes of the grid Ω_h at which the function $v^{(h)}$ assumes its maximum value, take one node (x_m, y_n) that has the maximum abscissa. According to our assumption, (x_m, y_n) is an interior node, i.e., $(x_m, y_n) \in \Omega_h^0$, and also $v_{mn} > v_{m+1,n}$. Then we have:

$$\Lambda_h v^{(h)}\Big|_{(x_m,y_n)} = \frac{(v_{m+1,n} - v_{mn}) + (v_{m-1,n} - v_{mn}) + (v_{m,n+1} - v_{mn}) + (v_{m,n-1} - v_{mn})}{h^2} < 0,$$

because the first term in the numerator is negative and the other three terms are non-positive. Contradiction with (12.9) proves the lemma. ∎

LEMMA 12.2
Let the function $v^{(h)} = \{v_{mn}\}$ *be defined on the grid* Ω_h, *and let it satisfy the inequality:*

$$\Lambda_h v^{(h)}\Big|_{(x_m,y_n)} \leq 0, \quad (x_m, y_n) \in \Omega_h^0, \tag{12.10}$$

at the interior nodes $(x_m, y_n) = (mh, nh)$ *of the grid. Then, the minimum value of* $v^{(h)}$ *on the grid* Ω_h *is attained at least at one point of the boundary* Γ_h.

PROOF The proof of Lemma 12.2 is similar to that of Lemma 12.1. ∎

THEOREM 12.1 (the maximum principle)
Let the function $v^{(h)} = \{v_{mn}\}$ *be defined on the grid* Ω_h, *and let it satisfy the finite-difference Laplace equation:*

$$\Lambda_h v^{(h)}\Big|_{(x_m,y_n)} = 0, \quad (x_m, y_n) \in \Omega_h^0, \tag{12.11}$$

at the interior nodes $(x_m, y_n) = (mh, nh)$ *of the grid. Then,* $v^{(h)}$ *assumes its maximum and minimum values at some nodes of the boundary* Γ_h.

PROOF The desired result is obtained as a combination of the results of Lemmas 12.1 and 12.2, because the function $v^{(h)}$ satisfies equation (12.11) if and only if it simultaneously satisfies inequalities (12.9) and (12.10). ∎

The discrete maximum principle, i.e., the property of the solutions of finite-difference equation (12.11) established by Theorem 12.1, is analogous to the continuous maximum principle for the solutions $v = v(x,y)$ of the Laplace equation

$v_{xx}+v_{yy}=0$. It is known that the solution $v(x,y)$ may only assume its maximum and minimum values at the boundary of the region where this solution is defined.

The maximum principle implies that the problem:

$$\boldsymbol{L}_h u^{(h)} = 0,$$

where

$$\boldsymbol{L}_h u^{(h)} = \begin{cases} \boldsymbol{\Lambda}_h u^{(h)}\big|_{(x_m,y_n)}, & (x_m,y_n)\in\Omega_h^0, \\ u^{(h)}\big|_{(x_m,y_n)}, & (x_m,y_n)\in\Gamma_h, \end{cases}$$

may only have a trivial solution, $u^{(h)}\equiv 0$, because the maximum and minimum values of this solution are attained at the boundary Γ_h and as such, are both equal to zero. Consequently, the finite-difference problem (12.5), (12.6) always has a unique solution for any given right-hand side $f^{(h)}$.

Let us now prove the stability estimate (12.8). According to formula (12.7), for an arbitrary polynomial of the second degree:

$$P(x,y) = ax^2+bxy+cy^2+dx+ey+f,$$

the following equality holds:

$$\boldsymbol{\Lambda}_h[P]_h = \frac{\partial^2 P}{\partial x^2}+\frac{\partial^2 P}{\partial y^2}, \tag{12.12}$$

because the fourth derivatives that appear in the remainder of Taylor's formula (12.7) are equal to zero. (Formula (12.12) is also true for cubic polynomials.)

Let $\varphi_{mn}\equiv\varphi(x_m,y_n)$ and $\psi_{mn}=\psi(s_{mn})$ be the right-hand sides of system (12.5), (12.6), and introduce an auxiliary grid function $P^{(h)}=\{P_{mn}\}$ defined on Ω_h:

$$P_{mn} = \frac{1}{4}[2-(x_m^2+y_n^2)]\cdot \max_{(x_k,x_l)\in\Omega_h^0}|\varphi_{kl}| + \max_{(x_k,x_l)\in\Gamma_h}|\psi_{kl}|.$$

As this function is obviously the trace of a quadratic polynomial on the grid Ω_h, formula (12.12) yields:

$$\boldsymbol{\Lambda}_h P^{(h)}\Big|_{(x_m,y_n)} = -\max_{(x_k,x_l)\in\Omega_h^0}|\varphi_{kl}|, \quad (x_m,y_n)\in\Omega_h^0.$$

Therefore, at the interior grid nodes, i.e., for $(x_m,y_n)\in\Omega_h^0$, the difference between the solution $u^{(h)}$ of problem (12.5), (12.6) and the function $P^{(h)}$ satisfies:

$$\boldsymbol{\Lambda}_h[u^{(h)}-P^{(h)}] = \boldsymbol{\Lambda}_h u^{(h)} - \boldsymbol{\Lambda}_h P^{(h)} = \varphi_{mn} + \max_{(x_k,x_l)\in\Omega_h^0}|\varphi_{kl}| \geq 0.$$

Then, according to Lemma 12.1, the difference $u^{(h)}-P^{(h)}$ assumes its maximum value at the boundary Γ_h. However, at the boundary this difference is non-positive:

$$\begin{aligned} u^{(h)}\Big|_{\Gamma_h} - P^{(h)}\Big|_{\Gamma_h} &= \psi_{mn}-P_{mn} \\ &= [\psi_{mn} - \max_{(x_k,x_l)\in\Omega_h}|\psi_{kl}|] + \frac{1}{4}[x_m^2+y_n^2-2]\cdot \max_{(x_k,x_l)\in\Omega_h^0}|\varphi_{kl}|, \end{aligned}$$

because everywhere on $\bar{\Omega}$ we have $x^2+y^2 \leq 2$ and consequently, both terms in square brackets on the right-hand side of the previous equality are non-positive. As such,

$$u_{mn} - P_{mn} \leq 0, \quad (x_m, y_n) \in \Omega_h,$$

i.e., $u^{(h)} \leq P^{(h)}$ everywhere on the grid Ω_h.

Likewise, for the sum $u^{(h)} + P^{(h)}$ we can prove that at the interior nodes $(x_m, y_n) \in \Omega_h^0$ it satisfies the inequality:

$$\mathbf{\Lambda}_h[u^{(h)} + P^{(h)}] \leq 0,$$

whereas at the boundary Γ_h the grid function $u^{(h)} + P^{(h)}$ is non-negative. Lemma 12.2 then implies:

$$u_{mn} + P_{mn} \geq 0, \quad (x_m, y_n) \in \Omega_h,$$

which means that $-P^{(h)} \leq u^{(h)}$ everywhere on the grid Ω_h.

Altogether, we then obtain:

$$|u_{mn}| \leq |P_{mn}| \leq \frac{1}{2} \cdot \max_{(x_k,x_l)\in\Omega_h^0} |\varphi_{kl}| + \max_{(x_k,x_l)\in\Gamma_h} |\psi_{kl}|$$

Consequently,

$$\|u^{(h)}\|_{U_h} = \max_{(x_m,y_n)\in\Omega_h} |u_{mn}| \leq \left(\max_{(x_k,x_l)\in\Omega_h^0} |\varphi_{kl}| + \max_{(x_k,x_l)\in\Gamma_h} |\psi_{kl}| \right) = \|f^{(h)}\|_{F_h},$$

which is equivalent to the stability estimate (12.8) with the constant $c = 1$.

12.1.3 Variable Coefficients

For a Dirichlet problem with variable coefficients:

$$\frac{\partial}{\partial x}\left(a\frac{\partial u}{\partial x}\right) + \frac{\partial}{\partial y}\left(b\frac{\partial u}{\partial y}\right) = \varphi(x,y), \quad (x,y) \in \Omega,$$
$$u\Big|_{\Gamma} = \psi(s),$$

where the functions $a = a(x,y) > 0$ and $b = b(x,y) > 0$ are sufficiently smooth on Ω, a finite-difference scheme can be constructed similarly to how it was done previously for constant coefficients. The expressions $\frac{\partial}{\partial x}\left(a\frac{\partial u}{\partial x}\right)$ and $\frac{\partial}{\partial y}\left(b\frac{\partial u}{\partial y}\right)$ can be approximately replaced by difference quotients according to the formulae:

$$\frac{\partial}{\partial x}\left(a(x,y)\frac{\partial u(x,y)}{\partial x}\right) \approx \tilde{\boldsymbol{\Lambda}}_{xx}u(x,y)$$
$$\stackrel{\text{def}}{=} \frac{1}{h}\left[a\left(x+\frac{h}{2},y\right)\frac{u(x+h,y)-u(x,y)}{h} - a\left(x-\frac{h}{2},y\right)\frac{u(x,y)-u(x-h,y)}{h}\right],$$
$$\frac{\partial}{\partial y}\left(b(x,y)\frac{\partial u(x,y)}{\partial y}\right) \approx \tilde{\boldsymbol{\Lambda}}_{yy}u(x,y)$$
$$\stackrel{\text{def}}{=} \frac{1}{h}\left[b\left(x,y+\frac{h}{2}\right)\frac{u(x,y+h)-u(x,y)}{h} - b\left(x,y-\frac{h}{2}\right)\frac{u(x,y)-u(x,y-h)}{h}\right].$$

Then, we obtain a scheme of type (12.5):

$$\boldsymbol{L}_h u^{(h)} = f^{(h)},$$

where

$$\boldsymbol{L}_h u^{(h)} = \begin{cases} \tilde{\boldsymbol{\Lambda}}_{xx}u^{(h)}\Big|_{(x_m,y_n)} + \tilde{\boldsymbol{\Lambda}}_{yy}u^{(h)}\Big|_{(x_m,y_n)}, & (x_m,y_n)\in\Omega_h^0, \\ u_{mn}, & (x_m,y_n)\in\Gamma_h, \end{cases}$$
$$f^{(h)} = \begin{cases} \varphi(x_m,y_n), & (x_m,y_n)\in\Omega_h^0, \\ \psi(s_{mn}), & (x_m,y_n)\in\Gamma_h. \end{cases} \tag{12.13}$$

Using the Taylor formula, one can make sure that this scheme is also second order accurate. Stability of this scheme can be proven by the methods similar to those of Section 12.1.2. In doing so, one would have to address some additional issues.

In practice, when solving real application-driven problems on the computer, one normally conducts theoretical analysis only for simple model formulations of the type outlined above. As for the actual error estimates, they are most often obtained experimentally, by comparing the results of computations on the grids with different values of the size h.

Exercises

1. Let the function $v^{(h)} = \{v_{mn}\}$ be defined on the grid Ω_h, and let it satisfy the finite-difference Laplace equation:

$$\boldsymbol{\Lambda}_h v^{(h)}\Big|_{(x_m,y_n)} = 0, \quad (x_m,y_n)\in\Omega_h^0,$$

at the interior nodes $(x_m,y_n) = (mh,nh)$ of the grid. Prove that either $v^{(h)} = \text{const}$ everywhere on Ω_h, or the maximum and minimum values of $v^{(h)}$ are not attained at any interior grid node (a strengthened maximum principle).

2. Prove that if the inequality $\boldsymbol{\Lambda}_n v^{(h)} \geq 0$ holds for all interior nodes of the grid Ω_h, and if this inequality becomes strict at least at one node from Ω_h^0, then $v^{(h)}$ does not attain its maximum value at any interior node.

3. Consider the following boundary value problem for the Poisson equation:

$$\frac{\partial^2 u}{\partial x^2}+\frac{\partial^2 u}{\partial y^2}=\varphi(x,y), \quad (x,y)\in\Omega,$$
$$u(x,y)=\psi_1(s), \quad x=1,\ y=0, \text{ or } y=1,$$
$$\frac{\partial u}{\partial x}=\psi_2(s), \quad x=0.$$

This problem can be approximated on the grid Ω_h by the scheme $\boldsymbol{L}_h u^{(h)}=f^{(h)}$, where:

$$\boldsymbol{L}_h u^{(h)}=\begin{cases}\Lambda_h u^{(h)}\Big|_{(x_m,y_n)} & (x_m,y_n)\in\Omega_h^0,\\ u_{mn}, & m=M,\ n=0, \text{ or } n=M,\\ \dfrac{u_{1,n}-u_{0,n}}{h}, & n=1,2,\dots,M-1,\end{cases}$$

$$f^{(h)}=\begin{cases}\varphi(x_m,y_n), & (x_m,y_n)\in\Omega_h^0,\\ \psi_1(s_{mn}), & m=M,\ n=0, \text{ or } n=M,\\ \psi_2(s_{0,n}), & n=1,2,\dots,M-1.\end{cases}$$

a) Prove that for any $\varphi(x_m,y_n)$, $\psi_1(s_{mn})$, and $\psi_2(s_{0,n})$, the system of finite-difference equations $\boldsymbol{L}_h u^{(h)}=f^{(h)}$ has a unique solution.

b)* Prove that if $\varphi(x_m,y_n)\geq 0$, $\psi_1(s_{mn})\leq 0$, and $\psi_2(s_{0,n})\geq 0$, then $u^{(h)}\leq 0$.

c)* Prove that for any $\varphi(x_m,y_n)$, $\psi_1(s_{mn})$, and $\psi_2(s_{0,n})$, the following estimate holds:

$$\max_{(x_m,y_n)\in\Omega_h}|u_{mn}|\leq c\left(\max_{(x_m,y_n)\in\Omega_h^0}|\varphi_{mn}|+\max_{m,n}|\psi_1(s_{mn})|+\max_{n}|\psi_2(s_{0,n})|\right),$$

where c is a constant that does not depend on h. Find the value of c.

12.2 The Notion of Finite Elements. Ritz and Galerkin Approximations

In this section, we introduce a popular and universal approach to the discretization of boundary value problems based on their alternative variational formulations. The resulting group of numerical techniques is commonly referred to as the method of finite elements. It facilitates construction of efficient and useful discretizations on irregular grids. Moreover, compared to the conventional finite-difference schemes, the finite element method typically imposes weaker constraints on the regularity of the solutions that are being approximated. The flexibility that exists in choosing the grid for finite elements allows one to refine this grid locally (if necessary). This capability may be crucial for resolving fine details of the solution in the regions where it is expected to undergo strong variations. Altogether, the method of finite elements may yield the same overall accuracy of approximation as a that of a traditional finite-difference scheme, but using (considerably) fewer nodes of the computational grid.

In our discussion, we first outline a standard variational method for solving boundary value problems for elliptic partial differential equations (Section 12.2.1). Its key idea is to equivalently replace the original problem by the problem of minimizing a specially chosen functional. Next, we describe the well-known Ritz method of building finite-dimensional approximations in the variational framework (Section 12.2.2). It turns out, however, that for some problems the corresponding functionals may have no minima. Accordingly, we need to use an alternative to the variational formulation along with a more general approximation technique that replaces the Ritz method; it is known as the Galerkin method (Section 12.2.3). Subsequently, we demonstrate how a particular choice of the basis for either Ritz method or Galerkin method leads to a specific realization of the finite element method (Section 12.2.4). Finally, we discuss the approximation properties of finite elements and illustrate the key components of the analysis of their convergence (Section 12.2.5).

The material of this section is more involved and can be skipped during the first reading. Overall, our account of variational formulations and finite elements is very brief. The reader can find more detail in the vast literature that exists on the subject. A classic text by Ciarlet [Cia02] can be a good starting point, followed by a more advanced text by Brenner and Scott [BS02]; we also recommend [AH05].

12.2.1 Variational Problem

Many boundary value problems for differential equations admit equivalent variational formulations. We begin with analyzing a simple example that still allows us to illustrate the key concepts. Namely, we consider a Dirichlet problem for the two-dimensional Poisson equation on the domain Ω with piecewise smooth boundary $\Gamma = \partial\Omega$.

Let W be a linear space of all real functions $w = w(x,y)$ that are continuous on $\bar{\Omega} = \Omega \cup \Gamma$ and also have bounded piecewise continuous first partial derivatives. (The derivatives may only undergo discontinuities of the first kind, or jumps, along no more than a finite collection of straight lines, which are different for different functions from W.) Introduce a norm on the space W:

$$\|w\|_W = \left[\iint\limits_{\Omega} w^2 dxdy + \iint\limits_{\Omega} \left(\frac{\partial w}{\partial x}\right)^2 dxdy + \iint\limits_{\Omega} \left(\frac{\partial w}{\partial y}\right)^2 dxdy\right]^{\frac{1}{2}} \tag{12.14}$$

Completion of the space W with respect to this norm (a Sobolev norm) yields the complete Sobolev space W_2^1, which is also denoted H^1.

Consider the following Dirichlet problem:

$$\begin{gathered}\Delta u \equiv \frac{\partial^2 u}{\partial x^2} + \frac{\partial^2 u}{\partial y^2} = \varphi(x,y), \quad (x,y) \in \Omega, \\ u\Big|_{\Gamma} = \psi(s),\end{gathered} \tag{12.15}$$

where s is the arc length along Γ, and $\varphi(x,y)$ and $\psi(s)$ are two given functions. We assume that the regularity of the data (i.e., that of the right-hand sides $\varphi(x,y)$ and

$\psi(s)$, as well as of the boundary Γ) is sufficient to ensure that the solution $u = u(x,y)$ of problem (12.15) exists and has continuous partial derivatives up to the second order everywhere on $\bar{\Omega} = \Omega \cup \Gamma$.

Before we can provide an alternative variational formulation of problem (12.15), we need to establish several auxiliary results. Let $v(x,y) \in W$ and $w(x,y) \in W$, then the first Green's formula holds (see, e.g., [TS63]):

$$\iint_{\Omega} v\Delta w dxdy = -\iint_{\Omega} \left(\frac{\partial v}{\partial x}\frac{\partial w}{\partial x} + \frac{\partial v}{\partial y}\frac{\partial w}{\partial y}\right) dxdy + \int_{\Gamma} v\frac{\partial w}{\partial n} ds, \tag{12.16}$$

where n is the direction of differentiation along the outward normal to Γ. Indeed, the integrand on the left-hand side can be recast as:

$$\begin{aligned} v\Delta w \equiv v\frac{\partial^2 w}{\partial x^2} + v\frac{\partial^2 w}{\partial y^2} &= \frac{\partial}{\partial x}\left(v\frac{\partial w}{\partial x}\right) + \frac{\partial}{\partial y}\left(v\frac{\partial w}{\partial y}\right) - \frac{\partial v}{\partial x}\frac{\partial w}{\partial x} - \frac{\partial v}{\partial y}\frac{\partial w}{\partial y} \\ &= \operatorname{div}(v \operatorname{grad} w) - \left(\frac{\partial v}{\partial x}\frac{\partial w}{\partial x} + \frac{\partial v}{\partial y}\frac{\partial w}{\partial y}\right), \end{aligned}$$

and then formula (12.16) is immediately obtained by the application of the Gauss theorem, see, e.g., [MF53, page 37]. Recall, this theorem says that the integral of the divergence of a vector field over the domain is equal to the flux of this field through the domain's boundary:

$$\iint_{\Omega} \operatorname{div} \boldsymbol{f} dxdy = \int_{\Gamma} (\boldsymbol{f}, \boldsymbol{n}) ds.$$

Moreover, by merely swapping the functions v and w and then subtracting the resulting equalities (12.16) from one another, we arrive at the second Green's formula:

$$\iint_{\Omega} (v\Delta w - w\Delta v) dxdy = \int_{\Gamma} \left(v\frac{\partial w}{\partial n} - w\frac{\partial v}{\partial n}\right) ds. \tag{12.17}$$

We will also need to introduce a standard scalar product on the space W:

$$(v,w) \stackrel{\text{def}}{=} \iint_{\Omega} vw dxdy, \quad v \;\&\; w \in W.$$

As all the functions in W are real, this scalar product is clearly commutative. Formula (12.16) can then be rewritten as:

$$(\Delta w, v) = -\iint_{\Omega} \left(\frac{\partial v}{\partial x}\frac{\partial w}{\partial x} + \frac{\partial v}{\partial y}\frac{\partial w}{\partial y}\right) dxdy + \int_{\Gamma} v\frac{\partial w}{\partial n} ds. \tag{12.18}$$

Finally, consider a linear subspace $\overset{\circ}{W} \subset W$ of all the functions from W that satisfy the homogeneous Dirichlet boundary condition at $\Gamma = \partial\Omega$: $\overset{\circ}{W} = \left\{w \,\middle|\, w \in W,\ w\big|_{\Gamma} = 0\right\}$.

Let $v = w \in \overset{\circ}{W}$, then we obtain:

$$(\Delta w, w) = -\iint_{\Omega} \left[\left(\frac{\partial w}{\partial x}\right)^2 + \left(\frac{\partial w}{\partial y}\right)^2 \right] dxdy. \tag{12.19}$$

This equality means that the operator $-\Delta$ is positive definite on the space $\overset{\circ}{W}$, i.e., for any $w \in \overset{\circ}{W}$, $w \not\equiv 0$, we have $-(\Delta w, w) > 0$.

THEOREM 12.2
Among all thc functions $w \in W$ that satisfy the Dirichlet boundary condition:

$$w\big|_{\Gamma} = \psi(s), \tag{12.20}$$

the solution $u = u(x,y)$ of problem (12.15) delivers a minimum numerical value to the functional:

$$I(w) = \iint_{\Omega} \left[\left(\frac{\partial w}{\partial x}\right)^2 + \left(\frac{\partial w}{\partial y}\right)^2 + 2\varphi w \right] dxdy. \tag{12.21}$$

PROOF Let $w = w(x,y)$ be a fixed function from the space W and let it satisfy boundary condition (12.20). Using formula (12.18), we can transform expression (12.21) into:

$$I(w) = -(\Delta w, w) + 2(\varphi, w) + \int_{\Gamma} \psi \frac{\partial w}{\partial n} ds. \tag{12.22}$$

Next, denote: $\xi(x,y) = w(x,y) - u(x,y)$. As $u(x,y)$ is twice continuously differentiable and $w \in W$, we also have $\xi \in W$ and moreover, $\xi\big|_{\Gamma} = 0$, i.e., $\xi \in \overset{\circ}{W}$. Then,

$$\begin{aligned}
I(w) &= I(u+\xi) = -(\Delta(u+\xi), u+\xi) + 2(\varphi, u+\xi) + \int_{\Gamma} \psi \frac{\partial (u+\xi)}{\partial n} ds \\
&= \underbrace{-(\Delta u, u) + 2(\varphi, u) + \int_{\Gamma} \psi \frac{\partial u}{\partial n} ds}_{=I(u) \text{ according to (12.22)}} - (\Delta u, \xi) - (\Delta \xi, u) - (\Delta \xi, \xi) \\
&\quad + 2(\varphi, \xi) + \int_{\Gamma} \psi \frac{\partial \xi}{\partial n} ds - \underbrace{\int_{\Gamma} \xi \frac{\partial u}{\partial n} ds}_{=0 \text{ as } \xi \in \overset{\circ}{W}} \\
&= I(u) - (\Delta \xi, \xi) + \underbrace{(\Delta u, \xi) - (\Delta \xi, u) - \int_{\Gamma} \xi \frac{\partial u}{\partial n} ds + \int_{\Gamma} \psi \frac{\partial \xi}{\partial n} ds}_{=0 \text{ according to (12.17)}},
\end{aligned}$$

where we substituted $u\big|_\Gamma = \psi$ and $\varphi = \Delta u$. As such, we see that $I(w) = I(u) - (\Delta(w-u), w-u)$. In other words, the value of the functional $I(w)$ for an arbitrary $w \in W$ is equal to its fixed value $I(u)$ on the solution of problem (12.15) plus a positive increment, because the operator $-\Delta$ is positive definite on $\overset{\circ}{W}$, see (12.19). Therefore, the minimum of the functional $I(w)$ is reached at the solution u of problem (12.15), which completes the proof. □

The statement inverse to that of Theorem 12.2 is also true. Namely, the function $w \in W$ that minimizes the functional (12.21), satisfies boundary condition (12.20), and is sufficiently smooth, also solves the Poisson equation of (12.15). In this context, the Poisson equation can be interpreted as the Lagrange-Euler equation for the functional $I(w)$.

Altogether, boundary value problem (12.15) admits the following variational formulation: *Among all the functions from the space W that satisfy boundary condition (12.20) find the one that minimizes the functional $I(w)$ given by formula (12.21).*

REMARK 12.1 Consider a particular case of the homogeneous boundary condition (12.20), i.e., $\psi(s) \equiv 0$. Then, formula (12.22) implies that

$$I(w) = -(\Delta w, w) + 2(\varphi, w). \tag{12.23}$$

According to Theorem 12.2, the functional $I(w)$ is minimized by the solution $u = u(x,y)$ of the Poisson equation $\Delta u = \varphi$ subject to the zero Dirichlet boundary condition: $u\big|_\Gamma = 0$. Hence there is a clear analogy with the considerations of Section 5.5, where the solution $\boldsymbol{x}$ of the linear system $\boldsymbol{A}\boldsymbol{x} = \boldsymbol{f}$, $\boldsymbol{A} > 0$, would deliver a minimum value to the quadratic function given by formula (5.79):

$$F(\boldsymbol{x}) = (\boldsymbol{A}\boldsymbol{x}, \boldsymbol{x}) - 2(\boldsymbol{f}, \boldsymbol{x}).$$

Indeed, we only need to substitute $\boldsymbol{A}$ for $-\Delta$ and $\boldsymbol{f}$ for φ. □

A standard approach to the numerical solution of the foregoing variational problem consists of building a sequence of admissible functions $w_N(x,y) \in W$, $w_N\big|_\Gamma = \psi(s)$, $N = 1, 2, \ldots$, such that

$$\lim_{N \to \infty} I(w_N) = I(u), \tag{12.24}$$

where $u = u(x,y)$ is the desired solution. This sequence is called a minimizing sequence. By choosing the term w_N with sufficiently large N, we can make the difference between $I(w_N)$ and $I(u)$ arbitrarily small. In its own turn, a small difference between $I(w_N)$ and $I(u)$ translates into a small difference between $w_N(x,y)$ and $u(x,y)$. This can be proved using the Friedrichs inequality:

$$\iint_\Omega w^2 dxdy \le C \cdot \left[\iint_\Omega \left(\frac{\partial w}{\partial x}\right)^2 dxdy + \iint_\Omega \left(\frac{\partial w}{\partial y}\right)^2 dxdy + \int_\Gamma w^2 ds \right], \tag{12.25}$$

which holds, in particular, for any $w \in W$. C is a constant that does not depend on w.

Let $\xi_N(x,y) = w_N(x,y) - u(x,y)$, which means $\xi_N\big|_\Gamma = 0$, $N = 1,2,\ldots$. Then, applying the Friedrichs inequality to the function $\xi_N(x,y)$ and using the equality $I(u+\xi_n) = I(u) - (\Delta\xi_N, \xi_N)$ along with (12.19), we obtain:

$$\begin{aligned}\|w_N - u\|_W^2 = \|\xi_N\|_W^2 &= \iint_\Omega \xi_N^2 dxdy + \iint_\Omega \left(\frac{\partial \xi_N}{\partial x}\right)^2 dxdy + \iint_\Omega \left(\frac{\partial \xi_N}{\partial y}\right)^2 dxdy \\ &\leq (C+1)\cdot\left[\iint_\Omega \left(\frac{\partial \xi_N}{\partial x}\right)^2 dxdy + \iint_\Omega \left(\frac{\partial \xi_N}{\partial y}\right)^2 dxdy\right] \\ &= (C+1)\cdot[I(w_N) - I(u)]. \end{aligned} \tag{12.26}$$

In other words, convergence of the minimizing sequence in the sense of (12.24) implies convergence of the sequence of functions: $w_N = w_N(x,y)$ to the solution $u = u(x,y)$ on the domain Ω as $N \longrightarrow \infty$ in the sense of the Sobolev norm (12.14).

12.2.2 The Ritz Method

In the previous section we saw that the functions $w \in W$, $w\big|_\Gamma = \psi(s)$, for which the values of the functional $I(w)$ are close to $I(u)$, also provide a good approximation to the solution u of problem (12.15) in the sense of the Sobolev norm (12.14). Ritz in 1908 proposed a special method for actually building those approximate solutions, which was later named after him. We illustrate the Ritz method using the Dirichlet problem (12.15) as an example and also assuming for simplicity that the boundary condition is homogeneous: $u\big|_\Gamma = 0$, i.e., $\psi(s) \equiv 0$.

As before, let $\overset{\circ}{W} \subset W$ be the linear subspace of all functions from W that satisfy the homogeneous Dirichlet boundary condition at $\Gamma = \partial\Omega$: $\overset{\circ}{W} = \left\{w \,\middle|\, w \in W,\ w\big|_\Gamma = 0\right\}$. Specify a positive integer number N and fix some N linearly independent functions:

$$w_n^{(N)} \in \overset{\circ}{W}, \quad n = 1,2,\ldots,N. \tag{12.27}$$

Consider now a linear span of the functions $w_n^{(N)}$ of (12.27), which we denote by $\overset{\circ}{W}{}^{(N)}$. It is an N-dimensional linear space of all possible linear combinations:

$$w_N(x,y,c_1,c_2,\ldots,c_N) = \sum_{n=1}^{N} c_n w_n^{(N)}$$

with real coefficients $c_1, c_2, \ldots, c_N$. Instead of the function $w \equiv u$ that minimizes the functional $I(w)$ of (12.21) on the space $\overset{\circ}{W}$, we will be looking for the function $w_N(x,y,c_1,c_2,\ldots,c_N)$ that minimizes the functional $I(w)$ on the N-dimensional subspace $\overset{\circ}{W}{}^{(N)}$. It is this function $w_N(x,y,c_1,c_2,\ldots,c_N) \equiv w_N(x,y)$ that will be taken as the approximate solution for the specific choice of the basis functions (12.27).

The problem of obtaining the approximate solution $w_N(x,y)$ is much simpler than the problem of obtaining the exact solution $u(x,y)$. Indeed, for the functional $I(w_N)$

we can write:

$$I(w_N) = \iint_{\Omega} \left[\left(\frac{\partial}{\partial x} \sum_{n=1}^{N} c_n w_n^{(N)} \right)^2 + \left(\frac{\partial}{\partial y} \sum_{n=1}^{N} c_n w_n^{(N)} \right)^2 \right] dxdy$$
$$+2 \iint_{\Omega} \varphi \left(\sum_{n=1}^{N} c_n w_n^{(N)} \right) dxdy, \tag{12.28}$$

and consequently, what we need to find is a set of N real numbers $\{c_1, c_2, \dots, c_N\}$ that would deliver a minimum value to the function $I(w_N(x,y,c_1,c_2,\dots,c_N))$ of N variables. The first term on the right-hand side of (12.28) is a quadratic form with respect to $c_1, c_2, \dots, c_N$. As the functions $w_n^{(N)} \in \overset{\circ}{W}$ of (12.27) are linearly independent, this quadratic form is positive definite. Positive definiteness immediately follows from application of the Friedrichs inequality (12.25) to a function represented in the general form: $w_N(x,y) = \sum_{n=1}^{N} c_n w_n^{(N)}$. Consequently, the quadratic function $I(w_N)$ given by (12.28) has a unique minimum (see Section 5.5).

The minimum of the function $I(w_N)$ is attained at its critical point, which is defined as a solution to the following system of linear algebraic equations:

$$\frac{\partial}{\partial c_m} I(w_N(x,y,c_1,c_2,\dots,c_N)) = 0, \quad m = 1,2,\dots,N.$$

By substituting the actual expression (12.28) for $I(w_N)$, this system is rewritten as:

$$\sum_{n=1}^{N} c_n \iint_{\Omega} \left[\frac{\partial w_m^{(N)}}{\partial x} \frac{\partial w_n^{(N)}}{\partial x} + \frac{\partial w_m^{(N)}}{\partial y} \frac{\partial w_n^{(N)}}{\partial y} \right] dxdy$$
$$= -\iint_{\Omega} \varphi w_m^{(N)} dxdy, \quad m = 1,2,\dots,N. \tag{12.29}$$

To obtain a more compact form of system (12.29), we we use formulae (12.18) & (12.19) and introduce an alternative scalar product on the space $\overset{\circ}{W}$:

$$(v,w)' \stackrel{\text{def}}{=} \iint_{\Omega} \left[\frac{\partial v}{\partial x} \frac{\partial w}{\partial x} + \frac{\partial v}{\partial y} \frac{\partial w}{\partial y} \right] dxdy = -(\Delta w, v), \quad v \,\&\, w \in \overset{\circ}{W}.$$

The norm $\|\cdot\|' \equiv \sqrt{(\cdot,\cdot)'}$ induced by this scalar product on the space $\overset{\circ}{W}$ is called the energy norm; by virtue of the Friedrichs inequality (12.25), it is equivalent to the Sobolev norm $\|\cdot\|_W$ (12.14), see inequality (12.26). Using the new scalar product, we can recast system (12.29) as follows:

$$\sum_{n=1}^{N} c_n \left(w_m^{(N)}, w_n^{(N)} \right)' = -\left(\varphi, w_m^{(N)} \right), \quad m = 1,2,\dots,N, \tag{12.30}$$

where $(\cdot,\cdot)$ is the original scalar product on W. The matrix of linear system (12.30):

$$A^{(N)} = \begin{bmatrix} \left(w_1^{(N)}, w_1^{(N)}\right)' & \cdots & \left(w_1^{(N)}, w_N^{(N)}\right)' \\ \vdots & \ddots & \vdots \\ \left(w_N^{(N)}, w_1^{(N)}\right)' & \cdots & \left(w_N^{(N)}, w_N^{(N)}\right)' \end{bmatrix} \tag{12.31}$$

is the Gram matrix of the system of functions (12.27). It is generally known that the Gram matrix of a linearly independent system of functions, such as system (12.27), is non-singular. In our particular case, it is easy to see directly that the matrix $A^{(N)}$ is symmetric and positive definite, because $(v,w)' = (w,v)'$ and for any N-dimensional vector $c = \begin{bmatrix} c_1 \\ \cdots \\ c_N \end{bmatrix} \neq 0$ we have:

$$\left(A^{(N)}c, c\right) = \left(\sum_{n=1}^{N} c_n w_n^{(N)}, \sum_{n=1}^{N} c_n w_n^{(N)}\right)' > 0.$$

Consequently, linear system (12.30) always has a unique solution. Its solution $\{c_1, c_2, \dots, c_n\}$ obtained for a given $\varphi = \varphi(x,y)$ determines the approximate solution $w_N(x,y) = \sum_{n=1}^{N} c_n w_n^{(n)}$ of problem (12.15) with $\psi(s) \equiv 0$.

In Section 12.2.4, we provide a specific example of the basis (12.27) that yields a particular realization of the finite element method based on the approach by Ritz. In general, it is clear that the quality of approximation by $w_N(x,y) = \sum_{n=1}^{N} c_n w_n^{(n)}$ is determined by the properties of the approximating space: $\overset{\circ}{W}{}^{(N)} = \text{span}\{w_1^{(N)}, w_2^{(N)}, \dots, w_N^{(N)}\}$ that spans the basis (12.27). The corresponding group of questions that also lead to the analysis of convergence for finite element approximations is discussed in Section 12.2.5.

12.2.3 The Galerkin Method

In Section 12.2.1 we saw that the ability to formulate a solvable minimization problem hinges on the positive definiteness of the operator $-\Delta$, see formula (12.19). This consideration gives rise to a concern that if the differential operator of the boundary value problem that we are solving is not positive definite, then the corresponding functionals may have no minima, which will render the variational approach of Section 12.2.1 invalid. A typical example when this indeed happens is provided by the Helmholtz equation:

$$\Delta u + k^2 u = \varphi. \tag{12.32a}$$

This equation governs the propagation of time-harmonic waves that have the wavenumber $k > 0$ and are driven by the sources $\varphi = \varphi(x,y)$. We will consider a Dirichlet boundary value problem for the Helmholtz equation (12.32a) on the domain Ω with the boundary $\Gamma = \partial\Omega$, and for simplicity we will only analyze the homogeneous boundary condition:

$$u\Big|_{\Gamma} = 0. \tag{12.32b}$$

Problem (12.32) has a solution $u = u(x,y)$ for any given right-hand side $\varphi(x,y)$. This solution, however, is not always unique. Uniqueness is only guaranteed if the quantity $-k^2$, see equation (12.32a), is not an eigenvalue of the corresponding Dirichlet problem for the Laplace equation. Henceforth, we will assume that this is not the case and accordingly, that problem (12.32) is uniquely solvable.

Regarding the eigenvalues of the Laplacian, it is known that all of them are real and negative and that the set of all eigenvalues is countable and has no finite limit points, see, e.g., [TS63]. It is also known that the eigenfunctions that correspond to different eigenvalues can be taken orthonormal.

Denote the Helmholtz operator by $\boldsymbol{L}$: $\boldsymbol{L} = \Delta + k^2\boldsymbol{I}$, and consider the following expression:

$$(\boldsymbol{L}w, w) \equiv (\Delta w, w) + k^2(w,w), \quad w \in \overset{\circ}{W}. \tag{12.33}$$

We will show that unlike in the case of a pure Laplacian, see formula (12.19), we can make this expression negative or positive by choosing different functions w. First, consider a collection of eigenfunctions[2] $\hat{v}_j = \hat{v}_j(x,y) \in \overset{\circ}{W}$ of the Laplacian: $\Delta\hat{v}_j = \lambda_j\hat{v}_j$, such that for the corresponding eigenvalues we have: $\sum_j \lambda_j + k^2 < 0$. Then, for $w = \sum_j \hat{v}_j$ the quantity $(\boldsymbol{L}w, w)$ of (12.33) becomes negative:

$$\begin{aligned}(\boldsymbol{L}w, w) &= \Big(\Delta\sum_j \hat{v}_j, \sum_j \hat{v}_j\Big) + k^2\Big(\sum_j \hat{v}_j, \sum_j \hat{v}_j\Big)\\ &= \sum_j \lambda_j(\hat{v}_j,\hat{v}_j) + k^2\sum_j(\hat{v}_j,\hat{v}_j) < 0,\end{aligned}$$

where we used orthonormality of the eigenfunctions: $(\hat{v}_i,\hat{v}_j) = 0$ for $i \neq j$ and $(\hat{v}_j,\hat{v}_j) = 1$. On the other hand, among all the eigenvalues of the Laplacian there is clearly the largest (smallest in magnitude): $\lambda_1 < 0$, and for all other eigenvalues we have: $\lambda_j < \lambda_1$, $j > 1$. Suppose that the wavenumber k is sufficiently large so that $\sum_{j=1}^{j_k} \lambda_j + k^2 > 0$, where $j_k \geq 1$ depends on k. Then, by taking w as the sum of either a few or all of the corresponding eigenfunctions, we will clearly have $(\boldsymbol{L}w, w) > 0$.

We have thus proved that the Helmholtz operator $\boldsymbol{L}$ is neither negative definite nor positive definite on $\overset{\circ}{W}$. Moreover, once we have found a particular $w \in \overset{\circ}{W}$ for which $(\boldsymbol{L}w,w) < 0$ and another $w \in \overset{\circ}{W}$ for which $(\boldsymbol{L}w,w) > 0$, then by scaling these functions (multiplying them by appropriate constants) we can make the expression $(\boldsymbol{L}w,w)$ of (12.33) arbitrarily large negative or arbitrarily large positive.

By analogy with (12.23), let us now introduce a new functional:

$$J(w) = -(\boldsymbol{L}w,w) + 2(\varphi,w), \quad w \in \overset{\circ}{W}. \tag{12.34}$$

[2] It may be just one eigenfunction or more than one.

Assume that $u = u(x,y)$ is the solution of problem (12.32), and represent an arbitrary $w \in \overset{\circ}{W}$ in the form $w = u + \xi$, where $\xi \in \overset{\circ}{W}$. Then we can write:

$$\begin{aligned}
J(u+\xi) &= -(\Delta(u+\xi), u+\xi) - k^2(u+\xi, u+\xi) + 2(\varphi, u+\xi) \\
&= \underbrace{-(\Delta u, u) - k^2(u,u) + 2(\varphi, u)}_{=J(u) \text{ according to (12.33), (12.34)}} - (\Delta u, \xi) - (\Delta\xi, u) - (\Delta\xi, \xi) \\
&\quad - k^2(u,\xi) - k^2(\xi, u) - k^2(\xi,\xi) + 2(\varphi, \xi) \\
&= J(u) - 2(\Delta u, \xi) - (\Delta\xi, \xi) - 2k^2(u,\xi) - k^2(\xi,\xi) + 2(\Delta u + k^2 u, \xi) \\
&= J(u) - (\boldsymbol{L}\xi, \xi),
\end{aligned}$$

where we used symmetry of the Laplacian on the space $\overset{\circ}{W}$: $(\Delta u, \xi) = (\Delta\xi, u)$, which immediately follows from the second Green's formula (12.17), and also substituted $\varphi = \Delta u + k^2 u$ from (12.32a).

Equality $J(u+\xi) = J(u) - (\boldsymbol{L}\xi, \xi)$ allows us to conclude that the solution $u = u(x,y)$ of problem (12.32) does not deliver an extreme value to the functional $J(w)$ of (12.34), because the increment $-(\boldsymbol{L}\xi, \xi) = -(\boldsymbol{L}(w-u), w-u)$ can assume both positive and negative values (arbitrarily large in magnitude). Therefore, the variational formulation of Section 12.2.1 does not apply to the Helmholtz problem (12.32), and one cannot use the Ritz method of Section 12.2.2 for approximating its solution.

The foregoing analysis does not imply, of course, that if one variational formulation does not work for a given problem, then others will not work either. The example of the Helmholtz equation does indicate, however, that it may be worth developing a more general method that would apply to problems for which no variational formulation (like the one from Section 12.2.1) is known. A method of this type was proposed by Galerkin in 1916. It is often referred to as the projection method.

We will illustrate the Galerkin method using our previous example of a Dirichlet problem for the Helmholtz equation (12.32a) with a homogeneous boundary condition (12.32b). Consider the same basis (12.27) as we used for the Ritz method. As before, we will be looking for an approximate solution to problem (12.32) in the form of a linear combination:

$$w_N(x,y) \equiv w_N(x,y,c_1,c_2,\dots,c_N) = \sum_{n=1}^{N} c_n w_n^{(N)}. \tag{12.35}$$

Substituting this expression into the Helmholtz equation (12.32a), we obtain:

$$\frac{\partial^2 w_N(x,y)}{\partial x^2} + \frac{\partial^2 w_N(x,y)}{\partial y^2} + k^2 w_N(x,y) - \varphi(x,y) = \delta_N(x,y),$$

where $\delta_N(x,y) \equiv \delta_N(x,y,c_1,c_2,\dots,c_N)$ is the corresponding residual. This residual can only be equal to zero everywhere on Ω if $w_N(x,y)$ were the exact solution. Otherwise, the residual is not zero, and to obtain a good approximation to the exact solution we need to minimize $\delta_N(x,y)$ in some particular sense.

If, for example, we were able to claim that $\delta_N(x,y)$ was orthogonal to all the functions in W in the sense of the standard scalar product $(\cdot,\cdot)$, then the residual would vanish identically on Ω, and accordingly, $w_N(x,y)$ would coincide with the exact solution. This, however, is an ideal situation that cannot be routinely realized in practice because there are too few free parameters $c_1, c_2, \dots, c_N$ available in formula (12.35). Instead, we will require that the residual $\delta_N(x,y)$ be orthogonal not to the entire space but rather to the same basis functions $w_m^{(N)} \in \overset{\circ}{W}$, $m = 1, 2, \dots, N$, as used for building the linear combination (12.35):

$$\left(\delta_N, w_m^{(N)}\right) = 0, \quad m = 1, 2, \dots, N.$$

In other words, we require that the projection of the residual δ_N onto the space $\overset{\circ}{W}{}^{(N)}$ be equal to zero. This yields:

$$\iint_\Omega \left(\frac{\partial^2 w_N}{\partial x^2} + \frac{\partial^2 w_N}{\partial y^2}\right) w_m^{(N)} dxdy + k^2 \iint_\Omega w_N w_m^{(N)} dxdy = \iint_\Omega \varphi w_m^{(N)} dxdy,$$
$$m = 1, 2, \dots, N.$$

With the help of the first Green's formula (12.16), and taking into account that all the functions are in $\overset{\circ}{W}$, the previous system can be rewritten as follows:

$$-\iint_\Omega \left(\frac{\partial w_N}{\partial x}\frac{\partial w_m^{(N)}}{\partial x} + \frac{\partial w_N}{\partial y}\frac{\partial w_m^{(N)}}{\partial y}\right) dxdy + k^2 \iint_\Omega w_N w_m^{(N)} dxdy$$
$$= -\sum_{n=1}^N c_n \iint_\Omega \left(\frac{\partial w_n^{(N)}}{\partial x}\frac{\partial w_m^{(N)}}{\partial x} + \frac{\partial w_n^{(N)}}{\partial y}\frac{\partial w_m^{(N)}}{\partial y}\right) dxdy + k^2 \sum_{n=1}^N c_n \iint_\Omega w_n^{(N)} w_m^{(N)} dxdy$$
$$= -\sum_{n=1}^N c_n \left(w_n^{(N)}, w_m^{(N)}\right)' + k^2 \sum_{n=1}^N c_n \left(w_n^{(N)}, w_m^{(N)}\right) = \left(\varphi, w_m^{(N)}\right), \quad m = 1, 2, \dots, N.$$

We therefore arrive at the following system of N linear algebraic equations with respect to the unknowns $c_1, c_2, \dots, c_N$:

$$\sum_{n=1}^N c_n \left[-\left(w_n^{(N)}, w_m^{(N)}\right)' + k^2 \left(w_n^{(N)}, w_m^{(N)}\right)\right] = \left(\varphi, w_m^{(N)}\right), \qquad (12.36)$$
$$m = 1, 2, \dots, N.$$

Substituting its solution into formula (12.35), we obtain an approximate solution to the Dirichlet problem (12.32). As in the case of the Ritz method, the quality of this approximation, i.e., the error as it depends on the dimension N, is determined by the properties of the approximating space: $\overset{\circ}{W}{}^{(N)} = \text{span}\{w_1^{(N)}, w_2^{(N)}, \dots, w_N^{(N)}\}$ that spans the basis (12.27). In this regard, it is important to emphasize that different bases can be selected in one and the same space, and the accuracy of the approximation does not depend on the choice of the individual basis, see Section 12.2.5 for more detail.

Let us also note that when the wavenumber $k = 0$, the Helmholtz equation (12.32a) reduces to the Poisson equation of (12.15). In this case, the linear system of the Galerkin method (12.36) automatically reduces to the Ritz system (12.30).

12.2.4 An Example of Finite Element Discretization

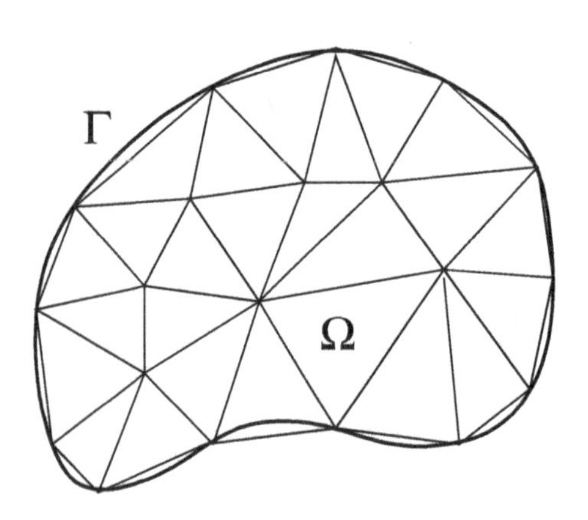

FIGURE 12.1: Unstructured triangular grid.

A frequently used approach to constructing finite element discretizations for two-dimensional problems involves unstructured triangular grids. An example of the grid of this type is shown in Figure 12.1. All cells of the grid have triangular shape. The grid is obtained by first approximating the boundary Γ of the domain Ω by a closed polygonal line with the vertexes on Γ, and then by partitioning the resulting polygon into a collection of triangles. The triangles of the grid never overlap. Each pair of triangles either do not intersect at all, or have a common vertex, or have a common side. Suppose that altogether there are N vertexes inside the domain Ω but not on the boundary Γ, where a common vertex of several triangles is counted only once. Those vertexes will be the nodes of the grid, we will denote them $Z_m^{(N)}$, $m = 1, 2, \dots, N$.

Once the triangular grid has been built, see Figure 12.1, we can specify the basis functions (12.27). For every function $w_n^{(N)} = w_n^{(N)}(x, y)$, $n = 1, 2, \dots, N$, we first define its values at the grid nodes:

$$w_n^{(N)}\left(Z_m^{(N)}\right) = \begin{cases} 1, & n = m, \\ 0, & n \neq m, \end{cases} \qquad n, m = 1, 2, \dots, N.$$

In addition, at every vertex that happens to lie on the boundary Γ we set $w_n^{(N)} = 0$. This way, the function $w_n^{(N)}$ gets defined at all vertexes of all triangles of the grid. Next, inside each triangle we define it as a linear interpolant, i.e., as a linear function of two variables: x and y, that assumes given values at all three vertexes of the triangle. Finally, outside of the entire system of triangles the function $w_n^{(N)}$ is set to be equal to zero. Note that according to our construction, the function $w_n^{(N)}$ is equal to zero on those triangles of the grid that do not have a given $Z_n^{(N)}$ as their vertex. On those triangles that do have $Z_n^{(N)}$ as one of their vertexes, the function $w_n^{(N)}$ can be represented as a fragment of the plane that crosses through the point in the 3D space that has elevation 1 precisely above $Z_n^{(N)}$ and through the side of the triangle opposite to the vertex $Z_n^{(N)}$. The finite elements obtained with the help of these basis functions are often referred to as piecewise linear.

In either Ritz or Galerkin method, the approximate solution is to be sought in the form of a linear combination:

$$w_N(x,y) = \sum_{n=1}^{N} c_n w_n^{(N)}(x,y).$$

In doing so, at a given grid node $Z_n^{(N)}$ only one basis function, $w_n^{(N)}$, is equal to one whereas all other basis functions are equal to zero. Therefore, we can immediately see that $c_n = w_N\left(Z_n^{(N)}\right)$. In other words, the unknowns to be determined when computing the solution by piecewise linear finite elements are the same as in the case of finite differences. They are the values of the approximate solution $w_N(x,y)$ at the grid nodes $Z_n^{(N)}$, $n = 1,2,\ldots,N$.

Once the piecewise linear basis functions have been constructed, the Ritz system of equations (12.30) for solving the Poisson problem (12.15) can be written as:

$$\sum_{n=1}^{N} w_N\left(Z_n^{(N)}\right)\left(w_m^{(N)}, w_n^{(N)}\right)' = -\left(\varphi, w_m^{(N)}\right), \quad m = 1,2,\ldots,N, \tag{12.37}$$

and the Galerkin system of equations (12.36) for solving the Helmholtz problem (12.32) can be written as:

$$\sum_{n=1}^{N} w_N\left(Z_n^{(N)}\right)\left[-\left(w_n^{(N)}, w_m^{(N)}\right)' + k^2\left(w_n^{(N)}, w_m^{(N)}\right)\right] = \left(\varphi, w_m^{(N)}\right), \quad m = 1,2,\ldots,N. \tag{12.38}$$

Linear systems (12.37) and (12.38) provide specific examples of implementation of the finite element method for two-dimensional elliptic boundary value problems.

The coefficients of systems (12.37) and (12.38) are given by scalar products [energy product $(\cdot,\cdot)'$ and standard product $(\cdot,\cdot)$] of the basis functions. It is clear that only those numbers $\left(w_n^{(N)}, w_m^{(N)}\right)'$ and $\left(w_n^{(N)}, w_m^{(N)}\right)$ can differ from zero, for which the grid nodes $Z_n^{(N)}$ and $Z_m^{(N)}$ happen to be vertexes of one and the same triangle. Indeed, if the nodes $Z_n^{(N)}$ and $Z_m^{(N)}$ are not neighboring in this sense, then the regions where $w_n^{(N)} \not\equiv 0$ and $w_m^{(N)} \not\equiv 0$ do not intersect, and consequently:

$$\left(w_m^{(N)}, w_n^{(N)}\right)' = \iint_{\Omega} \left[\frac{\partial w_m^{(N)}}{\partial x}\frac{\partial w_n^{(N)}}{\partial x} + \frac{\partial w_m^{(N)}}{\partial y}\frac{\partial w_n^{(N)}}{\partial y}\right] dxdy = 0 \tag{12.39a}$$

and

$$\left(w_m^{(N)}, w_n^{(N)}\right) = \iint_{\Omega} w_m^{(N)} w_n^{(N)} dxdy = 0. \tag{12.39b}$$

In other words, equation number m of system (12.37) [or, likewise, of system (12.38)] connects the value of the unknown function $w_N(Z_m^{(N)})$ with its values $w_N(Z_n^{(N)})$ only at the neighboring nodes $Z_n^{(N)}$. This is another similarity with conventional finite-difference schemes, for which every equation of the resulting system connects the values of the solution at the nodes of a given stencil.

The coefficients of systems (12.37) and (12.38) are rather easy to evaluate. Indeed, either integral (12.39a) or (12.39b) is, in fact, an integral over only a pair of grid triangles that have the interval $[Z_m^{(N)}, Z_n^{(N)}]$ as their common side. Moreover, the integral over each triangle is completely determined by its own geometry, i.e., by the length of its sides, but is not affected by the orientation of the triangle.

The integrand in (12.39a) is constant inside each triangle; it can be interpreted as the dot product of two gradient vectors (normals to the planes that represent the linear functions $w_m^{(N)}$ and $w_n^{(N)}$):

$$\frac{\partial w_m^{(N)}}{\partial x}\frac{\partial w_n^{(N)}}{\partial x} + \frac{\partial w_m^{(N)}}{\partial y}\frac{\partial w_n^{(N)}}{\partial y} = \left(\text{grad} w_m^{(N)}, \text{grad} w_n^{(N)}\right).$$

The value of the integral (12.39a) is then obtained by multiplying the previous quantity by the area of the triangle. Integral (12.39b) is an integral of a quadratic function inside each triangle, it can also be evaluated with no difficulties. This completes the construction of a finite element discretization for the specific choice of the grid and the basis functions that we have made.

12.2.5 Convergence of Finite Element Approximations

Consider the sequence of approximations $w_N(x,y) \equiv w_N(x,y,c_1,c_2,\dots,c_N) \in \overset{\circ}{W}$, $N = 1,2,\dots$, generated by the Ritz method. According to formula (12.26), this sequence will converge to the solution $u = u(x,y)$ of problem (12.15) with $\psi(s) \equiv 0$ in the Sobolev norm (12.14) provided that $I(w_N) \longrightarrow I(u)$ as $N \longrightarrow \infty$:

$$\|w_N - u\|_W^2 \leq \text{const} \cdot [I(w_N) - I(u)]. \tag{12.40}$$

The minuend on the right-hand side of inequality (12.40) is defined as follows:

$$I(w_N) = \inf_{w \in \overset{\circ}{W}^{(N)}} I(w),$$

where $\overset{\circ}{W}^{(N)}$ is the span of the basis functions (12.27). Consequently, we can write:

$$\|w_N - u\|_W^2 \leq \text{const} \cdot \inf_{w \in \overset{\circ}{W}^{(N)}} [I(w) - I(u)] \equiv \text{const} \cdot \inf_{w \in \overset{\circ}{W}^{(N)}} \left[\|w - u\|'\right]^2, \tag{12.41}$$

where $\|\cdot\|'$ is the energy norm on $\overset{\circ}{W}$. In other words, we conclude that convergence of the Ritz method depends on what is the best approximation of the solution u by the elements of the subspace $\overset{\circ}{W}{}^{(N)}$. In doing so, the accuracy of approximation is to be measured in the energy norm. A similar result also holds for the Galerkin method.

Of course, the question of how large the right-hand side of inequality (12.41) may actually be does not have a direct answer, because the solution $u = u(x,y)$ is not known ahead of time. Therefore, the best we can do for evaluating this right-hand side is first to assume that the solution u belongs to some class of functions $U \subset \overset{\circ}{W}$, and then try and narrow down this class as much as possible using all a priori information about the solution that is available. Often, the class U can be characterized in terms of smoothness, because the regularity of the data in the problem enable a certain degree of smoothness in the solution. For example, recall that the solution of problem (12.15) was assumed twice continuously differentiable. In this case, we can say that U is the class of all functions that are equal to zero at Γ and also have continuous second derivatives on $\bar{\Omega}$.

Once we have identified the maximally narrow class of functions U that contains the solution u, we can write instead of estimate (12.41):

$$\|w_N - u\|_W^2 \leq \text{const} \cdot \sup_{v \in U} \inf_{w \in \overset{\circ}{W}{}^{(N)}} \left[\|w - v\|' \right]^2. \tag{12.42}$$

Regarding this inequality, a natural expectation is that the narrower the class U, the closer the value on the right-hand side of (12.42) to that on the right-side of (12.41).

Next, we realize that the value on the right-hand side of inequality (12.42) depends on the choice of the approximating space $\overset{\circ}{W}{}^{(N)}$, and the best possible value therefore corresponds to:

$$\kappa_N(U, \overset{\circ}{W}) \stackrel{\text{def}}{=} \inf_{\overset{\circ}{W}{}^{(N)} \subset \overset{\circ}{W}} \sup_{v \in U} \inf_{w \in \overset{\circ}{W}{}^{(N)}} \|w - v\|'. \tag{12.43}$$

The quantity $\kappa_N(U, \overset{\circ}{W})$ is called the N-dimensional Kolmogorov diameter of the set U with respect to the space $\overset{\circ}{W}$ in the sense of the energy norm $\|\cdot\|'$. We have first encountered this concept in Section 2.2.4, see formula (2.39) on page 41. Note that the norm in the definition of the Kolmogorov diameter is not squared, unlike in formula (12.42). Note also that the outermost minimization in (12.43) is performed with respect to the N-dimensional space $\overset{\circ}{W}{}^{(N)}$ as a whole, and its result does not depend on the choice of a specific basis (12.27) in $\overset{\circ}{W}{}^{(N)}$. One and the same space can be equipped with different bases.

Kolmogorov diameters and related concepts are widely used in the theory of approximation, as well as in many other branches of mathematics. In the context of finite elements they provide optimal, i.e., unimprovable, estimates for the convergence rate of the method. Once the space $\overset{\circ}{W}$, the norm $\|\cdot\|'$, and the subspace $U \subset \overset{\circ}{W}$ have been fixed, the diameter (12.43) depends on the dimension N. Then, the optimal convergence rate of a finite element approximation is determined by how rapidly the diameter $\kappa_N(U, \overset{\circ}{W})$ decays as $N \longrightarrow \infty$. Of course, in every given case there is no guarantee that a particular choice of $\overset{\circ}{W}^{(N)}$ will yield the rate that would come anywhere close to the theoretical optimum provided by the Kolmogorov diameter $\kappa_N(U, \overset{\circ}{W})$. In other words, the rate of convergence for a specific Ritz or Galerkin approximation may be slower than optimal[3] as $N \longrightarrow \infty$. Consequently, it is the skill and experience of a numerical analyst that are required for choosing the approximating space $\overset{\circ}{W}^{(N)}$ is such a way that the actual quantity:

$$\sup_{v \in U} \inf_{w \in \overset{\circ}{W}^{(N)}} \|w - v\|'$$

that controls the right-hand side of (12.42) would not be much larger than the Kolmogorov diameter $\kappa_N(U, \overset{\circ}{W})$ of (12.43).

In many particular situations, the Kolmogorov diameters have been computed. For example, for the setup analyzed in this section, when the space $\overset{\circ}{W}$ contains all piecewise continuously differentiable functions on $\bar{\Omega} \subset \mathbb{R}^2$ equal to zero at $\Gamma = \partial\Omega$, the class U contains all twice continuously differentiable functions from $\overset{\circ}{W}$, and the norm is the energy norm, we have:

$$\kappa_N(U, \overset{\circ}{W}) = \mathcal{O}\left(\frac{1}{\sqrt{N}}\right) \quad \text{as} \quad N \longrightarrow \infty. \tag{12.44}$$

Under certain additional conditions (non-restrictive), it is also possible to show that the piecewise linear elements constructed in Section 12.2.4 converge with the same asymptotic rate $\mathcal{O}(N^{-1/2})$ when N increases, see, e.g., [GR87, § 39]. As such, the piecewise linear finite elements guarantee the optimal convergence rate of the Ritz method for problem (12.15), assuming that nothing else can be said about the solution $u = u(x,y)$ except that it has continuous second derivatives on $\bar{\Omega}$.

Note that as we have a total of N grid nodes on a two-dimensional domain Ω, for a regular uniform grid it would have implied that the grid size is $h = \mathcal{O}(N^{-1/2})$. In other words, the convergence rate of the piecewise linear finite elements appears to be $\mathcal{O}(h)$, and the optimal unimprovable rate (12.44) is also $\mathcal{O}(h)$. At a first glance, this seems to be a deterioration of convergence compared, e.g., to the standard central-difference scheme of Section 12.1, which converges with the rate $\mathcal{O}(h^2)$. This, however, is not the case. In Section 12.1, we measured the convergence in the norm that did not contain the derivatives of the solution, whereas in this section we are using the Sobolev norm (12.14) that contains the first derivatives.

Finally, recall that the convergence rates can be improved by selecting the approximating space that would be right for the problem, as well as by narrowing down the class of functions U that contains the solution u. These two strategies can be combined and implemented together in the framework of one adaptive procedure. Adaptive methods represent a class of rapidly developing, modern and efficient, approaches to finite element approximations. Both the size/shape of the elements, as

[3]Even though formula (12.42) is written as an inequality, the norms $\|\cdot\|_W$ and $\|\cdot\|'$ are, in fact, equivalent.

well as their order (beyond linear), can be controlled. Following a special (multigrid-type) algorithm, the elements are adapted dynamically, i.e., in the course of computation. For example, by refining the grid and/or increasing the order locally, these elements can very accurately approximate sharp variations in the solution. In other words, as soon as those particular areas of sharp variation are identified inside Ω, the class U becomes narrower, and at the same time, the elements provide a better, "fine-tuned," basis for approximation. Convergence of these adaptive finite element approximations may achieve spectral rates. We refer the reader to the recent monograph by Demkowicz for further detail [Dem06].

To conclude this chapter, we will compare the method of finite elements with the method of finite differences from the standpoint of how convergence of each method is established. Recall, the study of convergence for finite-difference approximations consists of analyzing two independent properties, consistency and stability, see Theorem 10.1 on page 314. In the context of finite elements, consistency as defined in Section 10.1.2 or 12.1.1 (small truncation error on smooth solutions) is no longer needed for proving convergence. Instead, we need approximation of the class $U \ni u$ by the functions from $\overset{\circ}{W}{}^{(N)}$, see estimate (12.42). Stability for finite elements shall be understood as good conditioning (uniform with respect to N) of the Ritz system matrix (12.31) or of a similar Galerkin matrix, see (12.36). The ideal case here, which cannot normally be realized in practice, is when the basis $w_n^{(N)}$, $n = 1, 2, \ldots, N$, is orthonormal. Then, the matrix $\boldsymbol{A}^{(N)}$ becomes a unit matrix. Stability still remains important when computing with finite elements, although not for justifying convergence, but for being able to disregard the small round-off errors.

Exercises

1. Prove that the quadratic form of the Ritz method, i.e., the first term on the right-hand side of (12.28), is indeed positive definite.

   Hint. Apply the Friedrichs inequality (12.25) to $w_N(x,y) = \sum_{n=1}^{N} c_n w_n^{(N)}$.

2. Consider a Dirichlet problem for the elliptic equation with variable coefficients:

$$\frac{\partial}{\partial x}\left(a(x,y)\frac{\partial u}{\partial x}\right) + \frac{\partial}{\partial y}\left(b(x,y)\frac{\partial u}{\partial y}\right) = \varphi(x,y), \quad (x,y) \in \Omega,$$

$$u\Big|_{\Gamma} = \psi(s), \quad \Gamma = \partial\Omega,$$

   where $a(x,y) \geq a_0 > 0$ and $b(x,y) \geq b_0 > 0$. Prove that its solution minimizes the functional:

$$I(w) = \iint_{\Omega} \left[a\left(\frac{\partial u}{\partial x}\right)^2 + b\left(\frac{\partial u}{\partial y}\right)^2 + 2\varphi w\right] dx dy$$

   on the set of all functions $w \in W$ that satisfy the boundary condition: $w\big|_{\Gamma} = \psi(s)$.

3. Let $\psi(s) \equiv 0$ in the boundary value problem of Exercise 2. Apply the Ritz method and obtain the corresponding system of linear algebraic equations.

4. Let $\psi(s) \equiv 0$ in the boundary value problem of Exercise 2. Apply the Galerkin method and obtain the corresponding system of linear algebraic equations.

5. Let the geometry of the two neighboring triangles that have common side $\left[Z_m^{(N)}, Z_n^{(N)}\right]$ be known. Evaluate the coefficients (12.39a) and (12.39b) for the finite element systems (12.37) and (12.38).

6. Consider problem (12.15) with $\psi(s) \equiv 0$ on a square domain Ω. Introduce a uniform Cartesian grid with square cells on the domain Ω. Partition each cell into two right triangles by the diagonal; in doing so, use the same orientation of the diagonal for all cells. Apply the Ritz method on the resulting triangular grid and show that it is equivalent to the central-difference scheme (12.5), (12.6).

Part IV

The Methods of Boundary Equations for the Numerical Solution of Boundary Value Problems

The finite-difference methods of Part III can be used for solving a wide variety of initial and boundary value problems for ordinary and partial differential equations. Still, in some cases these methods may encounter serious difficulties. In particular, finite differences may appear inconvenient for accommodating computational domains of an irregular shape. Moreover, finite differences may not always be easy to apply for the approximation of general boundary conditions, e.g., boundary conditions of different types on different portions of the boundary, or non-local boundary conditions. The reason is that guaranteeing stability in those cases may require a considerable additional effort, and also that the resulting system of finite-difference equations may be difficult to solve. Furthermore, another serious hurdle for the application of finite-difference methods is presented by the problems on unbounded domains, when the boundary conditions are specified at infinity.

In some cases the foregoing difficulties can be overcome by reducing the original problem formulated on a given domain to an equivalent problem formulated only on its boundary. In doing so, the original unknown function (or vector-function) on the domain obviously gets replaced by some new unknown function(s) on the boundary. A very important additional benefit of adopting this methodology is that the geometric dimension of the problem decreases by one (the boundary of a two-dimensional domain is a one-dimensional curve, and the boundary of a three-dimensional domain is a two-dimensional surface).

In this part of the book, we will discuss two approaches to reducing the problem from its domain to the boundary. We will also study the techniques that can be used for the numerical solution of the resulting boundary equations.

The first method of reducing a given problem from the domain to the boundary is the method of boundary integral equations of classical potential theory. It dates back to the work of Fredholm done in the beginning of the twentieth century. Discretization and numerical solution of the resulting boundary integral equations are performed by means of the special quadrature formulae in the framework of the method of boundary elements. We briefly discuss it in Chapter 13.

The second method of reducing a given problem to the boundary of its domain leads to boundary equations of a special structure. These equations contain projection operators first proposed in 1963 by Calderon [Cal63]. In Chapter 14, we describe a particular modification of Calderon's boundary equations with projections. It enables a straightforward discretization and numerical solution of these equations by the method of difference potentials. The latter was proposed by Ryaben'kii in 1969 and subsequently summarized in [Rya02]. The boundary equations with projections and the method of difference potentials can be applied to a number of cases when the classical boundary integral equations and the method of boundary elements will not work.

Hereafter, we will only describe the key ideas that provide a foundation for the method of boundary elements and for the difference potentials method. We will also identify their respective domains of applicability, and quote the literature where a more comprehensive description can be found.

Chapter 13

Boundary Integral Equations and the Method of Boundary Elements

In this chapter, we provide a very brief account of classical potential theory and show how it can help reduce a given boundary value problem to an equivalent integral equation at the boundary of the original domain. We also address the issue of discretization for the corresponding integral equations, and identify the difficulties that limit the class of problems solvable by the method of boundary elements.

13.1 Reduction of Boundary Value Problems to Integral Equations

To illustrate the key concepts, it will be sufficient to consider the interior and exterior Dirichlet and Neumann boundary value problems for the Laplace equation:

$$\Delta u \equiv \frac{\partial^2 u}{\partial x_1^2} + \frac{\partial^2 u}{\partial x_2^2} + \frac{\partial^2 u}{\partial x_3^2} = 0.$$

Let Ω be a bounded domain of the three-dimensional space $\mathbb{R}^3$, and assume that its boundary $\Gamma = \partial\Omega$ is sufficiently smooth. Let also Ω_1 be the complementary domain: $\Omega_1 = \mathbb{R}^3 \backslash \bar{\Omega}$. Consider the following four problems:

$$\Delta u = 0, \quad \boldsymbol{x} \in \Omega, \qquad u\big|_{\Gamma} = \varphi(\boldsymbol{x})\big|_{\boldsymbol{x}\in\Gamma}, \tag{13.1a}$$

$$\Delta u = 0, \quad \boldsymbol{x} \in \Omega, \qquad \frac{\partial u}{\partial n}\bigg|_{\Gamma} = \varphi(\boldsymbol{x})\big|_{\boldsymbol{x}\in\Gamma}, \tag{13.1b}$$

$$\Delta u = 0, \quad \boldsymbol{x} \in \Omega_1, \qquad u\big|_{\Gamma} = \varphi(\boldsymbol{x})\big|_{\boldsymbol{x}\in\Gamma}, \tag{13.1c}$$

$$\Delta u = 0, \quad \boldsymbol{x} \in \Omega_1, \qquad \frac{\partial u}{\partial n}\bigg|_{\Gamma} = \varphi(\boldsymbol{x})\big|_{\boldsymbol{x}\in\Gamma}, \tag{13.1d}$$

where $\boldsymbol{x} = (x_1, x_2, x_3) \in \mathbb{R}^3$, n is the outward normal to Γ, and $\varphi(\boldsymbol{x})$ is a given function for $\boldsymbol{x} \in \Gamma$. Problems (13.1a) and (13.1c) are the interior and exterior Dirichlet problems, respectively, and problems (13.1b) and (13.1d) are the interior and exterior Neumann problems, respectively. For the exterior problems (13.1c) and (13.1d), we also need to specify the desired behavior of the solution at infinity:

$$u(\boldsymbol{x}) \longrightarrow 0, \text{ as } |\boldsymbol{x}| \equiv (x_1^2 + x_2^2 + x_3^2)^{1/2} \longrightarrow \infty. \tag{13.2}$$

In the courses of partial differential equations (see, e.g., [TS63]), it is proven that the interior and exterior Dirichlet problems (13.1a) and (13.1c), as well as the exterior Neumann problem (13.1d), always have a unique solution. The interior Neumann problem (13.1b) is only solvable if

$$\int_{\Gamma} \varphi ds = 0, \tag{13.3}$$

where ds is the area element on the surface $\Gamma = \partial\Omega$. In case equality (13.3) is satisfied, the solution of problem (13.1d) is determined up to an arbitrary additive constant.

For the transition from boundary value problems (13.1) to integral equations, we use the fundamental solution of the Laplace operator, which is also referred to as the free-space Green's function:

$$G(\boldsymbol{x}) = -\frac{1}{4\pi}\frac{1}{|\boldsymbol{x}|}.$$

It is known that every solution of the Poisson equation:

$$\Delta u = f(\boldsymbol{x})$$

that vanishes at infinity, see (13.2), and is driven by a compactly supported right-hand side $f(\boldsymbol{x})$, can be represented by the formula:

$$u(\boldsymbol{x}) = \iiint G(\boldsymbol{x}-\boldsymbol{y})f(\boldsymbol{y})d\boldsymbol{y},$$

where the integration in this convolution integral is to be performed across the entire $\mathbb{R}^3$. In reality, of course, it only needs to be done over the region where $f(\boldsymbol{y}) \neq 0$.

As boundary value problems (13.1) are driven by the data $\varphi(\boldsymbol{x})$, $\boldsymbol{x} \in \Gamma$, the corresponding boundary integral equations are convenient to obtain using the single layer potential:

$$V(\boldsymbol{x}) = \int_{\Gamma} G(\boldsymbol{x}-\boldsymbol{y})\rho(\boldsymbol{y})ds_y \tag{13.4}$$

and the double layer potential:

$$W(\boldsymbol{x}) = \int_{\Gamma} \frac{\partial G(\boldsymbol{x}-\boldsymbol{y})}{\partial n_y}\sigma(\boldsymbol{y})ds_y \tag{13.5}$$

Both potentials are given by convolution type integrals. The quantity ds_y in formulae (13.4) and (13.5) denotes the area element on the surface $\Gamma = \partial\Omega$, while n_y is the normal to Γ that originates at $\boldsymbol{y} \in \Gamma$ and points toward the exterior, i.e, toward Ω_1. The functions $\rho = \rho(\boldsymbol{y})$ and $\sigma = \sigma(\boldsymbol{y})$, $\boldsymbol{y} \in \Gamma$, in formulae (13.4) and (13.5) are called densities of the single-layer potential and double-layer potential, respectively.

It is easy to see that the fundamental solution $G(\boldsymbol{x})$ satisfies the Laplace equation $\Delta G = 0$ for $\boldsymbol{x} \neq \boldsymbol{0}$. This implies that the potentials $V(\boldsymbol{x})$ and $W(\boldsymbol{x})$ satisfy the Laplace equation for $\boldsymbol{x} \notin \Gamma$, i.e., that they are are harmonic functions on Ω and on Ω_1. One can say that the families of harmonic functions given by formulae (13.4) and (13.5) are parameterized by the densities $\rho = \rho(\boldsymbol{y})$ and $\sigma = \sigma(\boldsymbol{y})$ specified for $\boldsymbol{y} \in \Gamma$.

Solutions of the Dirichlet problems (13.1a) and (13.1c) are to be sought in the form of a double-layer potential (13.5), whereas solutions to the Neumann problems (13.1b) and (13.1d) are to be sought in the form of a single-layer potential (13.4). Then, the the so-called Fredholm integral equations of the second kind can be obtained for the unknown densities of the potentials (see, e.g., [TS63]). These equations read as follows: For the interior Dirichlet problem (13.1a):

$$\sigma(\boldsymbol{x}) - \frac{1}{2\pi}\int_{\Gamma} \sigma(\boldsymbol{y}) \frac{\partial}{\partial n_y} \frac{1}{|\boldsymbol{x}-\boldsymbol{y}|} ds_y = -\frac{1}{2\pi}\varphi(\boldsymbol{x}), \quad \boldsymbol{x} \in \Gamma, \tag{13.6a}$$

for the interior Neumann problem (13.1b):

$$\rho(\boldsymbol{x}) - \frac{1}{2\pi}\int_{\Gamma} \rho(\boldsymbol{y}) \frac{\partial}{\partial n_y} \frac{1}{|\boldsymbol{x}-\boldsymbol{y}|} ds_y = -\frac{1}{2\pi}\varphi(\boldsymbol{x}), \quad \boldsymbol{x} \in \Gamma, \tag{13.6b}$$

for the exterior Dirichlet problem (13.1c):

$$\sigma(\boldsymbol{x}) + \frac{1}{2\pi}\int_{\Gamma} \sigma(\boldsymbol{y}) \frac{\partial}{\partial n_y} \frac{1}{|\boldsymbol{x}-\boldsymbol{y}|} ds_y = \frac{1}{2\pi}\varphi(\boldsymbol{x}), \quad \boldsymbol{x} \in \Gamma, \tag{13.6c}$$

and for the exterior Neumann problem (13.1d):

$$\rho(\boldsymbol{x}) + \frac{1}{2\pi}\int_{\Gamma} \rho(\boldsymbol{y}) \frac{\partial}{\partial n_y} \frac{1}{|\boldsymbol{x}-\boldsymbol{y}|} ds_y = \frac{1}{2\pi}\varphi(\boldsymbol{x}), \quad \boldsymbol{x} \in \Gamma. \tag{13.6d}$$

In the framework of the classical potential theory, integral equations (13.6a) and (13.6d) are shown to be uniquely solvable; their respective solutions $\sigma(\boldsymbol{x})$ and $\rho(\boldsymbol{x})$ exist for any given function $\varphi(x)$, $\boldsymbol{x} \in \Gamma$. The situation is different for equations (13.6b) and (13.6c). Equation (13.6b) is only solvable if equality (13.3) holds. The latter constraint reflects on the nature of the interior Neumann problem (13.1b), which also has a solution only if the additional condition (13.3) is satisfied. It turns out, however, that integral equation (13.6c) is not solvable for an arbitrary $\varphi(\boldsymbol{x})$ either, even though the exterior Dirichlet problem (13.1c) always has a unique solution. As such, transition from the boundary value problem (13.1c) to the integral equation (13.6c) appears inappropriate.

The loss of solvability for the integral equation (13.6c) can be explained. When we look for a solution to the exterior Dirichlet problem (13.1c) in the form of a double-layer potential (13.5), we essentially require that the solution $u(\boldsymbol{x})$ decay at infinity as $\mathcal{O}(|\boldsymbol{x}|^{-2})$, because it is known that $W(\boldsymbol{x}) = \mathcal{O}(|\boldsymbol{x}|^{-2})$ as $|\boldsymbol{x}| \longrightarrow \infty$. However, the original formulation (13.1c), (13.2) only requires that the solution vanish at infinity. In other words, there may be solutions that satisfy (13.1c), (13.2), yet they decay slower than $\mathcal{O}(|\boldsymbol{x}|^{-2})$ when $|\boldsymbol{x}| \longrightarrow \infty$ and as such, are not captured by the integral equation (13.6c). This causes equation (13.6c) to lose solvability for some $\varphi(\boldsymbol{x})$.

Using the language of classical potentials, the loss of solvability for the integral equation (13.6c) can also be related to the so-called resonance of the interior domain. The phenomenon of resonance manifests itself by the existence of a nontrivial (i.e., non-zero) solution to the interior Neumann problem (13.1b) driven by the zero boundary data: $\varphi(\boldsymbol{x}) \equiv 0$, $\boldsymbol{x} \in \Gamma$. Interpretation in terms of resonances is just another, equivalent, way of saying that the exterior Dirichlet problem (13.1c), (13.2) may have solutions that cannot be represented as a double-layer potential (13.5) and as such, will not satisfy equation (13.6c). Note that in this particular case equation (13.6c) can be easily "fixed," i.e., replaced by a slightly modified integral equation that will be fully equivalent to the original problem (13.1c), (13.2), see, e.g., [TS63].

Instead of the Laplace equation $\Delta u = 0$, let us now consider the Helmholtz equation on $\mathbb{R}^3$:

$$\Delta u + k^2 u = 0$$

that governs the propagation of time-harmonic waves with the wavenumber $k > 0$. The Helmholtz equation needs to be supplemented by the Sommerfeld radiation boundary conditions at infinity that replace condition (13.2):

$$u(\boldsymbol{x}) = \mathcal{O}(|\boldsymbol{x}|^{-1}), \quad \frac{\partial u(\boldsymbol{x})}{\partial |\boldsymbol{x}|} - iku(\boldsymbol{x}) = o(|\boldsymbol{x}|^{-1}), \quad |\boldsymbol{x}| \longrightarrow \infty. \tag{13.7}$$

The fundamental solution of the Helmholtz operator is given by:

$$G(\boldsymbol{x}) = -\frac{1}{4\pi}\frac{e^{ik|\boldsymbol{x}|}}{|\boldsymbol{x}|}.$$

The interior and exterior boundary value problems of either Dirichlet or Neumann type are formulated for the Helmholtz equation in the same way as they are set for the Laplace equation, see formulae (13.1). The only exception is that for the exterior problems condition (13.2) is replaced by (13.7). Both Dirichlet and Neumann exterior problems for the Helmholtz equation are uniquely solvable for any $\varphi(\boldsymbol{x})$, $\boldsymbol{x} \in \Gamma$, regardless of the value of k. The single-layer and double-layer potentials for the Helmholtz equation are defined as the same surface integrals (13.4) and (13.5), respectively, except that the fundamental solution $G(\boldsymbol{x})$ changes. Integral equations (13.6) that correspond to the four boundary value problems for the Helmholtz equation also remain the same, except that the quantity $\frac{1}{|\boldsymbol{x}-\boldsymbol{y}|}$ in every integral is replaced by $\frac{e^{ik|\boldsymbol{x}-\boldsymbol{y}|}}{|\boldsymbol{x}-\boldsymbol{y}|}$ (according to the change of G).

Still, even though, say, the exterior Dirichlet problem for the Helmholtz equation has a unique solution for any $\varphi(\boldsymbol{x})$, the corresponding integral equation of type (13.6c) is not always solvable. More precisely, it will be solvable for an arbitrary $\varphi(\boldsymbol{x})$ provided that the interior Neumann problem with $\varphi(\boldsymbol{x}) \equiv 0$ and the same wavenumber k has no solutions besides trivial. This, however, is not always the case. The situation is similar regarding the solvability of the integral equation of type (13.6d) that corresponds to the exterior Neumann problem for the Helmholtz equation. It is solvable for any $\varphi(\boldsymbol{x})$ only if the interior Dirichlet problem for the same k and $\varphi(\boldsymbol{x}) \equiv 0$ has no solutions besides trivial, which may or may not be true. We thus see that the difficulties in reducing boundary value problems for the Helmholtz equation to integral equations are again related to interior resonances, as in the case of the Laplace equation. These difficulties may translate into serious hurdles for computations.

Integral equations of classical potential theory have been explicitly constructed for the boundary value problems of elasticity, see, e.g., [MMP95], for the Stokes system of equations (low speed flows of viscous fluid), see [Lad69], and for some other equations and systems, for which analytical forms of the fundamental solutions are available.

Along with exploiting various types of surface potentials, one can build boundary integral equations using the relation between the values of the solution $u|_\Gamma$ and its normal derivative $\frac{\partial u}{\partial n}|_\Gamma$ at the boundary Γ of the region Ω. This relation can be obtained with the help of the Green's formula:

$$u(\boldsymbol{x}) = \int\limits_\Gamma \left(G(\boldsymbol{x}-\boldsymbol{y}) \frac{\partial u(\boldsymbol{y})}{\partial n_y} - \frac{\partial G(\boldsymbol{x}-\boldsymbol{y})}{\partial n_y} u(\boldsymbol{y}) \right) ds_y \tag{13.8}$$

that represents the solution $u(\boldsymbol{x})$ of the Laplace or Helmholtz equation at some interior point $\boldsymbol{x} \in \Omega$ via the boundary values $u|_\Gamma$ and $\frac{\partial u}{\partial n}|_\Gamma$. One gets the desired integral equation by passing to the limit $\boldsymbol{x} \longrightarrow \boldsymbol{x}_0 \in \Gamma$ while taking into account the discontinuity of the double-layer potential, i.e., the jump of the second term on the right-hand side of (13.8), at the interface Γ. Then, in the case of a Neumann problem, one can substitute a known function $\frac{\partial u}{\partial n}|_\Gamma = \varphi(\boldsymbol{x})$ into the integral and arrive at a Fredholm integral equation of the second kind that can be solved with respect to $u|_\Gamma$. The advantage of doing so compared to solving equation (13.6b) is that the quantity to be computed is $u|_\Gamma$ rather than the auxiliary density $\rho(\boldsymbol{y})$.

13.2 Discretization of Integral Equations and Boundary Elements

Discretization of the boundary integral equations derived in Section 13.1 encounters difficulties caused by the singular behavior of the kernels $\frac{\partial G}{\partial n_y}$ and G. In this section, we will use equation (13.6a) as an example and illustrate the construction of special quadrature formulae that approximate the corresponding integrals.

First, we need to triangulate the surface $\Gamma = \partial\Omega$, i.e., partition it into a finite number of non-overlapping curvilinear triangles. These triangles are called the boundary elements, and we denote them Γ_j, $j = 1, 2, \dots, J$. They may only intersect by their sides or vertexes. Inside every boundary element the unknown density $\sigma(\boldsymbol{y})$ is assumed constant, and we denote its value σ_j, $j = 1, 2, \dots, J$.

Next, we approximately replace the convolution integral in equation (13.6a) by the following sum:

$$\int\limits_{\Gamma} \sigma(\boldsymbol{y}) \frac{\partial}{\partial n_y} \frac{1}{|\boldsymbol{x}-\boldsymbol{y}|} ds_y = \sum_{j=1}^{J} \sigma_j \int\limits_{\Gamma_j} \frac{\partial}{\partial n_y} \frac{1}{|\boldsymbol{x}-\boldsymbol{y}|} ds_y.$$

Note that the integrals on the right-hand side of the previous equality depend only on Γ_j and on $\boldsymbol{x}$ as a parameter, but do not depend on σ_j. Let also $\boldsymbol{x}_k$ be some point from the boundary element Γ_k, and denote the values of the integrals as:

$$a_{kj} = \int\limits_{\Gamma_j} \frac{\partial}{\partial n_y} \frac{1}{|\boldsymbol{x}_k-\boldsymbol{y}|} ds_y.$$

Then, integral equation (13.6a) yields the following system of J linear algebraic equations with respect to the J unknowns σ_j:

$$2\pi\sigma_k - \sum_{j=1}^{J} a_{kj}\sigma_j = -\varphi(\boldsymbol{x}_k), \quad k = 1, 2, \dots, J. \tag{13.9}$$

When the number of boundary elements J tends to infinity and the maximum diameter of the boundary elements Γ_j simultaneously tends to zero, the solution of system (13.9) converges to the density $\sigma(\boldsymbol{y})$ of the double-layer potential that solves equation (13.6a). The unknown solution $u(\boldsymbol{x})$ of the corresponding interior Dirichlet problem (13.1a) is then computed in the form of a double-layer potential (13.5) given the approximate values of the density obtained by solving system (13.9).

Note that the method of boundary elements in fact comprises a wide variety of techniques and approaches. The computation of the coefficients a_{kj} can be conducted either exactly or approximately. The shape of the boundary elements Γ_j does not necessarily have to be triangular. Likewise, the unknown function $\sigma(\boldsymbol{y})$ does not necessarily have to be replaced by a constant inside every element. Instead, it can be taken as a polynomial of a particular form with undetermined coefficients. In doing so, linear system (13.9) will also change accordingly.

A number of established techniques for constructing boundary elements is available in the literature, along with the developed methods for the discretization of the boundary integral equations, as well as for the direct or iterative solution of the resulting linear algebraic systems. We refer the reader to the specialized textbooks and monographs [BD92, Hal94, Poz02] for further detail.

13.3 The Range of Applicability for Boundary Elements

Computer codes implementing the method of boundary elements are available today for the numerical solution of the Poisson and Helmholtz equations, the Lame system of equations that governs elastic deformations in materials, the Stokes system of equations that governs low speed flows of incompressible viscous fluid, the Maxwell system of equations for time-harmonic electromagnetic fields, the linearized Euler equations for time-harmonic acoustics, as well as for the solution of some other equations and systems, see, e.g., [Poz02]. An obvious advantage of using boundary integral equations and the method of boundary elements compared, say, with the method of finite differences, is the reduction of the geometric dimension by one. Another clear advantage of the boundary integral equations is that they apply to boundaries of irregular shape and automatically take into account the boundary conditions, as well as the conditions at infinity (if any). Moreover, the use of integral equations sometimes facilitates the construction of numerical algorithms that do not get saturated by smoothness, i.e., that automatically take into account the regularity of the data and of the solution and adjust their accuracy accordingly, see, e.g., [Bab86], [Bel89,BK01].

The principal limitation of the method of boundary elements is that in order to directly employ the apparatus of classical potentials for the numerical solution of boundary value problems, one needs a convenient representation for the kernels of the corresponding integral equations. Otherwise, no efficient discretization of these equations can be constructed. The kernels, in their own turn, are expressed through fundamental solutions, and the latter admit a simple closed form representation only for some particular classes of equations (and systems) with constant coefficients. We should mention though that the method of boundary elements, if combined with some special iteration procedures, can even be used for solving certain types of nonlinear boundary value problems.

Another limitation manifests itself even when the fundamental solution is known and can be represented by means of a simple formula. In this case, reduction of a boundary value problem to an equivalent integral equation may still be problematic because of the interior resonances. We illustrated this phenomenon in Section 13.1 using the examples of a Dirichlet exterior problem and a Neumann exterior problem for the Helmholtz equation.

Chapter 14

Boundary Equations with Projections and the Method of Difference Potentials

The method of difference potentials is a technique for the numerical solution of interior and exterior boundary value problems for linear partial differential equations. It combines some important advantages of the method of finite differences (Part III) and the method of boundary elements (Chapter 13). At the same time, it allows one to avoid certain difficulties that pertain to these two methods.

The method of finite differences is most efficient when using regular grids for solving problems with simple boundary conditions on domains of simple shape (e.g., a square, circle, cube, ball, annulus, torus, etc.). For curvilinear domains, the method of finite differences may encounter difficulties, in particular, when approximating boundary conditions. The method of difference potentials uses those problems with simple geometry and simple boundary conditions as auxiliary tools for the numerical solution of more complicated interior and exterior boundary value problems on irregular domains. Moreover, the method of difference potentials offers an efficient alternative to the difference approximation of boundary conditions.

The main advantage of the boundary element method is that it reduces the spatial dimension of the problem by one, and also that the integral equations of the method (i.e., integral equations of the potential theory) automatically take into account the boundary conditions of the problem. Its main disadvantage is that to obtain those integral equations, one needs a closed form fundamental solution. This requirement considerably narrows the range of applicability of boundary elements.

Instead of the integral equations of classical potential theory, the method of difference potentials exploits boundary equations with projections of Calderon's type. They are more universal and enable an equivalent reduction of the problem from the domain to the boundary, regardless of the type of boundary conditions. Calderon's equations do not require fundamental solutions. At the same time, they contain no integrals, and cannot be discretized using quadrature formulae. However, they can be approximated by difference potentials, which, in turn, can be efficiently computed.

A basic construction of difference potentials is given by difference potentials of the Cauchy type. It plays the same role for the solutions of linear difference equations and systems as the classical Cauchy type integral:

$$f(z) = \frac{1}{2\pi i} \oint_{\Gamma} \frac{\varphi(\zeta)}{\zeta - z} d\zeta$$

plays for the solutions of the Cauchy-Riemann system, i.e., for analytic functions.

In this chapter, we only provide a brief overview of the concepts and ideas related to the method of difference potentials. To illustrate the concepts, we analyze a number of model problems for the two-dimensional Poisson equation:

$$\frac{\partial^2 u}{\partial x^2} + \frac{\partial^2 u}{\partial y^2} = f(x,y). \tag{14.1}$$

These problems are very elementary themselves, but they admit broad generalizations discussed in the recent specialized monograph by Ryaben'kii [Rya02].

Equation (14.1) will be discretized using standard central differences on a uniform Cartesian grid:

$$\{(x_{m_1}, y_{m_2}) \,|\, (m_1 h, m_2 h),\ m_1 = 0, \pm 1, \pm 2, \dots,\ m_2 = 0, \pm 1, \pm 2, \dots\}. \tag{14.2}$$

The discretization is symbolically written as (we have encountered it previously, see Section 12.1):

$$\sum_{n \in N_m} a_{mn} u_n = f_m, \tag{14.3}$$

where N_m is the simplest five-node stencil:

$$N_m = \{(m_1, m_2), (m_1 \pm 1, m_2), (m_1, m_2 \pm 1)\}. \tag{14.4}$$

Note that we have adopted a multi-index notation for the grid nodes: $m = (m_1, m_2)$ and $n = (n_1, n_2)$. The coefficients a_{mn} of equation (14.3) are given by the formula:

$$a_{mn} = \begin{cases} -4h^{-2}, & \text{if} \quad n = m, \\ h^{-2}, & \text{if} \quad n = (m_1 \pm 1, m_2) \quad \text{or} \quad n = (m_1, m_2 \pm 1). \end{cases}$$

The material in the chapter is organized as follows. In Section 14.1, we formulate the model problems to be analyzed. In Section 14.2, we construct difference potentials and study their properties. In Section 14.3, we use the apparatus on Section 14.2 to solve model problems of Section 14.1 and thus demonstrate some capabilities of the method of difference potentials. Section 14.4 contains some general comments on the relation between the method of difference potentials and some other methods available in the literature; and Section 14.5 provides bibliographic information.

14.1 Formulation of Model Problems

In this section, we introduce five model problems, for which their respective approximate solutions will be obtained using the method of difference potentials.

14.1.1 Interior Boundary Value Problem

Find a solution $u = u(x,y)$ of the Dirichlet problem for the Laplace equation:

$$\Delta u = 0, \quad (x,y) \in D, \quad u\big|_\Gamma = \varphi(s), \tag{14.5}$$

where $D \subset \mathbb{R}^2$ is a bounded domain with the boundary $\Gamma = \partial D$, and $\varphi(s)$ is a given function the argument s, which is the arc length along Γ.

14.1.2 Exterior Boundary Value Problem

Find a bounded solution $u = u(x,y)$ of the Dirichlet problem:

$$\Delta u = 0, \quad (x,y) \in \mathbb{R}^2 \backslash D, \quad u\big|_\Gamma = \varphi(s). \tag{14.6}$$

The solution of problem (14.6) will be of interest to us only on some bounded neighborhood of the boundary Γ.

Let us immerse the domain D into a larger square D^0, see Figure 14.1, and instead of problem (14.6) consider the following modified problem:

$$\begin{aligned} &\Delta u = 0, \qquad (x,y) \in D^-, \\ &u\big|_\Gamma = \varphi(s), \quad u\big|_{\partial D_0} = 0, \end{aligned} \tag{14.7}$$

where $D^- \stackrel{\text{def}}{=} D^0 \backslash D$ is the region between $\Gamma = \partial D$ and the outer boundary of the square ∂D_0. The larger the size of the square D^0 compared to the size of $D \equiv D^+$, the better problem (14.7) approximates problem (14.6) on any finite fixed neighborhood of the domain D^+. Hereafter we will study problem (14.7) instead of problem (14.6).

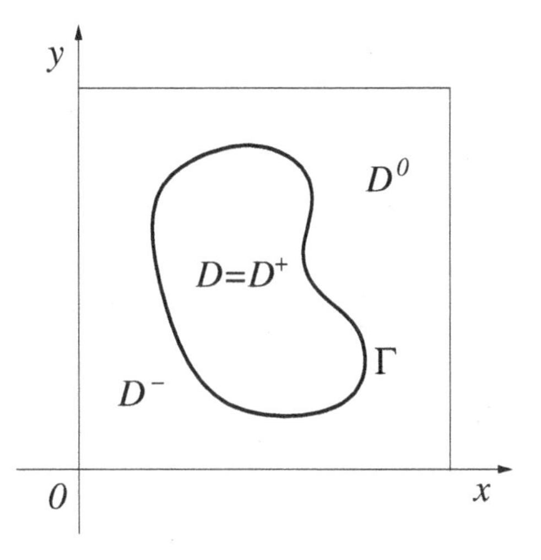

FIGURE 14.1: Schematic geometric setup.

14.1.3 Problem of Artificial Boundary Conditions

Consider the following boundary value problem on the square D^0:

$$Lu = f(x,y), \quad (x,y) \in D^0, \tag{14.8a}$$

$$u\big|_{\partial D^0} = 0, \tag{14.8b}$$

and assume that it has a unique solution $u = u(x,y)$ for any right-hand side $f(x,y)$. Suppose that the solution u needs to be computed not everywhere in D^0, but only on a given subdomain $D^+ \subset D^0$, see Figure 14.1. Moreover, suppose that outside D^+ the operator $\boldsymbol{L}$ transforms into a Laplacian, so that equation (14.8a) becomes:

$$\Delta u = 0, \quad (x,y) \in D^- = D^0 \setminus D^+. \tag{14.9}$$

Let us then define the artificial boundary as the the boundary $\Gamma = \partial D^+$ of the computational subdomain D^+. In doing so, we note that the original problem (14.8a), (14.8b) does not contain any artificial boundary. Having introduced Γ, we formulate the problem of constructing a special relation $lu\big|_\Gamma = 0$ on the artificial boundary Γ, such that for any $f(x,y)$ the solution of the new problem:

$$\boldsymbol{L}u = f(x,y), \quad (x,y) \in D^+, \tag{14.10a}$$

$$lu\big|_\Gamma = 0, \tag{14.10b}$$

will coincide on D^+ with the solution of the original problem (14.8a), (14.8b), (14.9). Condition (14.10b) is called an artificial boundary condition (ABC).

One can say that condition (14.10b) must equivalently replace the Laplace equation (14.9) outside the computational subdomain D^+ along with the boundary condition (14.8b) at the (remote) boundary ∂D^0 of the original domain. In our model problem, condition (14.8b), in turn, replaces the requirement that the solution of equation (14.8a) be bounded at infinity. One can also say that the ABC (14.10b) is obtained by transferring condition (14.8b) from the remote boundary of the domain D^0 to the artificial boundary ∂D^+ of the chosen computational subdomain D^+.

14.1.4 Problem of Two Subdomains

Let us consider the following Dirichlet problem on D^0:

$$\Delta u = f(x,y), \quad (x,y) \in D^0 = D^+ \cup D^-, \quad u\big|_{\partial D^0} = 0, \tag{14.11}$$

Suppose that the right-hand side $f(x,y)$ is unknown, but the solution $u(x,y)$ itself is known (say, it can be measured) on some neighborhood of the interface Γ between the subdomains D^+ and D^-, see Figure 14.1.

For an arbitrary domain $\Omega \subset \mathbb{R}^2$, define its indicator function:

$$\theta(\Omega) = \begin{cases} 1, & \text{if} \quad (x,y) \in \Omega, \\ 0, & \text{if} \quad (x,y) \notin \Omega. \end{cases}$$

Then, the overall solution can be represented as the sum of two contributions: $u = u^+ + u^-$, where $u^+ = u^+(x,y)$ is the solution due to the sources outside D^+:

$$\Delta u^+ = \theta(D^-) f(x,y), \qquad u^+\big|_{\partial D^0} = 0, \tag{14.12}$$

and $u^- = u^-(x,y)$ is the solution due to the sources inside D^+:

$$\Delta u^- = \theta(D^+) f(x,y), \qquad u^-\big|_{\partial D^0} = 0. \tag{14.13}$$

The problem is to find each term u^+ and u^- separately.

More precisely, given the overall solution and its normal derivative at the interface Γ:

$$\begin{bmatrix} u \\ \dfrac{\partial u}{\partial n} \end{bmatrix}_\Gamma = \begin{bmatrix} u^+ \\ \dfrac{\partial u^+}{\partial n} \end{bmatrix}_\Gamma + \begin{bmatrix} u^- \\ \dfrac{\partial u^-}{\partial n} \end{bmatrix}_\Gamma , \tag{14.14}$$

find the individual terms on the right-hand side of (14.14). Then, with no prior knowledge of the source distribution $f(x,y)$, also find the individual branches of the solution, namely:

$$u^+(x,y), \quad \text{for } (x,y) \in D^+,$$

which can be interpreted as the effect of D^- on D^+, see formula (14.12), and

$$u^-(x,y), \quad \text{for } (x,y) \in D^-,$$

which can be interpreted as the effect of D^+ on D^-, see formula (14.13).

14.1.5 Problem of Active Shielding

As before, let the square D^0 be split into two subdomains: D^+ and D^-, as shown in Figure 14.1. Suppose that the sources $f(x,y)$ are not known, but the values of the solution $u\big|_\Gamma$ and its normal derivative $\frac{\partial u}{\partial n}\big|_\Gamma$ at the interface Γ between D^+ and D^- are explicitly specified (e.g., obtained by measurements).

Along with the original problem (14.11), consider a modified problem:

$$\Delta w = f(x,y) + g(x,y), \quad (x,y) \in D^0, \quad w\big|_{\partial D^0} = 0, \tag{14.15}$$

where the additional term $g = g(x,y)$ on the right-hand side of equation (14.15) represents the active control sources, or simply active controls. The role of these active controls is to protect, or shield, the region $D^+ \subset D^0$ from the influence of the sources located on the complementary region $D^- = D^0 \backslash D^+$. In other words, we require that the solution $w = w(x,y)$ of problem (14.15) coincide on D^+ with the solution $v = v(x,y)$ of the problem:

$$\Delta v = \begin{cases} f(x,y), & (x,y) \in D^+, \\ 0, & (x,y) \in D^-, \end{cases} \qquad v = \big|_{\partial D^0} = 0, \tag{14.16}$$

which is obtained by keeping all the original sources on D^+ and removing all the sources on D^-. The problem of active shielding consists of finding all appropriate active controls $g = g(x,y)$. This problem can clearly be interpreted as a particular inverse source problem for the given differential equation (Poisson equation).

REMARK 14.1 Let us emphasize that we are not interested in finding the solution $u(x,y)$ of problem (14.11), the solution $w(x,y)$ of problem (14.15), or the solution $v(x,y)$ of problem (14.16). In fact, these solutions cannot even be obtained on D^0 because the right-hand side $f(x,y)$ is not known. Our objective is rather to find all $g(x,y)$ that would have a predetermined effect on the solution $u(x,y)$, given only the trace of the solution and its normal derivative on Γ. The desired effect should be equivalent to that of the removal of all sources on the subdomain D^-, i.e., outside D^+. ▯

An obvious particular solution to the foregoing problem of active shielding is given by the function:

$$g(x,y) \equiv \begin{cases} 0, & (x,y) \in D^+, \\ -f(x,y), & (x,y) \in D^-. \end{cases} \tag{14.17}$$

This solution, however, cannot be obtained explicitly, because the original sources $f(x,y)$ are not known. Even if they were known, active control (14.17) could still be difficult to implement in many applications.

14.2 Difference Potentials

In this section, we construct difference potentials for the simplest five-node discretization (14.3) of the Poisson equation (14.1). All difference potentials are obtained as solutions to specially chosen auxiliary finite-difference problems.

14.2.1 Auxiliary Difference Problem

Let the domain $D^0 \subset \mathbb{R}^2$ be given, and let M^0 be the subset of all nodes $m = (m_1 h, m_2 h)$ of the uniform Cartesian grid (14.2) that belong to $D^{(0)}$. We consider difference equation (14.3) on the set M^0:

$$\sum_{n \in N_m} a_{mn} u_n = f_m, \quad m \in M^0. \tag{14.18}$$

The left-hand side of this equation is defined for all functions $u_{N^0} = \{u_n\}$, $n \in N^0$, where $N^0 = \bigcup N_m$, $m \in M^0$, and N_m is the stencil (14.4). We also supplement equation (14.18) by an additional linear homogeneous (boundary) condition that we write in the form of an inclusion:

$$u_{N^0} \in U_{N^0}. \tag{14.19}$$

In this formula, U_{N^0} denotes a given linear space of functions on the grid N^0. The only constraint to satisfy when selecting a specific U_{N^0} is that problem (14.18), (14.19) must have a unique solution for an arbitrary right-hand side $f_{M^0} = \{f_m\}$, $m \in M^0$. If this constraint is satisfied, then problem (14.18), (14.19) is an appropriate auxiliary difference problem.

Note that a substantial flexibility exists in choosing the domain D^0, the corresponding grid domains M^0 and $N^0 = \bigcup N_m$, $m \in M^0$, and the space U_{N^0}. Depending on the particular choices made, the resulting auxiliary difference problem (14.18), (14.19) may or may not be easy to solve, and accordingly, it may or may not appear well suited for addressing a given application.

To provide an example of an auxiliary problem, we let D^0 be a square with its sides parallel to the Cartesian coordinate axes, and its width and height equal to kh, where k is integer. The corresponding sets M^0 and N^0 are shown in Figure 14.2. The set M^0 in Figure 14.2 consists of the black bullets, and the set $N^0 = \bigcup N_m$, $m \in M^0$, additionally contains the hollow bullets. The four corner nodes do not belong to either set.

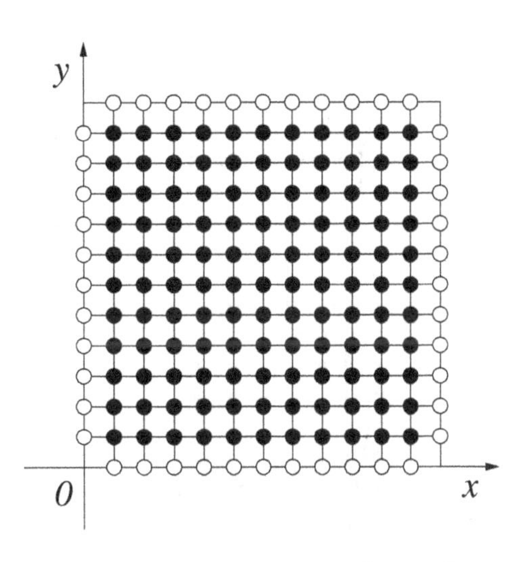

FIGURE 14.2: Grid sets M^0 and N^0.

Next, we define the space U_{N^0} as the space of all functions $u_{N^0} = \{u_n\}$, $n \in N^0$, that vanish on the sides of the square D^0, i.e., at the nodes $n \in N^0$ denoted by the hollow bullets in Figure 14.2. Then, the auxiliary problem (14.18), (14.19) is a difference Dirichlet problem for the Poisson equation subject to the homogeneous boundary condition. This problem has a unique solution for an arbitrary right-hand side $f_{M^0} = \{f_m\}$, see Section 12.1. The solution can be efficiently computed using separation of variables, i.e., the Fourier method of Section 5.7. In particular, if the dimension of the grid in one coordinate direction is a power of 2, then the computational complexity of the corresponding algorithm based on the fast Fourier transform (see Section 5.7.3) will only be $\mathcal{O}(h^{-2}|\ln h|)$ arithmetic operations.

14.2.2 The Potential $u^+ = P^+ v_\gamma$

Let D^+ be a given bounded domain, and let M^+ be the set of grid nodes m that belong to D^+. Consider the restriction of system (14.3) onto the grid M^+:

$$\sum_{n \in N_m} a_{mn} u_n = f_m, \quad m \in M^+. \tag{14.20}$$

The left-hand side of (14.20) is defined for all functions $u_{N^+} = \{u_n\}$, $n \in N^+$, where

$$N^+ = \bigcup N_m, \quad m \in M^+.$$

We will build difference potentials for the solutions of system (14.20). Let D^0 be a larger square, $D^+ \subset D^0$, and introduce an auxiliary difference problem of the type

(14.18), (14.19). Denote $D^- = D^0 \setminus \bar{D}^+$, and let M^- be the set of grid nodes that belong to D^-:

$$M^- = \{m \mid m \in D^-\} = M^0 \setminus M^+.$$

Along with system (14.20), consider the restriction of (14.3) onto the set M^-:

$$\sum_{n \in N_m} a_{mn} u_n = f_m, \quad m \in M^-. \tag{14.21}$$

The left-hand side of (14.21) is defined for all functions $u_{N^-} = \{u_n\}$, $n \in N^-$, where

$$N^- = \bigcup N_m, \quad m \in M^-.$$

Thus, system (14.18) gets split into two subsystems: (14.20) and (14.21), with their solutions defined on the sets N^+ and N^-, respectively.

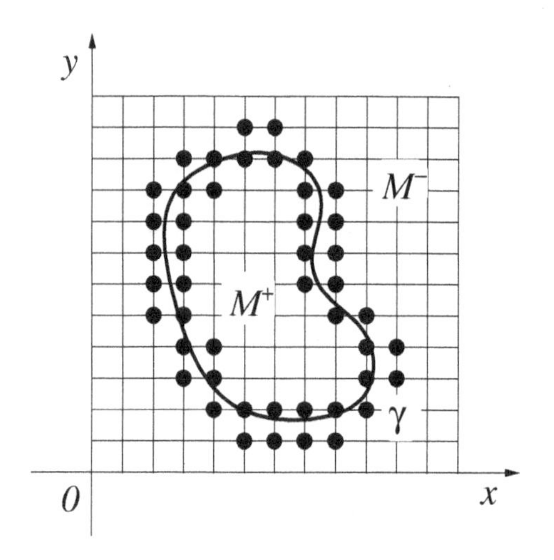

FIGURE 14.3: Grid boundary.

Let us now define the boundary γ between the grid domains N^+ and N^- (see Figure 14.3):

$$\gamma = N^+ \cap N^-.$$

It is a narrow fringe of nodes that straddles the continuous boundary Γ. We also introduce a linear space V_γ of all grid functions v_γ defined on γ. These functions can be considered restrictions of the functions $u_{N^0} \in U_{N^0}$ to the subset $\gamma \subset N^0$.

The functions $v_\gamma \in V_\gamma$ will be referred to as densities. We can now introduce the potential $u^+ = \boldsymbol{P}^+ v_\gamma$ with the density v_γ.

DEFINITION 14.1 *Consider the grid function:*

$$v_n = \begin{cases} v_\gamma\big|_n, & \textit{if } \; n \in \gamma, \\ 0, & \textit{if } \; n \notin \gamma, \end{cases} \tag{14.22}$$

and define:

$$f_m = \begin{cases} 0, & \textit{if } \; m \in M^+, \\ \sum_{n \in N_m} a_{mn} v_n, & \textit{if } \; m \in M^-, \end{cases} \tag{14.23}$$

Let $u_{N^0} = \{u_n\}$, $n \in N^0$, be the solution of the auxiliary difference problem (14.18), (14.19) driven by the right-hand side f_m of (14.23), (14.22). The difference potential $u^+ = \boldsymbol{P}^+ v_\gamma$ is the function $u_{N^+} = \{u_n\}$, $n \in N^+$, that coincides with u_{N^0} on the grid domain N^+.

Clearly, to compute the difference potential $u^+ = \boldsymbol{P}^+ v_\gamma$ with a given density $v_\gamma \in V_\gamma$, one needs to solve the auxiliary difference problem (14.18), (14.19) with the right-hand side (14.23), (14.22).

THEOREM 14.1
The difference potential $u^+ = u_{N^+} = \boldsymbol{P}^+ v_\gamma$ has the following properties.

1. *There is a function $u_{N^0} \in U_{N^0}$, such that $u_{N^+} = \boldsymbol{P}^+ v_\gamma = \theta(N^+) u_{N^0}\big|_{N^+}$, where $\theta(\cdot)$ denotes the indicator function of a set.*

2. *On the set M^+, the difference potential $u^+ = \boldsymbol{P}^+ v_\gamma$ satisfies the homogeneous equation:*

$$\sum_{n \in N_m} a_{mn} u_n = 0, \quad m \in M^+. \tag{14.24}$$

3. *Let $v_{N^0} \in U_{N^0}$, and let $v_{N^+} = \theta(N^+) v_{N^0}\big|_{N^+}$ be a solution to the homogeneous equation (14.24). Let the density v_γ be given, and let it coincide with v_{N^+} on γ: $v_\gamma = v_{N^+}\big|_\gamma$. Then the solution $v_{N^+} = \{v_n\}$, $n \in N^+$, is unique and can be reconstructed from its values on the boundary γ by the formula:*

$$v_{N^+} = \boldsymbol{P}^+ v_\gamma. \tag{14.25}$$

PROOF The first implication of the theorem follows immediately from Definition 14.1. The second implication is also true because the potential $\boldsymbol{P}^+ v_\gamma$ solves problem (14.18), (14.19), and according to formula (14.23), the right-hand side of this problem is equal to zero on M^+.

To prove the third implication, consider the function $v_{N^0} \in U_{N^0}$ from the hypothesis of the theorem along with another function $w_{N^0} = \theta(N^+) v_{N^0} \in U_{N^0}$. The function w_{N^0} satisfies equation (14.18) with the right-hand side given by (14.23), (14.22). Since the solution of problem (14.18), (14.19) is unique, its solution w_{N^0} driven by the right-hand side (14.23), (14.22) coincides on N^+ with the difference potential $v_{N^+} = \boldsymbol{P}^+ v_\gamma$. Therefore, the solution $v_{N^+} \equiv w_{N^0}\big|_{N^+}$ is indeed unique and can be represented by formula (14.25). ∎

Let us now introduce another operator, $\boldsymbol{P}^+_\gamma : V_\gamma \longmapsto V_\gamma$, that for a given density $\gamma \in V_\gamma$ yields the trace of the difference potential $\boldsymbol{P}^+ v_\gamma$ on the grid boundary γ:

$$\boldsymbol{P}^+_\gamma v_\gamma = \boldsymbol{P}^+ v_\gamma\big|_\gamma. \tag{14.26}$$

THEOREM 14.2
Let $v_\gamma \in V_\gamma$ be given, and let $v_\gamma = v_{N^+}\big|_\gamma$, where v_{N^+} is a solution to the homogeneous equation (14.24) *and $v_{N^+} = \theta(N^+) v_{N^0}\big|_{N^+}$, $v_{N^0} \in U_{N^0}$. Then,*

$$\boldsymbol{P}^+_\gamma v_\gamma = v_\gamma. \tag{14.27}$$

Conversely, if equation (14.27) holds for a given $v_\gamma \in V_\gamma$, then v_γ is the trace of some solution v_{N^+} of equation (14.24), *and* $v_{N^+} = \theta(N^+)v_{N^0}\big|_{N^+}$, $v_{N^0} \in U_{N^0}$.

PROOF The first (direct) implication of this theorem follows immediately from the third implication of Theorem 14.1. Indeed, the solution v_{N^+} of the homogeneous equation (14.24) that coincides with the prescribed v_γ on γ is unique and is given by formula (14.25). Considering the restriction of both sides of (14.25) onto γ, and taking into account the definition (14.26), we arrive at equation (14.27).

Conversely, suppose that for some v_γ equation (14.27) holds. Let us consider the difference potential $u^+ = \boldsymbol{P}^+ v_\gamma$ with the density v_γ. According to Theorem 14.1, u^+ is a solution of equation (14.24), and this solution can be extended from N^+ to N^0 so that the extension belongs to U_{N^0}. On the other hand, formulae (14.26) and (14.27) imply that

$$u^+\big|_\gamma = \boldsymbol{P}^+ v_\gamma\big|_\gamma = \boldsymbol{P}^+_\gamma v_\gamma = v_\gamma,$$

so that v_γ are the boundary values of this potential. As such, u^+ can be taken in the capacity of the desired v_{N^+}, which completes the proof. ∎

Theorems 14.1 and 14.2 imply that the operator $\boldsymbol{P}^+_\gamma$ is a projection:

$$(\boldsymbol{P}^+_\gamma)^2 = \boldsymbol{P}^+_\gamma.$$

In [Rya85], [Rya02], it is shown that this operator can be considered a discrete counterpart of Calderon's boundary projection, see [Cal63]. Accordingly, equation (14.27) is referred to as the boundary equation with projection. A given density $v_\gamma \in V_\gamma$ can be interpreted as boundary trace of some solution to the homogeneous equation (14.24) if and only if it satisfies the boundary equation with projection (14.27), or alternatively, if and only if it belongs to the range of the projection $\boldsymbol{P}^+_\gamma$.

14.2.3 Difference Potential $u^- = \boldsymbol{P}^- v_\gamma$

The difference potential $\boldsymbol{P}^- v_\gamma$ with the density $v_\gamma \in V_\gamma$ is a grid function defined on N^-. Its definition is completely similar and basically reciprocal to that of the potential $\boldsymbol{P}^+ v_\gamma$. To build the potential $\boldsymbol{P}^- v_\gamma$ and analyze its properties, one only needs to replace the superscript "+" by "−", and the superscript "−" by "+," on all occasions when these superscripts are encountered in Definition 14.1, as well as in Theorems 14.1 and 14.2 and their proofs.

14.2.4 Cauchy Type Difference Potential $w^{\pm} = \boldsymbol{P}^{\pm} v_\gamma$

DEFINITION 14.2 *The function $w^{\pm} = \boldsymbol{P}^{\pm} v_\gamma$ defined for $n \in N^0$ as:*

$$w_n^{\pm} = \begin{cases} w_n^{+} = \boldsymbol{P}^{+} v_\gamma\big|_n, & \text{if} \quad n \in N^{+}, \\ w_n^{-} = -\boldsymbol{P}^{-} v_\gamma\big|_n, & \text{if} \quad n \in N^{-}. \end{cases} \tag{14.28}$$

is called a Cauchy type difference potential $w^{\pm} = \boldsymbol{P}^{\pm} v_\gamma$ with the density v_γ.

This potential is, generally speaking, a two-valued function on γ, because each node $n \in \gamma$ simultaneously belongs to both N^+ and N^-.

Along with the Cauchy type difference potential, we will define an equivalent concept of the difference potential $w^{\pm} = \boldsymbol{P}^{\pm} v_\gamma$ with the jump v_γ. This new concept will be instrumental for studying properties of the potential $w^{\pm} = \boldsymbol{P}^{\pm} v_\gamma$, as well as for understanding a deep analogy between the Cauchy type difference potential and the Cauchy type integral from classical theory of analytic functions.

Moreover, the new definition will allow us to calculate all three potentials:

$$u^{+} = \boldsymbol{P}^{+} v_\gamma, \quad u^{-} = \boldsymbol{P}^{-} v_\gamma, \quad \text{and} \quad w^{\pm} = \boldsymbol{P}^{\pm} v_\gamma$$

at once, by solving only one auxiliary problem (14.18), (14.19) for a special right-hand side f_m, $m \in M^0$. On the other hand, formula (14.28) indicates that a straight-forward computation of $w^{\pm}$ according to Definition 14.2 would require obtaining the potentials $\boldsymbol{P}^{+} v_\gamma$ and $\boldsymbol{P}^{-} v_\gamma$ independently, which, in turn, necessitates solving two problems of the type (14.18), (14.19).

To define difference potential $w^{\pm} = \boldsymbol{P}^{\pm} v_\gamma$ with the jump v_γ, we first need to introduce piecewise regular functions. The functions u_{N^0} from the space U_{N^0} will be called regular. Given two arbitrary regular functions $u_{N^0}^{+}$ and $u_{N^0}^{-}$, we define a piecewise regular function $u_n^{\pm}$, $n \in N^0$, as:

$$u_n^{\pm} = \begin{cases} u_n^{+}, & \text{if} \quad n \in N^{+}, \\ u_n^{-}, & \text{if} \quad n \in N^{-}. \end{cases} \tag{14.29}$$

Let $U^{\pm}$ be a linear space of all piecewise regular functions (14.29). Any function (14.29) assumes two values u_n^{+} and u_n^{-} at every node n of the grid boundary γ. A single-valued function v_γ defined for $n \in \gamma$ by the formula:

$$v_\gamma\big|_n = [u^{\pm}]_n = u_n^{+} - u_n^{-}, \qquad n \in \gamma, \tag{14.30}$$

will be called a jump. Note that the space U_{N^0} of regular functions can be considered a subspace of the space $U^{\pm}$; this subspace consists of all functions $u^{\pm} \in U^{\pm}$ with zero jump: $v_\gamma = 0_\gamma$.

A piecewise regular function (14.29) will be called a piecewise regular solution of the problem:

$$\sum_{n \in N_m} a_{mn} u_n^{\pm} = 0, \quad u^{\pm} \in U^{\pm}, \tag{14.31}$$

if the functions u_n^+, $n \in N^+$, and u_n^-, $n \in N^-$, satisfy the homogeneous equations:

$$\sum_{n \in N_m} a_{mn} u_n^+ = 0, \quad m \in M^+, \tag{14.32a}$$

and

$$\sum_{n \in N_m} a_{mn} u_n^- = 0, \quad m \in M^-, \tag{14.32b}$$

respectively.

THEOREM 14.3
Problem (14.31) *has a unique solution for any given jump* $v_\gamma \in V_\gamma$. *This solution is obtained by the formula:*

$$u^\pm = v^\pm - u, \tag{14.33}$$

where the minuend on the right-hand side of (14.33) *is an arbitrary piecewise regular function* $v^\pm \in U^\pm$ *with the prescribed jump* v_γ, *and the subtrahend is a regular solution of the auxiliary difference problem* (14.18), (14.19) *with the right-hand side:*

$$f_m = \begin{cases} \sum_{n \in N_m} a_{mn} v_n^+, & m \in M^+, \\ \sum_{n \in N_m} a_{mn} v_n^-, & m \in M^-. \end{cases} \tag{14.34}$$

PROOF First, we notice that piecewise regular functions $v^\pm \in U^\pm$ with a given jump v_γ exist. Obviously, one of these functions is given by the formula:

$$v_n^\pm = \begin{cases} v_n^+, & n \in N^+, \\ v_n^-, & n \in N^-, \end{cases}$$

where

$$v_n^+ = \begin{cases} v_\gamma\big|_n, & \text{if} \quad n \in \gamma \subset N^+, \\ 0, & \text{if} \quad n \in N^+ \setminus \gamma. \end{cases}$$

$$v_n^- \equiv 0, \qquad \text{if} \quad n \in N^-.$$

The second term on the right-hand side of (14.33) also exists, since problem (14.18), (14.19) has one and only one solution for an arbitrary right-hand side f_m, $m \in M^0$, in particular, for the one given by formula (14.34).

Next, formula (14.33) yields:

$$[u^\pm] = [v^\pm] - [u] = v_\gamma - [u],$$

and since u is a regular function and has no jump:

$$[u]\Big|_n = 0, \quad n \in \gamma,$$

we conclude that the function $u^{\pm}$ of (14.33) is a piecewise regular function with the jump v_γ. Let us show that this function $u^{\pm}$ is, in fact, a piecewise regular solution. To this end, we need to show that the functions:

$$u_n^+ = v_n^+ - u_n, \quad n \in N^+, \tag{14.35a}$$

and

$$u_n^- = v_n^- - u_n, \quad n \in N^-, \tag{14.35b}$$

satisfy the homogeneous equations (14.32a) and (14.32b), respectively. First, substituting the function u_n^+ of (14.35a) into equation (14.32a), we obtain:

$$\sum_{n \in N_m} a_{mn} u_n^+ = \sum_{n \in N_m} a_{mn} v_n^+ - \sum_{n \in N_m} a_{mn} u_n, \quad m \in M^+. \tag{14.36}$$

However, $\{u_n\}$ is the solution of the auxiliary difference problem with the right-hand side (14.34). Hence, we have:

$$\sum_{n \in N_m} a_{mn} u_n = f_m = \sum_{n \in N_m} a_{mn} v_n^+, \quad m \in M^+.$$

Thus, the right-hand side of formula (14.36) cancels out and therefore, equation (14.32a) is satisfied. The proof that the function (14.35b) satisfies equation (14.32b) is similar.

It only remains to show that the solution to problem (14.31) with a given jump v_γ is unique. Assume that there are two such solutions. Then the difference between these solutions is a piecewise regular solution with zero jump. Hence it is a regular function from the space U_{N^0}. In this case, problem (14.31) coincides with the auxiliary problem (14.18), (14.19) driven by zero right-hand side. However, the solution of the auxiliary difference problem (14.18), (14.19) is unique, and consequently, it is equal to zero. Therefore, two piecewise regular solutions with the same jump coincide. $\square$

Theorem 14.3 makes the following definition valid.

DEFINITION 14.3 *A piecewise regular solution $u^{\pm}$ of problem* (14.31) *with a given jump $v_\gamma \in V_\gamma$ will be called a difference potential $u^{\pm} = \boldsymbol{P}^{\pm} v_\gamma$ with the jump v_γ.*

THEOREM 14.4

Definition 14.2 *of the Cauchy type difference potential $w^{\pm} = \boldsymbol{P}^{+} v_\gamma$ with the density $v_\gamma \in V_\gamma$ and Definition* 14.3 *of the difference potential $u^{\pm} = \boldsymbol{P}^{\pm} v_\gamma$ with the jump v_γ are equivalent, i.e., $w^{\pm} = u^{\pm}$.*

PROOF It is clear the formula (14.28) yields a piecewise regular solution of problem (14.31). Since we have proven that the solution of problem (14.31) with a given jump v_γ is unique, it only remains to show that the piecewise regular solution (14.28) has jump v_γ:

$$[w^\pm] = u_\gamma^+ - u_\gamma^- = \boldsymbol{P}_\gamma^+ v_\gamma - (-\boldsymbol{P}_\gamma^- v_\gamma) = \boldsymbol{P}_\gamma^+ v_\gamma + \boldsymbol{P}_\gamma^- v_\gamma = v_\gamma.$$

In the previous chain of equalities only the last one needs to be justified, i.e., we need to prove that

$$\boldsymbol{P}_\gamma^+ v_\gamma + \boldsymbol{P}_\gamma^- v_\gamma = v_\gamma, \quad v_\gamma \in V_\gamma. \tag{14.37}$$

Recall that $\boldsymbol{P}_\gamma^+ v_\gamma$ coincides on γ with the difference potential $u^+ = \boldsymbol{P}^+ v_\gamma$, which, in turn, coincides on N^+ with the solution of the auxiliary difference problem:

$$\sum_{n \in N_m} a_{mn} u_n^+ = f_m^+, \quad m \in M^0, \quad u_{N^0}^+ \in U_{N^0}, \tag{14.38}$$

driven by the right-hand side f_m of (14.23), (14.22):

$$f_m^+ = \begin{cases} 0, & \text{if} \quad m \in M^+, \\ \sum_{n \in N_m} a_{mn} v_n, & \text{if} \quad m \in M^-. \end{cases} \tag{14.39}$$

Quite similarly, the expression $\boldsymbol{P}_\gamma^- v_\gamma$ coincides on γ with the difference potential $u^- = \boldsymbol{P}^- v_\gamma$, which, in turn, coincides on N^- with the solution of the auxiliary difference problem:

$$\sum_{n \in N_m} a_{mn} u_n^- = f_m^-, \quad m \in M^0, \quad u_{N^0}^- \in U_{N^0}, \tag{14.40}$$

driven by the right-hand side:

$$f_m^- = \begin{cases} \sum_{n \in N_m} a_{mn} v_n, & m \in M^+, \\ 0, & m \in M^-. \end{cases} \tag{14.41}$$

Similarly to formula (14.39), the function v_n in formula (14.41) is determined by the given density v_γ according to (14.22).

Let us add equations (14.38) and (14.40). On the right-hand side we obtain:

$$f_m^+ + f_m^- = \sum_{n \in N_m} a_{mn} v_n, \quad m \in M^0.$$

Consequently, we have:

$$\sum_{n \in N_m} a_{mn} (u_n^+ + u_n^-) = \sum_{n \in N_m} a_{mn} v_n, \quad m \in M^0.$$

The function v_{N^0} determined by v_γ according to (14.22) belongs to the space U_{N^0}. Since the functions $u_{N^0}^+$ and $u_{N^0}^-$ are solutions of the auxiliary difference problems (14.38), (14.39) and (14.40), (14.41), respectively, each of them also belongs to U_{N^0}. Thus, their sum is in the same space: $z_{N^0} \equiv u^+_{N^0} + u^-_{N^0} \in U_{N^0}$, and it solves the following problem of the type (14.18), (14.19):

$$\sum_{n\in N_m} a_{mn} z_n = \sum_{n\in N_m} a_{mn} v_n, \quad m \in M^0, \quad z_{N^0} \in U_{N^0}.$$

Clearly, if we substitute v_{N^0} for the unknown z_{N^0} on the left-hand side, the equation will be satisfied. Then, because of the uniqueness of the solution, we have: $z_{N^0} = v_{N^0}$ or $z_n = u^+_n + u^-_n = v_n$, $n \in N^0$. This implies:

$$u^+_n + u^-_n = \boldsymbol{P}^+ v_\gamma\big|_n + \boldsymbol{P}^- v_\gamma\big|_n = v_\gamma\big|_n, \quad n \in \gamma \subset N^0,$$

which coincides with the desired equality (14.37). □

REMARK 14.2 Suppose that the potential $u^\pm = \boldsymbol{P}^\pm v_\gamma$ for a given jump $v_\gamma \in V_\gamma$ is introduced according to Definition 14.3. In other words, it is given by formula (14.33) that requires solving problem (14.18), (14.19) with the right-hand side (14.34). Thus, the functions u^+ and u^- are determined:

$$u^\pm_n = \begin{cases} u^+_n, & n \in N^+, \\ u^-_n, & n \in N^-. \end{cases}$$

Consequently, the individual potentials $\boldsymbol{P}^+ v_\gamma$ and $\boldsymbol{P}^- v_\gamma$ are also determined:

$$\begin{aligned} \boldsymbol{P}^+ v_\gamma\big|_n &= u^+_n, && \text{if} \quad n \in N^+, \\ \boldsymbol{P}^- v_\gamma\big|_n &= -u^-_n, && \text{if} \quad n \in N^-. \end{aligned}$$

by virtue of Theorem 14.4 and Definition 14.2. □

14.2.5 Analogy with Classical Cauchy Type Integral

Suppose that Γ is a simple closed curve that partitions the complex plane $z = x + iy$ into a bounded domain D^+ and the complementary unbounded domain D^-. The classical Cauchy type integral (where the direction of integration is counterclockwise):

$$u^\pm(z) = \frac{1}{2\pi i} \oint_\Gamma \frac{v_\Gamma(\zeta)}{\zeta - z} d\zeta, \tag{14.42}$$

can be alternatively defined as a piecewise analytic function that has zero limit at infinity and undergoes jump $v_\Gamma = [u^\pm]_\Gamma$ across the contour Γ. Here $u^+(z)$ and $u^-(z)$ are the values of the integral (14.42) for $z \in D^+$ and $z \in D^-$, respectively.

Cauchy type integrals can be interpreted as potentials for the solutions of the Cauchy-Riemann system:

$$\frac{\partial a}{\partial x} = \frac{\partial b}{\partial y}, \qquad \frac{\partial b}{\partial x} = -\frac{\partial a}{\partial y},$$

that connects the real part a and the imaginary part b of an analytic function $a+ib$.
Thus, the Cauchy type difference potential:

$$w^{\pm} = \boldsymbol{P}^{\pm} v_{\gamma}$$

plays the same role for the solutions of system (14.18) as the Cauchy type integral plays for the solutions of the Cauchy-Riemann system.

14.3 Solution of Model Problems

In this section, we use difference potentials of Section 14.2 to solve model problems of Section 14.1.

14.3.1 Interior Boundary Value Problem

Consider the interior Dirichlet problem (14.5):

$$\begin{aligned} \Delta u &= 0, \qquad (x,y) \in D, \\ u\big|_{\Gamma} &= \varphi(s), \qquad \Gamma = \partial D. \end{aligned} \tag{14.43}$$

Introduce a larger square $D^0 \supset D$ with its sides on the coordinate lines of the Cartesian grid (14.2). The auxiliary problem (14.18) (14.19) is formulated on D^0:

$$\begin{aligned} \sum_{n \in N_m} a_{mn} u_n &= f_m, \qquad m \in M^0, \\ u_{N^0}\big|_n &= 0, \qquad n \in \partial D^0. \end{aligned} \tag{14.44}$$

We will construct two algorithms for solving problem (14.43) numerically. In either algorithm, we will replace the Laplace equation of (14.43) by the difference equation:

$$\sum_{n \in N_m} a_{mn} u_n = 0, \qquad m \in M^+ \subset D. \tag{14.45}$$

Following the considerations of Section 14.2, we introduce the sets N^+, N^-, and $\gamma = N^+ \cap N^-$, the space V_γ of all functions defined on γ, and the potential: $u_N^+ = \boldsymbol{P}^+ v_\gamma$.

Difference Approximation of the Boundary Condition

In the first algorithm, we approximate the Dirichlet boundary condition of (14.43) by some linear difference boundary condition that we symbolically write as:

$$l u_\gamma = \varphi^{(h)}. \tag{14.46}$$

Theorem 14.2 implies that the boundary equation with projection:

$$u_\gamma = \boldsymbol{P}_\gamma^+ u_\gamma \tag{14.47}$$

is necessary and sufficient for $u_\gamma \in V_\gamma$ to be a trace of the solution to equation (14.45) on the grid boundary $\gamma \subset N$. Obviously, the trace u_γ of the solution to problem (14.45), (14.46) coincides with the solution of system (14.46), (14.47):

$$u_\gamma - \boldsymbol{P}_\gamma^+ u_\gamma = 0, \qquad l u_\gamma = \varphi^{(h)}. \tag{14.48}$$

Once the solution u_γ of system (14.48) has been computed, the desired solution u_{N^+} to equation (14.45) can be reconstructed from its trace u_γ according to the formula: $u_{N^+} = \boldsymbol{P}^+ u_\gamma$, see Theorem 14.1. Regarding the solution of system (14.48), iterative methods (Chapter 6) can be applied because the grid function $v_\gamma - \boldsymbol{P}_\gamma^+ v_\gamma$ can be easily computed for any $v_\gamma \in V_\gamma$ by solving the auxiliary difference problem (14.44). Efficient algorithms for solving system (14.48) are discussed in [Rya02, Part I].

Let us also emphasize that the reduction of problem (14.45), (14.46) to problem (14.48) drastically decreases the number of unknowns: from u_{N^+} to u_γ.

Spectral Approximation of the Boundary Condition

In the second algorithm for solving problem (14.43), we choose a system of basis functions $\psi_1(s), \psi_2(s), \dots$ on the boundary Γ of the domain D, and assume that the normal derivative $\frac{\partial u}{\partial n}\big|_\Gamma$ of the desired solution can be approximated by the sum:

$$\left.\frac{\partial u}{\partial n}\right|_\Gamma = \sum_{k=1}^{K} c_k \psi_k(s) \tag{14.49}$$

with any given accuracy. In formula (14.49), c_k are the coefficients to be determined, and K is a sufficiently large integer. We emphasize that the number of terms K in the representation (14.49) that is required to meet a prescribed tolerance will, generally speaking, depend on the properties of the approximating space: $\text{span}\{\psi_1(s), \dots, \psi_n(s), \dots\}$, as well as on the class of functions that the derivative $\frac{\partial u}{\partial n}\big|_\Gamma$ belongs to (i.e., on the regularity of the derivative, see Section 12.2.5).

For a given function $u\big|_\Gamma = \varphi(s)$ and for the derivative $\frac{\partial u}{\partial n}\big|_\Gamma$ taken in the form (14.49), we construct the function v_γ using the Taylor formula:

$$v_n = v_n(c_1, \dots, c_K) = \varphi(s_n) + \rho_n \sum_{k=1}^{K} c_k \psi_k(s_n), \quad n \in \gamma. \tag{14.50}$$

In formula (14.50), s_n denotes the arc length along Γ at the foot of the normal dropped from the node $n \in \gamma$ to Γ. The number ρ_n is the distance from the point $s_n \in \Gamma$ to the node $n \in \gamma$ taken with the sign "+" if n is outside Γ and the sign "−" if n is inside Γ.

We require that the function $v_\gamma = v_\gamma(c_1, \dots, c_K) \equiv v_\gamma(\boldsymbol{c})$ satisfy the boundary equation with projection (14.47):

$$v_\gamma(\boldsymbol{c}) - \boldsymbol{P}_\gamma^+ v_\gamma(\boldsymbol{c}) = 0. \tag{14.51}$$

The number of equations in the linear system (14.51) that has K unknowns: $c_1, c_2, \dots, c_K$, is equal to the number of nodes $|\gamma|$ of the boundary γ. For a fixed K

and a sufficiently fine grid (i.e., if $|\gamma|$ is large) this system is overdetermined. It can be solved by the method of least squares (Chapter 7). Once the coefficients $c_1, \dots, c_K$ have been found, the function $v_\gamma(\boldsymbol{c})$ is obtained by formula (14.50) and the solution u_{N^+} is computed as:

$$u_{N^+} \approx \boldsymbol{P}^+ v_\gamma. \tag{14.52}$$

A strategy for choosing the right scalar product for the method of least squares is discussed in [Rya02, Part I]. Efficient algorithms for computing the generalized solution $c_1, \dots, c_K$ of system (14.51), as well as the issue of convergence of the approximate solution (14.52) to the exact solution, are also addressed in [Rya02, Part I].

14.3.2 Exterior Boundary Value Problem

First, we replace the Laplace equation of (14.6) by the difference equation:

$$\sum_{n \in N_m} a_{mn} u_n = 0, \qquad m \in M^- = M^0 \setminus M^+. \tag{14.53}$$

Then, we introduce the auxiliary difference problem (14.44), the grid domains N^+ and N^-, the grid boundary γ, and the space V_γ, in the same way as we did for the interior problem in Section 14.3.1.

For the difference approximation of the boundary condition $u\big|_\Gamma = \varphi(s)$, we also use the same equation (14.46). The boundary equation with projection for the exterior problem takes the form [cf. formula (14.27)]:

$$u_\gamma - \boldsymbol{P}_\gamma^- u_\gamma = 0. \tag{14.54}$$

The function u_γ satisfies equation (14.54) if and only if it is the trace of a solution to system (14.53) subject to the homogeneous Dirichlet boundary conditions at ∂D^0, see (14.44). To actually find u_γ, one must solve equations (14.46) and (14.54) together. Then, one can reconstruct the solution u_{N^-} of the exterior difference problem from its boundary trace u_γ as $u_{N^-} = \boldsymbol{P}^- u_\gamma$.

Moreover, for the exterior problem (14.7) one can also build an algorithm that does not require difference approximation of the boundary condition. This is done similarly to how it was done in Section 14.3.1 for the interior problem. The only difference between the two algorithms is that instead of the overdetermined system (14.51) one will need to solve the system:

$$v_\gamma(\boldsymbol{c}) - \boldsymbol{P}_\gamma^- v_\gamma(\boldsymbol{c}) = 0,$$

and subsequently obtain the approximate solution by the formula

$$u_{N^-} \approx \boldsymbol{P}_\gamma^- v_\gamma \tag{14.55}$$

instead of formula (14.52).

14.3.3 Problem of Artificial Boundary Conditions

The problem of difference artificial boundary conditions is formulated as follows. We consider a discrete problem on D^0:

$$\sum_{n\in N_m} a_{mn}u_n = \begin{cases} f_m, & m\in M^+, \\ 0, & m\in M^-, \end{cases} \qquad u_{N^0}\in U_{N^0}, \tag{14.56}$$

and use the same M^0, M^+, M^-, N^+, N^-, γ, V_γ, and U_{N^0} as in Sections 14.3.1 and 14.3.2. We need to construct a boundary condition of the type $lv_\gamma = 0$ on the grid boundary $\gamma = N^+\cap N^-$ of the computational subdomain N^+, such that the solution $\{u_n\}$ of the problem:

$$\sum_{n\in N_m} a_{mn}u_n = f_m, \quad m\in M^+, \qquad lu_\gamma = 0, \tag{14.57}$$

will coincide on N^+ with the solution u_{N^0} of problem (14.56). The grid set γ and the boundary condition $lu_\gamma = 0$ are called an artificial boundary and an artificial boundary condition, respectively, because they are not a part of the original problem (14.56). They appear only because the solution is to be computed on the subdomain N^+ rather than on the entire N^0, and accordingly, the boundary condition $u_n = 0$, $n\in\partial D^0$, is to be transferred from ∂D^0 to γ.

One can use the exterior boundary equation with projection (14.54) in the capacity of the desired boundary condition $lu_\gamma = 0$. Indeed, relation (14.54) is necessary and sufficient for the solution u_{N^+} of the resulting problem (14.57) to admit an extension to the set N^- that would satisfy:

$$\sum_{n\in N_m} a_{mn}u_n = 0, \qquad m\in M^-,$$
$$u_n = 0, \quad \text{if} \quad n\in\partial D^0.$$

14.3.4 Problem of Two Subdomains

A finite-difference counterpart of the continuous problem formulated in Section 14.1.4 is the following. Suppose that $u_{N^0}\equiv\{u_n\}$, $n\in N^0$, satisfies:

$$\sum_{n\in N_m} a_{mn}u_n = f_m, \quad m\in M^0, \qquad u_n = 0, \quad \text{if} \quad n\in\partial D^0. \tag{14.58}$$

The solution to (14.58) is the sum of the solutions $u^+_{N^0}$ and $u^-_{N^0}$ of the problems:

$$\sum_{n\in N_m} a_{mn}u^+_n = \theta_{M^0}(M^-)f_m, \quad m\in M^0, \quad u^+_{N^0}\in U_{N^0}, \tag{14.59}$$

and

$$\sum_{n\in N_m} a_{mn}u_n^- = \theta_{M^0}(M^+)f_m, \quad m\in M^0, \quad u_{N^0}^- \in U_{N^0}, \tag{14.60}$$

respectively, where $\theta(\cdot)$ is the indicator function of a set.

Assume that the solution to problem (14.58) is known on γ, i.e., that we know u_γ. We need to find the individual terms in the sum:

$$u_\gamma = u_\gamma^+ + u_\gamma^-.$$

In doing so, the values of f_m, $m\in M^0$, are not known.

Solution of this problem is given by the formulae:

$$u_\gamma^+ = \boldsymbol{P}_\gamma^+ u_\gamma, \tag{14.61}$$

$$u_\gamma^- = \boldsymbol{P}_\gamma^- u_\gamma, \tag{14.62}$$

where

$$\boldsymbol{P}_\gamma^+ u_\gamma = \boldsymbol{P}^+ u_\gamma\big|_\gamma \quad \text{and} \quad \boldsymbol{P}_\gamma^- u_\gamma = \boldsymbol{P}^- u_\gamma\big|_\gamma.$$

Formulae (14.61) and (14.62) will be justified a little later.

Let $u^+ = \{u_n^+\}$, $n\in N^+$, and $u^- = \{u_n^-\}$, $n\in N^-$, be the restrictions of the solutions of problems (14.59) and (14.60) to the sets N^+ and N^-, respectively. Then,

$$u_n^+ = \boldsymbol{P}^+ u_\gamma\big|_n, \quad n\in N^+, \tag{14.63}$$

$$u_n^- = \boldsymbol{P}^- u_\gamma\big|_n, \quad n\in N^-. \tag{14.64}$$

To show that formulae (14.63) and (14.64) are true, we introduce the function:

$$w_n^\pm = \begin{cases} u_n^+, & n\in N^+, \\ -u_n^-, & n\in N^-. \end{cases} \tag{14.65}$$

Clearly, the function $w_n^\pm$, $n\in N^0$, is a piecewise regular solution of problem (14.58) with the jump u_γ on γ:

$$[w^\pm]_n = w_n^+ - w_n^- = u^+ + u_n^- = u_n, \quad n\in\gamma.$$

Then, according to Definition 14.3, the function (14.65) is a difference potential with the jump u_γ: $w^\pm = \boldsymbol{P}^\pm u_\gamma$. However, by virtue to Theorem 14.4, this difference potential coincides with the Cauchy type difference potential that has the same density v_γ. Consequently,

$$w_n^\pm = \begin{cases} w_n^+ = \boldsymbol{P}^+ u_\gamma, & n\in N^+, \\ w_n^- = -\boldsymbol{P}^- u_\gamma, & n\in N^-. \end{cases} \tag{14.66}$$

Comparing formula (14.66) with formula (14.65), we obtain formulae (14.63), (14.64). Moreover, if we consider relations (14.63) and (14.64) only at the nodes $n\in\gamma$, we arrive at formulae (14.61) and (14.62).

14.3.5 Problem of Active Shielding

Consider a difference boundary value problem on D^0:

$$\sum_{n\in N_m} a_{mn}u_n = f_m, \quad m \in M^0, \tag{14.67}$$
$$u_{N^0} \in U_{N^0}.$$

For this problem, neither the right-hand side f_m, $m \in M^0$, nor the solution u_n, $n \in N^0$, are assumed to be given explicitly. The only available data are provided by the trace u_γ of the solution u_{N^0} to problem (14.67) on the grid boundary γ.

Let us introduce an additional term g_m on the right-hand side of the difference equation from (14.67). This term will be called an active control source, or simply an active control. Problem (14.67) then transforms into:

$$\sum_{n\in N_m} a_{mn}w_n = f_m + g_m, \quad m \in M^0, \tag{14.68}$$
$$w_{N^0} \in V_{N^0}.$$

We can now formulate a difference problem of active shielding of the grid subdomain $N^+ \subset N^0$ from the influence of the sources f_m located in the subdomain $M^- \subset M^0$. We need to find all appropriate active controls g_m such that after adding them on the right-hand side of (14.67), the solution w_{N^0} of the new problem (14.68) will coincide on the protected region $N^+ \subset N^0$ with the solution v_{N^0} of the problem:

$$\sum_{n\in N_m} a_{mn}v_n = \begin{cases} f_m, & m \in M^+, \\ 0, & m \in M^-, \end{cases} \tag{14.69}$$
$$v_{N^0} \in V_{N^0}.$$

In other words, the effect of the controls on the solution on N^+ should be equivalent to eliminating the sources f_m on the complementary domain M^-. The following theorem presents the general solution for the control sources.

THEOREM 14.5

The solution w_{N^0} of problem (14.68) coincides on $N^+ \subset N^0$ with the solution v_{N^0} of problem (14.69) if and only if the controls g_m have the form:

$$g_m = \begin{cases} 0, & m \in M^+, \\ -\sum_{n\in N_m} a_{mn}z_n, & m \in M^-, \end{cases} \tag{14.70}$$

where $\{z_n\} = z_{N^0} \in U_{N^0}$ is an arbitrary grid function with the only constraint that it has the same boundary trace as that of the original solution, i.e., $z_\gamma = u_\gamma$.

PROOF First, we show that the solution $u_{N^0} \in U_{N^0}$ of the problem:

$$\sum_{n\in N_m} a_{mn}u_n = \varphi_m, \quad m \in M^0, \quad u_{N^0} \in U_{N^0}, \tag{14.71}$$

vanishes on N^+ if and only if φ_m has the form:

$$\varphi_m = \begin{cases} 0, & m \in M^+, \\ \sum_{n\in N_m} a_{mn}\xi_n, & m \in M^-, \end{cases} \tag{14.72}$$

where $\xi_{N^0} \in U_{N^0}$ is an arbitrary grid function that vanishes on γ: $\xi_\gamma = 0$.

Indeed, let $u_n = 0$, $n \in N^+$. The value of φ_m at the node $m \in M^+$ and the values of u_n, $n \in N_m \subset N^+ = \bigcup N_m$, $m \in M^+$, are related via (14.71). Consequently, $\varphi_m = 0$, $m \in M^+$. For the nodes $m \in M^-$, relation (14.72) also holds, because we can simply substitute the solution u_{N^0} of problem (14.71) for ξ_{N^0}. Conversely, let φ_m be given by formula (14.72), where $\xi_n = 0$, $n \in \gamma$. Consider an extension $\xi_{N^0} \in U_{N^0}$ of the grid function ξ_{N^-} from N^- to N^0, such that

$$\xi_{N^0}\Big|_{N^+} = 0. \tag{14.73}$$

Then, we can use a uniform expression for φ_m instead of (14.72):

$$\varphi_m = \sum_{n\in N_m} a_{mn}\xi_n, \quad m \in M^0.$$

Since the solution of problem (14.71) is unique, we have $u_{N^0} = \xi_{N^0}$. Therefore, according to (14.73), $u_n = 0$ for $n \in N^+$. In other words, the solution of (14.71) driven by the right-hand side (14.72) with $\xi_\gamma = 0$ is equal to zero on N^+.

Now we can show that formula (14.70) yields the general solution for active controls. More precisely, we will show that the solution w_{N^0} of problem (14.68) coincides with the solution v_{N^0} of problem (14.69) on the grid set N^+ if and only if the controls g_m are given by (14.70). Let us subtract equation (14.69) from equation (14.68). Denoting $z_{N^0} = w_{N^0} - v_{N^0}$, we obtain:

$$\sum_{n\in N_m} a_{mn}z_n = \begin{cases} g_m, & m \in M^+, \\ g_m + f_m, & m \in M^-. \end{cases} \tag{14.74}$$

Moreover, according to (14.67), we can substitute $f_m = \sum a_{mn}u_n$, $m \in M^-$, into (14.74). Previously we have proven that $z_n = 0$ for all $n \in N^+$ if and only if the right-hand side of (14.74) has the form (14.72), where $\xi_\gamma = 0$. Writing the right-hand side of (14.74) as:

$$\psi_m = \begin{cases} g_m, & m \in M^+, \\ g_m + \sum_{n\in N_m} a_{mn}u_n, & m \in M^-, \end{cases}$$

we conclude that this requirement is equivalent to the condition that g_m has the form (14.70), where $z_\gamma = u_\gamma$. $\square$

14.4 General Remarks

The method of difference potentials offers a number of unique capabilities that cannot be easily reproduced in the framework of other methods available in the literature. First and foremost, this pertains to the capability of taking only the boundary trace of a given solution and splitting it into two components, one due to the interior sources and the other one due to the complementary exterior sources with respect to a given region (see Sections 14.1.4 and 14.3.4). The resulting split is unambiguous; it can be realized with no constraints on the shape of the boundary and no need for a closed form fundamental solution of the governing equation or system. Moreover, it can be obtained directly for the discretization, and it enables efficient (numerical) solution of the problem of artificial boundary conditions (Sections 14.1.3 and 14.3.3) and the problem of active shielding (Sections 14.1.5 and 14.3.5).

The meaning, or physical interpretation, of the solutions that can be obtained and analyzed with the help of difference potentials, may vary. If, for example, the solution is interpreted as a wave field [governed, say, by the Helmholtz equation (12.32a)], then its component due to the interior sources represents outgoing waves with respect to a given region, and the component due to the exterior sources represents the incoming waves. Hence, the method of difference potentials provides a universal and robust procedure for splitting the overall wave field into the incoming and outgoing parts, and for doing that it only requires the knowledge of the field quantities on the interface between the interior and exterior domains. In other words, no knowledge of the actual waves' sources is needed, and no knowledge of the field beyond the interface is required either.

In the literature, the problem of identifying the outgoing and incoming waves has received substantial attention in the context of propagation of waves over unbounded domains. When the sources of waves are confined to a predetermined compact region in space, and the waves propagate away from this region toward infinity, numerical reconstruction of the wave field requires truncation of the overall unbounded domain and setting artificial boundary conditions that would facilitate reflectionless propagation of the outgoing waves through the external artificial boundary. Artificial boundary conditions of this type are often referred to as radiation or non-reflecting boundary conditions. Pioneering work on non-reflecting boundary conditions was done by Engquist and Majda in the late seventies — early eighties, see [EM77, EM79, EM81]. In these papers, boundary operators that enable one-way propagation of waves through a given interface (radiation toward the exterior) were constructed in the continuous formulation of the problem for particular classes of governing equations and particular (simple) shapes of the boundary. The approaches to their discrete approximation were discussed as well.

14.5 Bibliography Comments

The method of difference potentials was proposed by Ryaben'kii in his Habilitation thesis in 1969. The state-of-the-art in the development of the method as of the year 2001 can be found in [Rya02]. Over the years, the method of difference potentials has benefited a great deal from fine contributions by many colleagues. The corresponding acknowledgments, along with a considerably more detailed bibliography, can also be found in [Rya02].

For further detail on artificial boundary conditions, we refer the reader to [Tsy98]. The problem of active shielding is discussed in [Rya95, LRT01].

Let us also mention the capacitance matrix method proposed by Proskurowski and Widlund [PW76, PW80] that has certain elements similar to those of the method of difference potentials.

List of Figures

Referenced Books

[Ach92] N. I. Achieser. *Theory of Approximation.* Dover Publications Inc., New York, 1992. Reprint of the 1956 English translation of the 1st Russian edition; the 2nd augmented Russian edition is available, Moscow, Nauka, 1965.

[AH05] Kendall Atkinson and Weimin Han. *Theoretical Numerical Analysis: A Functional Analysis Framework*, volume 39 of *Texts in Applied Mathematics.* Springer, New York, second edition, 2005.

[Atk89] Kendall E. Atkinson. *An Introduction to Numerical Analysis.* John Wiley & Sons Inc., New York, second edition, 1989.

[Axe94] Owe Axelsson. *Iterative Solution Methods.* Cambridge University Press, Cambridge, 1994.

[Bab86] K. I. Babenko. *Foundations of Numerical Analysis [Osnovy chislennogo analiza].* Nauka, Moscow, 1986. [Russian].

[BD92] C. A. Brebbia and J. Dominguez. *Boundary Elements: An Introductory Course.* Computational Mechanics Publications, Southampton, second edition, 1992.

[Ber52] S. N. Bernstein. *Collected Works. Vol. I. The Constructive Theory of Functions [1905–1930].* Izdat. Akad. Nauk SSSR, Moscow, 1952. [Russian].

[Ber54] S. N. Bernstein. *Collected Works. Vol. II. The Constructive Theory of Functions [1931–1953].* Izdat. Akad. Nauk SSSR, Moscow, 1954. [Russian].

[BH02] K. Binder and D. W. Heermann. *Monte Carlo Simulation in Statistical Physics: An Introduction*, volume 80 of *Springer Series in Solid-State Sciences.* Springer-Verlag, Berlin, fourth edition, 2002.

[BHM00] William L. Briggs, Van Emden Henson, and Steve F. McCormick. *A Multigrid Tutorial.* Society for Industrial and Applied Mathematics (SIAM), Philadelphia, PA, second edition, 2000.

[Boy01] John P. Boyd. *Chebyshev and Fourier Spectral Methods.* Dover Publications Inc., Mineola, NY, second edition, 2001.

[Bra84] Achi Brandt. *Multigrid Techniques: 1984 Guide with Applications to Fluid Dynamics*, volume 85 of *GMD-Studien [GMD Studies].*

Gesellschaft für Mathematik und Datenverarbeitung mbH, St. Augustin, 1984.

[Bra93] James H. Bramble. *Multigrid Methods*, volume 294 of *Pitman Research Notes in Mathematics Series.* Longman Scientific & Technical, Harlow, 1993.

[BS02] Susanne C. Brenner and L. Ridgway Scott. *The Mathematical Theory of Finite Element Methods*, volume 15 of *Texts in Applied Mathematics.* Springer-Verlag, New York, second edition, 2002.

[But03] J. C. Butcher. *Numerical Methods for Ordinary Differential Equations.* John Wiley & Sons Ltd., Chichester, 2003.

[CdB80] Samuel Conte and Carl de Boor. *Elementary Numerical Analysis: An Algorithmic Approach.* McGraw-Hill, New York, third edition, 1980.

[CH06] Alexandre J. Chorin and Ole H. Hald. *Stochastic Tools in Mathematics and Science*, volume 1 of *Surveys and Tutorials in the Applied Mathematical Sciences.* Springer, New York, 2006.

[Che66] E. W. Cheney. *Introduction to Approximation Theory.* International Series in Pure and Applied Mathematics. McGraw-Hill Book Company, New York, 1966.

[CHQZ88] Claudio Canuto, M. Yousuff Hussaini, Alfio Quarteroni, and Thomas A. Zang. *Spectral Methods in Fluid Dynamics.* Springer Series in Computational Physics. Springer-Verlag, New York, 1988.

[CHQZ06] Claudio Canuto, M. Yousuff Hussaini, Alfio Quarteroni, and Thomas A. Zang. *Spectral Methods. Fundamentals in Single Domains.* Springer Series in Scientific Computation. Springer-Verlag, New York, 2006.

[Cia02] Philippe G. Ciarlet. *The Finite Element Method for Elliptic Problems*, volume 40 of *Classics in Applied Mathematics.* Society for Industrial and Applied Mathematics (SIAM), Philadelphia, PA, 2002. Reprint of the 1978 original [North-Holland, Amsterdam; MR0520174 (58 #25001)].

[dB01] Carl de Boor. *A Practical Guide to Splines*, volume 27 of *Applied Mathematical Sciences.* Springer-Verlag, New York, revised edition, 2001.

[DB03] Germund Dahlquist and Ake Björck. *Numerical Methods.* Dover Publications Inc., Mineola, NY, 2003. Translated from the Swedish by Ned Anderson, Reprint of the 1974 English translation.

[Dem06] Leszek Demkowicz. *Computing with hp-Adaptive Finite Elements. I. One- and Two-Dimensional Elliptic and Maxwell Problems.* Taylor & Francis, Boca Raton, 2006.

[Dör82] Heinrich Dörrie. *100 Great Problems of Elementary Mathematics.* Dover Publications Inc., New York, 1982. Their history and solution,

Reprint of the 1965 edition, Translated from the fifth edition of the German original by David Antin.

[DR84] Philip J. Davis and Philip Rabinowitz. *Methods of Numerical Integration.* Computer Science and Applied Mathematics. Academic Press Inc., Orlando, FL, second edition, 1984.

[D'y96] Eugene G. D'yakonov. *Optimization in Solving Elliptic Problems.* CRC Press, Boca Raton, FL, 1996. Translated from the 1989 Russian original, Translation edited and with a preface by Steve McCormick.

[Fed77] M. V. Fedoryuk. *The Saddle-Point Method [Metod perevala].* Nauka, Moscow, 1977. [Russian].

[FM87] L. Fox and D. F. Mayers. *Numerical Solution of Ordinary Differential Equations.* Chapman & Hall, London, 1987.

[GAKK93] S. K. Godunov, A. G. Antonov, O. P. Kiriljuk, and V. I. Kostin. *Guaranteed Accuracy in Numerical Linear Algebra*, volume 252 of *Mathematics and its Applications.* Kluwer Academic Publishers Group, Dordrecht, 1993. Translated and revised from the 1988 Russian original.

[Gan59] F. R. Gantmacher. *The Theory of Matrices. Vols. 1, 2.* Translated by K. A. Hirsch. Chelsea Publishing Co., New York, 1959.

[GKO95] Bertil Gustafsson, Heinz-Otto Kreiss, and Joseph Oliger. *Time Dependent Problems and Difference Methods.* Pure and Applied Mathematics (New York). John Wiley & Sons Inc., New York, 1995. A Wiley-Interscience Publication.

[GL81] Alan George and Joseph W. H. Liu. *Computer Solution of Large Sparse Positive Definite Systems.* Prentice-Hall Inc., Englewood Cliffs, N.J., 1981. Prentice-Hall Series in Computational Mathematics.

[Gla04] Paul Glasserman. *Monte Carlo Methods in Financial Engineering*, volume 53 of *Applications of Mathematics (New York).* Springer-Verlag, New York, 2004. Stochastic Modelling and Applied Probability.

[GO77] David Gottlieb and Steven A. Orszag. *Numerical Analysis of Spectral Methods: Theory and Applications.* Society for Industrial and Applied Mathematics, Philadelphia, Pa., 1977. CBMS-NSF Regional Conference Series in Applied Mathematics, No. 26.

[GR64] S. K. Godunov and V. S. Ryabenki. *Theory of Difference Schemes. An Introduction.* Translated by E. Godfredsen. North-Holland Publishing Co., Amsterdam, 1964.

[GR87] S. K. Godunov and V. S. Ryaben'kii. *Difference Schemes: An Introduction to Underlying Theory.* North-Holland, New York, Amsterdam, 1987.

[GVL89] Gene H. Golub and Charles F. Van Loan. *Matrix Computations*, volume 3 of *Johns Hopkins Series in the Mathematical Sciences*. Johns Hopkins University Press, Baltimore, MD, second edition, 1989.

[Hac85] Wolfgang Hackbusch. *Multigrid Methods and Applications*, volume 4 of *Springer Series in Computational Mathematics*. Springer-Verlag, Berlin, 1985.

[Hal94] W. S. Hall. *The Boundary Element Method*, volume 27 of *Solid Mechanics and its Applications*. Kluwer Academic Publishers Group, Dordrecht, 1994.

[Hen64] Peter Henrici. *Elements of Numerical Analysis*. John Wiley & Sons Inc., New York, 1964.

[HGG06] Jan S. Hesthaven, Sigal Gottlieb, and David Gottlieb. *Spectral Methods for Time-Dependent Problems*. Cambridge University Press, Cambridge, 2006.

[HJ85] Roger A. Horn and Charles R. Johnson. *Matrix Analysis*. Cambridge University Press, Cambridge, 1985.

[HNW93] E. Hairer, S. P. Nørsett, and G. Wanner. *Solving Ordinary Differential Equations I: Nonstiff Problems*, volume 8 of *Springer Series in Computational Mathematics*. Springer-Verlag, Berlin, second edition, 1993.

[HW96] E. Hairer and G. Wanner. *Solving Ordinary Differential Equations II: Stiff and Differential-Algebraic Problems*, volume 14 of *Springer Series in Computational Mathematics*. Springer-Verlag, Berlin, second edition, 1996.

[IK66] Eugene Isaacson and Herbert Bishop Keller. *Analysis of Numerical Methods*. John Wiley & Sons Inc., New York, 1966.

[Jac94] Dunham Jackson. *The Theory of Approximation*, volume 11 of *American Mathematical Society Colloquium Publications*. American Mathematical Society, Providence, RI, 1994. Reprint of the 1930 original.

[KA82] L. V. Kantorovich and G. P. Akilov. *Functional Analysis*. Pergamon Press, Oxford, second edition, 1982. Translated from the Russian by Howard L. Silcock.

[Kel95] C. T. Kelley. *Iterative Methods for Linear and Nonlinear Equations*, volume 16 of *Frontiers in Applied Mathematics*. Society for Industrial and Applied Mathematics (SIAM), Philadelphia, PA, 1995. With separately available software.

[Kel03] C. T. Kelley. *Solving Nonlinear Equations with Newton's Method*. Fundamentals of Algorithms. Society for Industrial and Applied Mathematics (SIAM), Philadelphia, PA, 2003.

[KF75] A. N. Kolmogorov and S. V. Fomin. *Introductory Real Analysis.* Dover Publications Inc., New York, 1975. Translated from the second Russian edition and edited by Richard A. Silverman, Corrected reprinting.

[Lad69] O. A. Ladyzhenskaya. *The Mathematical Theory of Viscous Incompressible Flow.* Second English edition, revised and enlarged. Translated from the Russian by Richard A. Silverman and John Chu. Mathematics and its Applications, Vol. 2. Gordon and Breach Science Publishers, New York, 1969.

[LeV02] Randall J. LeVeque. *Finite Volume Methods for Hyperbolic Problems.* Cambridge Texts in Applied Mathematics. Cambridge University Press, Cambridge, 2002.

[LG95] O. V. Lokutsievskii and M. B. Gavrikov. *Foundations of Numerical Analysis [Nachala chislennogo analiza].* TOO "Yanus", Moscow, 1995. [Russian].

[Liu01] Jun S. Liu. *Monte Carlo Strategies in Scientific Computing.* Springer Series in Statistics. Springer-Verlag, New York, 2001.

[Lor86] G. G. Lorentz. *Approximation of Functions.* Chelsea Publishing Company, New York, 1986.

[Mar77] A. I. Markushevich. *Theory of Functions of a Complex Variable. Vol. I, II, III.* Chelsea Publishing Co., New York, English edition, 1977. Translated and edited by Richard A. Silverman.

[MF53] Philip M. Morse and Herman Feshbach. *Methods of Theoretical Physics. 2 Volumes.* International Series in Pure and Applied Physics. McGraw-Hill Book Co., Inc., New York, 1953.

[MK88] K. M. Magomedov and A. S. Kholodov. *Difference-Characteristic Numerical Methods [Setochno-kharakteristicheskie chislennye metody].* Edited and with a foreword by O. M. Belotserkovskii. Nauka, Moscow, 1988. [Russian].

[MM05] K. W. Morton and D. F. Mayers. *Numerical Solution of Partial Differential Equations: An Introduction.* Cambridge University Press, Cambridge, second edition, 2005.

[MMP95] Solomon G. Mikhlin, Nikita F. Morozov, and Michael V. Paukshto. *The Integral Equations of the Theory of Elasticity*, volume 135 of *Teubner-Texte zur Mathematik [Teubner Texts in Mathematics].* B. G. Teubner Verlagsgesellschaft mbH, Stuttgart, 1995. Translated from the Russian by Rainer Radok [Jens Rainer Maria Radok].

[Nat64] I. P. Natanson. *Constructive Function Theory. Vol. I. Uniform Approximation.* Translated from the Russian by Alexis N. Obolensky. Frederick Ungar Publishing Co., New York, 1964.

[Nat65a] I. P. Natanson. *Constructive Function Theory. Vol. II. Approximation in Mean.* Translated from the Russian by John R. Schulenberger. Frederick Ungar Publishing Co., New York, 1965.

[Nat65b] I. P. Natanson. *Constructive Function Theory. Vol. III. Interpolation and Approximation Quadratures.* Translated from the Russian by John R. Schulenberger. Frederick Ungar Publishing Co., New York, 1965.

[OR00] J. M. Ortega and W. C. Rheinboldt. *Iterative Solution of Nonlinear Equations in Several Variables*, volume 30 of *Classics in Applied Mathematics.* Society for Industrial and Applied Mathematics (SIAM), Philadelphia, PA, 2000. Reprint of the 1970 original.

[Poz02] C. Pozrikidis. *A Practical Guide to Boundary Element Methods with the Software Library BEMLIB.* Chapman & Hall/CRC, Boca Raton, FL, 2002.

[PT96] G. M. Phillips and P. J. Taylor. *Theory and Applications of Numerical Analysis.* Academic Press Ltd., London, second edition, 1996.

[QSS00] Alfio Quarteroni, Riccardo Sacco, and Fausto Saleri. *Numerical Mathematics*, volume 37 of *Texts in Applied Mathematics.* Springer-Verlag, New York, 2000.

[RF56] V. S. Ryaben'kii and A. F. Filippov. *On Stability of Difference Equations.* Gosudarstv. Izdat. Tehn.-Teor. Lit., Moscow, 1956. [Russian].

[Riv74] Theodore J. Rivlin. *The Chebyshev Polynomials.* Wiley-Interscience [John Wiley & Sons], New York, 1974. Pure and Applied Mathematics.

[RJ83] B. L. Rozdestvenskii and N. N. Janenko. *Systems of Quasilinear Equations and Their Applications to Gas Dynamics*, volume 55 of *Translations of Mathematical Monographs.* American Mathematical Society, Providence, RI, 1983. Translated from the second Russian edition by J. R. Schulenberger.

[RM67] Robert D. Richtmyer and K. W. Morton. *Difference Methods for Initial-Value Problems.* Second edition. Interscience Tracts in Pure and Applied Mathematics, No. 4. Interscience Publishers John Wiley & Sons, Inc., New York-London-Sydney, 1967.

[Rud87] Walter Rudin. *Real and Complex Analysis.* McGraw-Hill Book Co., New York, third edition, 1987.

[Rya00] V. S. Ryaben'kii. *Introduction to Computational Mathematics [Vvedenie v vychislitel'nuyu matematiku].* Fizmatlit, Moscow, second edition, 2000. [Russian].

[Rya02] V. S. Ryaben'kii. *Method of Difference Potentials and Its Applications*, volume 30 of *Springer Series in Computational Mathematics.* Springer-Verlag, Berlin, 2002.

[Saa03] Yousef Saad. *Iterative Methods for Sparse Linear Systems.* Society for Industrial and Applied Mathematics, Philadelphia, PA, second edition, 2003.

[Sch02] Michelle Schatzman. *Numerical Analysis. A Mathematical Introduction.* Clarendon Press, Oxford, 2002.

[SN89a] Aleksandr A. Samarskii and Evgenii S. Nikolaev. *Numerical Methods for Grid Equations. Vol. I, Direct Methods.* Birkhäuser Verlag, Basel, 1989. Translated from the Russian by Stephen G. Nash.

[SN89b] Aleksandr A. Samarskii and Evgenii S. Nikolaev. *Numerical Methods for Grid Equations. Vol. II, Iterative Methods.* Birkhäuser Verlag, Basel, 1989. Translated from the Russian and with a note by Stephen G. Nash.

[Str04] John C. Strikwerda. *Finite Difference Schemes and Partial Differential Equations.* Society for Industrial and Applied Mathematics (SIAM), Philadelphia, PA, second edition, 2004.

[Tho95] J. W. Thomas. *Numerical Partial Differential Equations: Finite Difference Methods*, volume 22 of *Texts in Applied Mathematics.* Springer-Verlag, New York, 1995.

[Tho99] J. W. Thomas. *Numerical Partial Differential Equations: Conservation Laws and Elliptic Equations*, volume 33 of *Texts in Applied Mathematics.* Springer-Verlag, New York, 1999.

[TOS01] U. Trottenberg, C. W. Oosterlee, and A. Schüller. *Multigrid.* Academic Press Inc., San Diego, CA, 2001. With contributions by A. Brandt, P. Oswald and K. Stüben.

[TS63] A. N. Tikhonov and A. A. Samarskii. *Equations of Mathematical Physics.* Pergamon Press, Oxford, 1963.

[Van01] R. J. Vanderbei. *Linear Programming: Foundations and Extensions.* Kluwer Academic Publishers, Boston, 2001.

[vdV03] Henk A. van der Vorst. *Iterative Krylov Methods for Large Linear Systems*, volume 13 of *Cambridge Monographs on Applied and Computational Mathematics.* Cambridge University Press, Cambridge, 2003.

[Wen66] Burton Wendroff. *Theoretical Numerical Analysis.* Academic Press, New York, 1966.

[Wes92] Pieter Wesseling. *An Introduction to Multigrid Methods.* Pure and Applied Mathematics (New York). John Wiley & Sons Ltd., Chichester, 1992.

[WW62] E. T. Whittaker and G. N. Watson. *A Course of Modern Analysis. An Introduction to the General Theory of Infinite Processes and of Ana-*

lytic Functions: with an Account of the Principal Transcendental Functions. Fourth edition. Reprinted. Cambridge University Press, New York, 1962.

Referenced Journal Articles

[Ast71] G. P. Astrakhantsev. An iterative method of solving elliptic net problems. *USSR Comput. Math. and Math. Phys.*, 11(2):171–182, 1971.

[Bak66] N. S. Bakhvalov. On the convergence of a relaxation method with natural constraints on the elliptic operator. *USSR Comput. Math. and Math. Phys.*, 6(5):101–135, 1966.

[Bel89] V. N. Belykh. Algorithms without saturation in the problem of numerical integration. *Dokl. Akad. Nauk SSSR*, 304(3):529–533, 1989.

[BK01] Oscar P. Bruno and Leonid A. Kunyansky. Surface scattering in three dimensions: an accelerated high-order solver. *R. Soc. Lond. Proc. Ser. A Math. Phys. Eng. Sci.*, 457(2016):2921–2934, 2001.

[Bra77] Achi Brandt. Multi-level adaptive solutions to boundary-value problems. *Math. Comp.*, 31(138):333–390, 1977.

[Cal63] A. P. Calderon. Boundary-value problems for elliptic equations. In *Proceedings of the Soviet-American Conference on Partial Differential Equations at Novosibirsk*, pages 303–304, Moscow, 1963. Fizmatgiz.

[CFL28] R. Courant, K. O. Friedrichs, and H. Lewy. Über die partiellen Differenzengleichungen der mathematischen Physik. *Math. Ann.*, 100:32, 1928. [German].

[Col51] Julian D. Cole. On a quasi-linear parabolic equation occurring in aerodynamics. *Quart. Appl. Math.*, 9:225–236, 1951.

[CT65] James W. Cooley and John W. Tukey. An algorithm for the machine calculation of complex Fourier series. *Math. Comp.*, 19:297–301, 1965.

[EM77] Bjorn Engquist and Andrew Majda. Absorbing boundary conditions for the numerical simulation of waves. *Math. Comp.*, 31(139):629–651, 1977.

[EM79] Björn Engquist and Andrew Majda. Radiation boundary conditions for acoustic and elastic wave calculations. *Comm. Pure Appl. Math.*, 32(3):314–358, 1979.

[EM81] Björn Engquist and Andrew Majda. Numerical radiation boundary conditions for unsteady transonic flow. *J. Comput. Phys.*, 40(1):91–103, 1981.

[EO80] Björn Engquist and Stanley Osher. Stable and entropy satisfying approximations for transonic flow calculations. *Math. Comp.*, 34(149):45–75, 1980.

[Fed61] R. P. Fedorenko. A relaxation method for solving elliptic difference equations. *USSR Comput. Math. and Math. Phys.*, 1(4):922–927, 1961.

[Fed64] R. P. Fedorenko. On the speed of convergence of one iterative process. *USSR Comput. Math. and Math. Phys.*, 4(3):227–235, 1964.

[Fed67] M. V. Fedoryuk. On the stability in C of the Cauchy problem for difference and partial differential equations. *USSR Computational Mathematics and Mathematical Physics*, 7(3):48–89, 1967.

[Fed73] R. P. Fedorenko. Iterative methods for elliptic difference equations. *Russian Math. Surveys*, 28(2):129–195, 1973.

[Fil55] A. F. Filippov. On stability of difference equations. *Dokl. Akad. Nauk SSSR (N.S.)*, 100:1045–1048, 1955. [Russian].

[Hop50] Eberhard Hopf. The partial differential equation $u_t + uu_x = \mu u_{xx}$. *Comm. Pure Appl. Math.*, 3:201–230, 1950.

[Ise82] Arieh Iserles. Order stars and a saturation theorem for first-order hyperbolics. *IMA J. Numer. Anal.*, 2(1):49–61, 1982.

[Lax54] Peter D. Lax. Weak solutions of nonlinear hyperbolic equations and their numerical computation. *Comm. Pure Appl. Math.*, 7:159–193, 1954.

[LRT01] J. Loncarić, V. S. Ryaben'kii, and S. V. Tsynkov. Active shielding and control of noise. *SIAM J. Applied Math.*, 62(2):563–596, 2001.

[Lyu40] L. A. Lyusternik. The Dirichlet problem. *Uspekhi Matematicheskikh Nauk (Russian Math. Surveys)*, 8:115–124, 1940. [Russian].

[OHK51] George G. O'Brien, Morton A. Hyman, and Sidney Kaplan. A study of the numerical solution of partial differential equations. *J. Math. Physics*, 29:223–251, 1951.

[Ols95a] Pelle Olsson. Summation by parts, projections, and stability. I. *Math. Comp.*, 64(211):1035–1065, S23–S26, 1995.

[Ols95b] Pelle Olsson. Summation by parts, projections, and stability. II. *Math. Comp.*, 64(212):1473–1493, 1995.

[PW76] Wlodzimierz Proskurowski and Olof Widlund. On the numerical solution of Helmholtz's equation by the capacitance matrix method. *Math. Comp.*, 30(135):433–468, 1976.

[PW80] Wlodzimierz Proskurowski and Olof Widlund. A finite element-capacitance matrix method for the Neumann problem for Laplace's equation. *SIAM J. Sci. Statist. Comput.*, 1(4):410–425, 1980.

[Rya52] V. S. Ryaben'kii. On the application of the method of finite differences to the solution of Cauchy's problem. *Doklady Akad. Nauk SSSR (N.S.)*, 86:1071–1074, 1952. [Russian].

[Rya64] V. S. Ryaben'kii. Necessary and sufficient conditions for good definition of boundary value problems for systems of ordinary difference equations. *U.S.S.R. Comput. Math. and Math. Phys.*, 4:43–61, 1964.

[Rya70] V. S. Ryaben'kii. The stability of iterative algorithms for the solution of nonselfadjoint difference equations. *Soviet Math. Dokl.*, 11:984–987, 1970.

[Rya75] V. S. Ryaben'kii. Local splines. *Comput. Methods Appl. Mech. Engrg.*, 5:211–225, 1975.

[Rya85] V. S. Ryaben'kii. Boundary equations with projections. *Russian Math. Surveys*, 40:147–183, 1985.

[Rya95] V. S. Ryaben'kii. A difference screening problem. *Funct. Anal. Appl.*, 29:70–71, 1995.

[Str64] Gilbert Strang. Accurate partial difference methods. II. Non-linear problems. *Numer. Math.*, 6:37–46, 1964.

[Str66] Gilbert Strang. Necessary and insufficient conditions for well-posed Cauchy problems. *J. Differential Equations*, 2:107–114, 1966.

[Str94] Bo Strand. Summation by parts for finite difference approximations for d/dx. *J. Comput. Phys.*, 110(1):47–67, 1994.

[Tho49] L. H. Thomas. Elliptic problems in linear difference equations over a network. Technical report, Watson Scientific Computing Laboratory, Columbia University, New York, 1949.

[Tsy98] S. V. Tsynkov. Numerical solution of problems on unbounded domains. A review. *Appl. Numer. Math.*, 27:465–532, 1998.

[TW93] Christopher K. W. Tam and Jay C. Webb. Dispersion-relation-preserving finite difference schemes for computational acoustics. *J. Comput. Phys.*, 107(2):262–281, 1993.

[VNR50] J. Von Neumann and R. D. Richtmyer. A method for the numerical calculation of hydrodynamic shocks. *J. Appl. Phys.*, 21:232–237, 1950.

Index

C

G

N

Q

R

V

W